冯建逵 1986 年绘清东陵鸟瞰图（《清东陵与清西陵古建筑》未刊书稿插图）
A bird's-eye view of the Eastern Tombs in the Qing Dynasty, drawn by FENG Jiankui in 1986 (an unpublished illustration for the manuscript of *The Ancient Buildings of the Qing Eastern and Western Tombs* [*Qing Dongling yu Xiling Gujianzhu*])

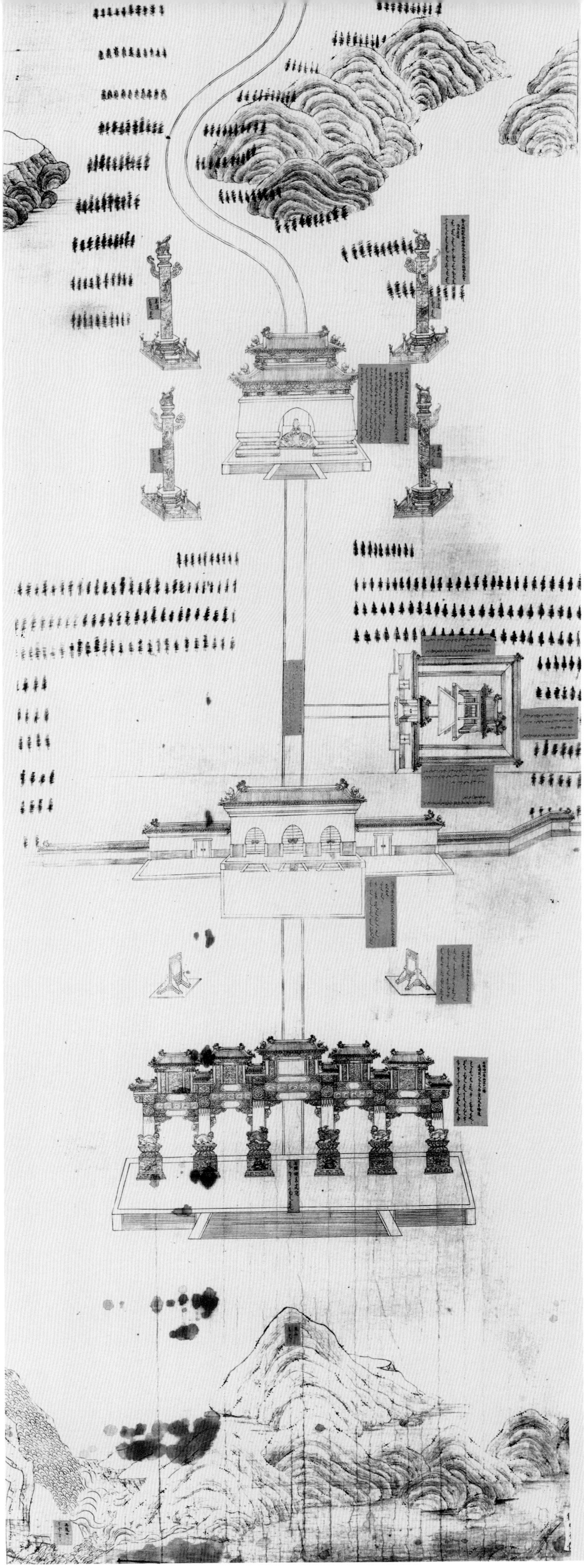

孝陵图（下段，截取时与孝陵图上段在龙凤门处有重合）。中国第一历史档案馆藏。康熙九年（1670年）工部绘制

Drawing of the Tomb of Piety (*Xiaoling tu*) (lower section, when the drawing was intercepted, overlapped with the upper section at the dragon and phoenix gate [*longfengmen*]), drawn by the Ministry of Engineering Works (Gongbu) in the ninth year of the Kangxi Emperor's reign (1670), in the collection of the First Historical Archives of China.

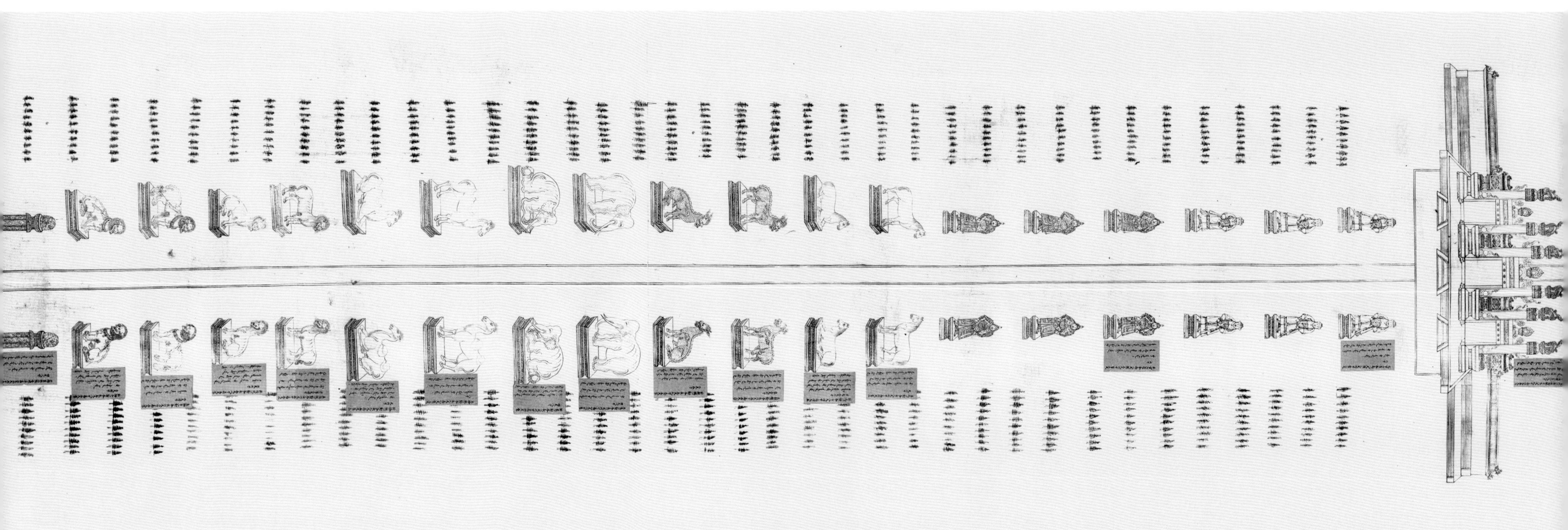

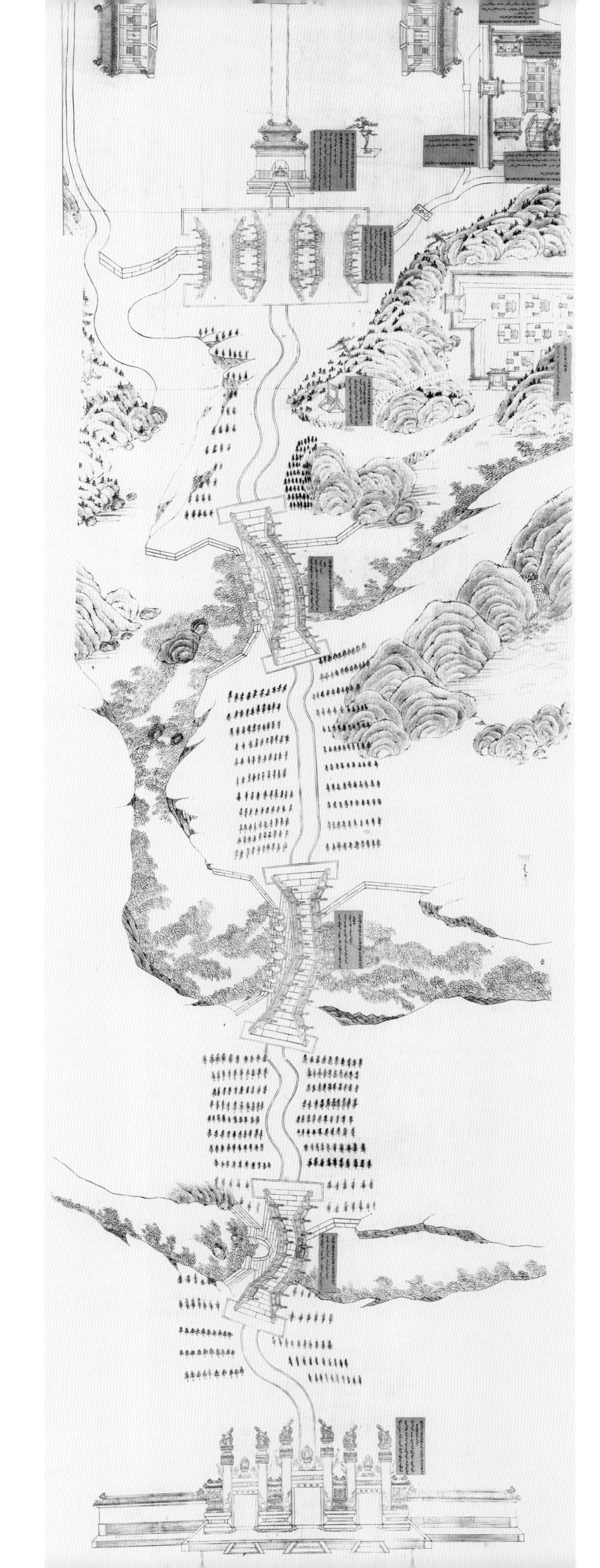

孝陵图（上段）。中国第一历史档案馆藏。康熙九年（1670年）工部绘制

Drawing of the Tomb of Piety (*Xiaoling tu*) (upper section), drawn by the Ministry of Engineering Works (Gongbu) in the ninth year of the Kangxi Emperor's reign (1670), in the collection of the First Historical Archives of China.

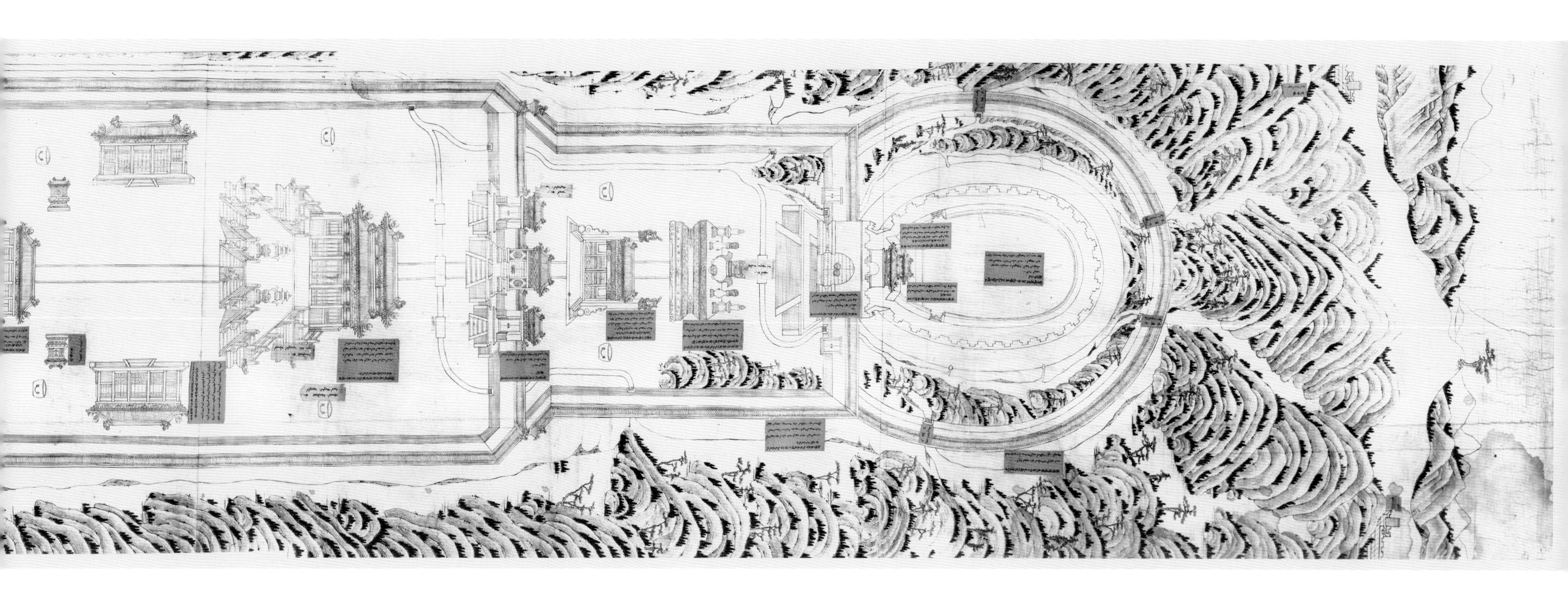

景陵图。中国第一历史档案馆馆藏。乾隆四十年（1775年）绘制

Drawing of the Tomb of Admiration (Jingling ling), drawn in the fortieth year of the Qianlong Emperor's reign (1775), in the collection of the First Historical Archives of China

裕陵图。中国第一历史档案馆馆藏。乾隆四十年（1775 年）绘制

Drawing of the Tomb of Prosperity (Yuling tu), drawn in the fortieth year of the Qianlong Emperor's reign (1775), in the collection of the First Historical Archives of China.

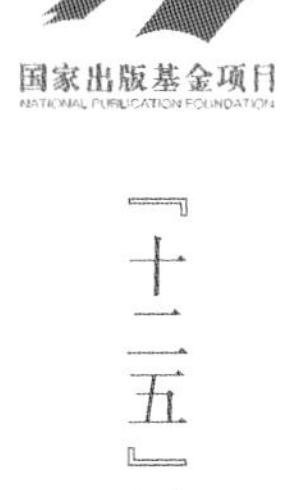

『十二五』国家重点图书出版规划项目

中国古建筑测绘大系·陵寝建筑

清东陵

天津大学建筑学院　清东陵文物管理处　编写

冯建逵　王其亨　编著

中国建筑工业出版社

Traditional Chinese Architecture Surveying and Mapping Series: Tomb Architecture

EASTERN QING TOMBS

Compiled by the School of Architecture, Tianjin University &
the Eastern Qing Tombs Cultural Relics Management Office
Edited by FENG Jiankui, WANG Qiheng

China Architecture & Building Press

『中国古建筑测绘大系』编委会

Editorial Board of the Traditional Chinese Architecture Surveying and Mapping Series

目录

Contents

导言

总述

世界文化遗产清东陵，位于北京迤东今河北省遵化市马兰峪镇西，是清朝入主中原后在关内开辟最早、规模最大的皇家陵区，东西宽约 20 公里，南北长约 125 公里，占地约 2500 平方公里。其中昌瑞山东西横迤，山南『前圈』约 40 平方公里为陵寝建筑区，东、南、西亘有风水墙；山北为风水禁地『后龙』，直抵承德境南雾灵山，利用山沟伐净草木辟出『火道』供巡防山火，旁立红桩，二十丈外设白桩，外十里又围青桩，往外另有 20 里宽的官山，严禁采樵伐木、采石取土和烧窑。该区域今已成为河北省最大的自然保护区（图 1）。

作为标举孝治天下的载体，清东陵的经营，肇自顺治八年（1651 年）顺治皇帝亲政后的『相度成规』；十年后，他生前未遑兴工的孝陵鼎建。继而直到光绪

Introduction

The Eastern Qing Tombs are located in Malanyu Town about 30 kms west of Zunhua City and 120 kms east of Beijing in Hebei Province. The tombs are the earliest and largest imperial tomb complex established by the Qing court after the Manchu rulers entered the central plains of China. The tombs cover an area of about 2,500 square kilometres: 20 kilometres across the east-west axis and 125 kilometres up and down the north-south axis. Within the tomb complex, Changrui Mountain lies along the east-west axis. The 40 square kilometre area where the tombs are built lies on the southern side of Changrui Mountain. This area is called the front field (*qianquan*). The 'front field' is enclosed with *fengshui* walls, marking its eastern, southern and western borders. On the northern side of Changrui Mountain is the back dragon (*houlong*) mountain, an area protected according to *fengshui* principles, and this extends north to Wuling Mountain south of Chengde. All the gullies encircling this protected area have been cleared for fire prevention. This fire break is marked by the red-coloured wooden posts along the edge of the cleared path. About 64 meters further out there are white coloured posts. About 5.76 kilometres further out there are cyan-coloured wooden posts and finally an 11.5 kilometre-wide government controlled mountain area. This whole area, where wood cutting, earth mining, and kiln firing were prohibited, has now become one of the largest natural reserves in Hebei Province (Fig.1).

Emblematic of 'governance through filiality', which was given high respect by the Qing dynasty, the establishment of the Eastern Qing Tombs was registered by close examination and recording of the area initiated by the Shunzhi Emperor during the eighth year of his reign (1651). Ten years later, after his death, work on his tomb, the Tomb of Filial Piety (Xiaoling), the main tomb of the Eastern Qing Tombs, began. The East Tomb of Stability (Ding Dongling) in Putuo Valley rebuilt for the Empress Dowager Cixi was completed in 1908. It marked the end of the 250-year-long construction of the Eastern Qing Tombs. The complex includes five emperors' tombs including the Tomb of Filial Piety (Xiaoling), the Tomb of Admiration (Jingling), the Tomb of Prosperity (Yuling), the Tomb of Stability (Dingling) and the Tomb of Benevolence (Huiling), four empresses' tombs including the West Tomb of Brightness (Zhao Xiling), the East Tomb of Filial Piety (Xiao Dongling), the East Tombs of Stability (Ding

图1 清东陵风水形势图，光绪元年（1875年）雷思起绘。日本东京大学东洋文化研究所藏。图中形象描绘了清东陵山水的来龙去脉。约2500平方公里的陵区外周树有红、白、青桩；昌瑞山东西迤逦，各陵寝荟萃山南，称为『前圈』，围有风水墙；山北为风水禁地『后龙』，具有生态环境保护性质

Fig.1 *Fengshui* topographic drawing of the Eastern Qing Tombs, drawn by LEI Siqi in the first year of the Guangxu Emperor's reign (1875), in the collection of the Institute for Advanced Studies on Asia, Tokyo University, Japan.

The drawing displays clearly the ins and outs of the overall mountain-river layout of the Eastern Qing Tombs. It covers a total area of about 2,500 square kilometres, with red, white and cyan-coloured wooden posts marking its different rings or edges. With Changrui Mountain lying along the east-west axis, various tombs are concentrated to the south of the mountain, forming the so-called 'front field' (*qianquan*). Surrounded by the *fengshui* walls, a prohibited zone is also formed on the northern side of the mountain, the so-called 'back dragon' (*houlong*), which has the quality of a natural reserve.

Dongling) in Puxiang Valley and Putuo Valley, and five imperial consorts' tombs affiliated with the five emperors' tombs. There were 157 people buried in the complex, including 5 emperors, 15 empresses, 115 consorts, 22 wives of princes, princesses and princes. Near the above-mentioned tombs, there were eleven barracks belonging to the Imperial Household Department (Neiwufu), and east to the *fengshui* wall nine barracks, which belonged to the Ministry of Rites (Libu), the Ministry of Engineering Works (Gongbu), and the Eight Banners (Baqi). These barracks host personnel in charge of rituals, engineering works and defence respectively. In addition, there are more than ten tombs for princes and princesses on both sides of the *fengshui* wall. Beside each emperor's tomb are either the tombs of the empresses or the consorts. The clear and coherent layout reflecting the strict system of ranks and privileges has become the distinctive feature of the Qing Tombs as opposed to that of the Ming Tombs (Fig.2).

The construction and management of the imperial tombs during the Qing Dynasty were known as the "Great Works of the Mountain Tomb", characterised by its large scale, high standards, and lengthy construction period. Continuing throughout the entire history of the Qing Dynasty, the construction of the Eastern Qing Tombs became the model with the greatest continuity. Tomb construction condensed the essence of architectural art, technology and project management system of the time. In conformity with the building artefacts, there are also extremely large numbers of project archival material which is well protected until today. The archetypal archives the Yangshi Lei Archives of the Qing Dynasty, have been listed in UNESCO's Memory of the World Programme in 2007. Known as the largest and richest existing source for ancient building design and presentation, there were some 4000 records related to the Tomb of Stability (Dingling), the East Tomb of Stability (Ding Dongling) and the Tomb of Benevolence (Huiling), which have documented in detail the process of site selection, survey, design, construction and management. These detailed records disclose in a vivid and subtle way the brilliant and elegant philosophy, procedure and methodology contained in Chinese traditional architectural design. First and foremost is the idea of taking the natural and cultural environment as a single holistic entity to organize the overall layout and exterior space of such large-scale building complexes. Additionally, the application of the module grid or *pingge* to survey, plan and design buildings amidst complicated topography, makes Chinese traditional architectural design unique and foremost even in today's world.

In 2012, to celebrate the 20-year anniversary of the UNESCO Memory of the World Programme, UNESCO selected 24 projects as archetypal cases and made them into plaques for exhibition at UNESCO headquarters in Paris. Among these 24 projects

三十四年（1908年）慈禧太后的菩陀峪定东陵重修告竣，历朝帝、后、妃陵寝的建设赓续两个半世纪，汇聚了孝陵、景陵、裕陵、定陵和惠陵共5座帝陵；昭西陵、孝东陵、普祥峪和菩陀峪定东陵共4座后陵；隶从景陵、裕陵、定陵和惠陵的5座妃园寝，共葬有5位皇帝，15位皇后，115位妃嫔，22位福晋、格格、皇子，计157人。各陵寝附近，还有11座内务府营房；风水墙东另有9座礼部、工部和八旗营房，住有祭祀、守护及维修人员。此外，风水墙两旁尚有十多座亲王、公主等的园寝。各帝陵左右，统绪分明地分布着后陵和妃园寝等，体制严整，显著区别于明代陵寝制度（图2）。

清代各陵寝建设规模大、规格高、工期长，号称山陵大工，贯穿了整个清史，衍为连续性最强的范例，凝聚了这时期建筑艺术、技术及工程管理体制的菁华。对应建筑实物，还传世海量工程档案。典型如2007年列为世界记忆遗产的清代样式雷图档，作为世界现知规模最大、内容最丰富的古代建筑设计图像资源，涉及定陵、定东陵和惠陵的就逾四千件，翔实记录了选址、勘测、设计、施工及管理等详情细节，细致生动地展现出中国传统建筑设计理念、程序和方法的高明与精湛，尤其是以自然和人文环境为本体的创作理念，大规模建筑组群的外部空间布局，对复杂地形运用『平格』模数网的勘测和规划设计方法，直至当代，仍然堪称独领风骚。

2012年联合国教科文组织巴黎总部庆祝『世界记忆工程』20周年，精选了24项最典型的世界记忆遗产，制成图版展出，唯以『中国清代样式雷图档』彰显古代建筑设计的高度智慧和杰出成就，凸显出其对人类文明史无与伦比的意义和价值。这也标志着，在世界建筑史上，曾长期存在有关中国古代建筑设计的『失语症』从此彻底终结（图3）。

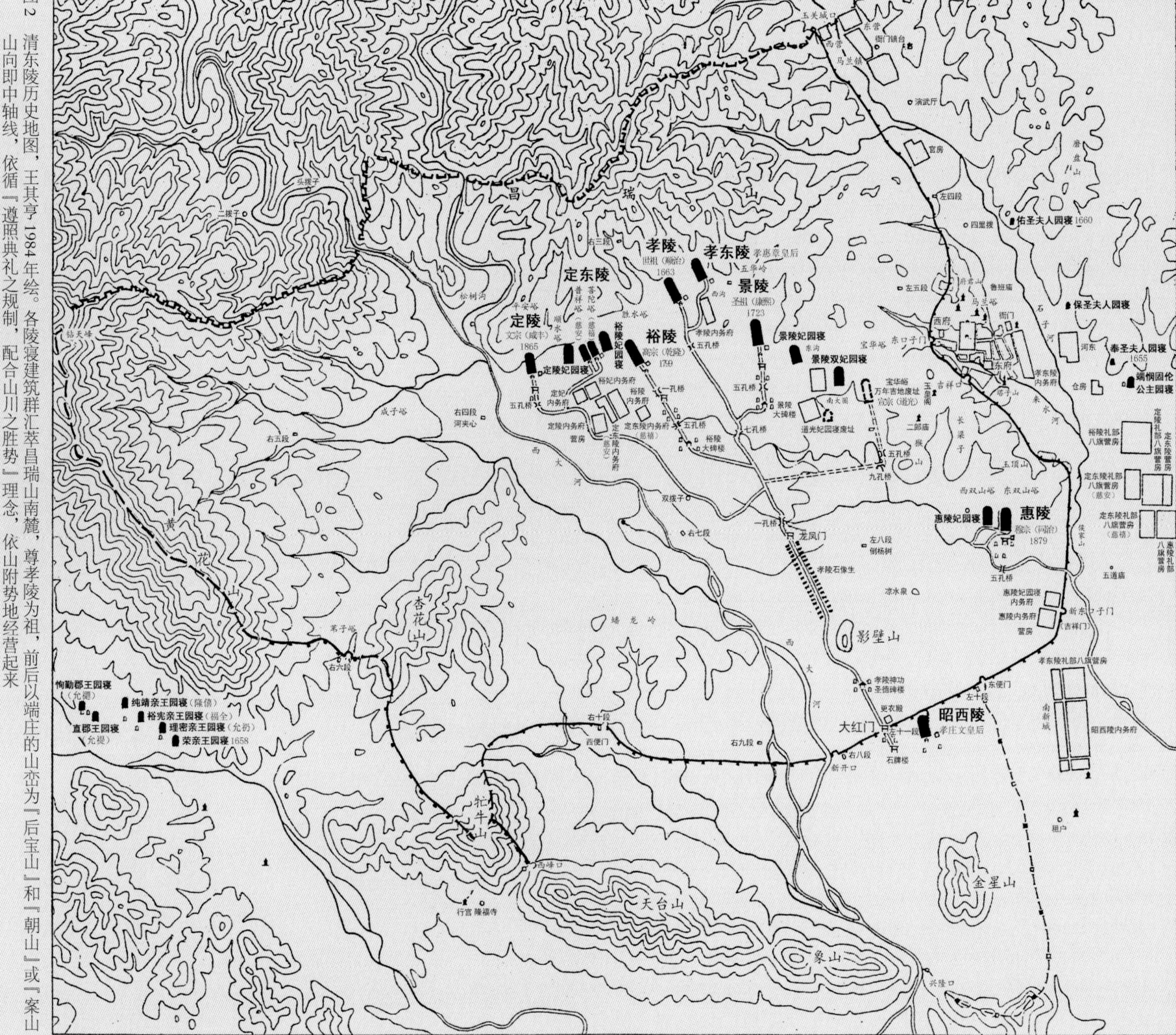

图 2 清东陵历史地图，王其亨 1984 年绘。各陵寝建筑群汇萃昌瑞山南麓，尊孝陵为祖，前后以端庄的山峦为『后宝山』和『朝山』或『案山』，山向即中轴线，依循『遵照典礼之规制，配合山川之胜势』理念，依山附势地经营起来

图 3 2012 年联合国教科文组织巴黎总部纪念『世界记忆工程』20 周年展版。该展共精选 24 个国家和地区最典型的世界记忆遗产项目，其中，辑录定陵、定东陵画样的『中国清代样式雷图档』，彰示了古代建筑设计智慧

Fig.2 Historic map of the Eastern Qing Tombs, drawn by WANG Qiheng in 1984.
All tomb building complexes are located on the southern side of Changrui Mountain. The Tomb of Filial Piety, regarded as the ancestor, is located in the centre, as determined by the dignified and elegant mountains: the back treasure mountain (*houbaoshan*) and the mountain that faces the tomb (*chaoshan*) or the mountain located opposite the tomb (*anshan*), which establish directions and formulate the central axis. According to the principle of "following the arrangements for ceremonial rituals and harmonising with the landscape", the whole tomb area was planned and constructed.

Fig.3 Plaque at the exhibition for the 20th-year anniversary of the Memory of the World Programme at UNESCO headquarters in Paris, 2012.
24 projects were selected as the archetypal cases representing their countries and areas respectively. Among these, the archival drawings of the Tomb of Stability (Dingling) and the East Tomb of Stability (Ding Dongling) from the Yangshi Lei Archives of the Qing Dynasty demonstrate the wisdom of traditional Chinese architectural design.

selected for the exhibition, the Yangshi Lei Archive of the Qing Dynasty was the only one to showcase the genius and extraordinary achievements of traditional Chinese architectural design, with its incomparable meaning and value dedicated to the history of human civilization. The exhibition also marked the end of the lack of understanding of traditional Chinese architectural design in the history of world architecture (Fig. 3).

I. The Foundation and Evolution for the Eastern Qing Tombs

In the Shunzhi Emperor's eighth regnal year (1651), the Emperor officially assumed power. He organised a hunting trip to Zunhua. During the trip the Sunzhi Emperor personally investigated Fengtai Mountain which, according to a rumour from late Ming-dynasty literati, had earlier been chosen as the tomb site for the Chongzhen Emperor of the Ming in 1640. As a result of the investigation, the southern side of the main peak of Fengtai Mountain was selected as the tomb site. The Shunzhi Emperor was still young, and there were still many important things to be done in the early years of the dynasty, including the frequent use of military force and unresolved problems in securing people's livelihood, both of which led to financial difficulties. Hence, construction on the tomb site was not really started except for the completion of the tombs for: Madam Fengsheng (Sunzhi's first wet nurse) which began in the 12th year of Shunzhi's reign; Prince Rong and Consort Dao in 1658; and Madam Yousheng (Shunzi's wet nurse) in 1660. These early initiatives could be considered as constituting the Shunzhi Emperor's putting his overall planning for the Eastern Qing Tombs into effect.

When the Kangxi Emperor ascended the throne, he began to build the Tomb of Filial Piety (Xiaoling) conforming with the regulations for imperial tombs during the Ming Dynasty for legitimacy between the end of 1661 and 1672. He also changed the name of Fengtai Mountain (meaning Phoenix Terrace) into Changrui Mountain (meaning Prosperity and Auspiciousness). The Kangxi Emperor continued building: the Tomb of Admiration (Jingling) for himself; the Imperial Consort Tombs affiliated with the Tomb of Admiration (Jingling Feiyuanqin); the temporary Offering Hall for the Empress Dowager Xiaozhuang and the East Tomb of Filial Piety for the Empress Dowager Xiaohui. These projects established the basic structure of the Eastern Qing Tombs. While the imperial tomb system of the Ming Dynasty was continued, reforms were also made, thus initiating the Qing imperial tomb systems for emperors, empresses and consorts, for which order and hierarchy dominate.

Later, the Yongzheng Emperor began to build the Tomb of Peace (Tailing) and its affiliated tombs for his consorts at the foot of Yongning Moutain in Yi County southwest of Beijing. In

一、清东陵建置沿革

顺治八年（1651年）顺治皇帝亲政，借校猎遵化，躬亲考察凤台山，在其主峰南选定了寿陵基址。此地明末文人传为崇祯十三年（1640年）曾选定为崇祯皇帝陵地，因顺治其时尚年轻，大清基业甫定，财用困难，未遑鸠工兴建。仅有顺治十二年始建的奉圣夫人园寝，顺治十五年落成的荣亲王园寝和悼妃园寝，顺治十七年竣工的佑圣夫人园寝，体现为他对清东陵总体规划的实施。

康熙皇帝践祚后，从顺治十八年末到康熙十一年（1672年）鼎建孝陵（图4、图5），规制参仿明陵以示正统，凤台山更名昌瑞山；接着他又在孝陵东南五华岭经营了景陵、景陵妃园寝（图6），以及孝庄皇后的暂安奉殿和孝惠皇后的孝东陵（图7），奠定了清东陵的基本架构，继承并改革明代陵寝制度，开创了有清一代统绪关系和等级秩序严整分明的帝陵、后陵和妃园寝制度。

嗣后，雍正皇帝在北京西南易县永宁山下兴建泰陵及妃园寝。乾隆皇帝为免泰陵疏离清东陵，策划了后嗣子、孙依次在昌瑞山和永宁山附近建陵的『东西陵昭穆制度』，率先选地孝陵西侧圣水峪经营其裕陵；该举措影响后世，最终形成了分别以孝陵、泰陵为主陵的清东陵和清西陵。其间，雍正三年（1725年）孝庄皇后的暂安奉殿改建为昭西陵，特建神道碑亭申明其遥隶盛京（沈阳）太宗皇太极的昭陵（图8）。雍正五年景陵添建圣德神功碑亭及四隅华表，亭内特立双碑以彰康熙帝『功德隆盛』。乾隆四年（1739年），乾隆皇帝感戴皇祖的贵妃佟佳氏与和妃瓜尔佳氏养育恩情，尊为太皇贵妃并『另建园寝皆有加于皇贵妃定制』，建起了各带宝城和方城明楼的景陵皇贵妃园寝（图9）。乾隆十三年，值裕陵营建（图10），景陵又添设了望柱、石像生和牌楼门。乾隆二十五年，落成已八年的裕陵妃园寝，又参仿景陵皇贵妃园寝增修东西配殿，为纯惠皇贵妃苏佳氏地宫添建宝城和方城明楼。

图 4　孝陵图。中国第一历史档案馆藏。康熙九年（1670 年）闰二月工部绘制。图中满、汉文黄签说明，石牌坊、大碑楼四隅擎天柱（华表）尚未建造；满文白签显示，该图曾用来安排当年九月康熙皇帝侍奉孝庄太皇太后、孝惠皇太后隆重举行他亲政后的首次上陵礼。

1. 全图。2. 上段：陵宫组群。3. 中段：望柱、石像生及龙凤门。4. 下段：石牌坊、下马牌、大红门、更衣殿、大碑楼及华表

Fig.4 *Drawing of the Tomb of Filial Piety* (*Xiaoling tu*) (Fig.4–1), drawn by the Ministry of Engineering Works (Gongbu) in the intercalary second month of the ninth year of the Kangxi Emperor's reign (1670), in the collection of the First Historical Archives of China.

According to the Manchurian-and-Chinese yellow tags on the drawing, the marble memorial gateway and the ornamental columns in the four corners of the great stela tower (The Pavilion for the Stela of Divine Merit and Sage Virtue) had not yet been built. And according to the Manchurian-written white tag, this drawing was used by Kangxi when he planned to assist his grandmother, Grand Empress Dowager Zhaosheng (posthumuous name Xiaozhuangwen) and his mother, Empress Dowager Renxian (posthumous name Xiaohuizhuang) in holding the first grand ceremonial visit to the tomb after he officially assumed power in the ninth month of the same year.

1. Overall site plan.
2. Upper section: the tomb palace complex.
3. Middle section: the ornamental columns (*wangzhu*), the stone statues (*shixiangsheng*) and the dragon and phoenix gate (*longfengmen*).
4. Lower section: the marble memorial gateway (*shipaifang*), the stela marking the place for dismounting from a horse (*xiamapai*), the great red gate building (*dahongmen*), the hall for changing clothes (*gengyidian*), the great stelae tower (*dabeilou*) and the ceremonial columns (*huabiao*).

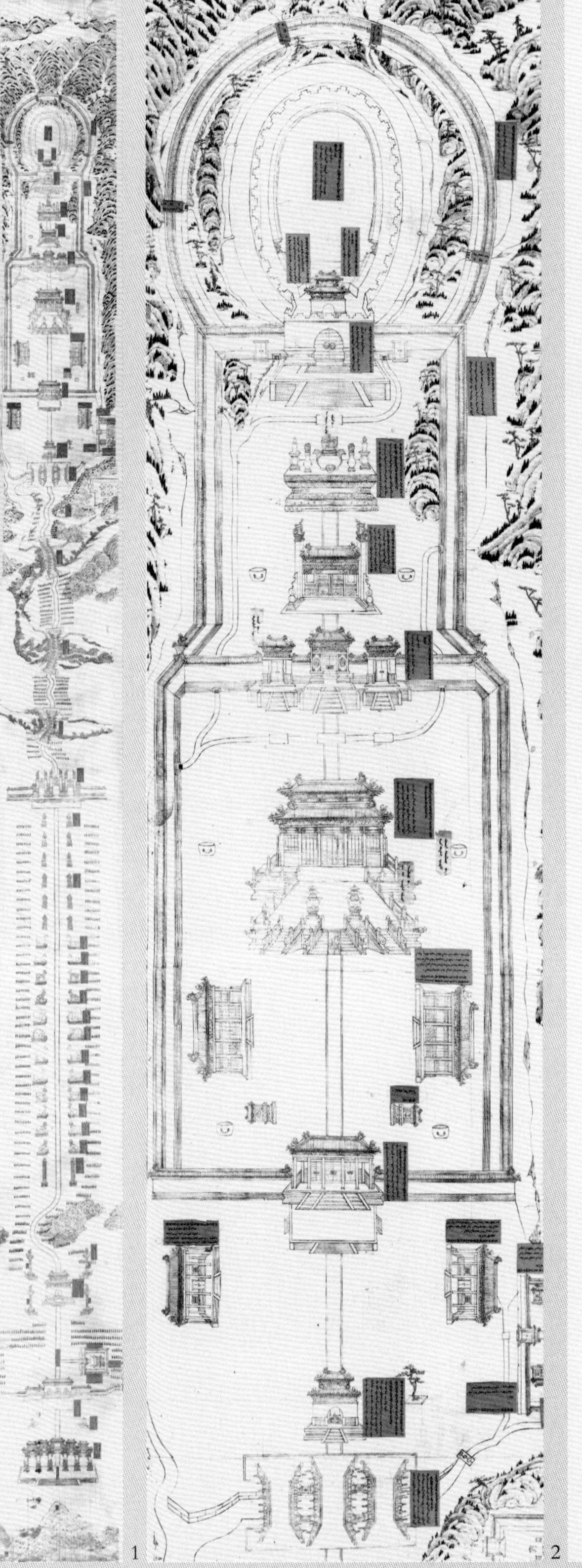

2

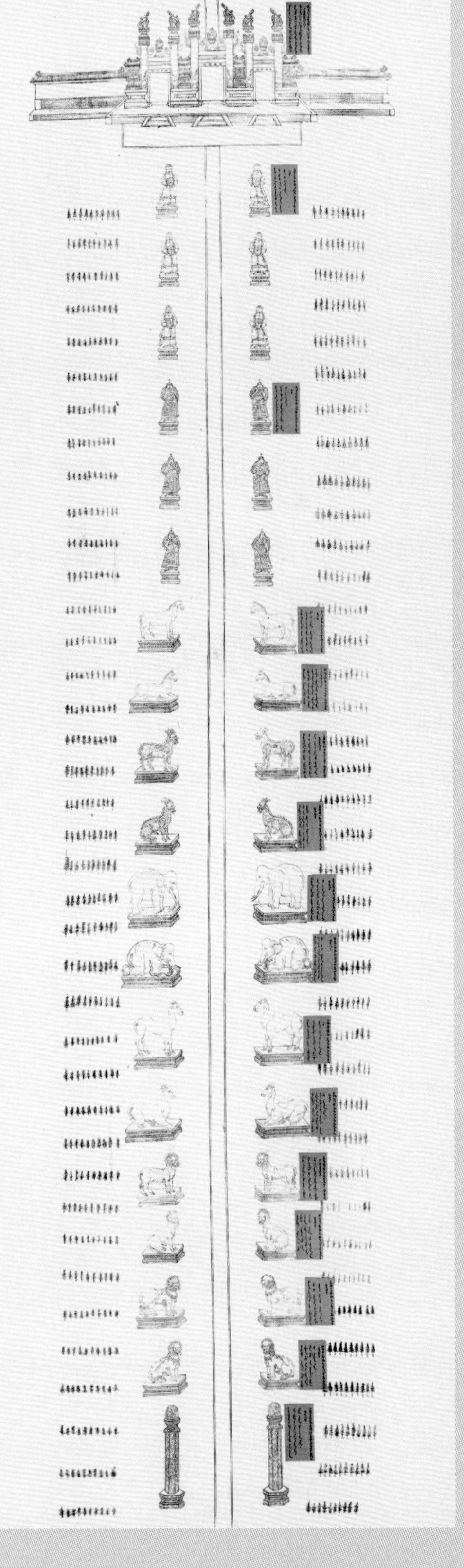

3

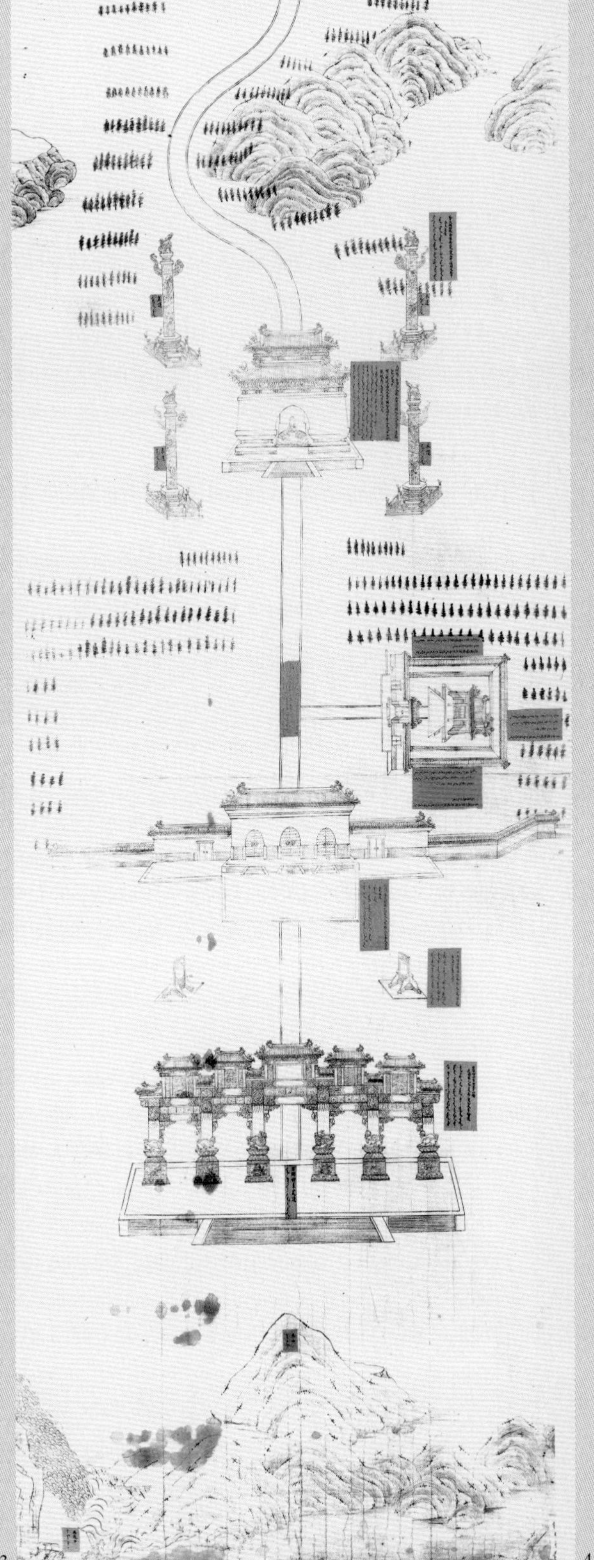

4

图6 景陵图。中国第一历史档案馆藏乾隆朝内务府舆图。康熙十五年（1676年）始建清东陵第二座帝陵，陵宫效仿孝陵，东南隅建景陵妃园寝。雍正、乾隆朝，景陵添建圣德神功碑亭、华表、望柱、石像生及牌楼门，皆明十三陵长陵外各帝陵所无。东面，乾隆又增建景陵皇贵妃园寝

图5 南怀仁 Ferdinand Verbiest《熙朝定案》插图。巴黎法国国家图书馆藏。康熙八年冬十月十二日，南怀仁用西法滑车牵引16轮大练车，装载数十吨重的孝陵大碑楼龟趺安全通过新修卢沟桥。康熙九年、十年冬，又用此法运过大碑楼四隅擎天柱（华表）和石牌坊等石构件，陆续在孝陵安位，保障了翌年孝陵工程完竣

Fig.5 Illustration in the *Final verdicts of the Gracious Reign* (*Xichao ding'an*), compiled by Ferdinand Verbiest (NAN Huairen, 1623-1688).
From the collection of the French National Library in Paris. In the winter of the twelfth day in the tenth month of the eighth year of the Kangxi Emperor's reign (1669), introduced by Ferdinand Verbiest, sixteen carriages based on western technology were dragged by the driving pulley to go through the newly built Lugou Bridge safely, bearing a turtle-shaped stela base (*guifu*) for the great stela tower (*dabeilou*) at the Tomb of Filial Piety (Xiaoling). In the winters of the ninth and tenth years of the Kangxi Emperor's reign, the same method was used to transport stone structures like the four ceremonial columns at the corners of the great stela tower and the marble memorial gateway, all of which were installed one after the other at the Tomb of Piety, thus ensuring completion of the Tomb of Filial Piety in the following year.

Fig.6 *Drawing of the Tomb of Admiration* (*Jingling tu*). Map from the Imperial Household Department (Neiwufu) during the Qianlong period, in the collection of the First Historical Archives of China. The Tomb of Admiration (Jingling) is the second imperial tomb built in the Eastern Qing Tombs. Its construction started in the 15th year of the Kangxi Emperor's reign (1676), following the example of the Tomb of Filial Piety (Xiaoling), with its imperial consort tombs in the southeast corner. During the Yongzheng and Qianlong periods, the Pavilion for the Stela of Sagely Virtue and Divine Merit (Shengde Shengong Beiting), the ceremonial columns (*huabiao*), the ornamental columns (*wangzhu*), the stone statues (*shixiangsheng*) and the memorial gateway (*pailoumen*) were added to the Tomb of Admiration (Jingling). Such additions had never been built for an imperial tomb except for the Long Tomb (Changling) located in the Thirteen Ming Tombs. To its east, the Qianlong Emperor later added the Imperial Noble Consorts' Tombs affiliated with the Tomb of Admiration (Jingling Huangguifei Yuanqin).

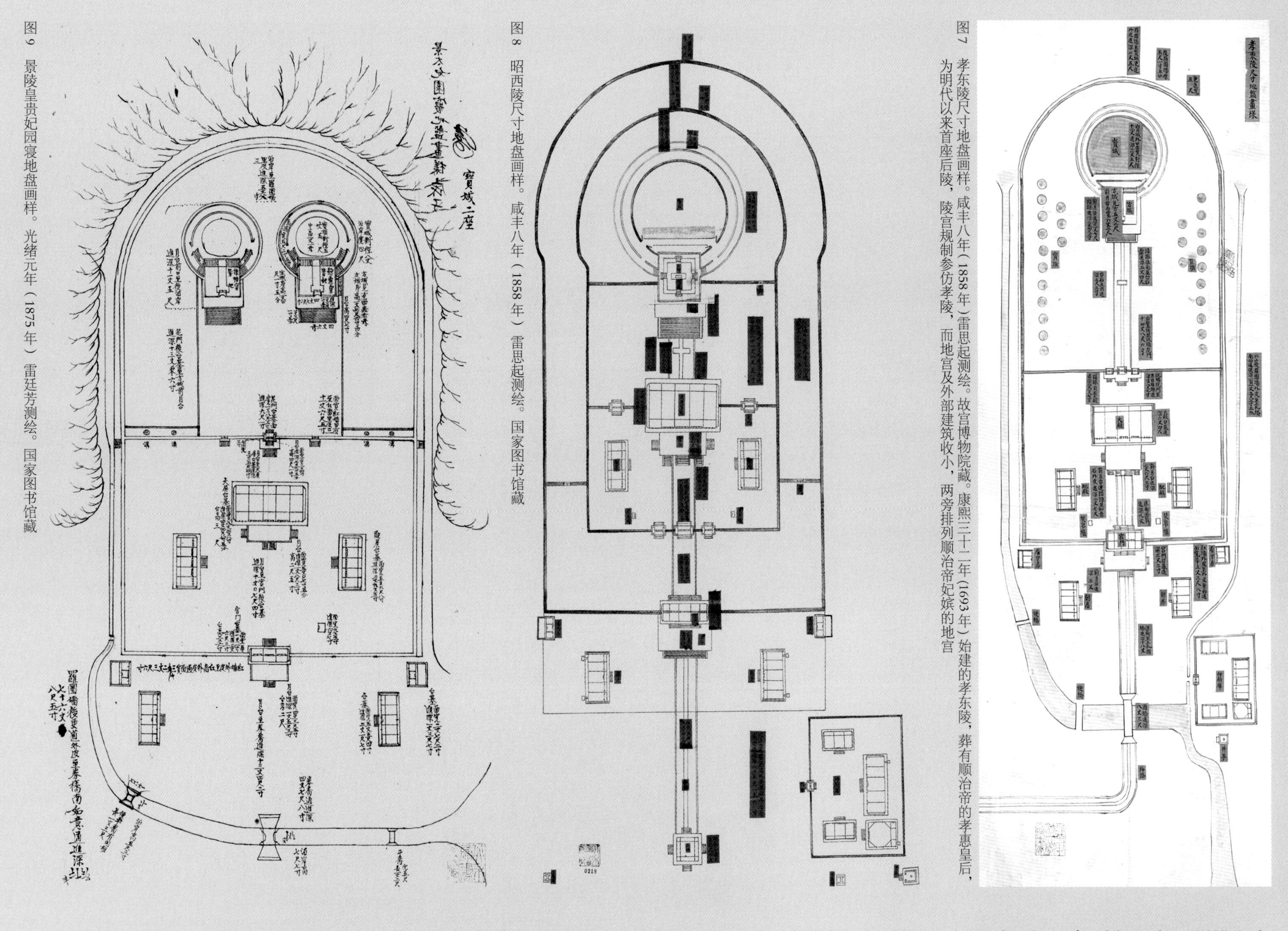

图7 孝东陵尺寸地盘画样。咸丰八年（1858年）雷思起测绘。故宫博物院藏。康熙三十二年（1693年）始建的孝东陵，葬有顺治帝的孝惠皇后，为明代以来首座后陵，陵宫规制参仿孝陵，而地宫及外部建筑收小，两旁排列顺治帝妃嫔的地宫

图8 昭西陵尺寸地盘画样。咸丰八年（1858年）雷思起测绘。国家图书馆藏

图9 景陵皇贵妃园寝地盘画样。光绪元年（1875年）雷廷芳测绘。国家图书馆藏

Fig.7 Drawing of the site plan showing the dimensions of the East Tomb of Filial Piety (Xiao Dongling). The site plan was surveyed and drawn by LEI Siqi in the Xianfeng Emperor's eighth regnal year (1858). In the collection of the Palace Museum.
In the Kangxi Emperor's 32rd regnal year, the construction started for the East Tomb of Filial Piety, in which Empress Xiaohui of the Shunzhi Emperor was buried. It was the first tomb for an empress built since the Ming Dynasty. It followed the example of the Tomb of Filial Piety (Xiaoling) by scaling down the size of both the underground palace as well as all the buildings above the ground. The tombs of other imperial consorts were built on both sides.

Fig.8 Drawing of the site plan showing the dimensions of the West Tomb of Brightness (Zhao Xiling). The site plan was surveyed and drawn by LEI Siqi in the Xianfeng Emperor's eighth regnal year (1858). In the collection of the National Library of China.

Fig.9 Drawing of the site plan of the Imperial Noble Consorts' Tombs affiliated with the Tomb of Admiration (Jingling Huangguifei Yuanqin), surveyed and drawn by LEI Tingfang in the Guangxu Emperor's 1st regnal year (1875). In the collection of the National Library of China.

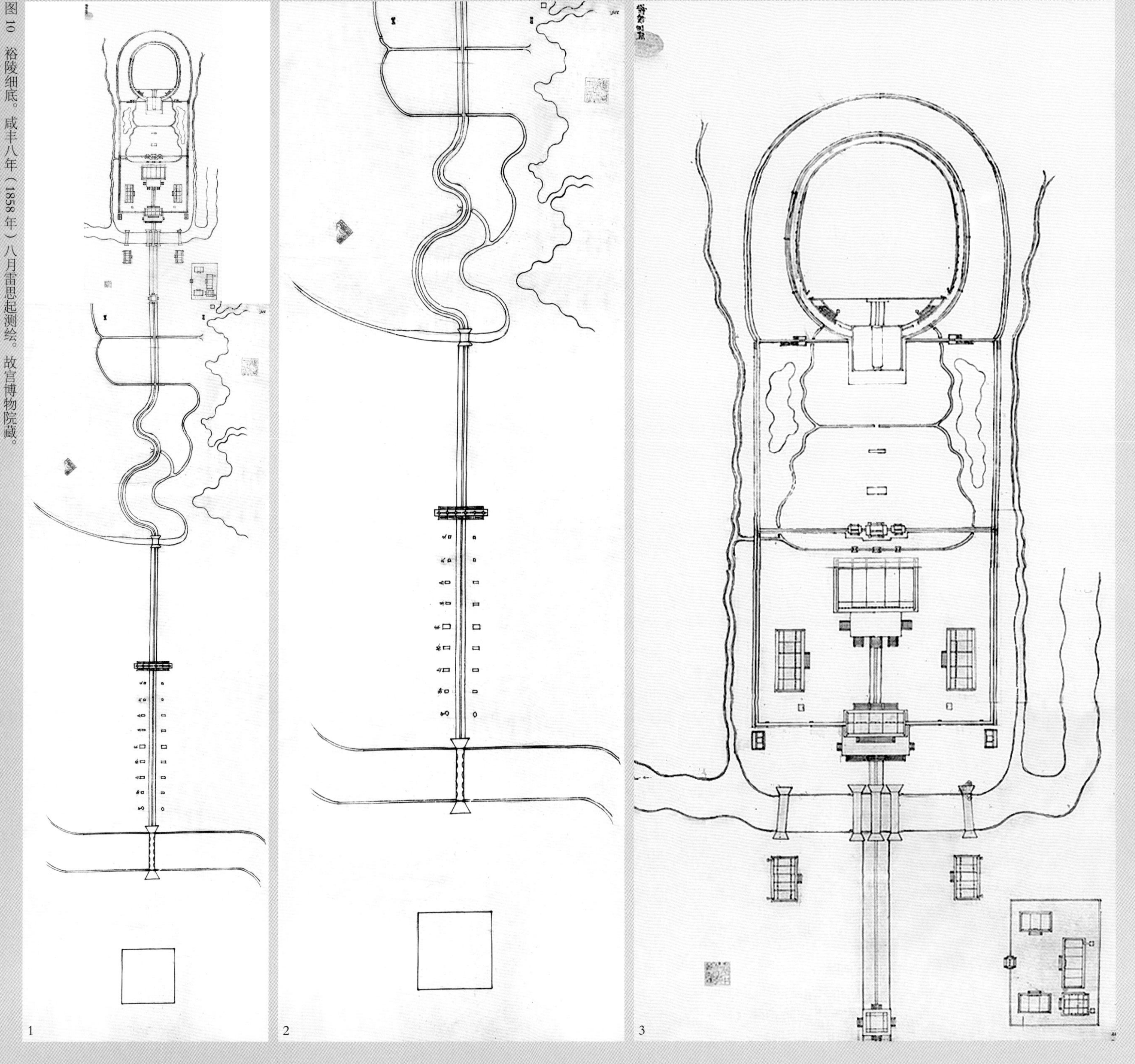

图 10 裕陵细底。咸丰八年（1858 年）八月雷思起测绘。故宫博物院藏。
1. 全图。2. 帖页：大碑楼、望柱、石像生及牌楼门的陵前引导空间。3. 本图：陵宫建筑群

Fig.10 Detailed plan of the Tomb of Prosperity (Yuling), surveyed and drawn by LEI Siqi in the Xianfeng Emperor's eighth regnal year (1858), in the collection of the Palace Museum.
1. Overall site plan.
2. Pasted page: the guiding space in front of the tomb, including the great stela tower (*dabeiyou*), the ornamental columns (*wangzhu*), the stone statues (*shixiangsheng*) and the memorial gateway (*pailoumen*).
3. Base drawing: the building complex of the Tomb of Prosperity.

嘉庆四年（1799年），在裕陵石像生群南，嗣皇帝嘉庆又依循成宪，为皇考添建了圣德神功碑亭和四隅华表。

到晚清，咸丰九年（1859年）咸丰皇帝参仿景陵及其妃园寝规制，择地裕陵西隅平安峪和顺水峪，兴工建造了定陵及其妃园寝，还刨用了父皇道光皇帝宝华峪吉地废址的大量旧料（图11）。同治十二年（1873年），他的遗孀慈安和慈禧太后，在定陵和裕陵妃园寝之间的普祥峪和菩陀峪，经营了规制划一、平行并列的两座后陵（图12）。不久，光绪元年（1875年），同治皇帝的惠陵和妃园寝，仿照定陵及其妃园寝规制，择地清东陵东南隅的东、西双山峪兴建，成为清东陵内终结性的陵寝（图13）。

清东陵的持续经营，形成了体制性的选址和荐名制度。孝陵以后，各帝陵多在皇帝登基后选址建造，称为『万年吉地』并冠以地名。在皇帝生前建成，先葬入已故皇后，暂以其谥号称名。临皇帝入葬，才由内阁会拟陵名，由嗣皇帝钦定；皇帝驾崩后经营的惠陵，则在兴工前已定陵名。皇帝奉安前去世的皇后，全都合葬帝陵，少数受宠的已故皇贵妃，也祔葬其中。

皇帝入葬后，在世皇后以卑不动尊为由，另在帝陵左右建陵，也称为『万年吉地』并冠以地名，如普祥峪、菩陀峪万年吉地；皇后逝后，才按帝陵左右方位冠以帝陵名，如昭西陵、孝东陵、普祥峪定东陵、菩陀峪定东陵。

隶从帝陵的妃园寝，向例选地帝陵左右，与帝陵同期兴工，其中按嫔妃不同等级分别建造地宫和宝顶，薨后陆续对位入葬。

清东陵各陵寝建置和安葬帝、后、妃，略可归纳为表1～表3。

order not to keep the Tomb of Peace apart from the Eastern Qing Tombs, the Qianlong Emperor planned and promoted a new *zhaomu* system①, instructing his successors to alternate between east and west when building their tombs near Changrui Mountain and Yongning Mountain respectively. As an example, he had chosen the Shengshui Valley west of the Tomb of Filial Piety (Xiaoling) to build his own tomb, the Tomb of Prosperity (Yuling). As a result, the Eastern Qing Tombs with the Tomb of Filial Piety (Xiaoling) as its main tomb and the Western Qing Tombs with the Tomb of Peace (Tailing) as its main tomb were finally defined in terms of shape and form. During this period, the temporary Offering Hall for the Empress Xiaozhuangwen was reconstructed as the West Tomb of Brightness (Zhao Xiling) in the third year of the Yongzheng Emperor's reign (1725), with a special construction of the pavilion for the stela on the spirit way, which, as declared on the stela, consonant with the Tomb of Brightness (Zhaoling) for the Taizong Emperor, Huangtaiji, in Shengjing (today's Shenyang) (Fig. 8). In the fifth year of Yongzheng's reign (1727), the Pavilion for the Stela of Sagely Virtue and Divine Merit (Shengde Shenggong Beiting), together with four ceremonial columns in each of its four corners, was added to the Tomb of Admiration (Jingling). In the pavilion, two stelae were erected to manifest the Kangxi Emperor's great achievements and virtues. In the fourth year of the Qianlong Emperor's reign (1739), in order to show his gratitude for the love and care received from Dowager Imperial Noble Consort, of the Tunggiya clan, and from Consort He, of the Guwalgiya clan, both consorts of his grandfather, Qianlong awarded both the title of 'Grand Imperial Noble Consort'. Their tombs were therefore built near the Tomb of Admiration (Jingling), each with an encircled realm of treasure (*baocheng*) and a square walled terrace with a memorial tower (*fangcheng minglou*), thus upgrading the existing commemorative system for imperial consorts (Fig. 9). In the thirteenth year of Qianlong's reign (1748), when the Tomb of Prosperity (Yuling) was under construction (Fig. 10), ceremonial columns (*huabiao*), stone statues (*shixiangsheng*) and a marble memorial gateway (*paifangmen*) were added to the Tomb of Admiration (Yuling). In the 25th year of Qianlong's reign (1760), following the example of the Imperial Consort Tombs affiliated with the Tomb of Admiration (Jingling Feiyuanqin), side halls were newly built and an encircled realm of treasure with a tumulus, and a square walled terrace with a memorial tower were added to the Tomb of Imperial Noble Consort Chunhui, of the Su clan, all located in the Imperial Consort Tombs affiliated with the Tomb of Prosperity (Yuling Feiyuanqin), eight years after its initial completion.

① The *zhaomu* system originated as a system for arranging spirit tablet in an ancestral temple. Dating from the early Western Zhou dynasty (11th century–770 BCE), it is usually described as an arrangement in which the main ancestor is placed in the middle, while younger ancestors were placed alternatively to the left and right. [translator's note]

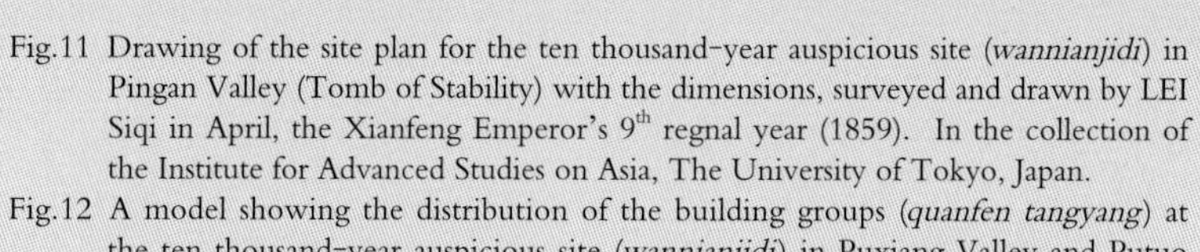

Fig.11 Drawing of the site plan for the ten thousand-year auspicious site (*wannianjidi*) in Pingan Valley (Tomb of Stability) with the dimensions, surveyed and drawn by LEI Siqi in April, the Xianfeng Emperor's 9th regnal year (1859). In the collection of the Institute for Advanced Studies on Asia, The University of Tokyo, Japan.

Fig.12 A model showing the distribution of the building groups (*quanfen tangyang*) at the ten thousand-year auspicious site (*wannianjidi*) in Puxiang Valley and Putuo Valley for the East Tomb of Stability (Ding Dongling), constructed by LEI Siqi in the Tongzhi Emperor's 12th regnal year (1873). In the collection of the Palace Museum.

Fig.13 The final draft of the overall layout of the Tomb of Benevolence (Huiling) area (Bird's-eye view after the completion of the Tomb of Benevolence [Huiling] and the Imperial Consorts' Tombs [Feiyuanqin], together with the barracks of the Imperial Household Department [Neiwufu], the Ministries of Rites and Engineering Works [Ligongbu] and the Eight Banners [Baqi]), drawn by LEI Tingfang in the Guangxu Emperor's fourth regnal year (1878). In the collection of the National Library of China.

图 12 普祥峪、菩陀峪万年吉地（定东陵）全分烫样。同治十二年（1873 年）五月雷思起制。故宫博物院藏

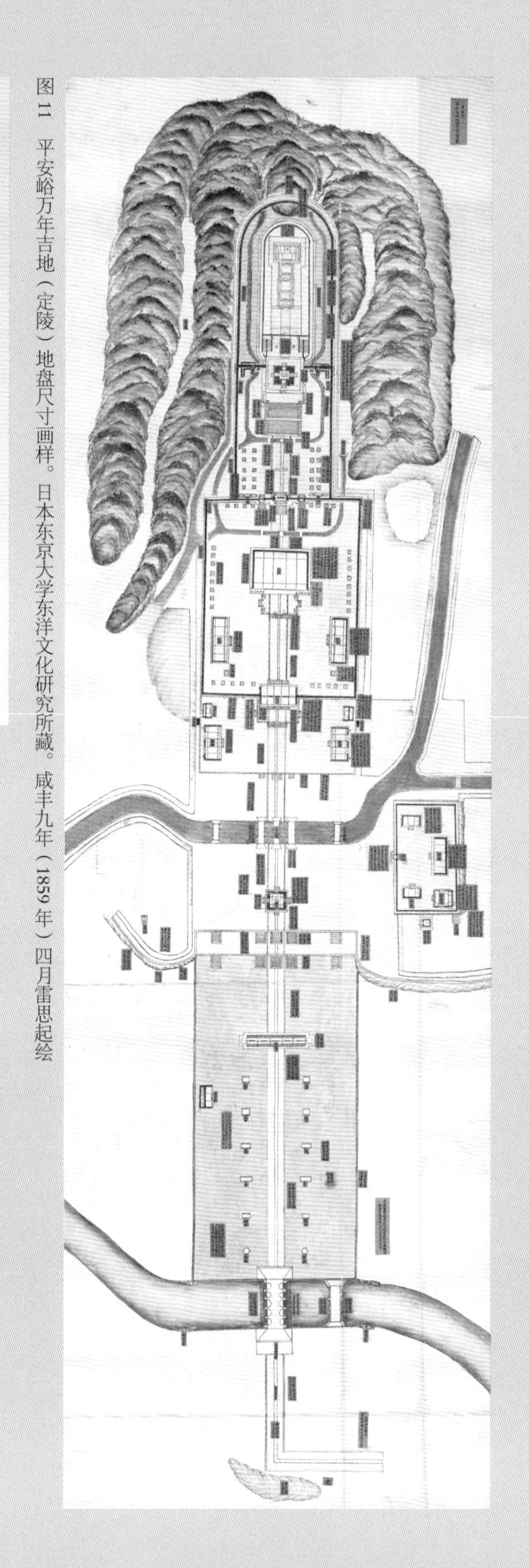

图 11 平安峪万年吉地（定陵）地盘尺寸画样。日本东京大学东洋文化研究所藏。咸丰九年（1859 年）四月雷思起绘

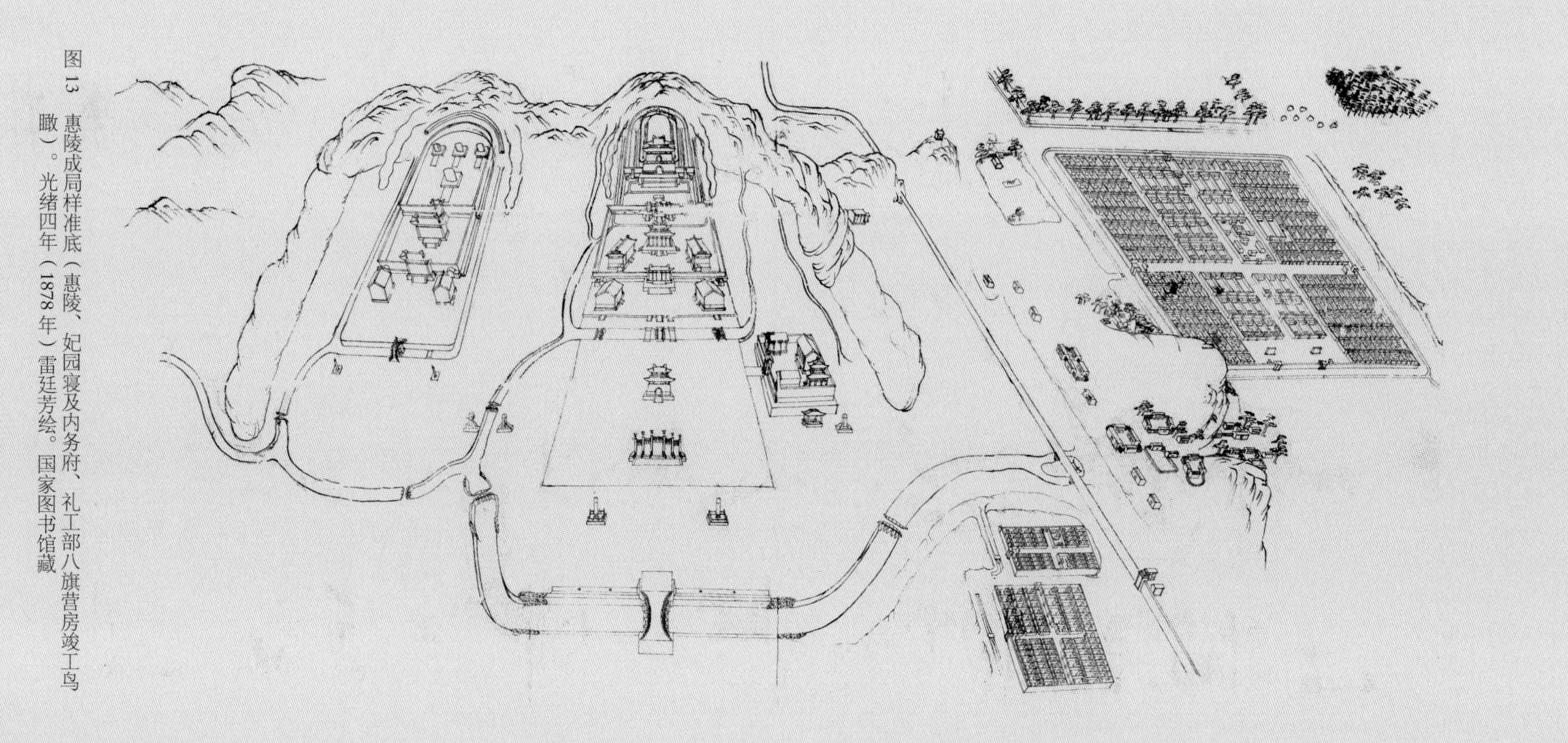

图 13 惠陵成局样准底（惠陵、妃园寝及内务府、礼工部八旗营房竣工鸟瞰）。光绪四年（1878 年）雷廷芳绘。国家图书馆藏

In the fourth year of the Jiaqing Emperor's reign, the successor emperor Jiaqing followed the example and system of the Tomb of Admiration (Jingling), in the area south of the stone statues of the Tomb of Prosperity (Yuling), added for his father the Pavilion for the Stela of Sagely Virtue and Divine Merit and the ceremonial columns at the four corners.

During the late Qing period, in the Xianfeng Emperor's ninth regnal year (1859), the Xianfeng Emperor followed the example and system of the Tomb of Admiration (Jingling) and its affiliated imperial consorts' tombs, and chose the sites of Ping'an Valley and Shunshui Valley west of the Tomb of Prosperity (Yuling) to built the Tomb of Stability (Dingling) for himself and his imperial consorts. In addition, he also reused a large number of old material from the abandoned site in Baohua Valley, once considered an auspicious place for the tomb of his father, the Daoguang Emperor (Fig. 11). In the Tongzhi Emperor's 12th rengal year (1873), the Xianfeng Emperor's two widows, Empress Dowager Ci'an and Empress Dowager Cixi decided to build their empress tombs in parallel and identically, with the sites in Puxiang Valley and Putuo Valley respectively, between the Tomb of Stability (Dingling) and the Imperial Consort Tombs affiliated to the Tomb of Prosperity (Yuling Feiyuanqin) (Fig. 12). Soon after, following the rules and regulations of the Tomb of Stability (Dingling) and its affiliated consorts' tombs (feiyuanqin), the Tomb of Benevolence (Huiling) for the Tongzhi Emperor and his imperial consort tombs were chosen to be built in east and west Shuangshan Valley, located in the southeast corner of the Eastern Qing Tombs, which were to be the last tombs built in this area (Fig. 13).

Following the continuous construction and management of the Eastern Qing Tombs, systematic arrangements were formulated for site selection and naming. After the completion of the Tomb of Filial Piety, most tomb sites for emperors were selected after their enthronement ceremony. The sites were often labelled by their local geographic name plus the unique term "ten thousand years auspicious site" (*wannian jidi*). Each tomb would then be constructed during the corresponding emperor's lifetime, and the deceased empress would be buried first, named after her posthumous title. When the emperor was to be buried, the formal name of the tomb would be discussed and proposed by the Cabinet, and then finalized and approved by the next emperor. The only exception is the Tomb of Benevolence (Huiling), which was constructed after the death of the Tongzhi emperor, and whose name had been decided before its construction. Empresses who died before the emperor would be buried in the same tomb with him, and a few deceased yet beloved imperial consorts would also be allowed to be buried in the same tomb with the emperor.

清东陵的帝陵建置　表 1

陵名	皇帝	年号	庙号	谥号	世系	在位时间	祔葬	始建时间	地点
孝陵	福临	顺治	世祖	章皇帝	皇太极九子	1644—1661 年	孝康皇后佟佳氏、孝献皇后董鄂氏	顺治十八年（1661 年）八月	清东陵昌瑞山主峰南麓
景陵	玄烨	康熙	圣祖	仁皇帝	福临三子	1661—1722 年	孝诚皇后赫舍里氏、孝昭皇后钮祜禄氏、孝懿皇后佟佳氏、孝恭皇后乌雅氏，敬敏皇贵妃章佳氏	康熙十五年（1676 年）二月	清东陵，孝陵东南隅五华岭
裕陵	弘历	乾隆	高宗	纯皇帝	胤禛四子	1735—1795 年	孝贤皇后富察氏、孝仪皇后魏氏，慧贤皇贵妃高氏、哲悯皇贵妃富察氏及淑嘉皇贵妃金氏	乾隆八年（1743 年）二月	清东陵，孝陵西侧胜水峪
定陵	奕詝	咸丰	文宗	显皇帝	旻宁四子	1850—1861 年	孝德皇后萨克达氏	咸丰九年（1859 年）四月	清东陵，裕陵以西平安峪
惠陵	载淳	同治	穆宗	毅皇帝	奕詝长子	1861—1874 年	孝哲皇后阿鲁特氏	光绪元年（1873 年）三月	清东陵东南双山峪

清东陵的后陵建置　表 2

陵名	安葬皇后	谥号	隶从	始建时间	地点	说明
昭西陵	博尔济吉特氏	孝庄文皇后	昭陵	康熙二十七年（1688 年）	清东陵东南大红门外	雍正三年（1725 年）添建地宫、宝城和方城明楼
孝东陵	博尔济吉特氏	孝惠章皇后	孝陵	康熙三十二年（1693 年）	清东陵孝陵东侧	方城明楼两翼还排列有顺治的 28 位妃嫔的宝顶
普祥峪定东陵	钮祜禄氏	孝贞显皇后	定陵	同治十二年（1873 年）	清东陵定陵东面普祥峪	
菩陀峪定东陵	叶赫那拉氏	孝钦显皇后	定陵	同治十二年（1873 年）	普祥峪东面菩陀峪	

清东陵的妃园寝建置　表 3

园寝名	安葬妃嫔	隶从	始建时间	地点
景陵妃园寝	温僖贵妃钮祜禄氏及其他妃、嫔、贵人、答应和常在共四十人，以及一位阿哥	景陵	康熙二十年　（1681 年）	景陵东侧
景陵双妃园寝	悫惠皇贵妃佟佳氏，惇怡皇贵妃瓜尔佳氏	景陵	乾隆四年　（1739 年）	景陵妃园寝东侧
裕陵妃园寝	废后乌喇那拉氏、纯惠皇贵妃苏佳氏及其他妃嫔共三十六人	裕陵	乾隆十二年　（1747 年）	裕陵西侧
定陵妃园寝	庄静皇贵妃他他拉氏等共十五位妃嫔	定陵	咸丰九年　（1859 年）	定陵东侧顺水峪
惠陵妃园寝	淑慎皇贵妃富察氏、恭肃皇贵妃阿鲁特氏、献哲皇贵妃赫舍里氏、荣惠皇贵妃西林觉罗氏	惠陵	光绪元年　（1875 年）	惠陵西隅西双山峪

After the burial of the emperor, the living empress should find a different site for her tomb next to the emperor's tomb. Since the empress is of a lower rank, her burial ceremony should not disturb the buried emperor. The naming of the tomb site for the empress follows the same arrangement, the geographic name plus the ten thousand-year auspicious site, for example, the Puxiang Valley and Putuo Valley ten thousand-year auspicious site. Only after the death of the empress was the tomb site to be formally named according to its orientation in relation to the emperor's tomb, for example, the West Tomb of Brightness (Zhao Xiling), the East Tomb of Filial Piety (Xiao Dongling), the East Tomb of Stability in Puxiang Valley (Puxiangyu Ding Dongling), East Tomb of Stability in Putuo Valley (Putuoyu Ding Dongling).

As for other imperial consorts, their tomb sites were often near the corresponding emperor's tomb site, which was built during the same period of time as the emperor's tomb. Their underground palaces and tumuli were built according to their place in the hierarchy and each of them was buried in the corresponding location once they died.

The overall information for the construction and arrangement of each tomb located in the Eastern Qing Tombs can be briefly summarized in the following Tab.1-Tab.3:

Tab.1 The inception of each emperor's tomb in the Eastern Qing Tombs

Tomb name	Emperor's name	Reign title	Temple title	Posthumous title	Bloodline	Duration period	Accompanied burials	Starting time for construction	Location
Tomb of Filial Piety (Xiaoling)	Fulin	Shunzhi	Shizu	Emperor Zhang	The ninth son of Huangtaiji	1644-1661	Empress Xiaokang, of the Tunggiya clan; Empress Xiaoxian, of the Donggo clan	The eighth month in the Shunzhi Emperor's 18th regnal year (1661)	Southern side of the main peak of Changrui Mountain
Tomb of Admiration (Jingling)	Xuanye	Kangxi	Shengzu	Emperor Ren	The third son of Fulin	1661-1722	Empress Xiaocheng, of the Heseri clan; Empress Xiaozhao, of the Niohuru clan; Empress Xiaoyi, of the Tunggiya clan; Empress Xiaogong, of the Uya clan; Imperial Noble Consort Jingmin, of the Janggiya clan	The second month of the Kangxi Emperor's 15th regnal year (1676)	Wuhua Ridge southeast of the Tomb of Filial Piety (Xiaoling)
Tomb of Prosperity (Yuling)	Hongli	Qianlong	Gaozong	Emperor Chun	The fourth son of Yinzhen	1735-1795	Empress Xiaoxian, of the Fuca clan; Empress Xiaoyi, of the Wei clan; Imperial Noble Consort Huixian, of the Gao clan; Imperial Noble Consort Zhemin, of the Fuca clan; Imperial Noble Consort Shujia, of the Jin clan	The second month of the Qianlong Emperor's 8th regnal year (1743)	Shengshui Ridge west of the Tomb of Filial Piety
Tomb of Stability (Dingling)	Yining	Xianfeng	Wenzong	Emperor Xian	The fourth son of Minning	1850-1861	Empress Xiaode, of the Sakda clan	The fourth month of the Xianfeng Emperor's 9th regnal year (1859)	Ping'an Valley west of the Tomb of Prosperity
Tomb of Benevolence (Huiling)	Zaichun	Tongzhi	Muzong	Emperor Yi	The first son of Yining	1861-1874	Empress Xiaozhe, of the Alute clan	The third month of the Guangxu Emperor's 1st regnal year (1873)	Shuangshan Valley in the southeast part of the Eastern Qing Tombs

Tab.2 The inception of each empress' tomb in the Eastern Qing Tombs

Tomb name	Empress' name	Posthumous title	Affiliation	Starting time for construction	Location	Remarks
West Tomb of Brightness (Zhao Xiling)	Borjigit Bumbutai	Empress Xiaozhuang-wen	Tomb of Brightness	The Kangxi Emperor's 27th regnal year (1688)	Outside the Great Red Gate, in the southeast part of the Eastern Qing Tombs	The underground palace, the encircled realm of treasure, and the square walled terrace with a memorial tower were added in the Yongzheng Emperor's 3rd regnal year (1725)
East Tomb of Filial Piety (Xiao Dong -ling)	Borjigit Alatan Qiqige	Empress Xiaohui-zhang	Tomb of Filial Piety	The Kangxi Emperor's 32nd regnal year (1693)	East of the Tomb of Filial Piety	28 tumuli of the Shunzhi emperor's imperial consorts' tombs lined up along the two sides of the square walled terrace with the memorial tower
East Tomb of Stability in Puxiang Valley (Puxiangyu Ding Dongling)	of the Niohuru Clan	Empress Xiaozhen-xian	Tomb of Stability	The Tongzhi emperor's 12th regnal year (1873)	Puxiang Valley east of the Tomb of Stability	
East Tomb of Stability in Putuo Valley (Putuoyu Ding Dongling)	Yehe Nara Xingzhen	Empress Xiaoqin-xian	Tomb of Stability	The Tongzhi Emperor's 12th regnal year (1873)	Putuo Valley east of the Puxiang Valley	

Tab.3 The inception of each imperial consort's tomb in the Eastern Qing Tombs

Tomb name	Buried consorts and concubines	Affiliation	Starting year of construction	Location
Imperial Consort Tombs affiliated with the Tomb of Admiration (Jingling Feiyuanqin)	Imperial Noble Consort Wenxi, of the Niohuru clan and other consorts, concubines, noble ladies, first-class and second-class female attendants, in total 40 women, as well as a young prince	Tomb of Admi-ration (Jingling)	The Kangxi Emperor's 20th regnal year (1681)	East of the Tomb of Admiration
Tombs of the Two Imperial Consorts buried side by side affiliated with the Tomb of Admiration (Jingling Shuangfeiyuanqin)	Imperial Noble Consort Quehui, of the Tunggiya clan; Imperial Noble Consort Chunyi, of the Guwalgiya clan	Tomb of Admira-tion (Jingling)	The Qianlong Emperor's 4th regnal year (1739)	East of the Imperial Consort Tombs affiliated with the Tomb of Admiration
Imperial Consort Tombs affiliated with the Tomb of Prosperity (Yuling Feiyuanqin)	Empress (demoted), of the Uranara clan; Imperial Noble Consort Chun-hui, of the Su clan and other consorts and concubines, in total 36	Tomb of Prosperity (Yuling)	The Qianlong Emperor's 12th regnal year (1747)	West of the Tomb of Prosperity
Imperial Consort Tombs affiliated with the Tomb of Stability (Dingling Feiyuanqin)	Imperial Noble Consort Zhuangjing, of the Tatala clan and 14 other consorts and concubines	Tomb of Stability (Dingling)	The Xianfeng Emperor's 9th regnal year (1859)	Shunshui Valley east of the Tomb of Stability
Imperial Consort Tombs affiliated with the Tomb of Benevolence (Huiling Feiyuanqin)	Imperial Noble Consort Shushen, of the Fuca clan; Imperial Noble Consort Gongsu, of the Arute clan; Imperial Consort Xianzhe, of the Heseri clan; Imperial Noble Consort Ronghui, of the Silin Gioro clan	Tomb of Benevo-lence (Huiling)	The Guangxu Emperor's 1st regnal year (1875)	West of the Tomb of Benevolence in the western part of Shuangshan Valley

二、清东陵的建筑规制

清东陵各陵寝中，孝陵被尊为祖陵，建筑规制最为完备。南部神道随山水向北延展，除桥座外，皆承袭明长陵，列置石牌坊、下马牌、大红门、具服殿、神功圣德碑亭、擎天柱（即华表）、望柱、石像生、龙凤门等，构成展谒引导空间。北倚昌瑞山的陵宫组群作为祭祀空间，仿照明晚期定型的帝陵，次第配置神道碑亭、神厨库、朝房、宫门、焚帛炉、配殿、享殿、陵寝门（即琉璃花门）、二柱门、石五供、方城明楼及哑巴院、月牙城、琉璃影壁和宝城、宝顶等，远比长陵节制。宫门、享殿由明代的祾恩门、祾恩殿改称为隆恩门、隆恩殿。

石牌坊而外，各建筑皆覆黄琉璃瓦，共六类屋顶式样：石牌坊五座顶楼、大红门、龙凤门四座琉璃墙顶为单檐庑殿顶；大碑楼、小碑亭、宰牲亭、隆恩殿、明楼皆重檐歇山顶；具服殿院门及正殿、神厨库院门、隆恩门、东西焚帛炉、东西配殿、琉璃花门、二柱门、月牙城前琉璃影壁俱单檐歇山顶；神厨、神库为单檐悬山顶；具服殿净房、东西朝房为单檐硬山顶；神厨库井亭为单檐盝顶。弥足珍贵的是，凡此形象，康熙九年《孝陵图》均有详实写照（图 14）。

应指出的是，孝陵鼎建，因时局掣肘，经营隆恩殿、神道碑亭等主体建筑，曾拆用明嘉靖十一年（1532 年）建在北京西苑的清馥殿组群的金丝楠木构材（图 14–19）。

II Building Arrangements for the Eastern Qing Tombs

Among all the Eastern Qing Tombs, the Tomb of Filial Piety (Xiaoling) for the Shunzhi Emperor is the most respected, being the first built before the surrounding tombs of his descendants. The Tomb of Filial Piety (Xiaoling) is the principle tomb and contains the most complete set of building arrangements. The spirit way (*shendao*) starts in the south and extends to the north, following the topography of the mountains and rivers. Except for the building arrangements for the bridges, all the building arrangements were inherited from the Long Tomb (Changling) of the Ming Dynasty. The arrangements advise on the spatial form of the entrance space: the marble memorial gateway (*shipaifang*), the stela marking the place for dismounting from one's horse (*xiamapai*), the great red gate building (*dahongmen*), the hall for ceremonial robes (*jufudian*), the Pavilion for the Stela of Divine Merit and Sagely Virtue (Shengong Shengde Beiting), the ceremonial columns (*huabiao*), the ornamental columns (*wangzhu*), the stone statues (*shixiangsheng*) and the dragon-phoenix gate (*longfengmen*). The building complex located at the southern end of Changrui Mountains is the major sacrificial space. Following the same arrangements as the imperial tomb arrangements established in the late Ming Dynasty, the Eastern Qing Tombs were planned seriatim, starting with the pavilion for the stela on the spirit way (*shendao beiting*), the culinary courtyard for sacrifice (*shenchuku*), the reception halls for court officials (*chaofang*), the main gate (*gongmen*), the sacrificial burners (*fenbolu*), the side halls (*peidian*), the main sacrificial hall (*xiangdian*), the gate of the tomb complex or the gate with glazed roof tiles (*lingqin men* or *liuli huamen*), the gate with two columns (*erzhumen*), the five stone sacrificial utensils (*shiwugong*), the square walled terrace and the memorial tower (*fangcheng minglou*), the courtyard of the mute (*yabayuan*), the crescent wall (*yueyacheng*), the screen wall with glazed tiles (*liuli yingbi*), and finally, the encircled realm of treasure and the tumulus (*baocheng baoding*). The Tomb of Filial Piety (Xiaoling) was built with greater moderation compared to the Long Tomb (Changling) of the Ming dynasty. The names of the main gate and the main sacrificial hall were renamed from *ling'enmen* and *ling'endian* (*ling'en* meaning 'divine auspiciousness and grace') to *long'enmen* and *long'endian* (*long'en* meaning 'monumental grace') respectively.

Except for the marble memorial gateway, all the buildings are covered with yellow glazed tiles. Altogether there are six categories of roofs: the five pavilions on top of the marble memorial gateway (*shipaifang*), the great red gate building (*dahongmen*), and the four glazed wall tops of the dragon and phoenix gate (*longfengmen*) are all built with single-eave hip-gable roofs; the great stela tower (*dabeilou*), the small stela pavilion (*xiaobeiting*), the ritual abattoir (*zaishengting*), the hall of monumental grace (*long'endian*), and the memorial tower

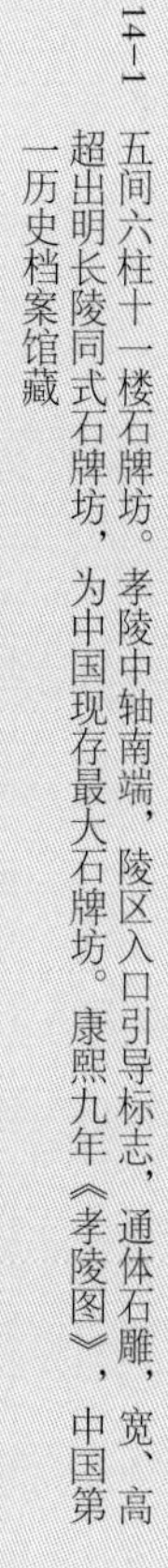

图14-1 五间六柱十一楼石牌坊。孝陵中轴南端，陵区入口引导标志，通体石雕，宽、高超出明长陵同式石牌坊，为中国现存最大石牌坊。康熙九年《孝陵图》，中国第一历史档案馆藏

图14-2 下马牌与大红门。大红门为孝陵暨清东陵入口，砖石结构，券门三，两翼连缀风水墙并设角门。大红门前对峙下马牌，石雕，竖刻满、蒙、汉文“官员人等在此下马”，告诫皇帝等谒陵者恭敬步行前往；该下马牌也配置在后区神道碑亭前。康熙九年《孝陵图》，中国第一历史档案馆藏

Fig.14-1 The marble memorial gateway with five bays, six columns, and eleven roofs (*wujian liuzhu shiyilou shipaifang*).

Located at the southern end of the central axis of the Tomb of Filial Piety (Xiaoling), the memorial gateway was built with stone sculptures as the landmark of the entrance area of the tomb. Being wider and higher than that of the similar styled memorial gateway of the Long Tomb (Changling) of the Ming Dynasty, the gateway is known as the largest existing marble memorial gateway in China.

Drawing of the Tomb of Filial Piety (*Xiaoling tu*), the Kangxi Emperor's 9th regnal year (1670). In the collection of The First Historical Archives of China.

Fig.14-2 The post marking the place for dismounting from horses and the great red gate building (*xiamapai* and *dahongmen*).

The great red gate building (*dahongmen*) is the entrance to the Tomb of Filial Piety (Xiaoling) and to all the Eastern Qing Tombs. It was built with brick and stone structures with three arches and winged by continuous *fengshui* walls with corner doors.

There are two posts marking the place for dismounting from horses (*xiamapai*) facing each other in front of the great red gate building. They are made of stone sculptures and inscribed with the sentence "Officials must dismount from their horses here" in Manchurian, Mongolian and Chinese respectively. The inscriptions remind those who come to pay homage, including the emperor, that they should all walk respectfully from here onwards. Similar posts are also found in front of the pavilions for the stelae on the spirit ways (*shendao beiting*).

Drawing of the Tomb of Filial Piety (*Xiaoling tu*), the Kangxi Emperor's 9th regnal year (1670). In the collection of The First Historical Archives of China.

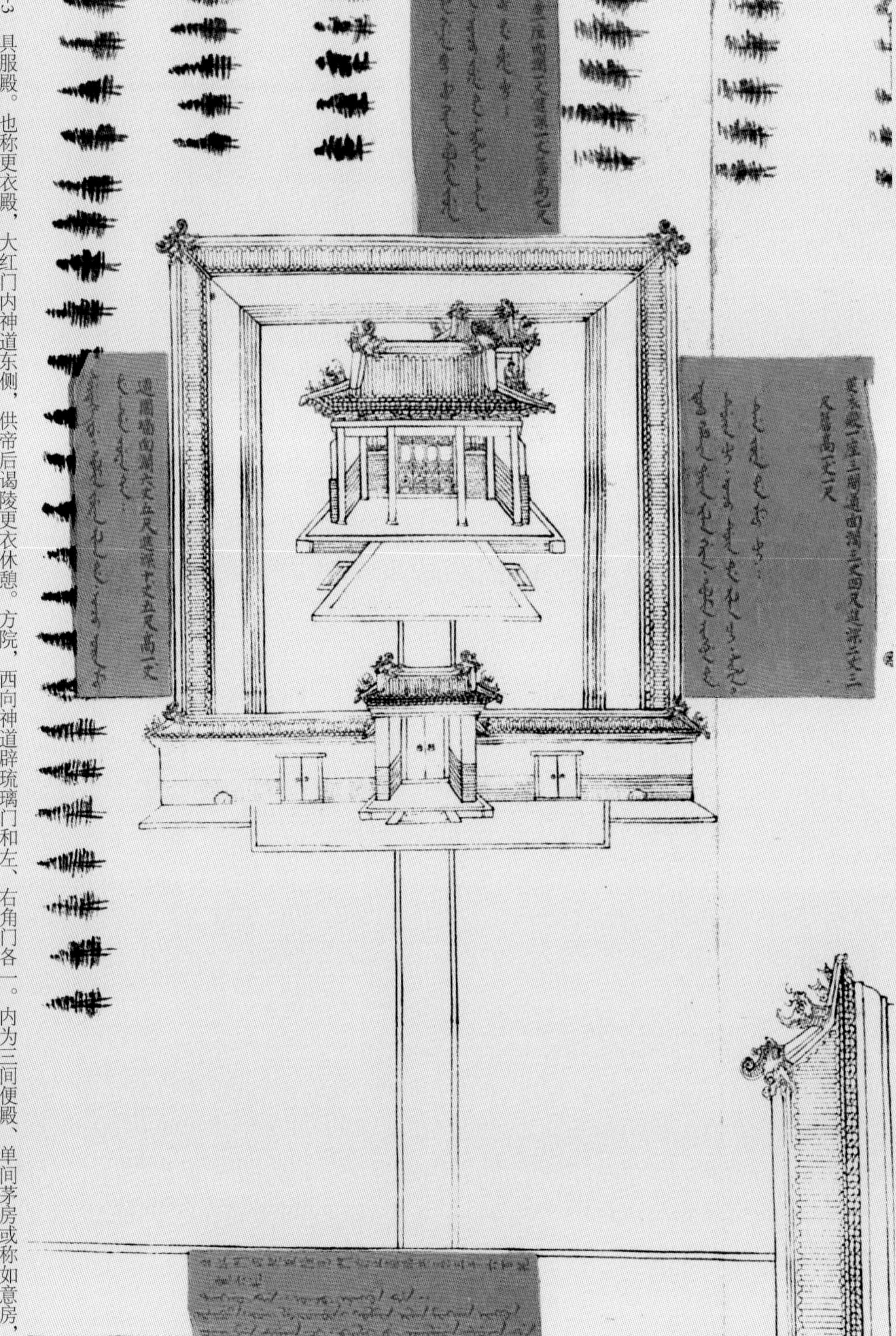

图 14-3　具服殿。也称更衣殿，大红门内神道东侧，供帝后谒陵更衣休憩。方院，西向神道辟琉璃门和左、右角门各一。内为三间便殿、单间茅房或称如意房，由明季五百余间缩简而来。康熙九年《孝陵图》，中国第一历史档案馆藏

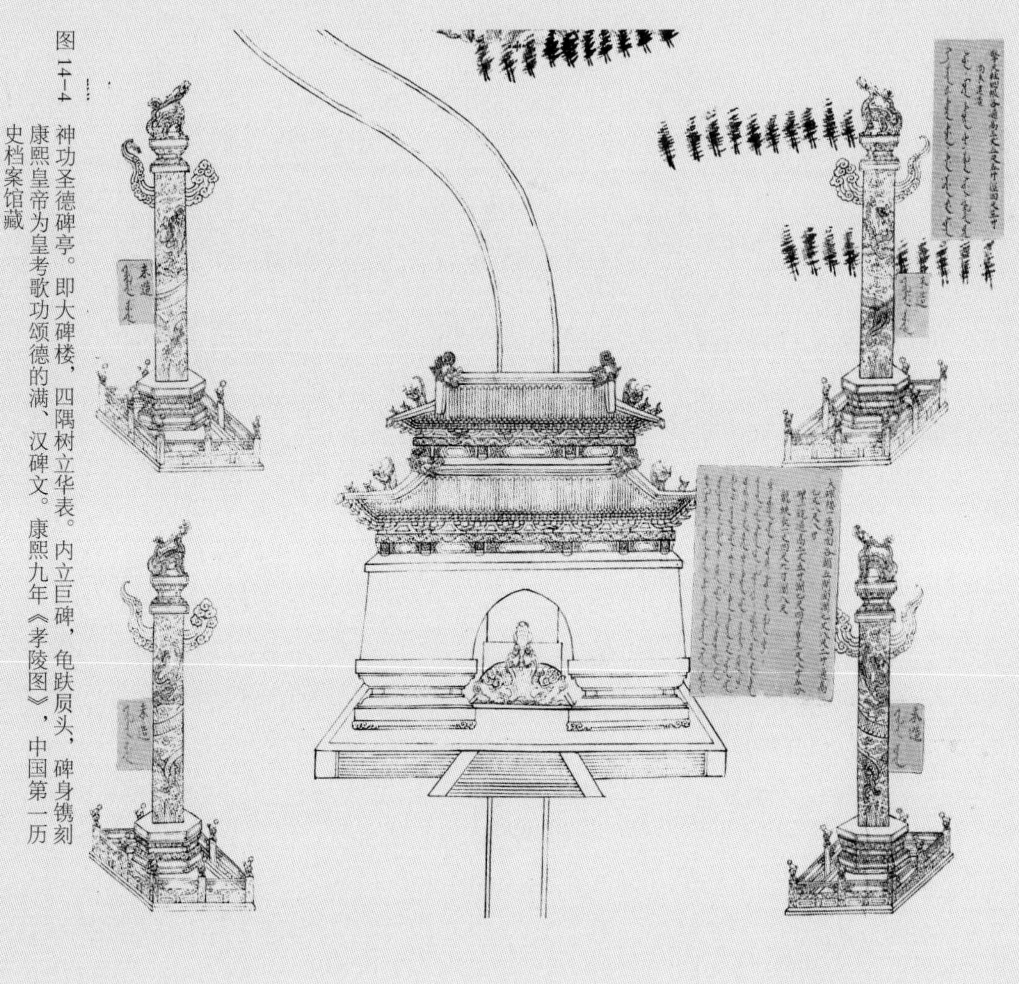

图 14-4　神功圣德碑亭。即大碑楼，四隅树立华表。内立巨碑，龟趺赑头，碑身镌刻康熙皇帝为皇考歌功颂德的满、汉碑文。康熙九年《孝陵图》，中国第一历史档案馆藏

Fig.14-3　The hall for ceremonial robes (*jufudian*).
Also known as hall for changing clothes (*gengyidian*), the hall for ceremonial robes is located inside the great red gate building (*dahongmen*) on the east side of the spirit way (*shendao*). Emperors and Empresses could change their clothes and take a rest there on the trip to pay homage at the tomb. The square courtyard around the hall opens on to the spirit way (*shendao*) through a glazed tiled gate (*liulimen*) in the middle with two corner doors (*jiaomen*) on each side. Inside the courtyard, there is a hall with three rooms and a single-room toilet (*maofang* or *ruyifang*), which is a simplified version of the five hundred rooms with similar functions during the Ming Dynasty.
Drawing of the Tomb of Filial Piety (*Xiaoling tu*), the Kangxi Emperor's 9th regnal year (1670). In the collection of The First Historical Archives of China.

Fig.14-4　The Pavilion for the Stela of Divine Merit and Sacred Virtue (Shengong Shengde Beiting), also known as the great stela tower (*dabeilou*), with ceremonial columns (*huabiao*) standing at all four corners. Inside the pavilion there is a huge stela mounted on a turtle-shaped base, with an inscription written by the Kangxi Emperor in Manchurian and Chinese languages. These inscriptions were to eulogize the deeds and virtues of his father, the Shunzhi Emperor.
Drawing of the Tomb of Filial Piety (*Xiaoling tu*), the Kangxi Emperor's 9th regnal year (1670). In the collection of The First Historical Archives of China.

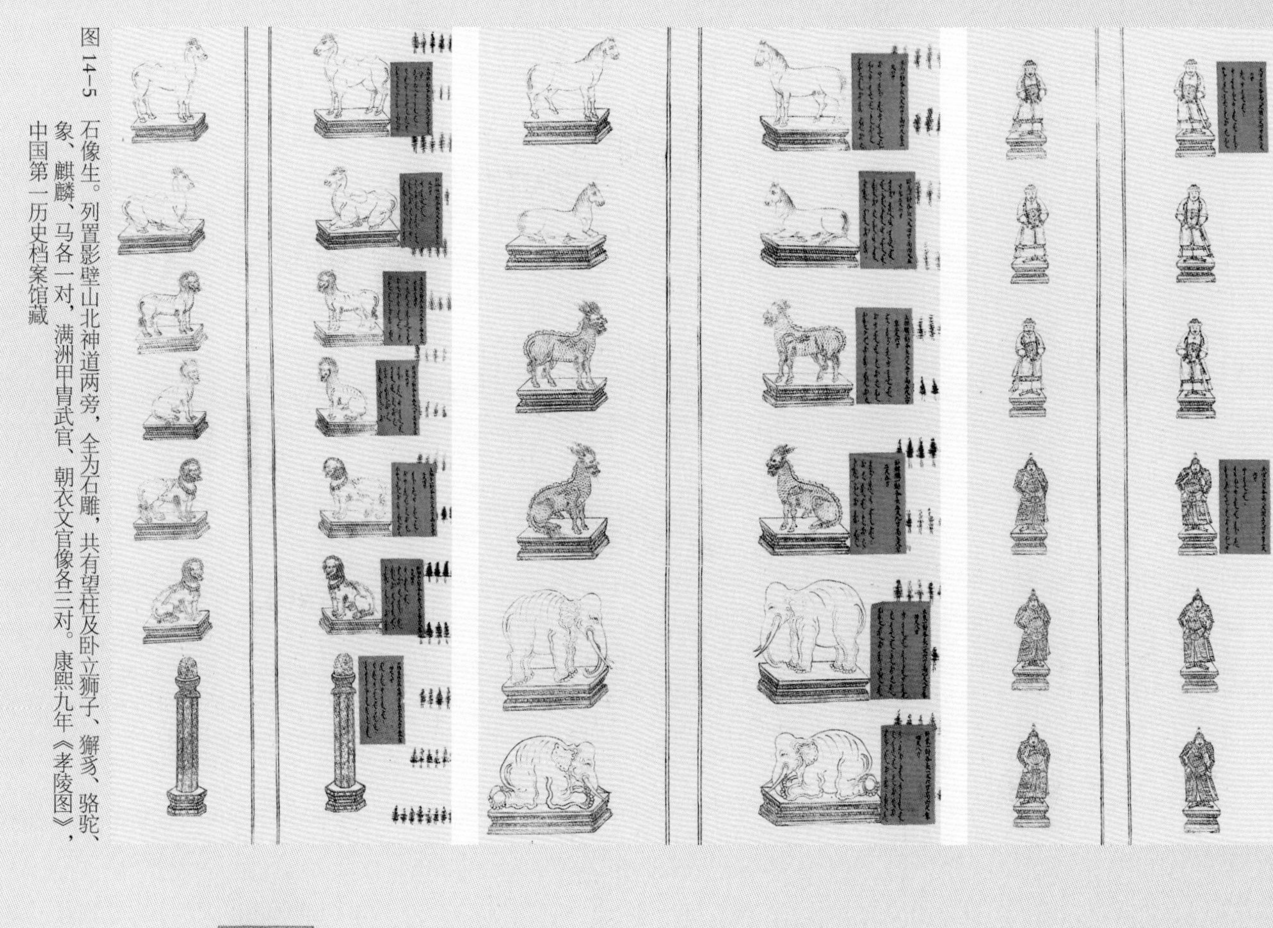

图 14-5 石像生。列置影壁山北神道两旁，全为石雕，共有望柱及卧立狮子、獬豸、骆驼、象、麒麟、马各一对，满洲甲胄武官、朝衣文官像各三对。康熙九年《孝陵图》，中国第一历史档案馆藏

Fig.14-5 The stone-sculptured statues (*shixiangsheng*).
Situated on both sides of the spirit way north of the Screen Wall Mountain (Yingbishan), all the statues are stone sculptured, including: a pair of ornamental columns, pairs of lions, *xiezhi*, camels, elephants, kylins and horses in resting and standing positions respectively, and three pairs of martial officials in armour and literary officials in court robes.
Drawing of the Tomb of Filial Piety (*Xiaoling tu*), the Kangxi Emperor's 9th regnal year (1670). In the collection of The First Historical Archives of China.

图 14-6 龙凤门。三座石构冲天式火焰牌坊，间缀琉璃照壁，两旁翊垣，为石像生群的底景，标志着陵区南部展谒引导空间的终结。康熙九年《孝陵图》，中国第一历史档案馆藏

Fig.14-6 The dragon and phoenix gate (*longfengmen*).
Composed of three stone-structured gateways decorated with sculptured flames soaring skywards (*chongtianshi huoyan pailou*), the dragon and phoenix gate has four screen walls of glazed titles between each two of the gateways and between the gateways and the roofed walls on both sides of the gate. The gate becomes the background of the stone statues and marks the end of the entrance space in the southern part of the tomb.
Drawing of the Tomb of Filial Piety (*Xiaoling tu*), the Kangxi Emperor's 9th regnal year (1670). In the collection of The First Historical Archives of China.

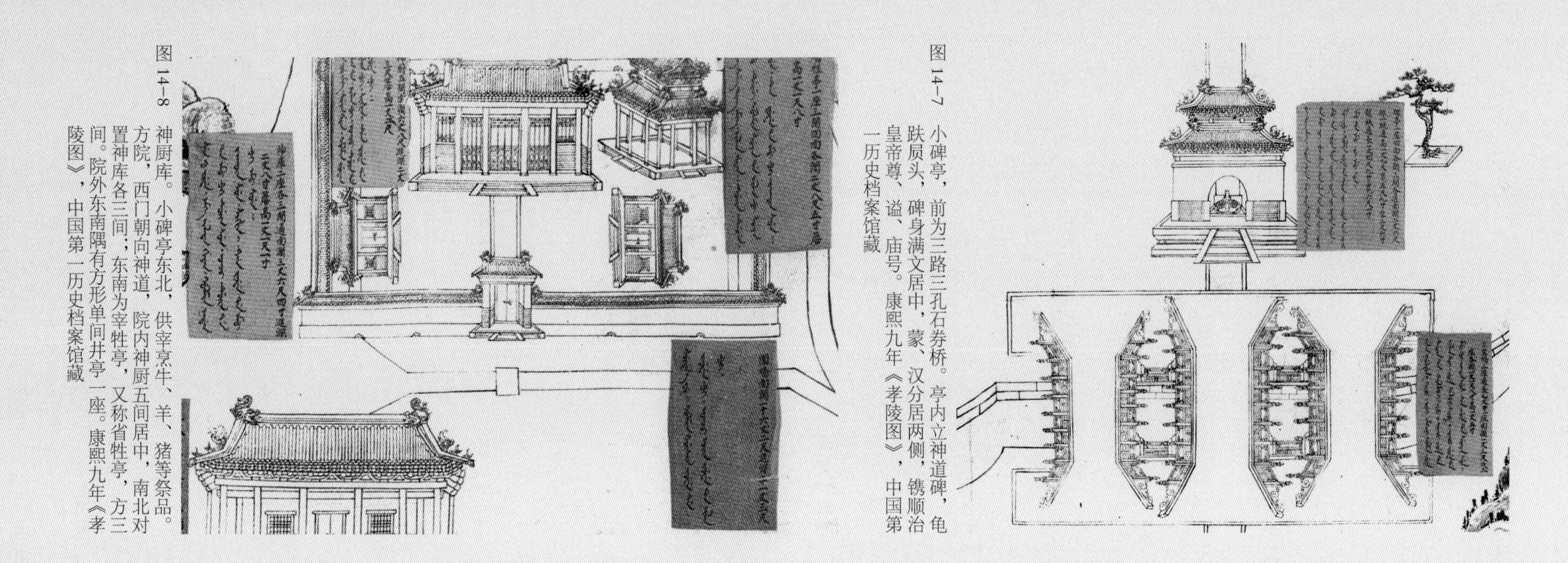

图 14-7 小碑亭，前为三路三孔石券桥。亭内立神道碑，龟趺赑头，碑身满文居中，蒙、汉分居两侧，镌顺治皇帝尊、谥、庙号。康熙九年《孝陵图》，中国第一历史档案馆藏

图 14-8 神厨库。小碑亭东北，供宰烹牛、羊、猪等祭品。方院，西门朝向神道，院内神厨五间居中，南北对置神库各三间；东南为宰牲亭，又称省牲亭，方三间。院外东南隅有方形单间井亭一座。康熙九年《孝陵图》，中国第一历史档案馆藏

Fig.14-7 The small stela pavilion (*xiaobeiting*).

With the three-way and three-arch stone bridge in the front, the pavilion houses a stela mounted on a turtle-shaped base, with the inscription of the regnal title, the posthumous title, and the temple title of the Shunzhi Emperor, with the Manchurian version in the middle and the Chinese and Mongolian versions on either side.

Drawing of the Tomb of Filial Piety (*Xiaoling tu*), the Kangxi Emperor's 9th regnal year (1670). In the collection of The First Historical Archives of China.

Fig.14-8 The culinary courtyard for sacrifice (*shenchuku*).

Northeast of the small stela pavilion, the culinary courtyard for sacrifice is a space where cows, sheep and pigs were slaughtered and cooked for sacrificial ceremonies. It is a square courtyard with its west gate facing the spirit way (*shendao*). Inside the courtyard, the five-room sacrificial kitchen (*shenchu*) is located in the center, with a three-room sacrificial storage house (*shenku*) on both sides facing south and north. A three-room square pavilion in the southeast corner is the ritual abattoir (*zaishengting*), or animal sacrifice pavilion (*xingshengting*). Outside the courtyard, a single-room square pavilion covering the well (*jingting*) is located in the southeast corner.

Drawing of the Tomb of Filial Piety (*Xiaoling tu*), the Kangxi Emperor's 9th regnal year (1670). In the collection of The First Historical Archives of China.

Fig.14-9 The reception halls for court officials (*chaofang*).

They are located on the east and west side and in front of the gate of monumental grace (*long'enmen*). The east *chaofang* is also called room for food and drink (*chafanfang*), and the west *chaofang* is called room for steamed bread (*bobofang*). Each hall is a five-room structure. They were where daily food, drinks and sacrificial offerings were prepared and where officials rest and slept overnight.

Drawing of the Tomb of Filial Piety (*Xiaoling tu*), the Kangxi Emperor's 9th regnal year (1670). In the collection of The First Historical Archives of China.

Fig.14-10 The gate of monumental grace (*long'enmen*).

As the entrance to the sacrificial space of the tomb, the gate of monumental grace is a five-room structure with a platform without sculptured railings in the front.

Drawing of the Tomb of Filial Piety (*Xiaoling tu*), the Kangxi Emperor's 9th regnal year (1670). In the collection of The First Historical Archives of China.

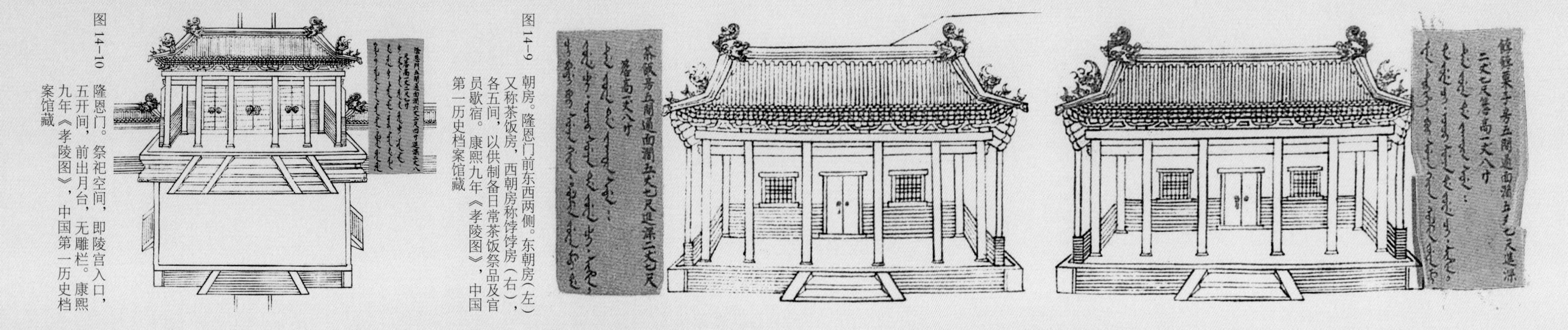

图 14-9 朝房。隆恩门前东西两侧。东朝房（左）又称茶饭房，西朝房称饽饽房（右），各五间，以供制备日常茶饭祭品及官员歇宿。康熙九年《孝陵图》，中国第一历史档案馆藏

图 14-10 隆恩门。祭祀空间，即陵宫入口，五开间，前出月台，无雕栏。康熙九年《孝陵图》，中国第一历史档案馆藏

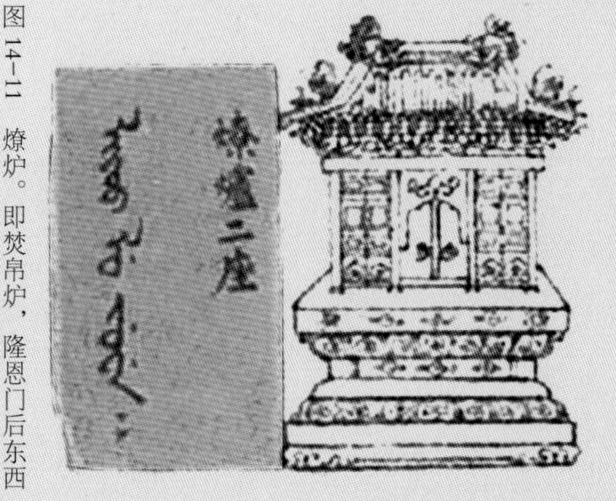

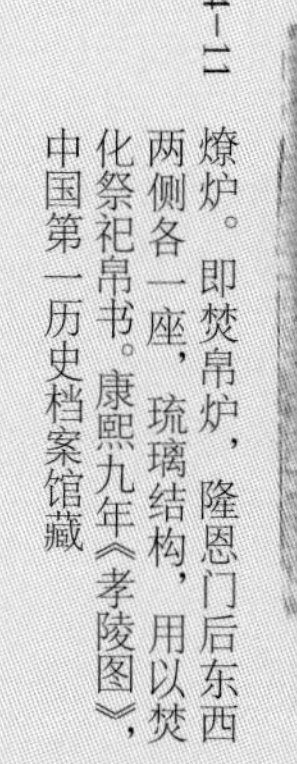

图 14-11　燎炉。即焚帛炉，隆恩门后东西两侧各一座，琉璃结构，用以焚化祭祀帛书。康熙九年《孝陵图》，中国第一历史档案馆藏

图 14-12　配殿。对称建在隆恩殿后东西两翼，皆五开间，用于祭祀事务。康熙九年《孝陵图》，中国第一历史档案馆藏

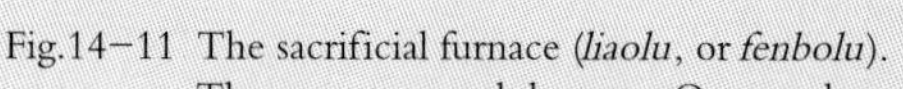

图 14-13　隆恩殿。陵寝主体建筑，举行各式祀典，五开间，前出周匝雕栏的须弥座大月台，殿内居中偏后三间暖阁内设神龛，供奉帝后神主牌位，统称神寝。康熙九年《孝陵图》，中国第一历史档案馆藏

Fig.14–11　The sacrificial furnace (*liaolu*, or *fenbolu*).
There are two such burners. One on the east and one on the west side behind the Gate of Monumental grace. The burners were built with a glazed-tile structure to burn silk manuscripts for the sacrificial ceremony.
Drawing of the Tomb of Filial Piety (*Xiaoling tu*), the Kangxi Emperor's 9th regnal year (1670). In the collection of The First Historical Archives of China.

Fig.14–12　The side hall (*peidian*).
The side halls are built in a symmetrical way on the east and west side behind the hall of monumental grace (*long'endian*). Both are five-room structures used for supporting sacrificial rituals.
Drawing of the Tomb of Filial Piety (*Xiaoling tu*), the Kangxi Emperor's 9th regnal year (1670). In the collection of The First Historical Archives of China.

Fig.14–13　The hall of monumental grace (*long'endian*).
The main building of the tomb complex, the hall of monumental grace is where all kinds of sacrificial ceremonies were held. It is a five-room building with a large platform in the front. The platform has a Sumeru base and is surrounded by sculptured railings. The back of the central three rooms is partitioned off from the interiors of the hall, forming a *nuange* (a section with a heating stove), where a shrine was set up, with the ancestral emperor's and empress(es)' memorial tablets for worship. The *nuange* is named *shenqin*, meaning the sleeping place of the ancestral spirits.
Drawing of the Tomb of Filial Piety (*Xiaoling tu*), the Kangxi Emperor's 9th regnal year (1670). In the collection of The First Historical Archives of China.

Fig.14–14　The gate of the sleeping spirits (*lingqinmen*).
The gate of the sleeping spirits is also named the gate of the tomb (*lingqingmen*) or the gate with glazed roof tiles (*liuli huamen*). The *lingqinmen* is composed of three gates with glazed roof tiles—a big gate in the middle and two small gates standing side by side. This form is similar to the entrance to the upper palace in the tomb complexes built in the Tang and Song dynasties. Only emperors and empresses paying homage at the tomb and officials who were responsible for daily cleaning, making offerings and patrolling were allowed to enter. Other people, even princes and officials with meritorious records were forbidden to enter the gate.
Drawing of the Tomb of Filial Piety (*Xiaoling tu*), the Kangxi Emperor's 9th regnal year (1670). In the collection of The First Historical Archives of China.

图 14-14　灵寝门。即陵寝门或琉璃花门。三座琉璃门一大两小并置，类同唐、宋陵寝上宫入口。除帝后谒陵和日常祭扫、巡护官员，严禁包括亲王、勋臣在内的其他人等进入。康熙九年《孝陵图》，中国第一历史档案馆藏

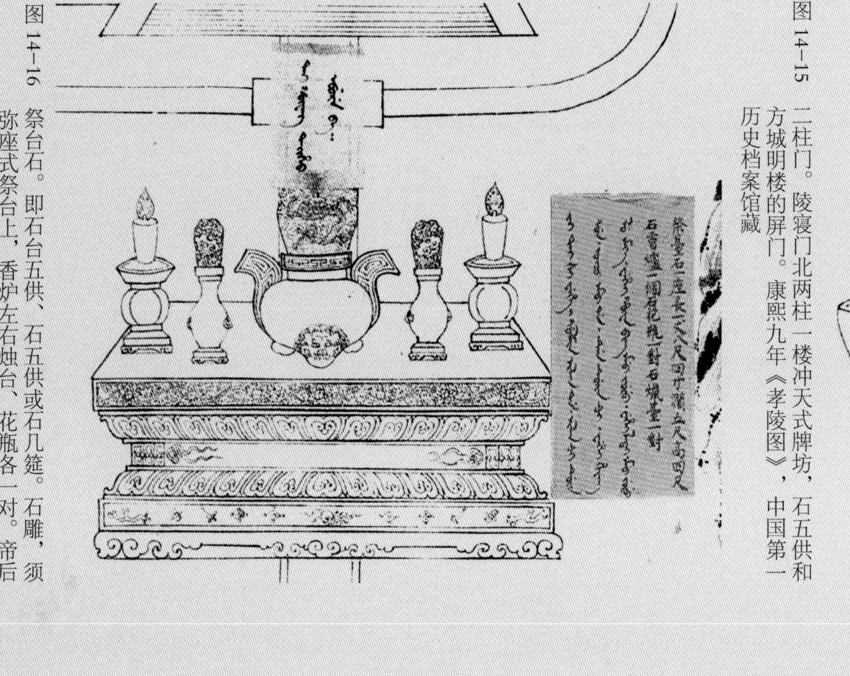

图 14-16 祭台石。即石台五供、石五供或石几筵。石雕，须弥座式祭台上，香炉左右烛台、花瓶各一对。帝后谒陵最隆重的祭礼就在此举行。康熙九年《孝陵图》，中国第一历史档案馆藏

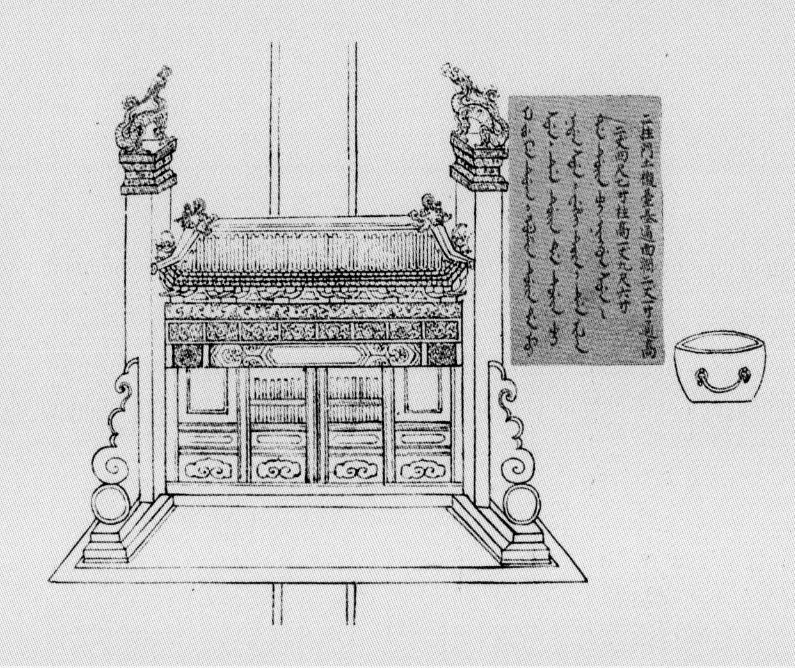

图 14-15 二柱门。陵寝门北两柱一楼冲天式牌坊，石五供和方城明楼的屏门。康熙九年《孝陵图》，中国第一历史档案馆藏

Fig.14-15 The two-columned gate (*erzhumen*).
North of the gate of the tomb (*lingqingmen*) is *erzhumen*, i.e. the soaring skywards gateway with two columns and one roof (*liangzhu yilou chongtianshi paifang*). It is the screen gate for the five stone sacrificial utensils (*shiwugong*) and the square walled terrace with the memorial tower (*fangcheng minglou*).
Drawing of the Tomb of Filial Piety (*Xiaoling tu*), the Kangxi Emperor's 9th regnal year (1670). In the collection of The First Historical Archives of China.

Fig.14-16 The stone sacrificial altar (*jitaishi*).
The stone sacrificial altar is also called the five stone ritual vessels (*shiwugong*), the five stone ritual vessels on the altar (*taishiwugong*) or the stone altar feast (*shijiyan*). The altar has a stone-sculptured Sumeru base. On top of the altar are: an incense burner in the middle, and a vase and a candle-holder on both sides. The grandest sacrificial ceremonies when emperors and empresses were paying homage to the tomb took place here.
Drawing of the Tomb of Filial Piety (*Xiaoling tu*), the Kangxi Emperor's 9th regnal year (1670). In the collection of The First Historical Archives of China.

Fig.14-17 The square walled terrace with the memorial tower (*fangcheng minglou*) are the main buildings at the end of the tomb complex. Beneath the square walled terrace (*fangcheng*) is a large platform reached by a ramp made of big bricks laid on edge (*jiangca*) in the front. On top of the walled terrace is the memorial tower (*minglou*), with a stela erected inside. The stela has an inscription incised and filled with gold, with the text, "Tomb of the Shizu Emperor Zhang (*Shizu Zhang Huangdi zhi Ling*)" , in Manchurian, Chinese and Mongolian languages. This inscription corresponds to a copper striped and gilded plaque inscribed with the text "Xiaoling" , also written in three languages, hanging over the upper eave of the southern facade. Right through the centre of the square walled terrace is the vault of the gateway (*mendongquan*), which leads to the courtyard of the mute (*yabayuan*). The courtyard is bounded by the glazed screen wall (*liuli yingbi*) and the crescent wall (*yueyacheng*) in the north. Next to the encircled realm of treasure (baocheng, i.e. the superstructure above the underground palace) are two curve stairways, through which one may reach the memorial tower (*minglou*) and the tumulus (*baoding*). In such a setting, the crescent wall, the encircled realm of treasure and the tumulus surround and guard the core of the tomb, the underground palace (*digong*).
Drawing of the Tomb of Filial Piety (*Xiaoling tu*), the Kangxi Emperor's 9th regnal year (1670). In the collection of The First Historical Archives of China.

图 14-17 方城明楼。陵寝终端的主体建筑，方城下设前出礓䃰的高大月台，城顶为明楼，内有阴刻填金满、汉、蒙文『世祖章皇帝之陵』的明楼碑，呼应明楼南面上檐下竖嵌铜条镀金的满、汉、蒙文『孝陵』斗匾。方城贯通门洞券，通入哑巴院，北为月牙城和琉璃影壁，两旁宝城壁设转向蹬道登临明楼和宝顶。月牙城、宝城和宝顶围护陵寝核心的地宫。康熙九年《孝陵图》，中国第一历史档案馆藏

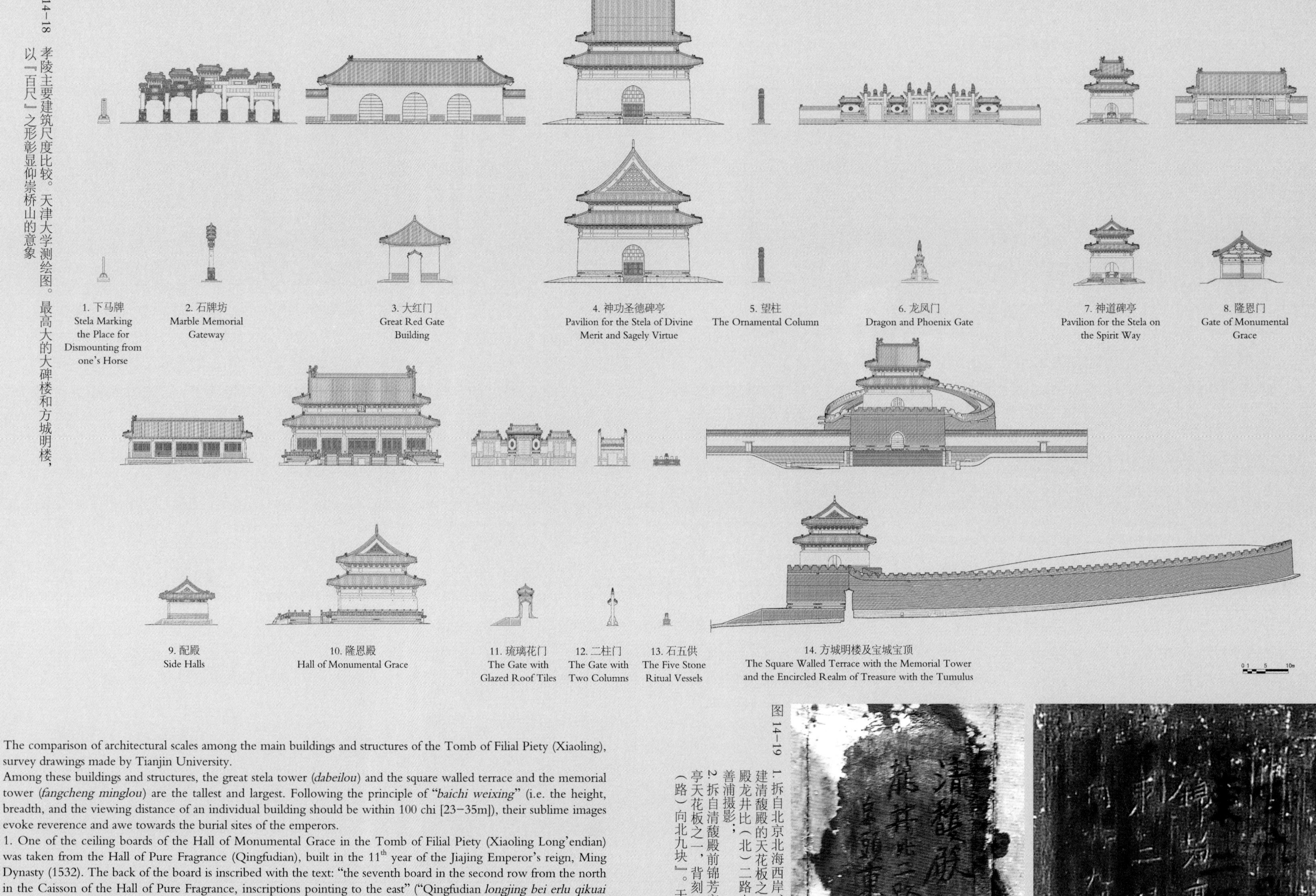

图14-18 孝陵主要建筑尺度比较。天津大学测绘图。最高大的大碑楼和方城明楼，以『百尺』之形彰显仰崇桥山的意象

Fig.14–18 The comparison of architectural scales among the main buildings and structures of the Tomb of Filial Piety (Xiaoling), survey drawings made by Tianjin University.
Among these buildings and structures, the great stela tower (*dabeilou*) and the square walled terrace and the memorial tower (*fangcheng minglou*) are the tallest and largest. Following the principle of "*baichi weixing*" (i.e. the height, breadth, and the viewing distance of an individual building should be within 100 chi [23–35m]), their sublime images evoke reverence and awe towards the burial sites of the emperors.

图14-19 1. 拆自北京北海西岸、明嘉靖十一年建清馥殿的天花板之一，背刻『清馥殿龙井比（北）二路七块字头东』于善浦摄影；2. 拆自清馥殿前锦芳亭的孝陵神道碑亭天花板之一，背刻『锦芳亭东二禄（路）向北九块』。于善浦摄影

Fig.14–19 1. One of the ceiling boards of the Hall of Monumental Grace in the Tomb of Filial Piety (Xiaoling Long'endian) was taken from the Hall of Pure Fragrance (Qingfudian), built in the 11th year of the Jiajing Emperor's reign, Ming Dynasty (1532). The back of the board is inscribed with the text: "the seventh board in the second row from the north in the Caisson of the Hall of Pure Fragrance, inscriptions pointing to the east" ("Qingfudian *longjing bei erlu qikuai zitoudong*"). Photo by YU Shanpu.
2. One of the ceiling boards of the Pavilion for the Stela on the Spirit Way in the Tomb of Filial Piety (Xiaoling Shendao Beiting) was taken from the Pavilion of Brightness and Beauty (Jinfangting) in front of the Hall of Pure Fragrance in the West Park of the Ming Dynasty. The back of the board is inscribed with the text: "the ninth board in the second row from the east in the Pavilion of Brightness and Beauty" ("Jinfangting *dong'erlu xiangbei jiukuai*"). Photo by YU Shanpu.

继孝陵后，康熙皇帝预建景陵，陵宫参照孝陵，宝城则由长圆式平面改为圆式。陵前神道衔接孝陵，裁去大碑楼、石像生和龙凤门等，以彰明统绪关联，凸显孝陵的主体地位。到雍正朝，景陵增建大碑楼和华表；乾隆皇帝又添设望柱、五对石像生以及五间六柱冲天式牌楼门。往后，乾隆皇帝的裕陵参仿孝陵、景陵，石像生却增至八对；隆恩殿东暖阁内又添设仙楼供奉佛像，也称佛楼，嗣后被各帝陵沿袭；陵寝门前玉带河三座单孔石拱桥，石雕望柱栒栏两端安设蹲龙，更远比孝陵三座石平桥奢华，成为清代陵寝绝响。

晚清，咸丰皇帝的定陵祖述前朝三陵，包括『量小为收』的五对石像生，但撤去二柱门，裁掉隆恩殿两山及后檐石雕望柱栏板；结合地势，琉璃花门以北院落内缩，宝城平面收窄为长圆式，方城前月台增为二层叠落，添设雕栏。最后，同治皇帝的惠陵依照定陵规制，却撤掉了石像生，仅留一对望柱，特地围以雕栏（图15）。

帝陵的陵宫建筑也成为后陵的蓝本，后陵的下马牌、神厨库、神道碑亭、朝房、隆恩门、隆恩殿、东西配殿、焚帛炉、陵寝门和石五供等，均类同帝陵而规模缩减，不设二柱门，环护地宫的宝城与方城明楼联为一体，没有哑巴院、月牙城和琉璃影壁。

其中，昭西陵隆恩殿为康熙二十七年（1688年）『皇上亲示画图』，迁建大内慈宁宫东的五间新宫，重檐庑殿顶，为清代陵寝孤例；此外，隆恩门和隆恩殿间隔墙，中建三座琉璃花门，也为特例。雍正三年（1725年）添建地宫、宝城、方城明楼和石五供，因地势局促，陵寝门建在隆恩殿两侧；陵前又仿效帝陵特建神道碑亭。

(*minglou*) are all equipped with double-eave hip-gable roofs. The gate to the yard of the hall for ceremonial robes (*jufudian*) and the hall itself, the gate to the culinary courtyard for sacrifices (*shenchuku*), the gate of monumental grace (*long'enmen*), the east and west sacrificial burners (*fenbolu*), the east and west side halls (*peidian*), the gate with glazed roof tiles (*liuli huamen*), the gate with two columns (*erzhumen*), and the screen wall of glazed tiles (*liuli yingbi*) in front of the crescent wall (*yueyacheng*) are all built with hip-gable roofs with a single eave. The kitchen for sacrifices (*shenchu*) and the storage house for sacrifices (*shenku*) are both built with single-eave suspension roofs. The storage room for keeping mobile toilets (*jingfang*) in the hall for ceremonial robes and the reception halls on the east and west sides are all single-eave gabled roofs. The well pavilion (*jingting*) in the culinary courtyard for sacrifices (*shenchuku*) is built with a flatted hip roof. What is so valuable is that all the images of all these buildings have been depicted and represented fully and accurately in the *Drawing of the Tomb of Filial Piety* (*Xiaoling tu*) completed in the Kangxi emperor's 9th regnal year (1670) (Fig.14).

During the early years of constructing of the Tomb of Filial Piety (Xiaoling), due to financial and material shortages, the major buildings such as the hall of monumental grace (*long'endian*) and the pavilion for the stela on the spirit way (*shendao beiting*) reused the *nanmu* made structural elements taken from the Hall of Pure Fragrance (Qingfudian) complex in the West Park (Xiyuan) in Beijing, which was built in the 11th year in the reign of Jiajing of the Ming Dynasty (1532) (Fig. 14-19).

After the completion of the Tomb of Filial Piety (Xiaoling), the Kangxi Emperor began construction of the Tomb of Admiration (Jingling), whose main structures followed those of the Tomb of Filial Piety (Xiaoling), except that the encircled realm of treasure (*baocheng*) was changed from an oval-shape to a round-shape. In addition, the spirit way (*shendao*) of the Tomb of Admiration (Jingling) joined with the spirit way of the Tomb of Filial Piety (Xiaoling), but without the great stela tower (*dabeilou*), the stone statues (*shixiangsheng*) and the dragon-phoenix gate (*longfengmen*), in order to clarify the unification and relationship between the two, and also to emphasise the dominant status of the Tomb of Filial Piety (Xiaoling). During the Yongzheng period, the great stela tower (*dabeilou*) and the ceremonial columns (*huabiao*) were added to the Tomb of Admiration (Jingling). The Qianlong Emperor further added ornamental columns (*wangzhu*), five pairs of stone statues (*shixiangsheng*) and a five-bay gateway with six columns soaring skywards (*wujian liuzhu chongtianshi pailoumen*). From then on, the Tomb of Prosperity (Yuling) for the Qianlong Emperor followed the arrangements of the Tomb of Filial Piety (Xiaoling) and the Tomb of Admiration (Jingling), but increased the number of stone statues (*shixiangsheng*) to eight pairs. In addition, an indoor tower of the

图 15 清东陵各帝陵的陵宫平面比较。天津大学测绘图

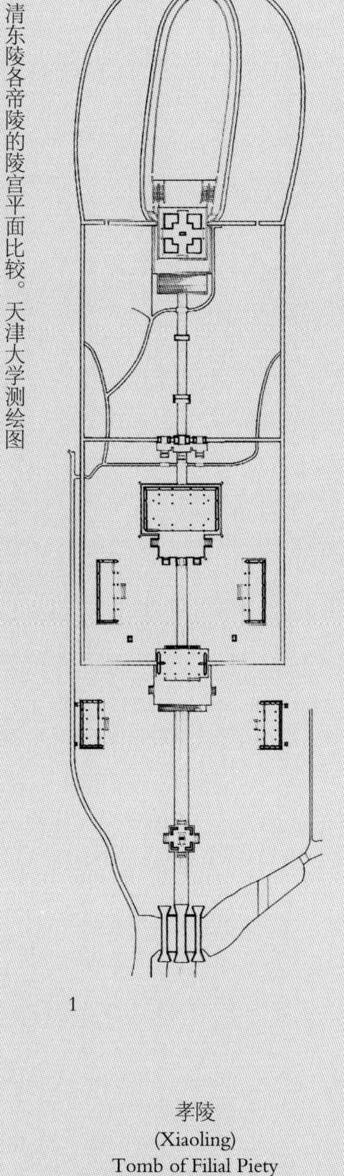

1

孝陵
(Xiaoling)
Tomb of Filial Piety
顺治十八年（1661）
The 18th year of the Shunzhi Emperor's reign

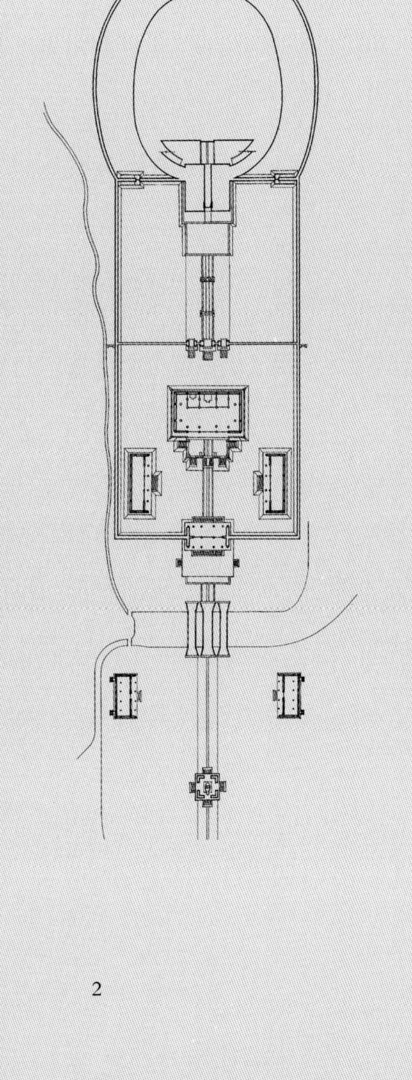

2

景陵
(Jingling)
Tomb of Admiration
康熙十五年（1676）
The 15th year of the Kangxi Emperor's reign

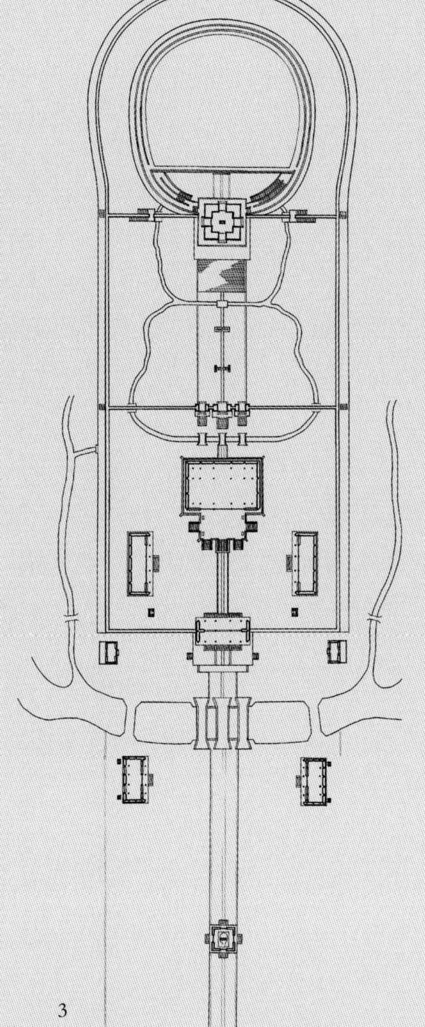

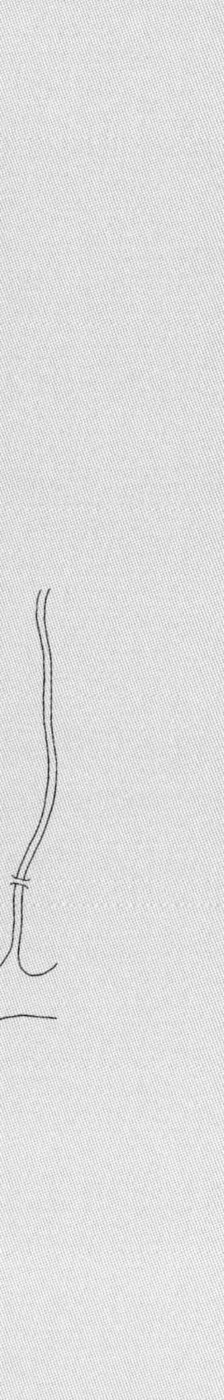

3

裕陵
(Yuling)
Tomb of Prosperity
乾隆八年（1743）
The 8th year of the Qianlong Emperor's reign

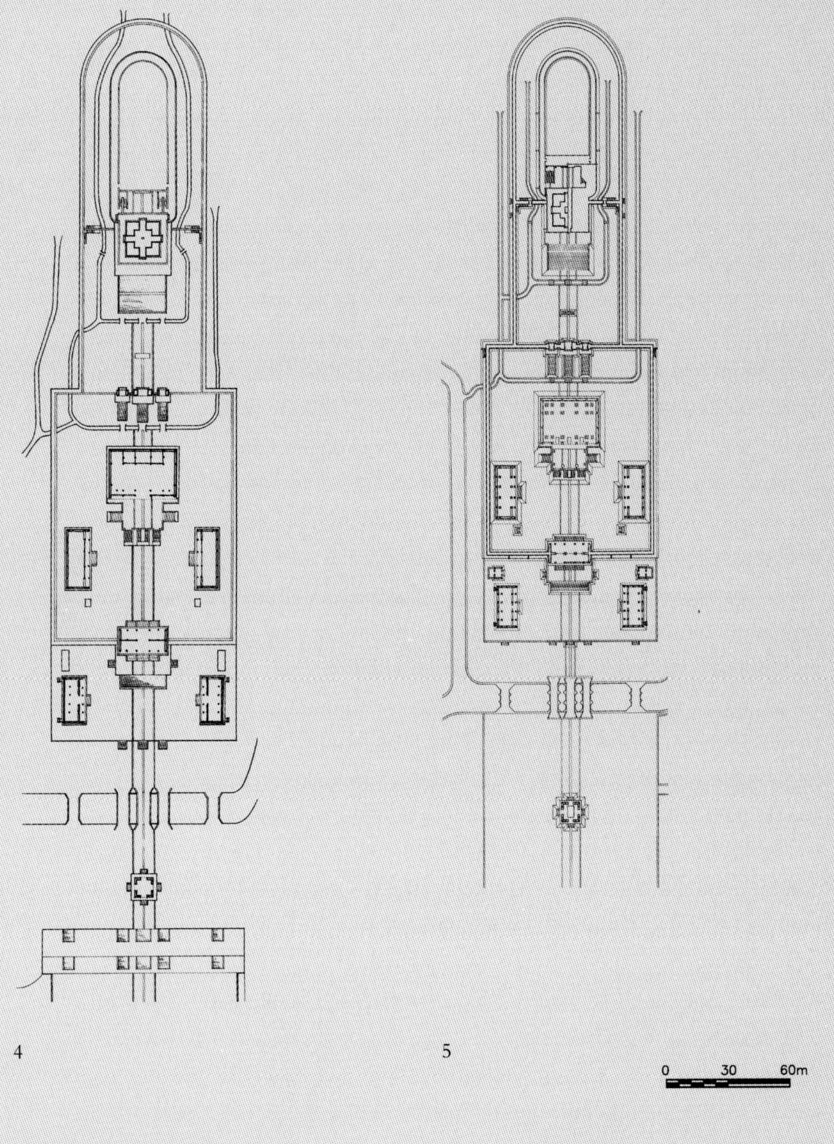

4

定陵
(Dingling)
Tomb of Stability
咸丰九年（1859）
The 9th year of the Xianfeng Emperor's reign

5

惠陵
(Huiling)
Tomb of Benevolence
光绪元年（1875）
The 1st year of the Guangxu Emperor's reign

Fig.15 The comparisons among the site plans for the emperors' tombs in the Eastern Qing Tombs. Survey drawings by Tianjin University.

immortal (*xianlou*), also called the tower of the Buddha (*folou*), where a statue of the Buddha was enshrined, was added to the east room with a heating stove (*dongnuange*) of the hall of monumental grace (*long'endian*). Similar *xianlou* or *folou* were followed in the later emperors' tombs. In front of the main gate of the Tomb of Prosperity (Yuling), there are three single-arch stone bridges over the Jade Ribbon River (Yudaihe); a stone squatting dragon (*dunlong*) is placed at each end of the stone-sculptured balusters and railings (*shidiao wangzhu xunlan*) on the bridge. The whole visual effect of these single-arch bridges is far more extravagant than the three flat stone bridges of the Tomb of Filial Piety (Xiaoling). They became the unique example of architectural art among all the imperial tombs of the Qing Dynasty.

In the late Qing period, the Tomb of Stability (Dingling) for the Xianfeng Emperor followed the building arrangements of the previous three emperors' tombs (Xiaoling, Jingling, and Yuling). The Tomb of Stability (Dingling) therefore included five pairs of reduced-sized stone statues (*shixiangsheng*), removing the gate with two columns (*erzhumen*), and the stone sculptured balusters and frieze panels (*wangzhu lanban*) on the left and right sides of the hall of monumental grace (*long'endian*) as well as those beneath its back eaves (*houyan*). As a response to the topography, the courtyard to the north of the screen wall with glazed tiles (*liuli huamen*) was reduced inwardly, and the plan of the encircled realm of treasure (*baocheng*) was squeezed into an oval shape. The front platform of the square walled terrace (*fangcheng*) was increased to two levels surrounded by sculptured railings and panels (*diaolan*). Finally, the Tomb of Benevolence (Huiling) for the Tongzhi Emperor, followed most of the guidance for the Tomb of Stability (Dingling), but removed all stone statues (*shixiangsheng*), except for a pair of ornamental columns (*wangzhu*) surrounded by sculptured railings (*diaolan*) (Fig. 15).

Building arrangements for the emperors' tombs were also taken as the blue prints for the empresses' tombs. The designs for buildings such as the dismounting stela (*xiamapai*), the culinary courtyard for sacrifice (*shenchuku*), the pavilion for the stela on the spirit way (*shendao beiting*), the halls for court officials (*chaofang*), the gate of monumental grace (*long'enmen*), the hall of monumental grace (*long'endian*), the east and west side halls (*peidian*), the sacrificial burners (*fenbolu*), the gate of the tomb complex (*lingqinmen*) and the five stone ritual vessels (*shiwugong*) were all like those for the emperors', but reduced in terms of size and scale. No gate with two columns (*er'zhumen*) was installed. The encircled realm of treasure (*baocheng*) surrounding and guarding the underground palace (*digong*) was joined with the square walled terrace and the memorial tower (*fangcheng minglou*), removing the courtyard of the mute (*yabayuan*), the crescent wall (*yueyacheng*)

在孝东陵，康熙五十七年（1718）安葬孝惠皇后，清东陵西侧顺治皇帝妃园寝已葬妃嫔等也全部迁入，大小不同的圆丘宝顶覆盖地宫，对称排列在方城明楼两侧，形成后陵、妃园寝兼具的特殊格局。

最晚营建的普祥峪和菩陀峪定东陵，其规制划一，平列东西，宝城作长圆式，陵垣前宽后窄，明显效法定陵。两组神厨库和井亭，南北序列地集中建在东侧大平台上。陵前效仿昭西陵建置神道碑亭，成为慈安、慈禧太后曾长期垂帘听政的特别标志（图16）。

隶从帝陵的妃园寝，均以康熙二十年（1681年）经营的景陵妃园寝为彝宪。其南端对置木下马桩，北为单孔石拱桥，单孔石平桥居东。桥北园寝入口称为宫门，三间单檐歇山顶，两翼红墙同北端的罗圈墙合抱。除宫门前东西置的班房各三间、厢房各五间为灰瓦硬山顶，各建筑和墙垣均覆绿琉璃瓦。宫门内，居中为五开间单檐歇山顶的享殿，殿东南安置一座燎炉，没有配殿。享殿后横贯面阔墙，中央拔起琉璃花门一座，左右辟随墙角门。门北的后院，序列月台宝顶覆盖的各妃嫔地宫。

乾隆四年（1739年）建景陵皇贵妃园寝，规制加尊。后院平行并列两座地宫，各覆圆式平面的宝城，前出方城明楼；前院享殿两翼添建配殿各五间，明楼和配殿皆为绿琉璃单檐歇山顶。乾隆二十六年（1761年）裕陵妃园寝改建，后院居中的纯惠皇贵妃地宫添建宝城方城明楼，前院增修东西配殿，皆如景陵皇贵妃园寝；由于原琉璃花门北场地局促，又仿昭西陵将琉璃花门及红墙改建在享殿两侧。清末，定陵和惠陵的妃园寝归复景陵妃园寝规制，最终使景陵皇贵妃园寝和裕陵妃园寝成为仅见于乾隆朝的特例（图17）。

图16 清东陵各后陵的陵宫平面比较。天津大学测绘图

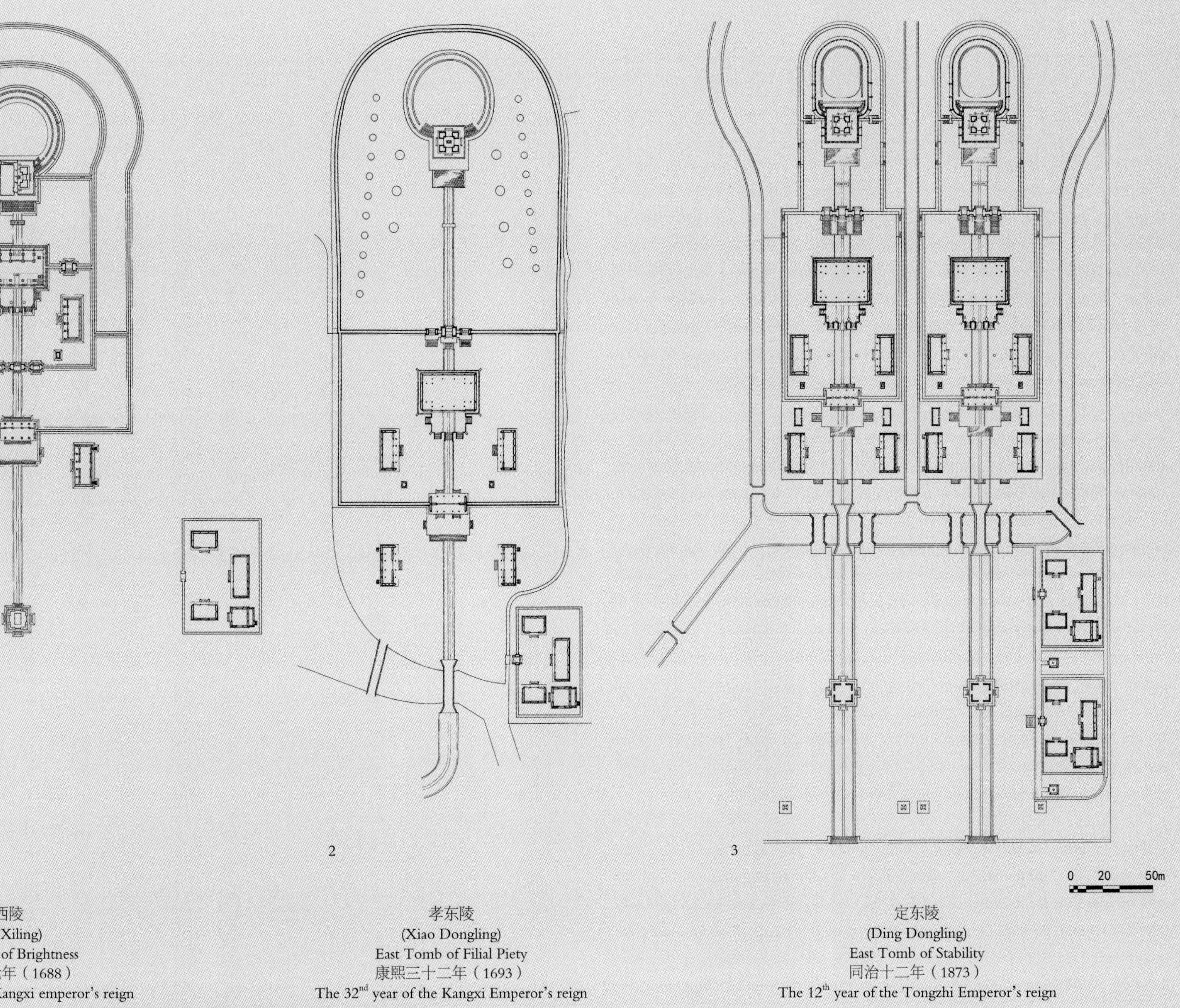

昭西陵
(Zhao Xiling)
West Tomb of Brightness
康熙二十七年（1688）
The 27th year of the Kangxi emperor's reign

孝东陵
(Xiao Dongling)
East Tomb of Filial Piety
康熙三十二年（1693）
The 32nd year of the Kangxi Emperor's reign

定东陵
(Ding Dongling)
East Tomb of Stability
同治十二年（1873）
The 12th year of the Tongzhi Emperor's reign

Fig.16 The comparisons among the site plans for the empresses' tombs in the Eastern Qing Tombs. Survey drawings by Tianjin University.

图 17 清东陵各妃园寝平面比较。天津大学测绘图

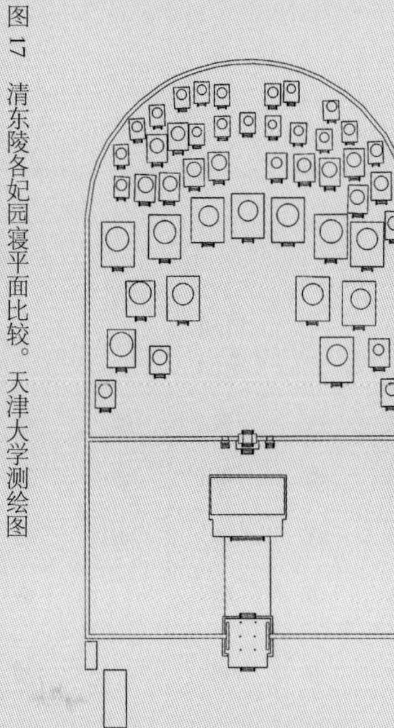

景陵妃园寝
(Jingling Feiyuanqin)
The Imperial Consorts' Tombs affiliated with the Tomb of Admiration
康熙二十年（1681）
The 20th year of the Kangxi Emperor's reign (1681)

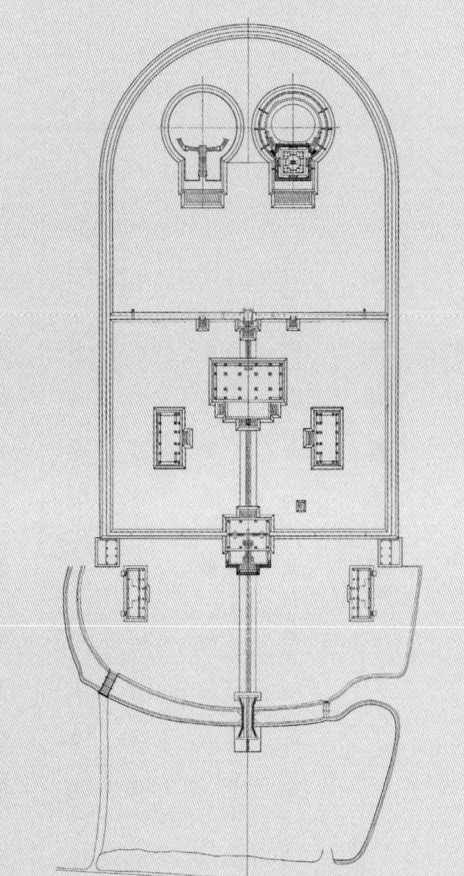

景陵双妃园寝（景陵皇贵妃园寝）
(Jingling Shuangfeiyuanqin [Jingling Huangguifei Yuanqin])
The Two Imperial Consorts' Tombs affiliated with the Tomb of Admiration
(The Imperial Noble Consort's Tombs affiliated with the Tomb of Admiration)
乾隆四年（1739）
The 4th year of the Qianlong Emperor's reign (1739)

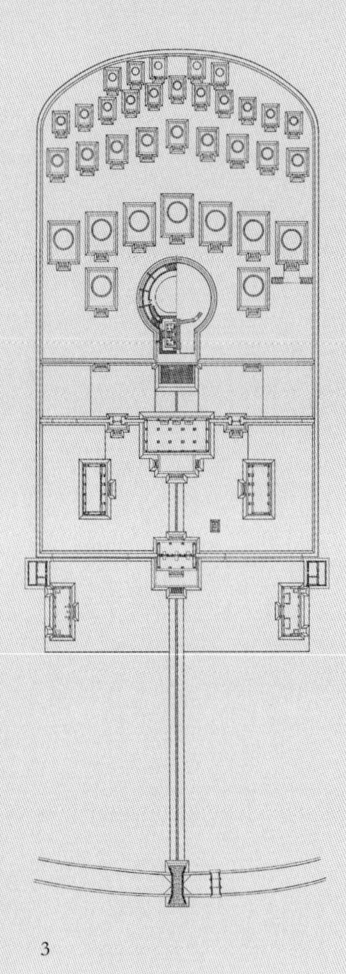

裕陵妃园寝
(Yuling Feiyuanqin)
The Imperial Consorts' Tombs affiliated with the Tomb of Prosperity
乾隆十二年（1747）
The 12th year of the Qianlong Emperor's reign (1747)

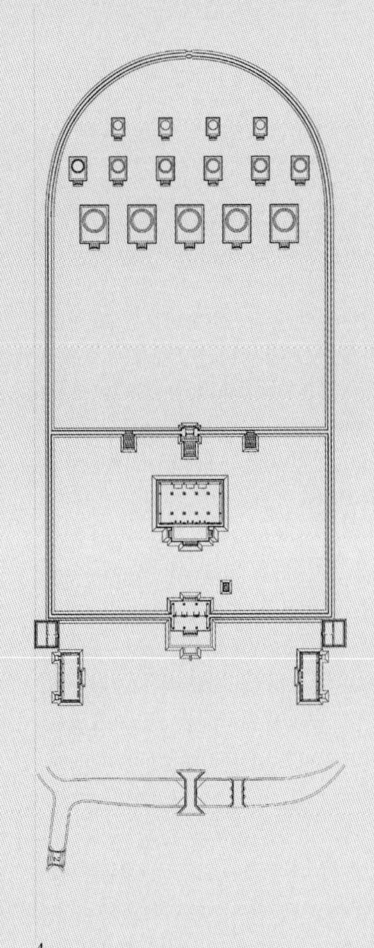

定陵妃园寝
(Dingling Feiyuanqin)
The Imperial Consorts' Tombs affiliated with the Tomb of Stability
咸丰九年（1859）
The 9th year of the Xianfeng Emperor's reign (1859)

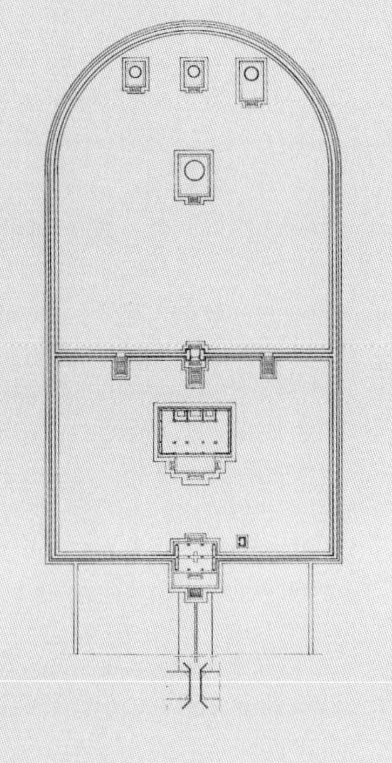

惠陵妃园寝
(Huiling Feiyuanqin)
The Imperial Consorts' Tombs affiliated with the Tomb of Benevolence
光绪元年（1875）
The 1st year of the Guangxu Emperor's reign (1875)

Fig.17 The comparisons among the site plans for the imperial consorts' tombs in the Eastern Qing Tombs. Survey drawings by Tianjin University.

and the screen wall of glazed tiles (*liuli yingbi*).

Among the empresses' tombs, the hall of monumental grace (*long'endian*) in the West Tomb of Brightness (Zhaoxiling) was an exception. With a hip roof and double eaves, it was built by relocating a five-room new building from the east of the Palace of Kindness and Tranquillity (Ci'ninggong) in the Forbidden City, by order of the Kangxi Emperor who made the architectural drawing in person in the 27th year of his reign (1688). In addition, the partition wall between the gate of monumental grace (*long'enmen*) and the hall of monumental grace (*long'endian*) as well as the three gates with glazed tiles (*liuli huamen*) in the middle of the wall were also exceptional. In the 3rd year of the Yongzhen Emperor's reign (1725), the underground palace (*digong*), the encircled realm of treasure (*baocheng*), the square walled terrace and the memorial tower (*fangcheng minglou*) and the five stone ritual vessels (*shiwugong*) were added. Due to topographical limitations, the gate of the tomb (*lingqinmen*) was built on both sides of the hall of monumental grace (*long'endian*). Following the example of emperors' tombs, a pavilion for the stela on the spirit way (*shendao beiting*) was built in the very front.

In the 57th year of the Kangxi Emperor's reign (1718), after the Empress Xiaohui was buried at the East Tomb of Filial Piety (Xiao Dongling), consorts and concubines of the Shunzi Emperor previously buried in the tomb for imperial consorts in the west of the Eastern Qing Tombs were re-interred in the East Tomb of Filial Piety (Xiao Dongling). Each consort and concubine had their own round tumulus (*yuanqiu baoding*) of different sizes covering their underground palaces. These tumuli were lined up symmetrically on both sides of the square walled terrace with the memorial tower (*fangcheng minglou*). This formed a special layout for the empress and the imperial consorts' tombs.

The East Tomb of Stability (Ding Dongling) in Puxiang Valley and the East Tomb of Stability (Ding Dongling) in Putuo Valley were the last-built of Eastern Qing Tombs. They were planned and built with the same layout and lay in parallel, each with an oval-shaped tumulus, wider in the front and narrower at the back, a feature that evidently follows that of the Tomb of Stability (Dingling). Two sets of culinary yards for sacrifice (*shenchuku*) and pavilions for the well (*jingting*) were gathered together along a south-north axis on a large platform at the east of the tomb. A pavilion for the stela on the spirit way (*shendao beiting*) was built right in front of each tomb, just as in the West Tomb of Brightness (Zhao Xiling), symbolising that the two empress Dowagers, Cian and Cixi, had held the real power ('reigning behind the curtain') for a long time (Fig. 16).

清东陵各陵寝建筑配置简表　　表 4

	陵寝名	大碑楼及华表	望柱、石像生	龙凤门	下马牌	小碑亭	朝房	宫门	焚帛炉	配殿	享殿	琉璃花门	二柱门	石五供
帝陵	孝陵	●	●	●	●	●	●	●	●	●	●	●	●	●
	景陵	●	●	❶	●	●	●	●	●	●	●	●	●	●
	裕陵	●	●	❶	●	●	●	●	●	●	●	●	●	●
	定陵		●	❶	●	●	●	●	●	●	●	●		●
	惠陵		设望柱，无石像生	❶	●	●	●	●	●	●	●	●		●
后陵	昭西陵				●	●	●	●	●	●	●	●		●
	孝东陵				●		●	●	●	●	●	●		●
	定东陵				●	●	●	●	●	●	●	●		●
妃园寝	景陵妃园寝				❷		❸	●	❹		●	❻ 1 座		
	景陵双妃园寝				❷		❸	●	❹	❺	●	❻ 2 座		
	裕陵妃园寝				❷		❸	●	❹	❺	●	❻ 2 座		
	定陵妃园寝				❷		❸	●	❹		●	❻ 1 座		
	惠陵妃园寝				❷		❸	●	❹		●	❻ 1 座		

（地宫地面部分详后。●为设置，部分建筑规制比较见图 18）

❶ 五间六柱冲天式牌楼门，道光朝统一改称龙凤门。

❷ 木质下马桩。

❸ 灰瓦屋顶。

❹ 帝后陵焚帛炉皆一对，妃园寝仅东面一座。

❺ 景陵双妃园寝、裕陵妃园寝设东配殿。

❻ 景陵双妃园寝和裕陵妃园寝设琉璃花门左右各一座，其余妃园寝仅居中设一座。

注：1. 帝后陵各建筑皆覆黄琉璃瓦，妃园寝除朝房外，皆覆绿琉璃瓦。

2. 帝后陵宫门称隆恩门，享殿称隆恩殿。

3. 朝房为硬山顶，余皆歇山顶。

4. 帝后陵碑亭、隆恩殿皆重檐，妃园寝各建筑皆单檐。

Tab. 4 Building arrangements of each tomb in the Eastern Qing Tombs

	Name of tombs	Great stela tower and ceremonial columns (*dabeilou ji huabiao*)	Ornamental columns and stone statues (*wangzhu shixiangsheng*)	Dragon and phoenix gate (*longfengmen*)	Dismounting stela (*xiamapai*)	Small stela pavilion (*xiaobeiting*)	Reception hall for court officials (*chaofang*)	Gate of the tomb (*gongmen*)	Sacrificial burner (*fenbolu*)	Side hall (*pei dian*)	Sacrificial hall (*xiangdian*)	Gate of glazed tiles (*liuli hua-men*)	Gate of two columns (*erzhu men*)	Five stone ritual vessels (*shi wugong*)
Emperors' tombs	Tomb of Filial Piety (Xiaoling)	●	●	●	●	●	●	●	●	●	●	●	●	●
	Tomb of Admiration (Jingling)	●	●	❶	●	●	●	●	●	●	●	●	●	●
	Tomb of Prosperity (Yuling)	●	●	❶	●	●	●	●	●	●	●	●	●	●
	Tomb of Stability (Dingling)		●	❶	●	●	●	●	●	●	●	●		●
	Tomb of Benevolence (Huiling)		with ornamental columns but without stone statues	❶	●	●	●	●	●	●	●	●		●
Empresses' tombs	The West Tomb of Brightness (Zhao Xiling)				●	●	●	●	●	●	●	●		●
	The East Tomb of Filial Piety (Xiao Dongling)				●		●	●	●	●	●	●		●
	The East Tomb of Stability (Ding Dongling)				●	●	●	●	●	●	●	●		●
Imperial consorts' tombs	Imperial Consorts' Tombs affiliated with the Tomb of Admiration (Jingling Feiyuanqin)				❷		❸	●	❹		●	❻		
	Two Imperial Consorts' Tombs affiliated with the Tomb of Admiration (Jingling Shuangfeiyuanqin)				❷		❸	●	❹	❺	●	❻		
	Imperial Consorts' Tombs affiliated with the Tomb of Prosperity (Yuling Feiyuanqin)				❷		❸	●	❹	❺	●	❻		
	Imperial Consorts' Tombs affiliated with the Tomb of Stability (Dingling Feiyuanqin)				❷		❸	●	❹		●	❻		
	Imperial Consorts' Tombs affiliated with the Tomb of Benevolence (Huiling Feiyuanqin)				❷		❸	●	❹		●	❻		

(●means being constructed. ●with a number indicates variations [see below]. Detailed information on the above-ground parts of the underground palace is discussed later. For comparisons among some building arrangements, see Fig. 18)

❶ The five-bay memorial gateway with six columns soaring skywards (*wujian liuzhu chongtianshi pailou men*) was renamed as the dragon and phoenix gate (*longfenmen*) during the Daoguang Emperor's reign.

❷ The wooden post for dismounting from a horse (*muzhi xiamazhuang*).

❸ The black-tiled roof.

❹ Each emperor and empress' tomb is equipped with a pair of sacrificial burners (*fenbolu*), while each imperial consort's tomb is equipped with a single sacrificial burner on its east side.

❺ Both the Two Imperial Consorts' Tombs affiliated with the Tomb of Admiration (Jingling Shuangfeiyuanqin) and the Imperial Consorts' Tombs affiliated with the Tomb of Prosperity (Yuling Feiyuanqin) have a side hall on the east.

❻ There are two gates with glazed tiles (*liuli huamen*) in the Two Imperial Consorts' Tombs affiliated with the Tomb of Admiration (Jingling Shuangfeiyuanqin) and the Imperial Consorts' Tombs affiliated with the Tomb of Prosperity (Yuling Feiyuanqin), while the other imperial consorts' tombs have only one gate with glazed tiles (*liuli huamen*) in the middle.

Note:

1. All buildings in the emperors' and empresses' tombs are covered with yellow glazed tiles. All buildings in the imperial consorts' tombs are covered with green glazed tiles except for the reception halls for court officials.

2.The main entrances for the emperors' and empresses' tombs are called the gate of monumental grace (*long'enmen*) and the sacrificial halls are called the hall of monumental grace (*long'endian*).

3.Only the reception halls for court officials are built with gabled roofs and the rest are built with hip-gabled roofs.

4.All the pavilions for the stela (*beiting*) and halls of monumental grace (*long'endian*) in emperors' and empresses' tombs are built with double eaves, while all the buildings in the imperial consorts' tombs are built with a single eave.

帝陵
Emperos'tombs

孝陵隆恩殿
The hall of monumental grace at the Tomb of Filial Piety (Xiaoling long'endian)
康熙十八年（1676年）始建
Construction began in the 18th year of the Shunzhi Emperor's reign (1661)

景陵隆恩殿
The hall of monumental grace at the Tomb of Admiration (Jingling long'endian)
康熙十八年（1676年）始建
Construction began in the 15th year of the Kangxi Emperor's reign (1676)

裕陵隆恩殿
The hall of monumental grace at the Tomb of Prosperity (Yuling long'endian)
乾隆八年（1743年）始建
Construction began in the 8th year of the Qianlong Emperor's reign (1743)

定陵隆恩殿
The hall of monumental grace at the Tomb of Stability (Dingling long'endian)
咸丰九年（1865年）始建
Construction began in the 9th year of the Xianfeng Emperor's reign (1860)

惠陵隆恩殿
The hall of monumental grace at the Tomb of Benevolence (Huiling long'endian)
光绪元年（1875年）始建
Construction began in the first year of the Guangxu Emperor's reign (1875)

后陵
Empresses' tombs

昭西陵隆恩殿（复原示意）
The hall of monumental grace at the West Tomb of Brightness (Zhao Xiling long' endian) (conjectural reconstruction drawings)
康熙二十七年（1688年）始建
Construction began in the 27th year of the Kangxi Emperor' s reign (1688)

孝东陵隆恩殿
The hall of immense grace at the East Tomb of Filial Piety (Xiao Dongling long' endian)
康熙三十二年（1693年）始建
Construction began in the 32th year of the Kangxi Emperor' s reign (1693)

定东陵隆恩殿
The hall of monumental grace at the East Tomb of Stability (Ding Dongling long' endian)
同治十二年（1873年）始建
Construction began in the 12th year of the Tongzhi Emperor' s reign (1873)

妃园寝
Imperial consorts' tombs

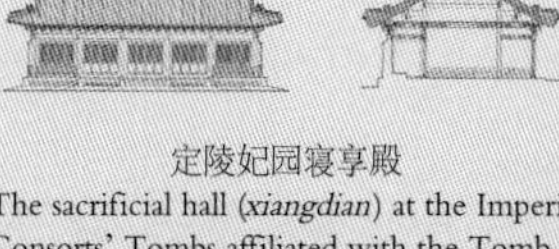

定陵妃园寝享殿
The sacrificial hall (*xiangdian*) at the Imperial Consorts' Tombs affiliated with the Tomb of Stability (Dingling Feiyuanqin)

惠陵妃园寝享殿
The sacrificial hall (*xiangdian*) at the Imperial Consorts'Tombs affiliated with the Tomb of Benevolence (Huiling Feiyuanqin)

1

帝陵
Emperor's tombs

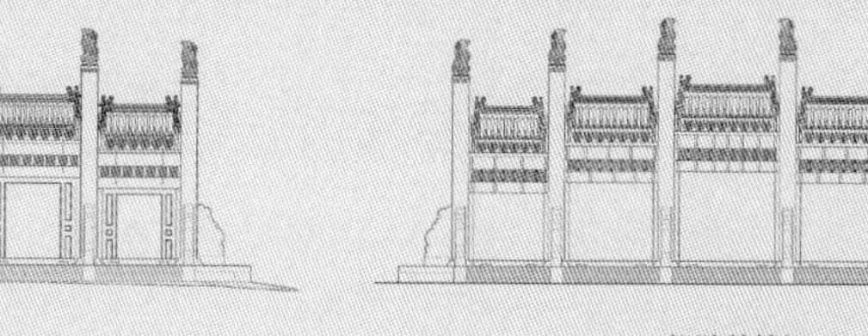
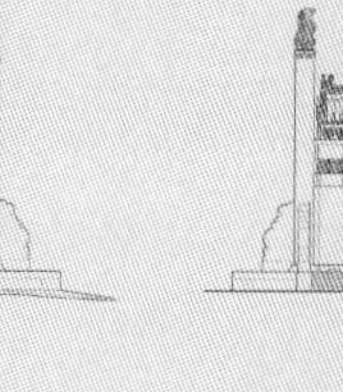

景陵牌楼门
The memorial gateway (*pailoumen*) of the Tomb of Admiration (Jingling)

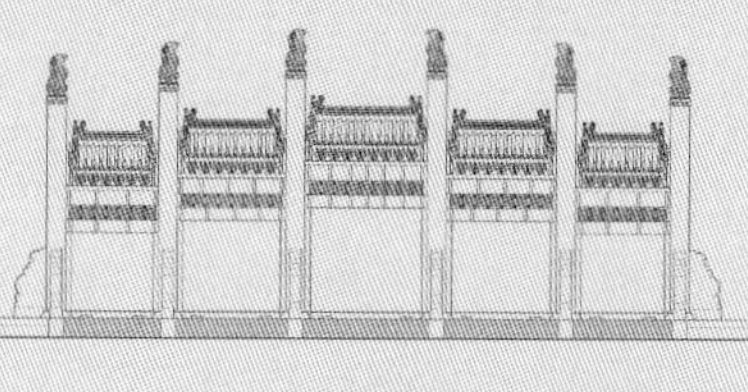
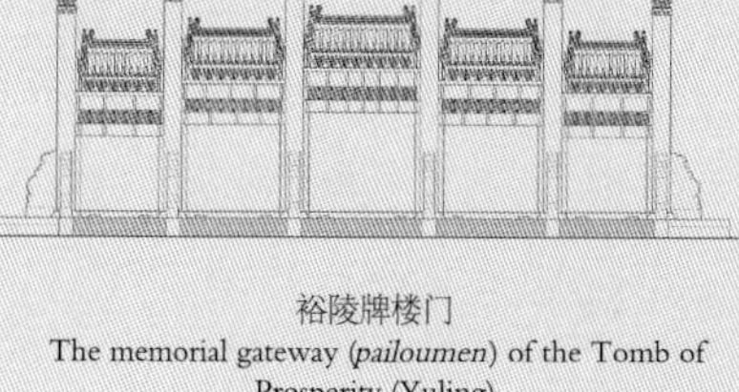

裕陵牌楼门
The memorial gateway (*pailoumen*) of the Tomb of Prosperity (Yuling)

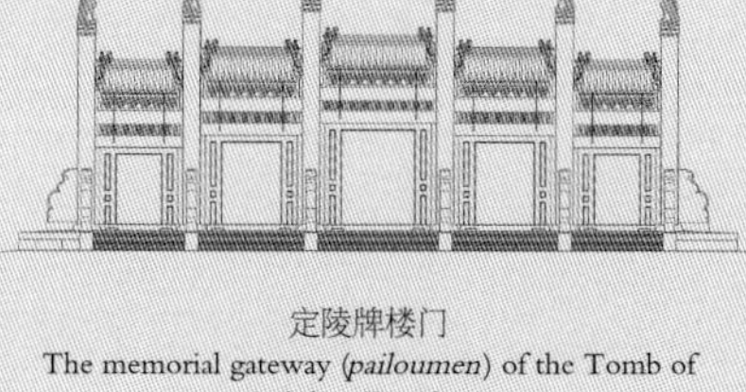

定陵牌楼门
The memorial gateway (*pailoumen*) of the Tomb of Stability (Dingling)

惠陵牌楼门
The memorial gateway (*pailoumen*) of the Tomb of Benevolence (Huiling)

2

帝陵
Emperor's tombs

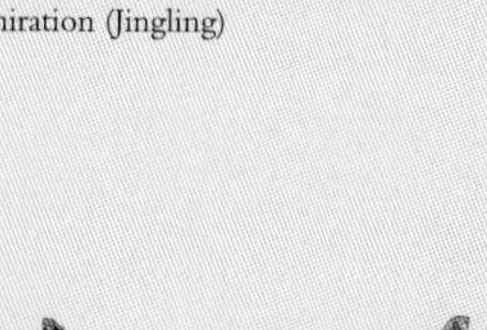

孝陵二柱门
The two-columned gate (*erzhumen*) of the Tomb of Filial Piety (Xiaoling)

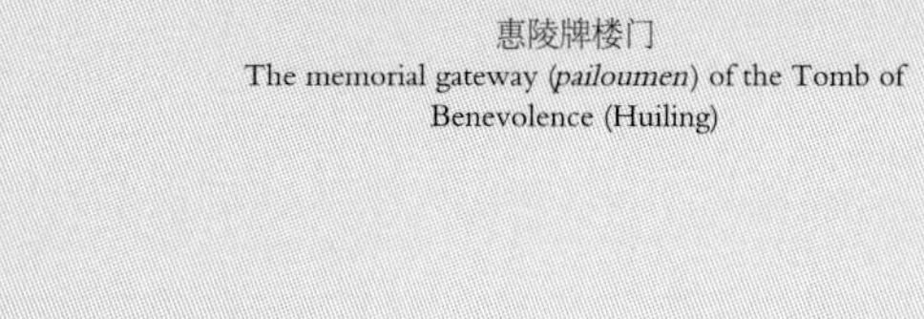

景陵二柱门
The two-columned gate (*erzhumen*) of the Tomb of Admiration (Jingling)

裕陵二柱门
The two-columned gate (*erzhumen*) of the Tomb of Prosperity (Yuling)

3

图18 建筑规制比较。1. 各帝后陵隆恩殿及妃园寝享殿。2. 各帝陵牌楼门。3. 各帝陵二柱门

Fig.18 The comparisons of building arrangements.
1. The halls of monumental grace (*long'endian*) in the emperors' and empresses' tombs and the sacrificial halls (*xiangdian*) in imperial consorts' tombs.
2. The memorial gateways (*pailoumen*) in the different emperors' tombs.
3. The gates with two columns (*erzhumen*) in the different emperors' tombs.
Survey drawings by Tianjin University.

三、清东陵的地宫制度

（一）地宫外部建筑

地宫为各陵寝核心，外部建造方城明楼、宝城和宝顶等，共有六种尊卑等级，见表5、图19–1、图19–2。

地宫外部建筑形制差别（表中●为设置） 表5

	帝	后	皇贵妃	贵妃、妃	嫔、贵人、常在	答应	备注
月台	●	●	●	●	●	●	
方城	●	●	●				
明楼	●	●	●				
哑巴院	●						
月牙城	●						
琉璃影壁	●						
宝城	●	●	●				
宝顶	●	●	●	●	●	●	
隧道	●	●	●	●	●		

The tombs of the imperial consorts' affiliated with their corresponding emperors all followed the building norms established in the construction of the imperial consorts' tombs affiliated with the Tomb of Admiration (Jingling) in the 20th year of the Kangxi Emperor's reign (1681). Wooden posts for dismounting from a horse (*xiamazhuang*) were set at its southern end, with a single-arch stone bridge to the north and a single-arch stone flat bridge to the east. Further north, the main entrance (*gongmen*) was built in the form of a three-room building with a single-eave and hip-gable roof. On each side of the gable walls of the main entrance was a red wall, which extended to the north and formed an enclosure with the wall surrounding the tumuli (*luoquanqiang*) in the north. The three-room duty houses for the guards (*banfang*) and the five-room side houses (*xiangfang*) on both the east and west sides in front of the main entrance were built in grey-tiled gable roofs. The other buildings and walls were roofed with green glazed tiles. Within the main entrance, a five-room sacrificial hall (*xiangdian*) with a single-eave and hip-gable roof stood in the middle. A sacrificial furnace was placed to the southeast of the hall, but with no side houses. Behind the sacrificial hall (*xiangdian*), there is a long wall traversing the whole site, with a gate with glaze tiles (*liuli huamen*) in the middle and two corner doors (*jiaomen*) on each side along the wall. Further beyond to the north lay the backyard, where the underground palaces of all imperial consorts and concubines were arranged in a hierarchical order, with platforms and tumuli on top.

In the fourth year of the Qianlong Emperor's reign (1739), the Imperial Noble Consorts' Tombs affiliated with the Tomb of Admiration (Jingling Huangguifei Yuanqin) were built with arrangements of a higher hiearchy. In the backyard of the tomb complex, two parallel underground palaces were built each with a round-shaped encircled realm of treasure (*baoding*) on top and a square walled terrace with a memorial tower (*fangcheng minglou*) in front. In the front yard, two five-room side halls (*peidian*) were added to the main hall (*xiangdian*) on each side. Both the memorial tower (*minglou*) and the side halls (*peidian*) were covered with a single-eave and hip-gable roof with green glazed tiles. When the Imperial Consorts' Tombs affiliated with the Tomb of Prosperity (Yuling Feiyuanqin) were reconstructed in the 26th year of the Qianlong emperor's reign (1760), the encircled realm of treasure, the square walled terrace and the memorial tower were added on top of the underground palace of the Imperial Noble Consort Chunhui's tomb which was in the centre of the backyard of the tomb complex. Two side halls to the east and west were added in the front yard of the tomb complex. All these reconstructions followed the arrangements for the Imperial Noble Consorts' Tombs affiliated with the Tomb of Admiration (Jingling Huangguifei Yuanqin) for the Kangxi emperor. Because of the site limitations for the original gate with glazed tiles (*liuli huamen*), a new gate with glazed tiles was rebuilt on

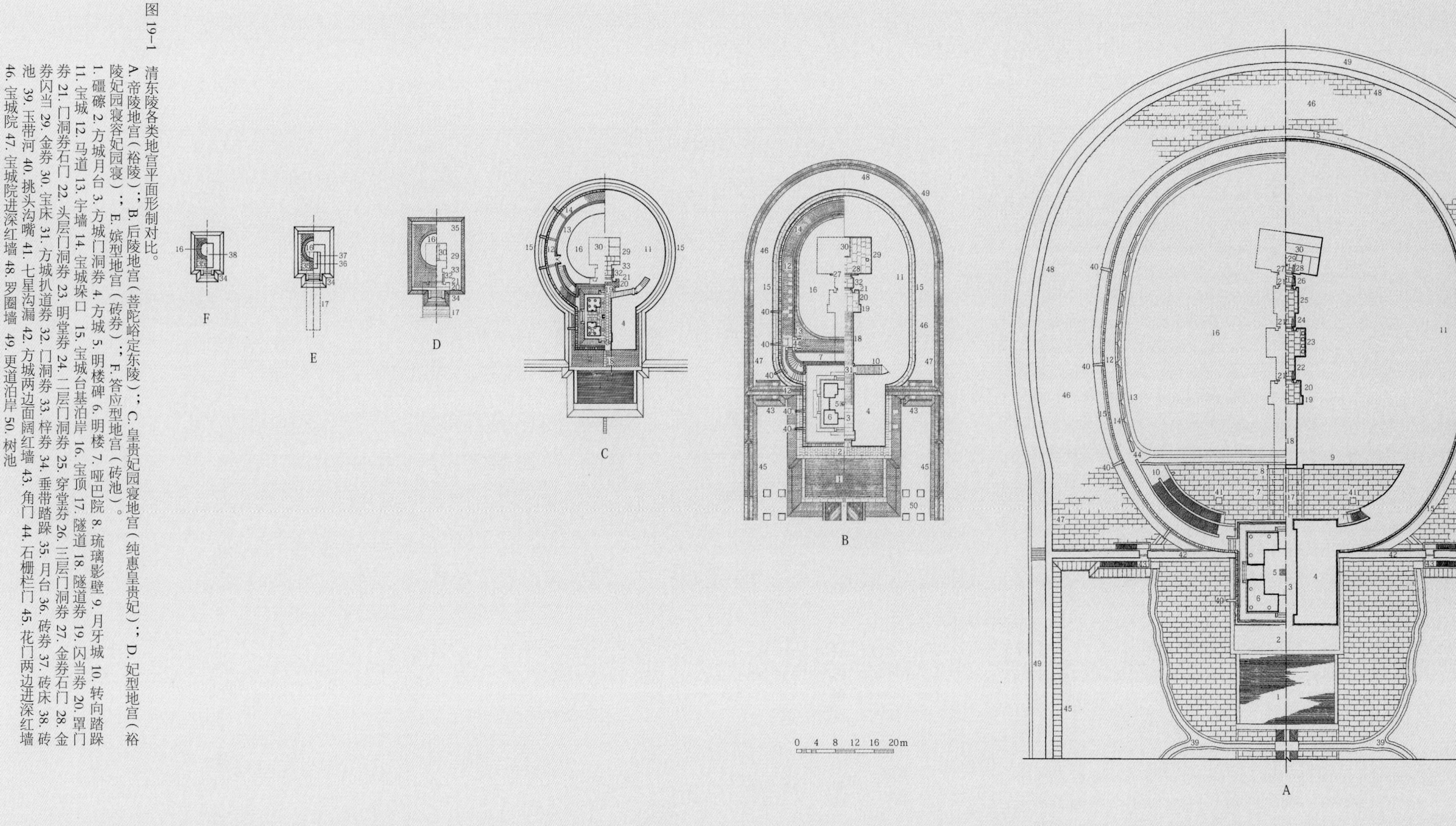

图19-1 清东陵各类地宫平面形制对比。

A. 帝陵地宫（裕陵）；B. 后陵地宫（普陀峪定东陵）；C. 皇贵妃园寝地宫（纯惠皇贵妃）；D. 妃型地宫（裕陵妃园寝容妃园寝）；E. 嫔型地宫（砖券）；F. 答应型地宫（砖池）。

1. 礓礤 2. 方城月台 3. 方城门洞券 4. 方城 5. 明楼碑 6. 明楼 7. 哑巴院 8. 琉璃影壁 9. 月牙城 10. 转向踏跺 11. 宝城 12. 马道 13. 宇墙 14. 宝城垛口 15. 宝城台基泊岸 16. 宝顶 17. 隧道 18. 隧道券 19. 闪当券 20. 罩门券 21. 门洞券石门 22. 头层门洞券 23. 明堂券 24. 二层门洞券 25. 穿堂券 26. 三层门洞券 27. 金券石门 28. 金券闪当 29. 金券 30. 宝床 31. 方城扒道券 32. 门洞券 33. 梓券 34. 垂带踏跺 35. 月台 36. 砖券 37. 砖床 38. 砖池 39. 玉带河 40. 挑头沟嘴 41. 七星沟漏 42. 方城两边面阔红墙 43. 角门 44. 石栅栏门 45. 花门两边进深红墙 46. 宝城院 47. 宝城院进深红墙 48. 罗圈墙 49. 更道泊岸 50. 树池

Fig.19-1 The comparisons among the plans of the various underground palaces in the Eastern Qing Tombs. Survey drawings by Tianjin University.

A. The underground palace of emperors (e.g. The Tomb of Prosperity [Yuling]).

B. The underground palace for empresses (e.g. The East Tomb of Stability [Ding Dongling] in Putuo Valley).

C. The underground palace for imperial noble consorts (e.g. Imperial Noble Consort Chunhui).

D. The underground palace for imperial consorts (e.g. The Tomb of Imperial Consort Rong at the Tombs of Imperial Consorts affiliated with the Tomb of Prosperity [Yuling Feiyuanqin Rongfei Yuanqin]).

E. The underground palace for concubines (in the style of the brick vault).

F. The underground palace for second-class female attendants (in the style of the brick chamber).

1. The ramp (*jiangca*) 2. The platform in front of the square walled terrace (*fangcheng yuetai*) 3. The vault of the passageway under the square walled terrace (*fangcheng mendongquan*) 4. The square walled terrace (*fangcheng*) 5. The stela in the memorial tower (*mingloubei*) 6. The memorial tower (*minglou*) 7. The courtyard of the mute (*yabayuan*) 8. The screen wall of glazed tiles (*liuli yingbi*) 9. The crescent wall (*yueyacheng*) 10. The curve stairway (*zhuanxiang taduo*) 11. The encircled realm of treasure (*baocheng*) 12. The paved path 13. The parapet wall (*yuqiang*) 14. crenellations of the encircled realm of treasure (*baocheng duokou*) 15. The base for the encircled realm of treasure (*baocheng taiji bo'an*) 16. The tumulus (*baoding*) 17. The tunnel 18. The tunnel vault (*suidaoquan*) 19. The recess vault (*shandangquan*) 20. The arched vault covering the main gate (*zhaomenquan*) 21. The stone doors of the vaults of doorways (*mendongquan shimen*) 22. The vault of the first doorway (*touceng mendongquan*) 23. The vault of the ceremonial hall (*mingtangquan*) 24. The vault of the second doorway (*erceng mendongquan*) 25. The vault of the hall of passage (*chuantangquan*) 26. The vault of the third doorway (*sanceng mendongquan*) 27. The stone doors of the golden vault (*jinquan shimeng*) 28. The recess of the golden vault (*jinquan shandang*) 29. The golden vault (*jinquan*) 30. The treasure bed (*baochuang*) 31. The vault that covers the staircase leading up to the square terrace (*fangcheng badaoquan*) 32. The vault of the doorway to the underground palace (*mendongquan*) 33. The transit vault between the vault of the doorway and the main vault (*ziquan*) 34. The stone steps with ramps on both sides (*chuidai taduo*) 35. The platform (*yuetai*) 36. The brick vault 37. The brick bed 38. The brick chamber 39. The jade ribbon river (*yudaihe*) 40. The water outlet spout (*tiaotou gouzui*) 41. The seven-hole stone drainage (*qixing goulou*) 42. The red east-west walls on either side of the square walled terrace 43. The corner gate (*jiaomen*) 44. The stone fence door (*shi zhalanmen*) 45. The two red, north-south walls perpendicular to the east-west wall on either side of the gates of glazed tiles 46. The courtyard between the encircled realm of treasure and the wall surrounding the tomb (*baochengyuan*) 47. The two red, north-south red walls perpendicular to the east-west walls on either side of the square walled terrace 48. The wall surrounding the tomb (*luoquanqiang*) 49. The outside base of the wall surrounding the tomb, along which the guards patrol at night (*gengdao bo'an*) 50. The tree box

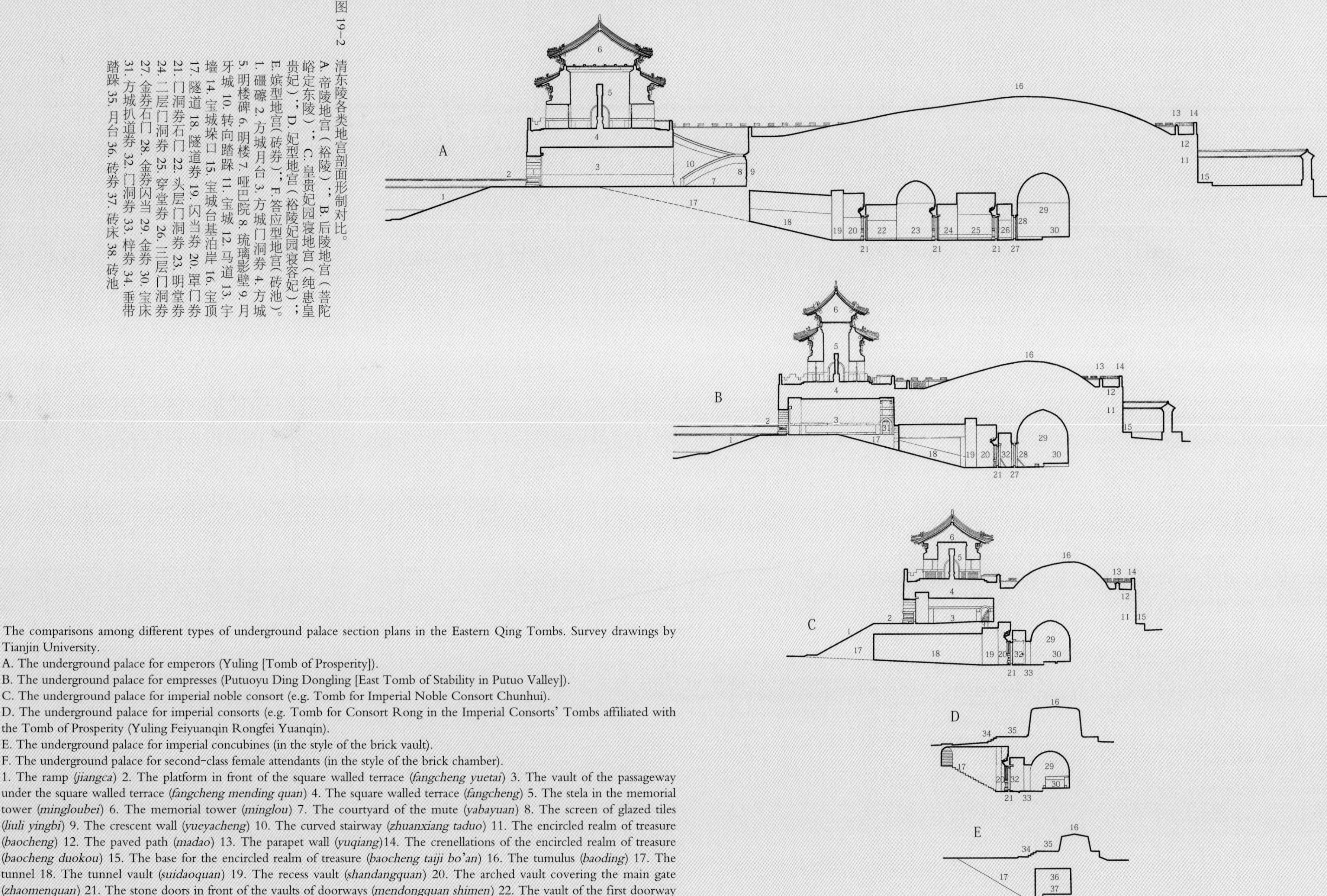

图 19–2 清东陵各类地宫剖面形制对比。A. 帝陵地宫（裕陵）；B. 后陵地宫（菩陀峪定东陵）；C. 皇贵妃园寝地宫（纯惠皇贵妃）；D. 妃型地宫（裕陵妃园寝容妃）；E. 嫔型地宫(砖券)；F. 答应型地宫(砖池)。
1. 礓礤 2. 方城月台 3. 方城门洞券 4. 方城 5. 明楼碑 6. 明楼 7. 哑巴院 8. 琉璃影壁 9. 月牙城 10. 转向踏跺 11. 宝城 12. 马道 13. 宇墙 14. 宝城垛口 15. 宝城台基泊岸 16. 宝顶 17. 隧道 18. 隧道券 19. 闪当券 20. 罩门券 21. 门洞券石门 22. 头层门洞券 23. 明堂券 24. 二层门洞券 25. 穿堂券 26. 三层门洞券 27. 金券石门 28. 金券闪当 29. 金券 30. 宝床 31. 方城扒道券 32. 门洞券 33. 梓券 34. 垂带踏跺 35. 月台 36. 砖券 37. 砖床 38. 砖池

Fig.19–2 The comparisons among different types of underground palace section plans in the Eastern Qing Tombs. Survey drawings by Tianjin University.
A. The underground palace for emperors (Yuling [Tomb of Prosperity]).
B. The underground palace for empresses (Putuoyu Ding Dongling [East Tomb of Stability in Putuo Valley]).
C. The underground palace for imperial noble consort (e.g. Tomb for Imperial Noble Consort Chunhui).
D. The underground palace for imperial consorts (e.g. Tomb for Consort Rong in the Imperial Consorts' Tombs affiliated with the Tomb of Prosperity (Yuling Feiyuanqin Rongfei Yuanqin).
E. The underground palace for imperial concubines (in the style of the brick vault).
F. The underground palace for second-class female attendants (in the style of the brick chamber).
1. The ramp (*jiangca*) 2. The platform in front of the square walled terrace (*fangcheng yuetai*) 3. The vault of the passageway under the square walled terrace (*fangcheng mending quan*) 4. The square walled terrace (*fangcheng*) 5. The stela in the memorial tower (*mingloubei*) 6. The memorial tower (*minglou*) 7. The courtyard of the mute (*yabayuan*) 8. The screen of glazed tiles (*liuli yingbi*) 9. The crescent wall (*yueyacheng*) 10. The curved stairway (*zhuanxiang taduo*) 11. The encircled realm of treasure (*baocheng*) 12. The paved path (*madao*) 13. The parapet wall (*yuqiang*)14. The crenellations of the encircled realm of treasure (*baocheng duokou*) 15. The base for the encircled realm of treasure (*baocheng taiji bo'an*) 16. The tumulus (*baoding*) 17. The tunnel 18. The tunnel vault (*suidaoquan*) 19. The recess vault (*shandangquan*) 20. The arched vault covering the main gate (*zhaomenquan*) 21. The stone doors in front of the vaults of doorways (*mendongquan shimen*) 22. The vault of the first doorway (*touceng mendongquan*) 23. The vault of the ceremonial hall (*mingtangquan*) 24. The vault of the second doorway (*erceng mendongquan*) 25. The vault of the hall of passage (*chuantangquan*) 26. The vault of the third doorway (*sanceng mendongquan*) 27. The stone doors of the golden vault (*jinquan shimeng*) 28. The recess of the golden vault (*jinquan shandang*) 29. The golden vault (*jinquan*) 30. The treasure bed (*baochuang*) 31. The vault that covers the staircase or ramp leading up to the square terrace (*fangcheng badaoquan*) 32. The vault for the doorway to the underground palace (*mendongquan*) 33. The transit vault between the vault of doorways and the main vault (*ziquan*) 34. Stone steps with ramps on both sides (*chuidai taduo*) 35. The Platform (*yuetai*) 36. The brick vault 37. The brick bed 38. The brick chamber

地宫外围建筑以帝陵等级最高，护卫地宫的同时，也具有重要典仪功能。梓宫入葬、帝后谒陵，均在方城明楼前行礼；宝顶则是嗣皇帝举行敷土礼的地方。

石五供北，循礓䃰往上，高大的月台上峙立方城，冠表陵宫，石雕须弥座下肩，城身和贯通南北的门洞券、城上宇墙及垛口皆砖砌。城顶明楼方三间，重檐歇山顶。四面砖墙辟券安隔扇门，内立石雕明楼碑，方须弥座，蟠龙顶，碑身南镌刻满、蒙、汉字皇帝庙号及谥号，填朱。楼外上檐下，南向悬木斗匾，镶嵌铜镀金的满、汉文陵名。

方城门洞券南口设青白石罩门券，内安朱漆实榻门。北口隔扇门外，方城两翼宝城合抱，北面横亘月牙城，围成哑巴院。正对方城门洞，月牙城前砌琉璃影壁，下为地宫入口。哑巴院两侧，各设排水七星地漏，以暗沟通出宝城外石砌泊岸；附于宝城内壁，各设转向踏跺或礓䃰以上达宝城和方城明楼。

宝城顶砖墁马道，两旁砖构宇墙，内侧覆黄琉璃顶，外侧起垛口；墙根石雕荷叶沟连通外泄雨水的挑沟嘴。内宇墙前部分设东西石栅栏门，内为覆盖地宫的丘状小夯灰土宝顶。

each side of the sacrificial hall (*xiangdian*) together with the red wall, following the example of the Tomb of Brightness (Zhaoxiling). Towards the end of the Qing Dynasty, Imperial Consorts' Tombs affiliated with the Tomb of Stability (Dingling Feiyuanqin) and the Tomb of Benevolence (Huiling Feiyuanqin) were again restored following the arrangements for the Imperial Consorts' Tombs affiliated with the Tomb of Admiration (Jingling Feiyuanqin). Thus the Imperial Noble Consort's Tombs affiliated with the Tomb of Admiration (Jingling Huangguifei Yuanqin) and the Imperial Consorts' Tombs affiliated with the Tomb of Prosperity (Yuling Huangguifei Yuanqin) became the only two imperial consorts' tombs with such unique arrangements during the Qianlong period (Fig. 17).

III Building Arrangements for the Underground Palaces in the Eastern Qing Tombs

1. Building above and surrounding the underground palaces

Above and surrounding the underground palace of each tomb is a building complex which includes the square walled terrace with a memorial tower (*fangcheng minglou*), the tumulus (*baoding*), and the encircled realm of treasure (*baocheng*), among others. According to the imperial hierarchy, these structures may be classified into six types (Tab. 5, Figs. 19-1, 19-2).

Tab. 5 Building arrangements of the six types of tombs in the Eastern Qing Tombs

	Emperor (*di*)	**Empress (*hou*)**	**Imperial noble consort (*huang guifei*)**	**Noble consort; Consort (*guifei, fei*)**	**Imperial concubine, fifth rank consort; Noble Lady, sixth rank consort; First-class female attendant (*pin, guiren, changzai*)**	**Second-class female attendant (*daying*)**
Platform (*yuetai*)	•	•	•	•	•	•
Square walled terrace (*fangcheng*)	•	•	•			
Memorial tower (*minglou*)	•	•	•			
Courtyard of the mute (*yabayuan*)	•					
Crescent wall (*yueyacheng*)	•					
Screen wall of glazed tiles (*liuli yingbi*)	•					
Encircled realm of treasure (*baocheng*)	•	•	•			
Tumulus (*baoding*)	•	•	•	•	•	•
Tunnel (*suidao*)	•	•	•	•	•	

Among the building complexes above the underground palaces, those of the emperors' rank highest. These building complexes not only safeguarded the underground palace, but were also the setting for important ceremonies. When the coffins of the emperors and empresses were buried, or when their descendent emperors and empresses came to pay homage at the tombs, ritual ceremonies were performed in front of the square walled terrace with the memorial tower. The tumulus (*baoding*) was where the successor emperor held the ceremony of adding fresh soil.

North of the five stone ritual vessels (*shiwugong*), one may ascend along the ramp, where one finds the lofty square walled terrace (*fangcheng*) standing on the giant platform, dominating and representing the site of the entire tomb complex. The Sumeru base is sculptured in stone. The wall of the square terrace, the vault of the north-south passageway under the square terrace, the parapet wall and the crenellations on top of the parapet wall are all made of bricks. On top of the square walled terrace is the three-bay memorial tower (*minglou*) with a square plan and a gable and hip roof with double eaves. On the four sides of the memorial tower were four brick walls with four arches, each fitted with partition doors. Inside the tower, there is a stone sculptured memorial stela (*mingloubei*), with a square-shaped Sumeru base, and dragon-shaped sculptures incised on the top of the stela. On the stela, the temple title and the posthumous title of the emperor who is buried there are both incised in Manchurian, Mongolian and Chinese languages respectively. The inscriptions are filled with cinnabar pigment. Outside the memorial tower and under the upper eave, there is a suspended wooden board facing southward. The board is inlaid with the name of the tomb in gold plated brass in Manchurian and Chinese.

On the southern side of the vault of the passageway under the square walled terrace (*fangcheng mendongquan*), there is another vault made of green-white stones covering the main gate (*zhaomenquan*). Under the vault covering the main gate are two red-lacquered solid doors. On the northern side of the vault of the passageway under the square walled terrace are partition doors. Beyond these partition doors and extending from both the east-west sides of the square walled terrace are two walls which connect with the encircled realm of treasure (*baocheng*). The north facade of the square walled terrace, the walls extending from the east-west sides of the square walled terrace, and the southern part of the encircled realm of treasure (baocheng) formed an enclosure called the courtyard of the mute (*yabayuan*). The southern part of the encircled realm of treasure inside the *yabayuan* is called the crescent wall (*yueyacheng*). Opposite the northern side of the passageway under the square walled terrace, the screen wall of glazed tiles (*liuli yingbi*) is built in

各后陵的地宫外围建筑，参仿帝陵而尺度缩减，没有哑巴院、月牙城及琉璃影壁。方城门洞券北端封以影壁墙，左右扒道券设踏跺，出上券门，为方城外东西转向平台，有排水沟眼下通吊井桶，折向宝城泊岸外侧水沟门。转向平台北旋设踏跺，往上，在定东陵，可达宝城马道；在其他后陵，则可上至围合宝顶的南部宇墙内，再穿出东西两翼的石栅栏门，达于宝城马道，南折即方城明楼。

各后陵地宫外围建筑尺度不一。方城门洞地面，定东陵呈水平，孝东陵上行斜坡，昭西陵为石踏跺。孝东陵、定东陵扒道券均与方城门洞垂直，昭西陵北斜。孝东陵宝顶中部，两侧分设吊井沟桶，通向宝城泊岸外侧水沟门，明显效法孝陵、景陵袭自明晚期帝陵做法，为清代后陵仅见。此外，宝城及宝顶平面，定东陵呈长圆式，其他后陵皆圆式；定东陵方城后部和宝顶间，以两道宇墙隔出一小哑巴院，也不同于其他后陵只隔一道宇墙。

皇贵妃型地宫外部，定制类同其他妃嫔，矩形月台上，偏北座以圆台形宝顶，与其他妃嫔比较，其尺度最大，前设青白石垂带踏跺五级。乾隆朝经营景陵皇贵妃和裕陵妃园寝三座皇贵妃地宫，外部参仿后陵添建宝城和方城明楼，尺度较后陵更小，明楼覆单檐绿琉璃歇山顶，属于特例（图 19–3）。

其他妃、嫔、答应地宫外均仅设宝顶、月台和踏跺，尺度递减，踏跺按四、三、二级递降。

图 19-3 清东陵各帝、后、皇贵妃方城明楼形制及尺度比较

帝陵
Emperor's tombs

孝陵方城明楼
The Square Walled terrace with the Memorial Tower in the Tomb of Filial Piety (Xiaoling Fangcheng Minglou)

景陵方城明楼
The Square Walled terrace with the Memorial Tower in the Tomb of Admiration (Jingling Fangcheng Minglou)

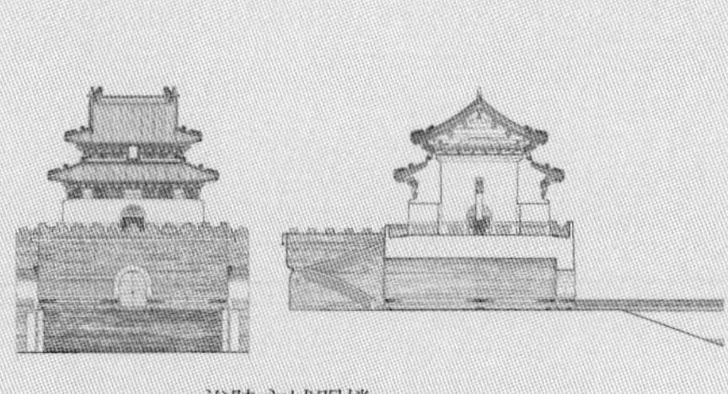

裕陵方城明楼
The Square Walled terrace with the Memorial Tower in the Tomb of Prosperity (Yuling Fangcheng Minglou)

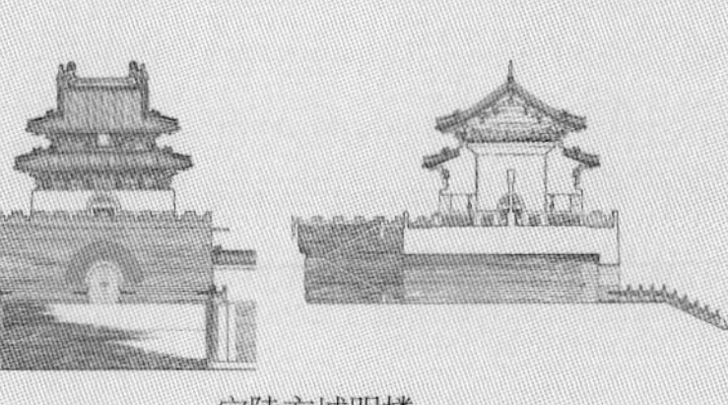

定陵方城明楼
The Square Walled terrace with the Memorial Tower in the Tomb of Stability (Dingling Fangcheng Minglou)

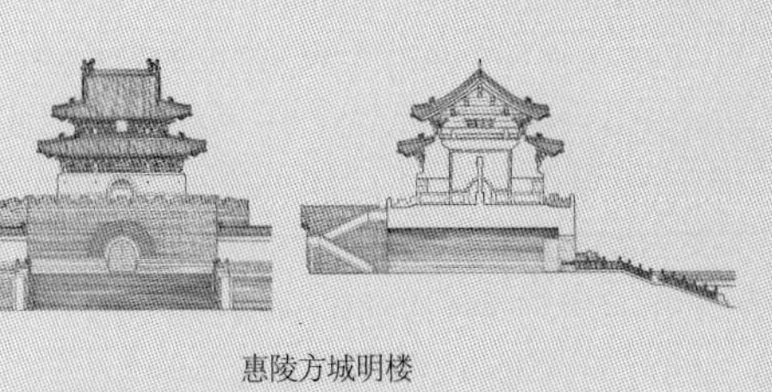

惠陵方城明楼
The Square Walled terrace with the Memorial Tower in the Tomb of Benevolence (Huiling Fangcheng Minglou)

后陵
Empresses' tombs

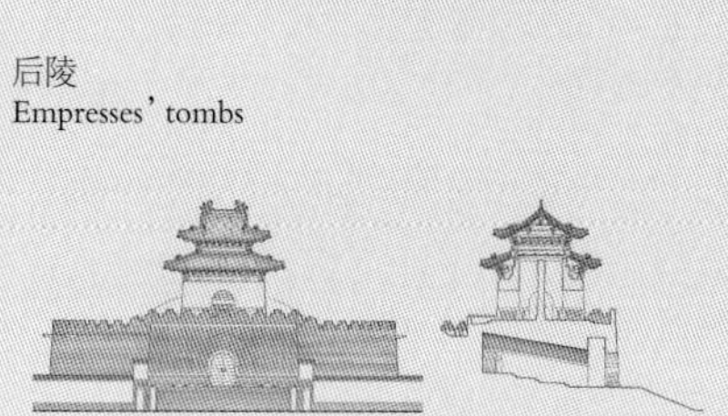

昭西陵方城明楼
The Square Walled terrace with the Memorial Tower in the West Tomb of Brightness (Zhao Xiling Fangcheng Minglou)

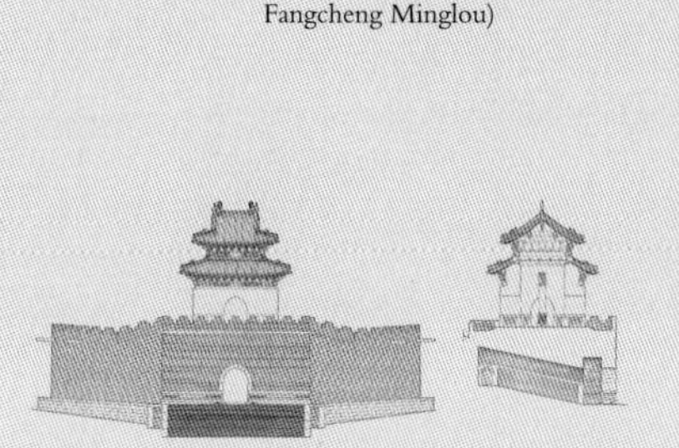

孝东陵方城明楼
The Square Walled terrace with the Memorial Tower in the East Tomb of Filial Piety (Xiao Dongling Fangcheng Minglou)

定东陵方城明楼
The Square Walled terrace with the Memorial Tower in the East Tomb of Stability (Ding Dongling Fangcheng Minglou)

妃园寝
Imperial consorts' tombs

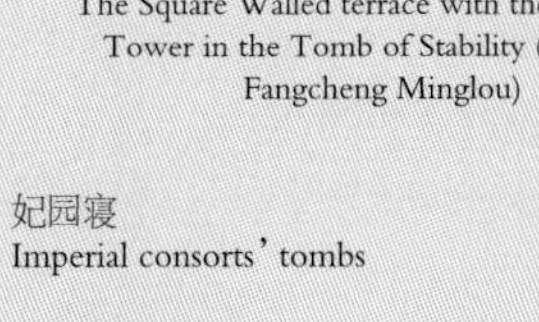

景陵双妃园寝方城明楼
The Square Walled terrace with the Memorial Tower in the Two Imperial Consorts' Tombs affiliated with the Tomb of Admiration (Jingling Shuangfeiyuanqin Fangcheng Minglou)

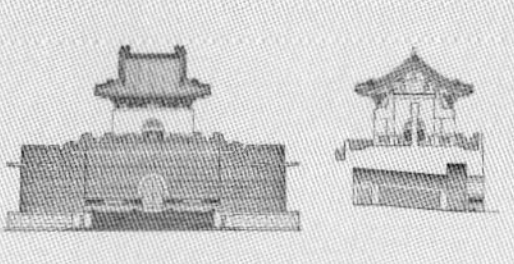

裕陵妃园寝方城明楼
The Square Walled terrace with the Memorial Tower in the Imperial Consorts' Tombs affiliated with the Tomb of Prosperity (Yuling Feiyuanqin Fangcheng Minglou)

Fig.19−3 The comparisons of the form and scale of the square walled terrace with the memorial tower in the tombs of emperors, empresses, and imperial noble consorts at the Eastern Qing Tombs.

front of the crescent wall. Underneath the screen wall of glazed tiles is the entrance to the underground palace. On both sides of the courtyard of the mute, seven-hole stone drains are installed. Under the courtyard, rain water is directed by a drainage ditch out of the stone base (*bo'an*) outside the encircled realm of treasure (*baocheng*). Attached to the inside of the encircled realm of treasure, the curved stairway (*zhuanxiang taduo*) or ramp (*jiangca*) guides people to reach the top of the encircled realm of treasure and the top of the square walled terrace and the memorial tower.

There is a brick-paved path on the top of the encircled realm of treasure (*baocheng*). Brick parapet walls (*yuqiang*) are built on both sides of the paved path. The inside parapet wall is covered by a yellow glazed tile roof, whereas the outside parapet wall is shaped with crenellations. At the foot of the outside parapet wall, there is a stone-sculptured lotus-leaf ditch with outlets that drain the water away. In the front part of the inside parapet wall, there are two stone fence-gates—one on the east and one on the west. Within the inside parapet wall is the tumulus (*baoding*) which is made of clay and covers the underground palace.

With the empresses' tombs, all the building complex above the underground palace followed the forms of the emperors' tombs, but all were scaled down. The empresses' tombs do not have the courtyard of the mute (*yabayuan*), the crescent wall (*yueyacheng*), and the screen of glazed tiles. The northern side of the vault of the passageway under the square walled terrace (*fangcheng mendongquan*) is blocked with a screen wall. On both the left (west) and right (east) side of the passageway is an arch-covered, curved stairway leading up to a gate known as the gate at the upper-end of the arch (*shangquanmen*). Through this gate is a landing platform on both the east and west sides of the square walled terrace. On the platform there is a drainage channel connected with the vertical drainage channel (*diaojingtong*) under the square terrace. The vertical drainage channel turns and directs water to outlets on the outside of the base of the encircled realm of treasure (*baocheng bo'an*). The landing platform turns north, with several steps going up to the top of the encircled realm of treasure. At the East Tomb of Stability (Ding Dongling), following these steps, one may reach the paved path on top of the encircled realm of treasure. At the other empresses' tombs, one may ascend to the tumulus enclosed by the inside parapet wall. One may then exit the inside parapet wall through the two stone-fence gates at the east and west, thus reaching the paved path on top of the encircled realm of treasure. When turning south, one arrives at the top of the square terrace with the memorial tower (*fangcheng minglou*).

（二）地宫内部形制

同外部建筑对应，六类地宫内部形制主要差别如表6：

地宫内部形制差别（●为设置） 表6

	帝	后	皇贵妃	贵妃、妃	嫔、贵人、常在	答应
挡券墙	●	●	●	●	●	
隧道券	●	●	●			
闪当券	●	●	●			
罩门券	●	●	●			
头层门洞券	●	●	●			
头层石门	●	●	●			
明堂券	●					
二层门洞券	●					
二层石门	●					
穿堂券	●					
三层门洞券	●					
三层石门	●					
金券闪当	●	●	梓券	梓券		
金券石门	●	●				
金券	●	●	正券	石券	砖券	砖池
宝床	●	●	石床	石床	砖床	

在清东陵，等级最高、规模最大的各帝陵地宫规制趋同，按晚明帝陵『前殿、中殿、后殿，重门相隔』意向，自方城门洞中央砖砌隧道北下，穿哑巴院，从月牙城底通入地宫。往里，顺次贯联隧道券、闪当券、罩门券、头层门洞券、明堂券、二层门洞券、穿堂券、三层门洞券、金券；四重对开的石门分设在各门洞券和凹入金券南壁的闪当内。

三座门洞券石门前，均对置雕作须弥座及马蹄柱的门对，覆枋子瓦片即雕出门簪、瓦垄和脊饰的单檐庑殿顶，背后横贯黄铜大管扇，与门枕管束门扇转动。金券石门无门对和枋子瓦片，大黄铜管扇南面显作门上槛。

With regard to the empresses' tombs: the scales of the building complex above the underground palace of the empresses' tombs vary. The construction of the ground of the passageway under the square walled terrace (*fangcheng mendong*) provides examples: at the East Tomb of Stability (Dingling), the ground is level; at the East Tomb of Filial Piety (Xiao Dongling), the ground is an ascending sloping path; at the West Tomb of Brightness (Zhao Xiling), the ground takes the form of several stone steps. At the East Tomb of Filial Piety (Xiao Dongling) and the East Tomb of Stability (Ding Dongling), the *padaoquan* (the vault that covers the staircase leading up to the square terrace) is perpendicular to the passageway under the square walled terrace; at the West Tomb of Brightness (Zhao Xiling), the *padaoquan* inclines towards the north. Across the middle of the tumulus of the East Tomb of Filial Piety (Xiao Dongling), there is a vertical underground ditch on the east side and the west side. These vertical underground ditches connect with the drainage outlets outside the base of the encircled realm of treasure. This is the only example of the empresses' tombs that followed the conventions of the emperors' tombs as in the Tomb of Filial Piety (Xiaoling) and the Tomb of Admiration (Jingling), both of which inherited the conventions of the late Ming emperors' tombs. The plans of the encircled realm of treasure and the tumulus at the East Tomb of Stability (Ding Dongling) are both oval, but for the other empresses' tombs, they are round. Another variation is that at the East Tomb of Stability (Ding Dongling), there is a small courtyard of the mute (*yabayuan*) between the two layers of parapet walls, whereas there is only one layer of parapet wall and no courtyard of the mute (*yabayuan*) at the other empresses' tombs.

Regarding the imperial noble consorts' tombs, the form of the building complex above the underground palace is similar to that of the other consorts' and concubines' tombs. On top of the rectangular platform, the round-terrace shaped tumuli of imperial consorts' tombs are in the north. Their sizes are the largest among all the tumuli of the consorts and concubines' tombs. In front of the platform are five green-white stone steps with stone ramps on both sides. During the Qianlong period, when constructing the underground palaces for the three imperial noble consorts' tombs in the Imperial Noble Consorts' Tombs affiliated with the Tomb of Admiration (Jingling Huangguifei Yuanqin) and the Imperial Consorts' Tombs affiliated with the Tomb of Prosperity (Yuling Feiyuanqin), the encircled realm of treasure (*baocheng*) and the square walled terrace with the memorial tower (*fangcheng minglou*) were added to the building complexes above the underground palaces. The forms of these additional structures follow those of the empresses' tombs', but at a smaller scale; the memorial tower (*minglou*) is covered by a roof with green glazed-tiles with a single eave. All these are singular examples (Fig. 19.3).

As for the tombs of the imperial consorts (*fei*), concubines (*pin*), and *daying* (second-class female attendants), there are usually only the tumulus, the platform, and the steps above the underground palace. The scale is reduced accordingly, corresponding to their place in the hierarchy, e.g. the number of steps at the tombs of *fei*, *pin*, and *daying* would be four, three, and two respectively.

2. The building arrangements of the underground palaces

Corresponding to the six types of building complex above and surrounding the underground palaces, there are six types of underground palaces. Their main differences are shown in the Tab. below (Tab. 6).

Tab. 6 The Building arrangements of the six types of underground palaces in the Eastern Qing Tombs

	Emperor (*di*)	**Empress (*hou*)**	**Imperial noble consort (*huang guifei*)**	**Noble consort; Consort (*guifei, fei*)**	**Imperial concubine, fifth rank consort; Noble Lady, sixth rank consort; First -class female attendant (*pin, guiren, changzai*)**	**Second-class female attendant (*daying*)**
Wall at the end of the vaulted arch (*dangquan qiang*)	•	•	•	•	•	
Tunnel vault (*suidaoquan*)	•	•	•			
Recess vault (*shandangquan*)	•	•	•			
Vault covering the main gate (*zhaomenquan*)	•	•	•			
Vault of the first doorway (*touceng mendongquan*)	•	•	•			
First stone doors (*touceng shimen*)	•	•	•			
Vault of the ceremonial hall (*mingtang quan*)	•					
Vault of the second doorway (*erceng mendongquan*)	•					
Second stone doors (*erceng shimen*)	•					
Vault of the hall of the passage (*chuantangquan*)	•					
Vault of the third doorway (*sanceng mendongquan*)	•					
Third stone doors (*sanceng shimen*)	•					
Recess of the golden vault (*jinquan shandang*)	•	•	Transit vault between the vault of doorway and the main vault (*ziquan*)	Transit vault between the vault of doorway and the main vault (*ziquan*)		
Stone doors of the golden vault (*jinquan shimen*)	•	•				
Golden vault (*jinquan*)	•	•	Main vault (*zhengquan*)	Stone vault (*shiquan*)	Brick vault (*zhuanquan*)	Brick chamber (*zhuanchi*)
Treasure bed (*baochuang*)	•	•	Stone bed (*shichuang*)	Stone bed (*shichuang*)	Brick bed (*zhuanchuang*)	

地宫各券皆筒拱，双心圆券形；隧道券、闪当券砖构，其他皆石券，均覆黄琉璃顶，定陵以后改为砖砌蓑衣顶；再筑小夯灰土宝顶掩蔽。俟皇帝葬毕，地宫石门掩闭，隧道券南封砌挡券墙和挡券影壁，隧道填为甬路，倚月牙城筑琉璃影壁，掩蔽地宫入口。

象征前朝后寝的明堂券、金券拱轴垂直于中轴线。明堂券两侧，列置方形须弥座式石雕宝、册座，安木匣专盛帝后的宝、册，皆檀香木制。金券后部，石雕棺床呈凹字形平面，北面称正面宝床，东西为垂手宝床，床沿雕作仰覆莲须弥座。宝床正中凿留圆孔称为金井，上置皇帝梓宫，两旁分列皇后和皇贵妃灵柩，各棺椁四角均卡以精雕彩绘的龙山石（图20）。惠陵地宫仅葬一帝一后，故只设正面宝床，宝、册座改置金券两旁。

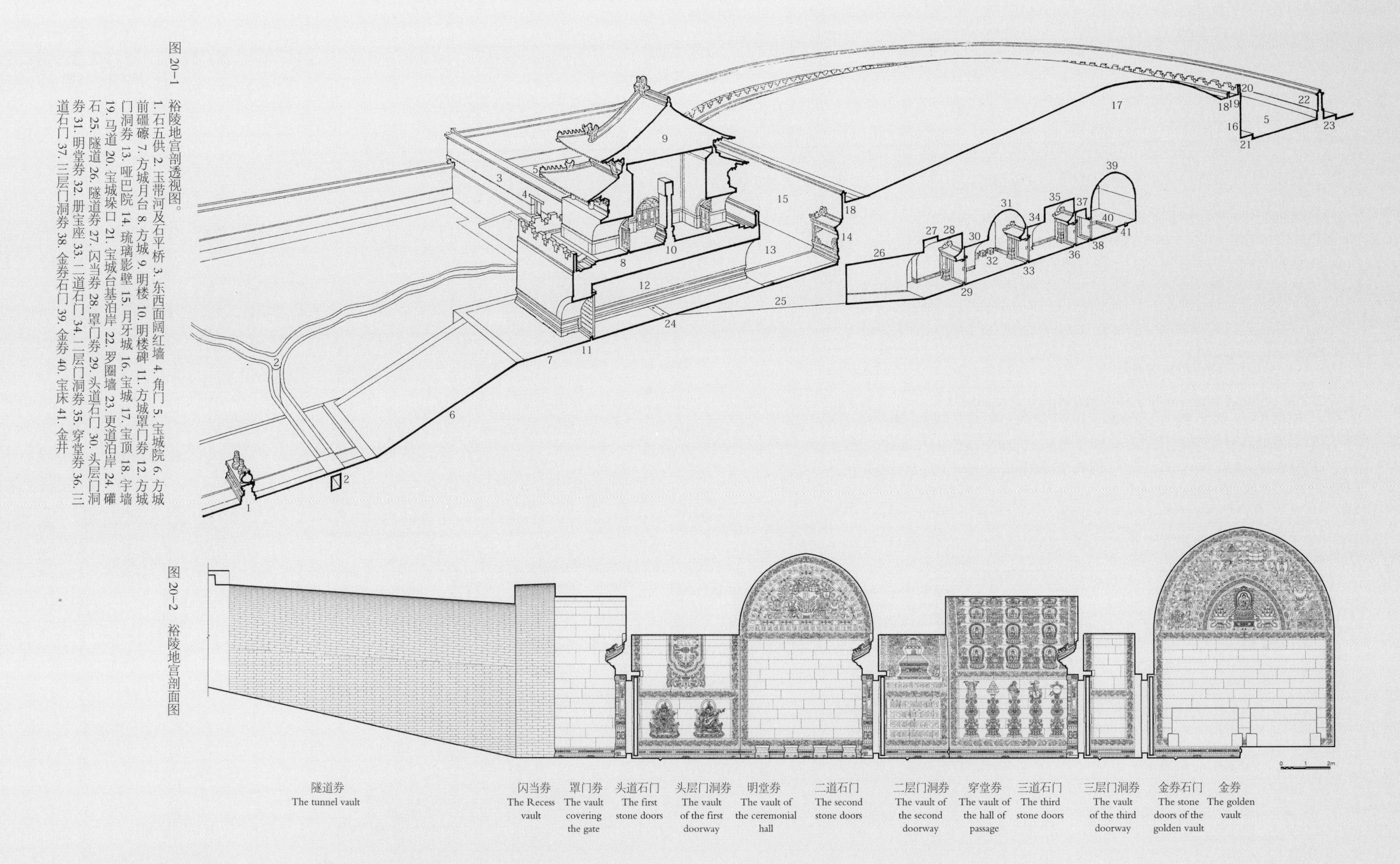

图 20−1 裕陵地宫剖透视图。

1. 石五供 2. 玉带河及石平桥 3. 东西面阔红墙 4. 角门 5. 宝城院 6. 方城前礓礤 7. 方城月台 8. 方城 9. 明楼 10. 明楼碑 11. 方城罩门券 12. 方城门洞券 13. 哑巴院 14. 琉璃影壁 15. 月牙城 16. 宝城 17. 宝顶 18. 宇墙 19. 马道 20. 宝城垛口 21. 宝城台基泊岸 22. 罗圈墙 23. 更道泊岸 24. 礓石 25. 隧道 26. 隧道券 27. 闪当券 28. 罩门券 29. 头道石门 30. 头层门洞券 31. 明堂券 32. 册宝座 33. 二道石门 34. 二层门洞券 35. 穿堂券 36. 三道石门 37. 三层门洞券 38. 金券石门 39. 金券 40. 宝床 41. 金井

图 20−2 裕陵地宫剖面图

Fig.20−1 Perspective Section of the Underground Palace of the Tomb of Prosperity (Yuling Digong).

1. The five stone ritual vessels (*shiwugong*) 2. The jade belt river (*yudaihe*) and a stone flat bridge (*shipingqiao*) 3. The east−west red walls alongside the square walled terrace 4. The corner gates 5. (*baochengyuan*) 6. The ramp in front of the square walled terrace (*fangcheng jiangca*) 7. The platform in front of the square walled terrace (*fangcheng yuetai*) 8. The square walled terrace (*fangcheng*) 9. The memorial tower (*minglou*) 10. The stelae in the memorial tower (*mingloubei*) 11. The vault covering the main gate of the square walled terrace (*fangcheng zhaomenquan*) 12. The vault of the passageway under the square walled terrace (*fangcheng mendongquan*) 13. The courtyard of the mute (*yabayuan*) 14. The screen of glazed tiles (*liuli yingbi*) 15. The crescent wall (*yueyacheng*) 16. The encircled realm of treasure (*baocheng*) 17. The tumulus (*baoding*) 18. The parapet wall (*yuqiang*) 19. The paved path (*madao*) 20. The crenellations of the encircled realm of treasure (*baocheng duokou*) 21. The base of the encircled realm of treasure (*baocheng taiji bo'an*) 22. The surrounding wall of the tomb (*luoquanqiang*) 23. The outside base of the wall surrounding the tomb, along which the guards patrol at night (*gengdao bo'an*) 24. The threshold stone (*guanshi*) 25. The tunnel 26. The tunnel vault (*suidaoquan*) 27. The recess vault (*shandangquan*) 28. The vault covering the gate (*zhaomenquan*) 29. The first stone doors (*toudao shimen*) 30. The vault of the first doorway (*touceng mendongquan*) 31. The vault of the ceremonial hall (*mingtangquan*) 32. The pedestals for the imperial seal and the edict documents (*cebaozuo*) 33. The second stone doors (*erdao shimen*) 34. The vault of the second doorway (*erceng mengdongquan*) 35. The vault of the hall of passage (*chuantangquan*) 36. The third stone doors (*sandao shimen*) 37. The vault of the third doorway (*sanceng mendongquan*) 38. The stone doors of the golden vault (*jinquan shimen*) 39. The golden vault (*jingquan*) 40. The treasure bed (*baochuang*) 41. The golden well (*jinjing*)

Fig.20−2 Section of the Underground Palace of the Tomb of Prosperity (Yuling Digong).

At the Eastern Qing Tombs, the forms of the emperors' underground palaces, which are the highest in hierarchy and the largest in scale, are very similar. They follow the concept of 'front hall, middle hall, and back hall, each separated by double gates' implemented in the imperial tombs of the Ming dynasty. Starting from the passageway under the square walled terrace and following a brick tunnel through the centre of the passageway, one proceeds downwards and northtowards; one moves through the courtyard of the mute (*yabayuan*) and the end of the crescent wall (*yueyacheng*), entering the underground palace. There, one comes across a series of vaults: the tunnel vault, the recess vault, the vault covering the main gate, the vault of the first doorway, the vault of the ceremonial hall, the vault of the second doorway, the vault of the hall of the passage, the vault of the third doorway, and the golden vault (the vault of the main chamber). Four layers of stone double doors are installed in each vault and in the recess of the southern wall of the golden vault.

On both sides of the three stone doors in front of the three vaults of the three doorways, there are a pair of stone-sculptured ornamental columns (*mendui*). Each *mendui* consists of two horseshoe columns and a Sumeru pedestal. Over the stone doors there is the so-called "*fangziwapian*" structure, that is, a hip roof with a single eave (*danyan wudian*) covered with round eave tiles (*walong*) and supported with beams; the door is decorated with door pegs (*menzan*) and the roof has ridge decorations (*jishi*). On the back of the door frame is a large brass lintel (*huangtong daguanshan*). On each end of the brass lintel is an upper pivot seat. Together with the stone hinge on the ground or the lower pivot seat (*menzhen*), they control the rotation of the door. In the golden vault, there are no *mendui* and *fangziwapian*. The southern (front) side of the large brass lintel is visible and serves as the top frame of the door.

All the vaults in the underground palaces are barrel vaults – the shape of the vault consists of two arcs which approximate a semi-circle. The tunnel vaults and recess vaults are made of bricks. The other vaults are all made of stones and are covered with yellow glazed tiles. This arrangement changed with the construction of the Tomb of Stability (Dingling), where all the vaults are built with bricks with a straw raincoat top (*suoyiding*), on top of which is then built the tumulus made of rammed marl as a covering structure. After the burial of the emperor, the stone doors of the underground palaces were closed. A wall and screen were then built at the south end of the tunnel vault. The tunnel would then be filled up to become a path. A screen with glazed tiles would be built against the crescent wall so as to cover the entrance to the underground palace.

Emblematic of the layout of the 'front court and back bedchamber' (*qianchao houqin*) in the imperial palaces, the vault of the memorial hall (*mingtangquan*) and the golden vault

裕陵地宫已开放，精丽堂皇，前无古人。八扇石门浮雕八大菩萨，诸石券壁、顶凸布四大天王、三十五佛、五方佛及狮子、宝塔、五供、八宝、佛花等藏传佛教图案，配饰阴刻经咒番（藏）字29464个，梵（古印度）字647个。图样及经文缜密布局，分件预制，经文按钦准木板字体镌刻，券石专门在倒扣阴模的样制券坑预制。经精心装配，浑然一体，宛若西天梵境，彰显了乾隆作为曼殊师利大皇帝，安藏辑蕃，定国家清平之基于永久的治国理念（图21）。

清代后陵仅菩陀峪定东陵地宫开放，方城门洞中央设隧道北下，迄隧道券入地宫，往北纵贯闪当券、罩门券、门洞券石门、金券石门。隧道券和闪当券砖砌，余皆石构。金券北安设石雕须弥座宝床，中央凿金井，慈禧灵柩安奉其上；床前对置宝、册座。地宫两侧设龙须沟，将渗水排向陵外马槽沟（图22）。

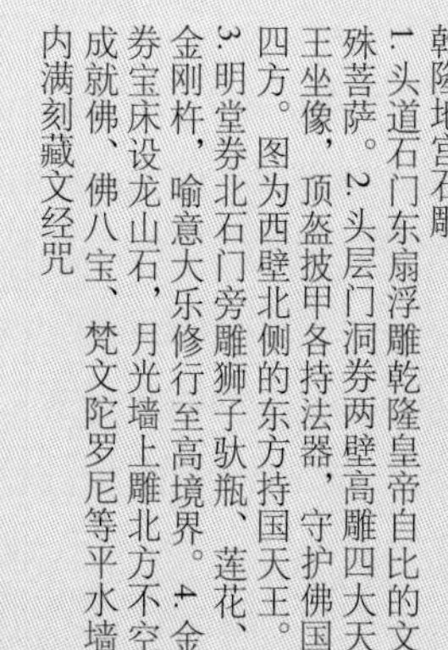

图 21 乾隆地宫石雕。
1. 头道石门东扇浮雕乾隆皇帝自比的文殊菩萨。2. 头层门洞券两壁高雕四大天王坐像，顶盔披甲各持法器，守护佛国四方。图为西壁北侧的东方持国天王。3. 明堂券北石门旁雕狮子驮瓶、莲花、金刚杵，喻意大乐修行至高境界。4. 金券宝床设龙山石，月光墙上雕北方不空成就佛、佛八宝、梵文陀罗尼等平水墙内满刻藏文经咒

图 22 定东陵菩陀峪地宫剖透视图。
1. 礓磜 2. 方城月台 3. 方城罩门券 4. 方城南门洞券 5. 方城北门洞券 6. 扒道券下券门 7. 金刚墙 8. 方城 9. 明楼碑 10. 明楼 11. 宝城城身 12. 马道 13. 宝城垛口 14. 宇墙 15. 宝城台基泊岸 16. 宝顶 17. 隧道 18. 隧道券 19. 闪当券 20. 罩门券 21. 门洞券石门 22. 门洞券 23. 金券石门 24. 金券闪当 25. 金券 26. 宝床及金井 27. 花门院进深红墙 28. 方城两边面阔墙 29. 宝城院进深墙 30. 宝城院 31. 罗圈墙 32. 更道泊岸

Fig.21 The stone sculptures in the Underground Palace of the Qianlong Emperor.
1. The door on the east side of the first stone doors is decorated in relief with the Manjusri bodhisattva, a self-image of the Qianlong Emperor. 2. The two side walls of the vault over the first doorway are both engraved with the Four Heavenly Guardians in a seated position. The Guardians wear helmets and armoury, holding their different musical instruments while guarding the four orientations of the Buddhist realm. The illustration shows the Upholder of the Nation on the north of the western wall guarding the East orientation. 3. The walls on either side of the stone doors north of the vault over the ceremonial hall (*mingtangquan*) are each engraved with the combined pattern of a lion carrying a vase, a lotus flower, and a Vajra (a club with a ribbed spherical head), together indicating that one has reached the Land of Ultimate Bliss through upholding Buddhist practices. 4. There are dragon-mountain stone bolts (*longshanshi*) on the treasure bed under the golden vault (*jinquan*). On the crescent arch with lustrous stones (*yueguangqiang*) are engraved: Amoghasiddhi (one of the Five Wisdom Buddhas), the Buddhist Eight Treasures, and mantras in Sanskrit. The walls beneath the crescent arch with lustrous stones are carved with Tibetan Buddhist incantations.

Fig.22 Perspective Section of the Underground Palace of the East Tomb of Stability in Putuo Valley (Ding Dongling Putuoyu).
1. The ramp (*jiangca*) 2. The platform in front of the square walled terrace (*fangcheng yuetai*) 3. The vault covering the gate of the square walled terrace (*fangcheng zhaomenquan*) 4. The south end of the vault of the passageway under the square walled terrace (*fangchengnan mendongquan*) 5. The north end of the vault of the passageway under the square walled terrace (*fangchengbei mendongquan*) 6. The lower arched gate of the vault that covers the staircase leading up to the square terrace (*padaoquan xiaquanmen*) 7. The wall of the vajra (*jingangqiang*) 8. The square walled terrace (*fangcheng*) 9. The stela in the memorial tower (*mingloubei*) 10. The memorial tower (*minglou*) 11. The retaining wall of the encircled realm of treasure (*baocheng chengshen*) 12. The paved path (*madao*) 13. The crenellations of the encircled realm of treasure (*baocheng duokou*) 14. The parapet wall (*yuqiang*) 15. The base for the encircled realm of treasure (*baocheng taiji bo'an*) 16. The tumulus (*baoding*) 17. The tunnel (*suidao*) 18. The tunnel vault (*suidaoquan*) 19. The recess vault (*shandangquan*) 20. The vault covering the doors (*zhaomenquan*) 21. The stone doors of the vault of the doorway (*mendongquan shimen*) 22. The vault of the doorway (*mendongquan*) 23. The stone doors of the golden vault (*jinquan shimen*) 24. The recess of the golden vault (*jinquan shandang*) 25. The golden vault (*jinquan*) 26. The treasure bed and the golden well (*baochuang ji jinjing*) 27. The two red, north-south walls perpendicular to the east-west wall on either side of the gates of glazed tiles 28. The red, east-west walls on either side of the square walled terrace 29. The two north-south walls in the courtyard between the encircled realm of treasure and the wall surrounding the tomb (*baochengyuan jinshenqiang*) 30. The courtyard between the encircled realm of treasure and the wall surrounding the tomb (*baochengyuan*) 31. The wall surrounding the tomb (*luoquanqiang*) 32. The outside base of the wall surrounding the tomb, along which the guards patrol at night (*gengdao bo'an*)

(*jinquan*) organized along an axis perpendicular to the central axis. On both sides of the vault of the memorial hall, there are square-planned Sumeru pedestals for the imperial seals of treasure (*bao*) and the edict document (*ce*), both of which are placed in specialised sandalwood boxes on top of the Sumeru pedestals. In the back of the golden vault (*jinquan*), there is a stone-sculptured coffin-bed in the form of "凹". To the north is the main treasure-bed (*zhengmian baochuang*); on the east and west are the attending treasure-beds (*chuishou baochuang*). The borders of the beds are sculpted in the shape of Sumeru pedestals decorated with overlapping lotus-petals. In the centre of the treasure-bed is a chiseled round small hole called "the golden well" (*jinjing*), upon which is placed the emperor's coffin (*zigong*, or home chamber). On both sides of the emperor's coffin are the coffins of the empresses and the imperial noble consorts. Each of the four corners of all coffins is fixed to the treasure-bed with a dragon-mountain stone bolt (*longshanshi*) with exquisite carved patterns and painted colours (Fig. 20). In the underground palace of the Tomb of Benevolence (Huiling), only one emperor and one empress are buried. Therefore there is only the main treasure-bed (*zhengmian baochuang*) and instead pedestals for the imperial seals of treasure (*bao*) and the edict document (*ce*) on each side of the golden vault.

The underground palace of the Tomb of Prosperity (Yuling) is open to the public. Its elegance and magnificence are unprecedented. All eight stone gates are decorated in relief with the Eight Great Bodhisattvas of Tibetan Buddhism. All the walls and the areas under the vaults are also decorated in relief with Tibetan Buddhist patterns—the Four Heavenly Guardians, the thirty-five Buddhas, the Five Wisdom Buddhas, the lion, the pagoda, the five ritual vessels, the Eight Treasures, and the Buddhist flowers. These patterns in relief are supplemented with an engraved Tibetan Buddhist incantation consisting of 29,464 Tibetan characters and 647 Brahmi characters. With a thoughtful design, both the patterns in relief and the engraved incantations, consisting of pre-made, separate panels, are all fully integrated. The incantations follow the imperial standard font for woodcut prints. Stones that constitute the vault (*quanshi*) were pre-made in a specialised mould pit for vaults. The whole interior décor is so meticulously assembled that it appears as an integrated whole, creating an illusion of the Buddhist paradise. The décor reveals the Qianlong emperor as the Great Manjusri Emperor and his governance ideal – to pacify Tibet and calm other minority ethnic tribes, thus ensuring the foundation for a peaceful empire (Fig. 21).

Of all the Qing empresses' tombs, only the underground palace of the Empress Dowager Cixi, the Eastern Tomb of Stability at Putuo Valley (Putuoyu Ding Dongling) is open to the public. One enters the tomb through a vaulted passageway under the square walled terrace. Then following

两道石门为雕饰重点，门扇铺首皆兽面仰月。门洞券石门类同帝陵，唯门对上覆单檐歇山顶，檐下出三踩斗栱及麻叶云头，栱眼刻火焰宝珠，椽头镌寿字、『卍』字，勾头镂祥云，滴水雕蝙蝠。金券石门上横黄铜管扇，南面露明为门上槛，前出镂刻祥云龙凤的门簪；管扇上的月光石精雕海水江崖流云和游龙翔凤。精美如此，远胜帝陵。

皇贵妃型地宫开放实例仅见裕陵妃园寝纯惠皇贵妃地宫，由方城月台前隧道券通入，纵贯闪当券、罩门券、门洞券石门、梓券和正券；前三券用砖，后四券石构。石门类同裕陵门洞券石门，雕饰较简。正券北设须弥座式石雕棺床，中央气土眼上置皇贵妃棺椁，前后安掐棺石，东为乾隆废后乌喇那拉氏灵柩。葬后石门掩闭，月台前封砌马尾礓礤（图 23）。

妃型地宫实例，仅见已开放的裕陵妃园寝容妃地宫，由月台南敞口隧道通入，往北配置罩门券、门洞券石门、门洞券、梓券和石券，罩门券砖构，其余均用石，石门形制如纯惠皇贵妃地宫。石券后部通设石雕须弥座式棺床，中央气土眼上安奉容妃棺柩，卡以掐棺石（图 24）。

嫔型地宫多称砖券，无开放实例。按样式雷相关图样及工程档案，月台前预留隧道斜入，无门单券，靠北砖砌须弥座式棺床，中央气土眼上安奉棺椁。葬毕，券外砌挡券砖，建踏跺，隧道填平（图 25）。

答应型地宫见于样式雷相关画样及工程档案，直称砖池或天落池，实际是掘地为竖穴，砖砌敞口方池，无棺床，地面中央留气土眼以安放棺椁。葬后，池口覆盖棚盖石，平砌城砖数层，夯以提溜盖面黄土，再建置月台、宝顶、踏跺和四围散水（图 26）。

Fig.23 Perspective Section of an example underground palace for imperial noble consorts.
1. The tunnel vault (*suidaoquan*) 2. The recess vault (*shandangquan*) 3. The vault covering the main gate (*zhaomenquan*) 4. The stone doors (*shimen*) 5. The vault of the doorway (*mendongquan*) 6. The transit vault (*ziquan*) 7. The main vault (*zhengquan*) 8. The vital energy-earth hole (*qituyan*)

Fig.24 Perspective Section of an example underground palace for imperial consorts.
1. The vault covering the main gate (*zhaomenquan*) 2. The stone doors (*shimen*) 3. The vault of the doorway (*mendongquan*) 4. The transit vault (*ziquan*) 5. The stone vault (*shiquan*) 6. The vital energy-earth hole (*qituyan*)

Fig.25 Perspective Section of an example underground palace for imperial concubines.
1. The tumulus (*baoding*) 2. The tunnel (*suidao*) 3. The brick vault (*zhuanquan*) 4. The brick bed (*zhuanchuang*) 5. The steps with ramps on both sides (*chuidai taduo*) 6. The platform (*yuetai*)

Fig.26 Perspective Section of an example underground palace for second-class female attendants.
1. The tumulus (*baoding*) 2. The brick chamber (*zhuanchi*) 3. The steps with ramps on both sides (*chuidai taduo*) 4. The platform (*yuetai*)

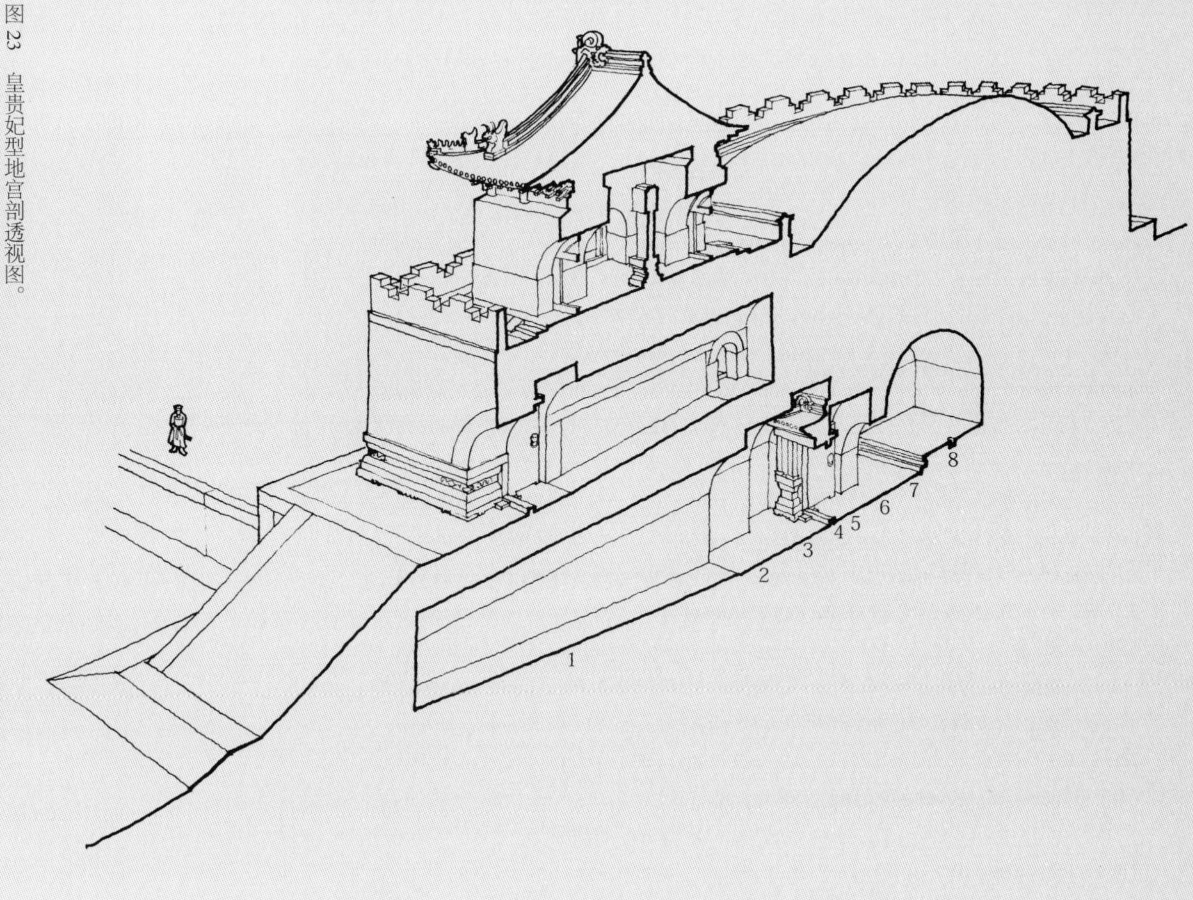

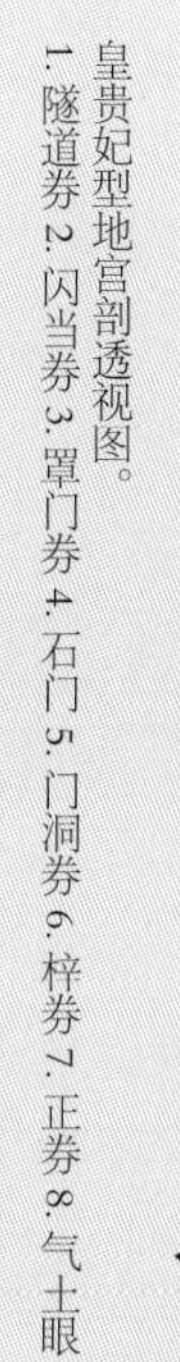

图23 皇贵妃型地宫剖透视图。
1. 隧道券 2. 闪当券 3. 罩门券 4. 石门 5. 门洞券 6. 梓券 7. 正券 8. 气土眼

图24 妃园寝妃型地宫剖透视图。
1. 罩门券 2. 石门 3. 门洞券 4. 梓券 5. 石券 6. 气土眼

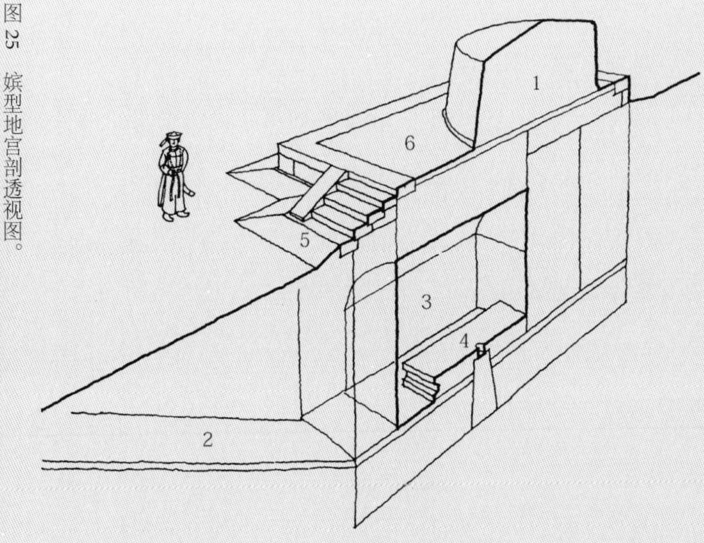

图25 嫔型地宫剖透视图。
1. 宝顶 2. 隧道 3. 砖券 4. 砖床 5. 垂带踏跺 6. 月台

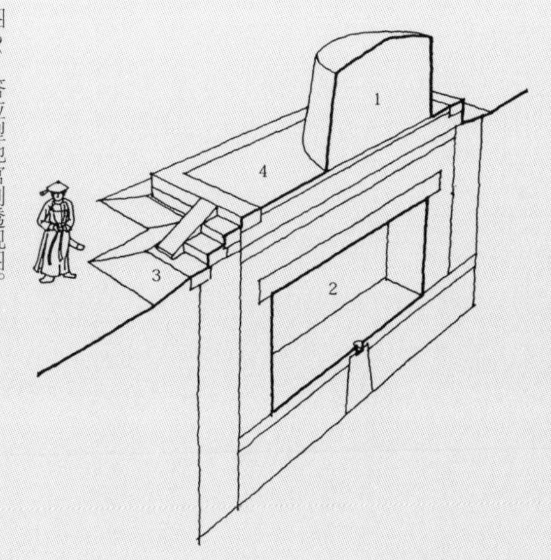

图26 答应型地宫剖透视图。
1. 宝顶 2. 砖池 3. 垂带踏跺 4. 月台

a brick tunnel connected with the passageway, one proceeds downwards and northwards. After passing through the tunnel vault (*suidaoquan*), one enters the underground palace and comes across, first the recess vault, then the vault covering the main gate, then the stone doors of the vault of the doorway, and then the stone doors of the golden vault. Except for the tunnel vault and the recess vault made of bricks, the others are all made of stones. North of the golden vault is the stone-sculptured treasure bed (*baochuang*) with a Sumeru base. In the centre of the treasure bed is the golden well, upon which is laid the coffin of the Empress Dowager Cixi. In front of the treasure bed are two pedestals, one beneath the imperial seal (*bao*) and the other beneath the edict documents (*ce*). On both sides of the underground palace were dragon-whisker drainage channels (*longxugou*), by which the water that seeped through holes in the ground could be directed to the horse-trough ditch (*macaogou*) outside the tomb area (Fig. 22).

The stone doors of the vault of the doorway and the stone doors of the golden vault are the focus of the sculpted-décor. The design of the door knocker (*pushou*) is in the form of an animal head with a ring in its mouth – known as "the animal's head admiring the moon." The stone doors of the vault of the doorway are similar to those in the emperors' tombs. However, the degree of magnificence and elegance well exceeds that of the emperors' tombs. The ornamental columns are covered with a gable and hip roof with a single eave. Under the hip roof there are three brackets (*sancai dougong*). The protruding ends of the beams above the top brackets are shaped in a cloud pattern (*maye yuntou*). In the space between each bracket (*gongyan*) the pattern of "treasure beads in flame" (*huoyan baozhu*) is engraved. At the end of each rafter the character of " 壽 " (*shou*, meaning longevity) or " 卍 " (*wan*, meaning auspiciousness) is engraved. The *goutou* (round tiles at the end of the eave) are engraved with the pattern of auspicious clouds, whereas the *dishui* (tiles hanging beneath the end of the intervening channel) are engraved with the pattern of bats. Across the stone doors of the golden vault is a brass lintel (*huangtong guanshan*), the southern side of which is free from décor, forming the upper frame of the gate, from which there are protruding decorative door pegs (*menzan*), engraved with auspicious clouds, dragons and phoenixes; above the *guanshan* is the crescent arch with lustrous stones (*yueguangshi*), which are engraved with the patterns of sea waves, great rivers and mountains, floating clouds, roaming dragons and flying phoenixes.

Of the underground palaces of the imperial noble consorts, the only one open to the public is that of the Imperial Noble Consort Chunhui. Entering from the tunnel vault in front of the platform of the square walled terrace (*fangcheng*), one passes through a series of vaults: the recess vault (*shandangquan*), the vault covering the main gate (*zhaomenquan*), the stone gate of the vault of the passage (*mendongquan shimen*), the transit vault (*ziquan*) and the main vault (*zhengquan*). The first

四、清东陵的设计意匠

康熙中叶后，各陵寝建设作为国家重大工程，耗银动辄数百万两，按例由皇帝钦派承修大臣组建工程处——又称钦工处，负责规划设计和施工。工程预算另有钦派勘估处审计，奏准后支领经费招商董修。工程处下设样式房和算房，择优录用样子匠（即建筑师）和算手（即会计师）。样式房负责设计，主持人称『掌案』，康熙朝以后主要出自雷氏世家，被世人誉称『样式雷』。

各工程中，凡选勘风水地势，测绘既有建筑以资参考，拟定规划设计和施工方案，制作烫样（即模型），绘制已做现做活计图，记录工程进展，等等，皆须样式雷率同样式房匠人完成，包括大量相关草图（即糙底）、正式图样的细底、准底（即底本），以及精工制作的进呈烫样或画样，还对应有各类文书。清代陵寝的设计程序、理念和方法，根据『世界记忆遗产』的样式雷相关图档，可以空前清晰地展现。

清代陵寝的规划设计，素重『遵照典礼之规制，配合山川之胜势』。各陵寝组群布局和单体形制，要恪遵宗法礼制的统绪关系和尊卑等级，更要追求高山仰止、景行行止的审美意境。这样，样式雷必须随董工官员等赴现场察勘风水，全面调查并统筹工程地质、生态、四至景观及环境容量等要素，最终确定基址（即吉地），进而缜密协同其山水格局，展开规划设计，力使山川自然美与建筑人文美臻向高度统一。

其中，要顾及前行展谒时可以看到建筑组群在其底景（即后宝山）的映托下有序展开，返回时也须有相应对景（即案山或朝山）；陵寝中轴线即山向，就由此优选确定。同时，如何因地制宜，保障地基坚固，便利防洪排水，优化景观与生态，减少土石方量，缩短工期，节省人力和财力；凡此，促成选址和规划设计的密切结合，甚至反复交叉进行（图27）。

这样，地形测量（即『抄平子』）就成为要务。届时，根据山向，用白灰从基址中心（即穴中）向四面划出方格网，按规划设计精度需求，方格尺度20丈、10丈或5丈。然后测量灰线网格上各交点的标高，形成计量性描述地形的文本或图样，即『平子单』『格子本』和『平格样』（图28）。

three vaults were built with bricks, and the following four vaults were built with stone. The shape of the stone gate/door is similar to that of the Tomb of Prosperity, although its decoration is simpler. North of the main vault is the stone-sculptured coffin-bed upon a Sumeru pedestal. At the centre of the coffin-bed is the hole of vital energy-earth (*qituyan*) upon which was laid the imperial noble consort's coffin fixed to the coffin-bed with stone bolts (*qiaguanshi*) in the front and the back. East of the consort's coffin is the coffin of the demoted empress from the Ula Nara clan. After the burial, the stone gate was closed, and a ramp built in front of the platform (Fig. 23).

Of the underground palace of consorts, the only one that is open to the public is that of the Consort Rong in the Imperial Consorts' Tombs affiliated with the Tomb of Prosperity (Yuling Feiyuanqin). Entering the open tunnel at the south of the platform and heading north, one passes through the vault covering the main gate (*zhaomenquan*), the stone door of the vault of the doorway (*mendongquan shimen*), the vault of the doorway (*mendongquan*), the transit vault (*ziquan*) and the stone vault (*shiquan*). The vault covering the main gate was made of bricks, whereas others were made of stone. The form of the stone door is similar to that in the underground palace of the Imperial Noble Consort Chunhui. At the back of the stone vault is a stone-sculptured coffin-bed with a Sumeru pedestal. At the centre of the coffin-bed is the hole of vital energy-earth (*qituyan*) upon which was laid Consort Rong's coffin fixed to the coffin-bed with stone bolts (*qiaguan shi*) (Fig. 24).

Imperial Concubines' underground palaces mostly use brick vaults. None of these are open to the public. According to related drawings in the Yangshi Lei Archive and documentation of engineering works, in front of the platform there is a sloping tunnel that goes down to a single-vault tomb chamber without doors. In the north of the chamber is a brick coffin-bed with a Sumeru base. In the centre of the coffin-bed is the vital energy-earth hole (*qituyan*) upon which was laid the coffin. After the burial, a wall was built outside the vaulted tomb chamber to close it off. The tunnel was then filed up and steps were built in front of the platform (Fig. 25).

As recorded in drawings in the Yangshi Lei Archive and documentation of engineering works, in front of the underground palaces of the second-class female attendants (*daying*) are called brick chambers (*zhuanchi*) or sky drop-down chambers (*tianluochi*). The pattern is: to dig a vertical shaft first, then build an open rectangular chamber with bricks. There is no coffin bed in the chamber. In the centre of the bottom of the chamber is reserved a vital energy-earth hole (*qituyan*) upon which the coffin is laid. After the burial, the chamber will be covered by a large tomb-covering stone (*penggaishi*) upon which a few layers of wall-bricks will be laid. On the surface of the bricks will be rammed yellow earth. Finally, a platform, a tumulus, steps and sloping brickworks to disperse water (*sanshui*) will be built (Fig. 26).

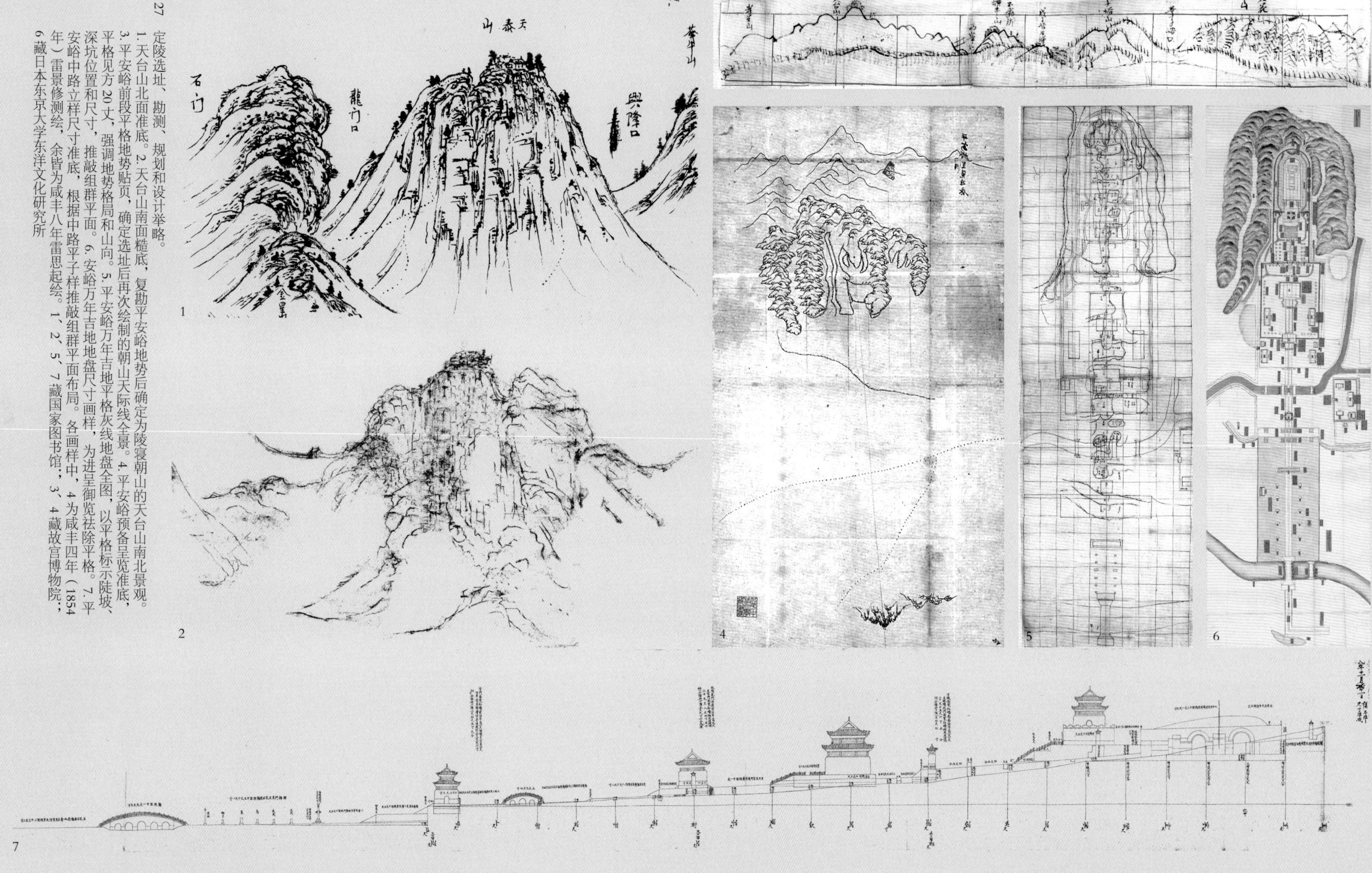

图27 定陵选址、勘测、规划和设计举略。1. 天台山北面准底。2. 天台山南面糙底，复勘平安峪地势后确定为陵寝朝山的天台山南北景观。3. 平安峪前段平格地势贴页，确定选址后再次绘制的朝山天际线全景。4. 平安峪预备呈览准底，平格见方20丈，强调地势格局和山向。5. 平安峪万年吉地平格灰线地盘全图，以平格标示陡坡、深坑位置和尺寸，推敲组群平面。6. 安峪万年吉地地盘尺寸画样，为进呈御览祛除平格。7. 平安峪中路立样尺寸准底，根据中路平子样推敲组群平面布局。各画样中，4为咸丰四年（1854年）雷景修测绘，余皆为咸丰八年雷思起绘。1、2、5、7藏国家图书馆，3、4藏故宫博物院，6藏日本东京大学东洋文化研究所

Fig.27 Tomb of Stability (Dingling) complex-site selection, surveying, planning and design:

1. The *zhundi* (designs drawn to scale with instruments, including detailed measurements and instructions for construction) for the north side of Tiantai Mountain.
2. After surveying the Ping'an Valley landforms, this *caodi* (primary drawing) confirms the south side of Tiantai Mountain as the *chaoshan* (the mountain that the tomb faces) for the tomb complex.
3. One page with a *pingge* (grid plan) showing both the terrain of the front part of Ping'an Valley and a panoramic outline of the *chaoshan* (the mountain that the tomb faces) after the selection of the site was confirmed.
4. *Zhundi* (designs drawn to scale with instruments, including detailed measurements and instructions for construction) for Ping'an Valley awaiting submission to the emperor; the size of the grid is 20 *zhang* (c. 66.6m); the drawing emphasizes the topography and the *shanxiang* (central axis of the tomb complex).
5. Master *pingge* (grid plan) of the construction site in Ping'an Valley; the *pingge* (grid plan) marked the locations and measurements of the mounds and pits, helping to consider the layout of the group of buildings.
6. *Huayang* (architectural drawing) of the construction site in Ping'an Valley; the *pingge* (grid plan) was removed [from the drawing] for presentation to the emperor.
7. *Zhundi* (designs drawn to scale with instruments, including detailed measurements and instructions for construction) of elevations of the group of buildings along the central axis in Ping'an Valley. Based on the master plan of the central axis, this detailed drawing helps to consider the layout of the group of buildings.

All these design drawings were by LEI Siqi in the Xianfeng Emperor's 8th regnal year (1858), except Item 4 which was measured and drawn by LEI Jingxiu in the Xianfeng Emperor's 4th regnal year (1854).

Items 1, 2, 5 and 7 were collected by the National Library of China; Items 3 and 4 were collected by the Palace Museum, China; Item 6 was collected by the Institute for Advanced Studies on Asia, University of Tokyo, Japan.

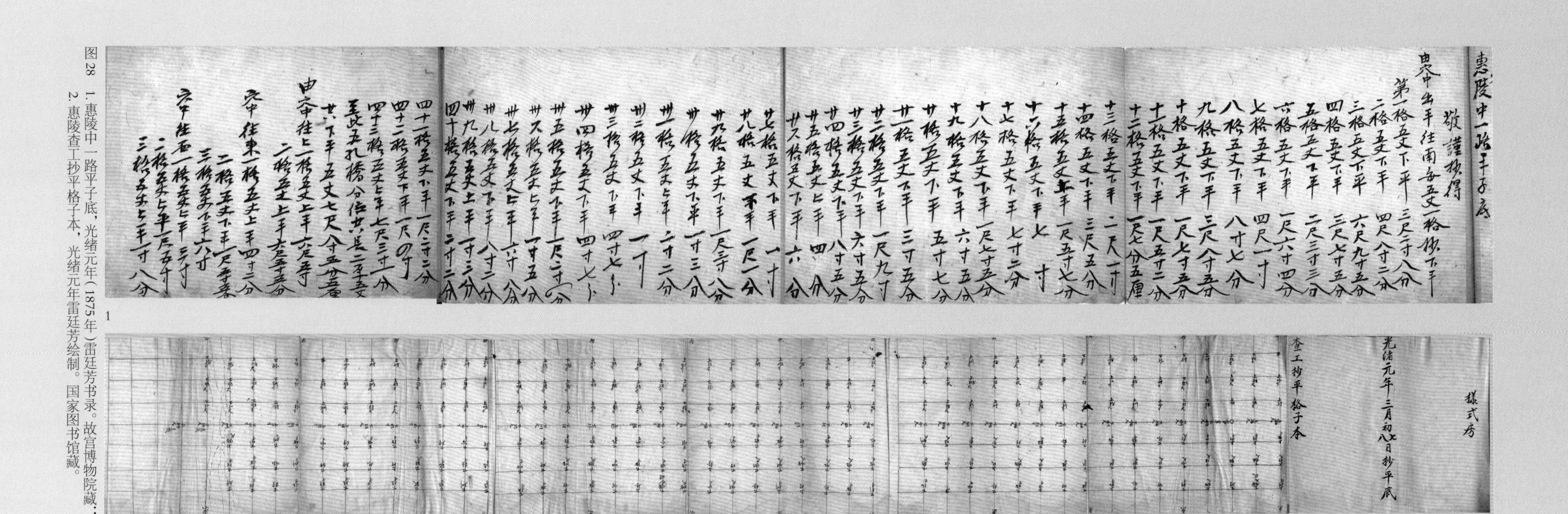

图 28　1. 惠陵中一路平子底，光绪元年（1875 年）雷廷芳书录。故宫博物院藏；
2. 惠陵查工抄平格子本，光绪元年雷廷芳绘制。国家图书馆藏。

Fig.28　1. *Pingzidi* of the first route in the middle area in the Tomb of Benevolence (Huiling) complex, authored by LEI Tingfang in the Guangxu Emperor's 1st regnal year (1875), collected by the Palace Museum.
2. *Geziben* for topographic surveying by the project examiner for the Tomb of Benevolence, authored by LEI Tingfang in the Guangxu Emperor's 1st regnal year (1875), collected by the National Library of China.

IV. The Art of Designing the Eastern Qing Tombs

After the middle of his reign, the Kangxi Emperor placed considerable value on the construction of imperial tombs, costing millions of *taels* (c.50 grams) of silver. Normally, the emperor ordered the ministers responsible for imperial construction to set up a Construction Office, also known as the Qingongchu, whose duties included planning, design and construction supervision. The project budget had first to be examined by auditors and then calls were made for investment after getting approval from the emperor. Under the Construction Office were the Design Office (Yangshifang) and the Construction-cost-control Office (Suanfang) and they had different responsibilities: the former selected the architects, while the latter chose the accountants. The Design Office was responsible for design; the chief architect of the Design Office was called the *zhang'an*. After the latter part of the Kangxi Emperor's reign, the chief architects were mainly from the family of LEI and, therefore, held the honorific title Yangshi Lei.

In projects, when determining the *fengshui* and the topography, surveying and drawing the existing architecture was the first step, followed by designing the masterplans and construction plans, making architectural models (*tangyang*), charts of the already-completed and ongoing design plans (*yizuo xianzuo huoji*), etc. All these were in charge of Yangshi Lei and the Design Office who prepared a variety of primary design drafts (*caodi*), more refined architectural drawings – *xidi* (designs drawn to scale with instruments) and *zhundi* (designs drawn to scale with instruments, including detailed measurements and instructions for construction), as well as working out detailed models for the emperor's inspection with the corresponding documents. The design procedures, concepts and techniques for the Qing imperial tombs are now clearly demonstrated by the relevant records of the Yangshi Lei listed in the Memory of the World Register.

With regard to the planning and design principles for tombs during the Qing Dynasty, great attention was paid to 'following the arrangements for ceremonial rituals and harmonising with the landscape'. The distribution of various building complexes, as well as the form of each individual building, had to follow the conventions and hierarchies of the imperial ritual system and also aspire to pursue both the sublimity and dynamics of the viewing experience. Under these requirements, Yangshi Lei and other officers had to conduct fieldwork in advance to explore the *fengshui* relations of the proposed sites, systematically investigating and comprehensively considering aspects of the geology, ecology, surrounding landscape views, environmental capacity, etc, to eventually

按平子单或平格样的数字，可精准绘出地表断面，完全契合当代数字地面高程模型（DEM），具有高度科学性，用于建筑群的平面布局和竖向设计，也便于核算工程量及控制施工。建筑外部空间设计运用平格，融通传统风水『形势』说，包括同山水胜景有机联系的『百尺为形』的单体和『千尺为势』的组群尺度控制，以『过白』景框『于小者近者之外求其远者大者』，强化组群空间序列的内在关联，前瞻后顾时各建筑间精妙对话，构成『至哉！形势之相异也，远近行止之不同，心目之大观也』。

结合实地勘测完成的组群布局和主体建筑设计方案，要精工绘制并贴签注说，恭呈御览，直至钦准，随即进入陵寝建筑设计的关键步骤——制作烫样。俟御览钦准，才能据以绘制施工设计图、编制《工程做法》并核算工料钱粮。

陵寝烫样，包括地势烫样、全分烫样（即建筑组群布局烫样）、个样（即建筑单体烫样）和细样（即装修烫样）。制作烫样，各分件及其组装，事先要进行设计，绘制分件草图，在样式房内安排样子匠依样制作（图 29）。

在烫样完成前，还要绘制表现图，供皇帝、太后及承修大臣和管理官员等审阅。俟制成呈览，还要提供相关揭看说明，以便按序观览（图 30）。

烫样而外，在纸或布帛上用界尺、毛笔、炭条、颜料绘制图样——也叫画样或样式，是样式雷表达其设计理念、指导施工最主要的手段，存量巨大，图学表现方法丰富多彩，技艺精湛。主要有地盘（即平面）、立样（即立面）、剖面等投影图，包括契合现代理念的旋转剖视图、阶梯剖视图乃至等高线图，以及各类透视表现图。各相关画样每按统一比例绘制，注重尺寸注说，以精准表述建筑的三维空间构成，完全契合现代建筑设计图（图 31 ~ 图 33）。

determine the selected site of the tomb. Afterwards, planning and design procedures began, carefully assessing the landscape pattern in order to ensure the harmonisation of the natural beauty of the landscape and the architectural beauty of the humanity.

The central axis of the tombs, also known as the *shanxiang* (the location of the tomb according to *fengshui*) was determined through considering the views when entering and exiting the site: various groups of buildings had to be well-organised to ensure that they are viewed in a certain sequence: the view on entering the site when the emperor came to visit and pray for the blessings of his ancestors focuses on the building complex against the background of *houbaoshan* (the back of treasure mountain), while the view on exiting the site focuses on the building complex against the background provided by the *anshan* (the mountain located opposite the tomb) or *chaoshan* (the mountain that the tomb faces). Meanwhile, design elements should also be arranged to ensure that the foundations were selected according to local conditions, flood control and drainage, landscape views and ecological conditions, reducing the amount of stone and earthwork required and shortening the construction time in order to save labour and reduce the budget. All these considerations led to site selection and planning design being closely related (Fig. 27).

Therefore, topographic surveying, also known as *chaopingzi*, became a prioritised task. In reference to the central axis of the tomb complex, the surveyors would use lime to draw the grid outwards from the centroid of the site. The grid should be scaled either at 5, 10, or 20 *zhang* (one *zhang*≈3.33m), according to the planned precision of the design. Afterwards, the elevation of each intersection in the grid was measured and marked, thereby formulating texts and images for a qualitative description of the topography, which were called *pingzidan*, *geziben* and *pinggeyang* (Fig. 28).

According to the elevation data from the *pingzidan* or *pinggeyang*, the section plan of the site can be illustrated, thus perfectly fitting, or rather anticipating, the principles of a Digital Elevation Model (DEM). The section plan can be used for building distribution, design elevation, labour quantity evaluation and construction control. The use of *pingge* in designing exterior spaces is based on the principle of '*xing-shi*' in feng shui theory. In order to achieve harmony between buildings and the surrounding landscape, the *pingge* (gridded elevation model) helps to control the design of both the individual buildings and groups of buildings. The principle of "*baichi wei xing*" is used to control the dimensions of individual buildings (i.e. the height, breadth, and the viewing distance of an individual building should be within 100 *chi* [c.23-35m]). The principle of "*qianchi wei shi*" is used

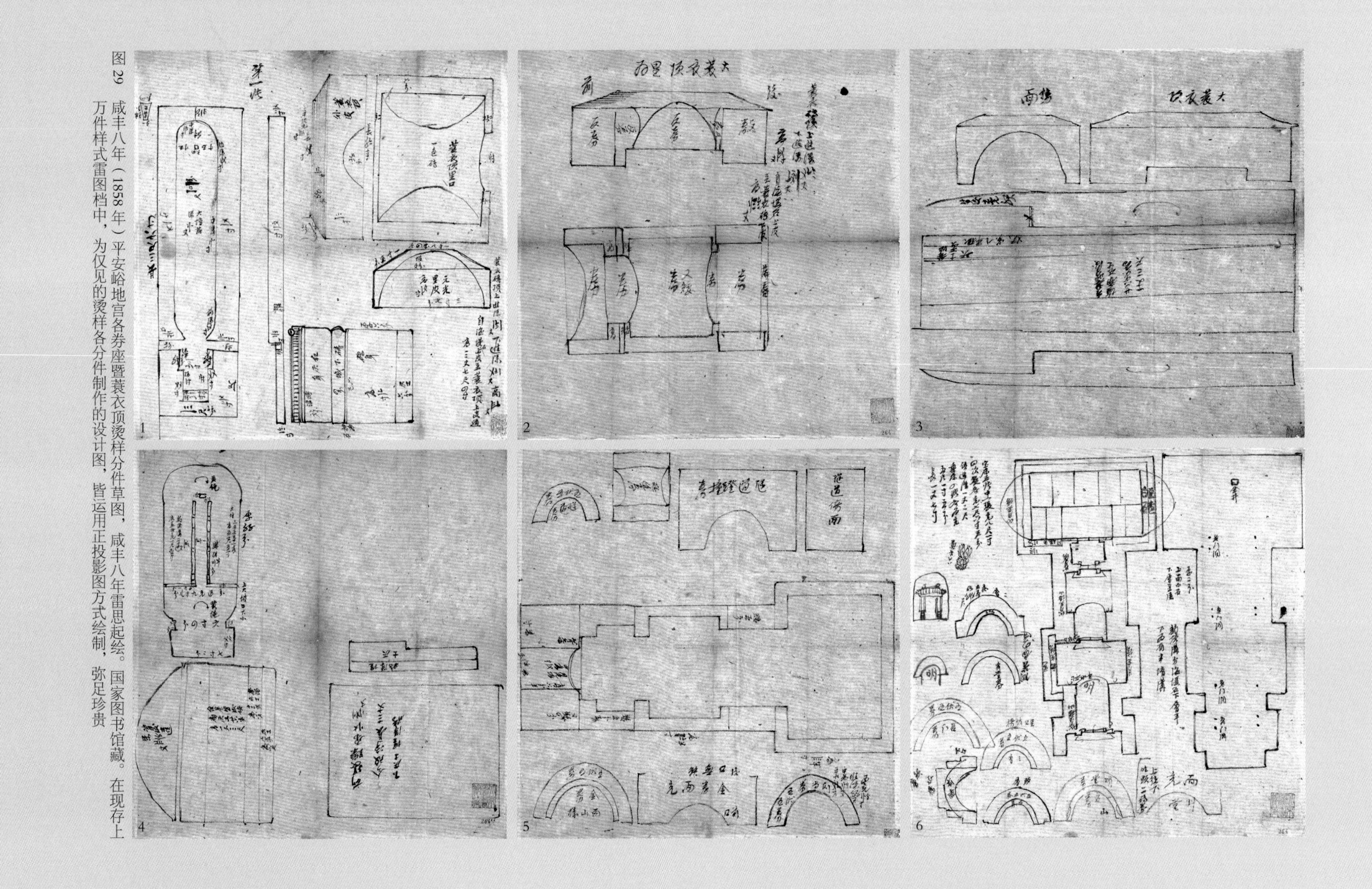

图 29 咸丰八年（1858 年）平安峪地宫各券座暨蓑衣顶烫样分件草图，咸丰八年雷思起绘。国家图书馆藏。在现存上万件样式雷图档中，为仅见的烫样各分件制作的设计图，皆运用正投影图方式绘制，弥足珍贵

Fig.29 Draft drawings of the models for the components of the supports and the straw raincoats of the vaults, authored by LEI Siqi in the Xianfeng Emperor's 8th regnal year, in the National Library of China collection. Among the thousands of drawings in the Yangshi Lei archives, this is the only *tangyang* where the design of each component was drawn on an orthographic projection.

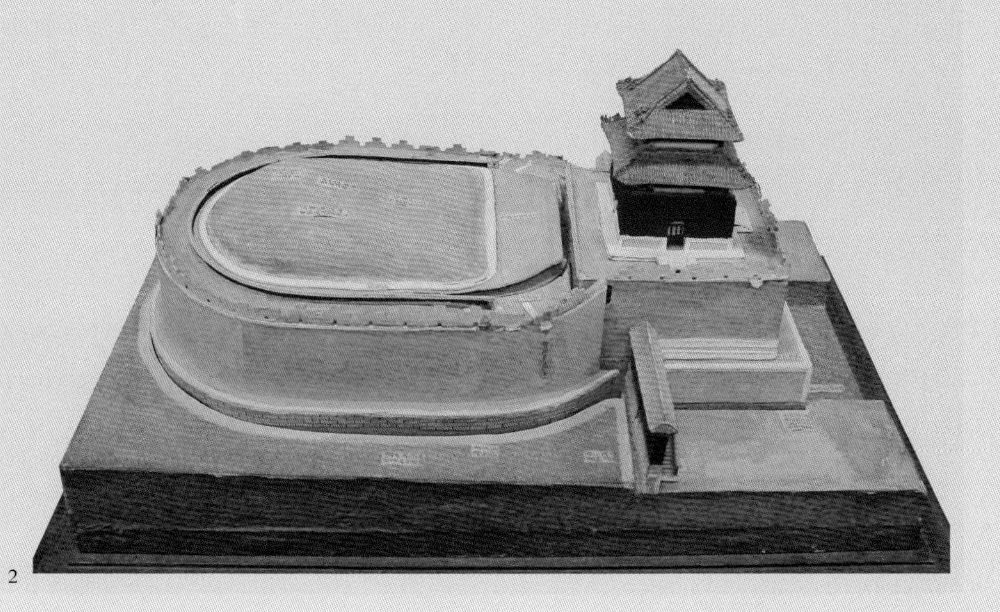

稟

地宮九道券由上往下

先開寶頂盖一層

二開填廂灰土一層

三開蓑衣磚頂一層 此層計三塊

四開各道磚券一層

五開各道石券一層

六係平水牆代石門一層

七係平水背後磚一層

八係石床代海墁石一層

九係灰土代龍鬚溝一層

十至大槽底

由南往北起

頭道踏跺一座 代石欄杆

二道方城明楼一座

三道啞叭院代隧道 轉向踏跺二座

四道琉璃影壁月牙城隧道內背磚

地宮九道券由南往北起

第一層隧道券 上代蹬撞券磚一層

第二層閃當磚券 上 蓑衣磚一層

第三層罩門石券 上 磚券一層

第四層頭層門洞券石券 上 券磚一層

第五層明堂石券 上磚券一層 係奉安冊寶

第六層二層門洞券 上磚券一層

第七層穿堂石券 上 磚券一層

第八層頭層門洞券 上磚券一層

第九層金券 內石床 上磚券一層

平水石九道一連 代石門四層

背後磚九道一連

背後土內 背後一塊 兩邊二塊

外寶城泊岸一道

以上各款均貼簽分斷一至九號

图 30 1. 普祥峪菩陀峪万年吉地宝城明楼西面立样图，故宫博物院藏，同治十二年（1873年）雷思起绘；2. 普祥峪定东陵烫样，清华大学建筑学院藏，雷思起等制；3. 平安峪《地宫烫样揭看次序》，咸丰八年（1858年）十月雷思起拟，国家图书馆藏

Fig.30 1. The elevation for the western side of the *baocheng* (the superstructure of the tomb chamber) and the *minglou* (the memorial tower) on the construction sites inside Puxiang Valley and Putuo Valley, authored by LEI Siqi in the Tongzhi Emperor's 12th regnal year (1873), in the collection of the Palace Museum.
2. The model of the Eastern Tomb of Stability (Ding Dongling) in Puxiang Valley, authored by LEI Siqi, in the collection of the School of Architecture, Tsinghua University.
3. The instructions for the model of the tombs, authored by LEI Siqi in the tenth month of the Xianfeng Emperor's 8th regnal year (1858), in the collection of the National Library of China.

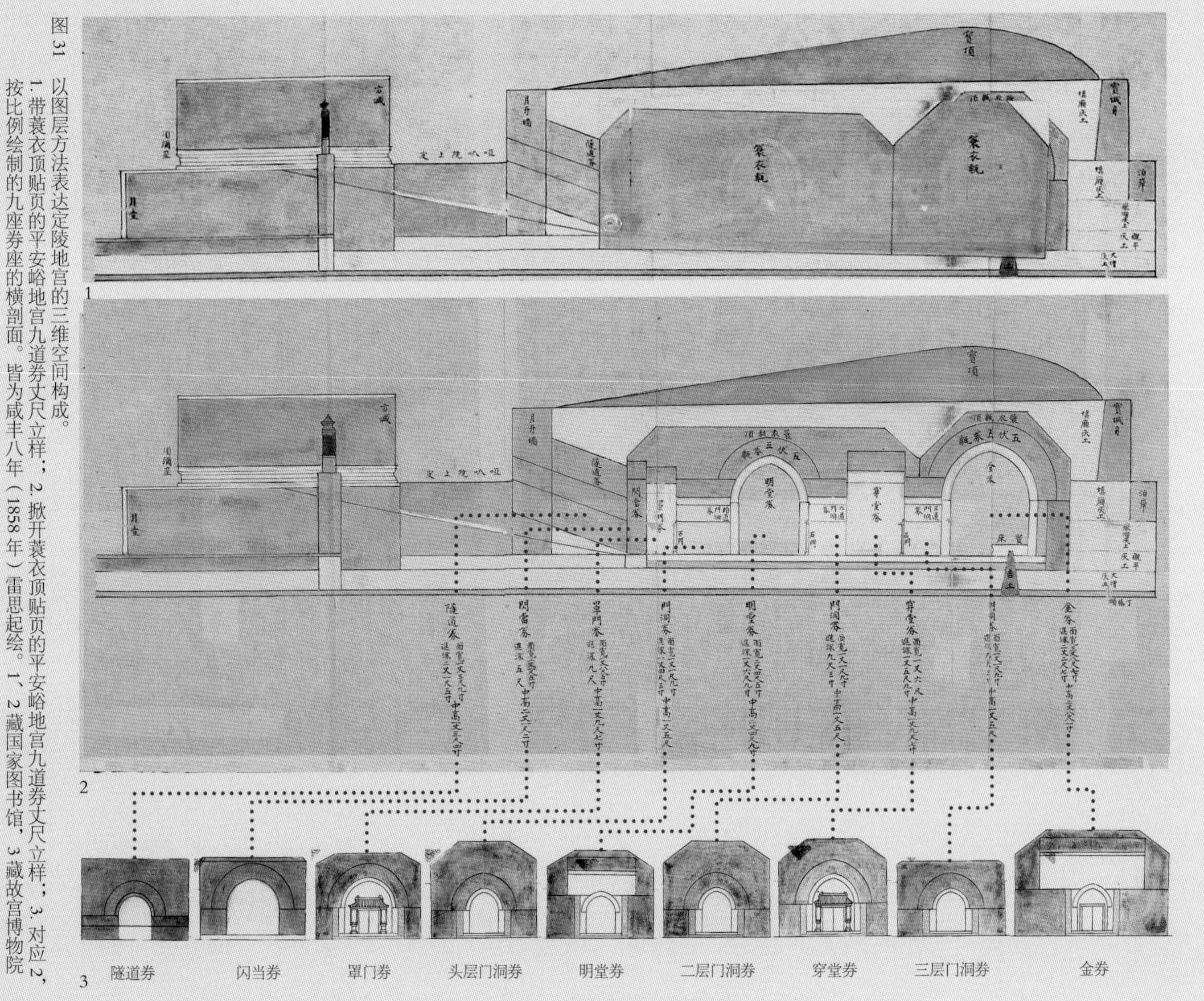

图 31 以图层方法表达定陵地宫的三维空间构成。1. 带蓑衣顶贴页的平安峪地宫九道券丈尺立样；2. 掀开蓑衣顶贴页的平安峪地宫九道券丈尺立样；3. 对应 2，按比例绘制的九座券座的横剖面。皆为咸丰八年（1858 年）雷思起绘。1、2 藏国家图书馆，3 藏故宫博物院

图 32 惠陵五孔石券桥做法层次画样（层层细样），光绪二年（1876 年）雷廷芳绘。国家图书馆藏。以图层方法全面表达了自桥基打桩到桥面望柱雕栏的做法尺寸

Fig.31 A layered method is used to express the three-dimensions of the layout of the Underground Palace, Tomb of Stability:
1. Sections of the nine vaults in sequence in the Underground Palace in Ping'an Valley, with an attached page showing the straw raincoated (*suoyiding*) vaults;
2. Sections of the nine vaults in sequence in the Underground Palace in Ping'an Valley, with the attached page which may be removed to reveal the next layer;
3. Details corresponding to item 2; the cross sections of the nine vaults in sequence drawn to scale.
All items were drawn by LEI Siqi in the Xianfeng Emperor's 8th regnal year (1858). Items 1 and 2 are in the collection of the National Library of China and Item 3 is in the collection of the Palace Museum.

Fig.32 The *huayang* (architectural drawings) of the five-arch stone bridge in the Tomb of Benevolence (Huiling) (multiple layers showing the construction of different levels), drawn by LEI Tingfang in the Guangxu Emperor's 2nd regnal year (1876), in the collection of the National Library of China. The layer method is used to express the size and assembly of the bridge construction from the pile foundations to the carved pillars on the bridge deck.

图 33 定陵隆恩殿景泰蓝五供立样准底。1. 花瓶；2. 香炉；3. 烛台及蜡烛。同治二年（1863 年）雷思起绘。国家图书馆藏

Fig.33 The *zhundi* (refined drawings) of five cloisonné ritual vessels in the Hall of Monumental Grace in the Tomb of Stability (Dingling Long'endian).
1. Vase
2. Incense burner
3. Candlesticks and candles
Drawn by LEI Siqi in the Tongzhi Emperor's 2nd regnal year, in the collection of the National Library of China.

to control the layout of a group of buildings (i.e. the measurement across each of the key spaces, such as a courtyard, and the viewing distance for the next enclosure should be within 1000 *chi* [c.230-350m]). The *pingge* also helps to determine the position of the gates to buildings so that the frame of each gate acts as a viewing frame (for the next building along the central axis). The use of these *pingge* based methods ensures that the groups of buildings form an organic whole with changing spatial relationships. Each individual building has a carefully thought out spatial relationship with adjacent buildings. Together the buildings and groups of buildings in the surrounding landscape afford visitors a multiplicity of ever changing unique experiences on their pilgrimages through the tomb complex.

The field-work-based plan of building distribution and design, carefully elaborated with clear notes, had to be submitted to the emperor for review and approval. Afterwards, the project was able to move to the key stage of tomb design – making *tangyang* or models. Only when such models were approved by the emperor, the following stages of making construction drawings, compiling *Construction Methods* (*Gongcheng Zuofa*), and accounting material and monetary expenditures could proceed.

Models (*tangyang*) of a tomb include topographical models (*dishi tangyang*), models showing the distribution of the building groups (*quanfen tangyang*), models of individual building entities (*geyang*) and models of interior decoration (*xiyang*). In building a model, the components and corresponding assembly should be planned and based on drafts of the components. They were finally manufactured by craftsmen in the Design Office (Fig. 29).

Before the completion of the *tangyang* models, architectural renderings were to be drawn for review by the emperor, the empress dowager, the ministers in charge of the project and the administrators. These drawings were to be submitted together with related instructions to ensure that they were examined in a particular sequence (Fig. 30).

Apart from the model (*tangyang*), numerous delicate drawings, called *huayang* or *yangshi* (architectural drawings), were drawn on paper, cloth or silk using wooden rulers, brushes, charcoal and painting colours. These drawings were some of the most important methods expressing Yangshi Lei's ideas of design and providing instructions for the construction of the project. This immense trove of drawings exhibits a multiplicity of perfected skills of graphical expression in *dipan* (layout plans) and *liyang* (sections and elevations). Besides, the rotated section views, laddered section views and contour maps congruent

作为超大型工程。陵寝营建动辄数年，每年要避开隆冬酷暑，分两个工期展开；木、石、瓦、搭彩、彩画诸匠及夯夫等工种要合理衔接；动土、每期开工、立基、供梁、竖柱、上梁、合龙等，要隆重行礼，以示尊重并激励工匠。为实时掌控进程，样式雷要跟踪记录，除了文档，还以成局图（即鸟瞰图）方式（图 34），按相同比例，分期绘制已做现做活计图，即施工进程图（图 35）。

今天，把这些已做现做活计图稍加合成处理，可获得动画效果，生动反映施工全过程各环节的详情细节：如殿宇如何开挖地基，奠定基础，继而竖柱、上梁、挂瓦、合龙、油饰彩画等；复杂的地宫还展示了发券，砌筑蓑衣顶，夯筑宝顶，抹饰包金土等。从世界范围比较，在没有摄影或录像技术的时代，样式雷有关施工进程的图像记录，显然堪称大型工程管理史上的奇绩。

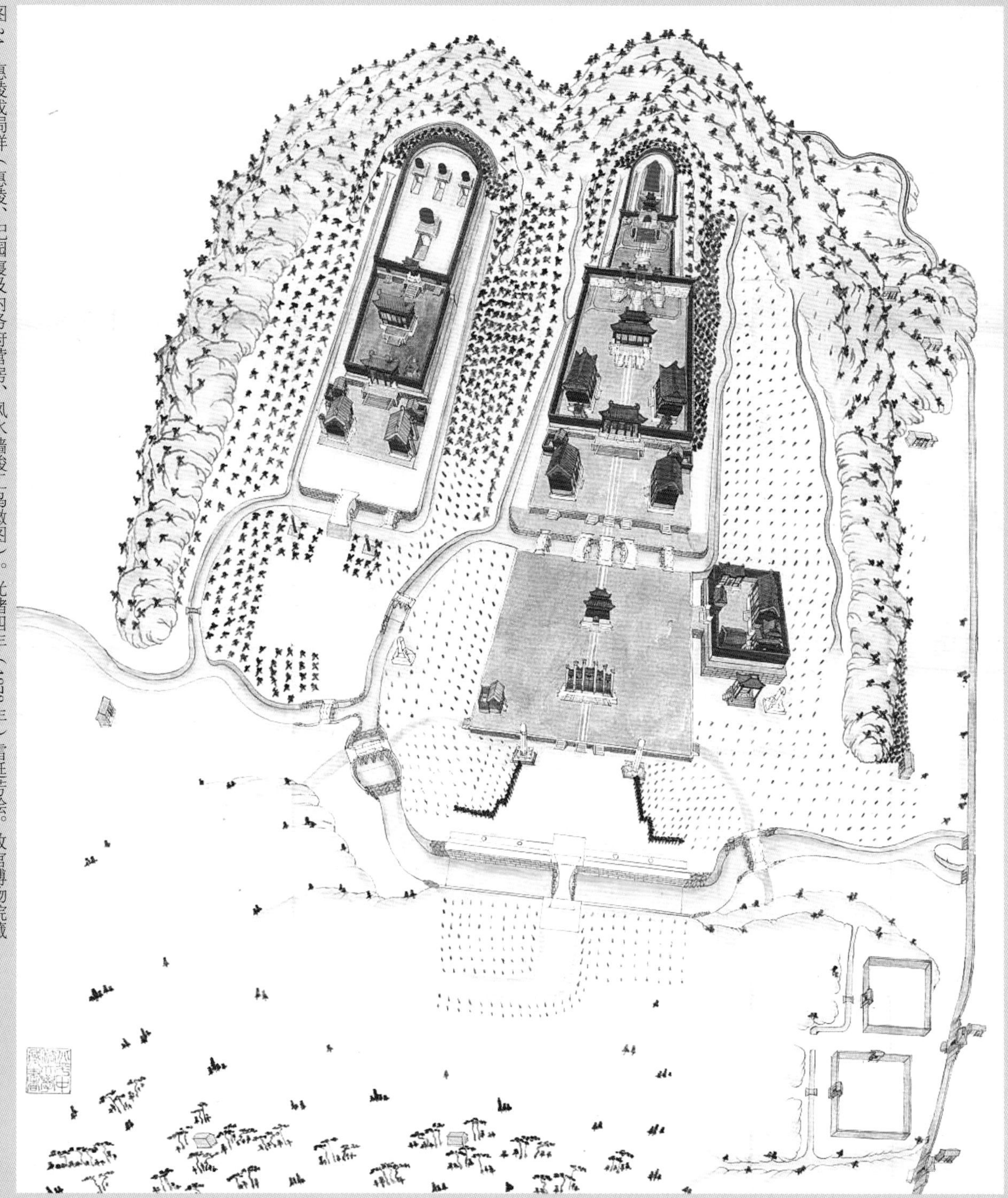

图 34 惠陵成局样（惠陵、妃园寝及内务府营房、风水墙竣工鸟瞰图）。光绪四年（1878 年）雷廷芳绘。故宫博物院藏

Fig.34 A bird's–eye view of the Tomb of Benevolence complex (Huiling *chengjuyang*) (Tomb of Benevolence [Huiling], Imperial Consorts' Tombs [Feiyuanqin], Imperial Household Department [Neiwufu]'s Barracks and the *fengshui* wall), drawn by LEI Tingfang in the Guangxu Emperor's 4th regnal year (1878), in the collection of the Palace Museum.

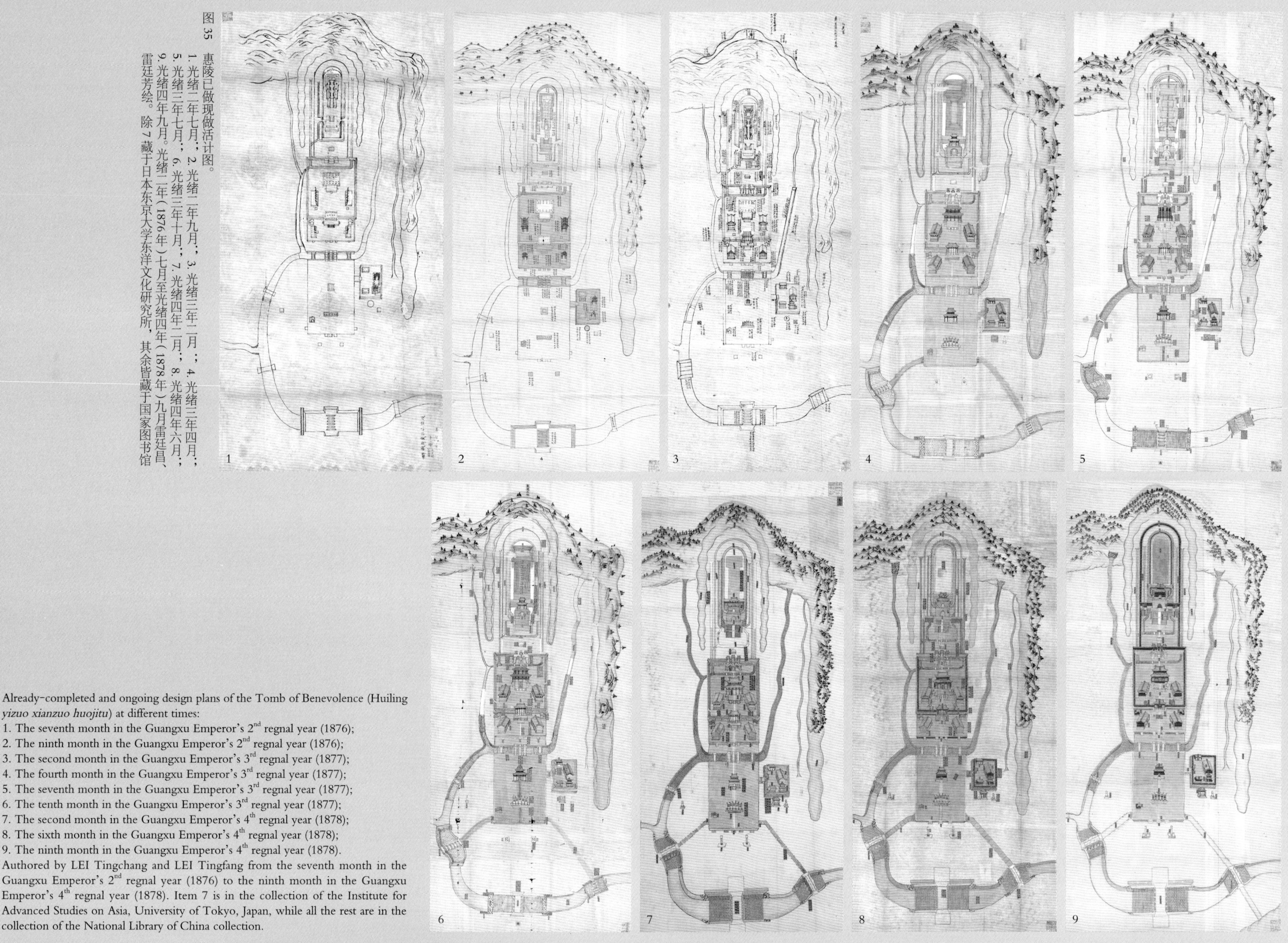

图 35　惠陵已做现做活计图。
1. 光绪二年七月；2. 光绪二年九月；3. 光绪三年二月；4. 光绪三年四月；
5. 光绪三年七月；6. 光绪三年十月；7. 光绪四年二月；8. 光绪四年六月；
9. 光绪四年九月。光绪二年（1876 年）七月至光绪四年（1878 年）九月雷廷昌、雷廷芳绘。除 7 藏于日本东京大学东洋文化研究所，其余皆藏于国家图书馆

Fig.35 Already-completed and ongoing design plans of the Tomb of Benevolence (Huiling *yizuo xianzuo huojitu*) at different times:
1. The seventh month in the Guangxu Emperor's 2nd regnal year (1876);
2. The ninth month in the Guangxu Emperor's 2nd regnal year (1876);
3. The second month in the Guangxu Emperor's 3rd regnal year (1877);
4. The fourth month in the Guangxu Emperor's 3rd regnal year (1877);
5. The seventh month in the Guangxu Emperor's 3rd regnal year (1877);
6. The tenth month in the Guangxu Emperor's 3rd regnal year (1877);
7. The second month in the Guangxu Emperor's 4th regnal year (1878);
8. The sixth month in the Guangxu Emperor's 4th regnal year (1878);
9. The ninth month in the Guangxu Emperor's 4th regnal year (1878).
Authored by LEI Tingchang and LEI Tingfang from the seventh month in the Guangxu Emperor's 2nd regnal year (1876) to the ninth month in the Guangxu Emperor's 4th regnal year (1878). Item 7 is in the collection of the Institute for Advanced Studies on Asia, University of Tokyo, Japan, while all the rest are in the collection of the National Library of China collection.

五、清东陵的保护和研究

有清一代，清东陵保护严密，有工部衙门专掌建筑养护岁修（图 36）；陵寝仪树、前圈海树和后龙山树则由内务府树户、八旗及绿营严管，恢宏的陵寝建筑融入苍郁林海，一派胜境。

清末，清东陵见知世界。1902 年德国驻华公使穆默（Alfons von Mumm）的《穆默的摄影日记》出版，刊出不少当年 4 月在清东陵拍摄的珍贵影像（图 37）。不久，德国政府特派杰出建筑师柏世曼（Ernst Boerschmann），1907 年 3 月专赴清东陵摄影测绘，这里『天才般的设计』使之叹服：『神圣的树林与隐现其中的大理石建筑、琉璃屋顶散发着不可思议的魔力，宁静的陵寝建筑在巨大山峦的环抱中显现着独立而松散的自由——每一名观者都会沉迷于这种由内在结构所营造的奇妙空间之中』（图 38）。

中国建筑以自然环境作为创作本体，与西方建筑观念完全异趣，引发了斯宾格勒（Oswald Spengler）、李约瑟（Joseph Terence Montgomery Needham）、培根（Edmund N. Bacon）等欧美大学者对中国陵墓艺术的交口赞颂。

1924 年以后，燕京大学社会学系创建者之一、美国学者甘博（Sidney David Gamble）也曾留下清东陵的珍贵影像（图 39）。

with today's design ideas, as well as other perspective drawings are also observed. All relevant drawings were painted in the same ratio with measurements and notes in order to carefully reflect the constitution of three-dimensional architectural space, in conformity with modern design drawings. (Figs. 31, 32 and 33).

The construction of a tomb, as a large-scale project, could last several years. There were two periods of construction each year, avoiding the extreme climatic conditions in winter and summer. Different craftsmen: carpenters, masons, bricklayers, scaffolders, painters and hammerers, were arranged sequentially as the construction process requires. Magnificent ceremonies to motivate and honour the craftsmen were held at important stages of construction, viz: ground-breaking, the beginning of the construction period, foundation laying, scaffolding, column erection, beam installation and joining two sections in a structure such as an arched vault. To supervise the project, Yangshi Lei kept records, both in texts and *chengjuytu*, which are aerial view maps (Fig. 34) by drawing both ongoing and already-completed project plans at the same scale (Fig. 35).

Today, an animation effect may be obtained after we digitise these already-completed and ongoing design plans and show them in sequence. The animation vividly shows the project details at different stages. For example, excavating the foundations,

foundation laying, column erection, tiling, *helong* (joining two sections in a structure such as a bridge), oil painting and colouring, the elaborate underground palaces also show examples of creating inverted vaults, manufacturing straw raincoated vaults (*suoyiding*), ramming the tumulus and painting the yellow marl (*baojintu*). The Yangshi Lei records are exceptional in the history of project management throughout the world, remembering that photo and video technology did not exist at that time.

V. Conservation and Research on the Eastern Qing Tombs

There were strict rules for the protection of the Eastern Qing Tombs during the Qing Dynasty. The preservation works were supervised by different departments. The Ministry of Engineering Works (Gongbu) was responsible for the annual maintenance and repair of buildings (Fig. 36), while the tree registry (*shuhu*) of the Imperial Household Department (Neiwufu), the Eight Banners (Baqi) and the Green Banner Army (Luying) had strict regulations for the management of the *yi* trees which were planted in designated positions inside the tombs for ritual purposes, the *hai* trees which were planted naturally and

Fig.36 "The Seal of the Ministry of Engineering Works for the Eastern Qing Tombs Administration (Dongling Gongbu Banli Shiwu Guanfang)", made by the Ministry of Construction for the Eastern Qing Tombs, in the Qianlong Emperor's 12th regnal year (1747). This ministry, previously known as the Administration of Tombs Management and Construction (Ligong Guanli Xiujian Shiwu Gongbu), was responsible for the annual repair of the Eastern Qing Tombs. They were issued a seal in the Kangxi Emperor's 25th regnal year (1686).

Fig.37 A rare photograph by Alfons von Mumm showing the Empress Dowager Cixi's Putuo Valley site which was being renovated:
1. In the foreground of the photograph is the square walled terrace (*fangcheng*) surrounded by scaffolding; the memorial tower (*minglou*) had been demolished for renovation. In the background is the encircled realm of treasure (*baocheng*) covered by a canopy for renovation.
2. The hall of monumental grace (*long'endian*) with the canopy under renovation.
3. The renovation of the truss of the hall of monumental grace (*long'endian*).

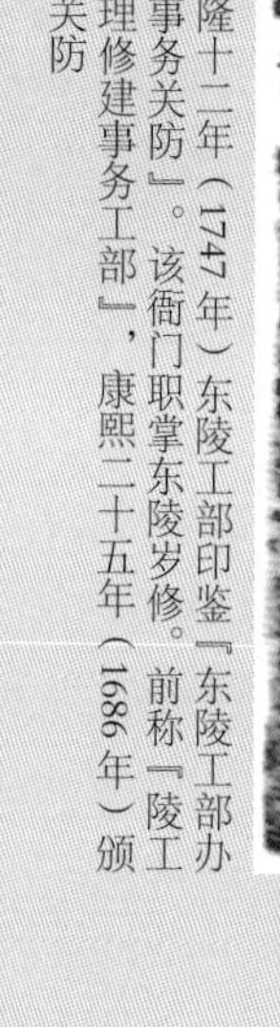

图 36　乾隆十二年（1747 年）东陵工部印鉴『东陵工部办理事务关防』。该衙门职掌东陵岁修。前称『陵工管理修建事务工部』，康熙二十五年（1686 年）颁给关防

图 37　慈禧太后菩陀峪吉地重修时的罕见场景：1. 前为方城，围合架木，明楼已拆待修，后为重修宝城的大罩棚；2. 重修隆恩殿的大罩棚；3. 重修隆恩殿内的梁架。1902 年 4 月德国公使穆默拍摄

图 38　1. 柏世曼 1907 年 3 月摄孝陵石牌坊；2. 柏世曼 1912 年绘清东陵全图

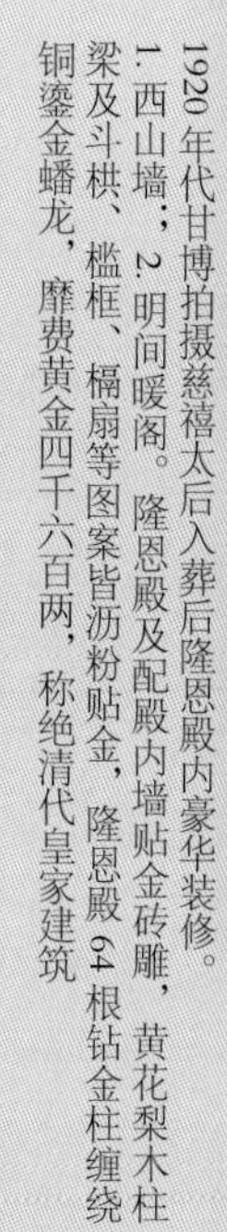
图 39　1920 年代甘博拍摄慈禧太后入葬后隆恩殿内豪华装修。1. 西山墙；2. 明间暖阁。隆恩殿及配殿内墙贴金砖雕，黄花梨木柱梁及斗栱、槛框、槅扇等图案皆沥粉贴金，隆恩殿 64 根钻金柱缠绕铜鎏金蟠龙，靡费黄金四千六百两，称绝清代皇家建筑

Fig.38 1. Photo of the Marble Memorial Gateway, Tomb of Filial Piety, taken by Ernst Boerschmann in March 1907.
2. The overall layout of the Eastern Qing Tombs, drawn by Ernst Boerschmann in 1912.

Fig.39 Photographs of the luxurious decorative interior of the hall of monumental grace (*long'endian*) in the Empress Dowager Cixi's Tomb, taken by Sidney Gamble in the 1920s.
1. Western interior wall.
2. A smaller chamber with a heating stove inside the central chamber. Bricks with gold foil were tiled in the hall of monumental grace and its side halls. The wooden pillars and beams, made from fragrant rosewood (*dalbergia odorifera*), as well as the brackets, door frames and partition boards were, all made using gold foil paint and embossed painting. There are 24 pillars (called *zuanjinzhu*) all decorated with a gold-plated bronze dragon motif. This building, costing 4,600 *taels* (one taels ≈ 50 grams) of gold, was an extraordinary example of imperial architecture in the Qing dynasty.

1931年5月末，日本著名建筑史家关野贞一行调查清东陵，竹岛卓一、荒木清三测量绘图上百幅，可惜在1945年煨于战火；关野贞摄158幅照片却幸存至今。

实地考察外，德国民族学博物馆于1921、1927年先后购藏惠陵妃园寝全分烫样、地宫烫样，以及定陵地宫烫样（图40）；1927年末成立的北京人文科学研究所，也收藏不少清东陵文档和样式雷画样（图41），今存中国科学图书馆；荒木清三自清东陵返京后，立即搜购了20幅样式雷画样，到1933年达277件，还有陵寝工程文书逾万件，今存日本东京大学东洋文化研究所（如图1、图11、图35-7等）。

① 埃德阿德·科戈尔：恩斯特·伯希曼与中国宗教建筑（1906—1931）[M]. 柏林：德格若伊特，2015：86，90.

randomly in the front field (*qianquan*) and the *shan* trees which were planted in the back dragon (*houlong*) mountains. The magnificent tomb buildings were integrated with the luxuriant forest surrounding the tomb complex.

The Eastern Qing Tombs were not known to the world until the late Qing Dynasty. In 1902, Alfons von Mumm, a diplomat from the German Empire posted to the Qing Empire, published *Ein Tagebuch in Bildern* containing the precious photographs he took in April of the same year (Fig. 37). In 1907, the German Empire assigned the famous architect Ernst Boerschmann to make a photogrammetric analysis of the Eastern Qing Tombs. To Boerschmann, "the around fifteen single burial sites were ingenious," and he commented: "The unbelievable magic of the sacred grove with its marble buildings and coloured glazed roof, the calm of the grave temples in the valleys, and the sense of freedom and solitude in front of the mountains – one enjoys the unbelievable sense of space around this structure" (Fig. 38).①

Chinese architectural design emphasises a sense of harmony between buildings and their surrounding natural environment, thus differing from Western architectural views. This distinction of Chinese architecture drew appreciation and admiration from Western scholars such as Oswald Spengler, Joseph Needham and Edmund Bacon.

After 1924, Sidney Gamble, one of the founders of the Sociology Department of Yenching University, took rare photographs of the Eastern Qing Tombs (Fig. 39).

During late May 1931, Tadashi Sekino, the famous Japanese architectural historian, conducted a field study of the Eastern Qing Tombs. His team members, Takuichi Takeshima and Araki Seizo, made hundreds of architectural drawings and maps, all of which were destroyed during the war in 1945. However, 158 photographs taken by Tadashi Sekino have survived until the present day.

Apart from fieldwork, the Ethnological Museum of Berlin successively purchased the topographic model (*quanfen tangyang*) and the model of the underground palace (*digong tangyang*) of the Imperial Consorts' Tombs affiliated with the Tomb of Benevolence (Huiling Feiyuanqin), as well as a model of the underground palace of the Tomb of

① Eduard Kögel. The Grand Documentation: Ernst Boerschmann and Chinese Religious Architecture (1906–1931) [M]. Berlin: De Gruyter, 2015: 86, 90.

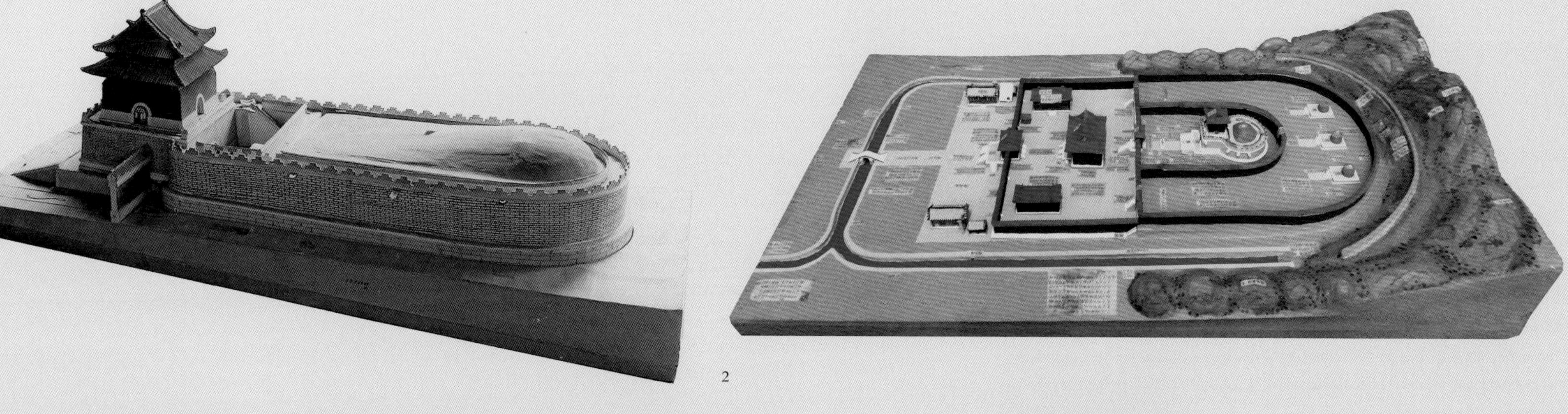

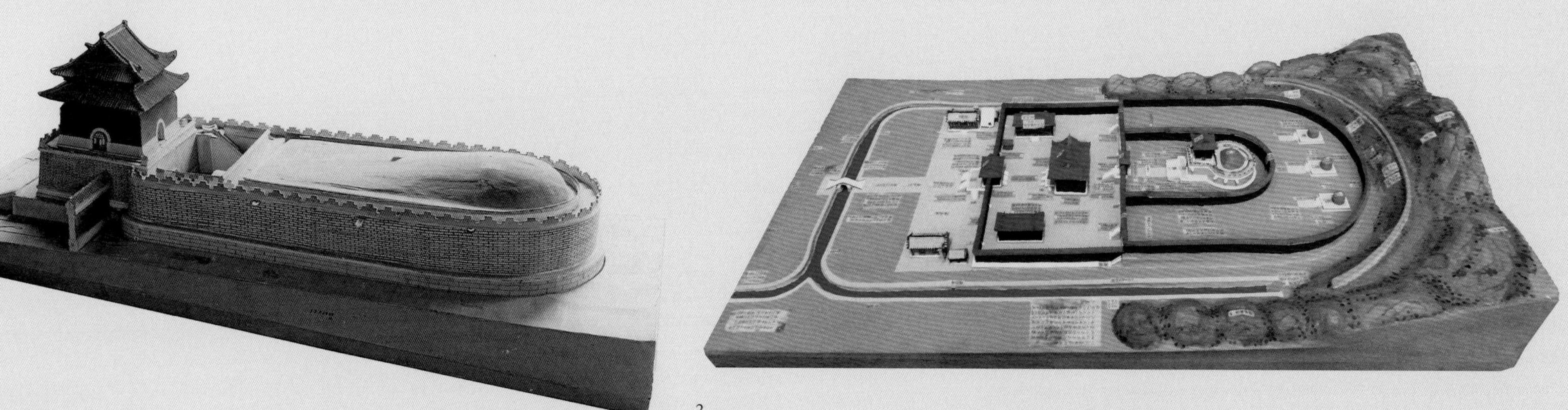

图 40 1 德国民族学博物馆藏烫样。1.1927 年购藏定陵地宫烫样，咸丰八年（1858 年）十一月雷思起制；2.1921 年购藏惠陵妃园寝全分烫样，光绪元年（1875 年）五月雷思起制

图 41 中国科学图书馆藏：1.《平安峪万年吉地工程备要》；2.《平安峪万年吉地工程续要》；3.《菩陀峪万年吉地工程备要》卷 24；4.《普祥峪万年吉地工程备要》卷 1

Fig.40 Models (*tangyang*) of the Qing tombs collected by the Ethnological Museum of Berlin.
1. Model of the Underground Palace (*digong tangyang*), Tomb of Stability (Dingling), bought in 1927, made by LEI Siqi in the eleventh month of the Xianfeng Emperor's 8th regnal year (1858);
2. Topographic model (*quanfen tangyang*) of the Imperial Concubine Tombs affiliated with the Tomb of Benevolence (Huiling Feiyuanqin), bought in 1921, made by LEI Siqi in the fifth month of the Guangxu Emperor's first regnal year (1876).

Fig.41 The following are all collected by the National Science Library, Chinese Academy of Sciences:
1. *Documentation for the Construction Work for the Ping'an Valley Site* (Ping'anyu *wannian jidi gongcheng beiyao*);
2. *Continued Documentation for the Construction Work for Ping'an Valley Site* (Ping'anyu *wannian jidi gongcheng xuyao*);
3. *Documentation for the Construction work on the Putuo Valley Site* (Putuoyu *wannian jidi gongcheng beiyao*), Volume 24;
4. *Documentation for the Construction work on the Puxiang Valley Site* (Puxiangyu *wannian jidi gongcheng beiyao*), Volume 1.

Stability (Dingling) in 1921 and 1927 respectively (Fig. 40). Also, the Beijing Institute of Human Sciences, established in late 1927, collected a large number of documents and Yangshi Lei's architectural drawings (*huayang*) of the Eastern Qing Tombs (Fig. 41). These historical documents are currently stored in the National Science Library of the Chinese Academy of Sciences. After returning to Beijing from the Eastern Qing Tombs, Araki Seizo purchased and collected twenty of Yangshi Lei's architectural drawings. By 1933, his collection had 277 items, together with more than 10,000 tombs project documents. Araki Seizo's collection is now preserved in the Institute of Advanced Studies on Asia, University of Tokyo (Figs.1, 11, 35–7).

Chinese scholars' studies of the Eastern Qing Tombs have a relatively shorter history, but they have taken the lead in more systematic and in-depth research. ZHU Qiqian, the founder of the Society for the Study of Chinese Architecture (SSCA) noted: "The Qing Ministry of Engineering Works (Gongbu)'s *Construction Methods* (*Gongcheng Zuofa*) only had regulations but no illustrations. Therefore, we organized craftsmen to help add illustrations and annotations to existing documentation as inferences to remedy any shortcomings in the original work. This work laid the foundations for the SSCA." When reprinting the Song-dynasty *Treatise on Architectural Methods or State Building Standards* (*Yingzao Fashi*) imitating the Song style from 1919-1925, the SSCA organised experienced craftsmen to create drawings following the Qing official construction methods including annotations about the terms. These drawings were included as an appendix to the reprinted *Treatise on Architectural Methods or State Building Standards* (*Yingzao Fashi*) (Figs. 42-1, 42-2). In March 1932, LIANG Sicheng, the head of the Department of Building Standards of SSCA started to read and annotate the Qing Ministry of Engineering Works' *Construction Methods* (*Gongcheng Zuofa*) and *Architectural Material Estimates* (*Yingzao Suanli*), leading to the publication of the *Qing Structural Regulations* (*Qingshi Yingzo Zeli*) in June 1934. Sicheng Liang's writings solved problems in understanding Qing official architecture construction methods and terms, later becoming a classic example of Chinese architectural research (Figs. 42-3, 42-4).

In May 1930, ZHU Qiqian sent a letter to the China Foundation for the Promotion of Education and Culture, responsible for managing the Boxer Indemnity, stating that: "*the craftsmanship inherited by the Lei Family and the design drawings they kept, must be regarded as the art of our ancestors. We can use them to investigate the relics … We can also use the drawings to corroborate the relics. As for the underground palaces of the*

中国学者研究稍晚，却率先系统深入实质。中国营造学社创始人朱启钤先生强调：『以清工部《工程做法》，有法无图，复纠集匠工，依例推求，补绘图释，以匡原著不足，中国营造学社之基，于兹成立。』1919—1925年仿宋重刊《营造法式》时，中国营造学社特聘老工匠按清代官式建筑做法绘图并标明相关术语作为重刊《营造法式》的附图（图42-1、图42-2），开中国营造学社相关研究先河。到1932年3月，学社法式部主任梁思成解读《工程做法》及匠籍《营造算例》的《清式营造则例》脱稿，1934年6月出版，解决了清代官式建筑做法和术语的解读，成为中国建筑史学经典（图42-3、图42-4）。

同期，朱启钤1930年5月专函管理美国退还庚款的中华文化基金会：『雷氏世守之工……所保存之图样，亦不得不视为前民艺术之表现。……可以考求遗迹；……并可与实物互相印证；至陵寝地宫向守秘密，今乃借此为公开研究；实于营造学、考古学均有重要之价值』。上万件样式雷图档迅即抢救性入藏国立北平图书馆，得到系统整理和研究利用。

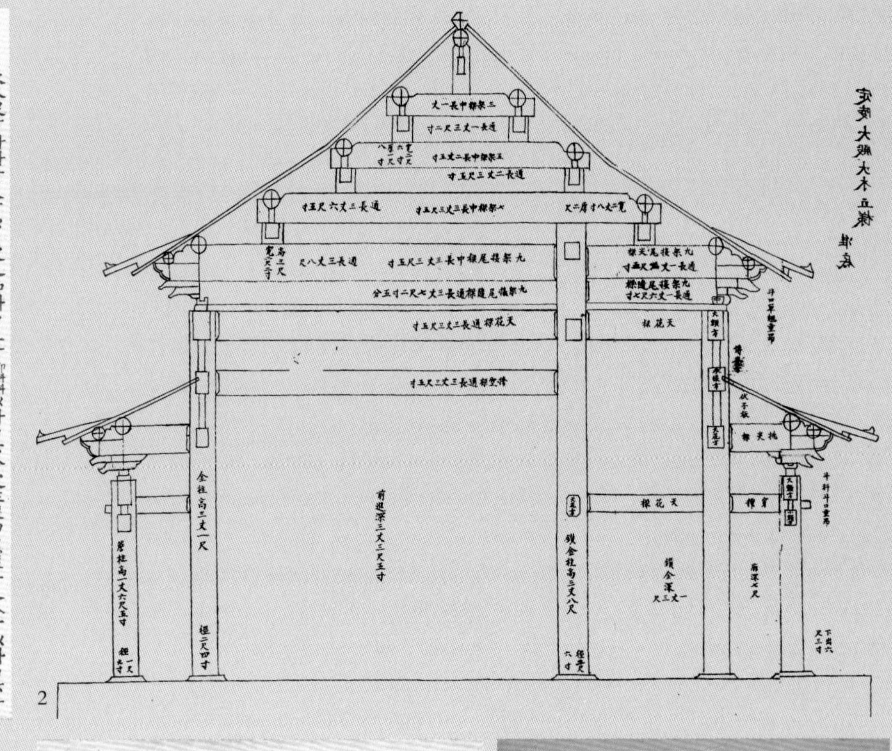

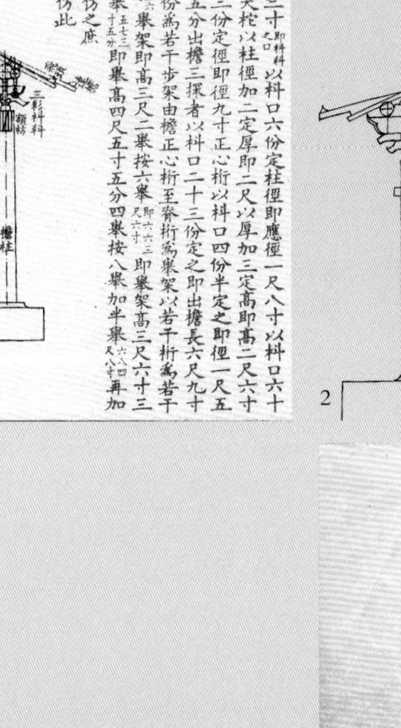

图 42 1.1925 年仿宋本《营造法式》卷 30 附，贺新赓、秦渭滨等绘；2. 国家图书馆 187-2-24 定陵大殿大木立样，雷思起绘；3. 梁思成著《清式营造则例》；4. 梁思成编订《营造算例》

Fig.42 1. Reprint of the *Treatise on Architectural Methods or State Building Standards* (*Yingzao Fashi*) (1925) imitating the Song style, Volume 30, characters in red show the Qing official architectural construction methods and terms, annotated by HE Xingeng and QIN Weibin;
2. Section drawings of the wooden structure of the main hall of the Tomb of Stability, drawn by LEI Siqi, collected by the National Library of China, 187-2-24;
3. *Qing Structural Regulations* (*Qingshi Yingzao Zeli*), authored by LIANG Sicheng;
4. *Architectural Material Estimates* (*Yingzao Suanli*), edited by LIANG Sicheng.

基于此，1934年9月学社文献部主任刘敦桢偕莫宗江、陈明达调查清西陵，翌年3月发表《易县清西陵》，融通文献、雷氏图档和实物测绘，前所未有地揭橥了关内外各陵寝建置沿革、术语名词、地宫构造（图43）。社员王璧文接踵发表《清官式石桥做法》《清官式石闸及石涵洞做法》。凡此，无不展现了中国建筑史学和文化遗产保护事业一代前贤的非凡功力。

自清室逊位，旧制崩解，清东陵饱经战乱和匪患，到1928年，陵寝建筑大多毁损，上百万仪树、海树无存。当年7月，乾隆裕陵和慈禧定东陵地宫惨遭军阀孙殿英盗掘，国民政府内政部才于8月30日成立东陵管理处。然而此后时局动荡，战争频仍，清东陵匪患更嚣，陵寝大多被盗。

Qing tombs, they have been kept secret. Today our study of these drawings is more open research which will allow the findings to exert great influence on future architecture and archaeology studies." Afterwards, over ten thousand of the Yangshi Lei architectural archives were rescued by the National Beiping Library, and systematically catalogued for research use.

Based on these works, in September 1934, LIU Dunzhen, director of the Archives Department of SSCA, together with MO Zongjiang and CHEN Mingda, conducted a field study on the Western Qing Tombs. In March of the following year, the outcome of their work, combining documents, the Yangshi Lei archives and mapping surveys, was published as *The Western Qing Tombs in Yi County* (*Yixian Qing Xiling*), which unprecedentedly revealed the building arrangements, evolutions, terms and structures of various tombs inside and outside Shanhaiguan[①] (Fig. 43). Then, Biwen Wang, a member of the SSCA, published *The Governmental Construction Methods of Stone Bridges in the Qing Dynasty* (*Qing guanshi shiqiao zuofa*) and *The Governmental Construction Methods of Stone Lock Gates and Stone Culverts* (*Qing guanshi shizha ji shi handing zuofa*). All these show the extraordinary efforts and accomplishments of the previous generation of Chinese architectural historians in cultural heritage conservation.

Since the abdication of the Qing imperial family, the instability of Chinese society, along with war and banditry, had led to serious destruction and looting of the Eastern Qing Tombs. By 1928, most buildings in the Eastern Qing Tombs were destroyed and millions of *yi* trees and *hai* trees were felled. In July of that year, the underground palaces in the Tomb of Prosperity (Yuling) of the Qianlong Emperor and the Eastern Tomb of Stability (Ding Dongling) of the Empress Dowager Cixi suffered from looting led by SUN Dianying. This prompted the Ministry of Internal Affairs (Neizhengbu) of the National Government of the Republic of China to establish the Eastern Tombs Management Office (Dongling Guanlichu) on 30th August 1928. However, the turbulent period, with frequent wars, resulted in the continued looting of the Eastern Qing Tombs.

Under the new sovereignty of the People's Republic of China since 1949, bandits and thieves have been severely punished by the government. The preservation of the Eastern Qing Tombs, along with the work of re-afforestation, were temporarily operated and

① Shanhaiguan, also known as Shanhai Pass, located in the north east of Beijing, is one of the major passes through the Great Wall of China (translator's note).

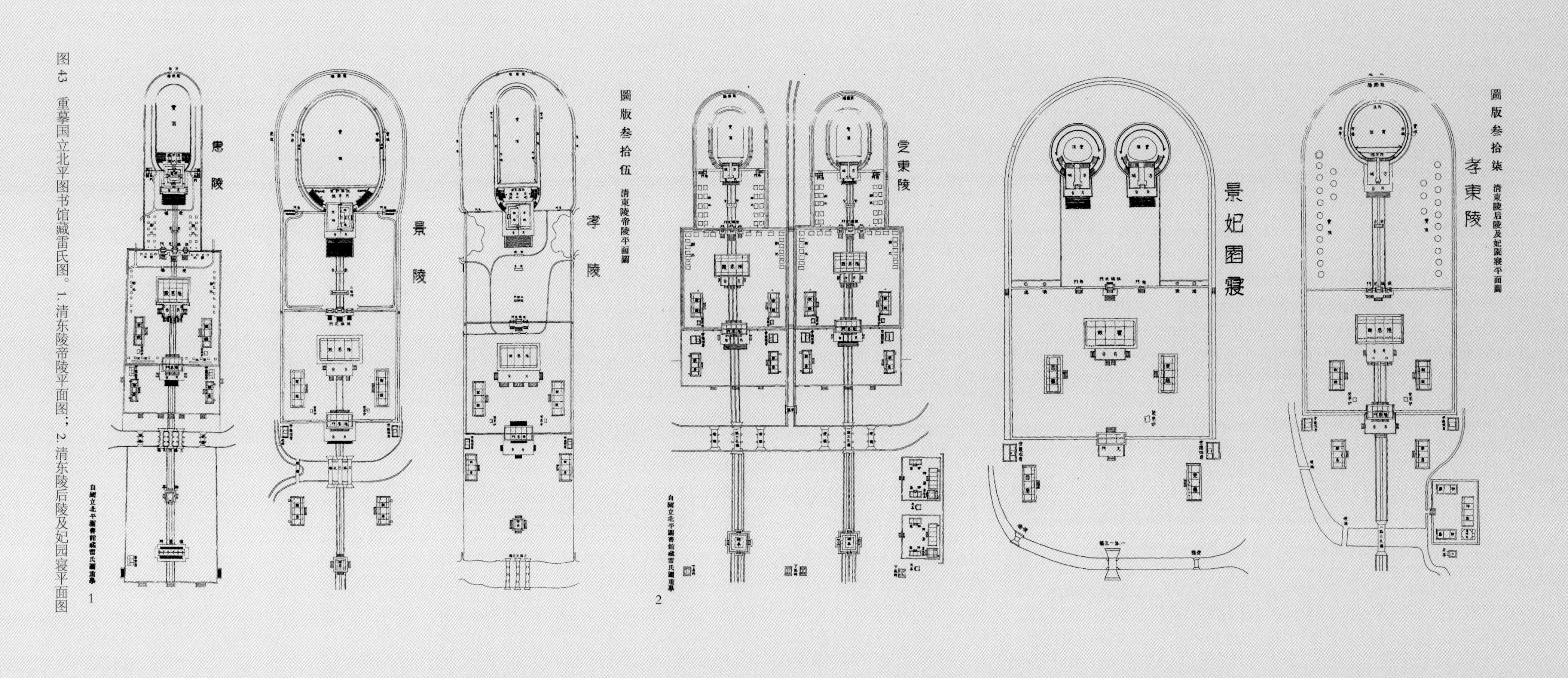

图 43 重摹国立北平图书馆藏雷氏图。1. 清东陵帝陵平面图；2. 清东陵后陵及妃园寝平面图

Fig.43 Drawings copied from the Yangshi Lei archival drawings collected in the National Beiping Library.
1.The drawing for the layout of the emperors' tombs in the Eastern Qing Tombs;
2.The layout of the tombs of empresses and the imperial consorts among the Eastern Qing Tombs.

managed by the Afforestation Bureau of Hebei Province. In July 1952, the great tower of the stela (*dabeilou*) of the Tomb of Admiration (Jingling) was destroyed by thunder, which prompted the establishment of the Office of Cultural Relics in the Eastern Qing Tombs (Qing Dongling Wenwu Baoguansuo). Preservation measures were implemented immediately, including blocking the holes used for robbing the tombs, refurbishing the damaged buildings and installing lightning rods on the main buildings. In 1956, the Eastern Qing Tombs were listed as major protected historical and cultural sites of Hebei Province and, later, in 1961, they were among the first batch of major historical and cultural sites protected at the national level (Fig. 44).

After the Cultural Revolution, the Chinese government invested heavily in the conservation of ancient buildings. Since January 1978, the tombs, after careful preservation and repair, have been opened for Chinese and international visitors. In terms of research, after Jinhua Yu published *Eastern and Western Qing Tombs* (*Qing Dongling yu Xiling*) based on the Qing imperial archives, YU Shanpu, led all the staff in the Cultural Relics Office of the Eastern Qing Tombs to publish numerous works, based on research on the imperial archives. These publications benefitted from an intimate and long-lasting collaboration with the Palace Museum, the First Historical Archives of China and the Chinese Academy of Heritage (Fig. 45). *The Great Tower for the Stela in the Qing Tomb of Filial Piety* (*Qing Xiaoling Da Beilou*) by GUO Wanxiang, published in 2009, is an important work providing a rigorous record of the conservation practices for cultural relics.

In 1952, Professor LU Sheng, a former member of the SSCA, initiated the use of the Eastern Qing Tombs as an important research and teaching field site for Tianjin University. Subsequently, Professor FENG Jiankui and his student WANG Qiheng, maintained the tradition of using the Qing tombs as a field site for the university. Since 1979, thousands of teachers and students have been organised to conduct large-scale field surveys there, prompting the use of digital technologies. Over five thousand pieces of Yangshi Lei archives have been carefully identified (Fig. 46). Their fruitful work has significantly promoted the conservation and research on the Eastern Qing Tombs. This has become a leading example of integrating academic research with heritage conservation practices, winning national awards several times.

Owing to the well-preserved building relics, the rich original documents and design drawings, the fruitful modern survey work, as well as the comprehensive management

1949年中华人民共和国成立，政府严惩盗匪，清东陵保护结合植树造林，由河北省造林局暂管。1952年7月景陵大碑楼毁于雷火，清东陵文物保管所迅即成立，封堵各陵寝盗洞，修补残损建筑，主体建筑安装避雷针。1956年，清东陵列为河北省重点文物保护单位，1961年成为首批全国重点文物保护单位（图44）。

『文革』结束后，国家投入巨资修缮古建筑，自1978年1月，悉心整治后的各陵寝陆续向国内外游人开放。研究管理方面，继1978年俞进化依据清宫档案撰著《清东陵与西陵》，清东陵文管所发挥同故宫博物院、中国第一历史档案馆、中国文物研究所等单位长期密切合作的优势传统，于善浦先生带领文管所员工，爬疏宫廷档案，涌出大量论著（图45）。2009年出版的郭万祥《清孝陵大碑楼》，则是翔实记录文物保护工程实践的重要著作。

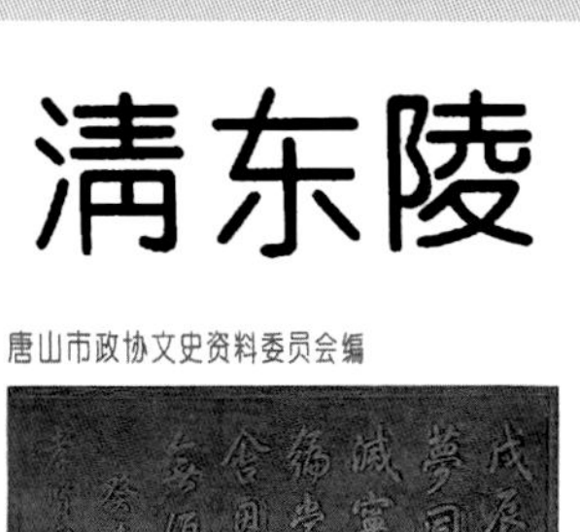

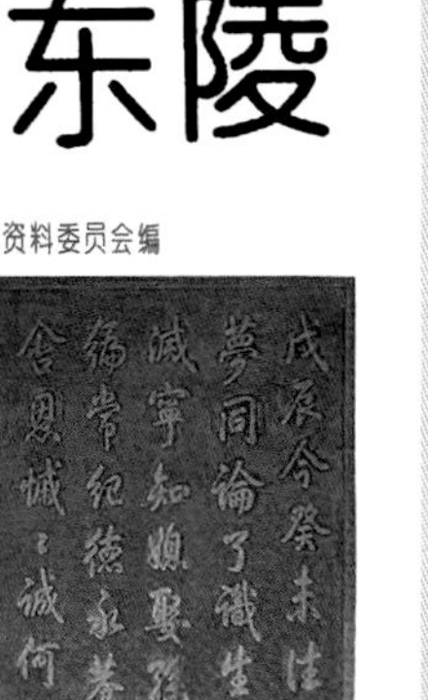

图45 改革开放以来爆发性涌出有关清东陵的著作举略

图44 全国重点文物保护单位碑——清东陵

Fig.44 The Stela declaring the Eastern Qing Tombs as being a major national historical and cultural conservation site.

Fig.45 Some studies of the Eastern Qing Tombs which emerged after the Chinese reform and opening up policy.

从1952年原中国营造学社成员卢绳教授开始，清东陵还成为天津大学的重要教学科研基地。冯建逵教授及弟子王其亨秉承这一传统，1979年至今，组织数千师生展开清代陵寝的大规模测绘，实时推广数字化技术。系统挖掘相关工程档案，细致鉴识总量逾五千件陵寝样式雷图档，成果丰硕（图46），有力推进了清东陵的保护研究，成为高校科研教学和遗产保护实践相结合的成功样板，屡获国家奖项。

由于实物遗存完好，原始档案及设计图样丰富，现代测绘研究成果充实，保护管理机制完备，2000年初联合国教科文组织专家实地考核验收，11月30日世界遗产委员会缔约国大会全票通过清东陵列为世界文化遗产，被评价为『代表了人类创造精神的杰作』。2007年6月20日『中国清代样式雷建筑图档』列入世界记忆名录，其中清东陵图档竟达3800件，该世界遗产的历史、艺术和科学的多重价值更为彰著（图47）。

为了更好保护清东陵，展示其古建筑菁华，唐山清东陵保护区管委会和天津大学建筑学院密切合作，精选历年测绘图纸，按统序组织为312页图版，辅以珍稀历史图像，简要阐述其建置沿革、建筑规制、设计意匠和保护研究历程，编辑成书，奉献给公众。

and conservation system, this assemblage has attracted international institutions. After an on-site assessment by UNESCO experts in early 2000, the Eastern Qing Tombs were unanimously voted to be a World Heritage site on 30th November 2000. The Eastern Qing Tombs were assessed as being: "a masterpiece of the creative genius of humanity." Furthermore, the Qing Dynasty Yangshi Lei Archives were inscribed in the Memory of the World Programme on 20th June 2007. The archives include up to 3,800 items from the Eastern Qing Tombs, thus revealing the historical, artistic and scientific significance of this world heritage site (Fig. 47).

To better protect the Eastern Qing Tombs and show the essence of their architecture, the Eastern Qing Tombs Protection Committee in Tangshan and the School of Architecture in Tianjin University have collaborated closely, selecting survey and mapping drawings over the years, and organising a 312-page publication with rare historical photographs. The book provides a brief account of the foundation and evolution, building arrangements, art of design, as well as the history of preservation and research on the Eastern Qing Tombs.

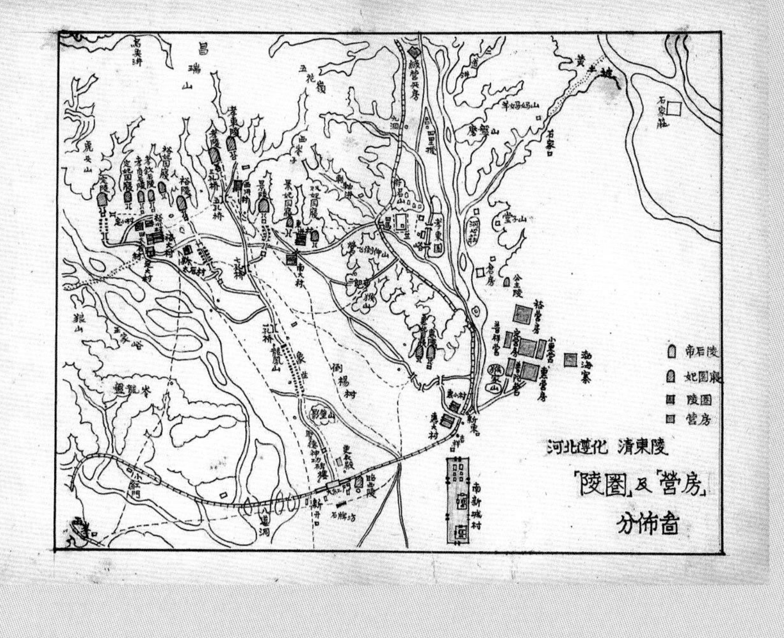

图 46 天津大学建筑学院的研究举略。
1. 卢绳 1954 年 4 月摄普陀峪、普祥峪定东陵远景；2. 卢绳 1954 年 4 月绘清东陵陵圈（内务府营房）及（礼工部八旗）营房分布图清东陵鸟瞰图；3. 冯建逵 1986 年绘清东陵鸟瞰图（《清东陵与清西陵古建筑》未刊书稿插图）；4. 王其亨 2003 年著《中国建筑艺术全集——清代陵墓建筑》

Fig.46 Examples of research by the School of Architecture, Tianjin University:
1. Putuo Valley and the Eastern Ding Tombs inside Puxiang Valley, photographed by LU Sheng;
2. A map showing the distribution of the representative houses of the Imperial Household Department (Neiwufu) and the barracks of the Ministries of the Eastern Qing Tombs (Ligongbu) and Eight Banners (Baqi), drawn by LU Sheng in April 1954;
3. A bird's-eye view of the Eastern Tombs in the Qing Dynasty, drawn by FENG Jiankui in 1986 (an unpublished illustration for the manuscript of *The Ancient Buildings of the Eastern Qing Tombs and Western Qing Tombs* [*Qing Dongling yu Xiling Gujianzhu*]);
4. The book cover of *The Complete Collection of Chinese Architectural Art: Tomb Architecture in the Qing Dynasty* (*Zhongguo Jianzhu Yishu Quanji: Qingdai Lingmu Jianzhu*), authored by WANG Qiheng.

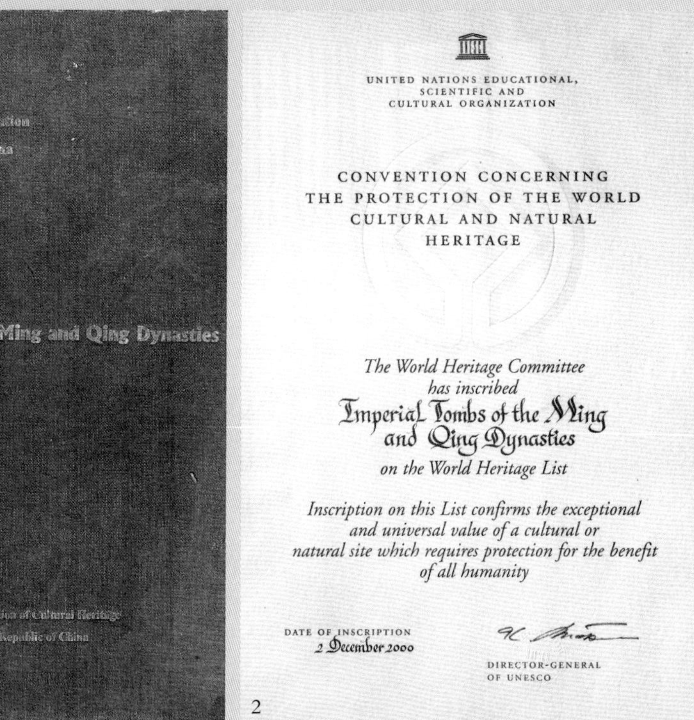
UNITED NATIONS EDUCATIONAL, SCIENTIFIC AND CULTURAL ORGANIZATION

CONVENTION CONCERNING THE PROTECTION OF THE WORLD CULTURAL AND NATURAL HERITAGE

The World Heritage Committee has inscribed

Imperial Tombs of the Ming and Qing Dynasties

on the World Heritage List

Inscription on this List confirms the exceptional and universal value of a cultural or natural site which requires protection for the benefit of all humanity

DATE OF INSCRIPTION 2 December 2000

DIRECTOR-GENERAL OF UNESCO

2

UNESCO

United Nations Educational, Scientific and Cultural Organization

MEMORY OF THE WORLD

Certifies the inscription of

The Qing Dynasty Yangshi Lei Archives

National Library of China

(institution)

Beijing (town) *People's Republic of China* (country)

on

the Memory of the World Register

Date 07 AOUT 2007

Koïchiro Matsuura
Director-General, UNESCO

Information on the website : http://www.unesco.org/webworld - Click on the logo "Memory of the world" down the screen.

3

图 47 天津大学数十年持续研究成果，成为两项世界遗产的申报文本的主体。1. 明清皇家陵寝（明显陵、清东陵、清西陵）申报世界文化遗产文本，国家文物局 1989 年；2. 世界文化遗产证书，联合国教科文组织 2000 年；3. 世界记忆遗产证书，联合国教科文组织 2007 年

Fig.47 Research conducted by Tianjin University became the main document resource for the application to have the Eastern Qing Tombs considered as a world heritage site:
1. Application documents for the Imperial Tombs of the Ming and Qing Dynasties to be listed in the World Cultural Heritage Convention, The National Cultural Heritage Administration, China, 1989;
2. Certificate of World Cultural Heritage Site, UNESCO, 2000;
3. Certificate of Memory of the World Programme, UNESCO, 2007.

图版

Figures

昭西陵
West Tomb of Brightness (Zhao Xiling)

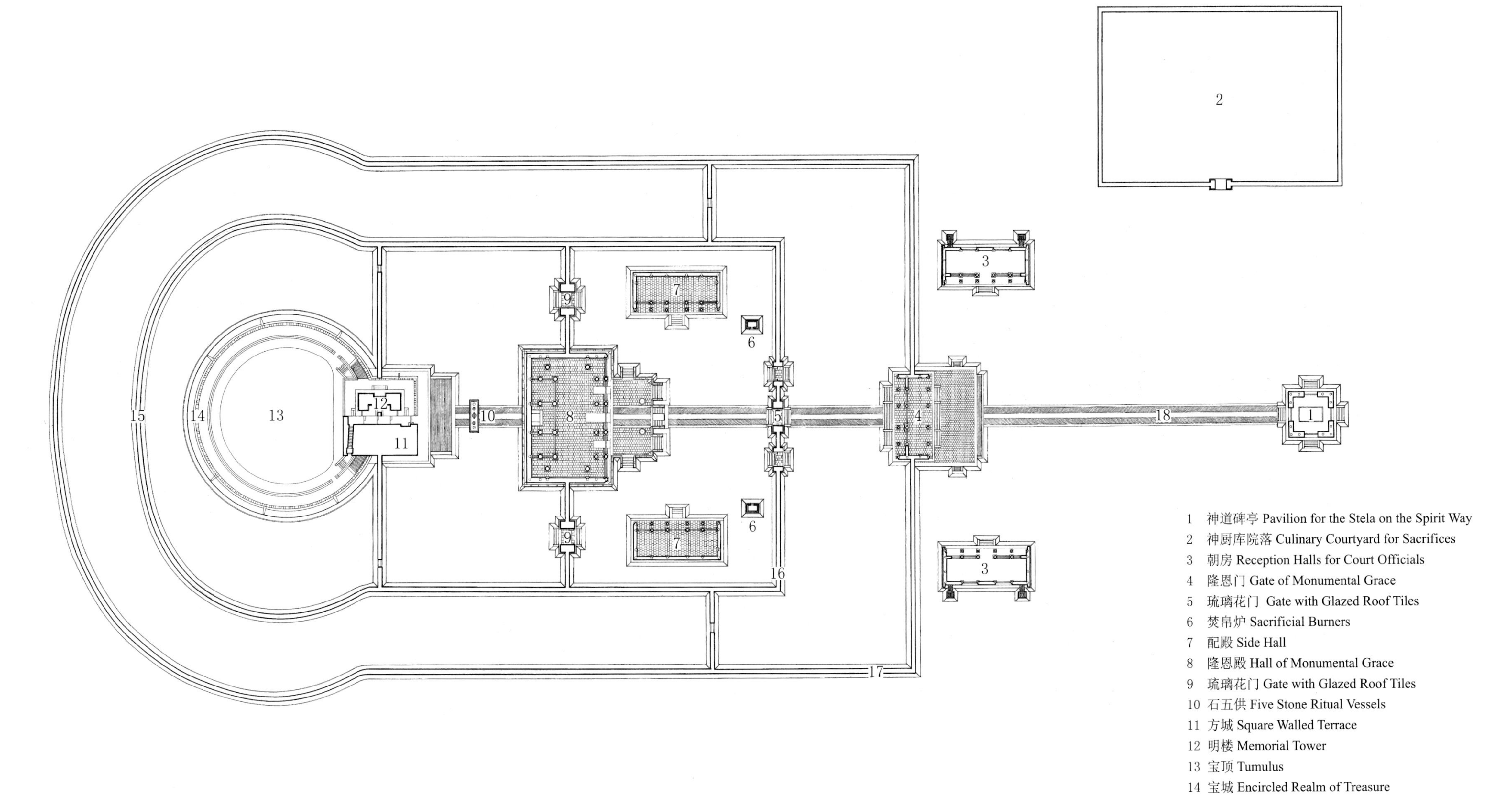

1 神道碑亭 Pavilion for the Stela on the Spirit Way
2 神厨库院落 Culinary Courtyard for Sacrifices
3 朝房 Reception Halls for Court Officials
4 隆恩门 Gate of Monumental Grace
5 琉璃花门 Gate with Glazed Roof Tiles
6 焚帛炉 Sacrificial Burners
7 配殿 Side Hall
8 隆恩殿 Hall of Monumental Grace
9 琉璃花门 Gate with Glazed Roof Tiles
10 石五供 Five Stone Ritual Vessels
11 方城 Square Walled Terrace
12 明楼 Memorial Tower
13 宝顶 Tumulus
14 宝城 Encircled Realm of Treasure
15 罗圈墙 Wall Surrounding the Tomb
16 内围墙 Inner Tomb Wall
17 外围墙 Outer Tomb Wall
18 神道 Spirit Way

0 10 40m

N

昭西陵总平面图
Site plan of the West Tomb of Brightness

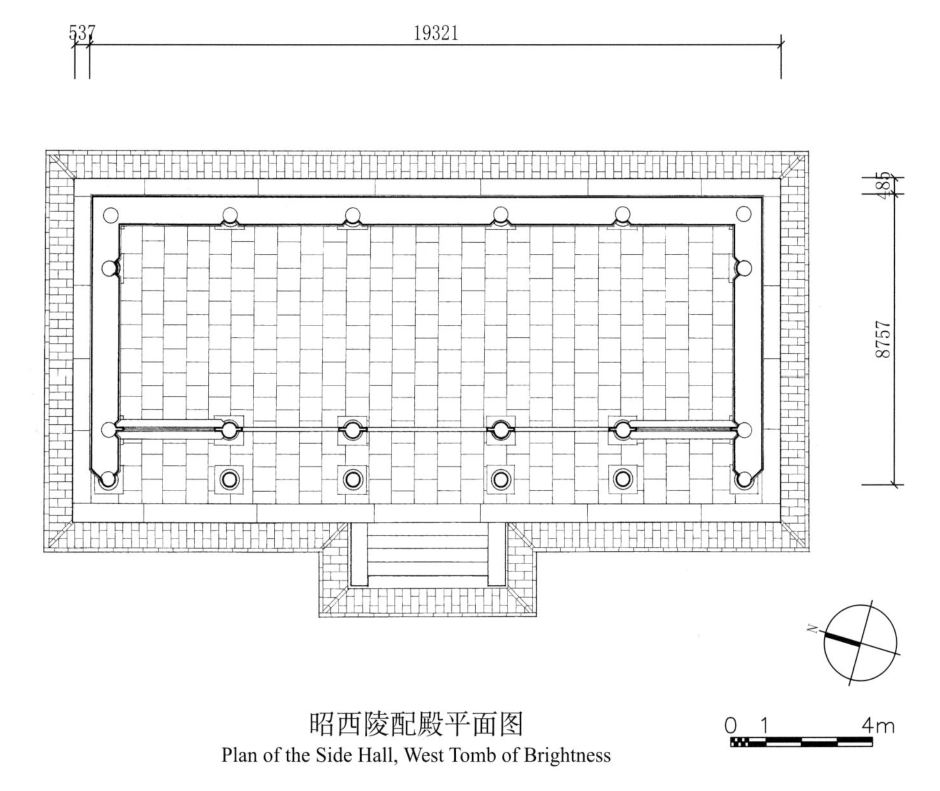

昭西陵配殿平面图
Plan of the Side Hall, West Tomb of Brightness

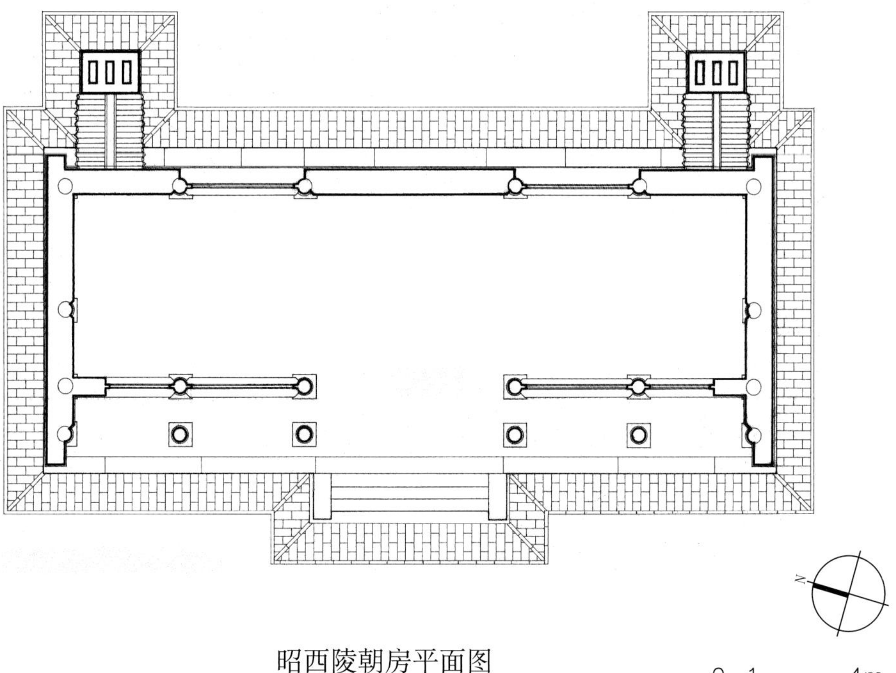

昭西陵朝房平面图
Plan of the Reception Hall for Court Officials, West Tomb of Brightness

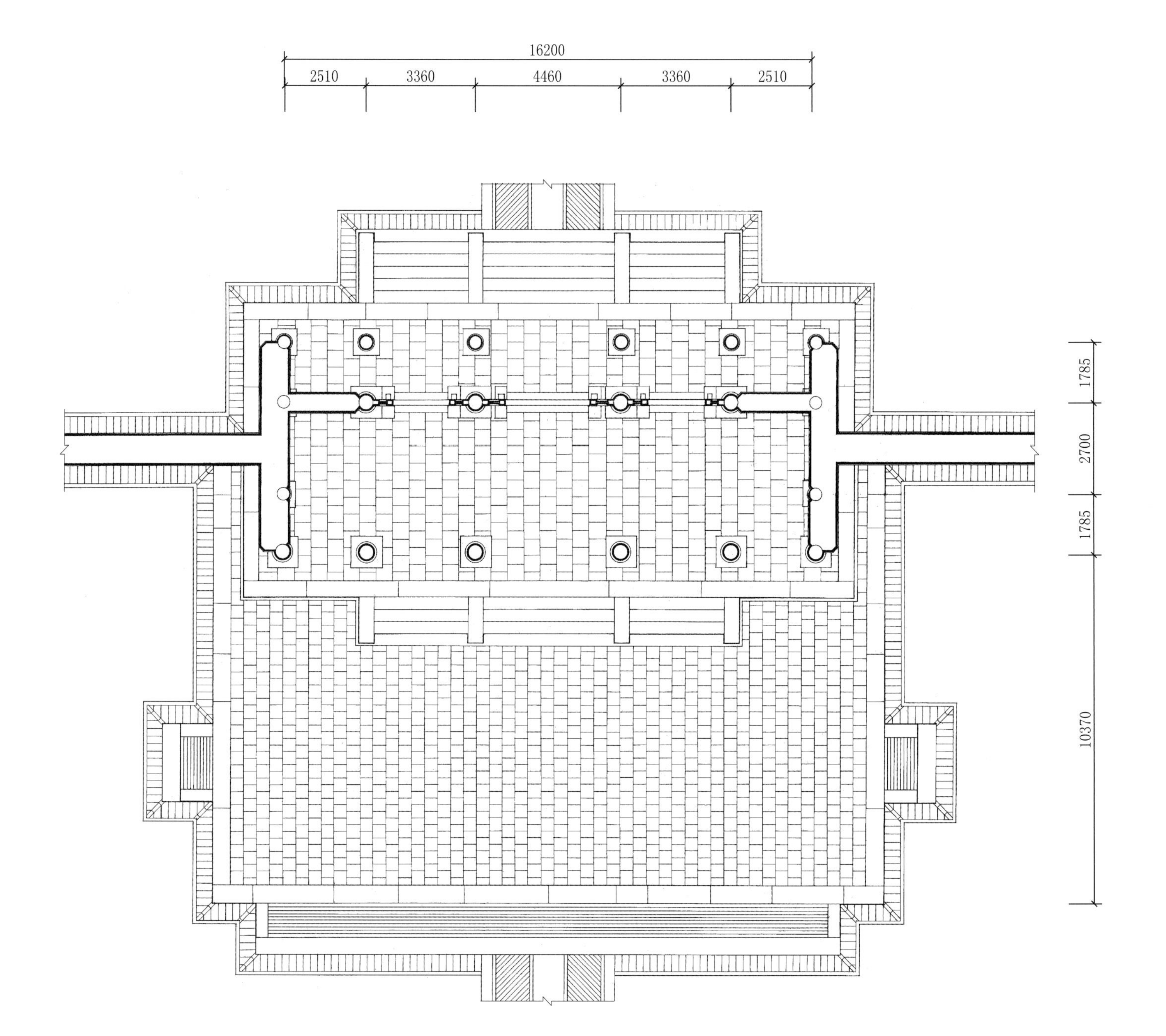

昭西陵隆恩门平面图
Plan of the Gate of Monumental Grace, West Tomb of Brightness

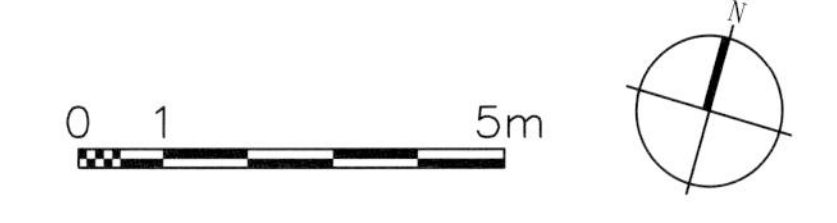

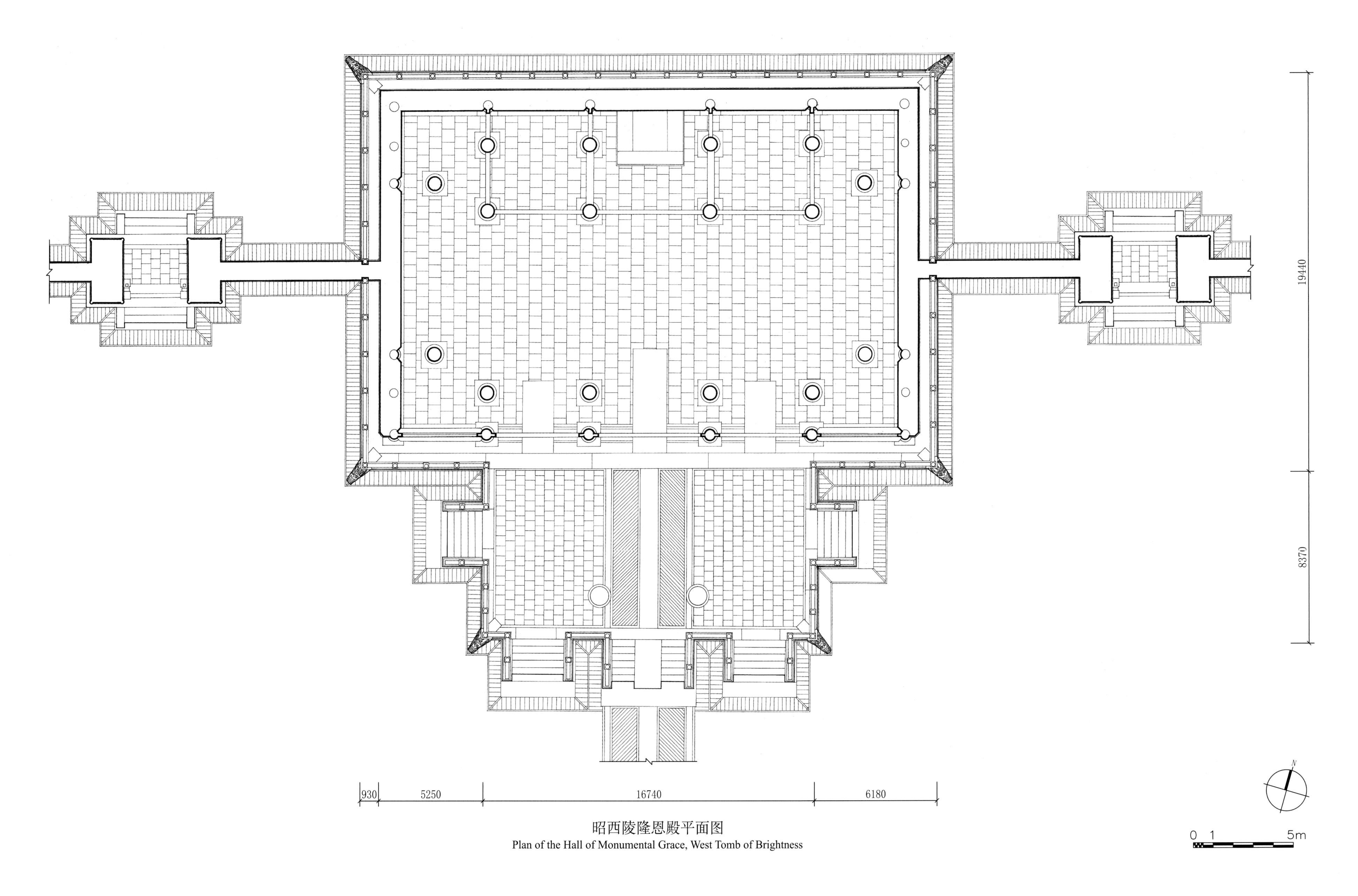

昭西陵隆恩殿平面图
Plan of the Hall of Monumental Grace, West Tomb of Brightness

昭西陵方城明楼正立面图
Front elevation of the Square Walled Terrace and the Memorial Tower, West Tomb of Brightness

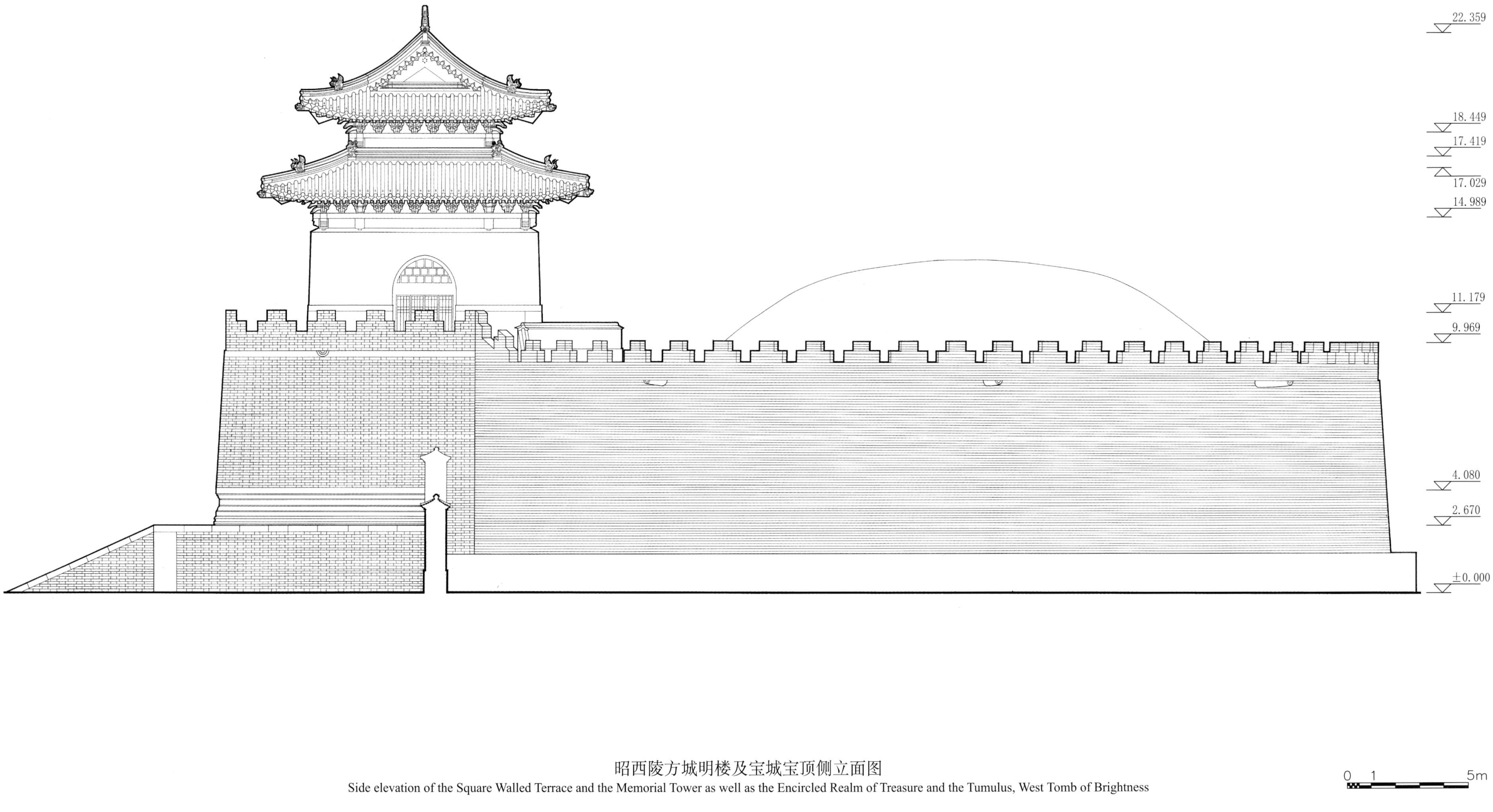

昭西陵方城明楼及宝城宝顶侧立面图

Side elevation of the Square Walled Terrace and the Memorial Tower as well as the Encircled Realm of Treasure and the Tumulus, West Tomb of Brightness

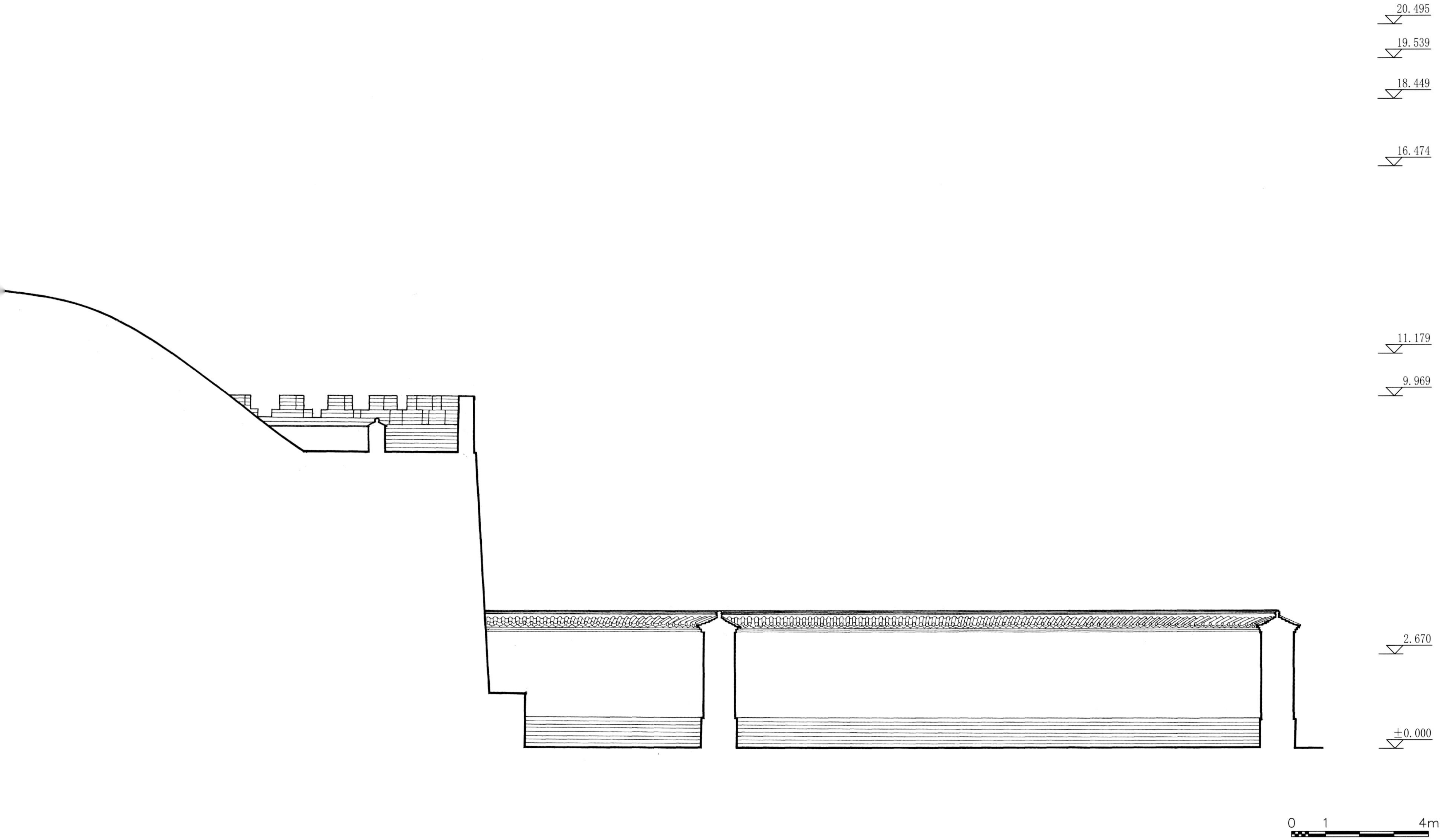

20.495
19.539
18.449
16.474
11.179
9.969
2.670
±0.000

0 1 4m

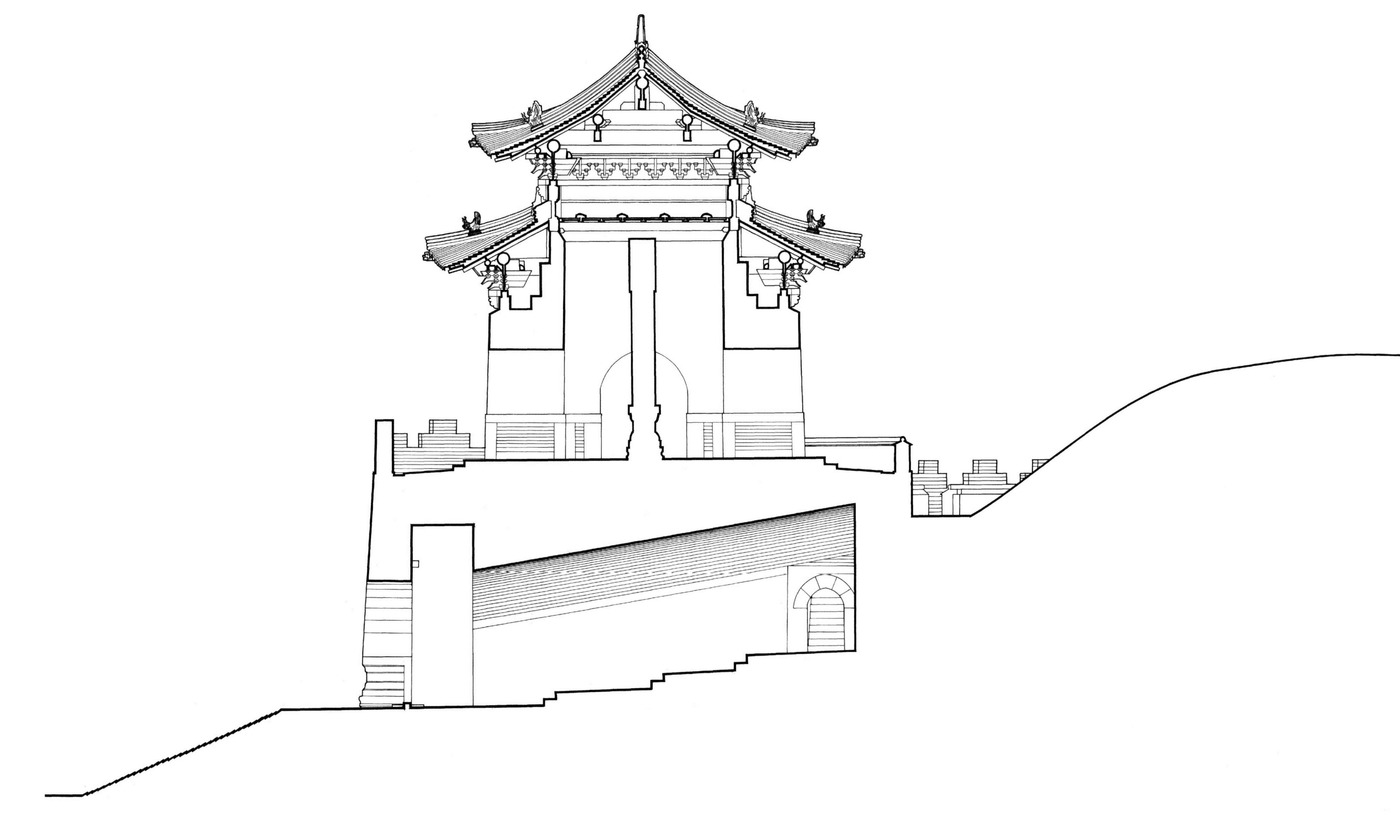

昭西陵方城明楼及宝城宝顶横剖面图
Cross section of the Square Walled Terrace and the Memorial Tower as well as the Encircled Realm of Treasure and the Tumulus, West Tomb of Brightness

孝陵

Tomb of Filial Piety
(Xiaoling)

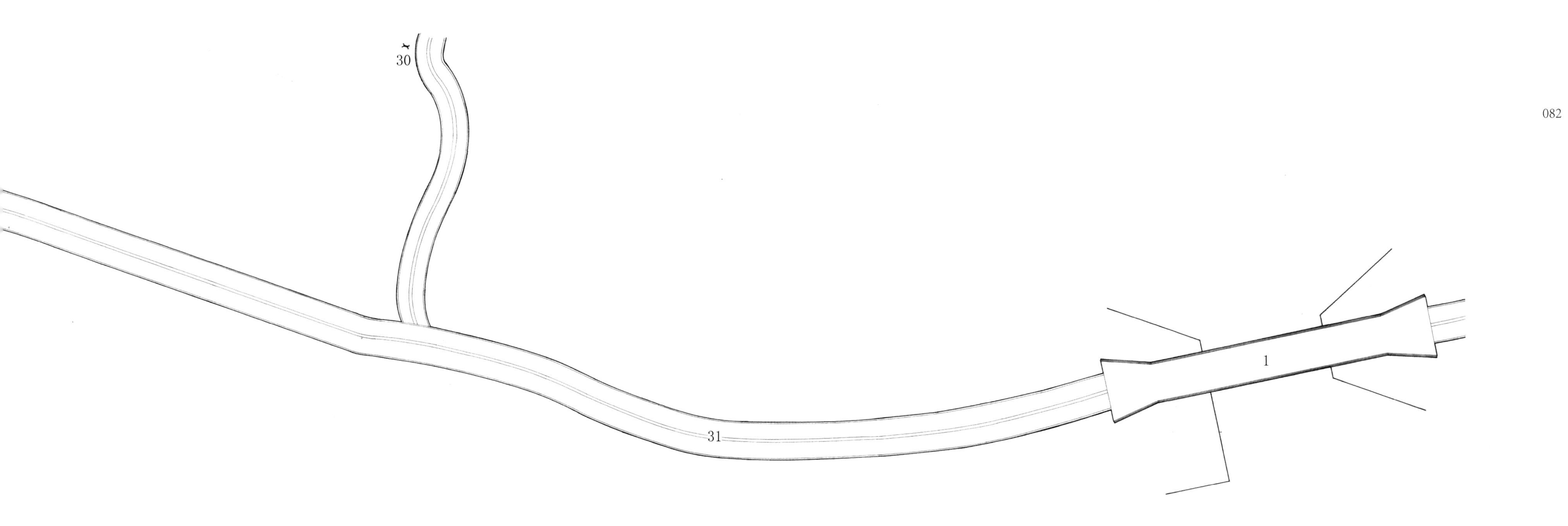

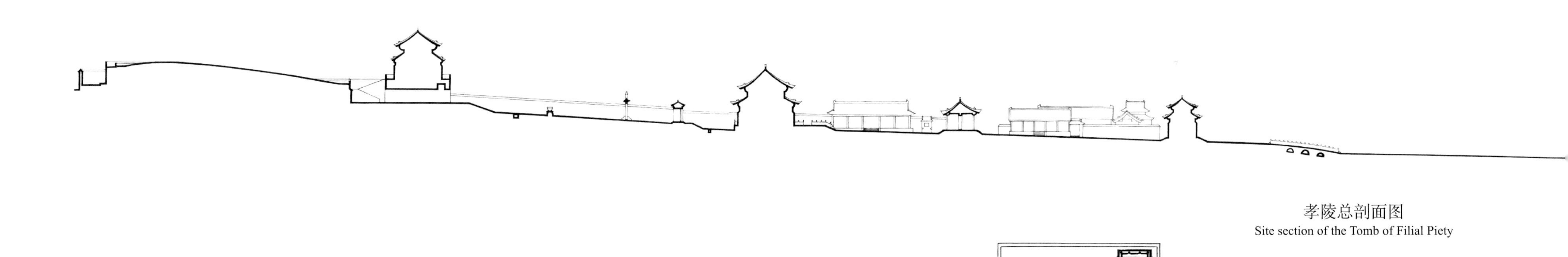

孝陵总剖面图
Site section of the Tomb of Filial Piety

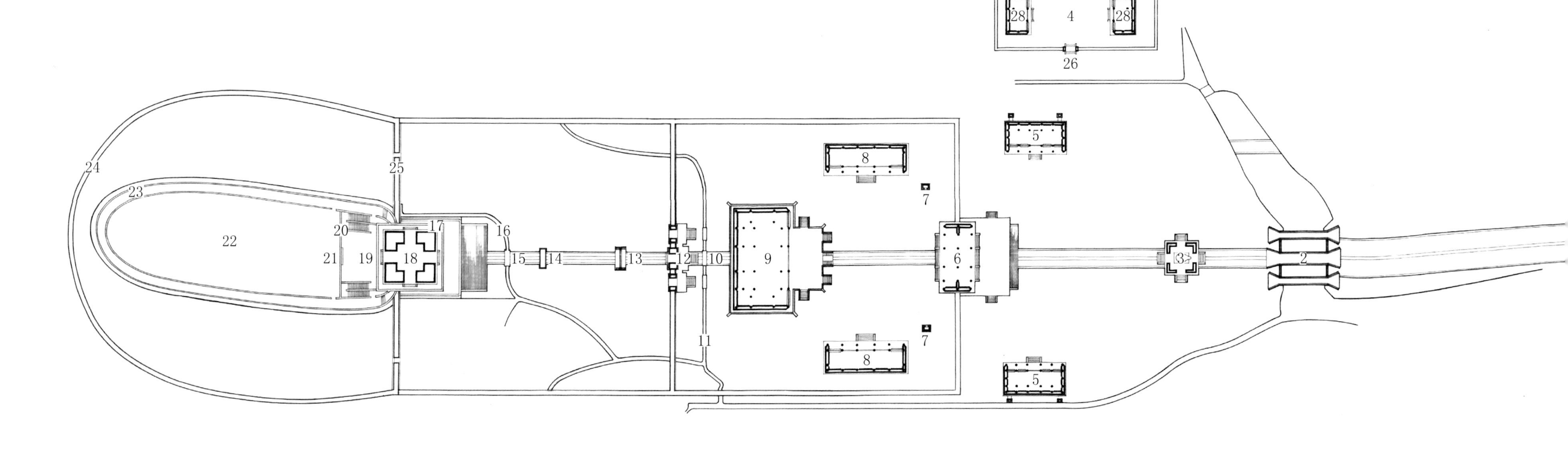

1 五孔石券桥 Five-Arch Stone Bridge
2 三路三孔桥 Three-Arch Stone Bridge
3 神道碑亭 Pavilion for the Stela on the Spirit Way
4 神厨库院落 Culinary Courtyard for Sacrifices
5 朝房 Reception Halls for Court Officials
6 隆恩门 Gate of Monumental Grace
7 焚帛炉 Sacrificial Burners
8 配殿 Side Hall
9 隆恩殿 Hall of Monumental Grace
10 石平桥 Flat Stone Bridge
11 玉带河 Jade Ribbon River
12 琉璃花门 Gate with Glazed Roof Tiles
13 二柱门 Gate with Two Columns
14 石五供 Five Stone Ritual Vessels
15 石平桥 Flat Stone Bridge
16 月牙桥 Crescent Bridge
17 方城 Square Walled Terrace
18 明楼 Memorial Tower
19 哑巴院 Courtyard of the Mute
20 月牙城 Crescent Wall
21 琉璃影壁 Screen Wall of Glazed Tiles
22 宝顶 Tumulus
23 宝城 Encircled Realm of Treasure
24 罗圈墙 Wall Surrounding the Tomb
25 卡子墙 Spacer Wall
26 神厨库院门 Culinary Courtyard Gate for Sacrifices
27 神厨 Sacrificial Kitchen
28 神库 Sacrificial Storehouse
29 宰牲亭 Ritual Abattoir
30 下马牌 Stela Marking the Place for Dismounting from one's Horse
31 神道 Spirit Way

孝陵总平面图
Site plan of the Tomb of Filial Piety

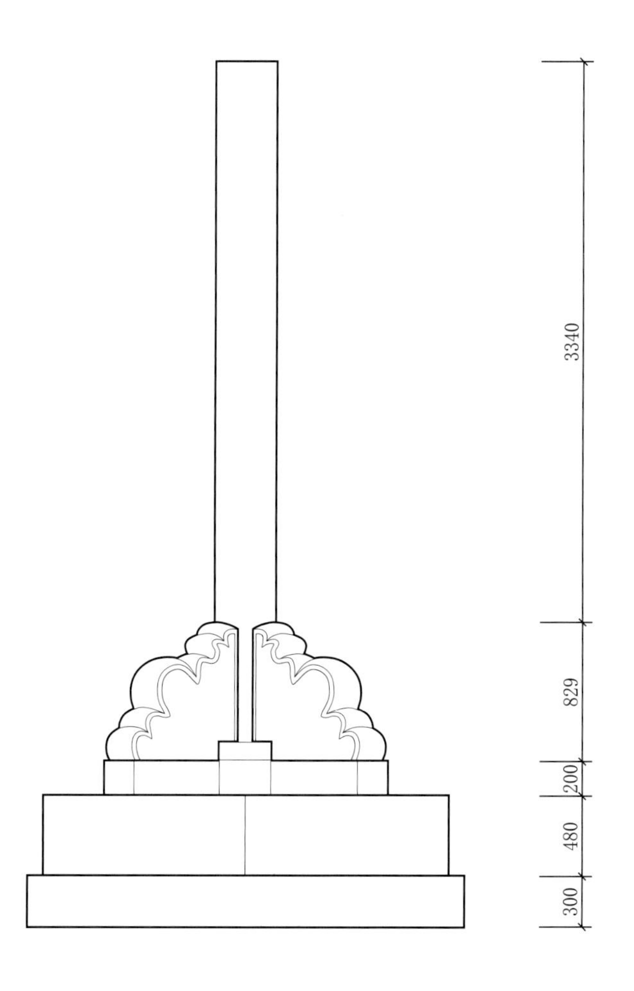

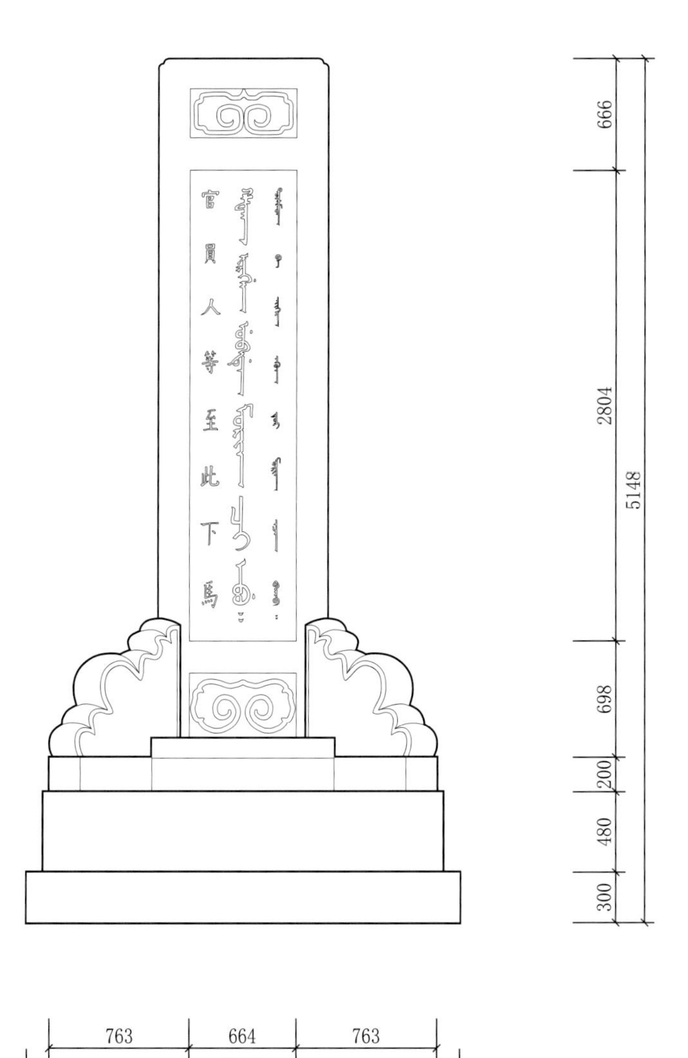

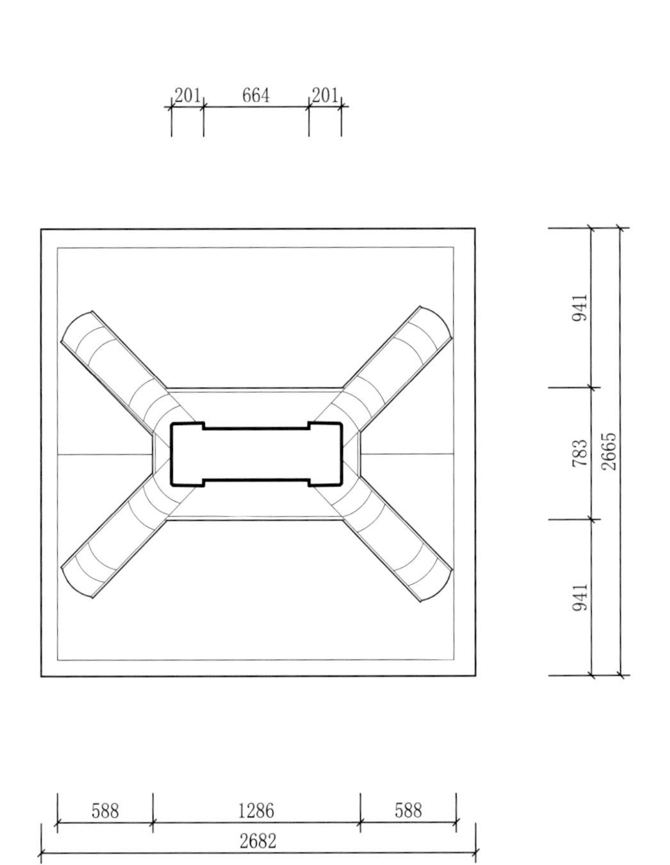

孝陵下马牌大样图

Detailed drawing of the Stela marking the place for dismounting from one's horse, Tomb of Filial Piety

0 0.2 1m

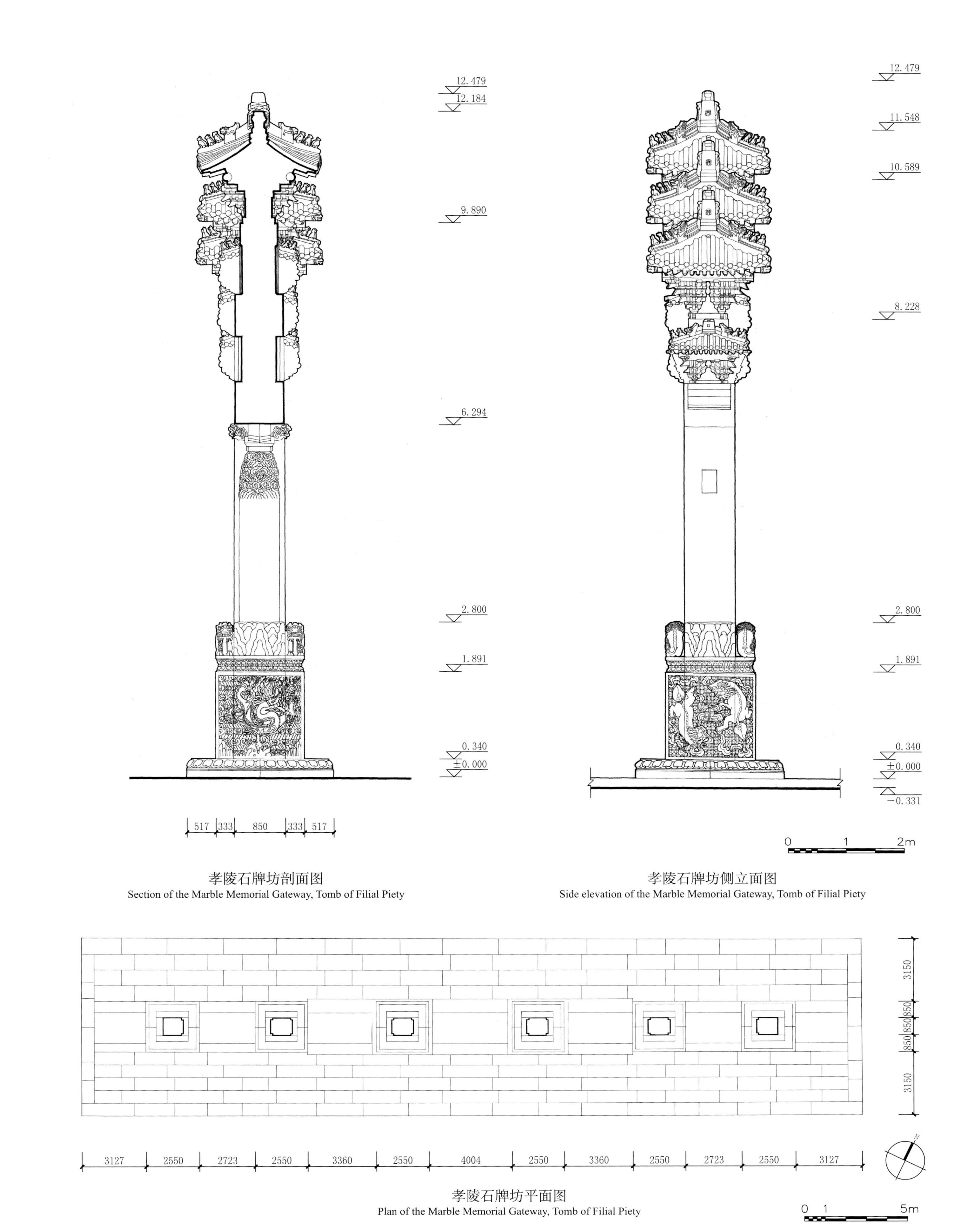

孝陵石牌坊剖面图
Section of the Marble Memorial Gateway, Tomb of Filial Piety

孝陵石牌坊侧立面图
Side elevation of the Marble Memorial Gateway, Tomb of Filial Piety

孝陵石牌坊平面图
Plan of the Marble Memorial Gateway, Tomb of Filial Piety

12.184
11.253
10.249
9.890
9.067
8.222
6.294
5.548
4.765
1.891
±0.000
−0.331
5910
5288
4387
0
1
2m

孝陵石牌坊正立面图
Front elevation of the Marble Memorial Gateway, Tomb of Filial Piety

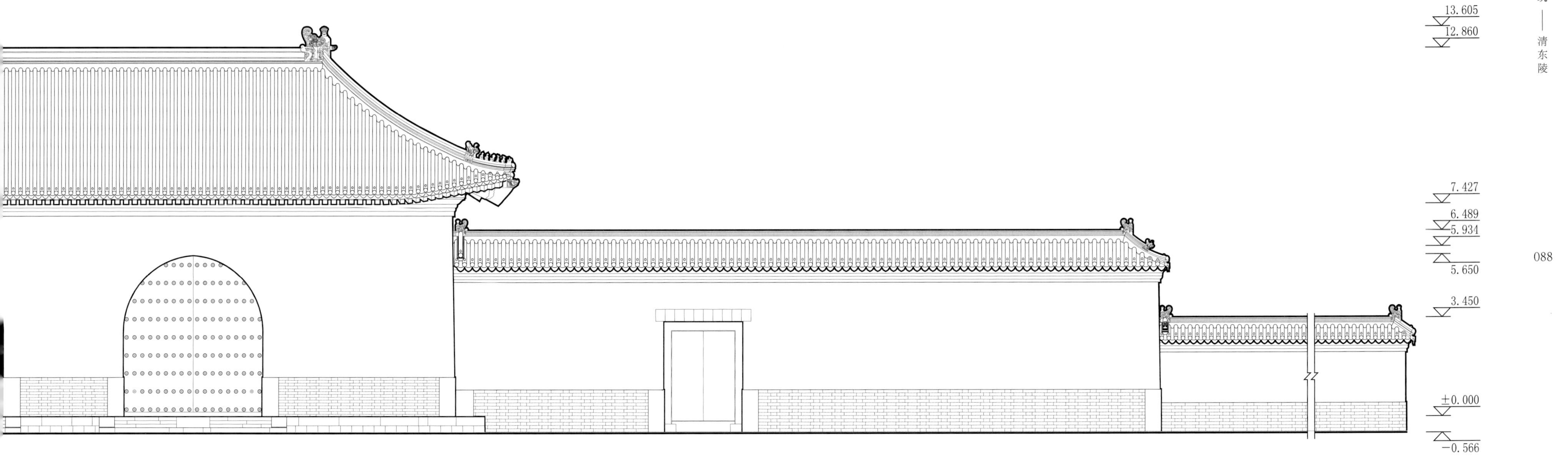
13.605
12.860
7.427
6.489
5.934
5.650
3.450
±0.000
−0.566
4337
5032
6887
1050
6517
2855
15123
0
1
5m

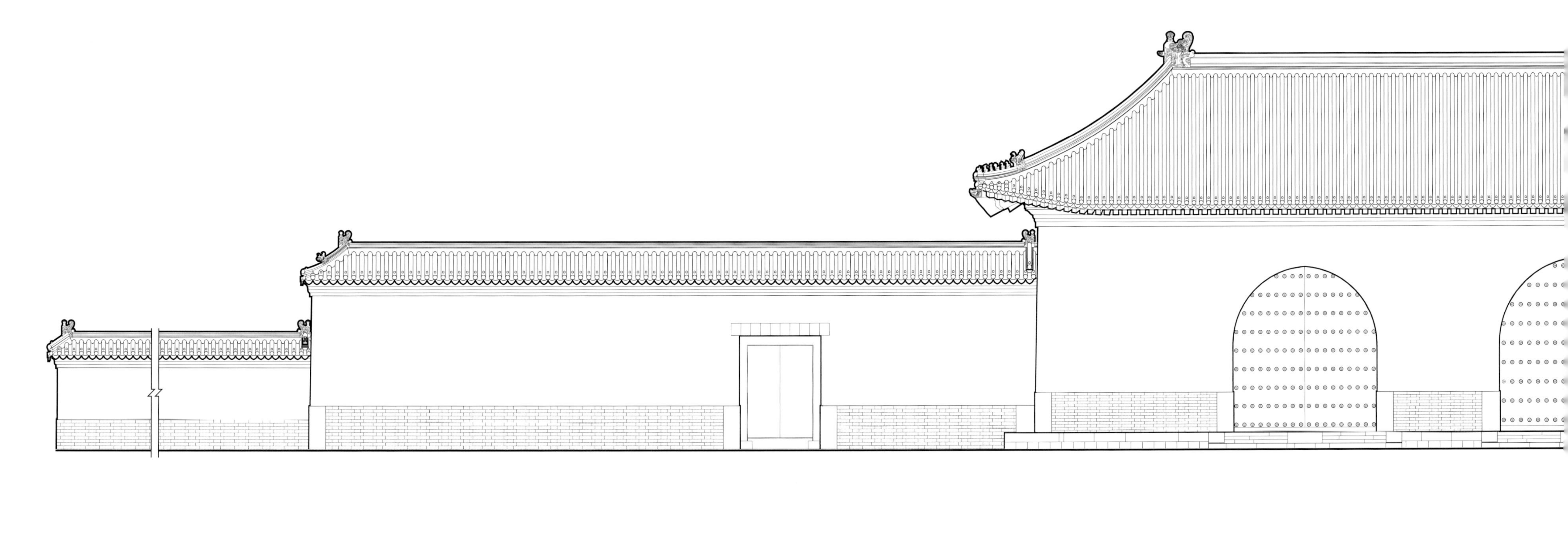

孝陵大红门正立面图
Front elevation of the Great Red Gate Building, Tomb of Filial Piety

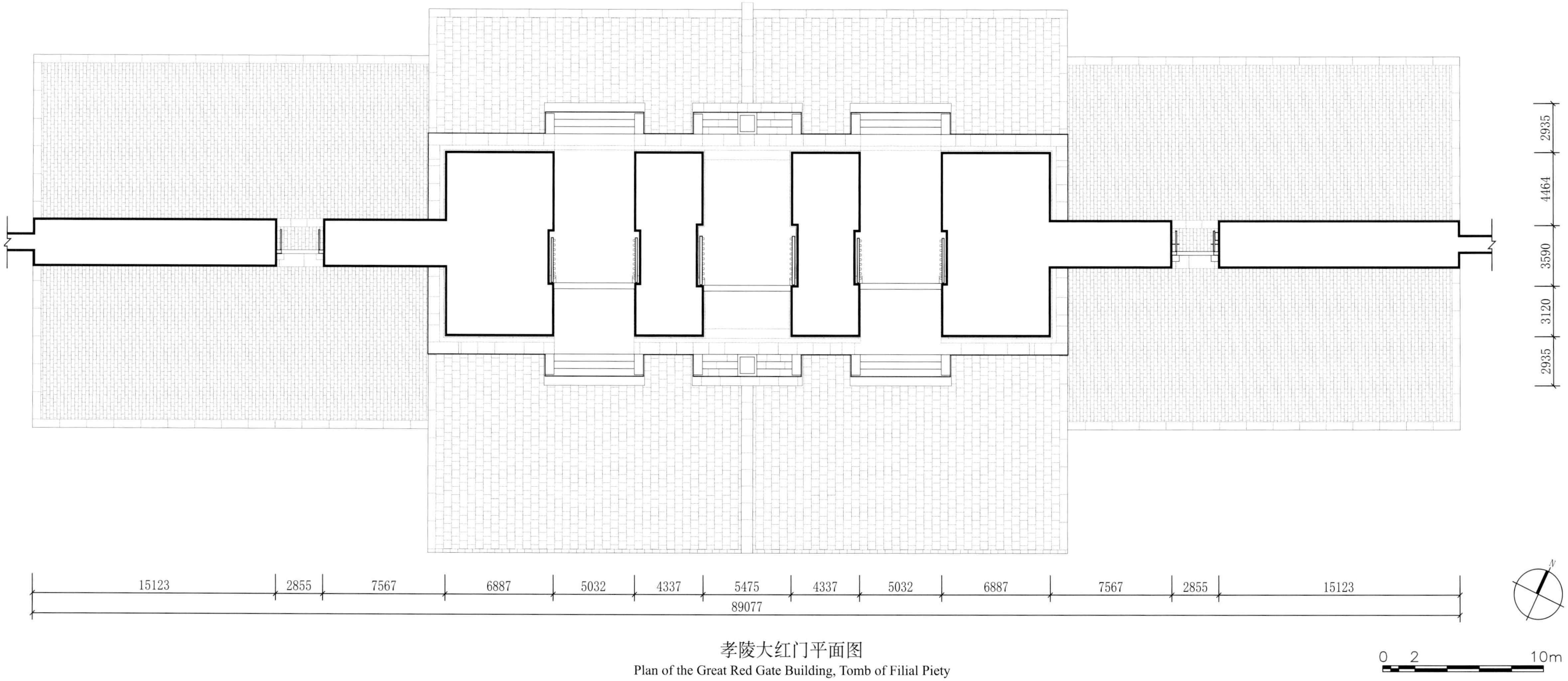

孝陵大红门平面图
Plan of the Great Red Gate Building, Tomb of Filial Piety

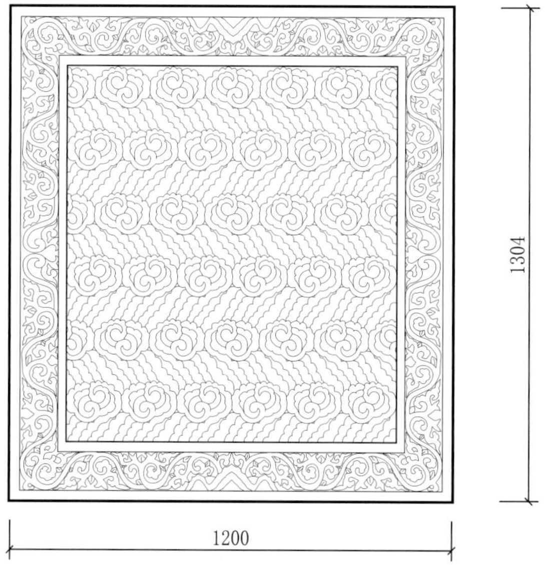

孝陵大红门丹陛大样图
Detailed drawing of the marble carving on the steps to the Great Red Gate, Tomb of Filial Piety

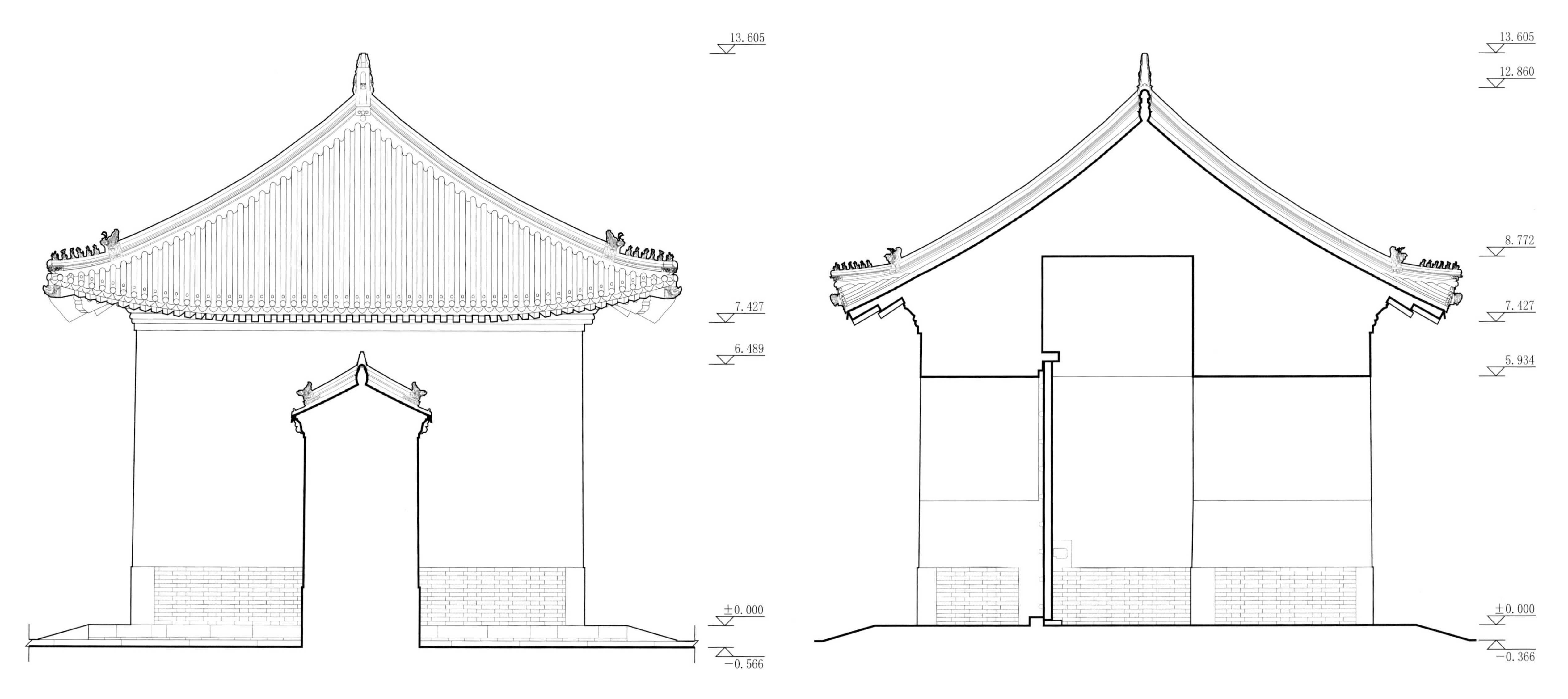

孝陵大红门侧立面图
Side elevation of the Great Red Gate Building, Tomb of Filial Piety

孝陵大红门横剖面图
Cross section of the Great Red Gate, Tomb of Filial Piety

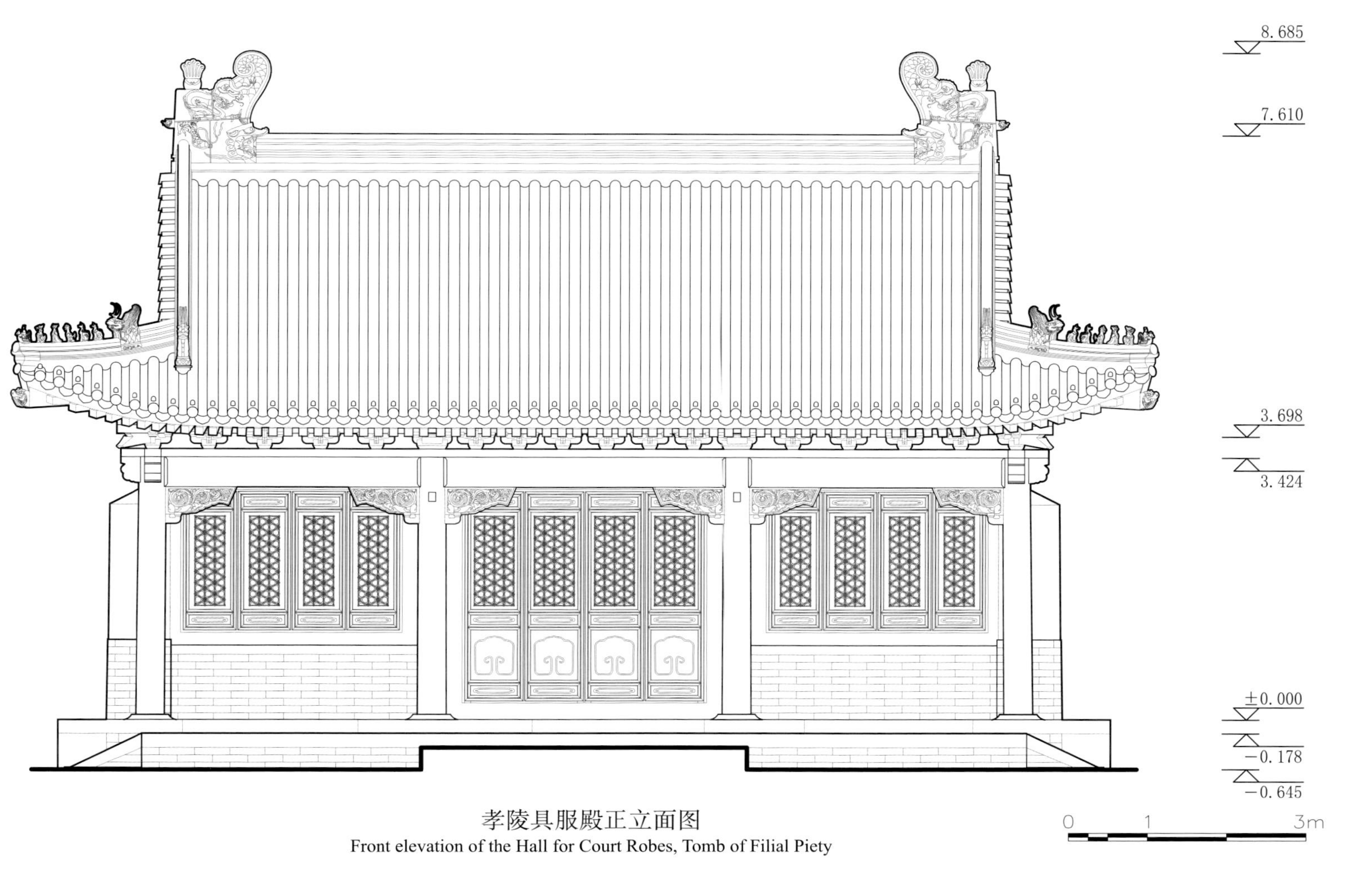

孝陵具服殿正立面图
Front elevation of the Hall for Court Robes, Tomb of Filial Piety

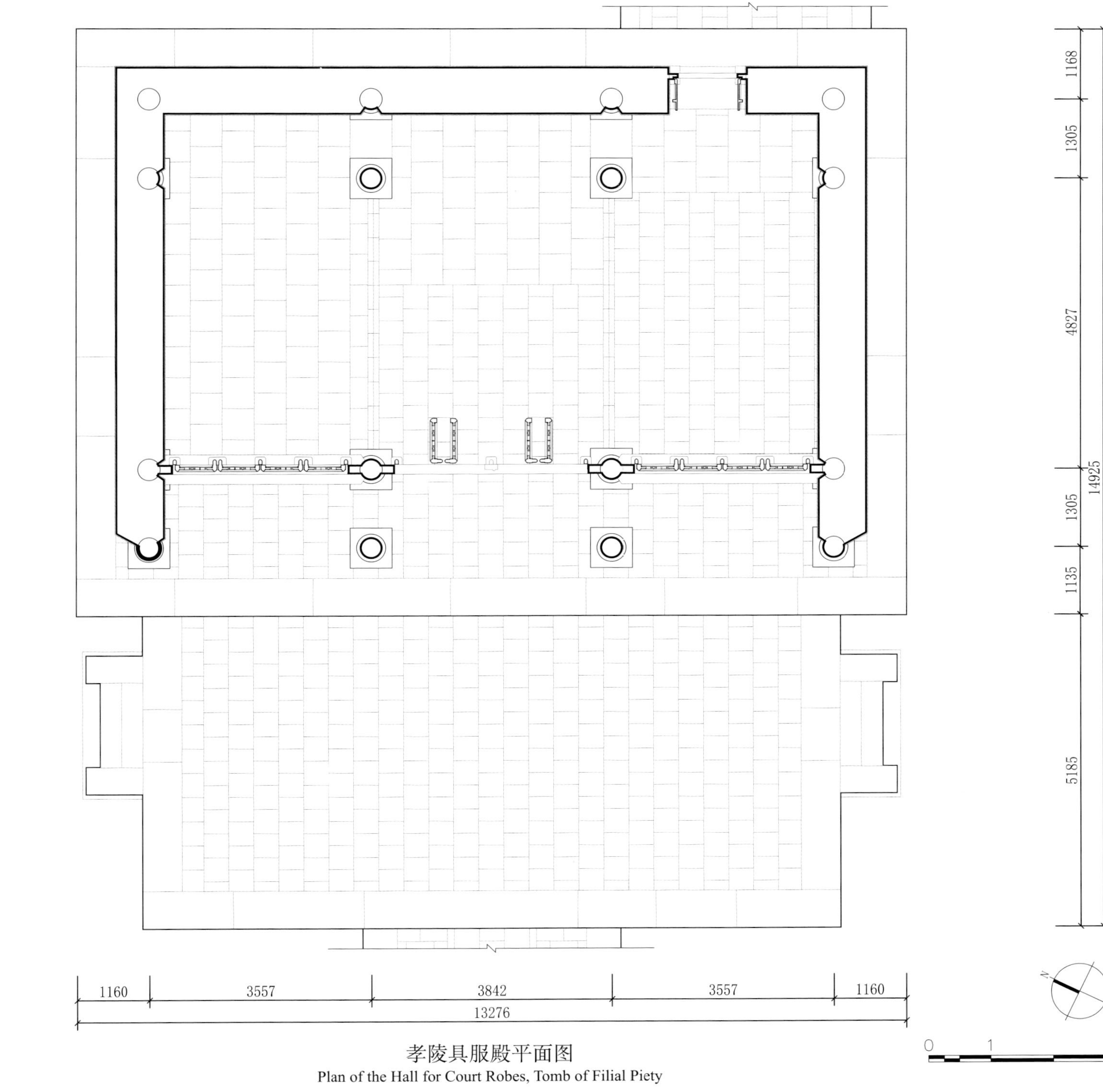

孝陵具服殿平面图
Plan of the Hall for Court Robes, Tomb of Filial Piety

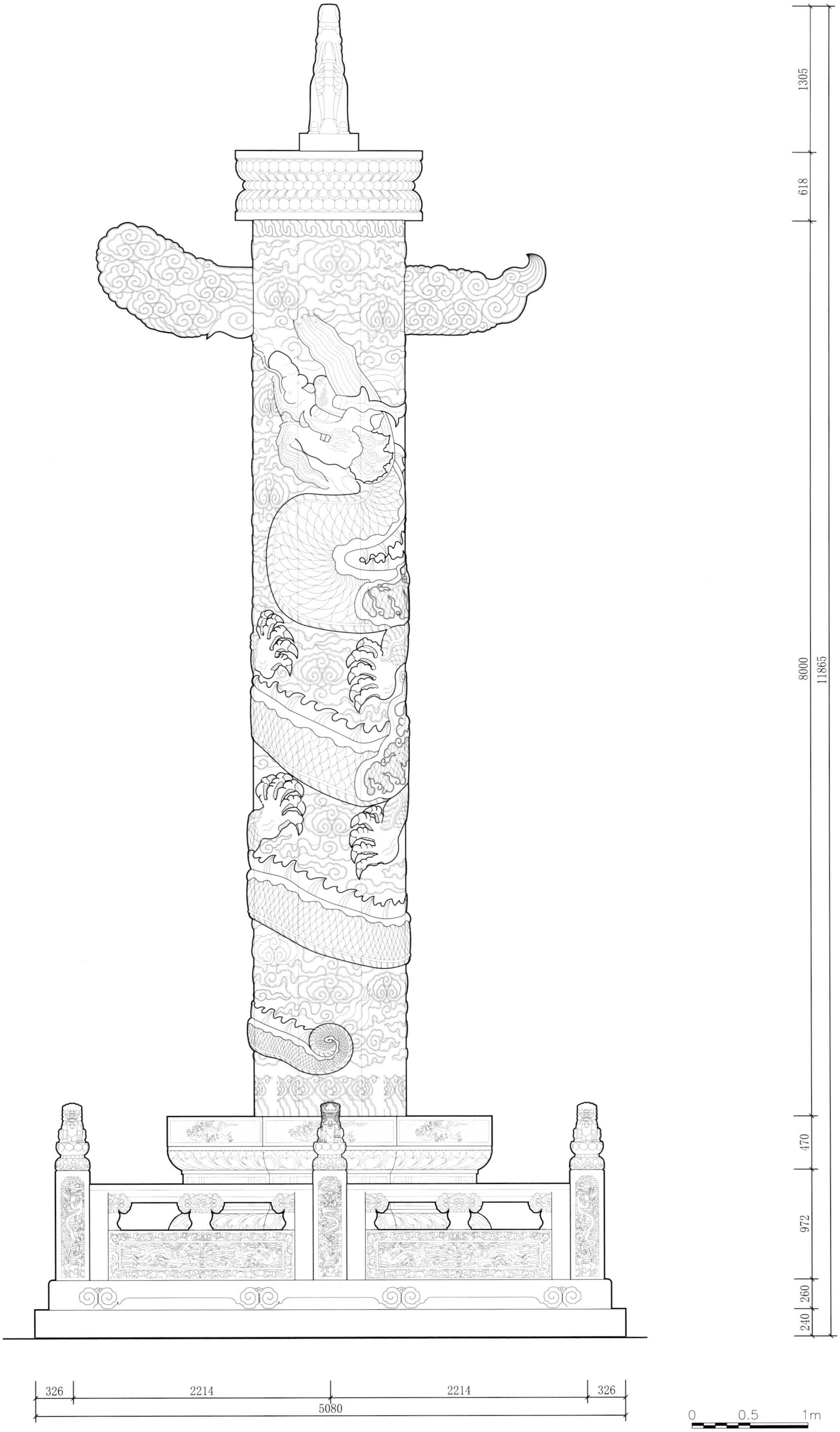

孝陵神功圣德碑亭东南华表南立面图

South elevation of the Ceremonial Column, southeast of the Pavilion for the Stelae of Divine Merit and Sage Virtue, Tomb of Filial Piety

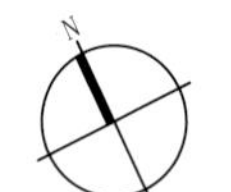

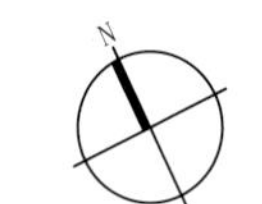

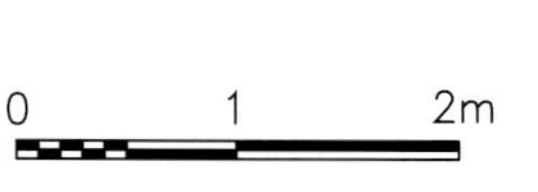

孝陵神功圣德碑亭东南华表平面图
Plan of the Ceremonial Column, southeast to the Pavilion for the Stelae of Divine Merit and Sage Virtue, Tomb of Filial Piety

孝陵神功圣德碑亭总平面图
General plan of the Pavilion for the Stelae of Divine Merit and Sage Virtue, Tomb of Filial Piety

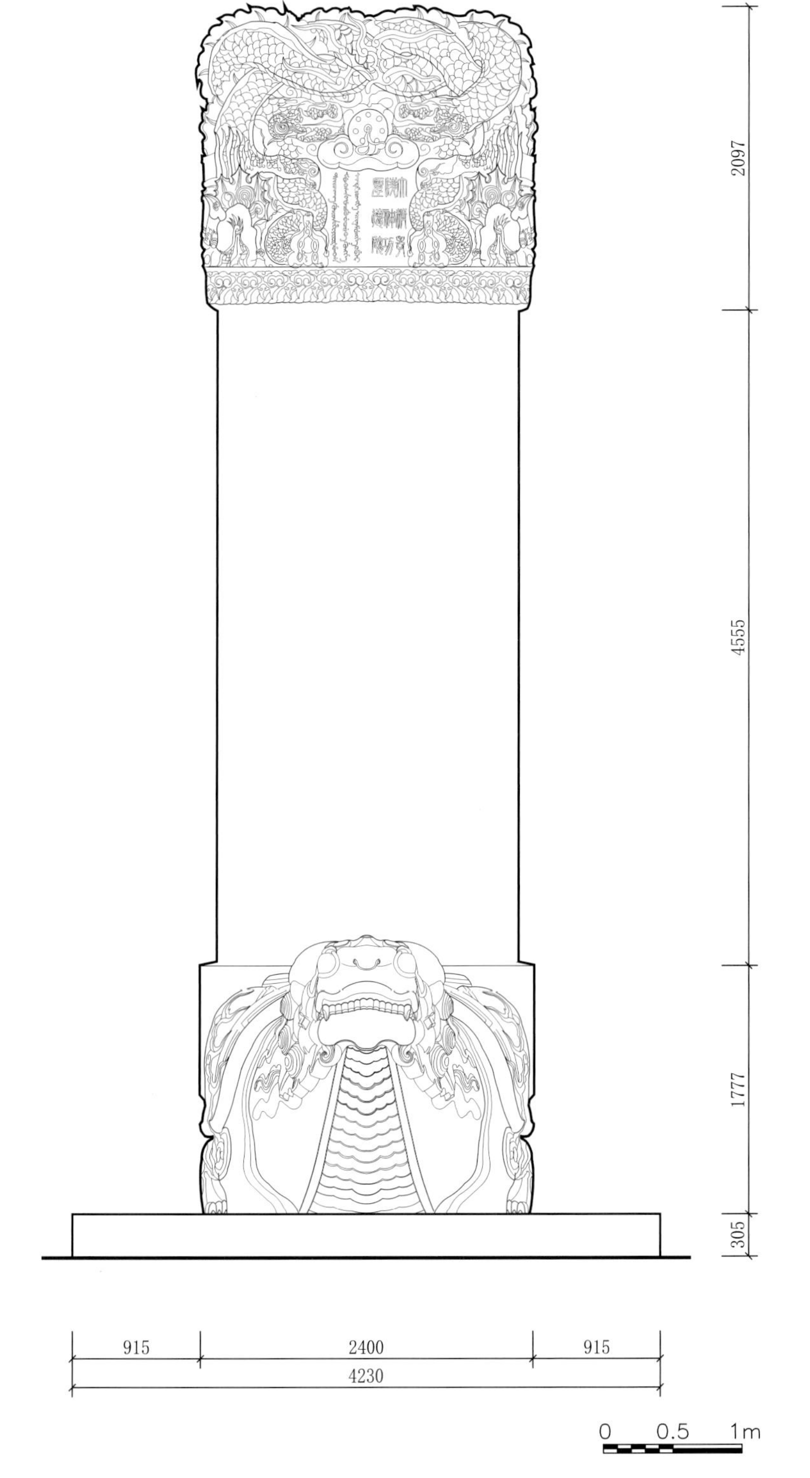

孝陵神功圣德碑亭石碑正立面图
Front elevation of the Stelae of Divine Merit and Sage Virtue in the Pavilion, Tomb of Filial Piety

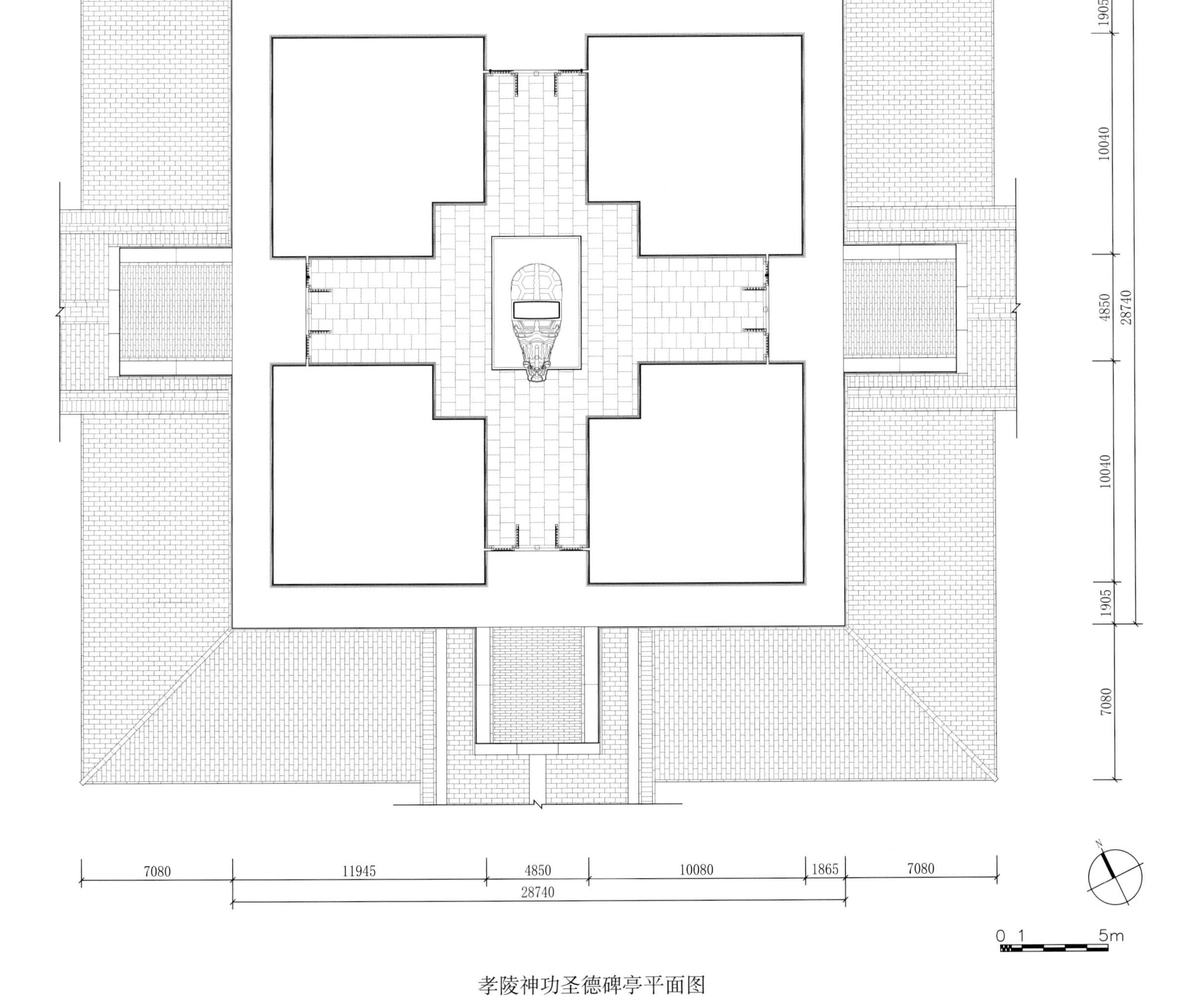

孝陵神功圣德碑亭平面图
Plan of the Pavilion for the Stelae of Divine Merit and Sage Virtue, Tomb of Filial Piety

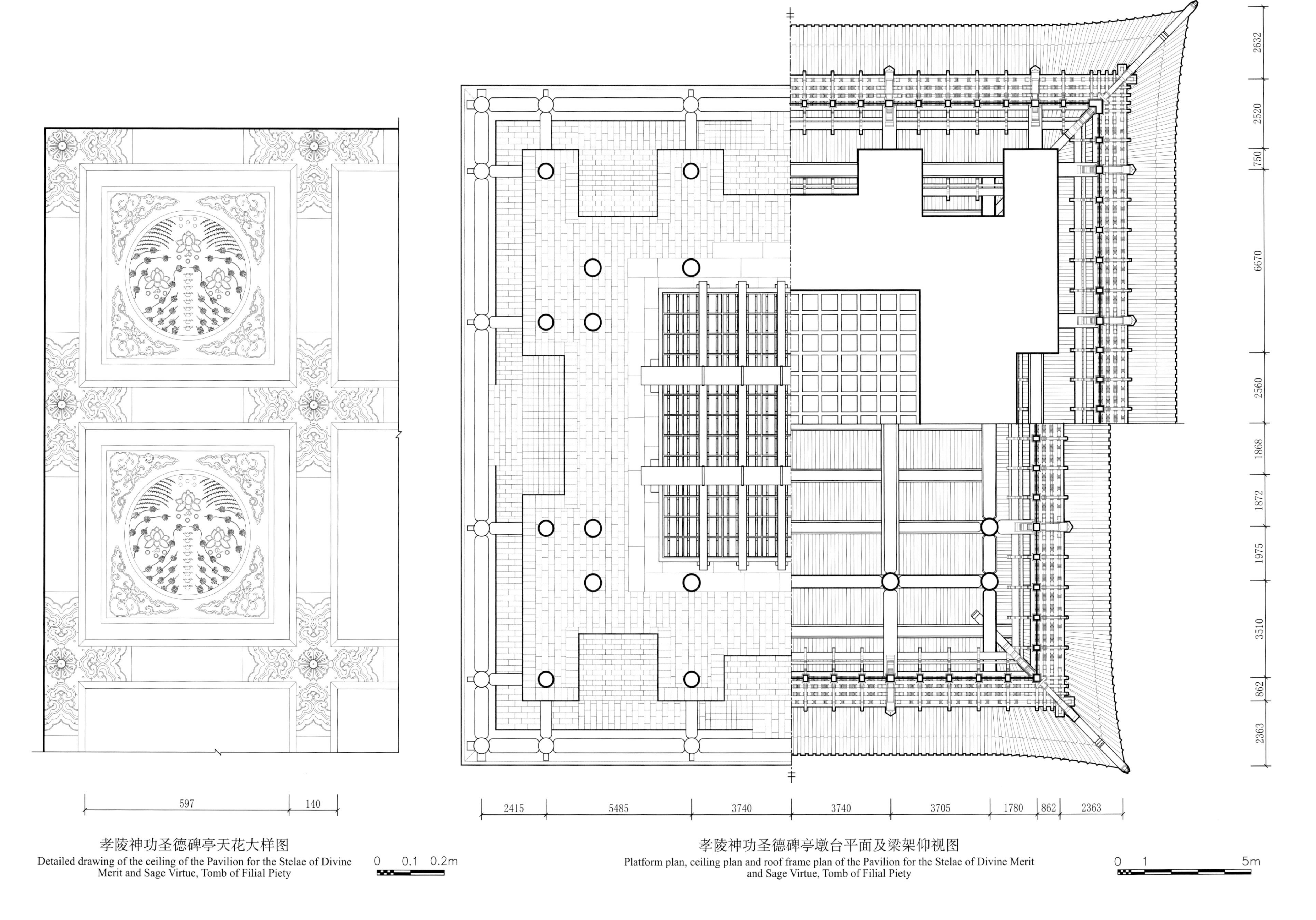

孝陵神功圣德碑亭天花大样图
Detailed drawing of the ceiling of the Pavilion for the Stelae of Divine Merit and Sage Virtue, Tomb of Filial Piety

孝陵神功圣德碑亭墩台平面及梁架仰视图
Platform plan, ceiling plan and roof frame plan of the Pavilion for the Stelae of Divine Merit and Sage Virtue, Tomb of Filial Piety

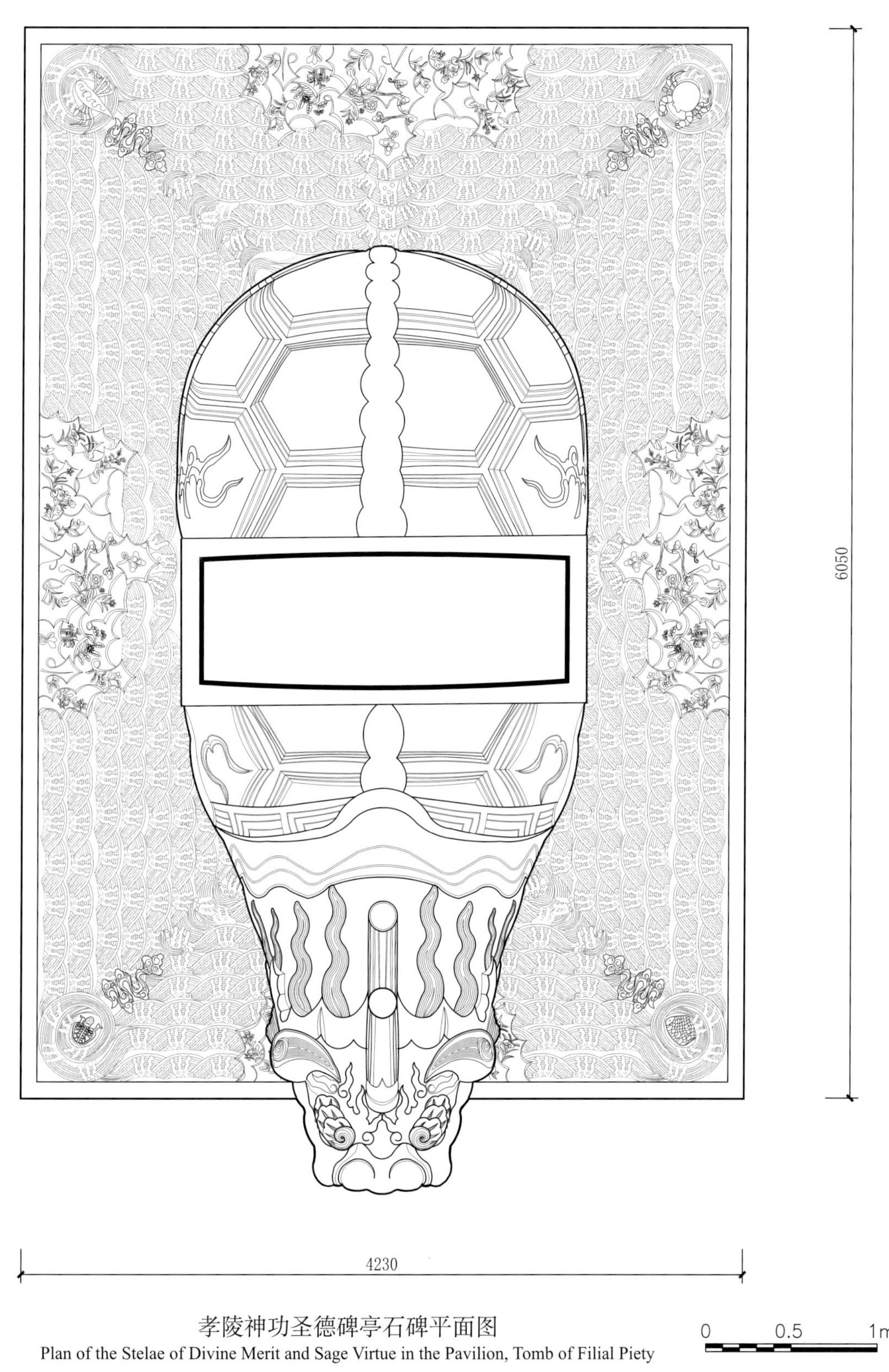

孝陵神功圣德碑亭石碑平面图
Plan of the Stelae of Divine Merit and Sage Virtue in the Pavilion, Tomb of Filial Piety

2097
4555
1765
300
0 0.5 1m

孝陵神功圣德碑亭石碑侧立面图
Side elevation of the Stelae of Divine Merit and Sage Virtue in the Pavilion, Tomb of Filial Piety

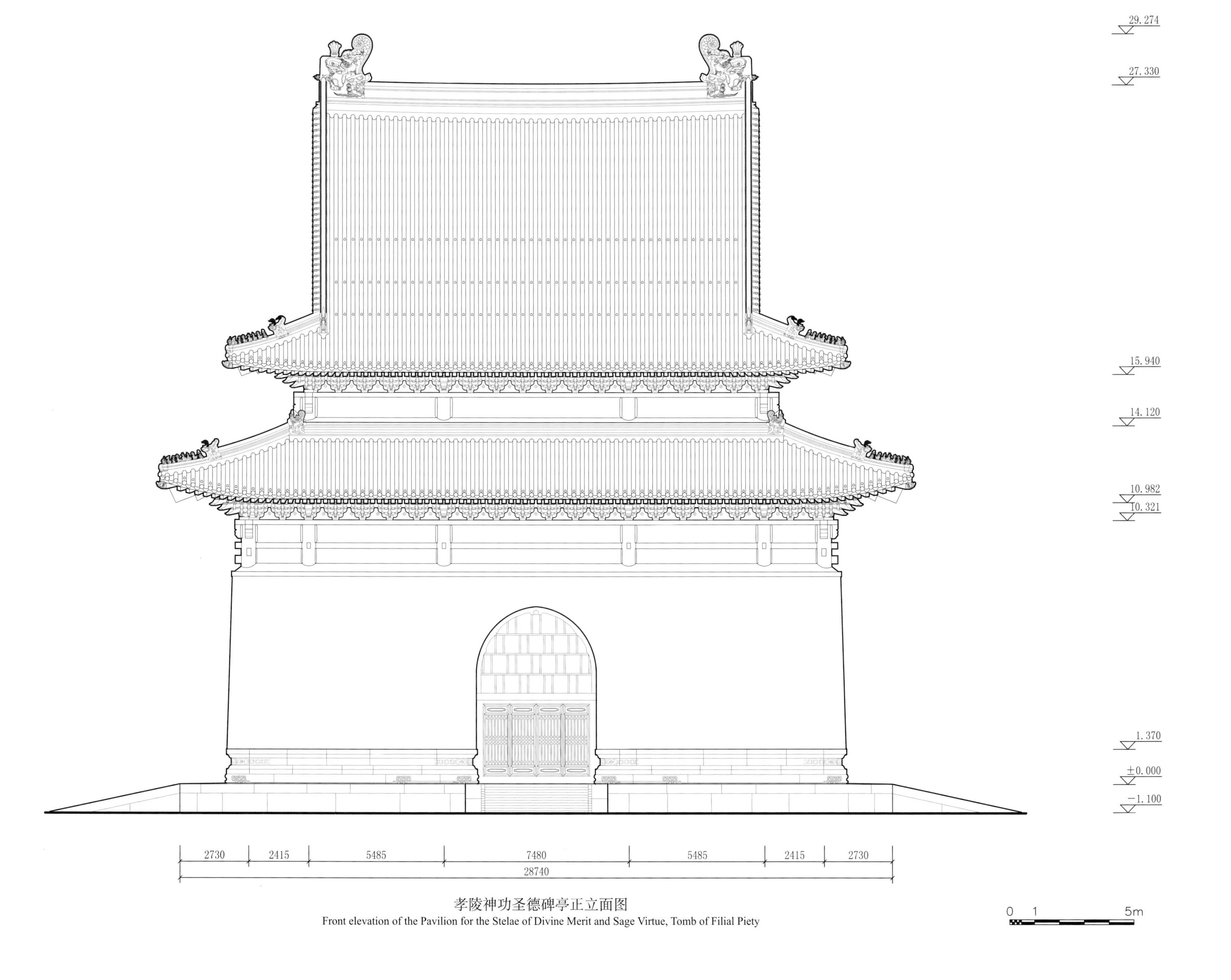

孝陵神功圣德碑亭正立面图
Front elevation of the Pavilion for the Stelae of Divine Merit and Sage Virtue, Tomb of Filial Piety

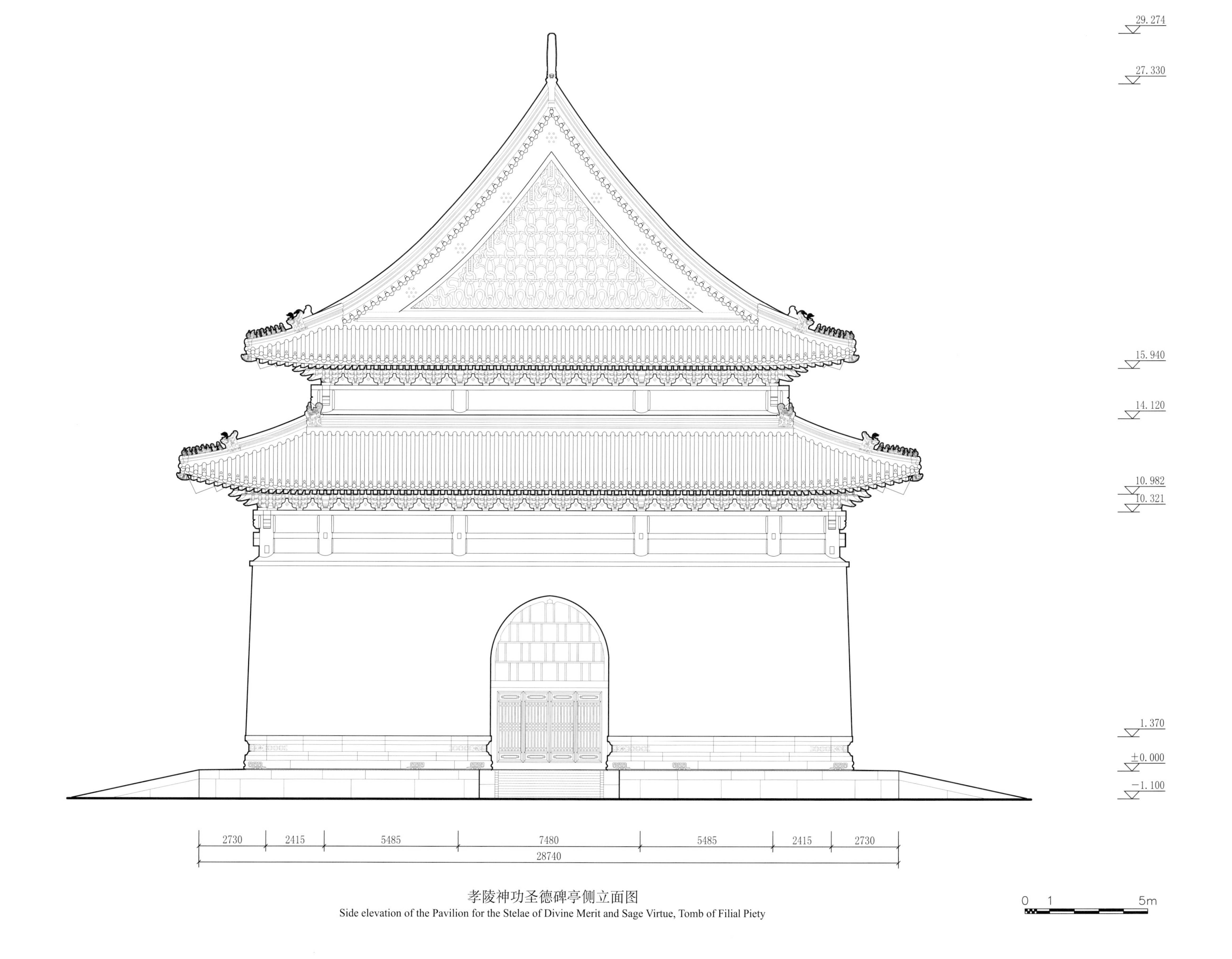

孝陵神功圣德碑亭侧立面图
Side elevation of the Pavilion for the Stelae of Divine Merit and Sage Virtue, Tomb of Filial Piety

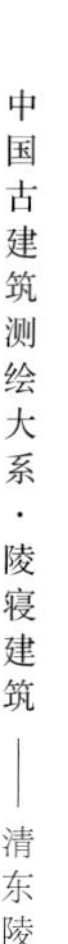

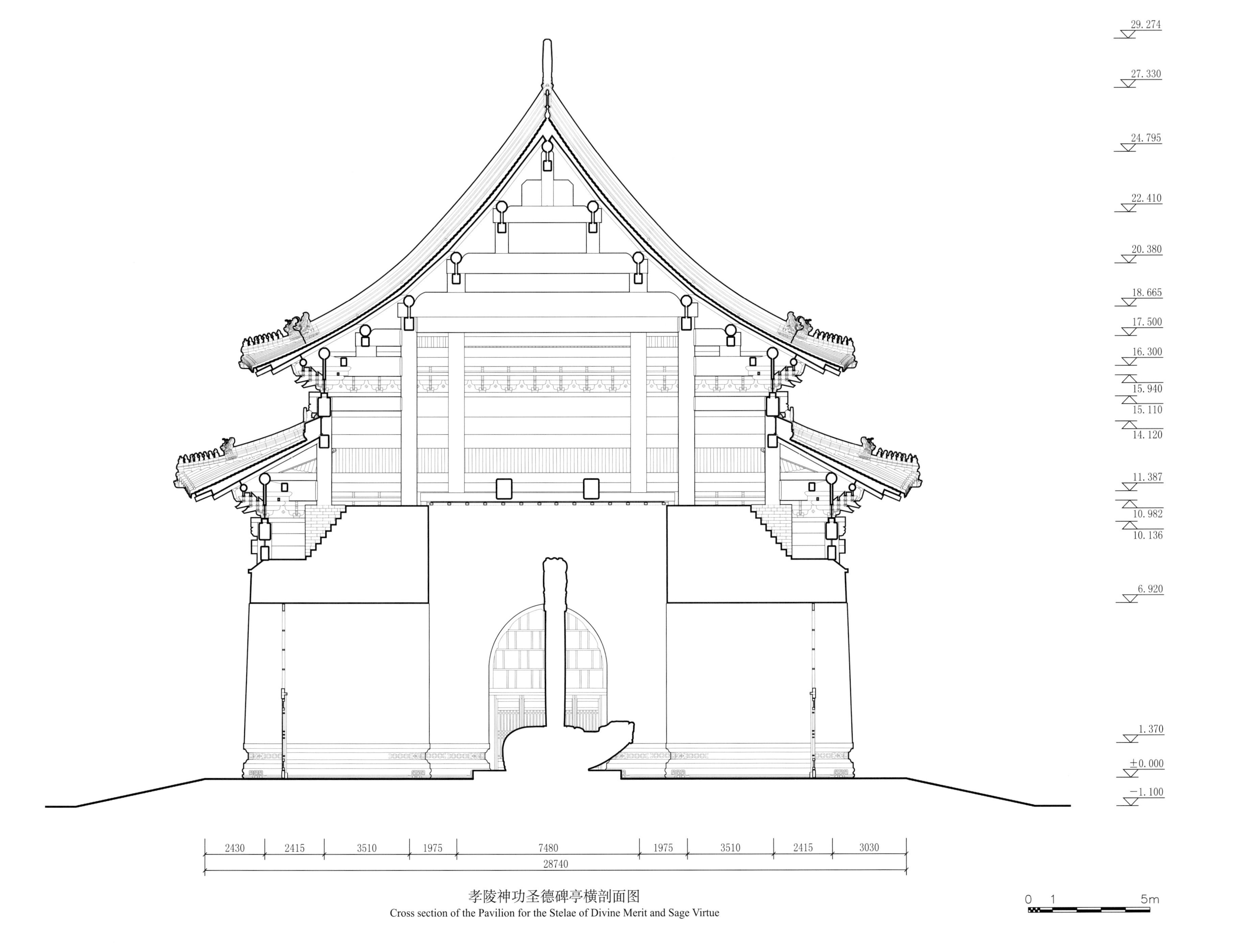

孝陵神功圣德碑亭横剖面图

Cross section of the Pavilion for the Stelae of Divine Merit and Sage Virtue

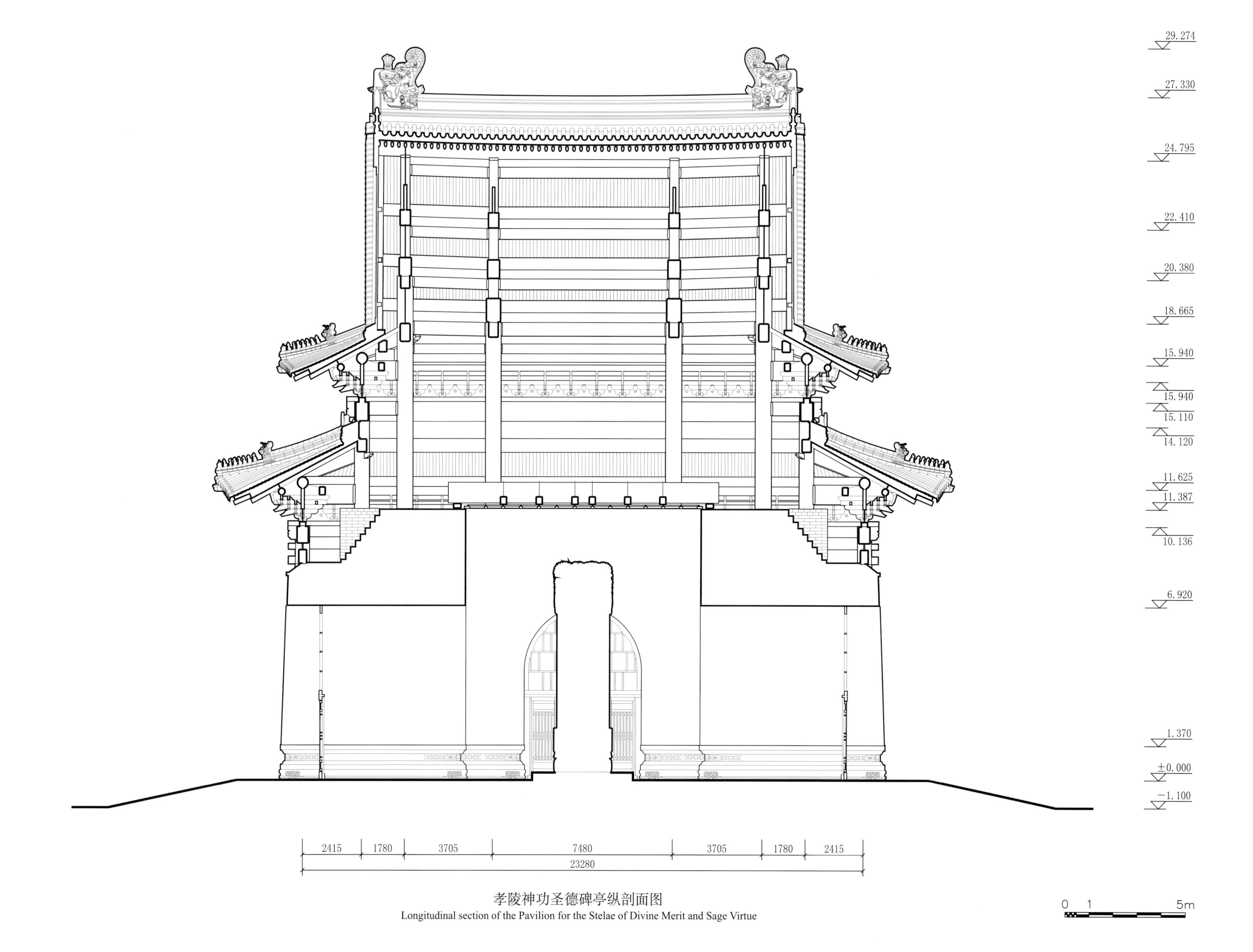

孝陵神功圣德碑亭纵剖面图
Longitudinal section of the Pavilion for the Stelae of Divine Merit and Sage Virtue

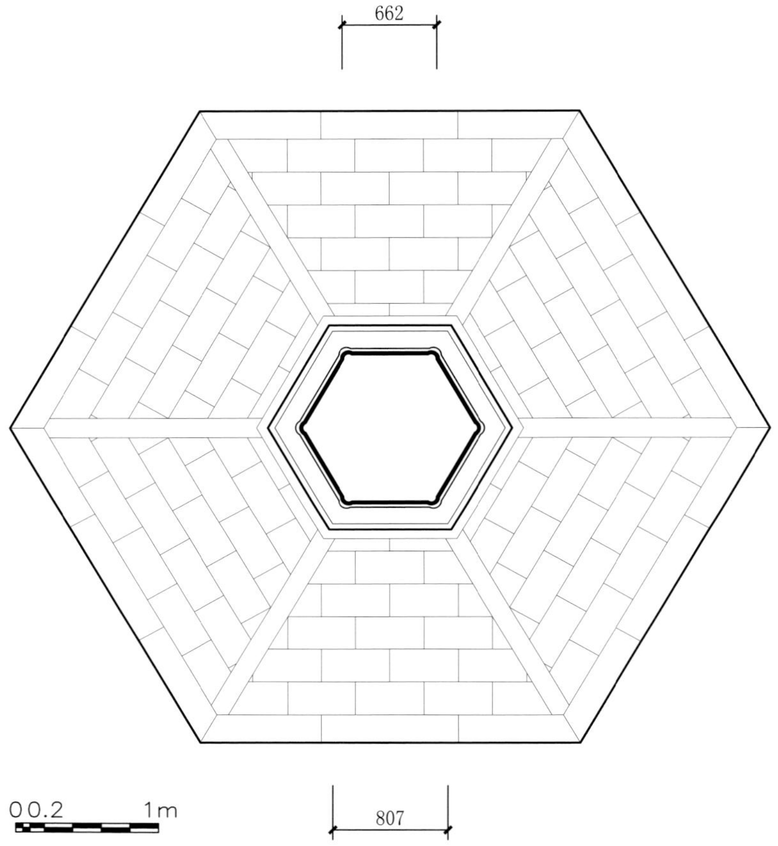

6.585

5.305

4.360

±0.000

−0.860

−1.017

0 0.3 0.9m

孝陵神道望柱立面图
Elevation of the Ornamental Column of the Spirit Way, Tomb of Filial Piety

662

807

孝陵神道望柱平面图
Plan of the Ornamental Column of the Spirit Way, Tomb of Filial Piety

0 0.2 1m

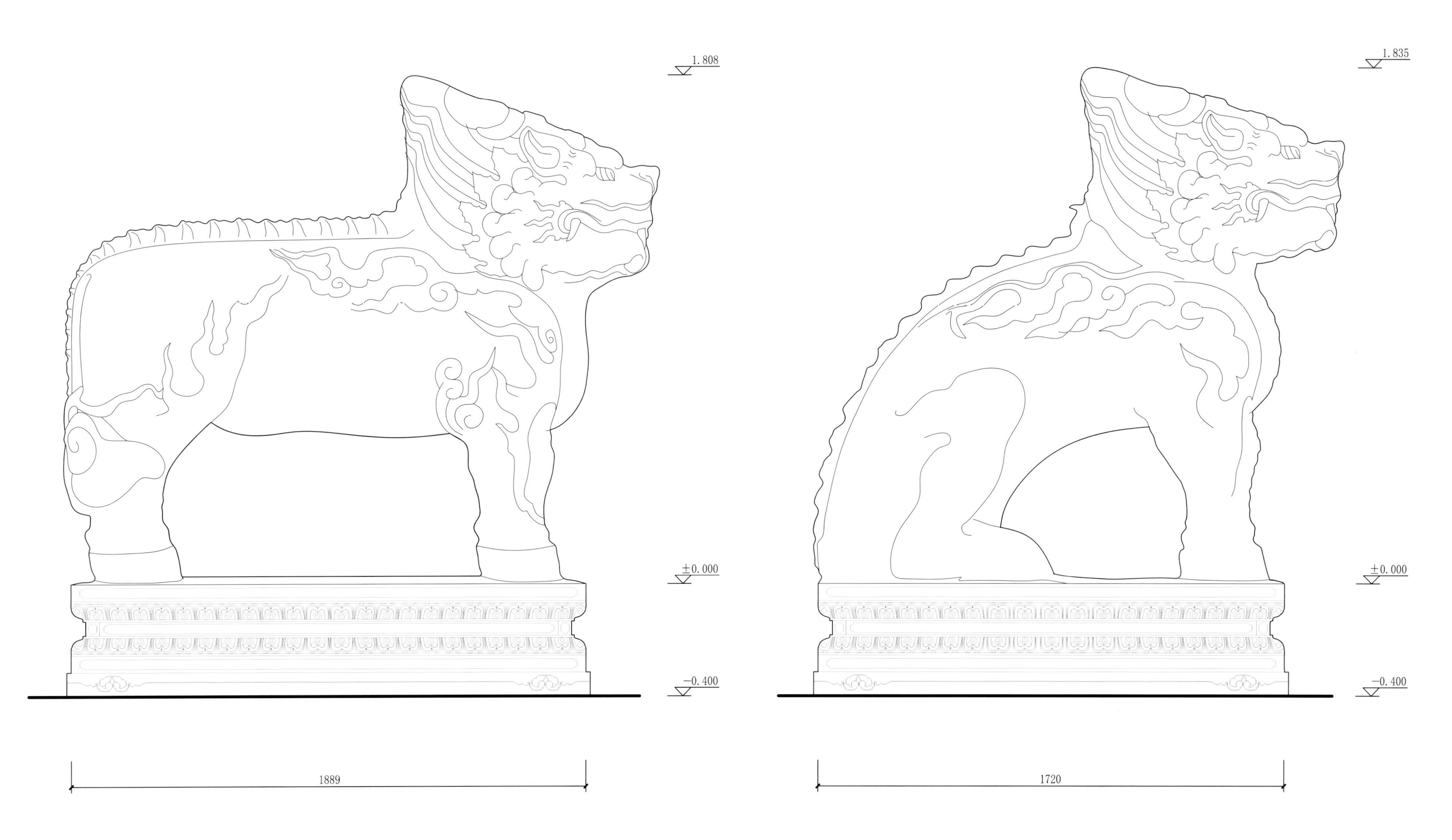

孝陵神道石像生立麒麟大样图
Detailed drawing of one of the Standing Qilins among the Stone Statues along the Spirit Way, Tomb of Filial Piety

孝陵神道石像生卧麒麟大样图
Detailed drawing of one of the Sitting Qilins among the Stone Statues along the Spirit Way, Tomb of Filial Piety

1.642

±0.000

−0.418

1728

孝陵神道石像生立狮子大样图
Detailed drawing of one of the Standing Lions among the Stone Statues along the Spirit Way , Tomb of Filial Piety

1.642

±0.000

−0.418

1728

孝陵神道石像生卧狮子大样图
Detailed drawing of one of the Sitting Lions among the Stone Statues along the Spirit Way, Tomb of Filial Piety

孝陵神道石像生立獬豸大样图
Detailed drawing of one of the Standing Xiezhis, among the Stone Statues along the Spirit Way, Tomb of Filial Piety

孝陵神道石像生卧獬豸大样图
Detailed drawing of one of the Sitting Xiezhis, among the Stone Statues along the Spirit Way, Tomb of Filial Piety

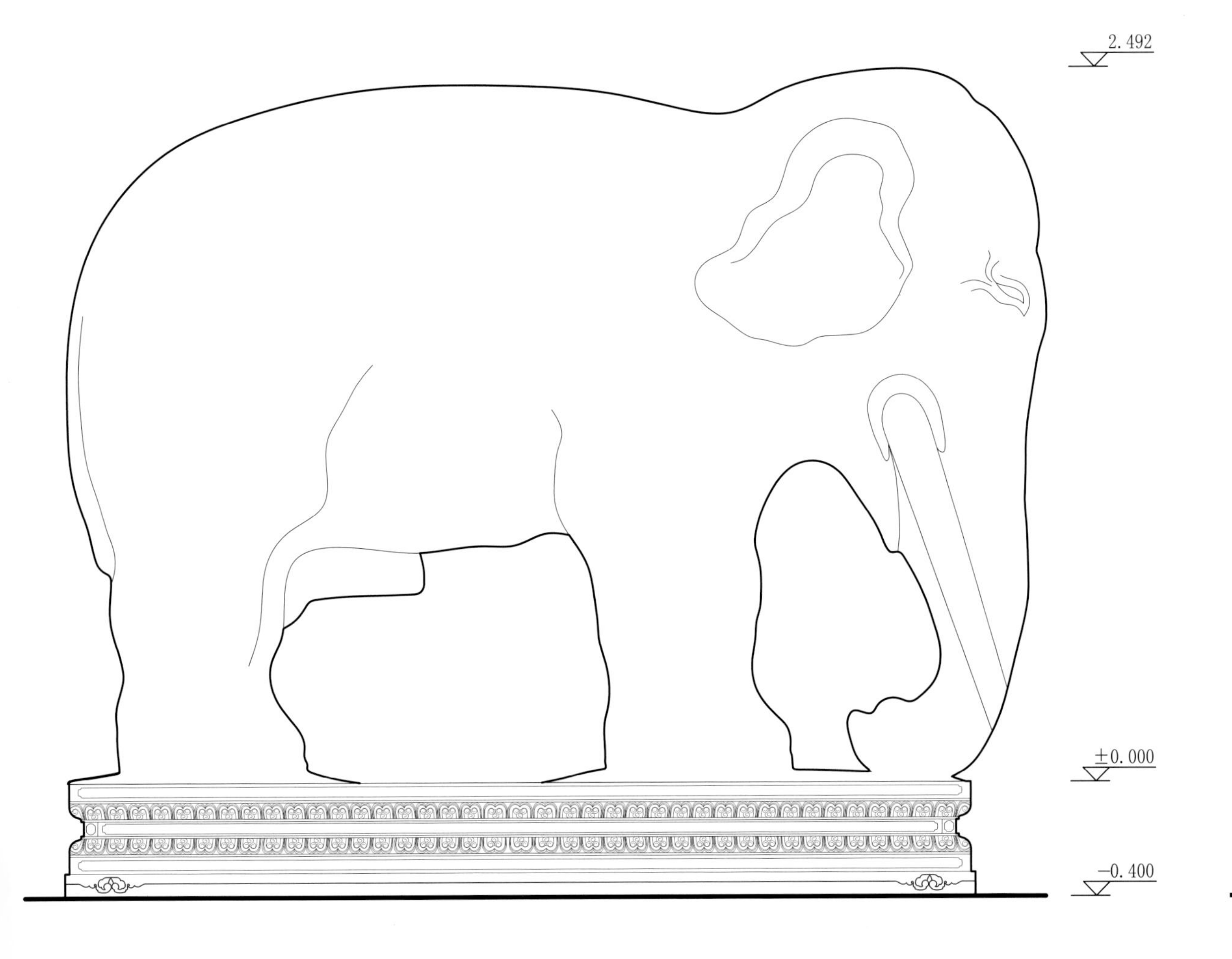

3262

孝陵神道石像生立象大样图

Detailed drawing of one of the Standing Elephants, among the Stone Statues along the Spirit Way, Tomb of Filial Piety

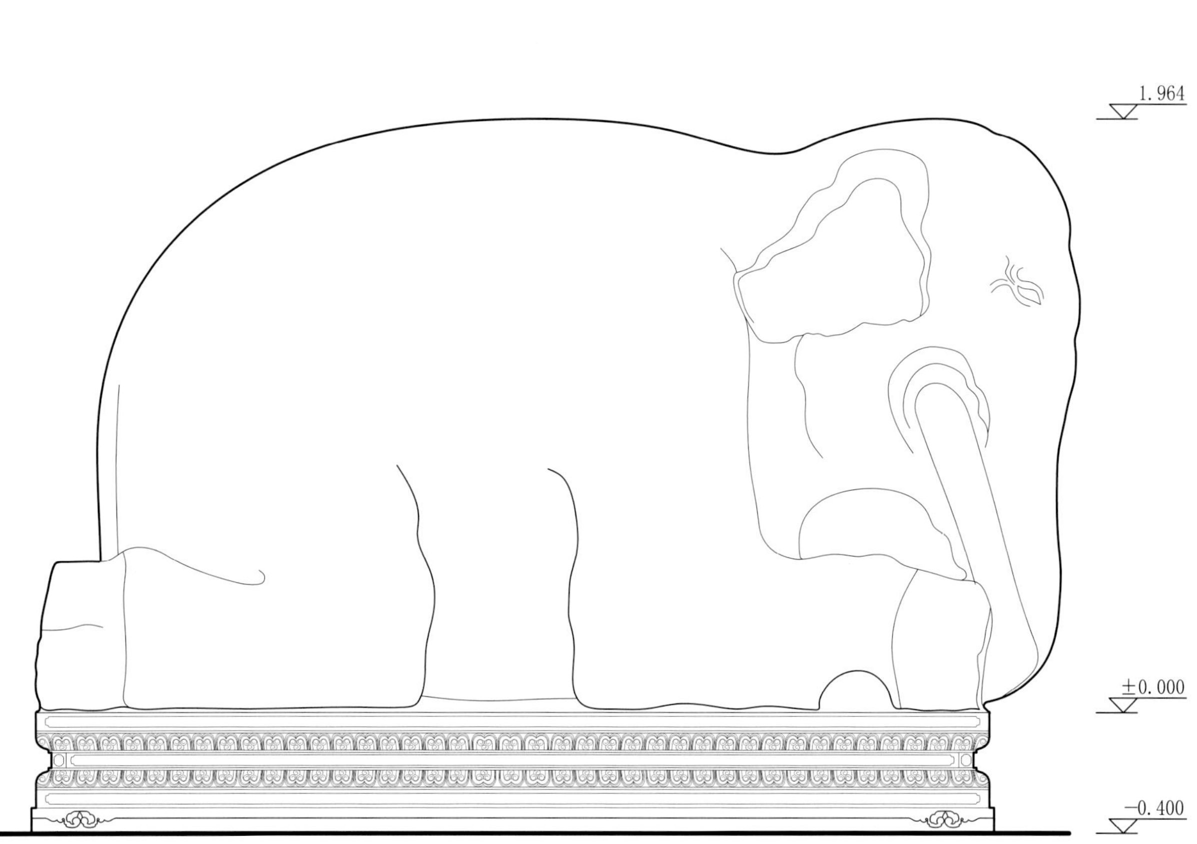

3262

孝陵神道石像生卧象大样图

Detailed drawing of one of the Sitting Elephants, among the Stone Statues along the Spirit Way, Tomb of Filial Piety

2.263

±0.000

−0.400

2473

孝陵神道石像生立骆驼大样图

Detailed drawing of one of the Standing Camels, among the Stone Statues along the Spirit Way, Tomb of Filial Piety

孝陵神道石像生卧骆驼大样图

Detailed drawing of one of the Sitting Camels, among the Stone Statues along the Spirit Way, Tomb of Filial Piety

0 0.1 0.8m

孝陵神道石像生文臣大样图
Detailed drawing of one of the Literary Officials, among the Stone Statues along the Spirit Way, Tomb of Filial Piety

孝陵神道石像生武将大样图
Detailed drawing of one of the Martial Officials, among the Stone Statues along the Spirit Way, Tomb of Filial Piety

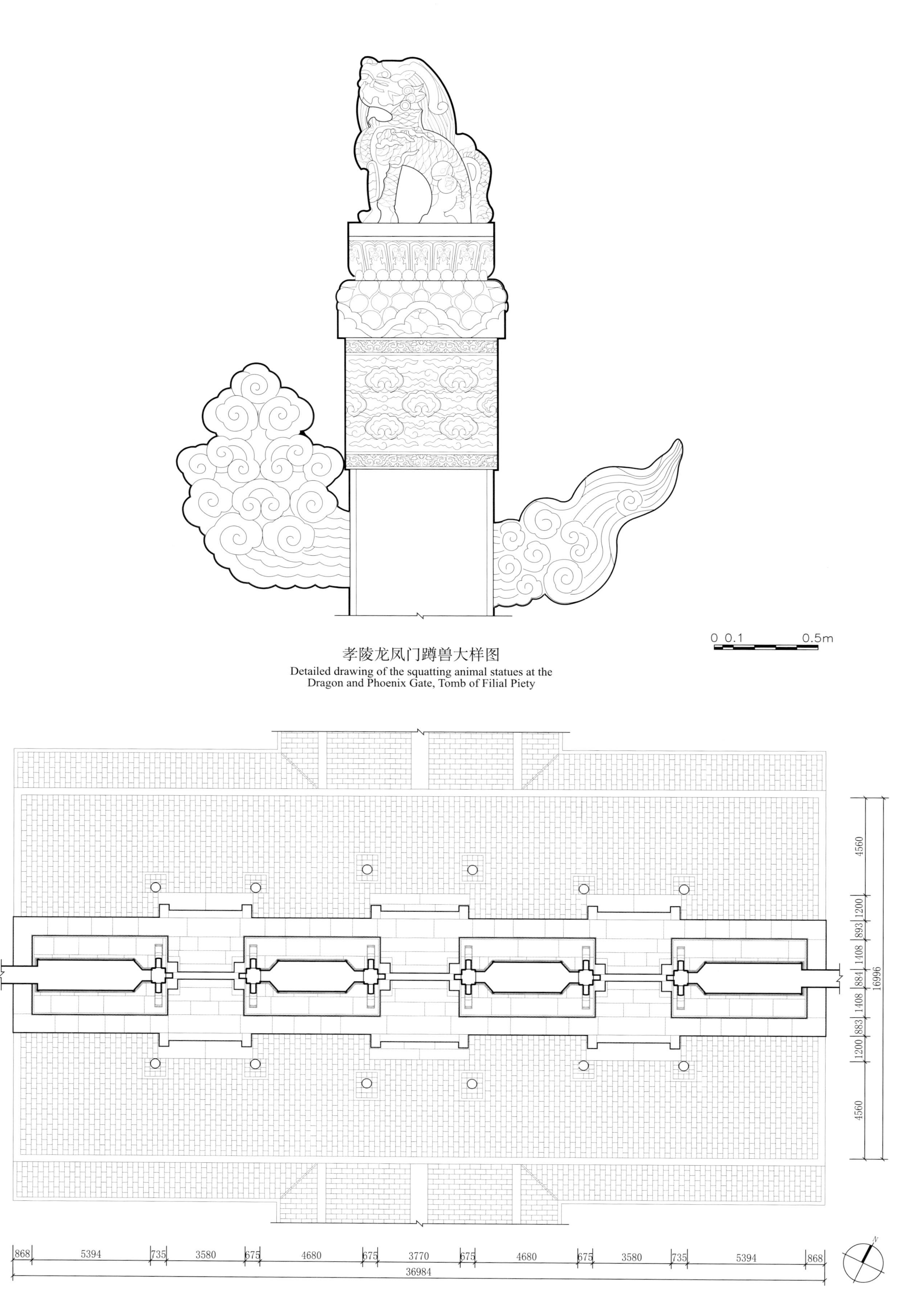

孝陵龙凤门蹲兽大样图
Detailed drawing of the squatting animal statues at the Dragon and Phoenix Gate, Tomb of Filial Piety

孝陵龙凤门平面图
Plan of the Dragon and Phoenix Gate, Tomb of Filial Piety

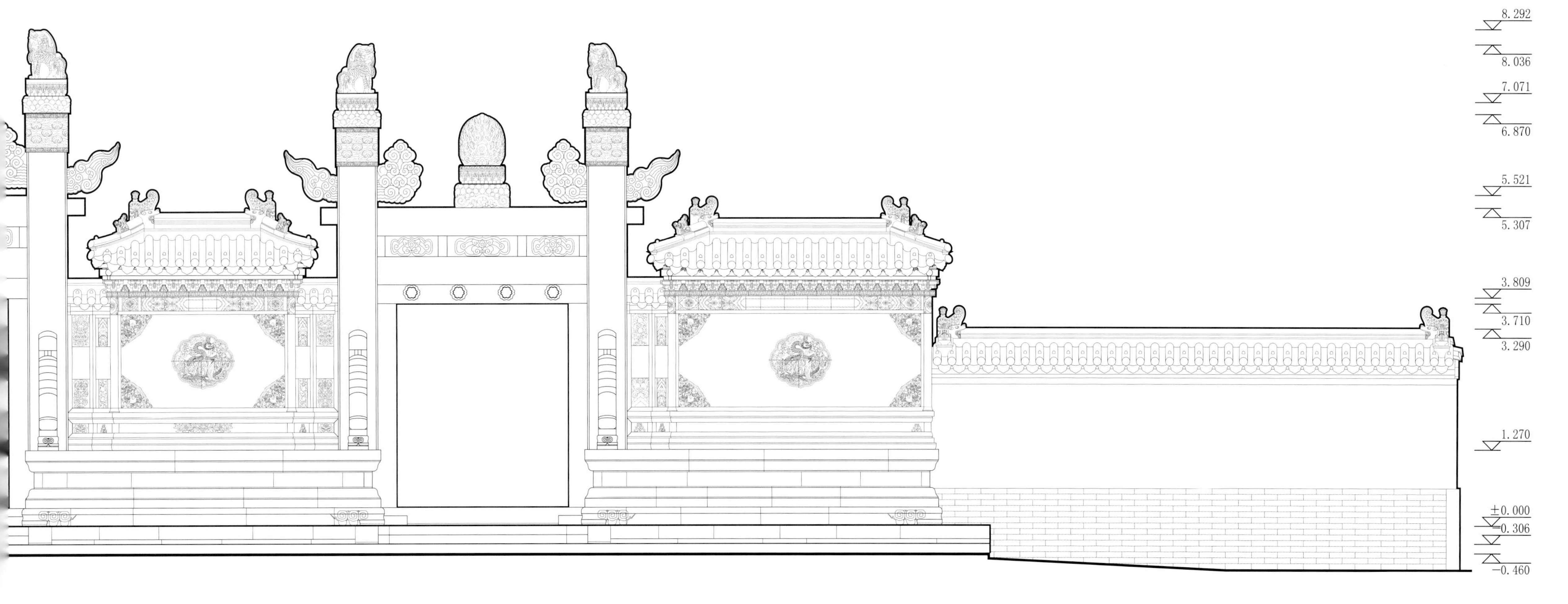
8.292
8.036
7.071
6.870
5.521
5.307
3.809
3.710
3.290
1.270
±0.000
-0.306
-0.460
670
1147
3068
1147
670
2920
670
1117
4435
1060
8100
0
1
2m

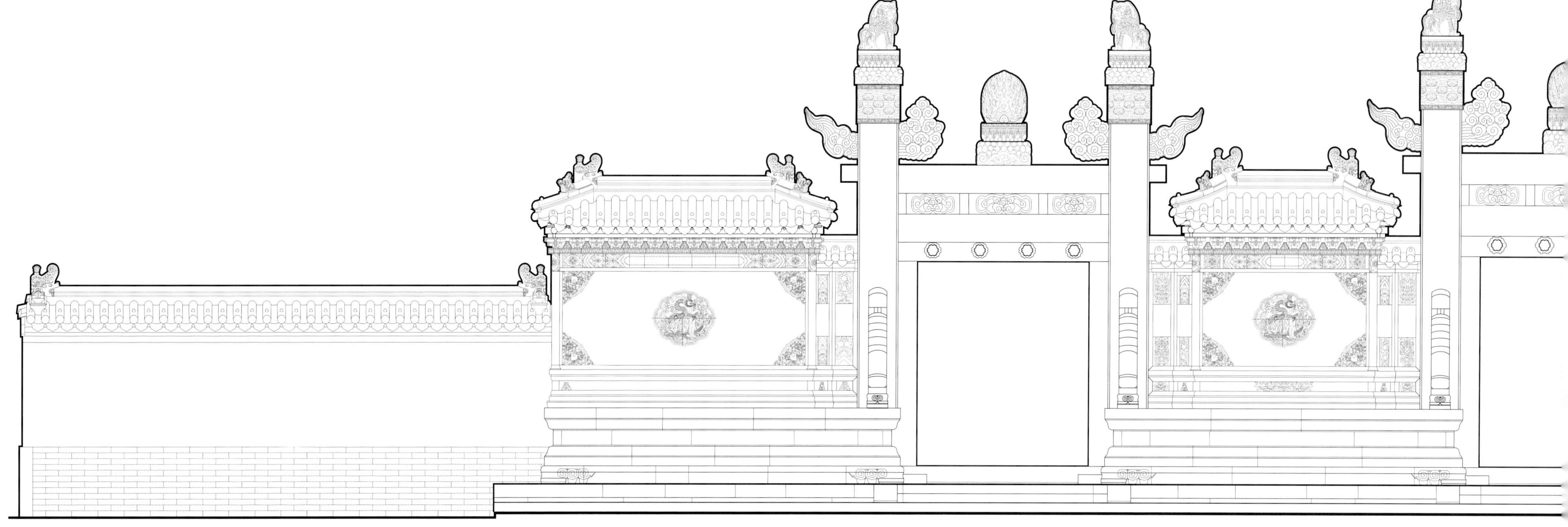

孝陵龙凤门正立面图
Front elevation of the Dragon and Phoenix Gate,
Tomb of Filial Piety

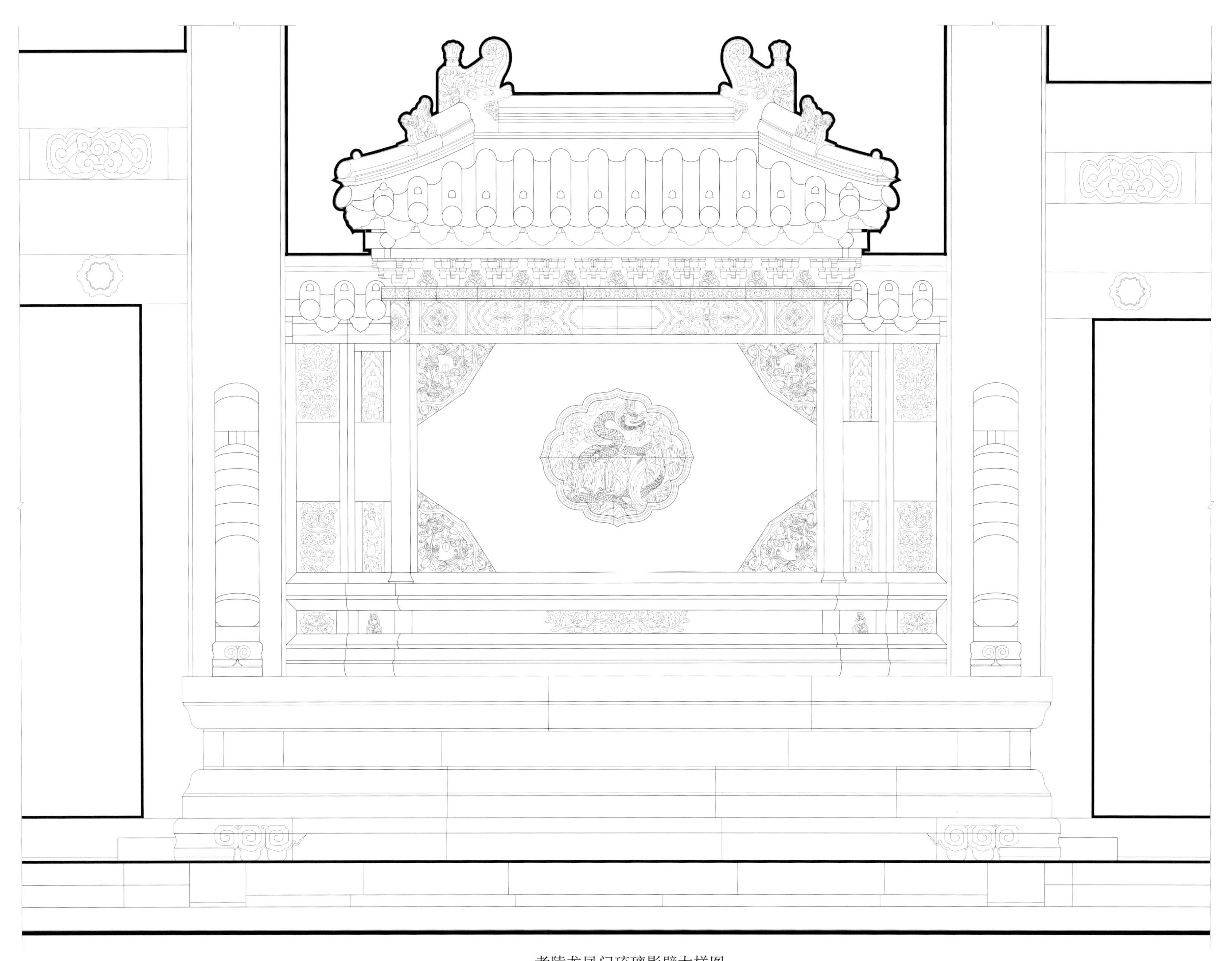

孝陵龙凤门琉璃影壁大样图

Detailed drawing of the Screen Wall of Glazed Tiles at the Dragon and Phoenix Gate, Tomb of Filial Piety

0 0.2 1m

8.292
8.036
5.441
3.911
3.657
1.270
±0.000

孝陵龙凤门侧立面图
Side elevation of the Dragon and Phoenix Gate, Tomb of Filial Piety

8.292
7.481
7.093
5.521
3.911
3.809
1.270
0.305
±0.000
−0.306

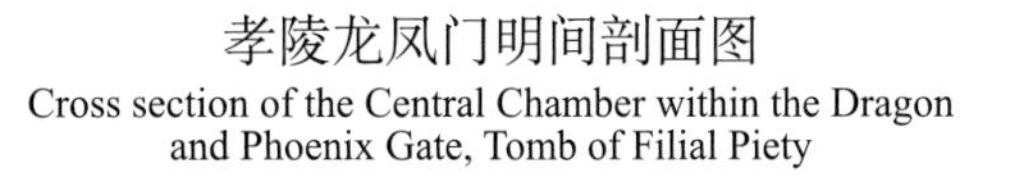

孝陵龙凤门明间剖面图
Cross section of the Central Chamber within the Dragon and Phoenix Gate, Tomb of Filial Piety

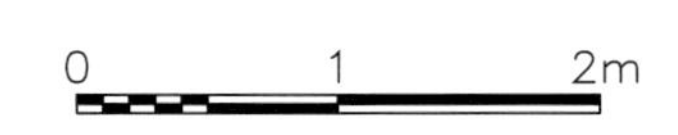

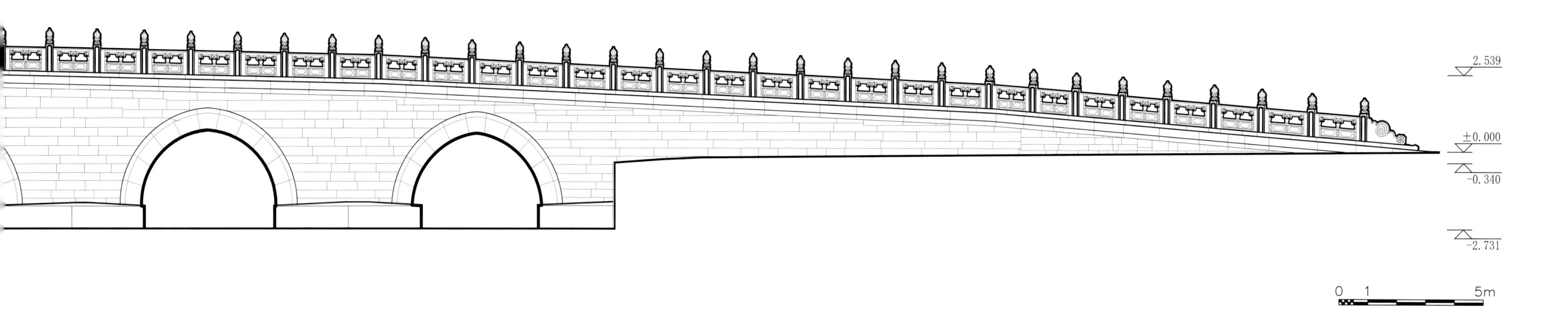
2.539
±0.000
-0.340
-2.731
0
1
5m

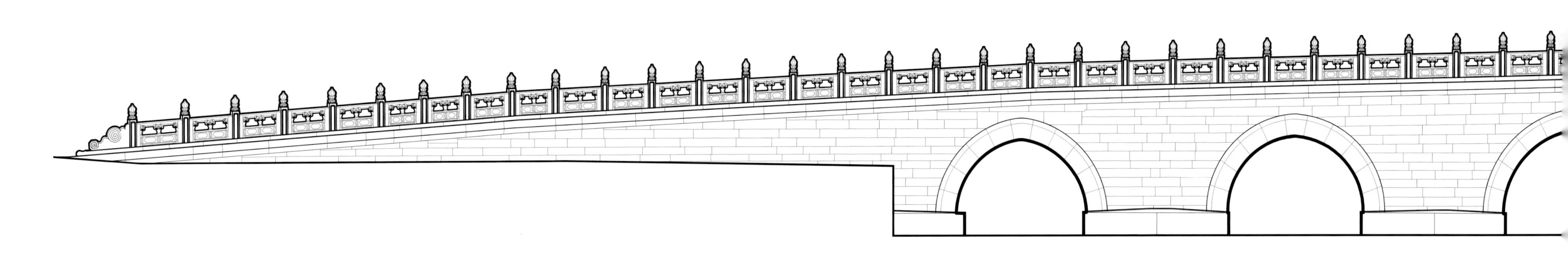

孝陵五孔石券桥东立面图
East elevation of the Five-Arched Stone Bridge, Tomb of Filial Piety

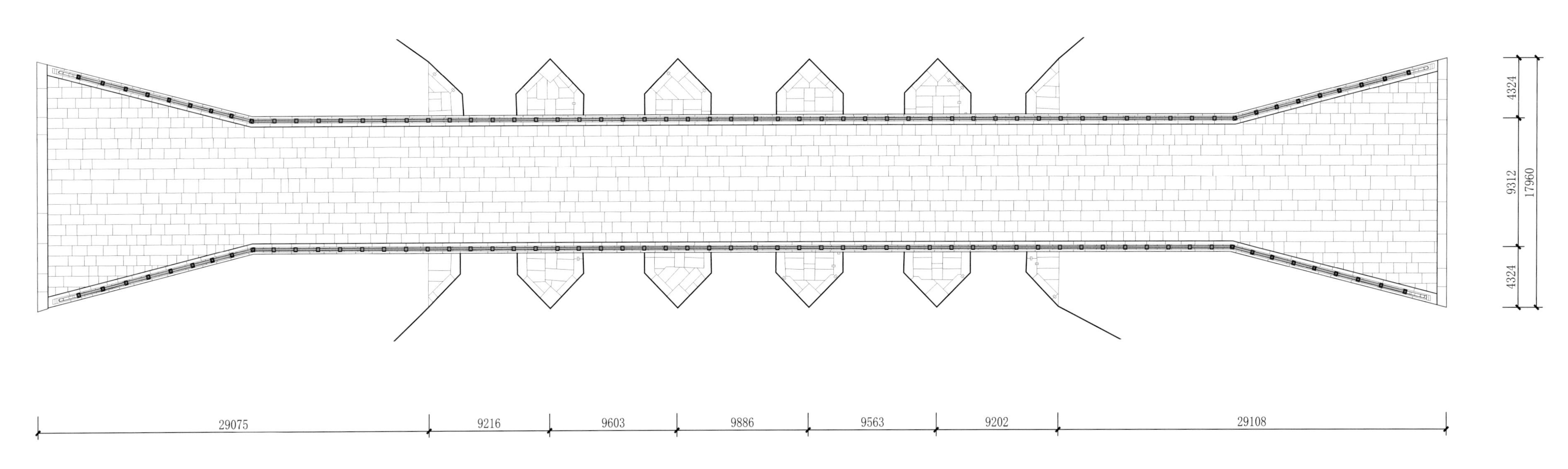

孝陵五孔石券桥平面图
Plan of the Five-Arched Stone Bridge, Tomb of Filial Piety

1050
483
567
567
483
1050
4200
560 560 2200 560 560
4440
0 0.25 1m

孝陵神厨库院门平面图
Plan of the Gate in the Culinary Courtyard for Sacrifices, Tomb of Filial Piety

4.862
3.453
2.680
2.100
±0.000
−0.410
560 560 2200 560 560
4440
0 0.25 1m

孝陵神厨库院门正立面图
Elevation of the Gate in the Culinary Courtyard for Sacrifices, Tomb of Filial Piety

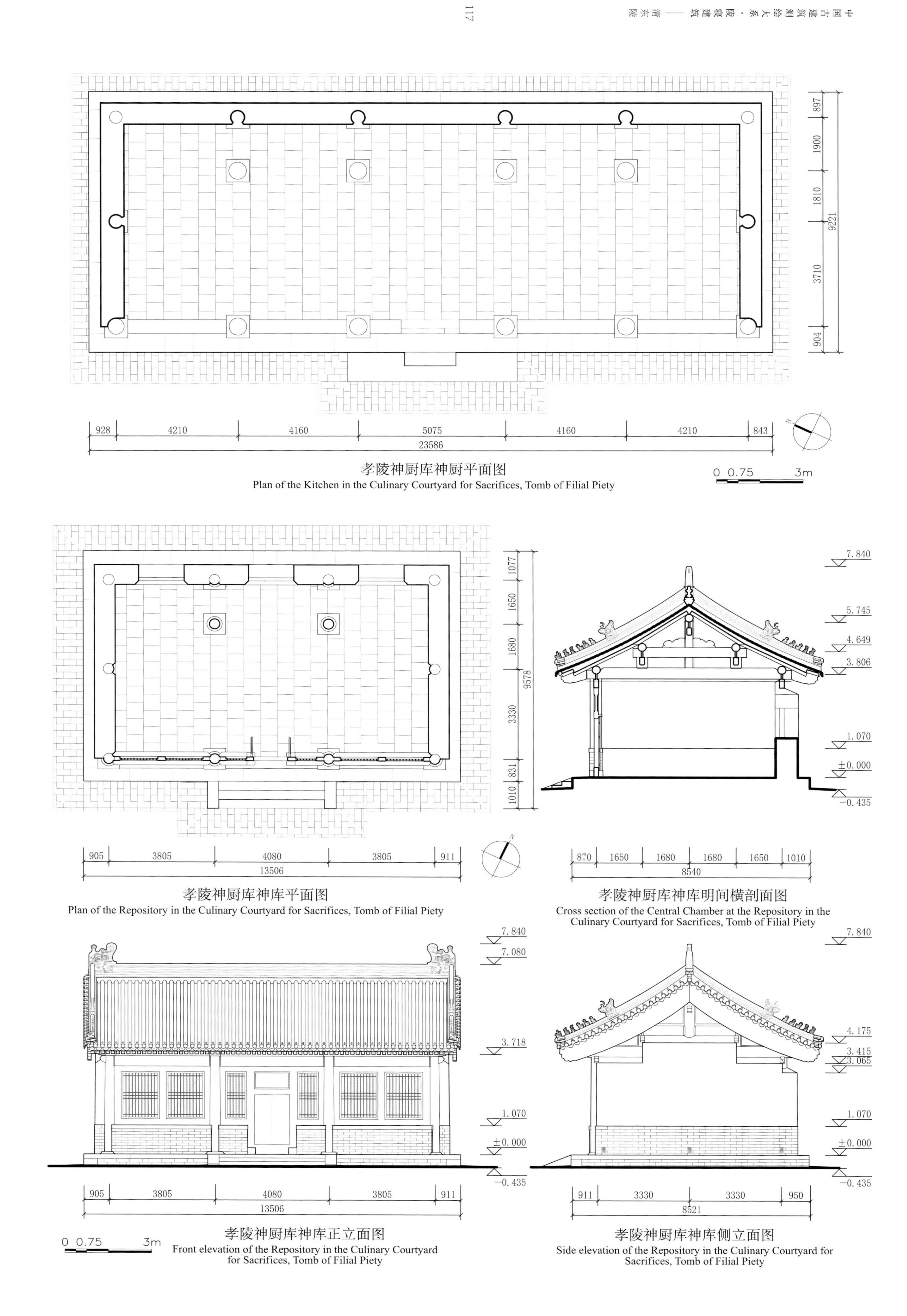

孝陵神厨库神厨平面图
Plan of the Kitchen in the Culinary Courtyard for Sacrifices, Tomb of Filial Piety

孝陵神厨库神库平面图
Plan of the Repository in the Culinary Courtyard for Sacrifices, Tomb of Filial Piety

孝陵神厨库神库明间横剖面图
Cross section of the Central Chamber at the Repository in the Culinary Courtyard for Sacrifices, Tomb of Filial Piety

孝陵神厨库神库正立面图
Front elevation of the Repository in the Culinary Courtyard for Sacrifices, Tomb of Filial Piety

孝陵神厨库神库侧立面图
Side elevation of the Repository in the Culinary Courtyard for Sacrifices, Tomb of Filial Piety

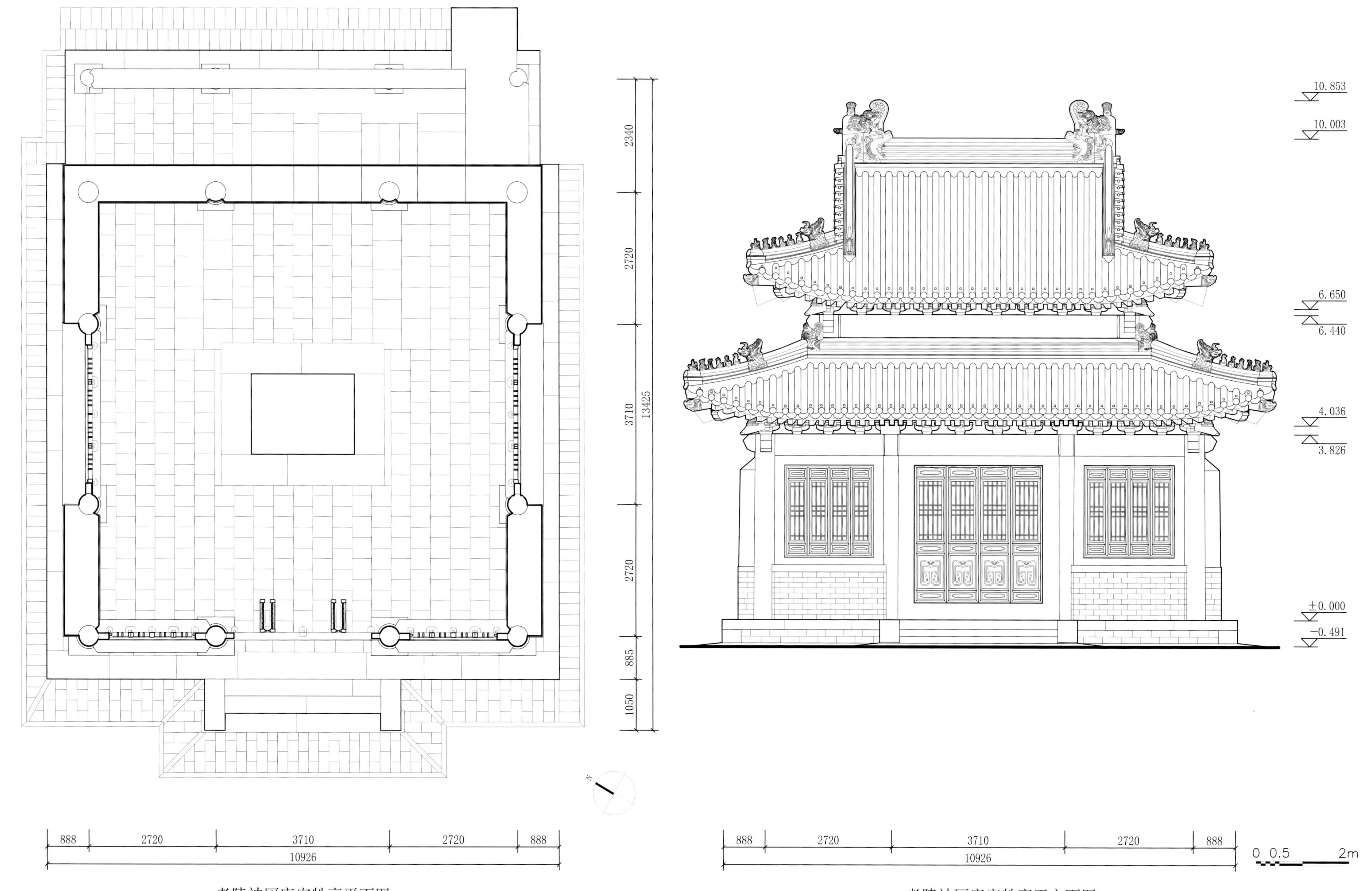

孝陵神厨库宰牲亭平面图
Plan of the Ritual Abattoir in the Culinary Courtyard for Sacrifices, Tomb of Filial Piety

孝陵神厨库宰牲亭正立面图
Front elevation of the Ritual Abattoir in the Culinary Courtyard for Sacrifices, Tomb of Filial Piety

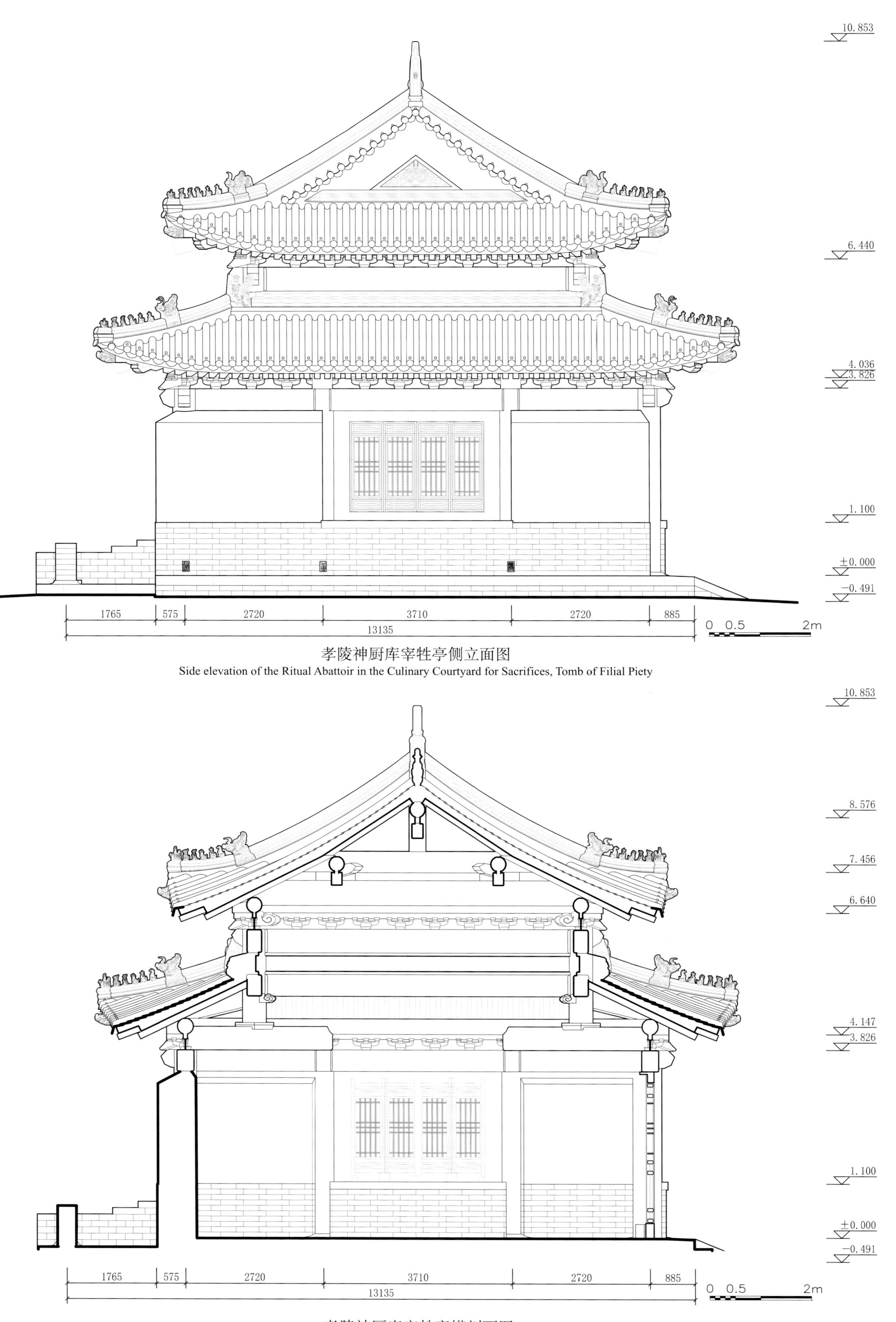

孝陵神厨库宰牲亭侧立面图
Side elevation of the Ritual Abattoir in the Culinary Courtyard for Sacrifices, Tomb of Filial Piety

孝陵神厨库宰牲亭横剖面图
Cross section of Ritual Abattoir in the Culinary Courtyard for Sacrifices, Tomb of Filial Piety

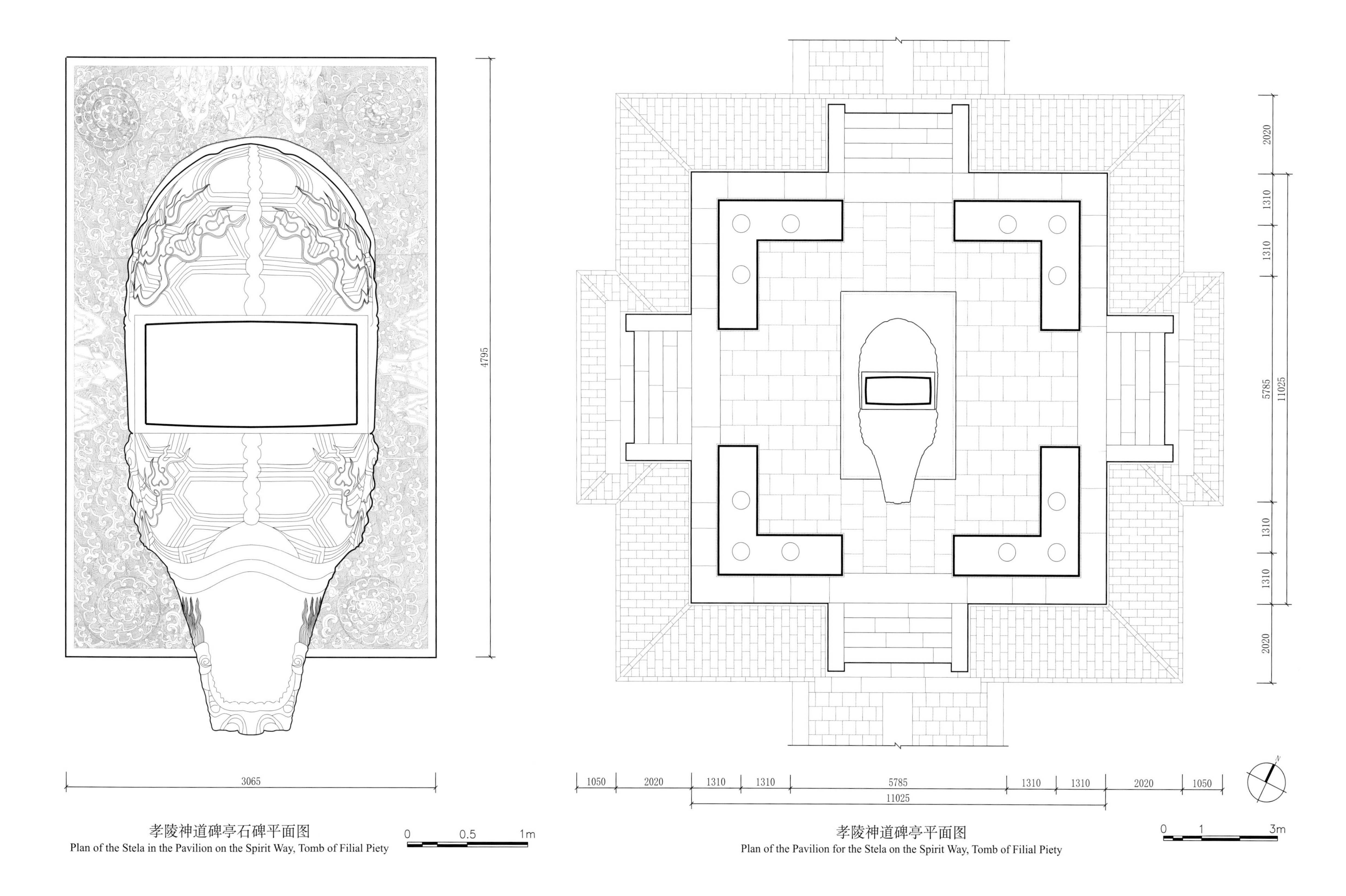

孝陵神道碑亭石碑平面图
Plan of the Stela in the Pavilion on the Spirit Way, Tomb of Filial Piety

孝陵神道碑亭平面图
Plan of the Pavilion for the Stela on the Spirit Way, Tomb of Filial Piety

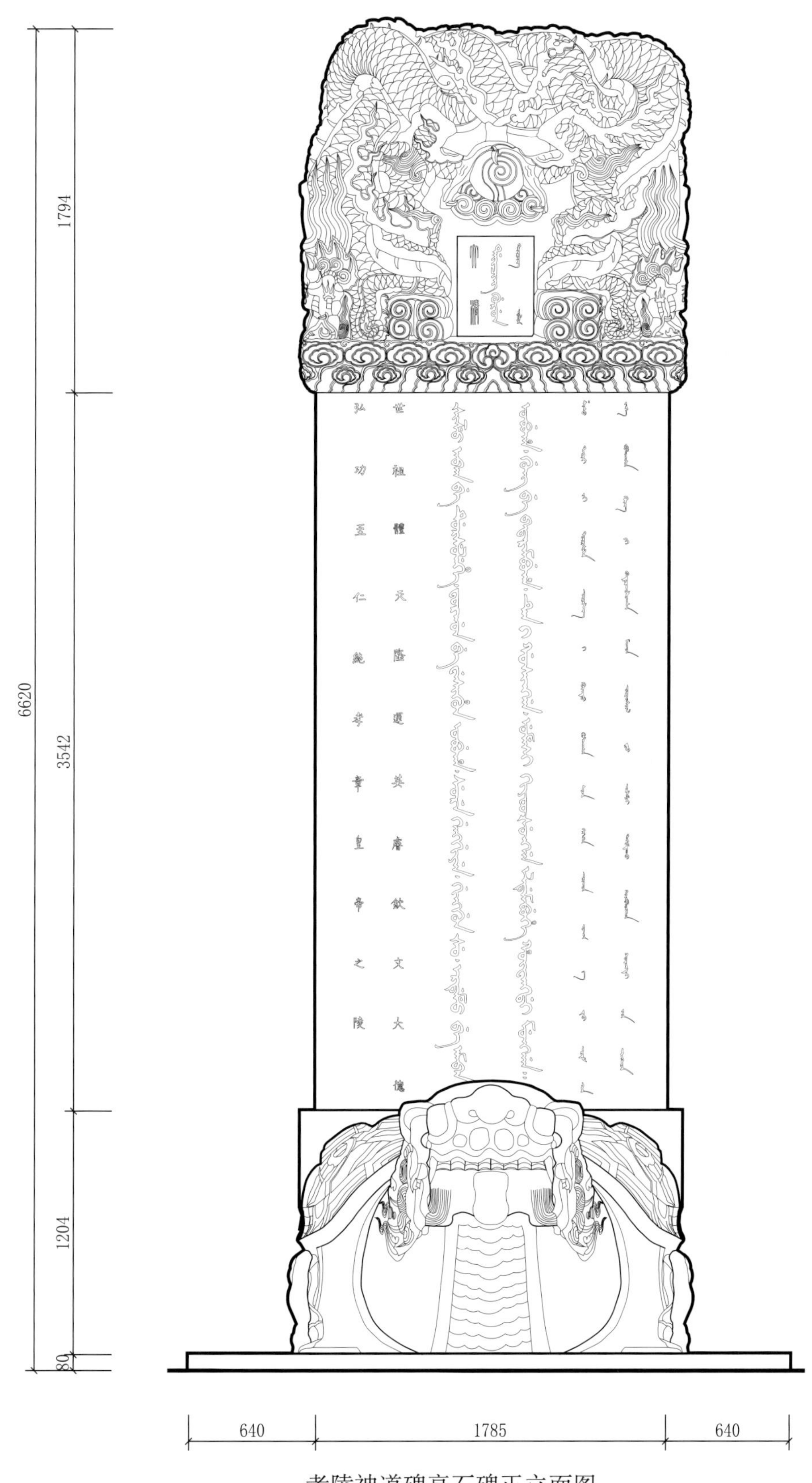

孝陵神道碑亭石碑正立面图
Front elevation of the Stela in the Pavilion on the Spirit Way, Tomb of Filial Piety

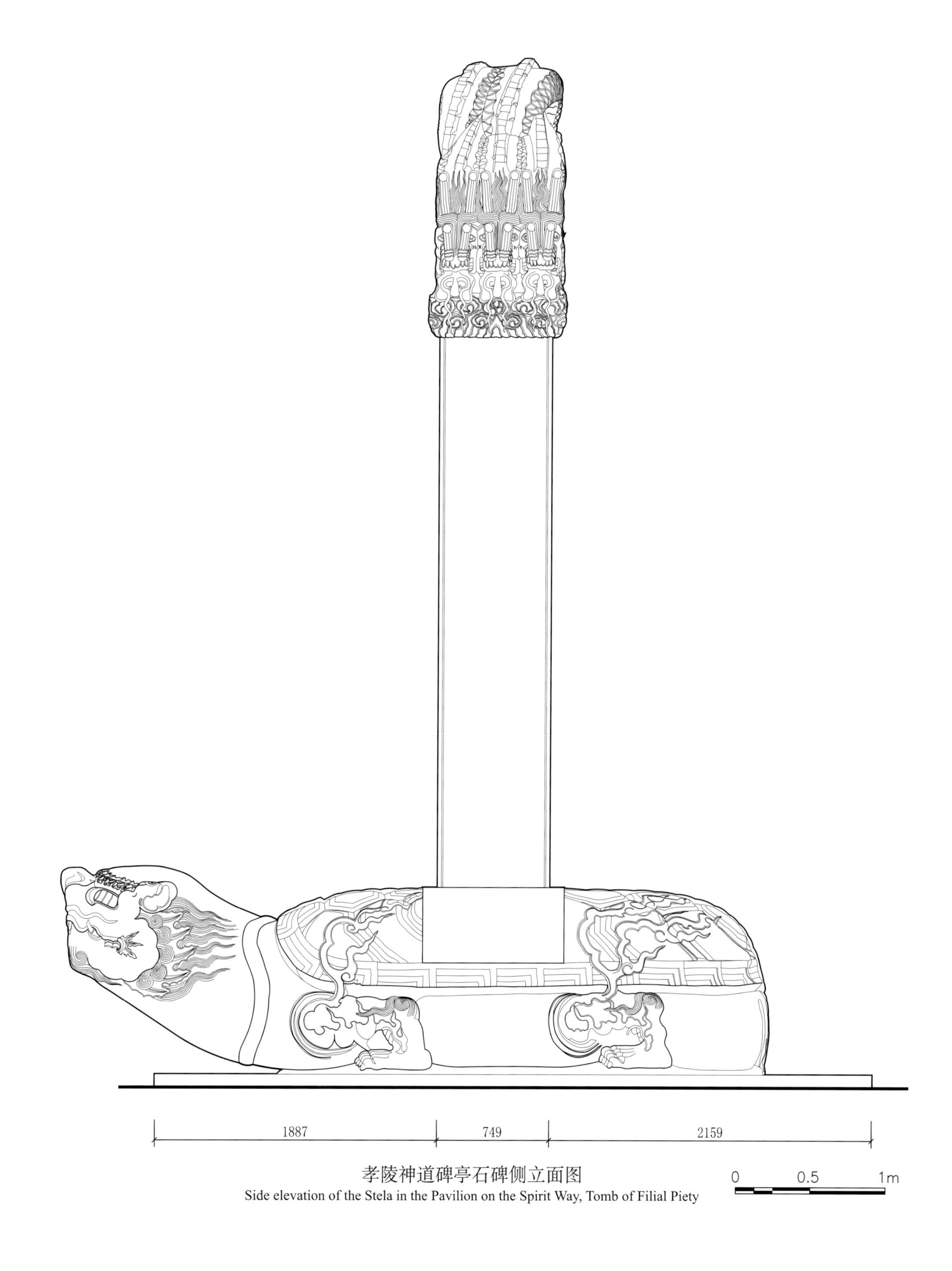

孝陵神道碑亭石碑侧立面图
Side elevation of the Stela in the Pavilion on the Spirit Way, Tomb of Filial Piety

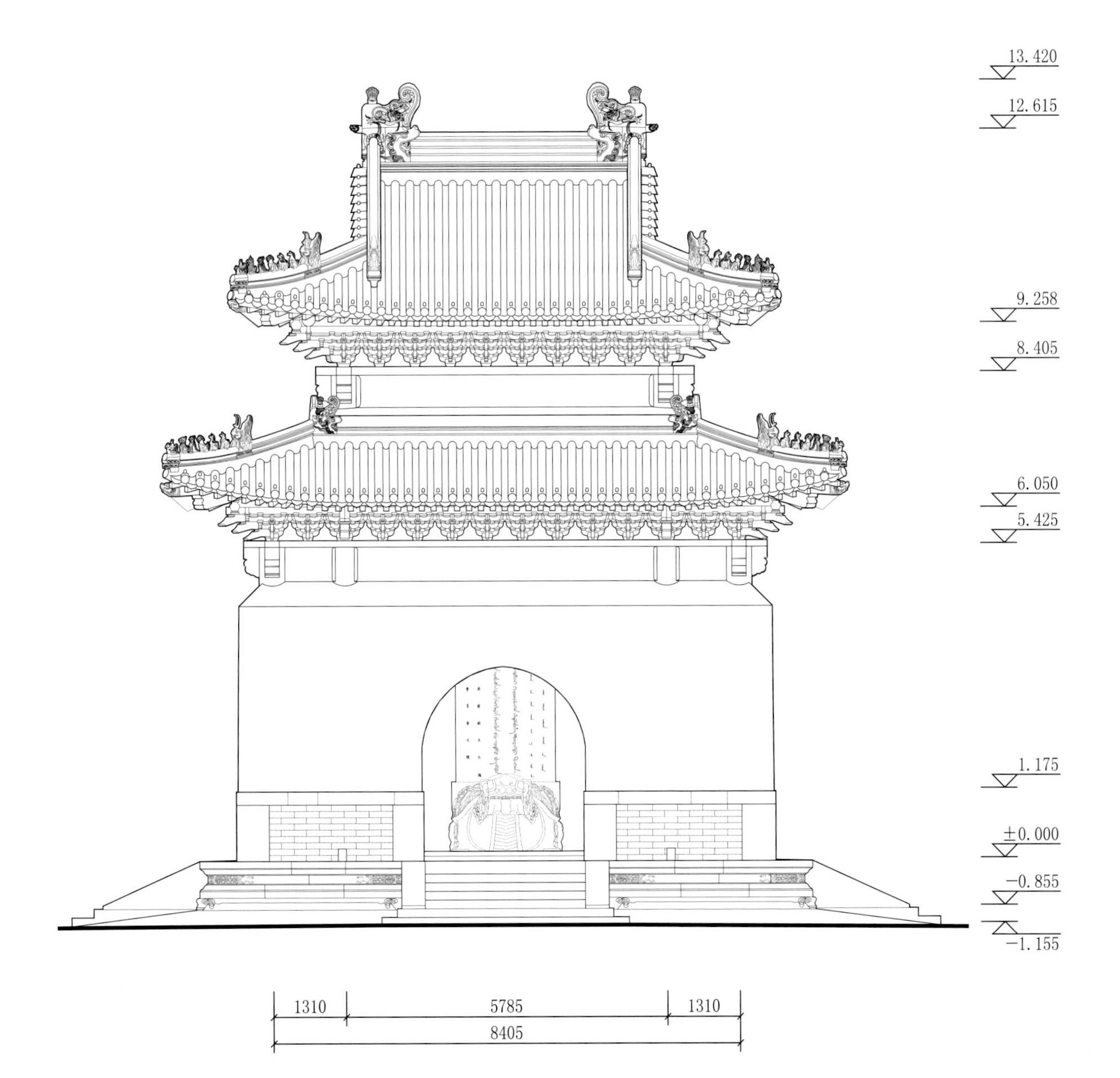

孝陵神道碑亭正立面图
Front elevation of the Pavilion for the Stela on the Spirit Way, Tomb of Filial Piety

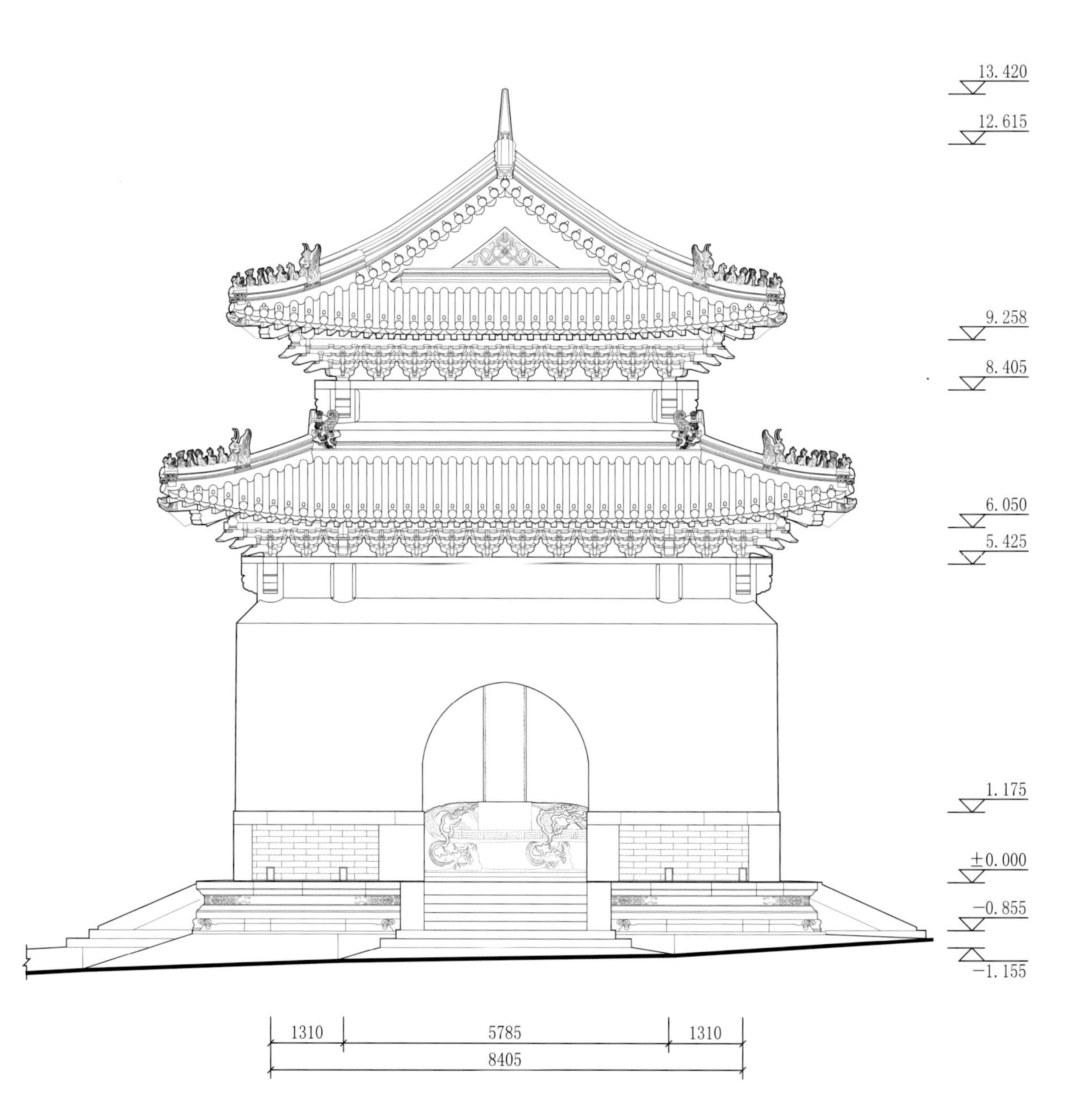

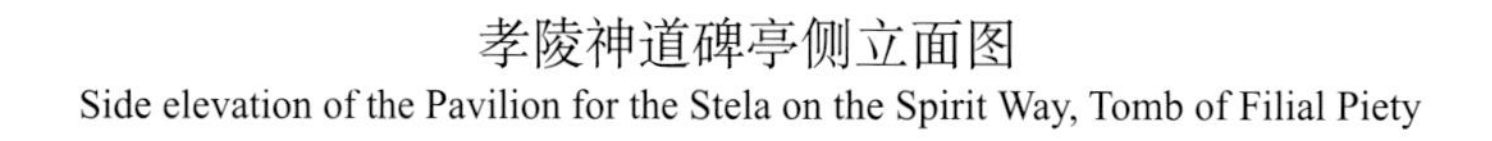

孝陵神道碑亭侧立面图
Side elevation of the Pavilion for the Stela on the Spirit Way, Tomb of Filial Piety

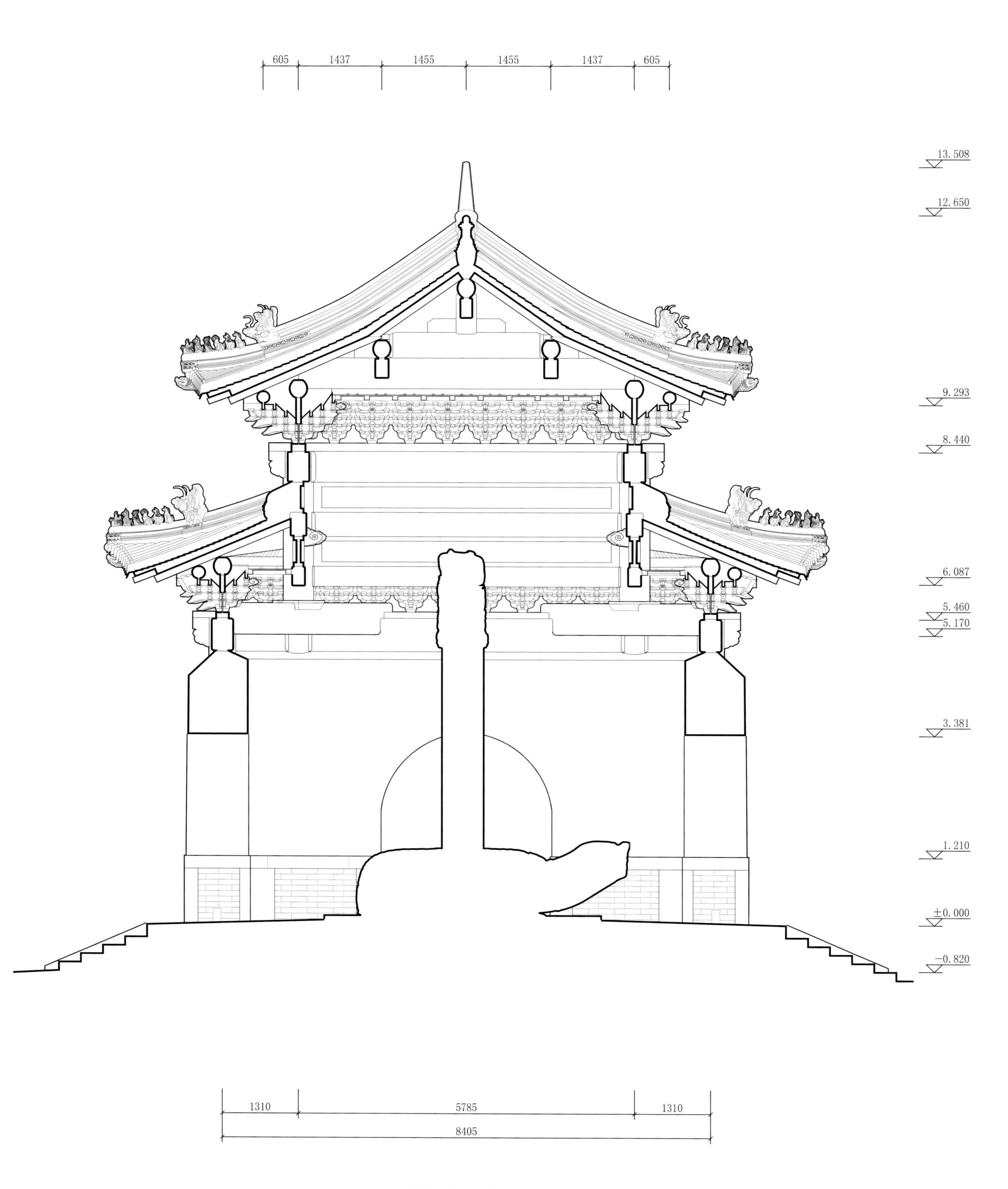

孝陵神道碑亭横剖面图
Cross section of the Pavilion for the Stela on the Spirit Way, Tomb of Filial Piety

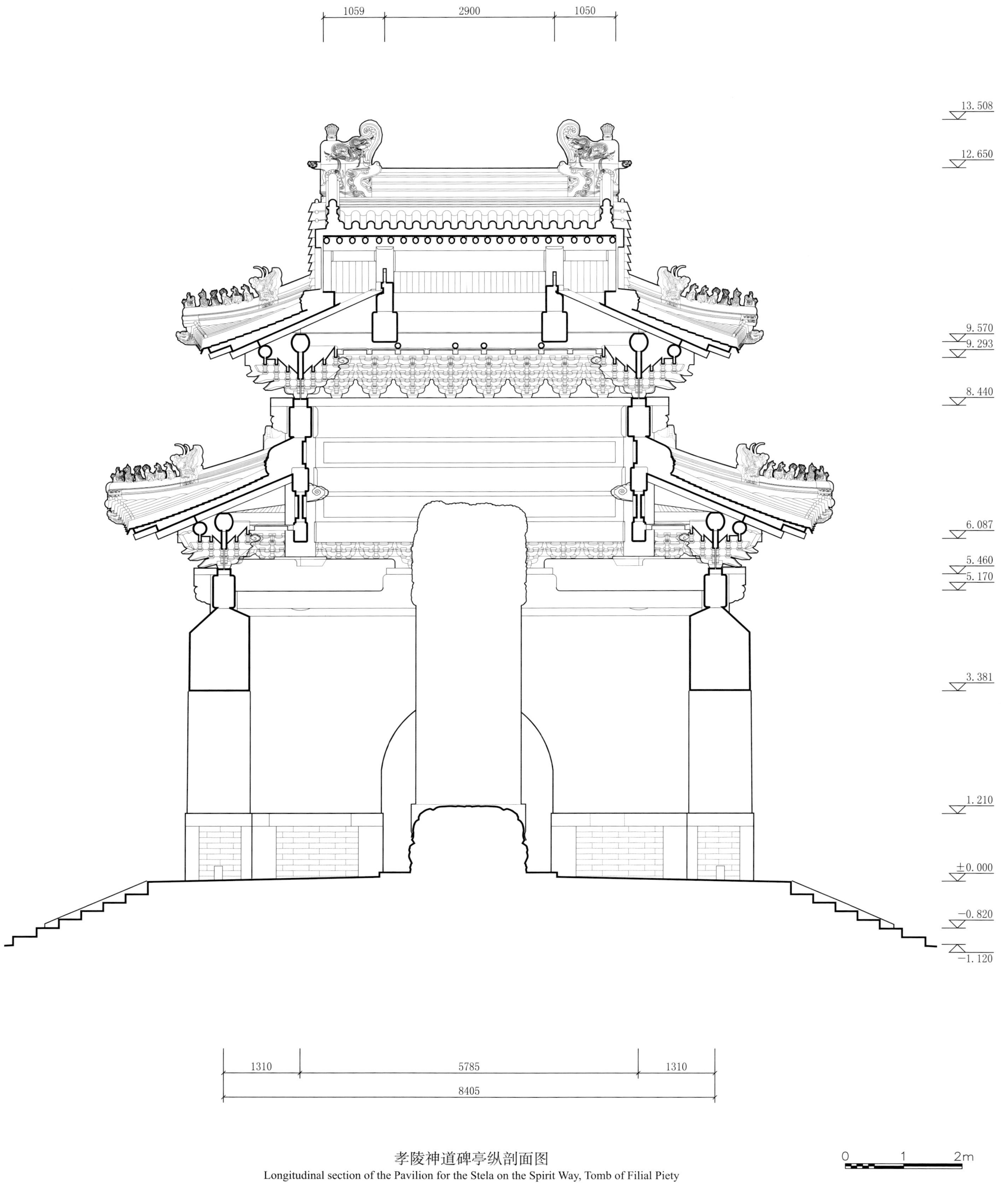

孝陵神道碑亭纵剖面图
Longitudinal section of the Pavilion for the Stela on the Spirit Way, Tomb of Filial Piety

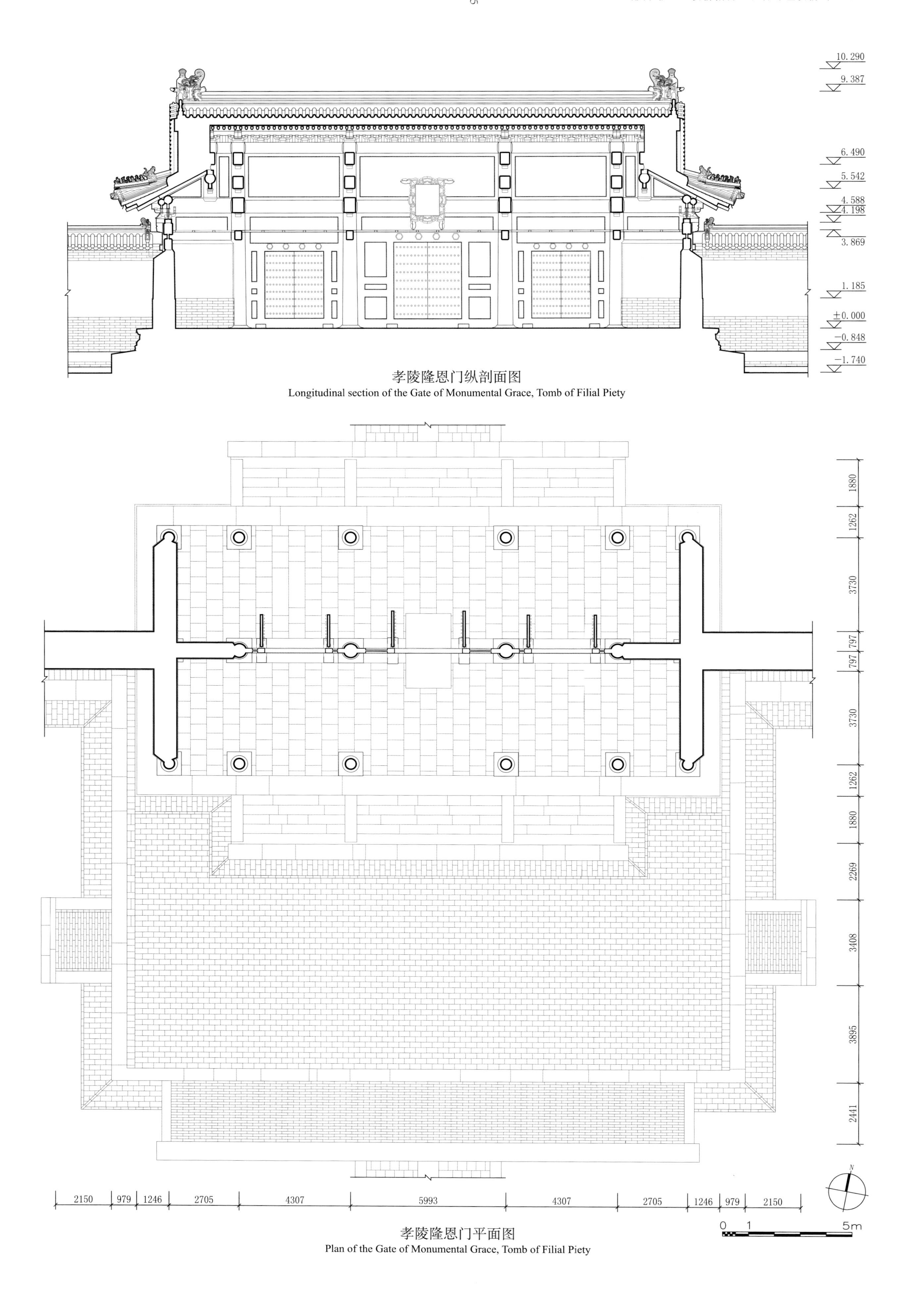

孝陵隆恩门纵剖面图
Longitudinal section of the Gate of Monumental Grace, Tomb of Filial Piety

孝陵隆恩门平面图
Plan of the Gate of Monumental Grace, Tomb of Filial Piety

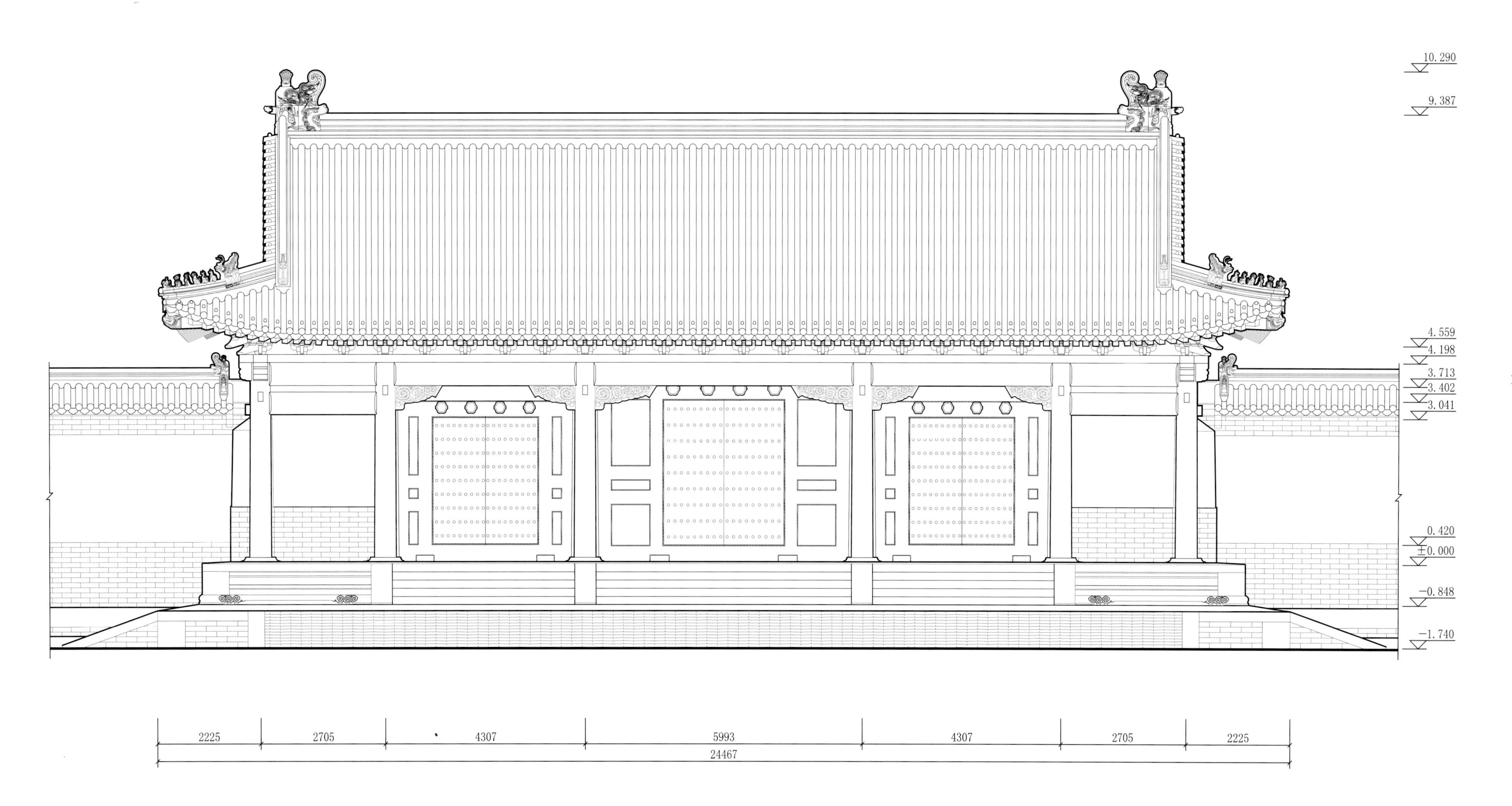

孝陵隆恩门正立面图

Front elevation of the Gate of Monumental Grace, Tomb of Filial Piety

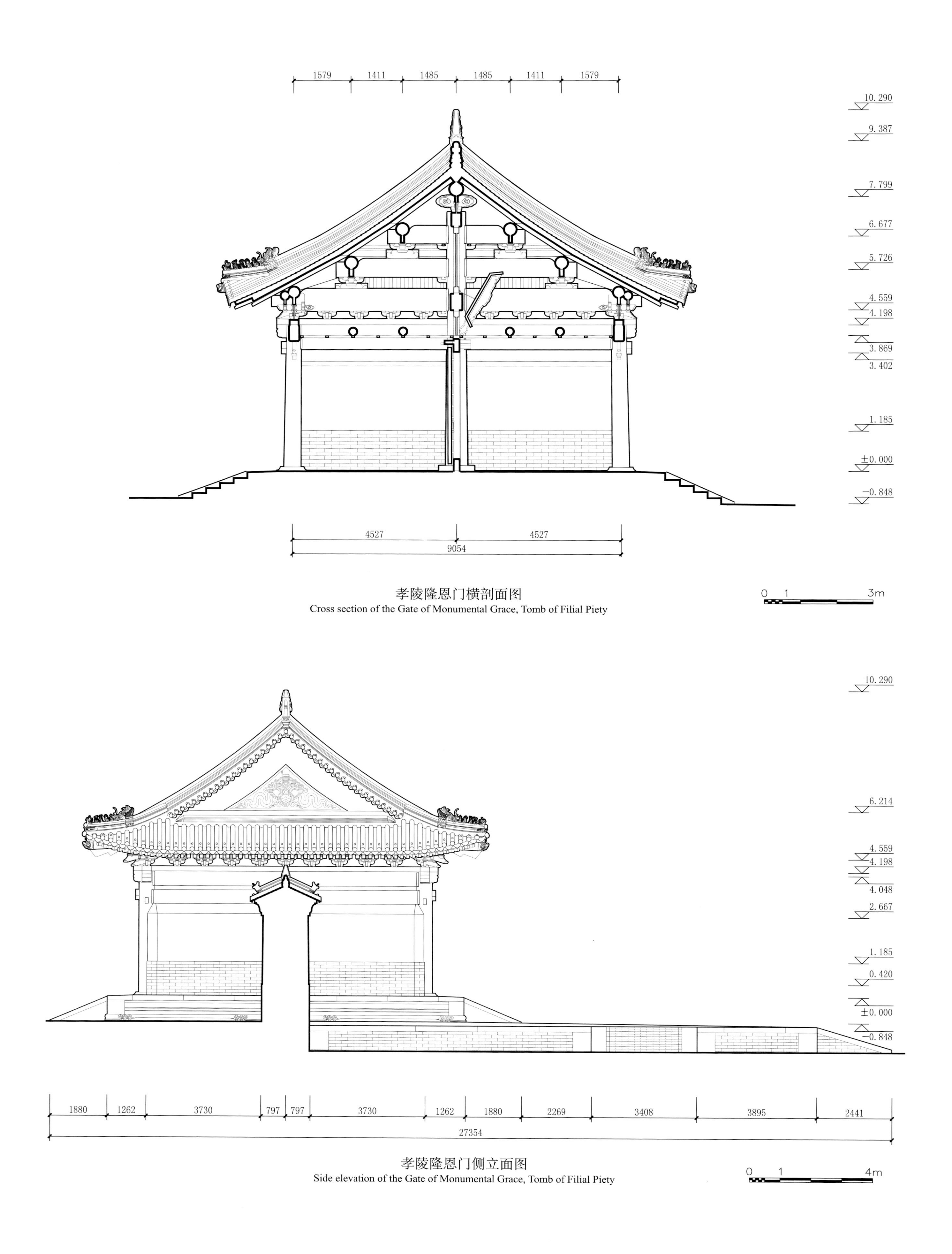

孝陵隆恩门横剖面图
Cross section of the Gate of Monumental Grace, Tomb of Filial Piety

孝陵隆恩门侧立面图
Side elevation of the Gate of Monumental Grace, Tomb of Filial Piety

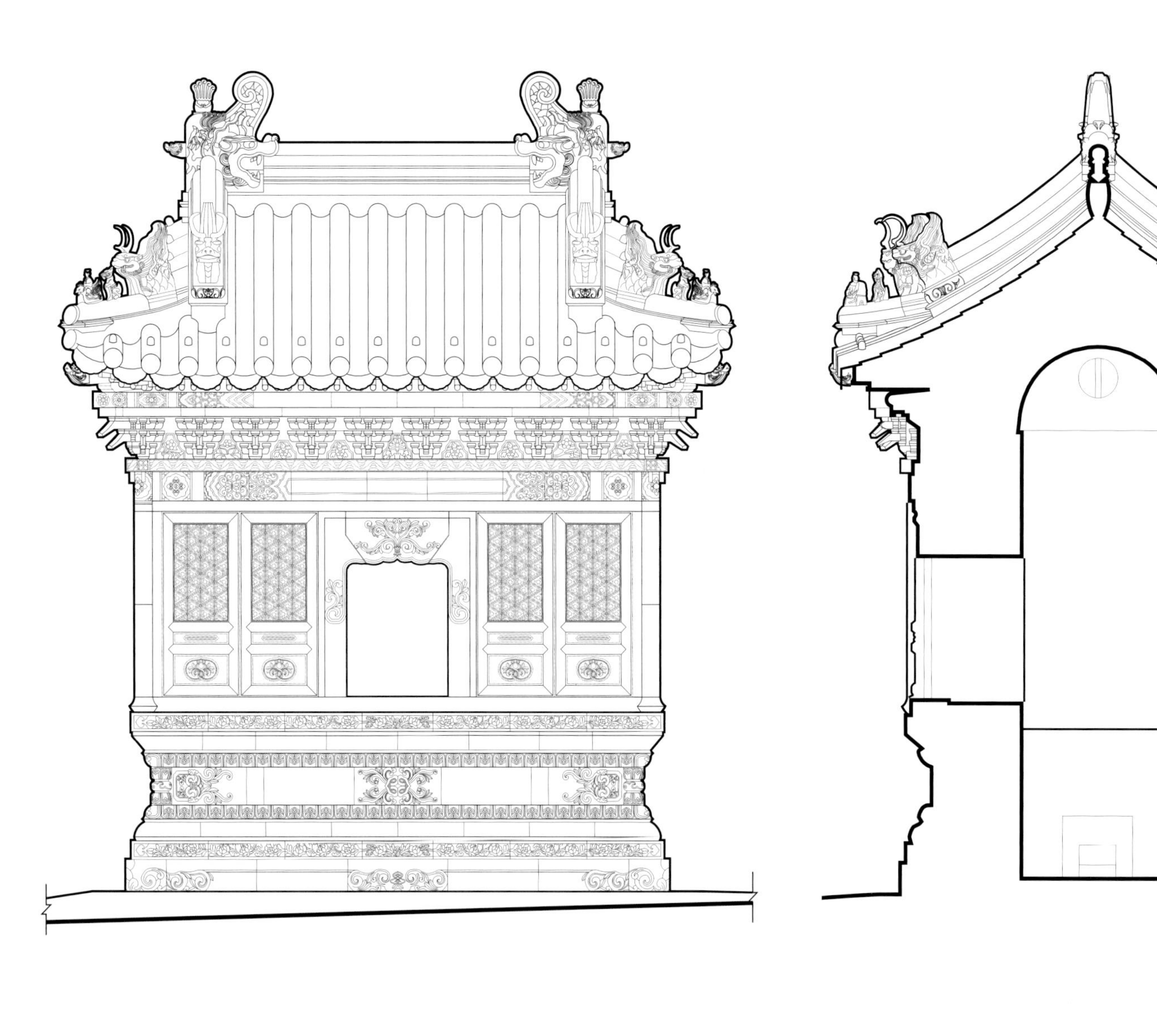

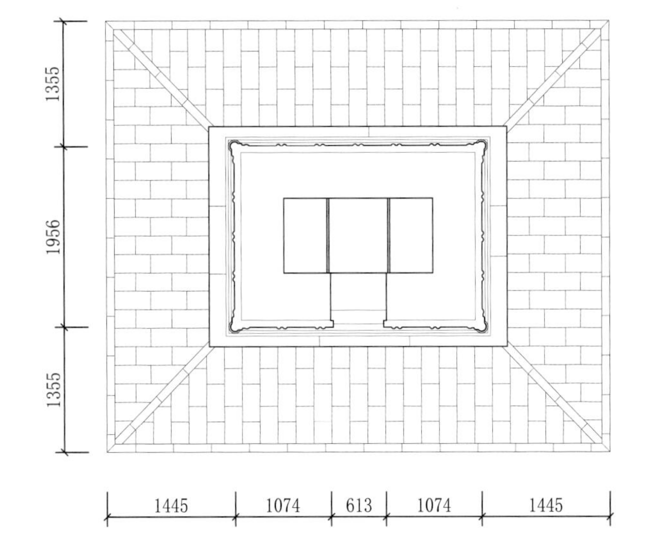

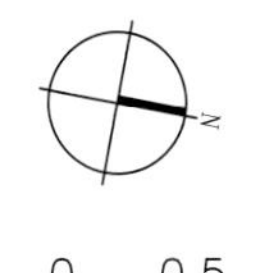

0 0.5 1m

孝陵西焚帛炉大样图

Detailed drawing of the Western Sacrificial Burner, Tomb of Filial Piety

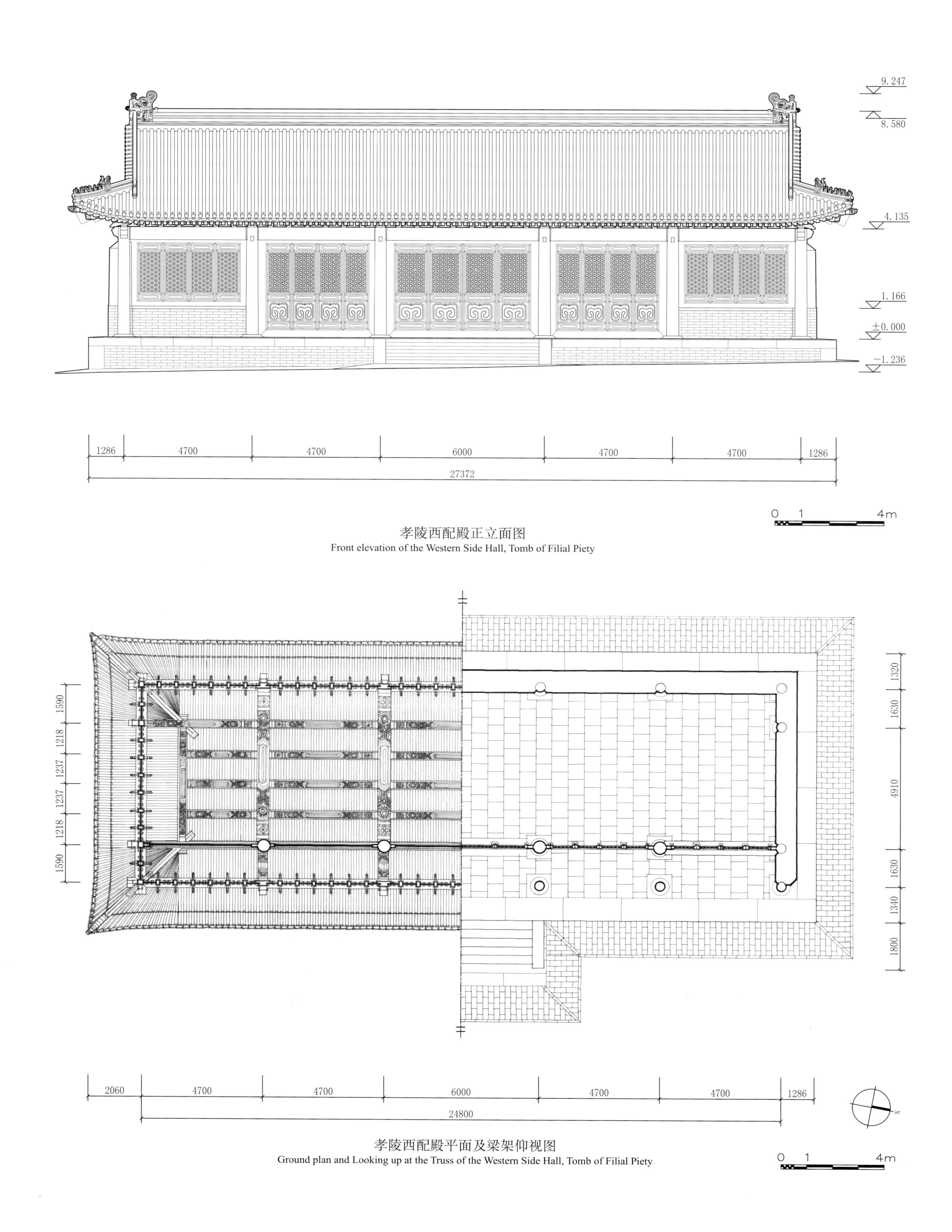

孝陵西配殿正立面图
Front elevation of the Western Side Hall, Tomb of Filial Piety

孝陵西配殿平面及梁架仰视图
Ground plan and Looking up at the Truss of the Western Side Hall, Tomb of Filial Piety

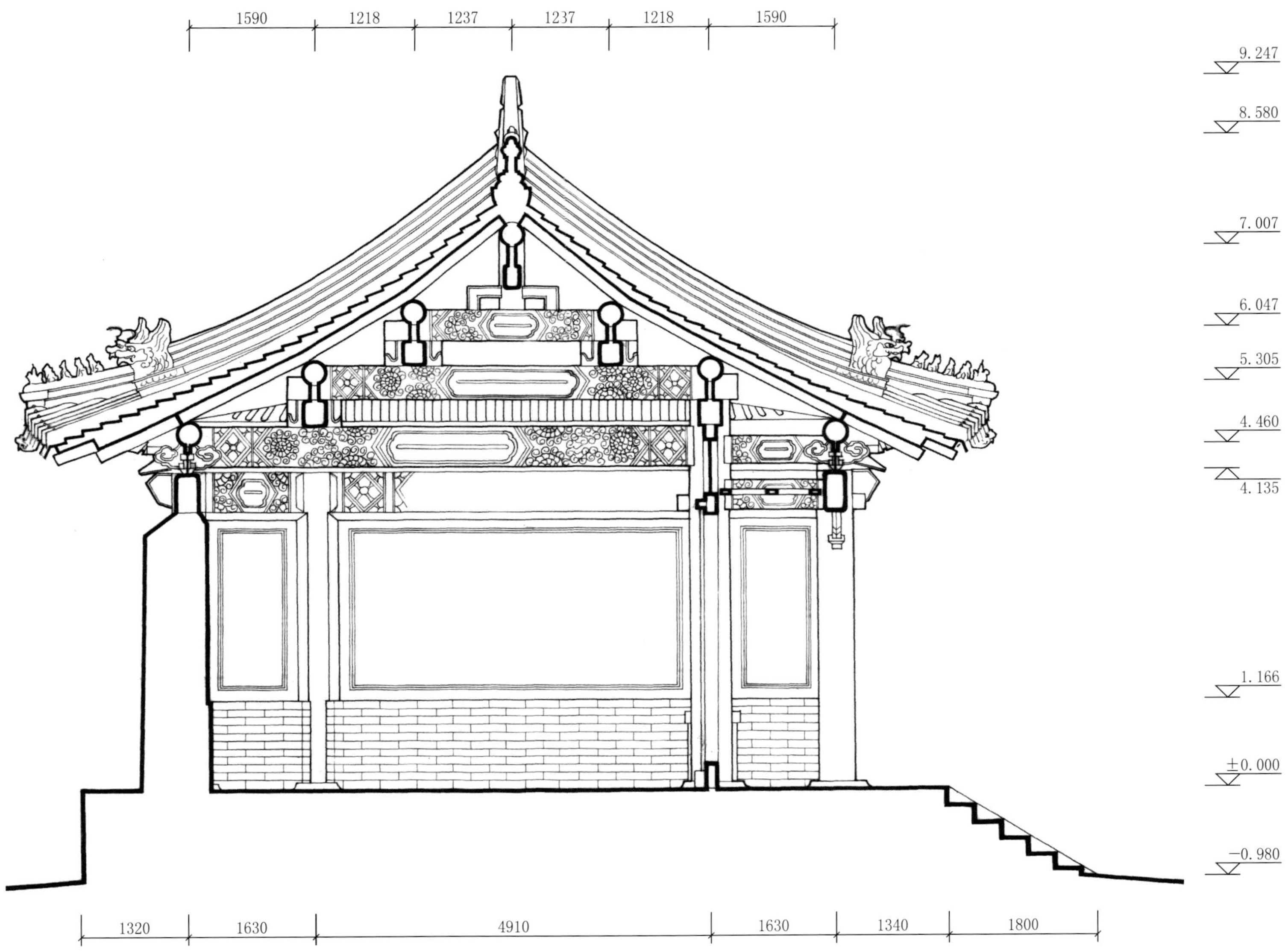

孝陵西配殿横剖面图
Cross section of the Western Side Hall, Tomb of Filial Piety

孝陵西配殿侧立面图
Side elevation of the Western Side Hall, Tomb of Filial Piety

5253
12590
12590
3090

孝陵隆恩殿梁架仰视图
Looking up at the Truss of the Hall of Monumental Grace, Tomb of Filial Piety

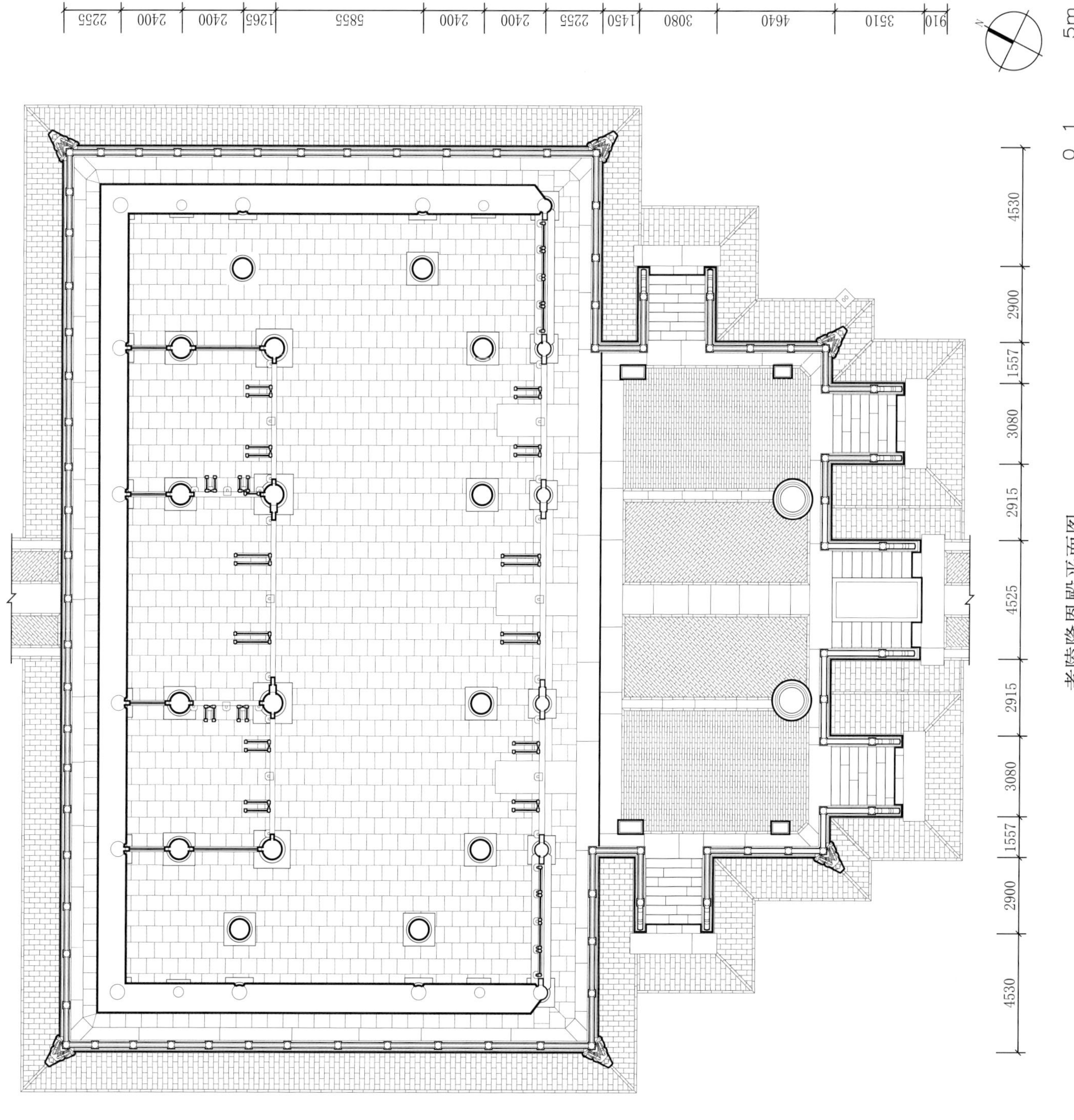

孝陵隆恩殿平面图
Plan of the Hall of Monumental Grace, Tomb of Filial Piety

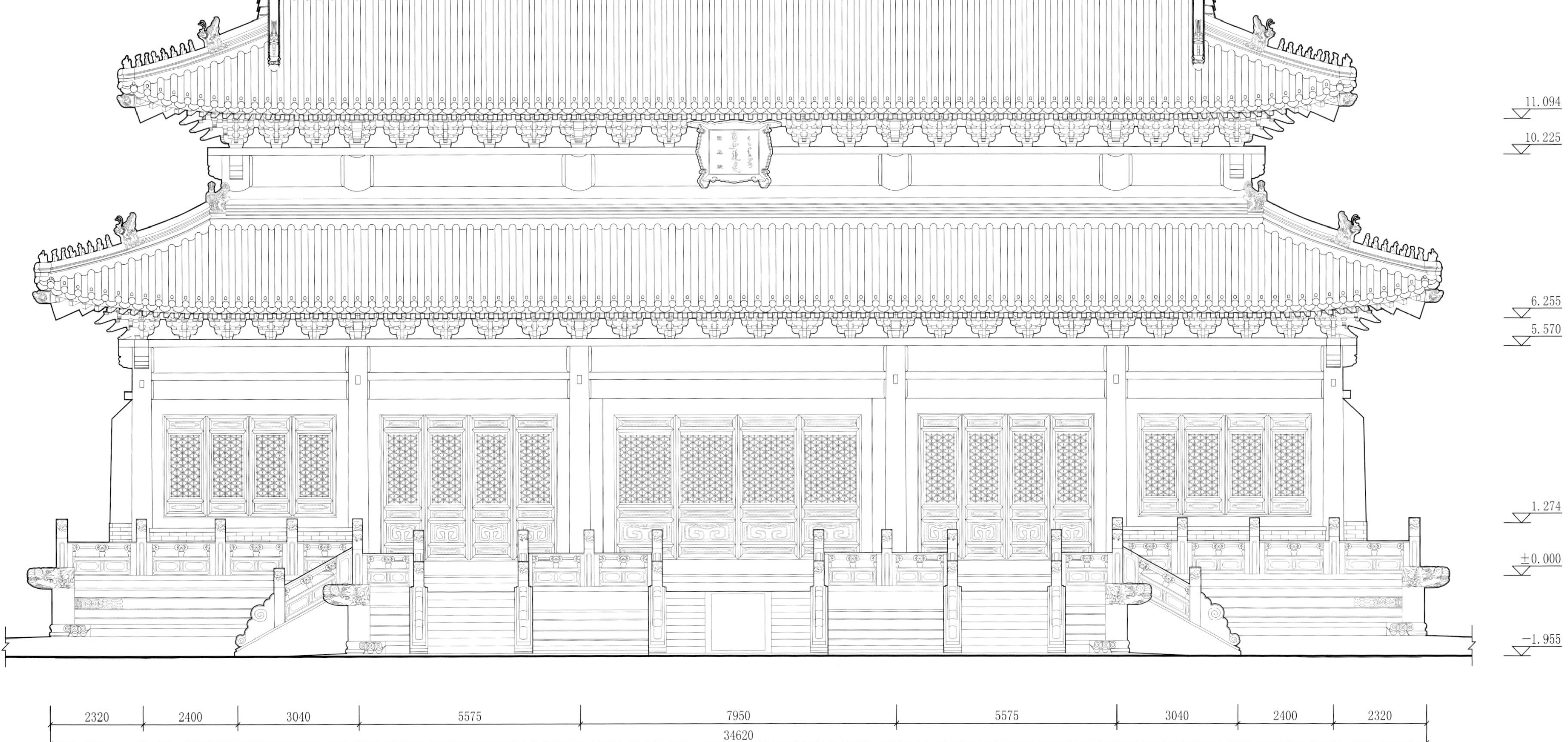

孝陵隆恩殿正立面图

Front elevation of the Hall of Monumental Grace, Tomb of Filial Piety

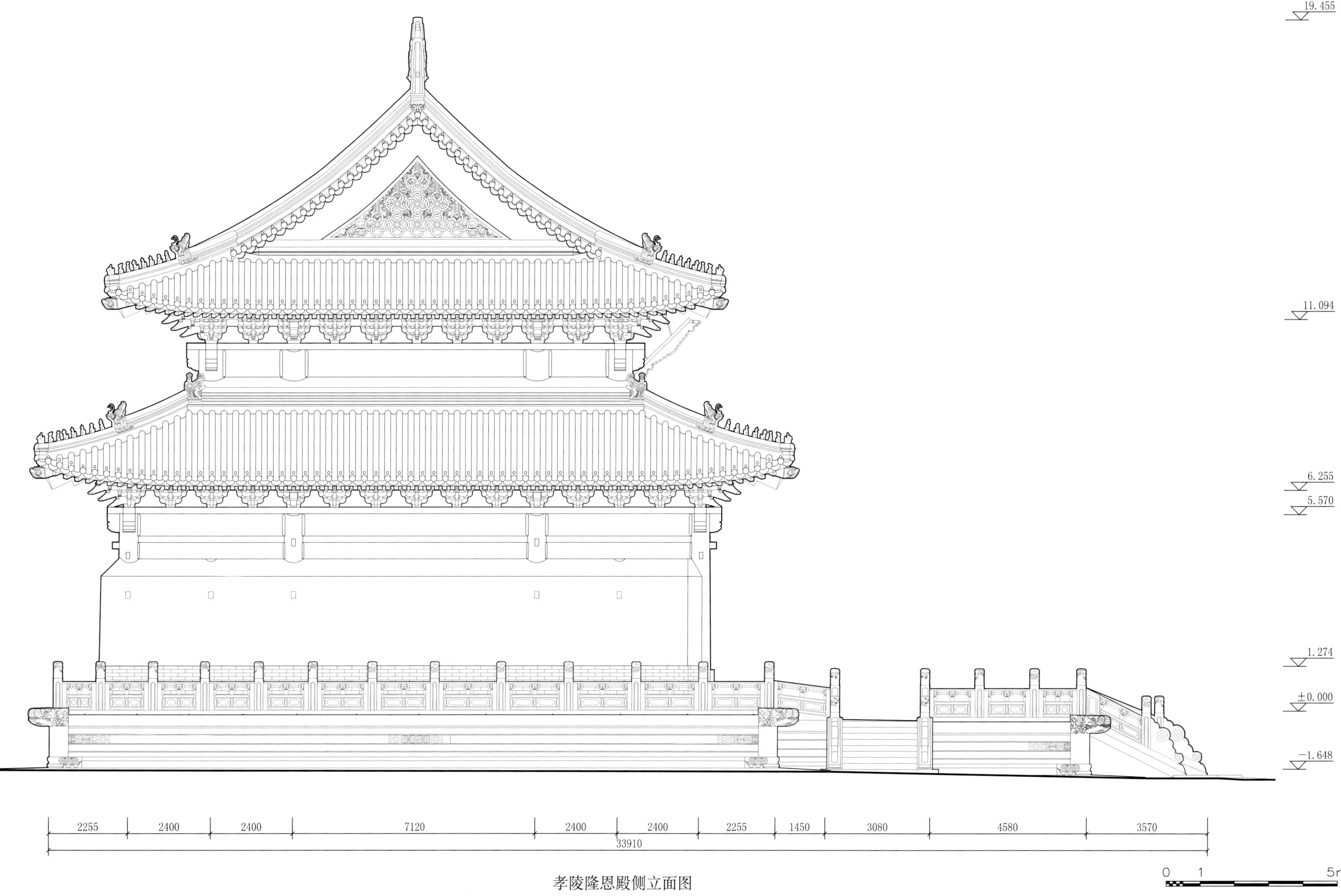

孝陵隆恩殿侧立面图

Side elevation of the Hall of Monumental Grace, Tomb of Filial Piety

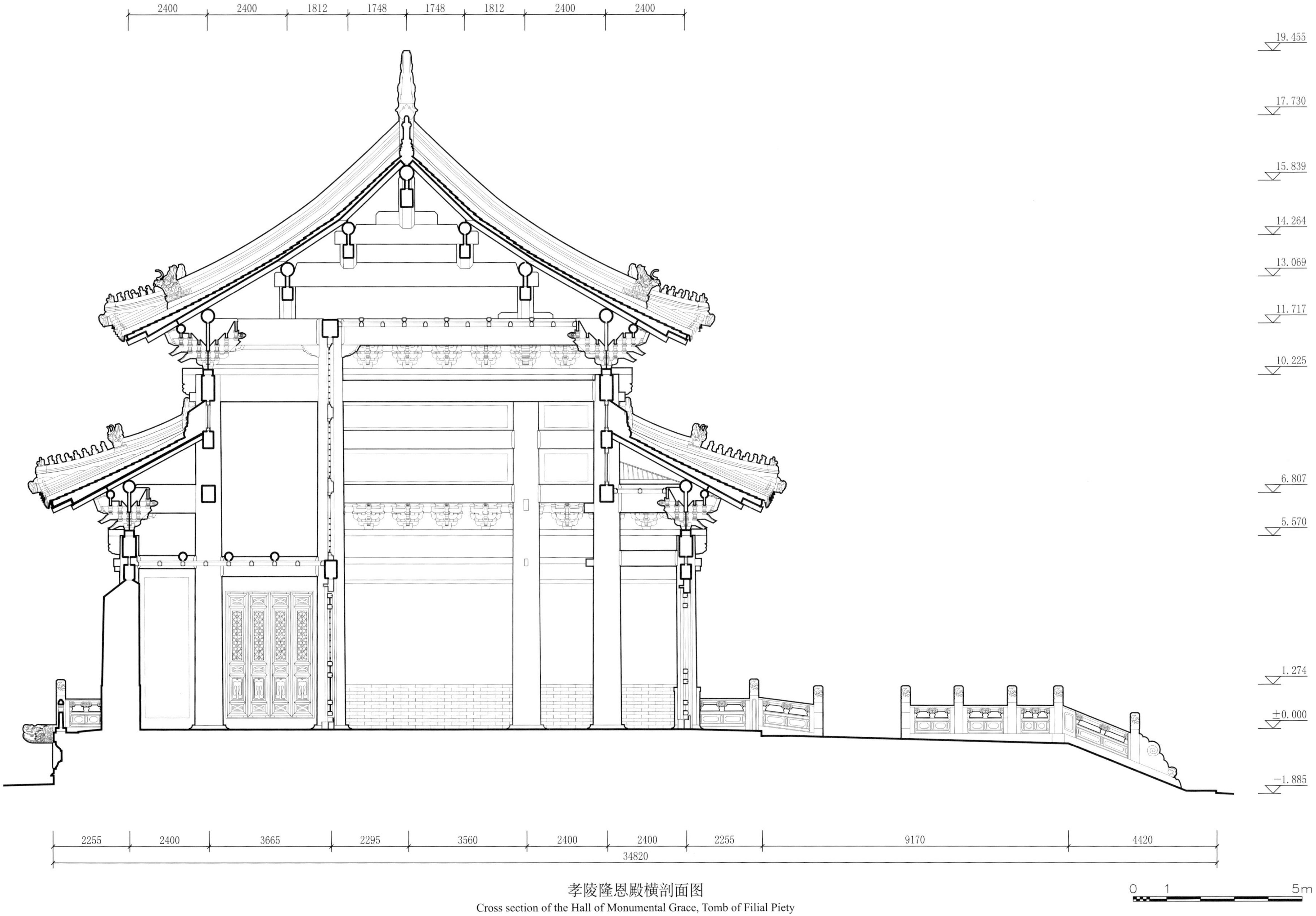

2400 2400 1812 1748 1748 1812 2400 2400

19.455
17.730
15.839
14.264
13.069
11.717
10.225
6.807
5.570
1.274
±0.000
−1.885

2255 2400 3665 2295 3560 2400 2400 2255 9170 4420
34820

0 1 5m

孝陵隆恩殿横剖面图
Cross section of the Hall of Monumental Grace, Tomb of Filial Piety

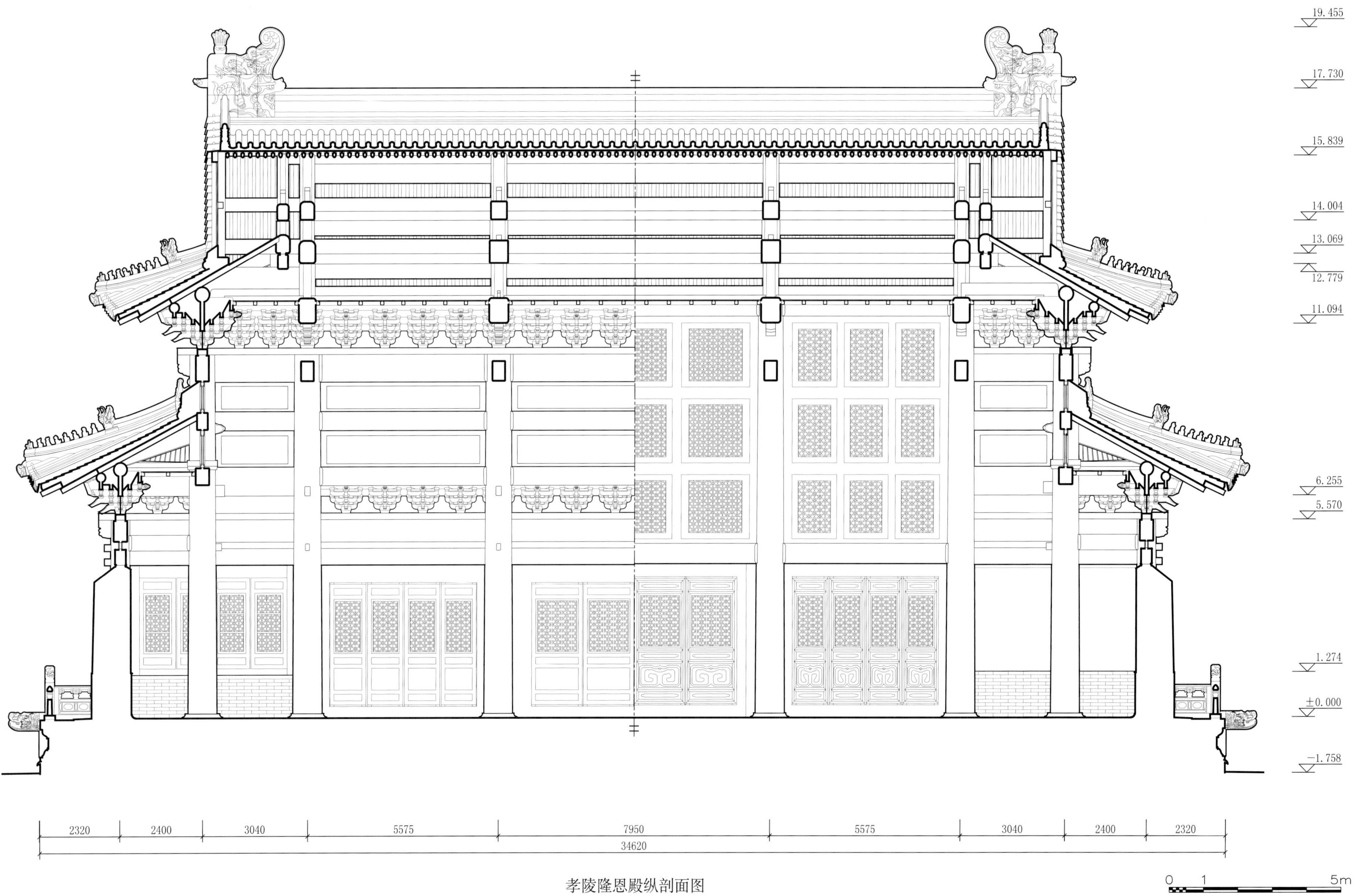

孝陵隆恩殿纵剖面图

Longitudinal section of the Hall of Monumental Grace, Tomb of Filial Piety

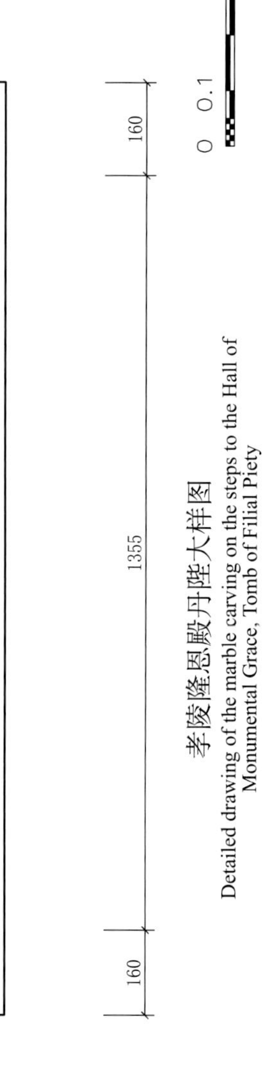

孝陵隆恩殿丹陛大样图
Detailed drawing of the marble carving on the steps to the Hall of Monumental Grace, Tomb of Filial Piety

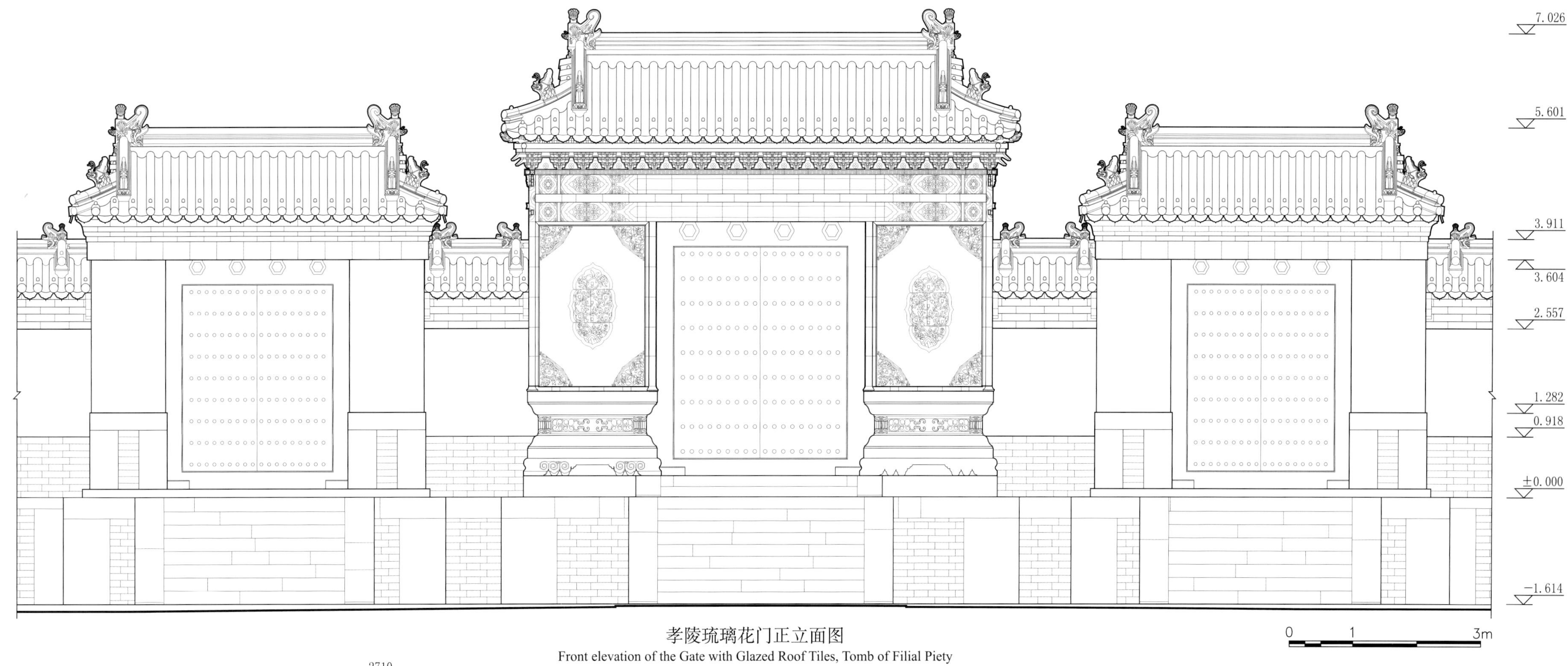

孝陵琉璃花门正立面图
Front elevation of the Gate with Glazed Roof Tiles, Tomb of Filial Piety

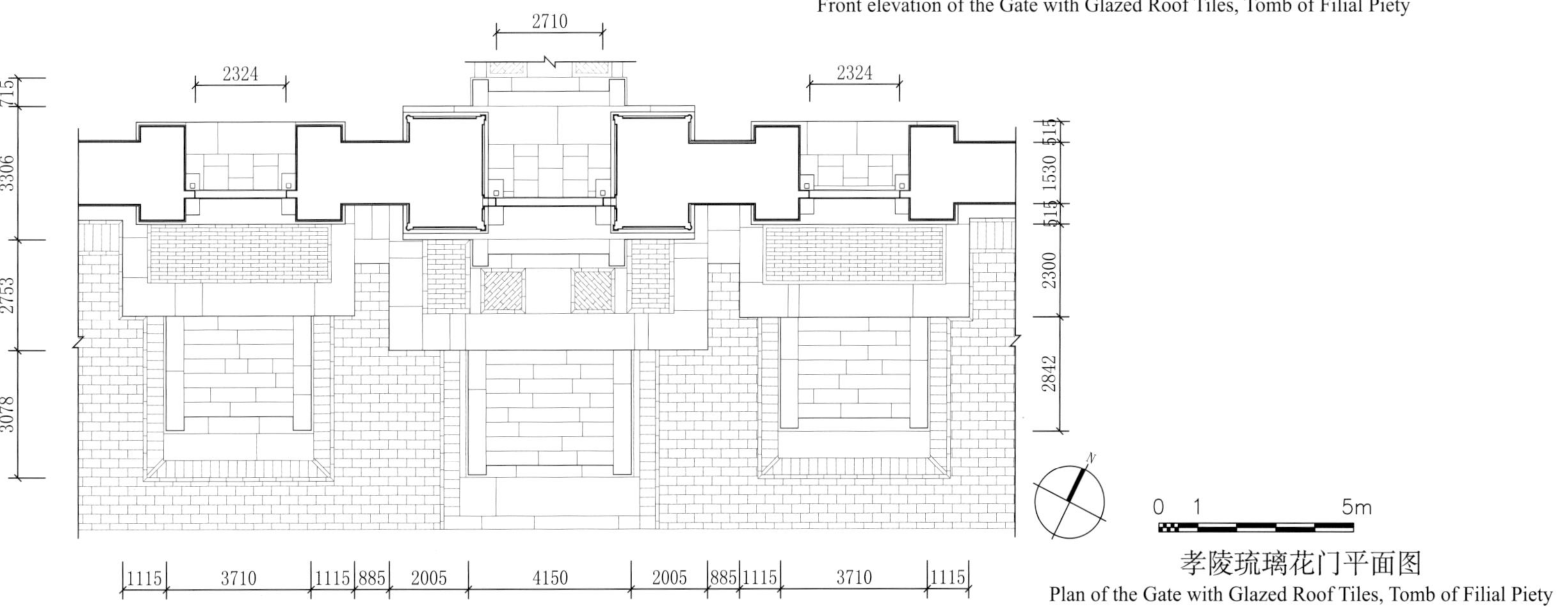

孝陵琉璃花门平面图
Plan of the Gate with Glazed Roof Tiles, Tomb of Filial Piety

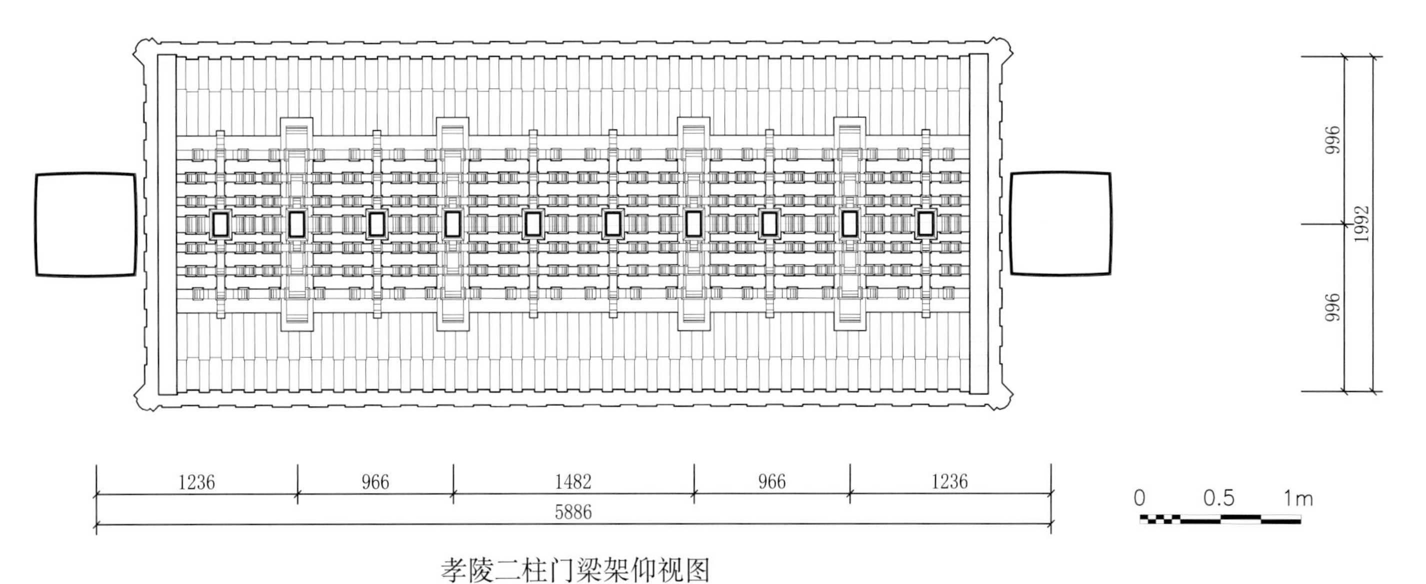

孝陵二柱门梁架仰视图
Looking up at the Truss (*liangjia*) of the Gate with Two Columns, Tomb of Filial Piety

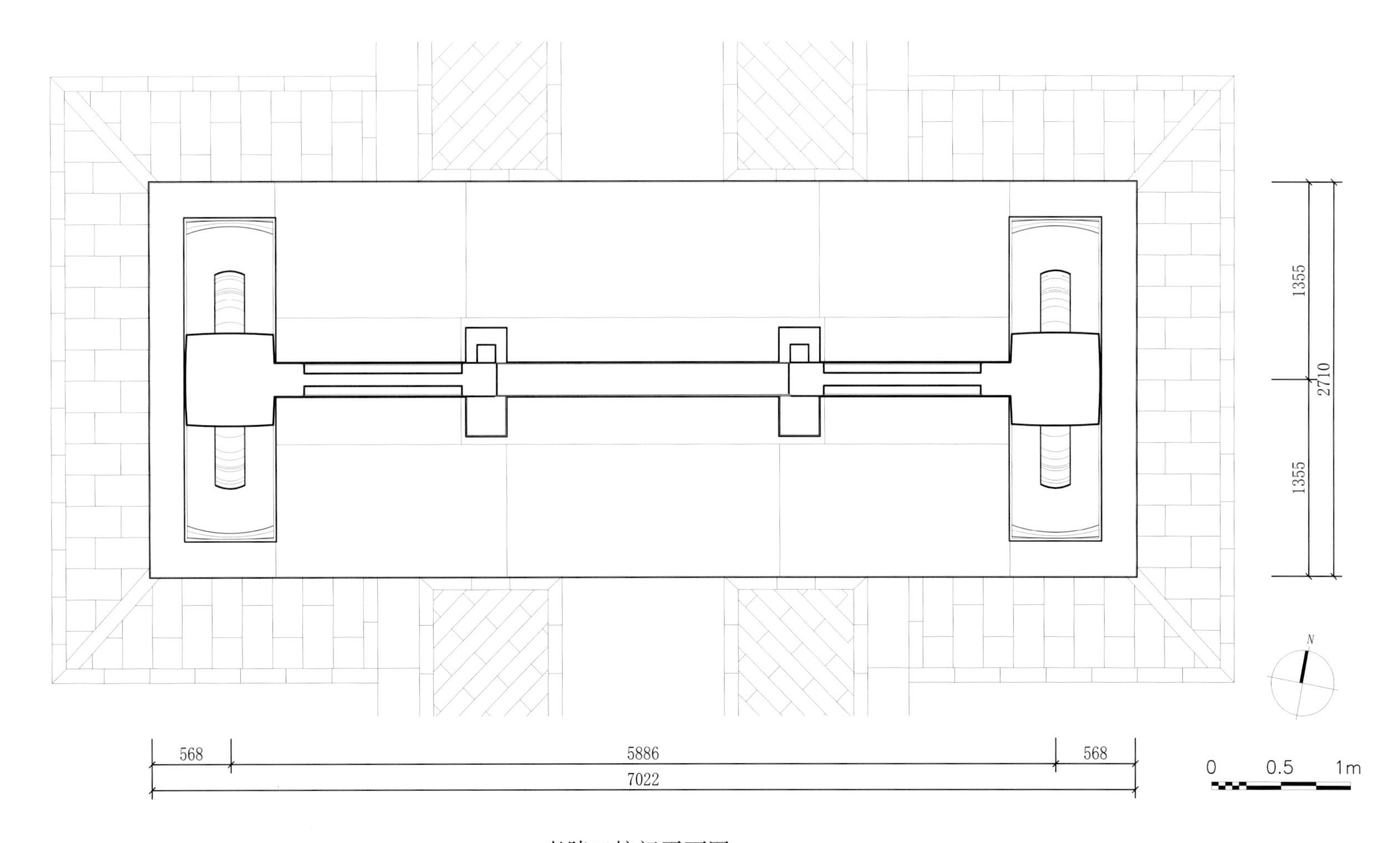

孝陵二柱门平面图
Plan of the Gate with Two Columns, Tomb of Filial Piety

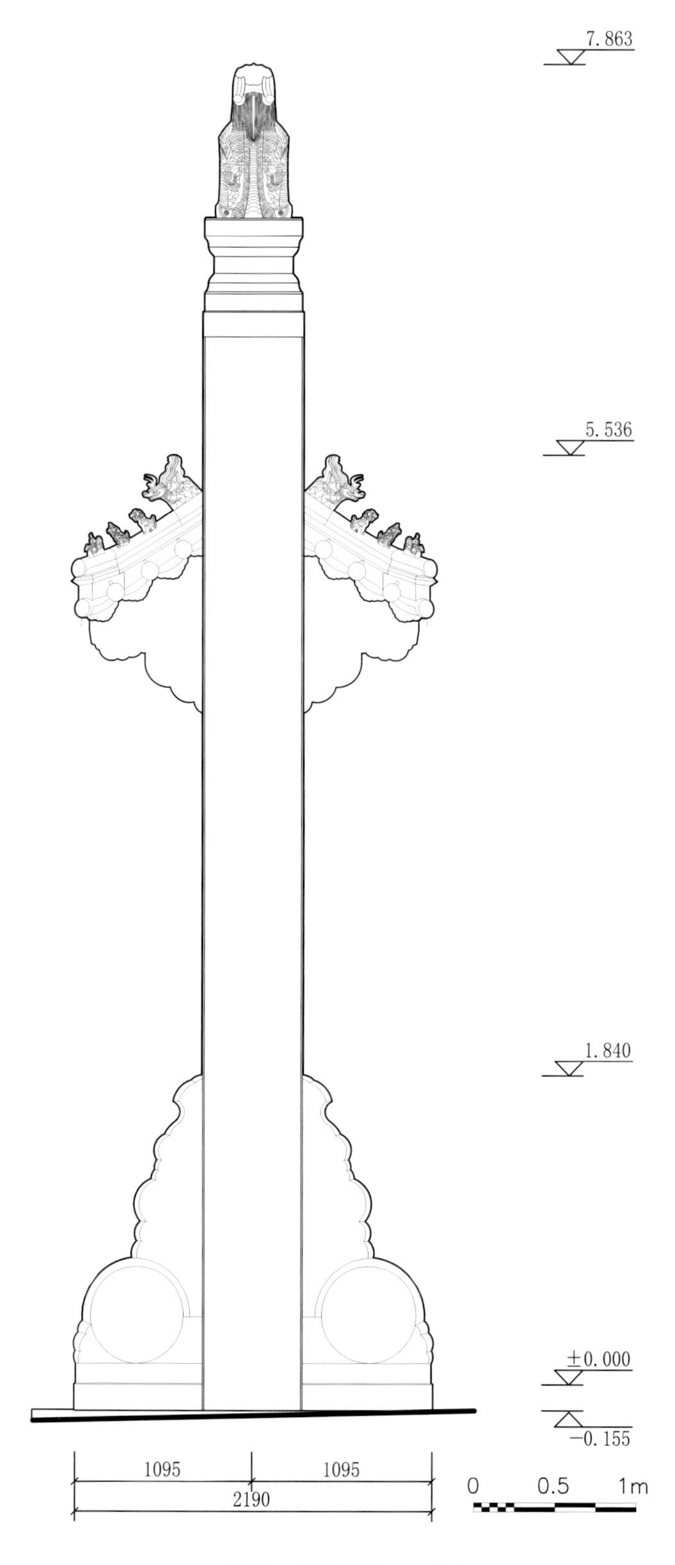

孝陵二柱门侧立面图
Side elevation of the Gate with Two Columns, Tomb of Filial Piety

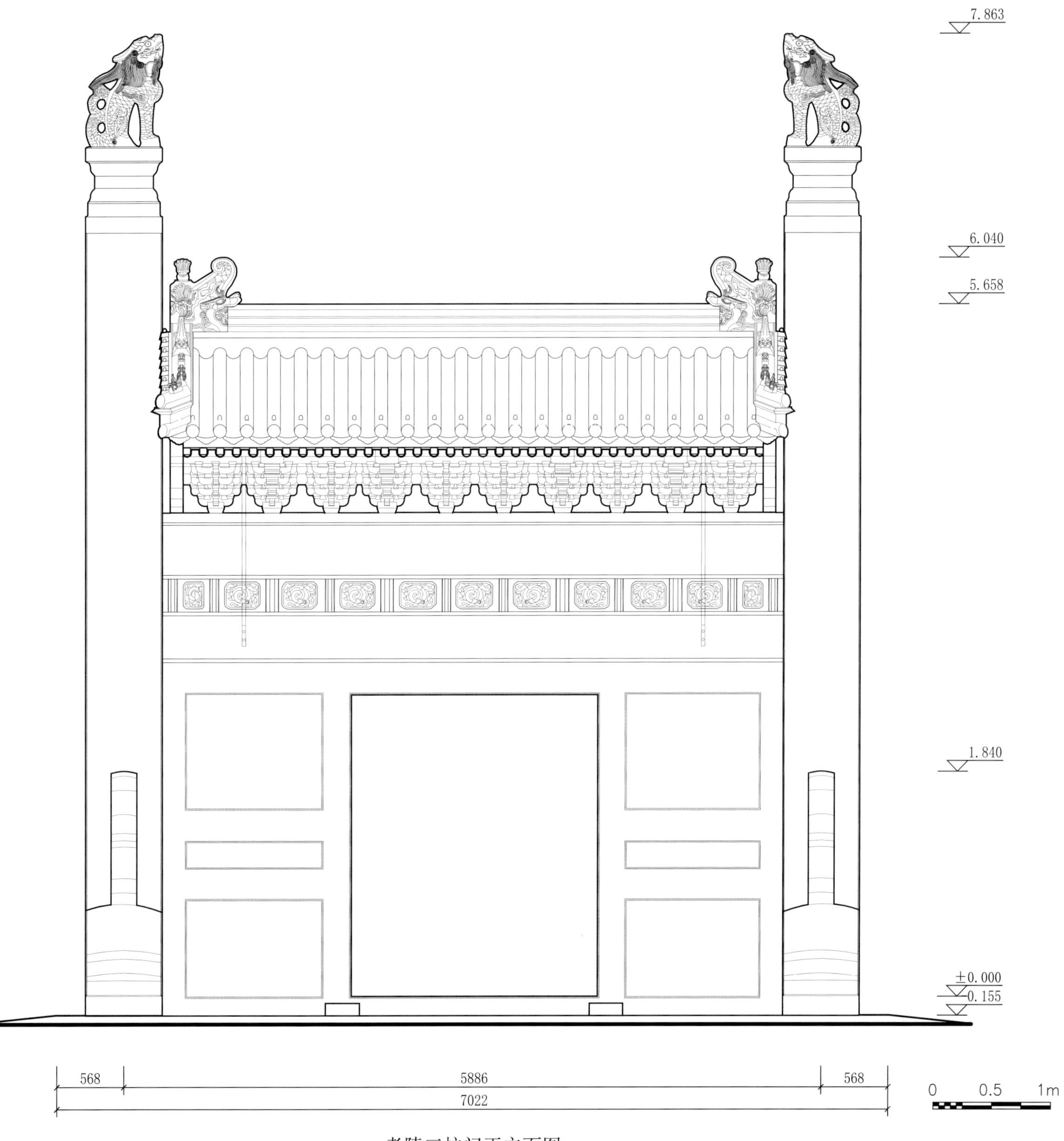

孝陵二柱门正立面图
Front elevation of the Gate with Two Columns, Tomb of Filial Piety

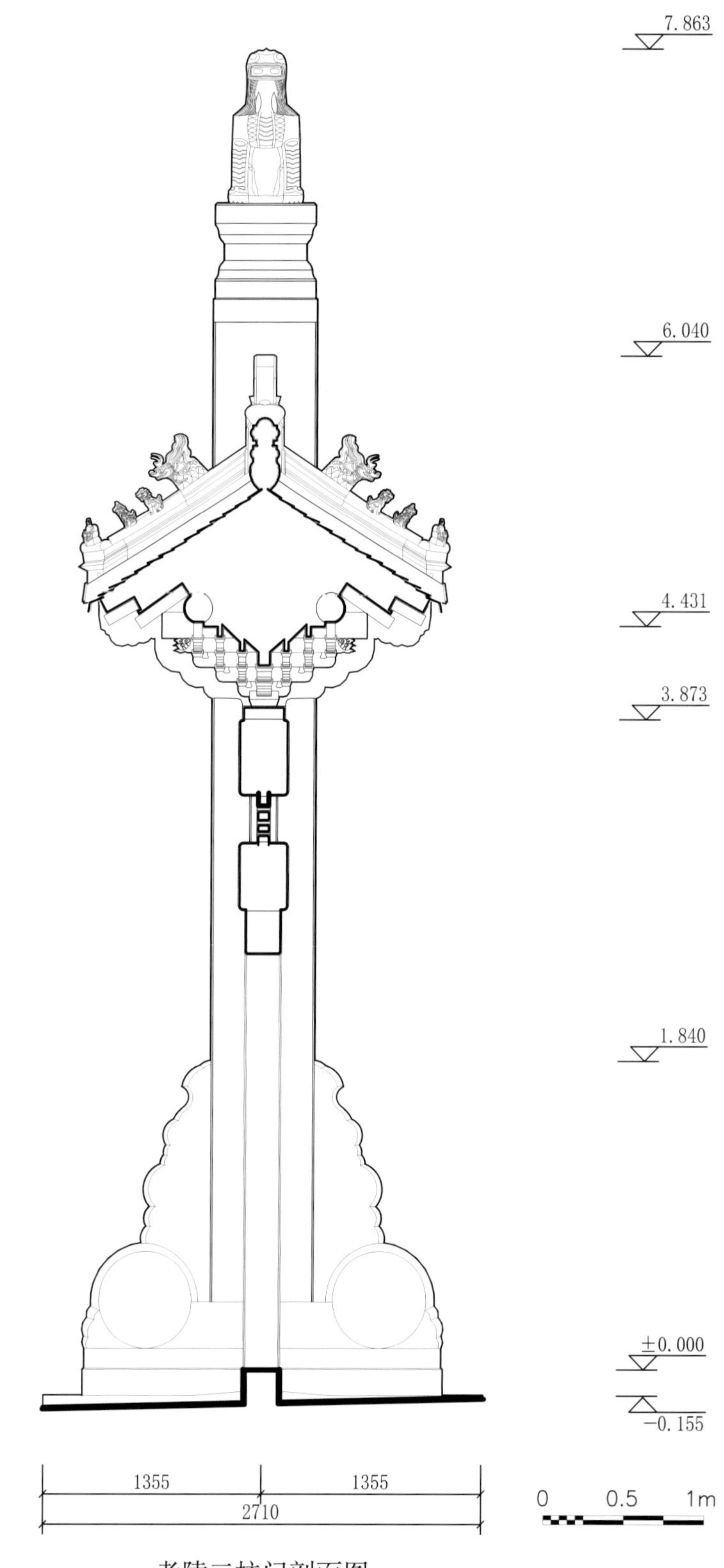

孝陵二柱门剖面图
Section of the Gate with Two Columns, Tomb of Filial Piety

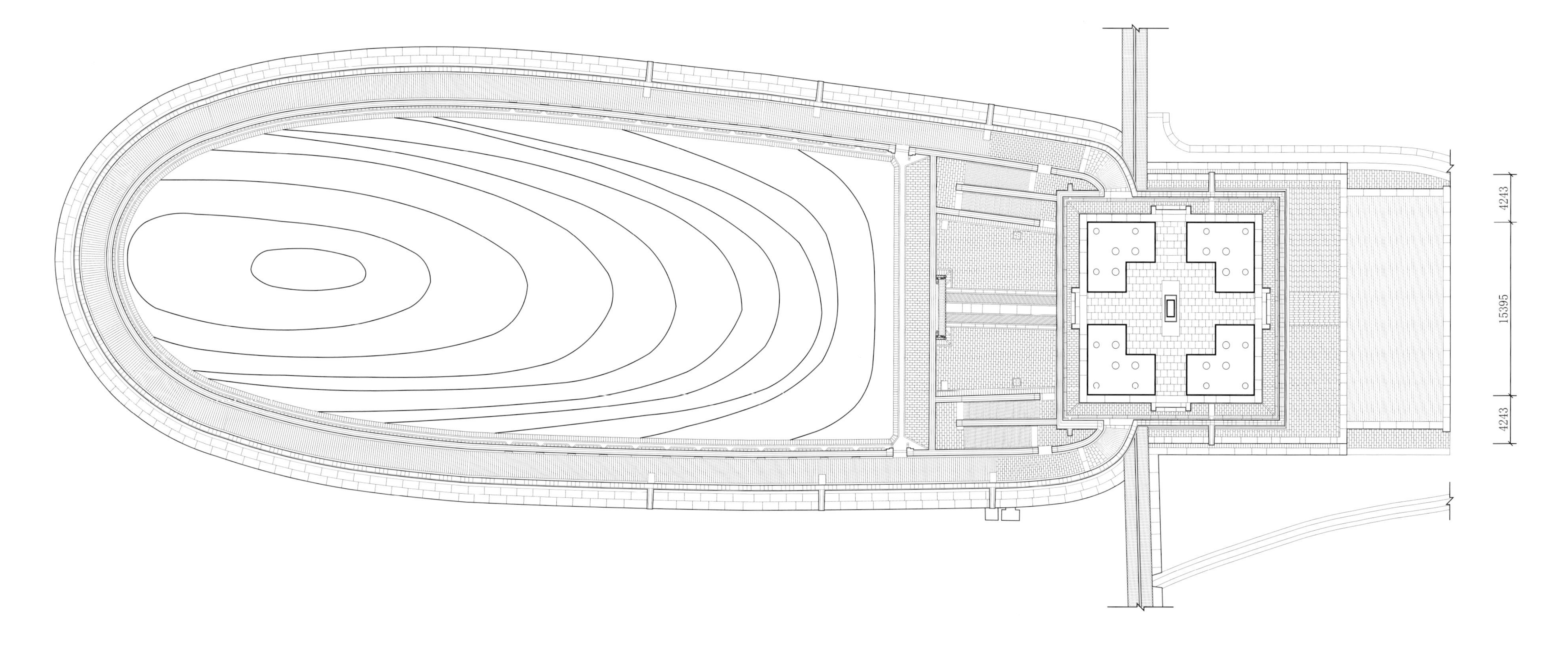

孝陵方城明楼及宝城宝顶平面图

Plan of the Square Walled Terrace and the Memorial Tower as well as the Encircled Realm of Treasure and the Tumulus, Tomb of Filial Piety

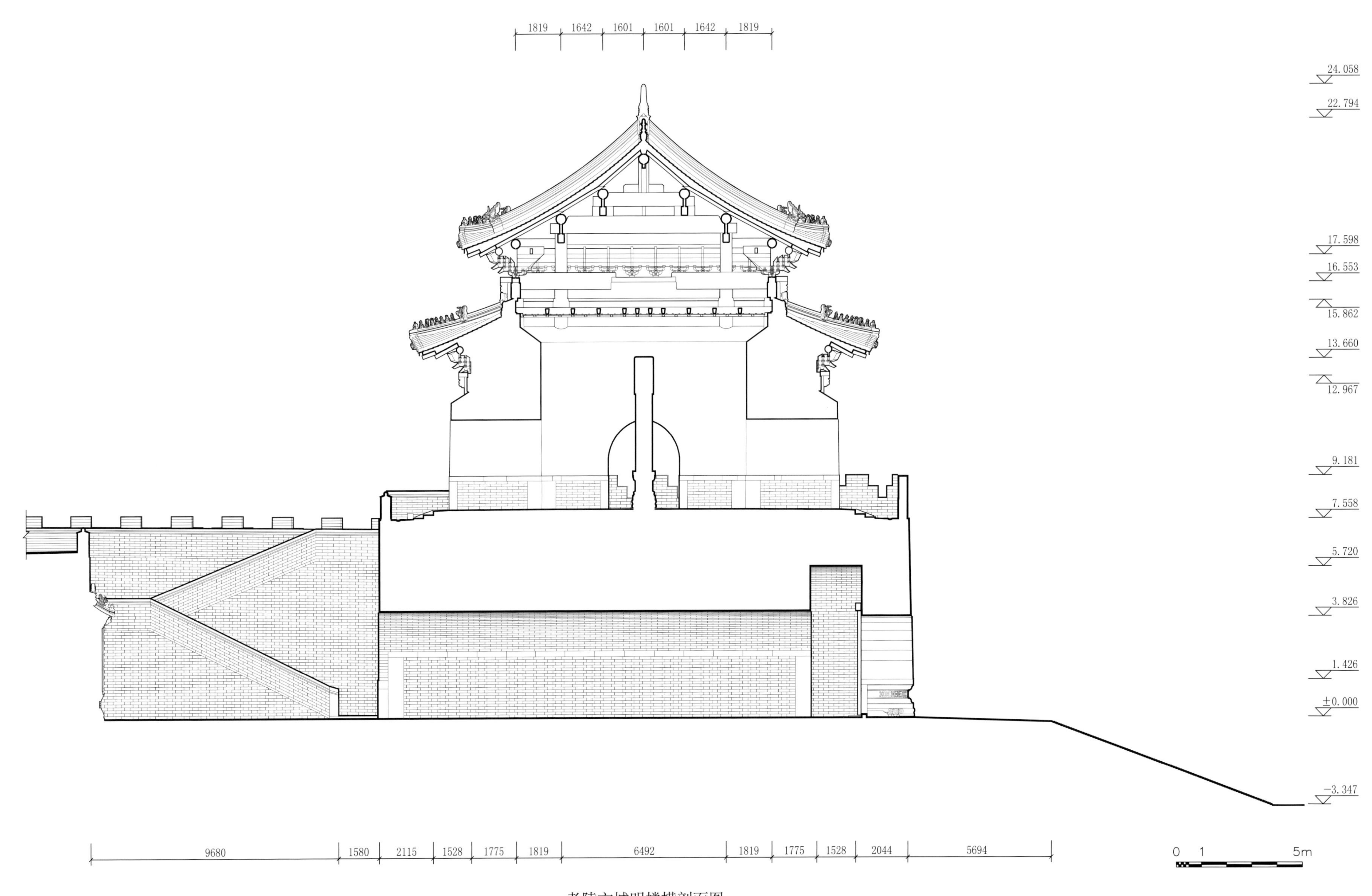

孝陵方城明楼横剖面图

Cross section of the Square Walled Terrace and the Memorial Tower, Tomb of Filial Piety

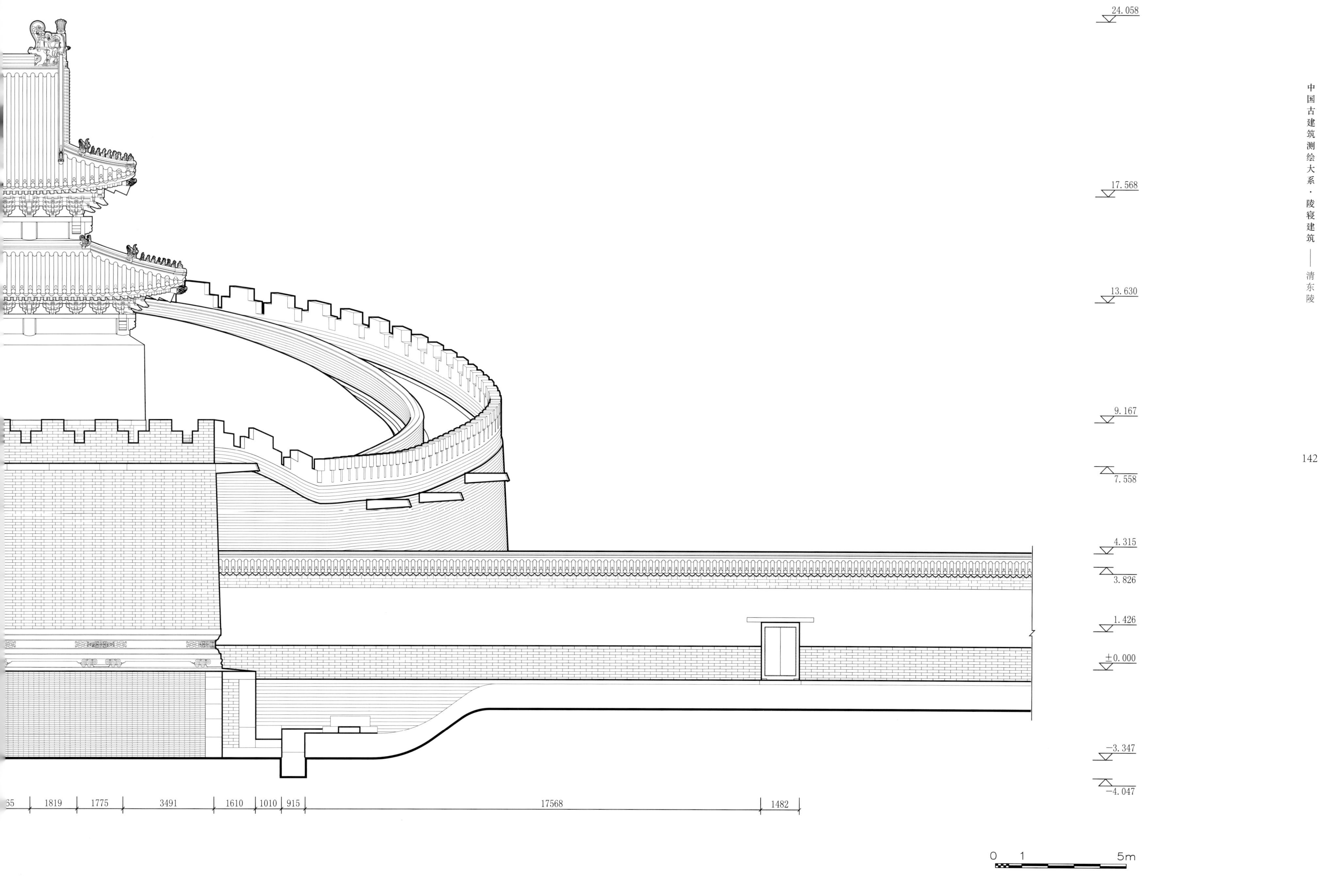
24.058
17.568
13.630
9.167
7.558
4.315
3.826
1.426
±0.000
-3.347
-4.047
65
1819
1775
3491
1610
1010
915
17568
1482
0
1
5m

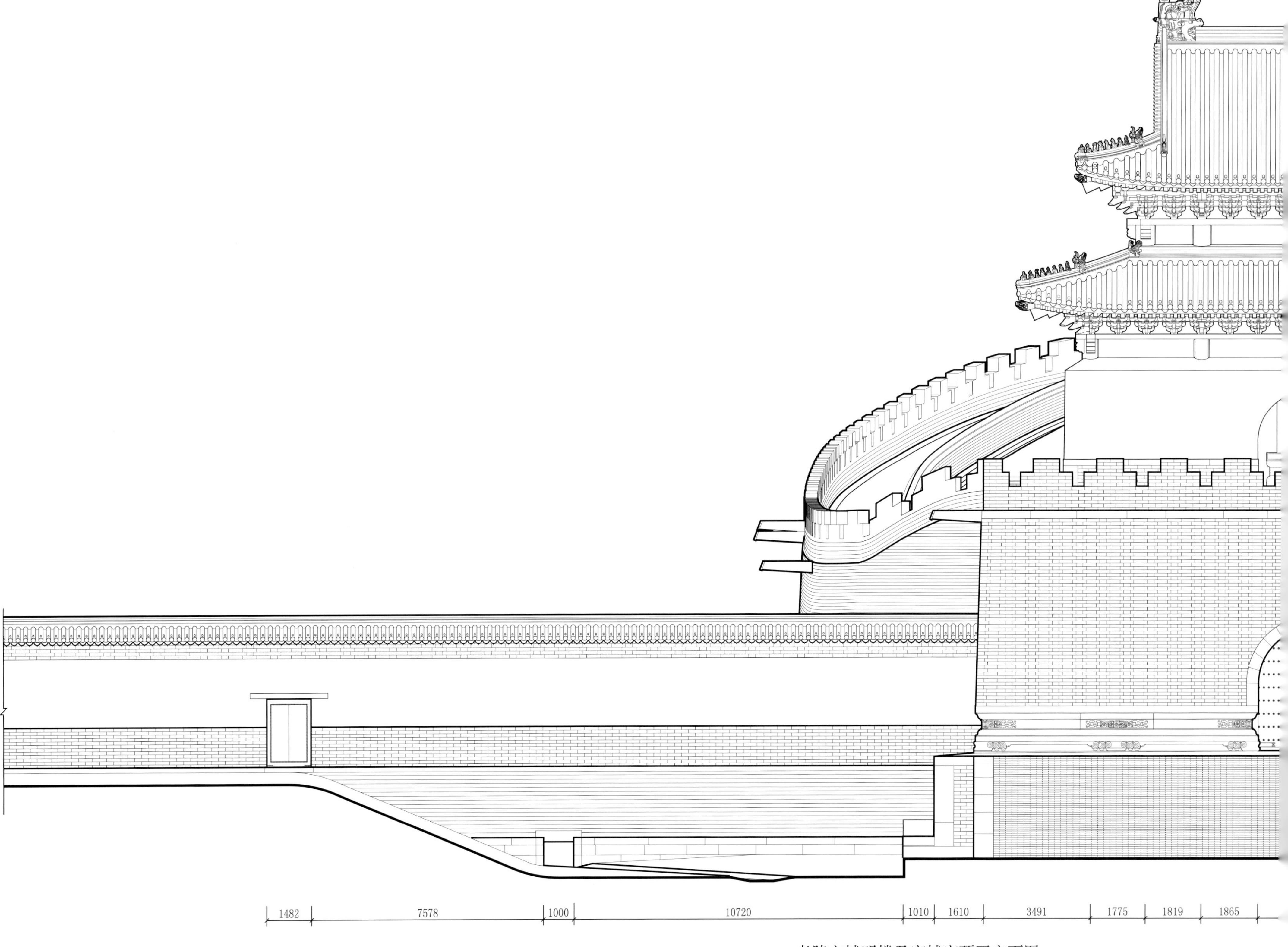

孝陵方城明楼及宝城宝顶正立面图

Front elevation of the Square Walled Terrace and the Memorial Tower as well as the Encircled Realm of Treasure and the Tumulus, Tomb of Filial Piety

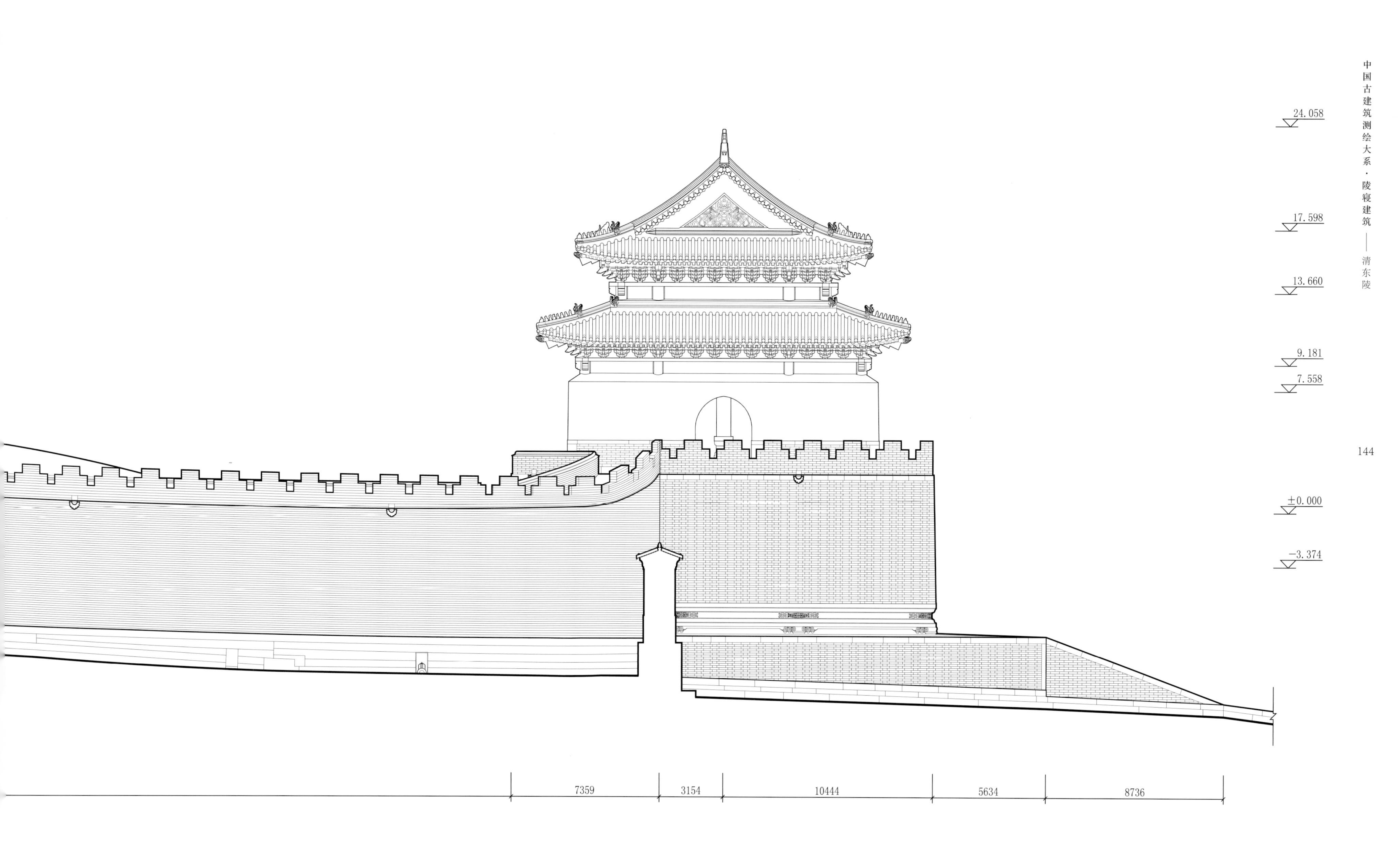
24.058
17.598
13.660
9.181
7.558
±0.000
−3.374
7359
3154
10444
5634
8736

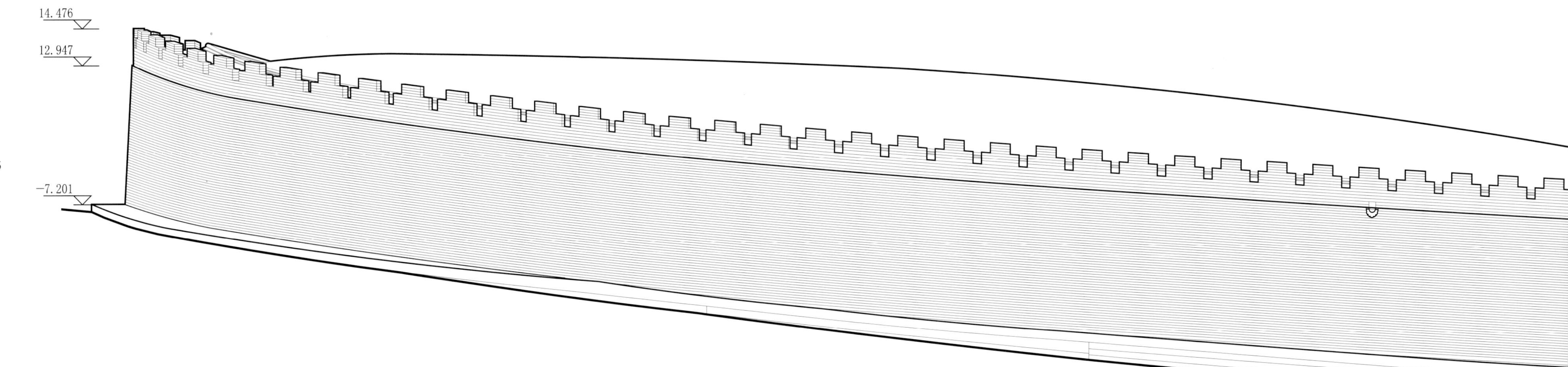

孝陵方城明楼及宝城宝顶侧立面图
Side elevation of the Square Walled Terrace and the Memorial Tower as well as the Encircled Realm of Treasure and the Tumulus, Tomb of Filial Piety

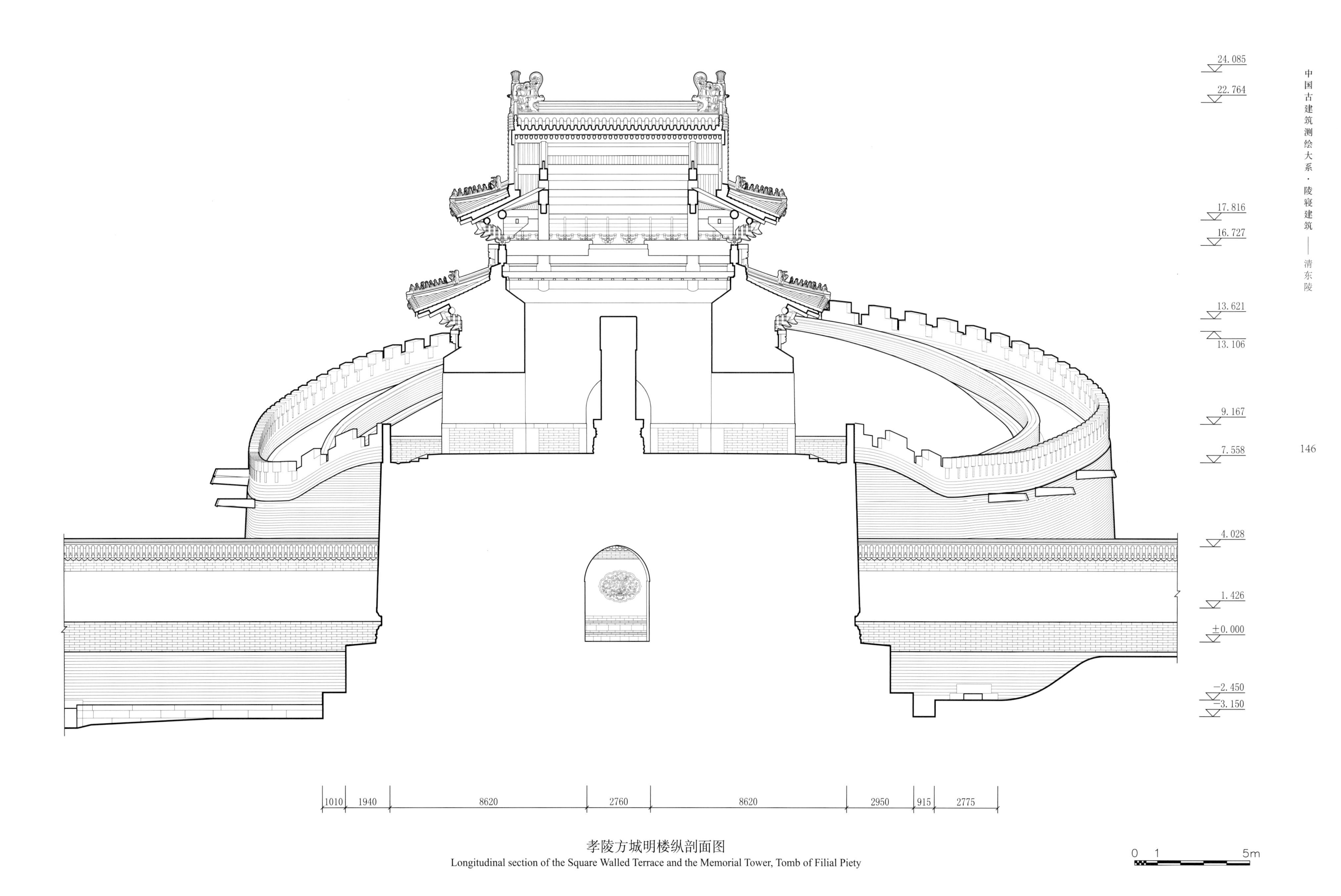

孝陵方城明楼纵剖面图
Longitudinal section of the Square Walled Terrace and the Memorial Tower, Tomb of Filial Piety

孝陵明楼石碑大样图
Detailed drawing of the Marble Stela in the Memorial Tower, Tomb of Filial Piety

0 0.5 1m

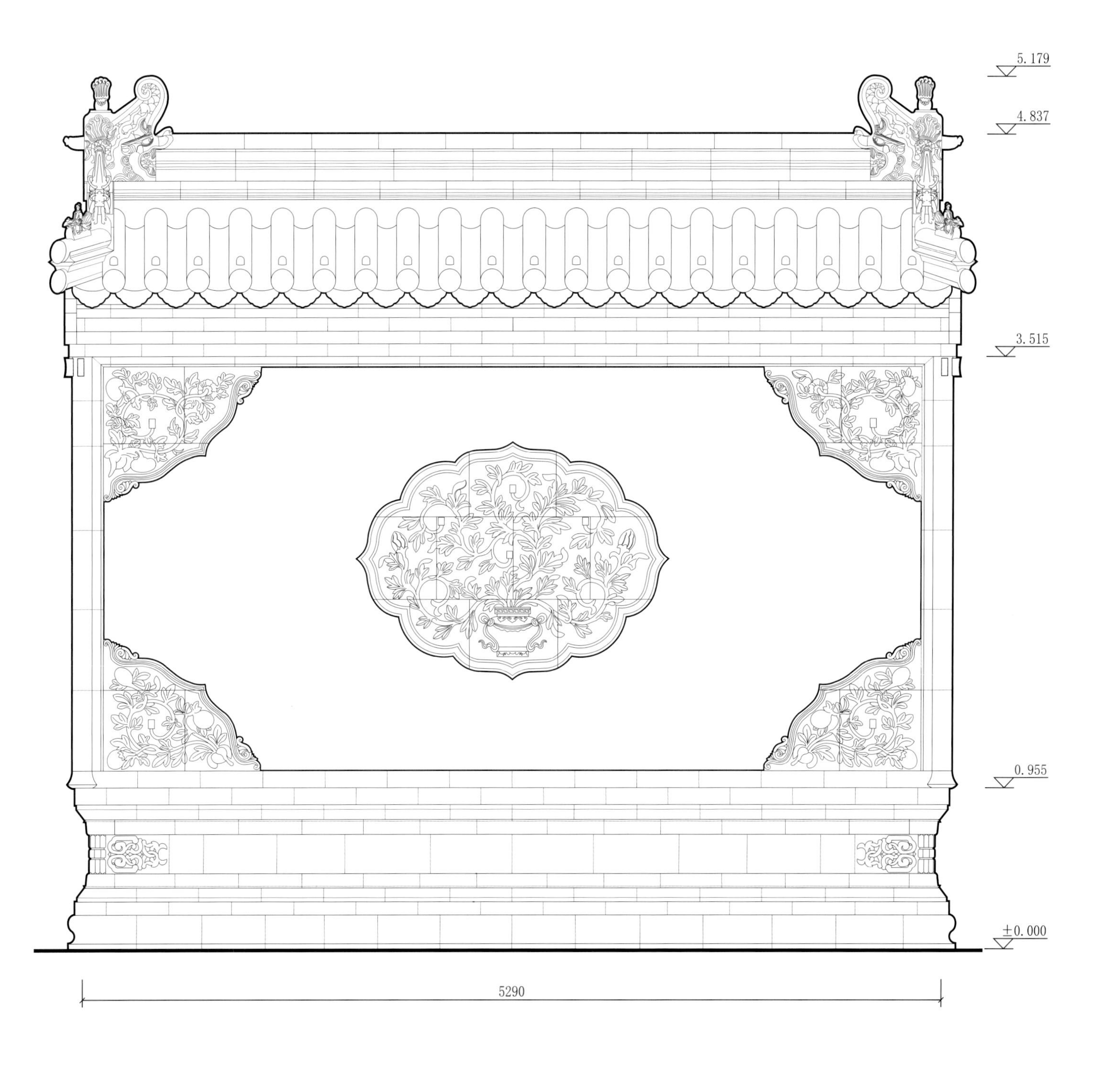

孝陵月牙城琉璃影壁正立面图
Front elevation of the Wall with Glazed Tiles in the Crescent Wall, Tomb of Filial Piety

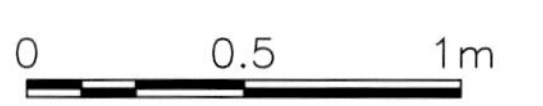

孝东陵
East Tomb of Filial Piety (Xiao Dongling)

孝东陵总剖面图
Site section of the East Tomb of Filial Piety

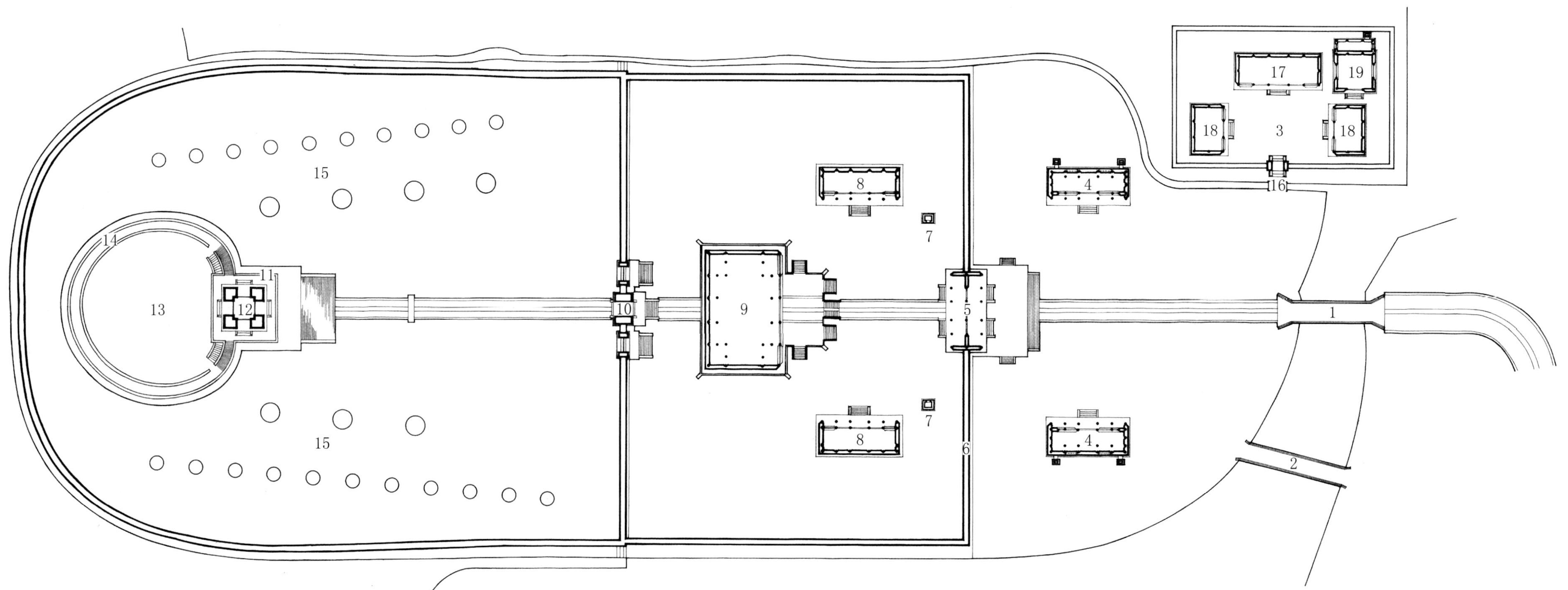

1 三孔石券桥 Three-Arch Stone Bridge
2 石平桥 Flat Stone Bridge
3 神厨库院落 Culinary Courtyard for Sacrifices
4 朝房 Reception Halls for Court Officials
5 隆恩门 Gate of Monumental Grace
6 陵墙 Tomb Wall
7 焚帛炉 Sacrificial Burners
8 配殿 Side Hall
9 隆恩殿 Hall of Monumental Grace
10 琉璃花门 Gate with Glazed Roof Tiles
11 方城 Square Walled Terrace
12 明楼 Memorial Tower
13 宝顶 Tumulus
14 宝城 Encircled Realm of Treasure
15 妃嫔宝顶 Concubines' Tumulus
16 神厨库院门 Culinary Courtyard Gate for Sacrifices
17 神厨 Sacrificial Kitchen
18 神库 Sacrificial Storehouse
19 宰牲亭 Ritual Abattoir

孝东陵总平面图
Site plan of the East Tomb of Filial Piety

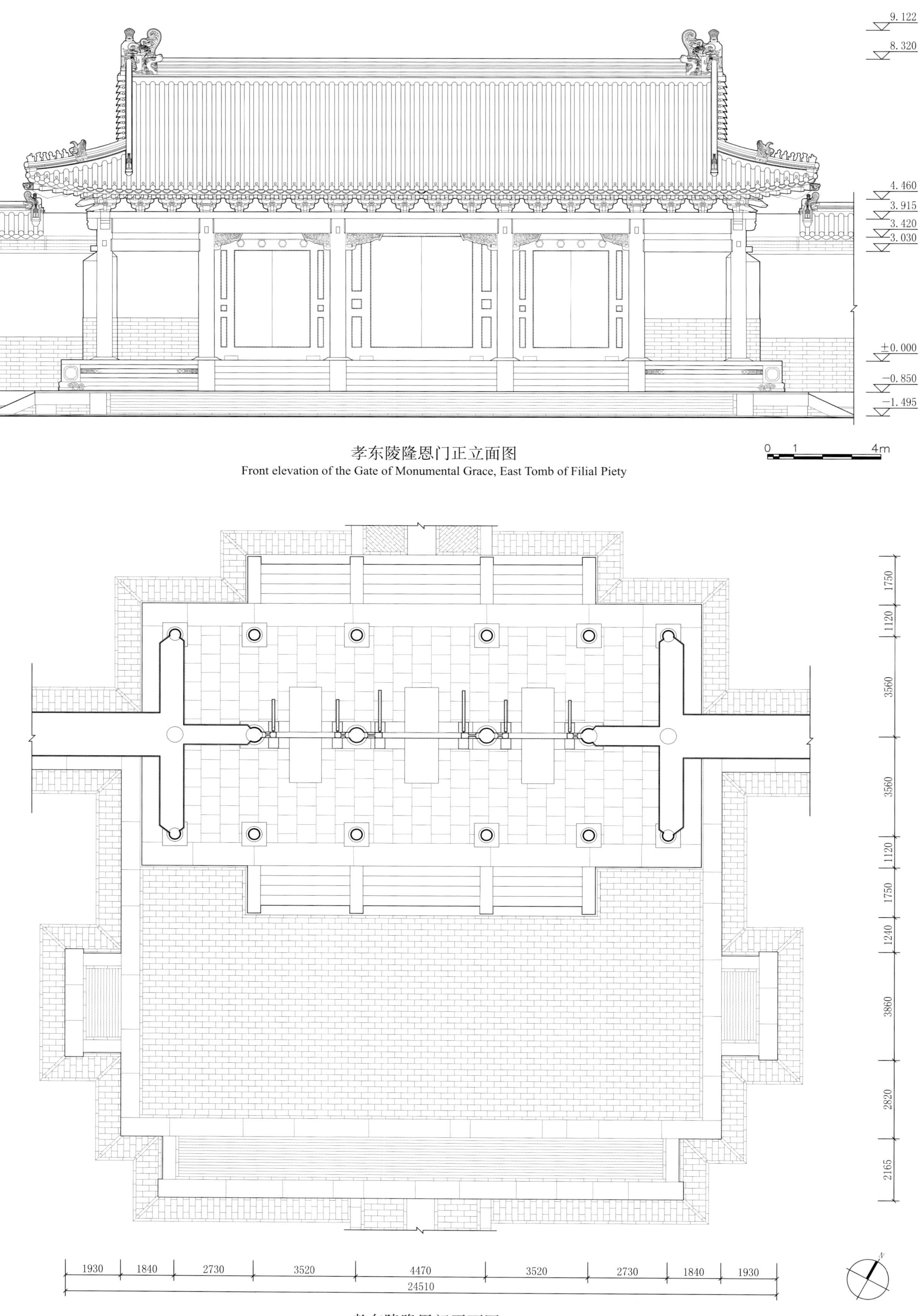

孝东陵隆恩门正立面图
Front elevation of the Gate of Monumental Grace, East Tomb of Filial Piety

孝东陵隆恩门平面图
Plan of the Gate of Monumental Grace, East Tomb of Filial Piety

9.122
8.320
6.865
5.820
5.410
4.460
4.330
3.915
1.170
±0.000
−0.850
−1.283

2730 3520 4470 3520 2730
16970

0 1 3m

孝东陵隆恩门纵剖面图
Longitudinal section of the Gate of Monumental Grace, East Tomb of Filial Piety

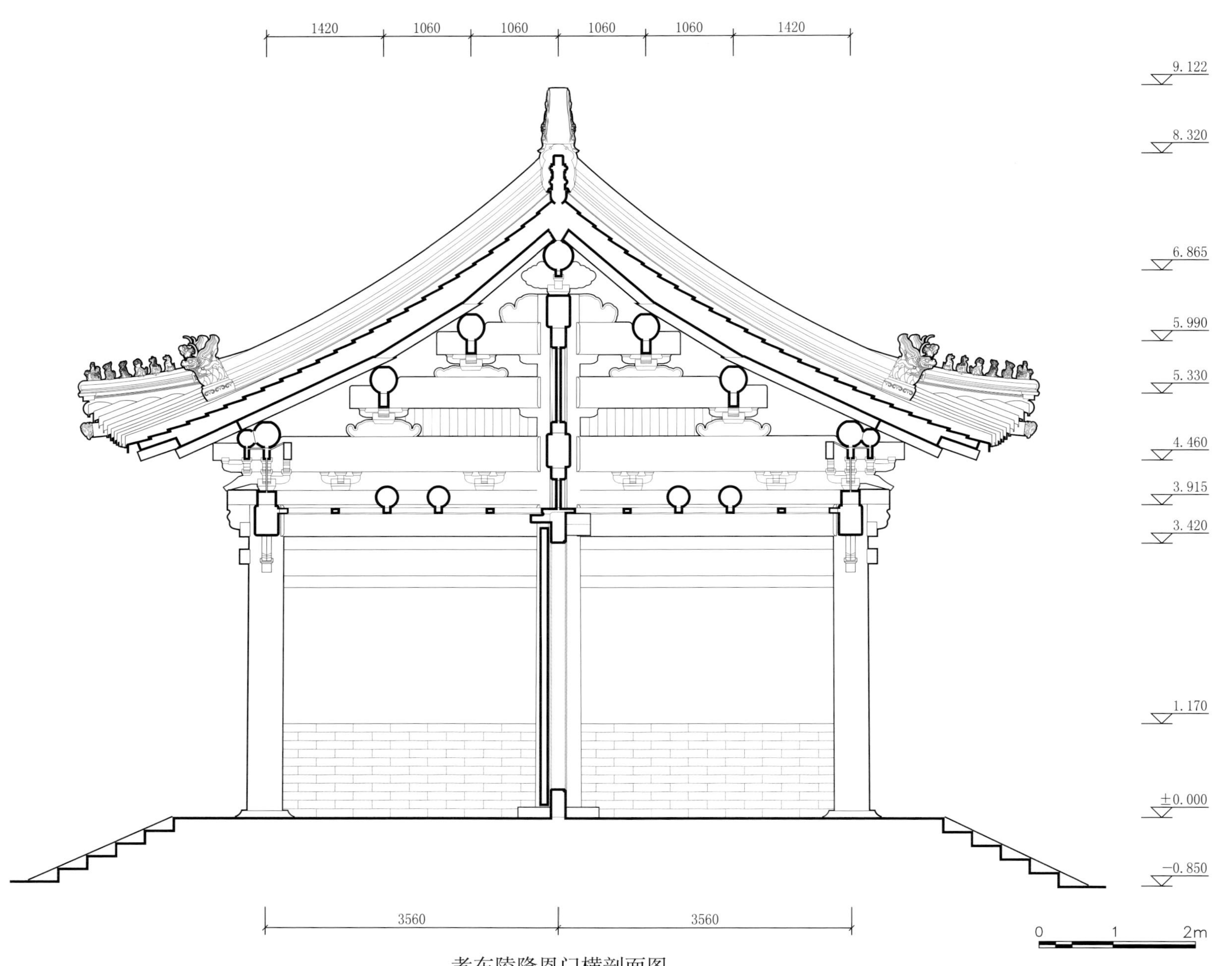

孝东陵隆恩门横剖面图
Cross section of the Gate of Monumental Grace, East Tomb of Filial Piety

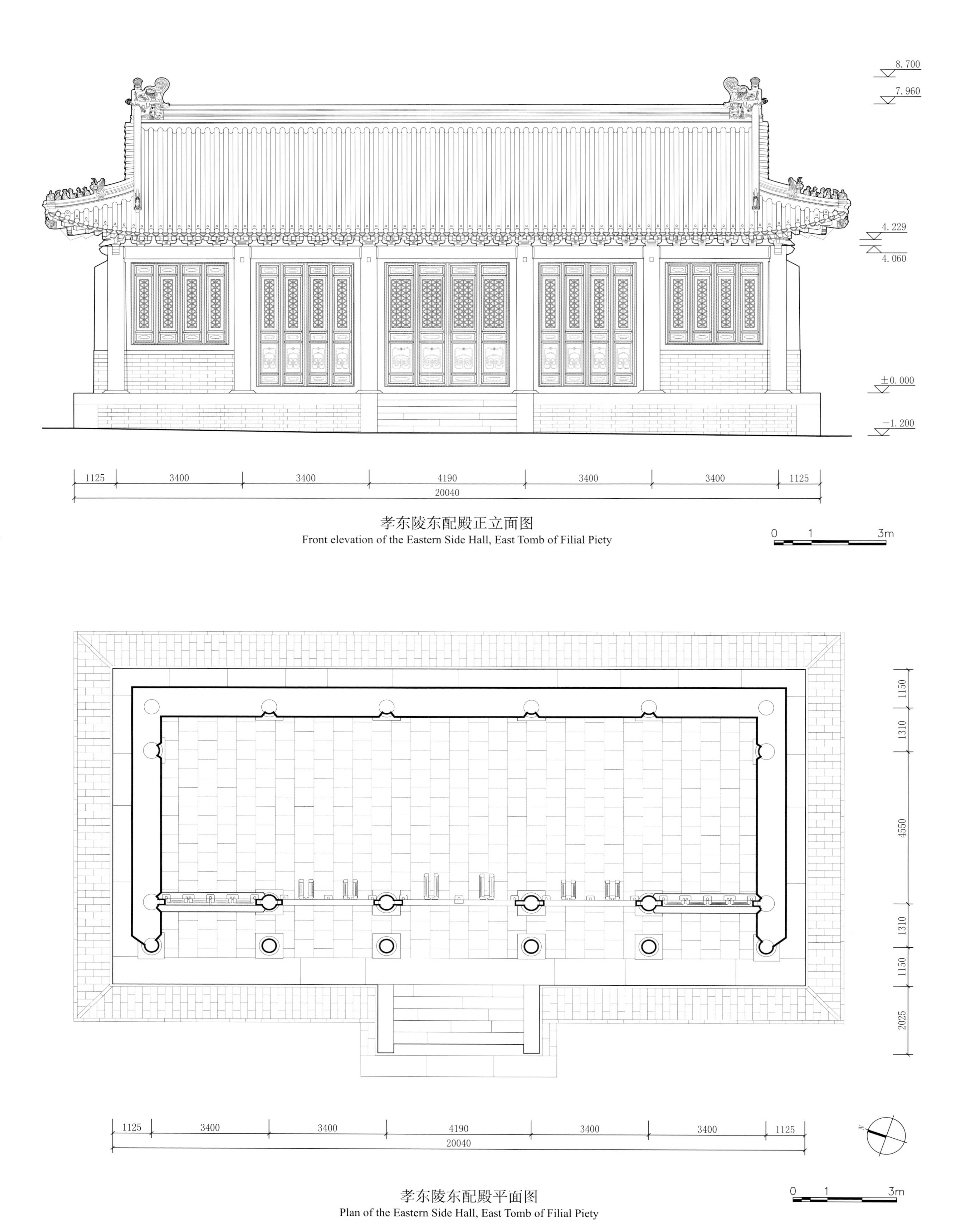

孝东陵东配殿正立面图
Front elevation of the Eastern Side Hall, East Tomb of Filial Piety

孝东陵东配殿平面图
Plan of the Eastern Side Hall, East Tomb of Filial Piety

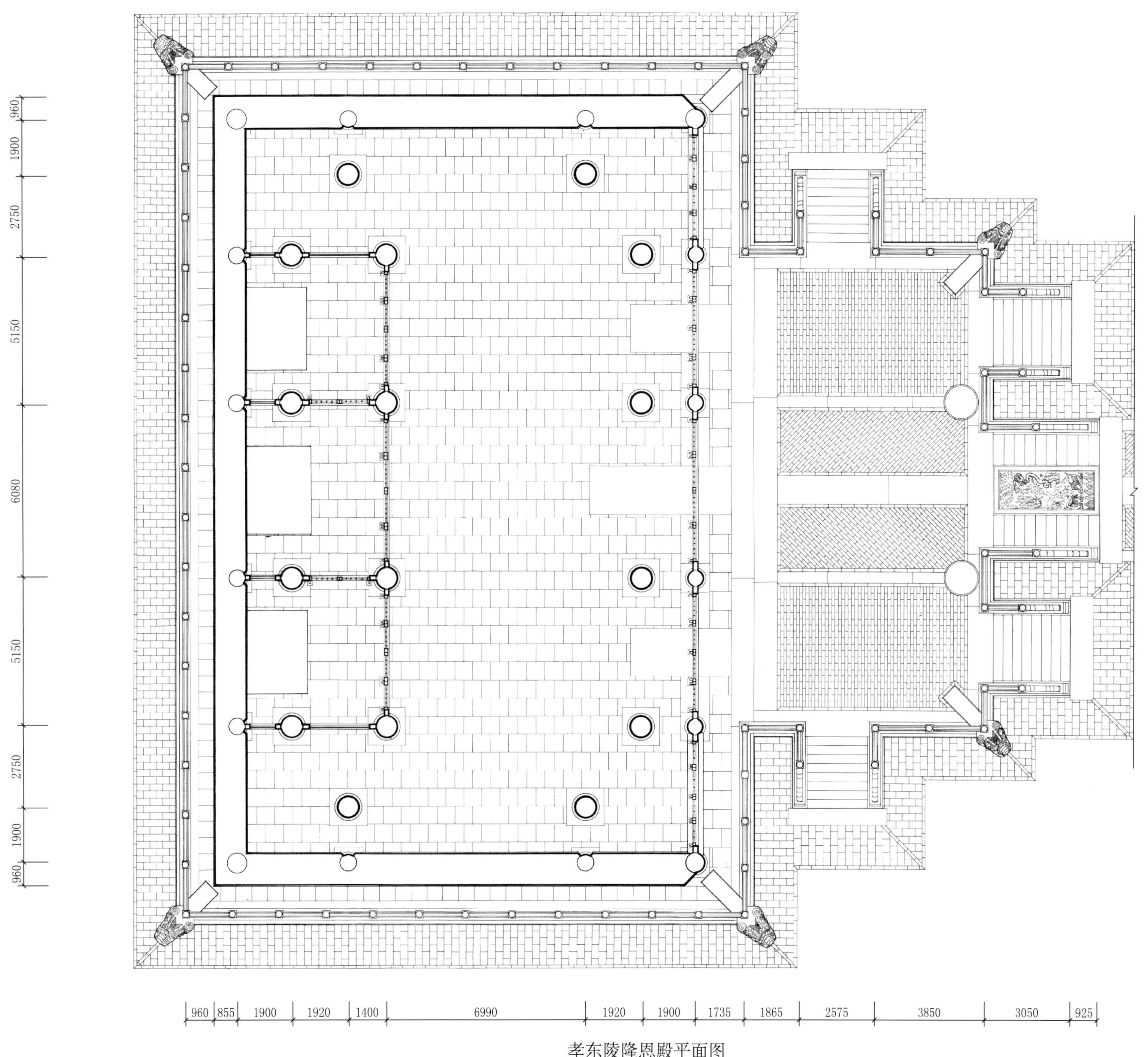

孝东陵隆恩殿平面图

Plan of the Hall of Monumental Grace, East Tomb of Filial Piety

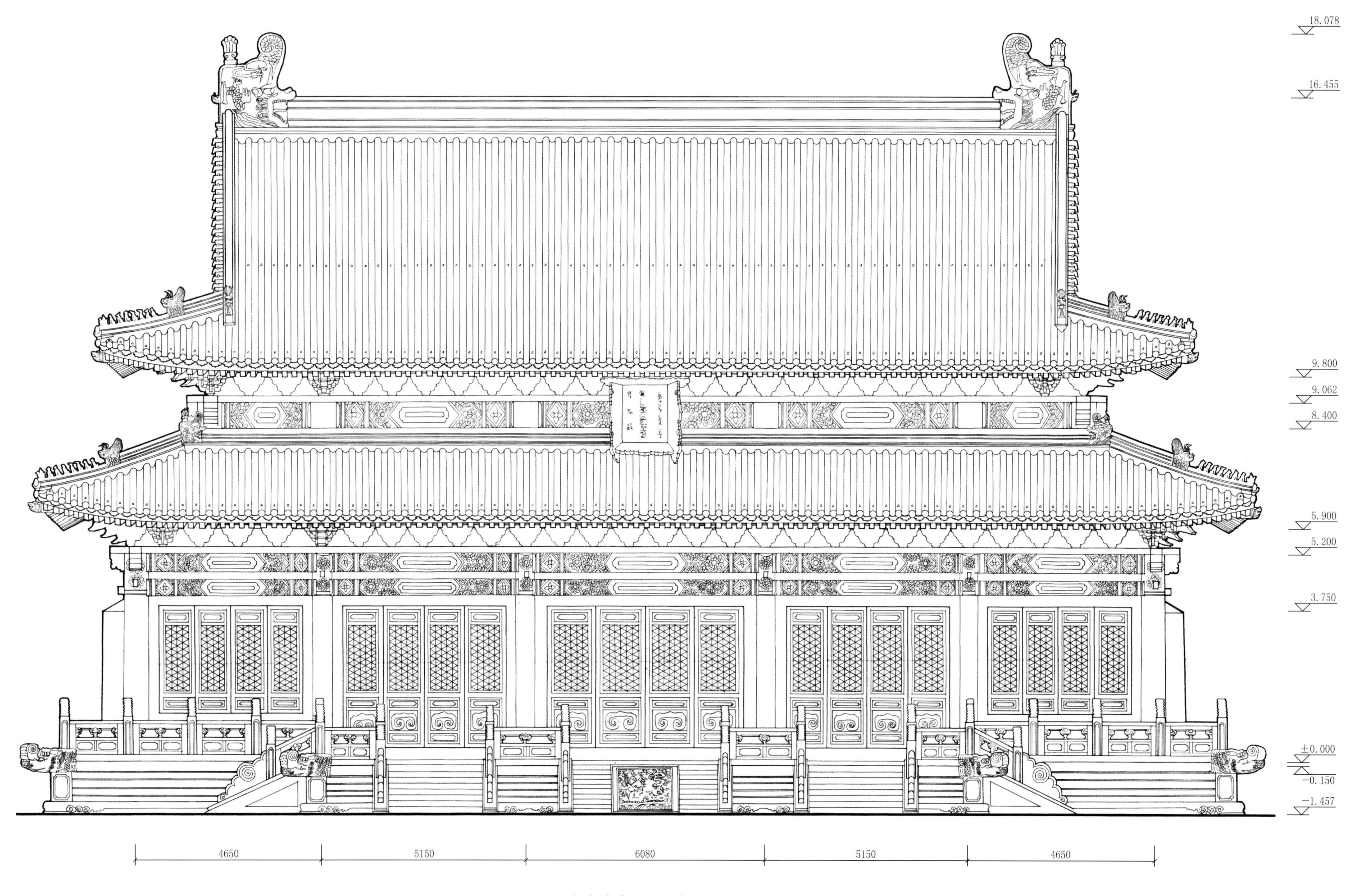

孝东陵隆恩殿正立面图

Front elevation of the Hall of Monumental Grace, East Tomb of Filial Piety

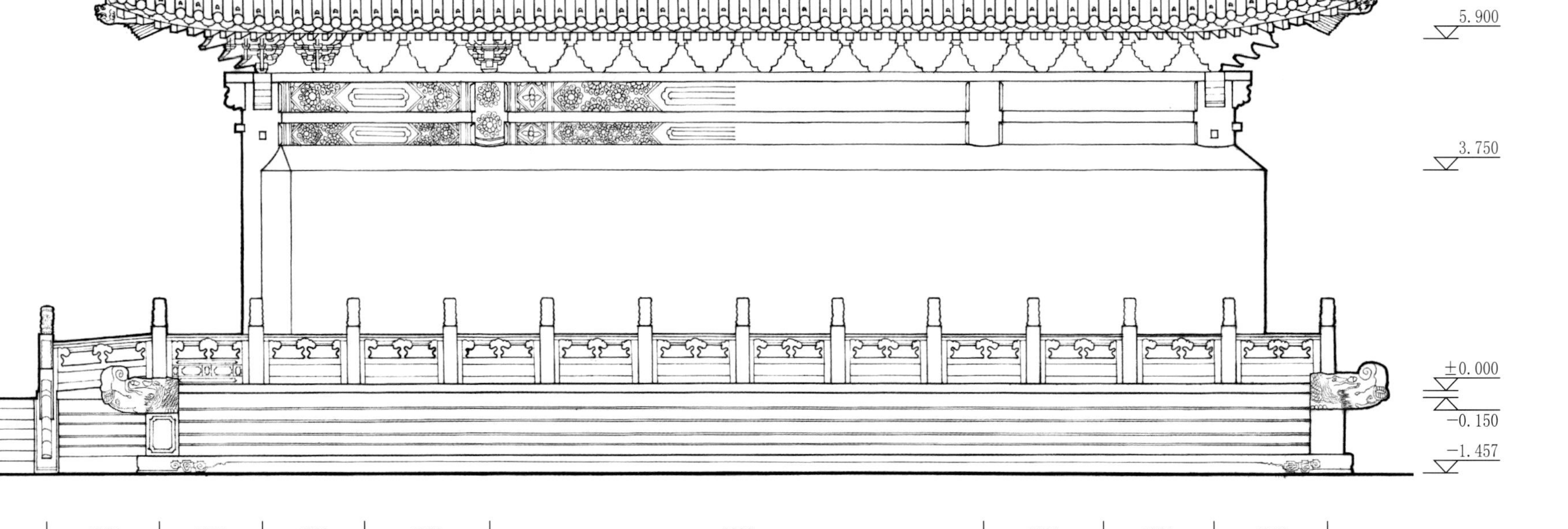

孝东陵隆恩殿侧立面图

Side elevation of the Hall of Monumental Grace, East Tomb of Filial Piety

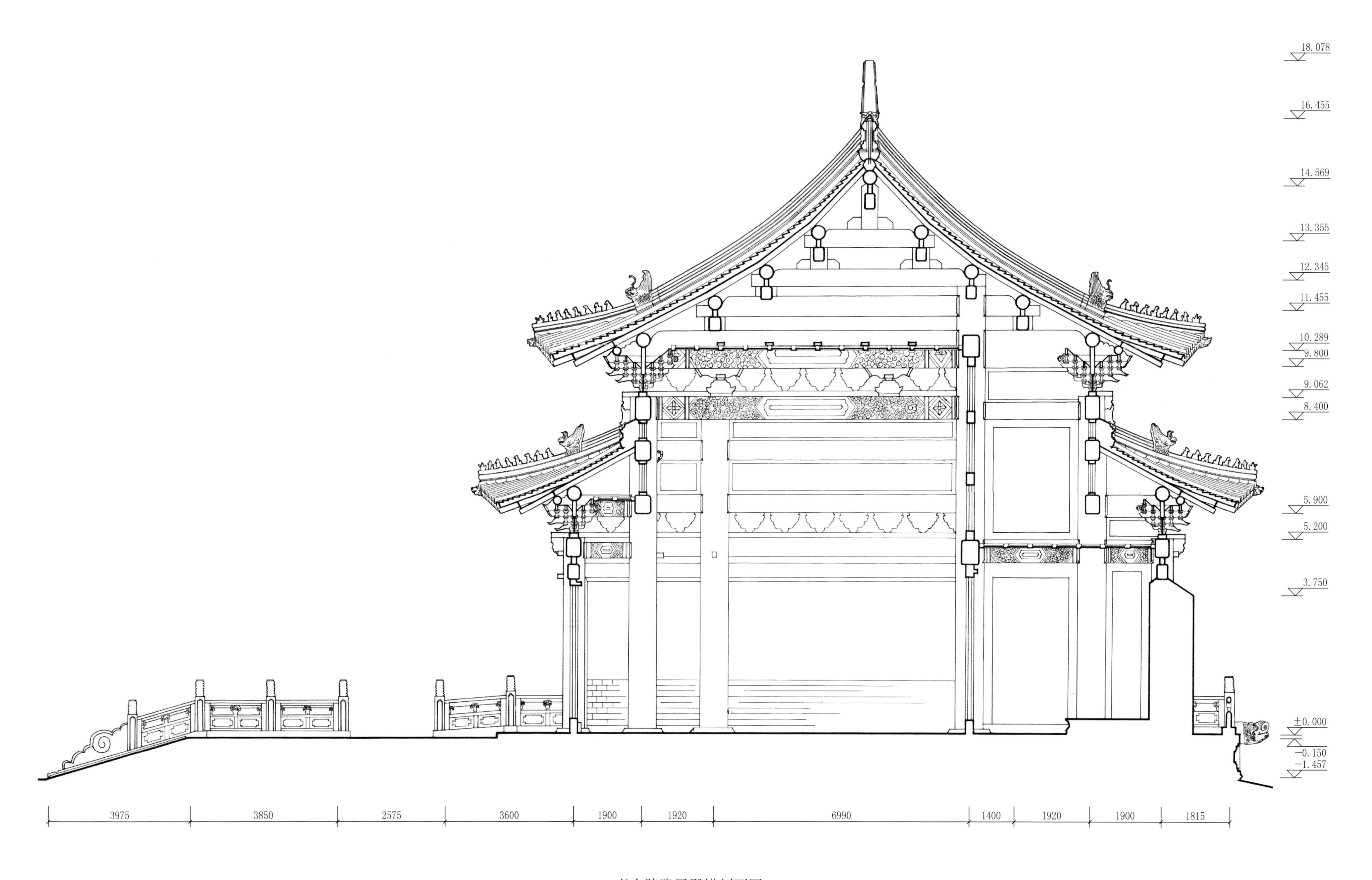

孝东陵隆恩殿横剖面图

Cross section of the Hall of Monumental Grace, East Tomb of Filial Piety

孝东陵隆恩殿纵剖面图

Longitudinal section of the Hall of Monumental Grace, East Tomb of Filial Piety

4180

1760

0 0.1 0.4m

孝东陵隆恩殿丹陛石大样图

Detailed drawing of the marble carving on the steps to the Hall of Monumental Grace, East Tomb of Filial Piety

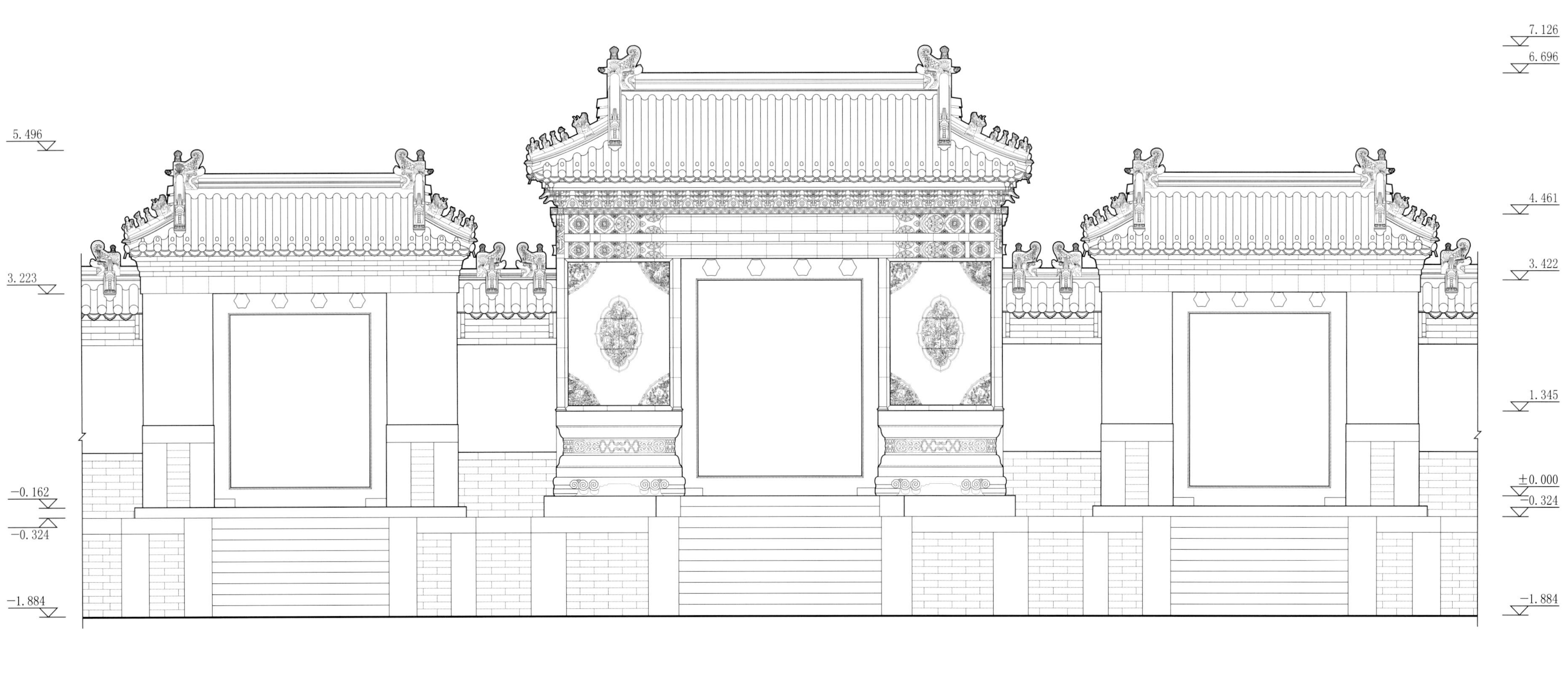

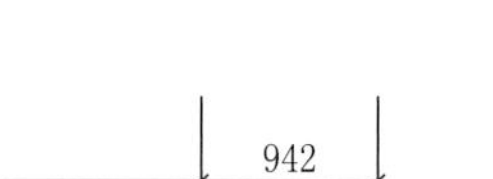

孝东陵琉璃花门正立面图

Front elevation of the Gate with Glazed Roof Tiles, East Tomb of Filial Piety

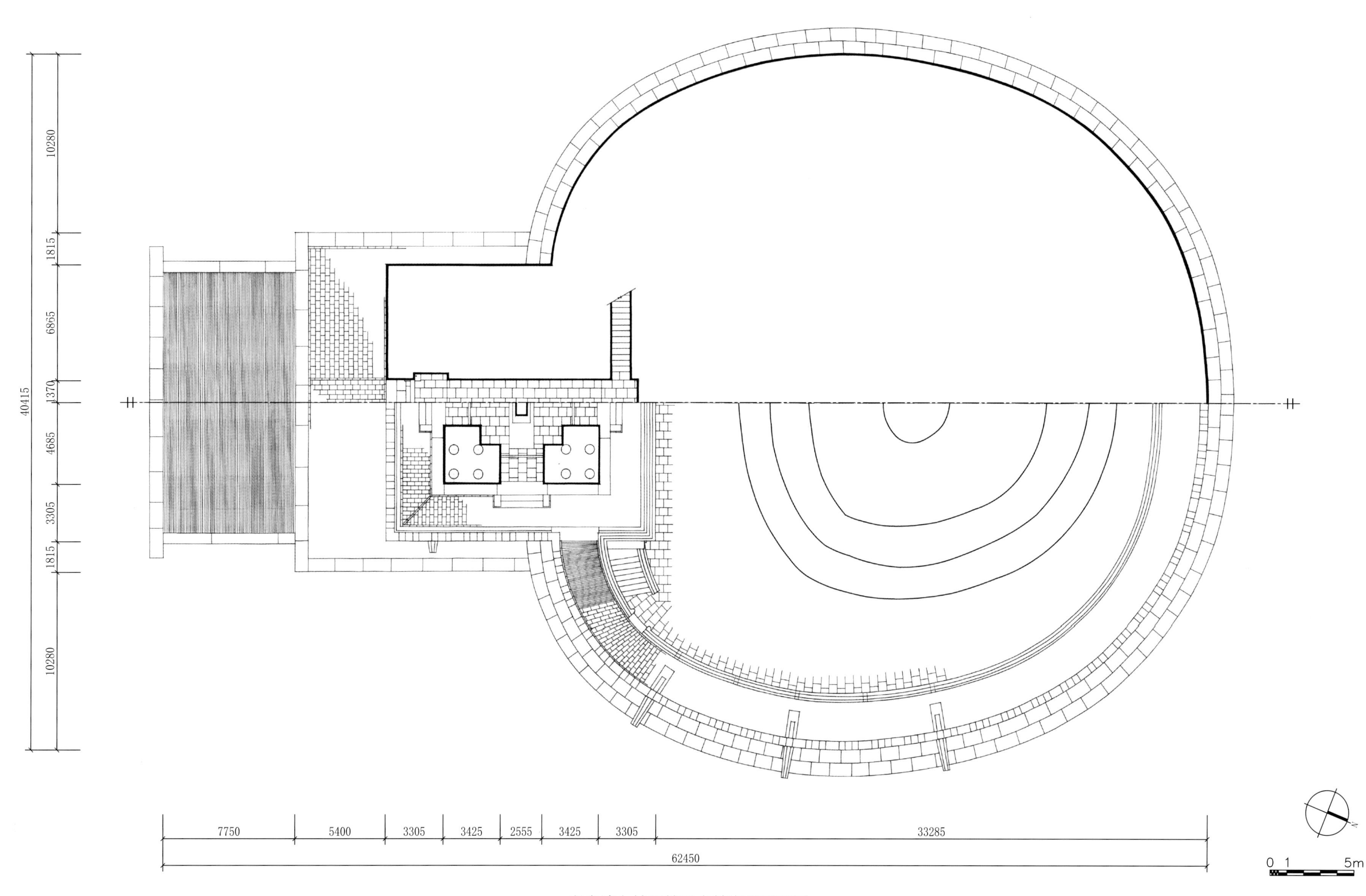

孝东陵方城明楼及宝城宝顶平面图

Plan of the Square Walled Terrace and the Memorial Tower as well as the Encircled Realm of Treasure and the Tumulus, East Tomb of Filial Piety

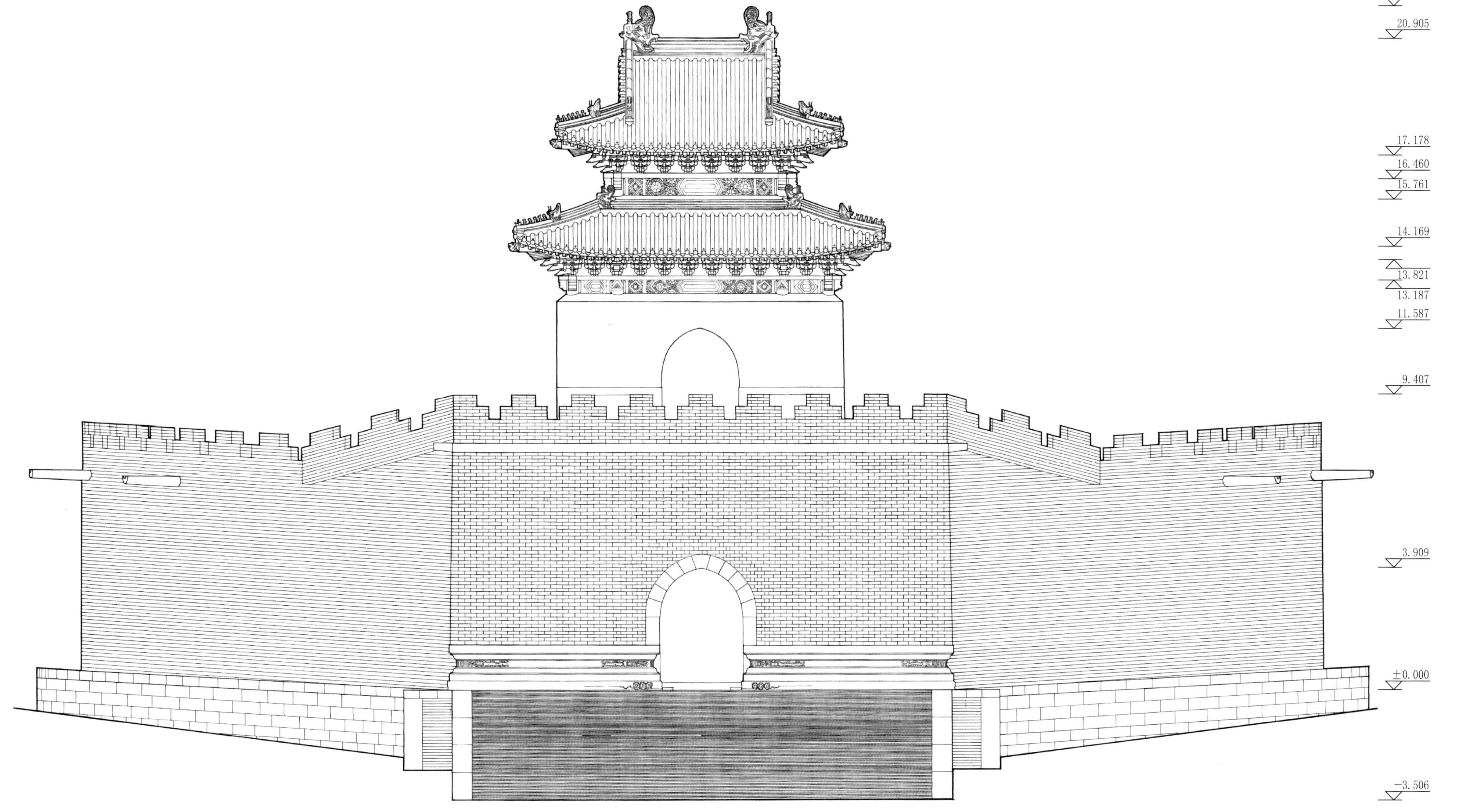

孝东陵方城明楼及宝城宝顶正立面图

Front elevation the Square Walled Terrace and the Memorial Tower as well as the Encircled Realm of Treasure and the Tumulus, East Tomb of Filial Piety

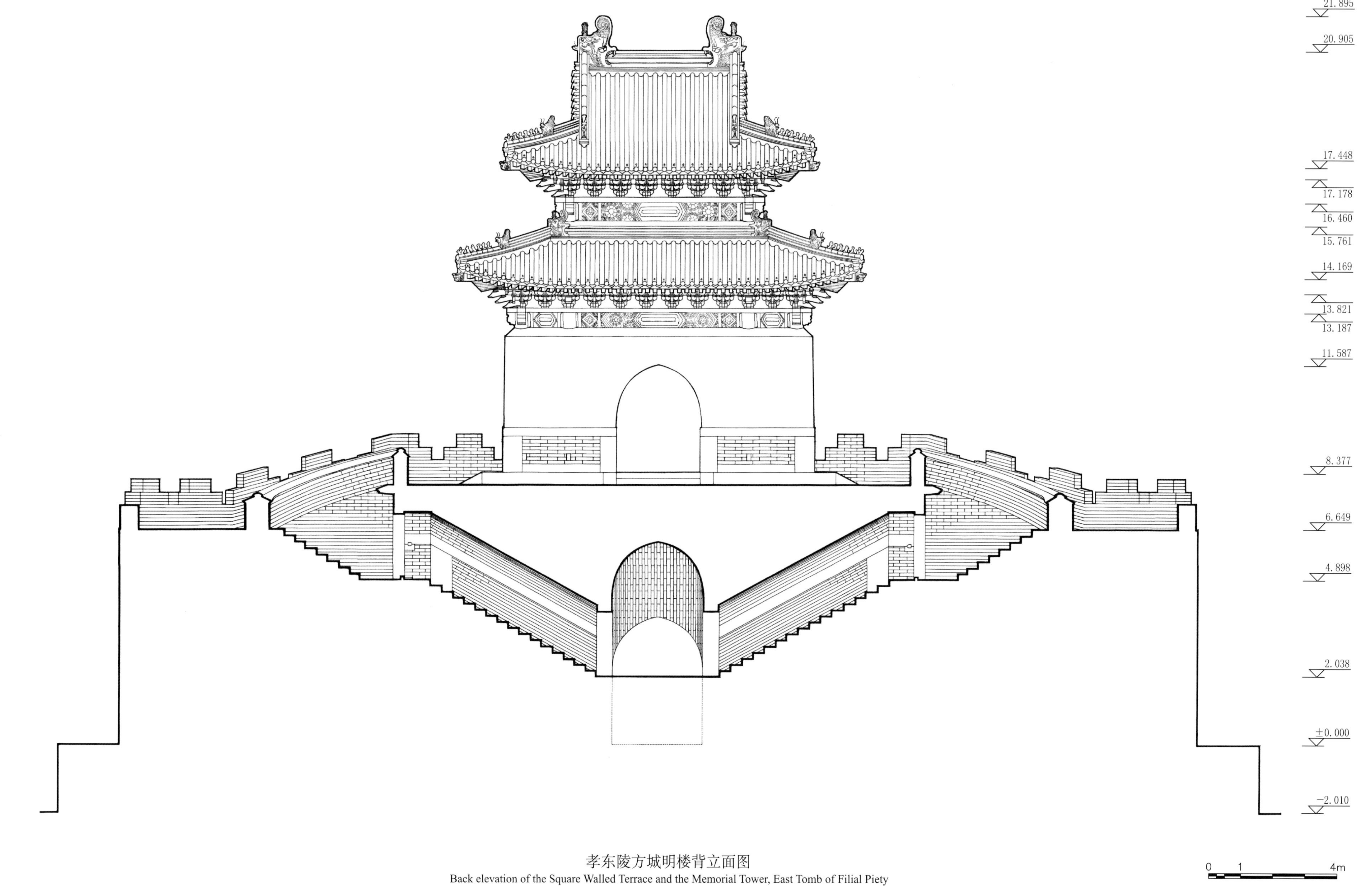

孝东陵方城明楼背立面图

Back elevation of the Square Walled Terrace and the Memorial Tower, East Tomb of Filial Piety

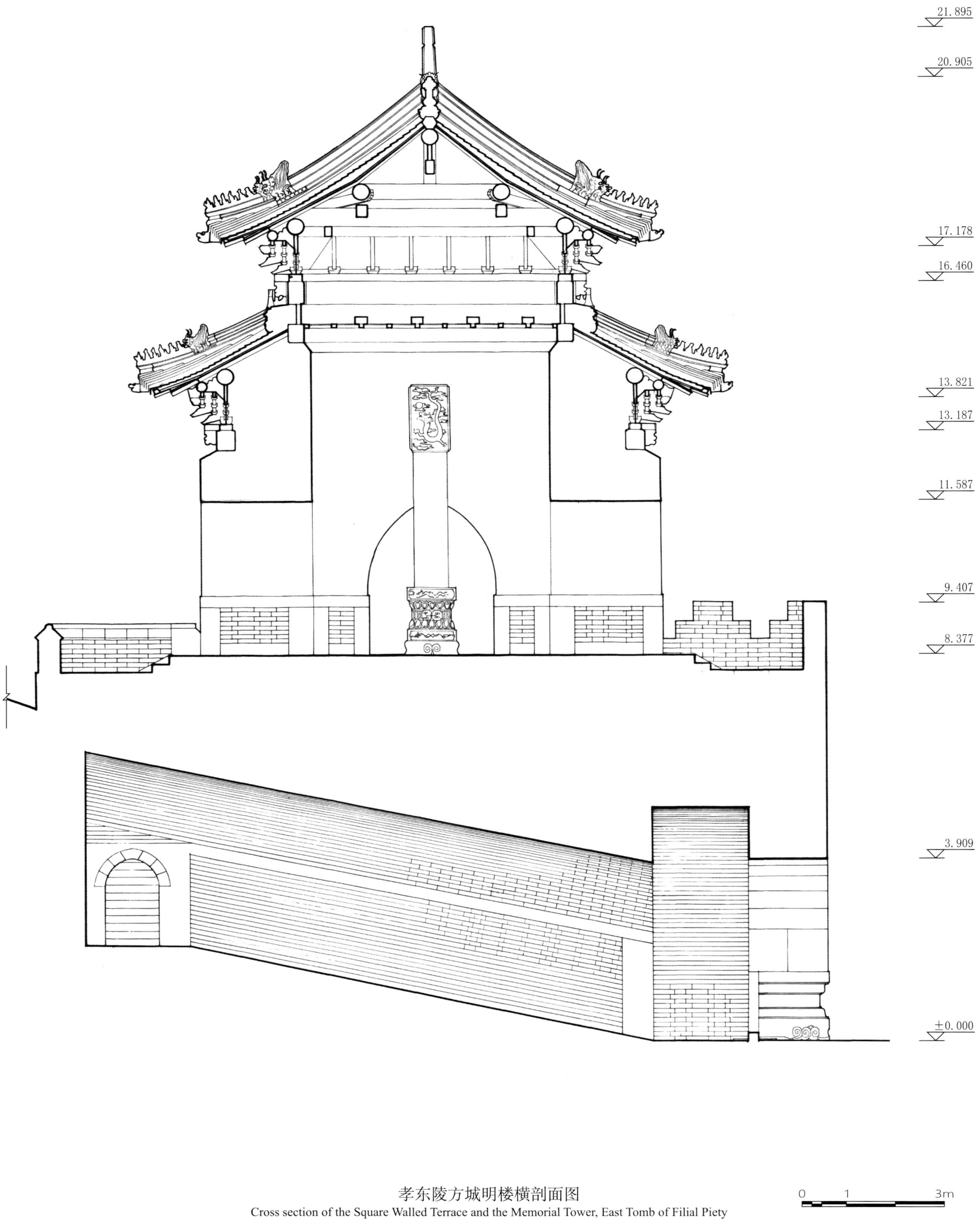

孝东陵方城明楼横剖面图

Cross section of the Square Walled Terrace and the Memorial Tower, East Tomb of Filial Piety

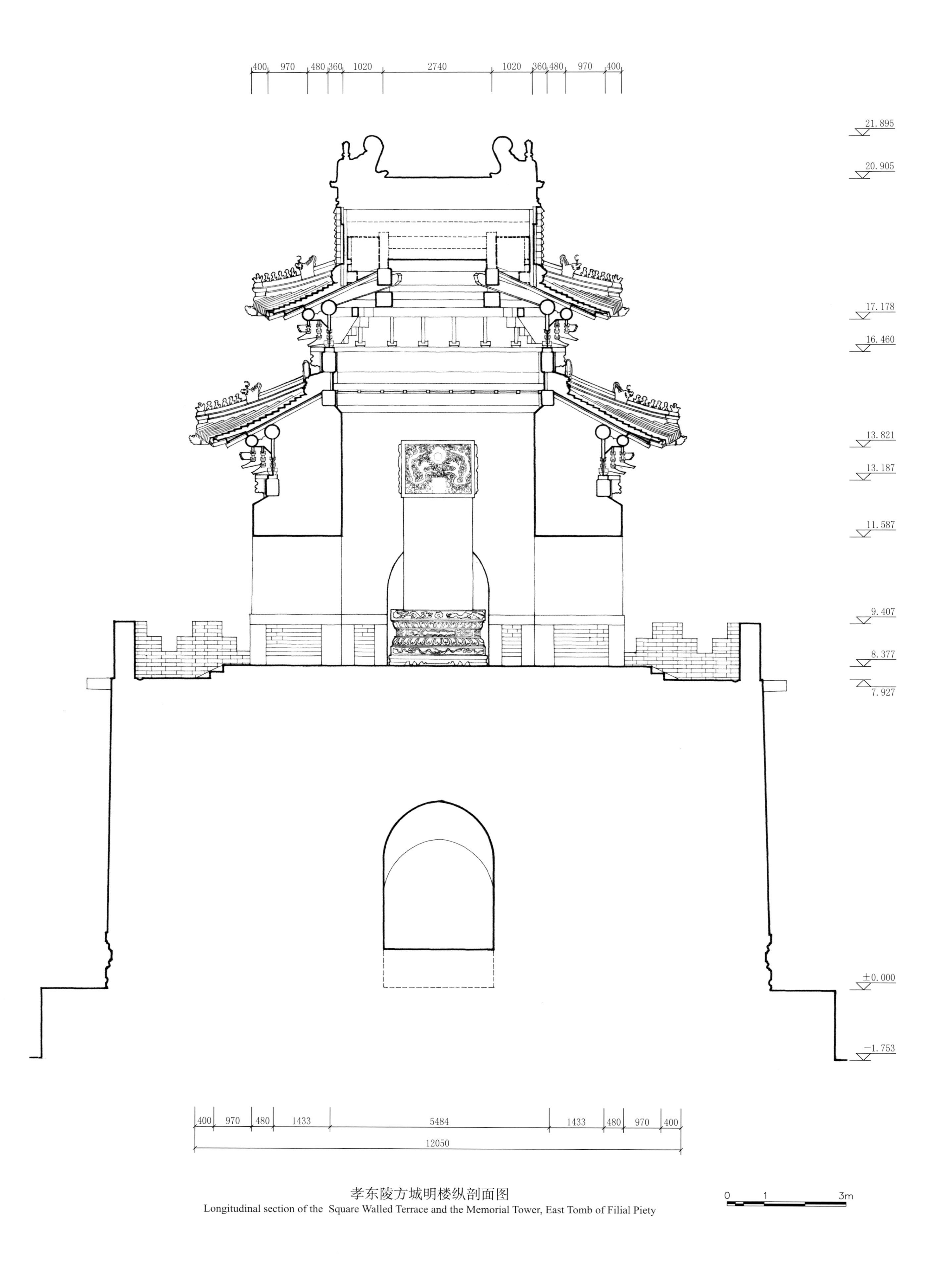

孝东陵方城明楼纵剖面图
Longitudinal section of the Square Walled Terrace and the Memorial Tower, East Tomb of Filial Piety

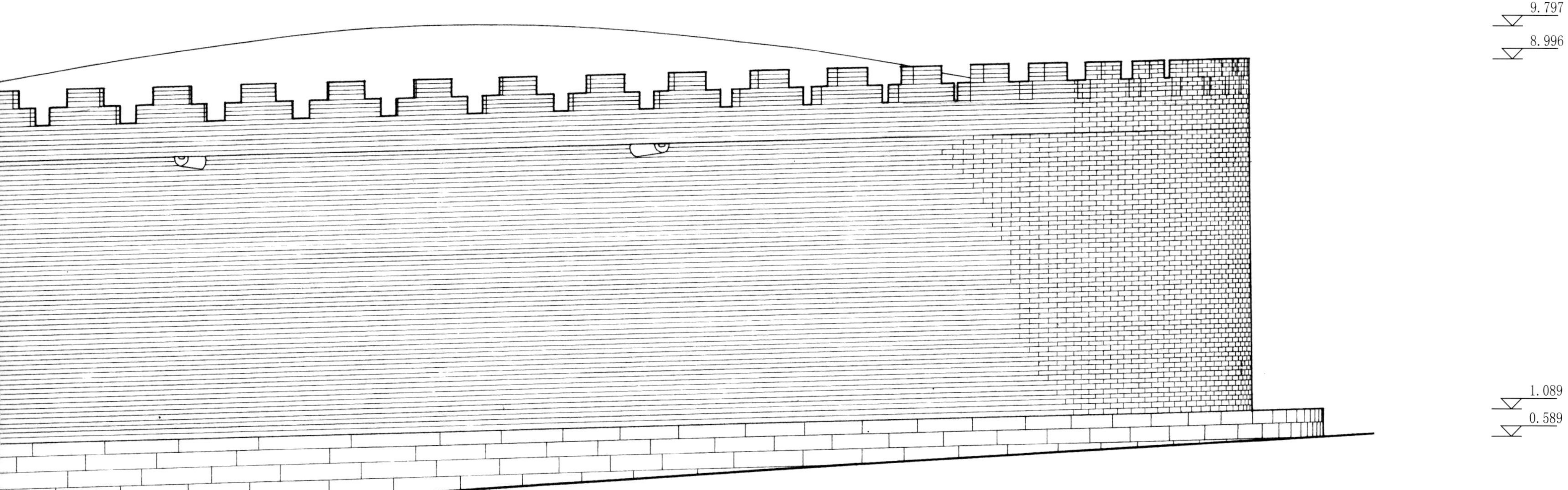
9.797
8.996
1.089
0.589
0 1 3m

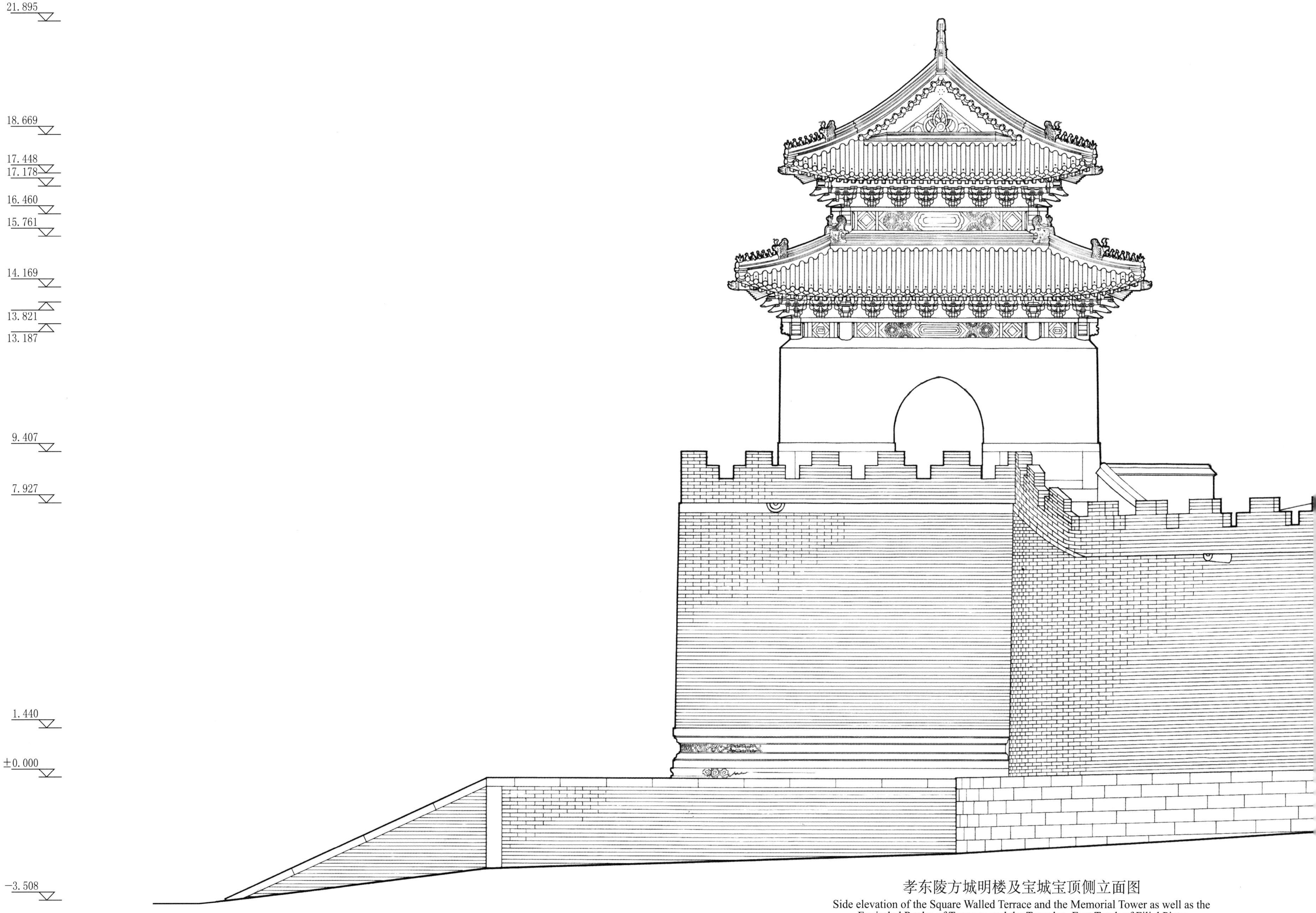

21.895
18.669
17.448
17.178
16.460
15.761
14.169
13.821
13.187
9.407
7.927
1.440
±0.000
−3.508

孝东陵方城明楼及宝城宝顶侧立面图

Side elevation of the Square Walled Terrace and the Memorial Tower as well as the Encircled Realm of Treasure and the Tumulus, East Tomb of Filial Piety

景陵

Tomb of Admiration (Jingling)

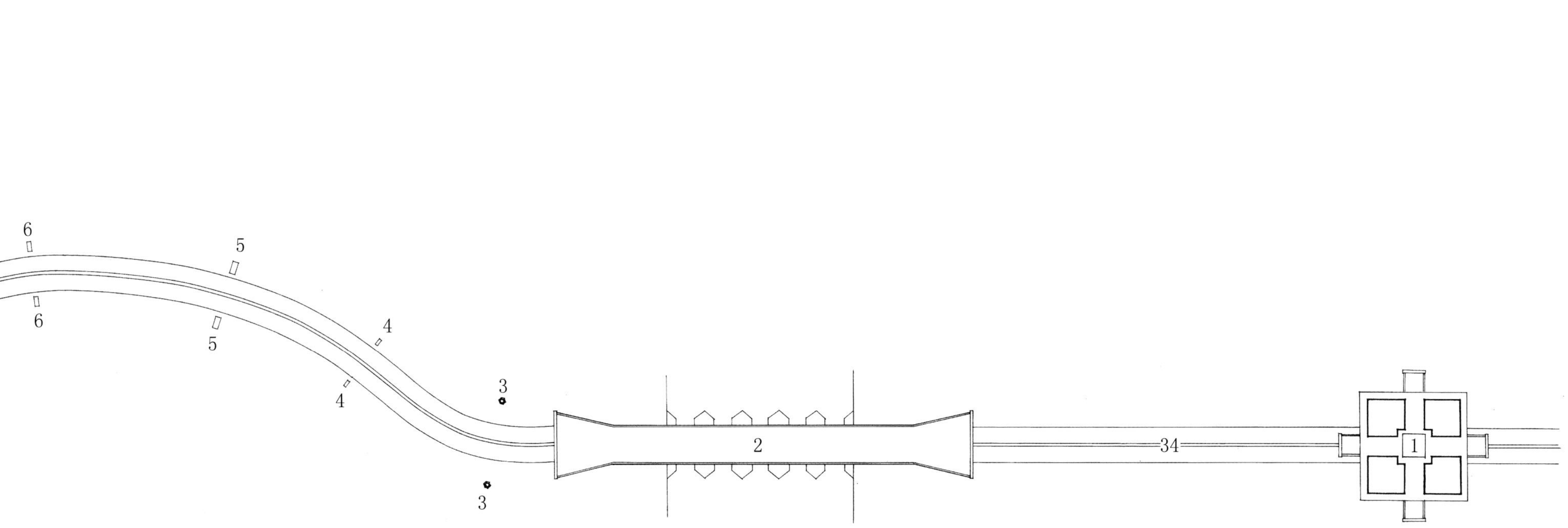

1 圣德神功碑亭 Pavilion for the Stela of Sagely Virtue and Divine Merit
2 五孔石券桥 Five-Arch Stone Bridge
3 望柱 Ornamental Column
4 立狮 Standing Lion
5 立象 Standing Elephant
6 立马 Standing Horse
7 武将 Martial Official
8 文臣 Literary Official
9 下马牌 Stela Marking the Place for Dismounting from one's Horse
10 神厨库院落 Culinary Courtyard for Sacrifices
11 牌坊 Marble Memorial Gateway
12 神道碑亭 Pavilion for the Stela on the Spirit Way
13 朝房 Reception Halls for Court Officials
14 三路三孔桥 Three-Way Three-Arch Bridges
15 隆恩门 Gate of Monumental Grace
16 配殿 Side Hall
17 隆恩殿 Hall of Monumental Grace
18 琉璃花门 Gate with Glazed Roof Tiles
19 二柱门 Gate with Two Columns
20 石五供 Five Stone Ritual Vessels
21 方城 Square Walled Terrace
22 明楼 Memorial Tower
23 哑巴院 Courtyard of the Mute
24 月牙城 Crescent Wall
25 琉璃影壁 Screen Wall of Glazed Tiles
26 宝顶 Tumulus
27 宝城 Encircled Realm of Treasure
28 罗圈墙 Wall Surrounding the Tomb
29 卡子墙 Spacer Wall
30 神厨库院门 Culinary Courtyard Gate for Sacrifices
31 神厨 Sacrificial Kitchen
32 神库 Sacrificial Storehouse
33 宰牲亭 Ritual Abattoir
34 神道 Spirit Way

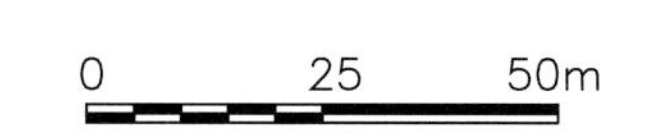

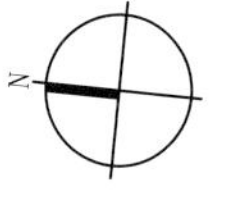

景陵总剖面图
Site section of the Tomb of Admiration

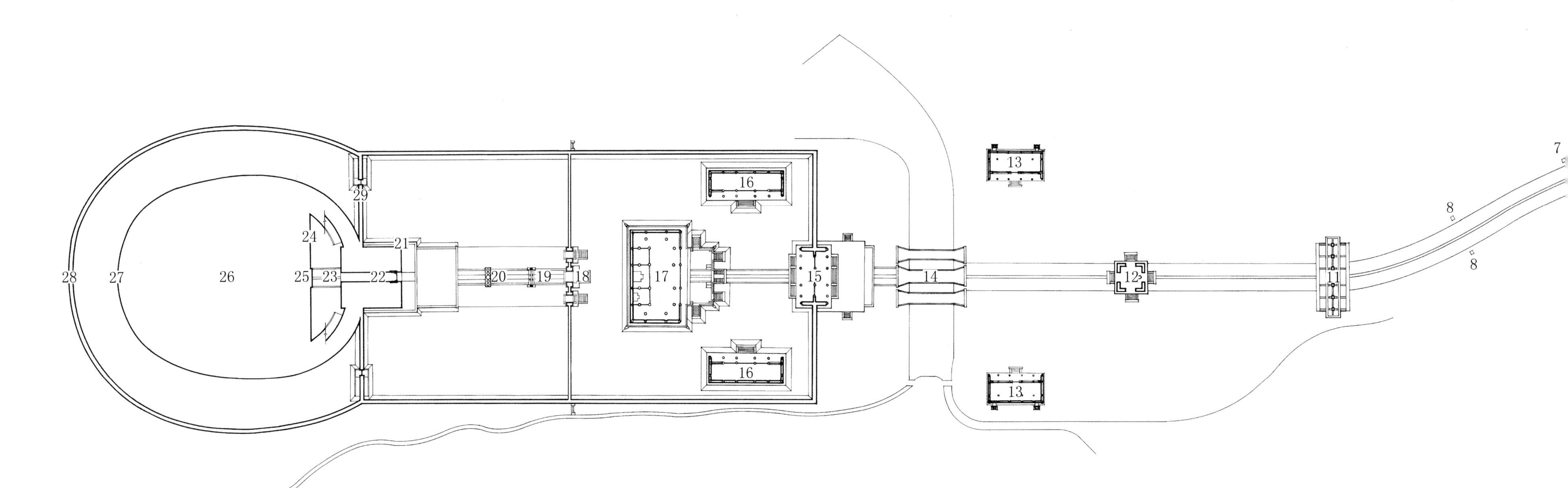

景陵总平面图
Site plan of the Tomb of Admiration

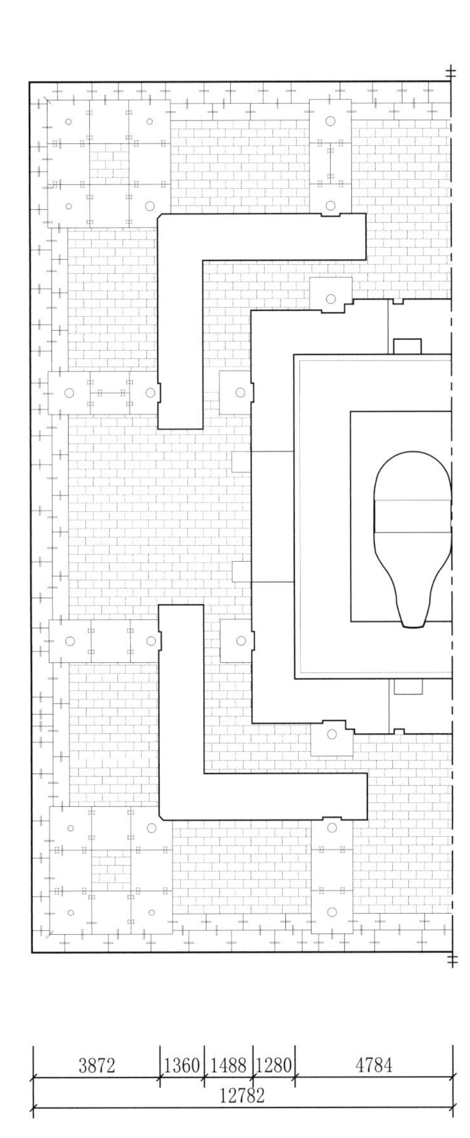

景陵圣德神功碑亭墩台平面图
Plan of the platform beneath the Pavilion for the Stela of Sage Virtue and Divine Merit, Tomb of Admiration

840 4750 1735 8136 2016 5216 2016 8136 1735 4750 840

38490

0 1 5m

景陵圣德神功碑亭底层平面图
Ground plan of the Pavilion for the Stela of Sage Virtue and Divine Merit, Tomb of Admiration

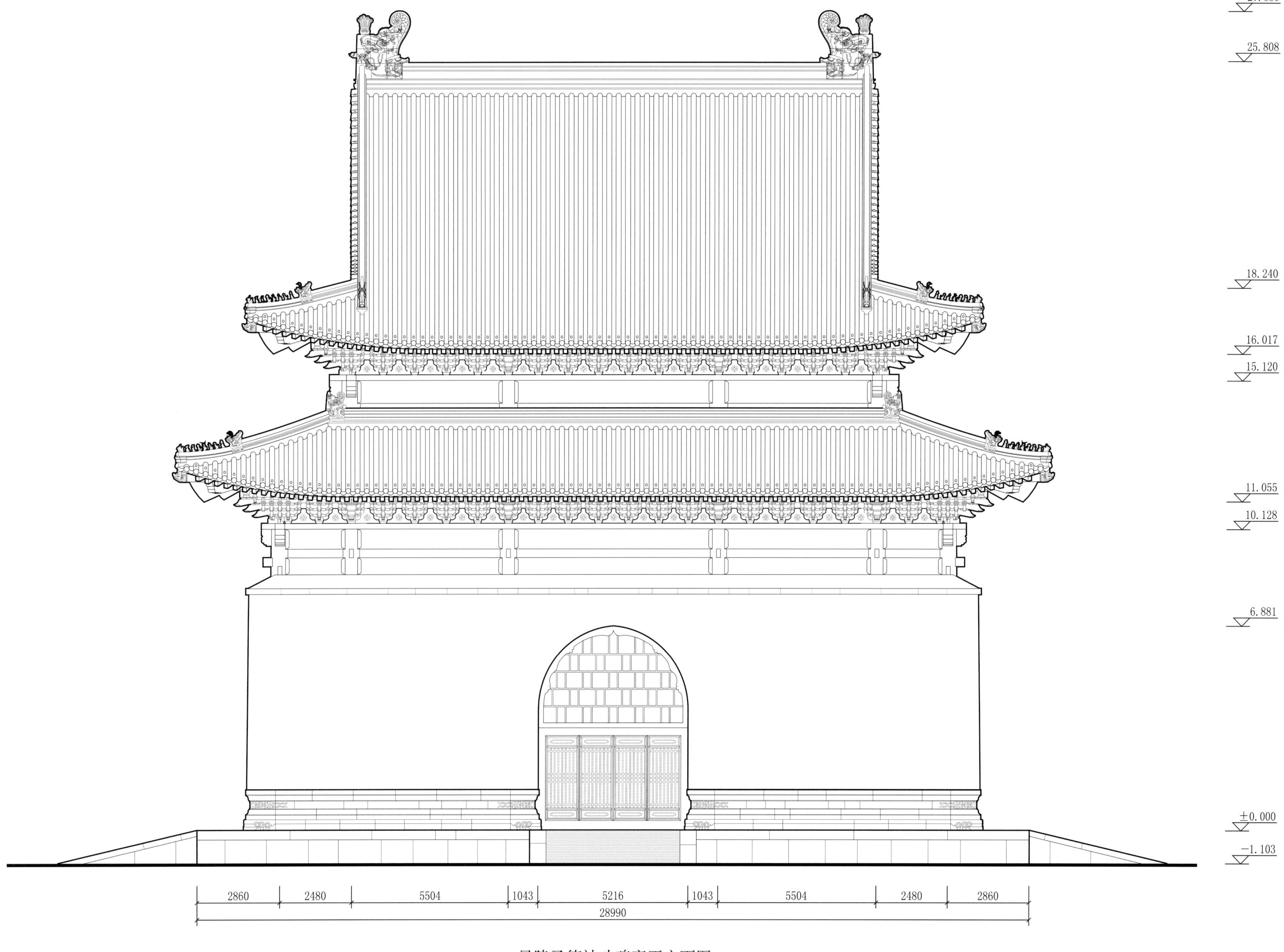

景陵圣德神功碑亭正立面图

Front elevation of the Pavilion for the Stela of Sage Virtue and Divine Merit, Tomb of Admiration

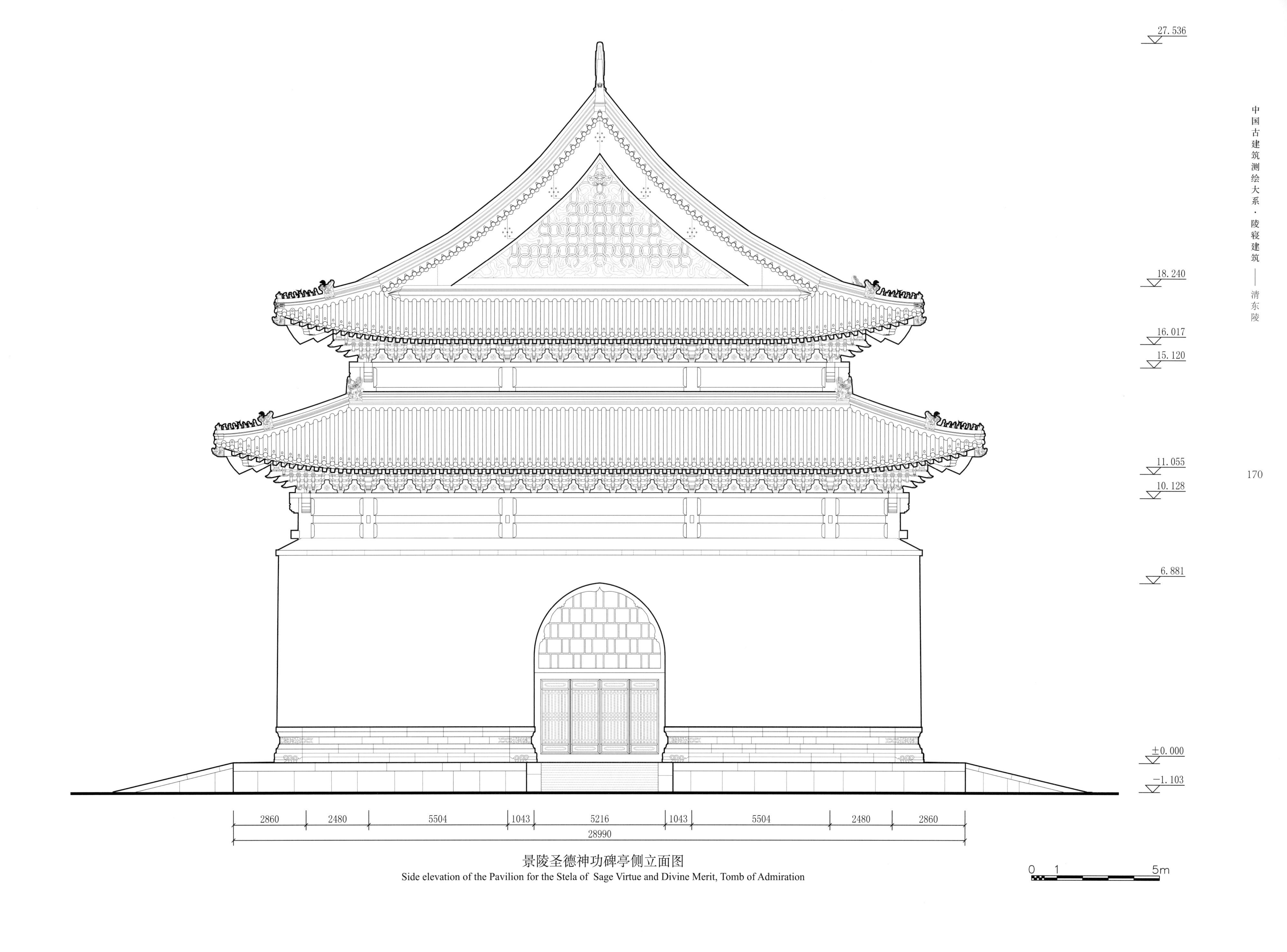

景陵圣德神功碑亭侧立面图

Side elevation of the Pavilion for the Stela of Sage Virtue and Divine Merit, Tomb of Admiration

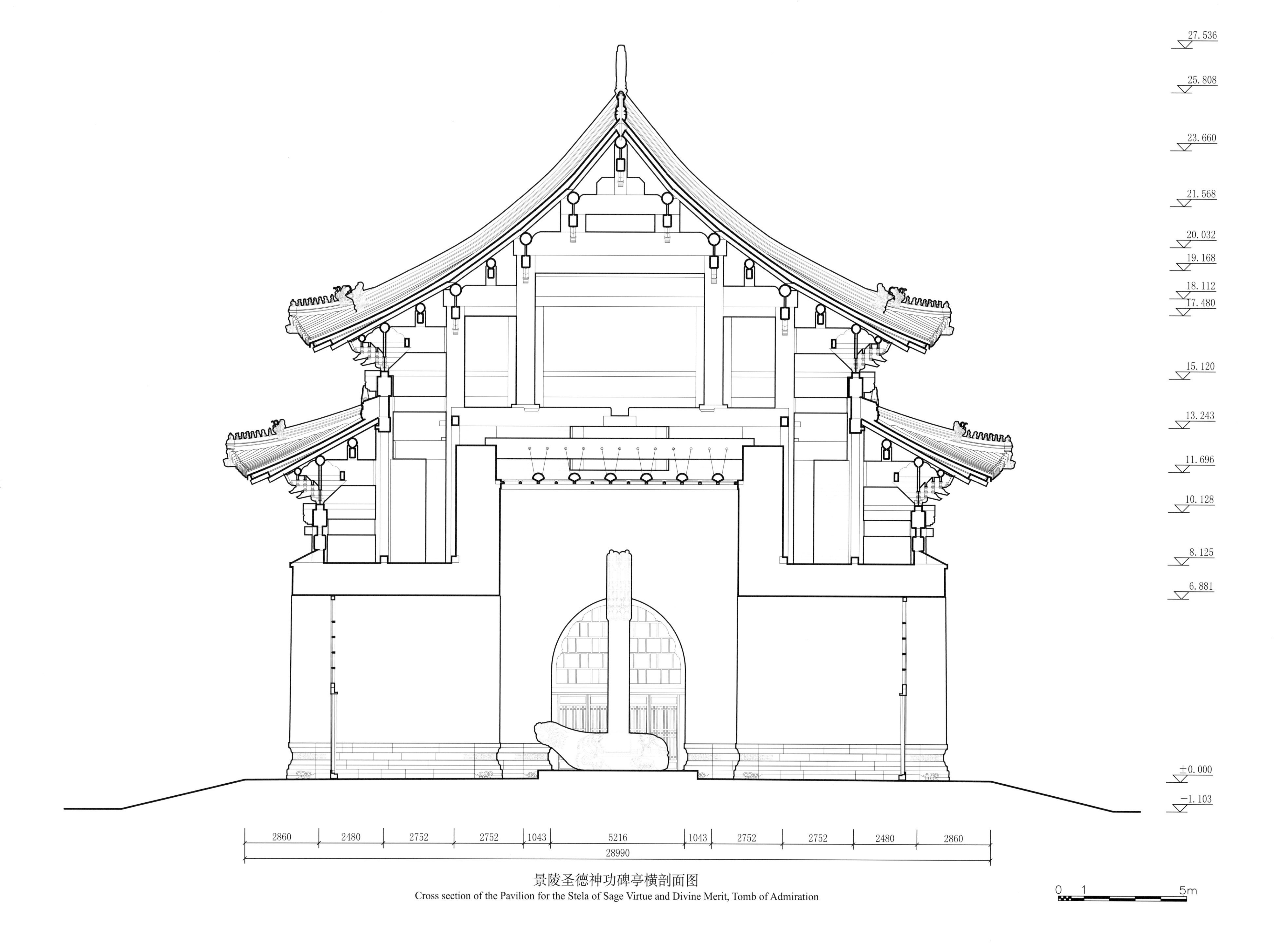

景陵圣德神功碑亭横剖面图

Cross section of the Pavilion for the Stela of Sage Virtue and Divine Merit, Tomb of Admiration

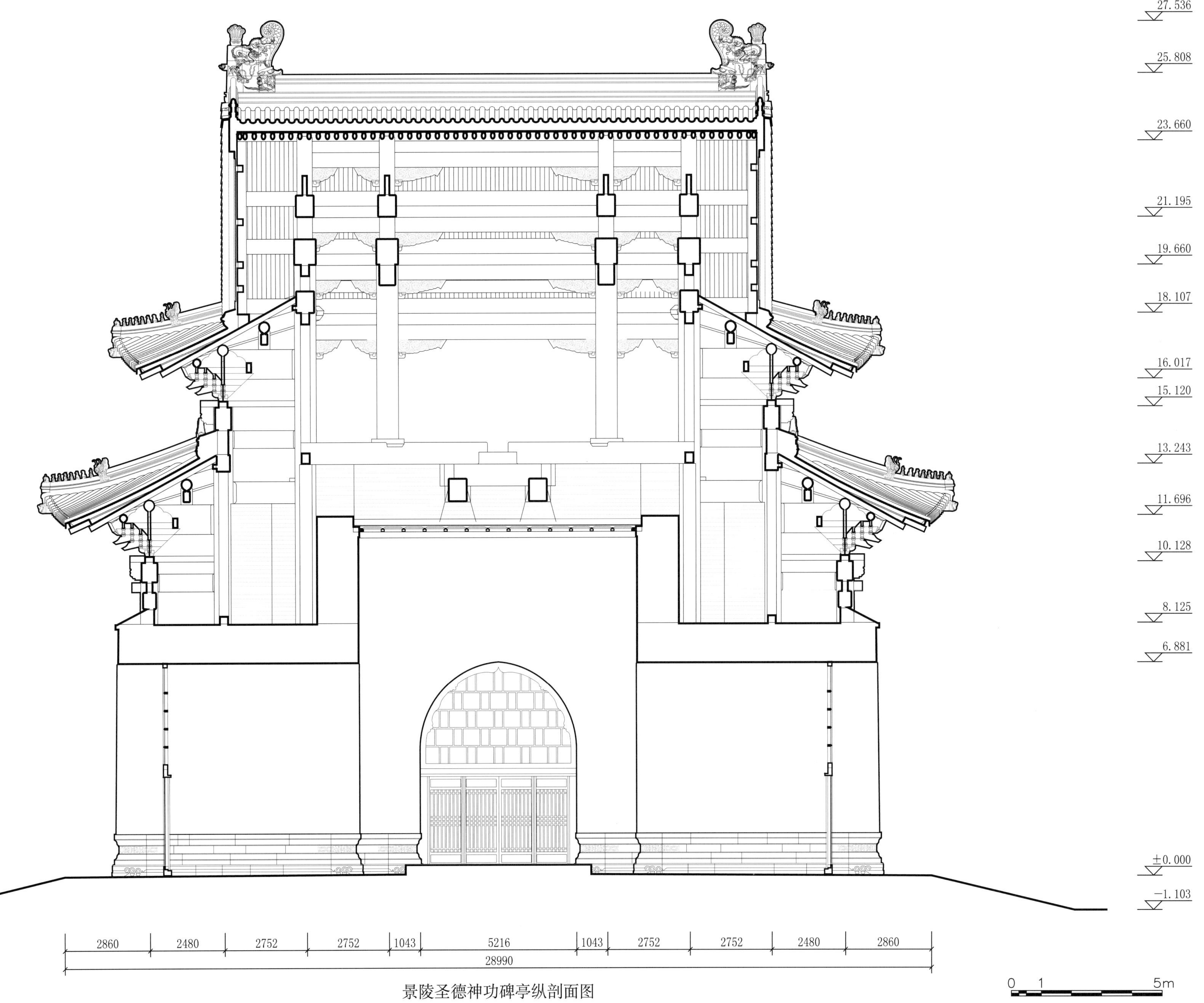

景陵圣德神功碑亭纵剖面图

Longitudinal section of the Pavilion for the Stela of Sage Virtue and Divine Merit, Tomb of Admiration

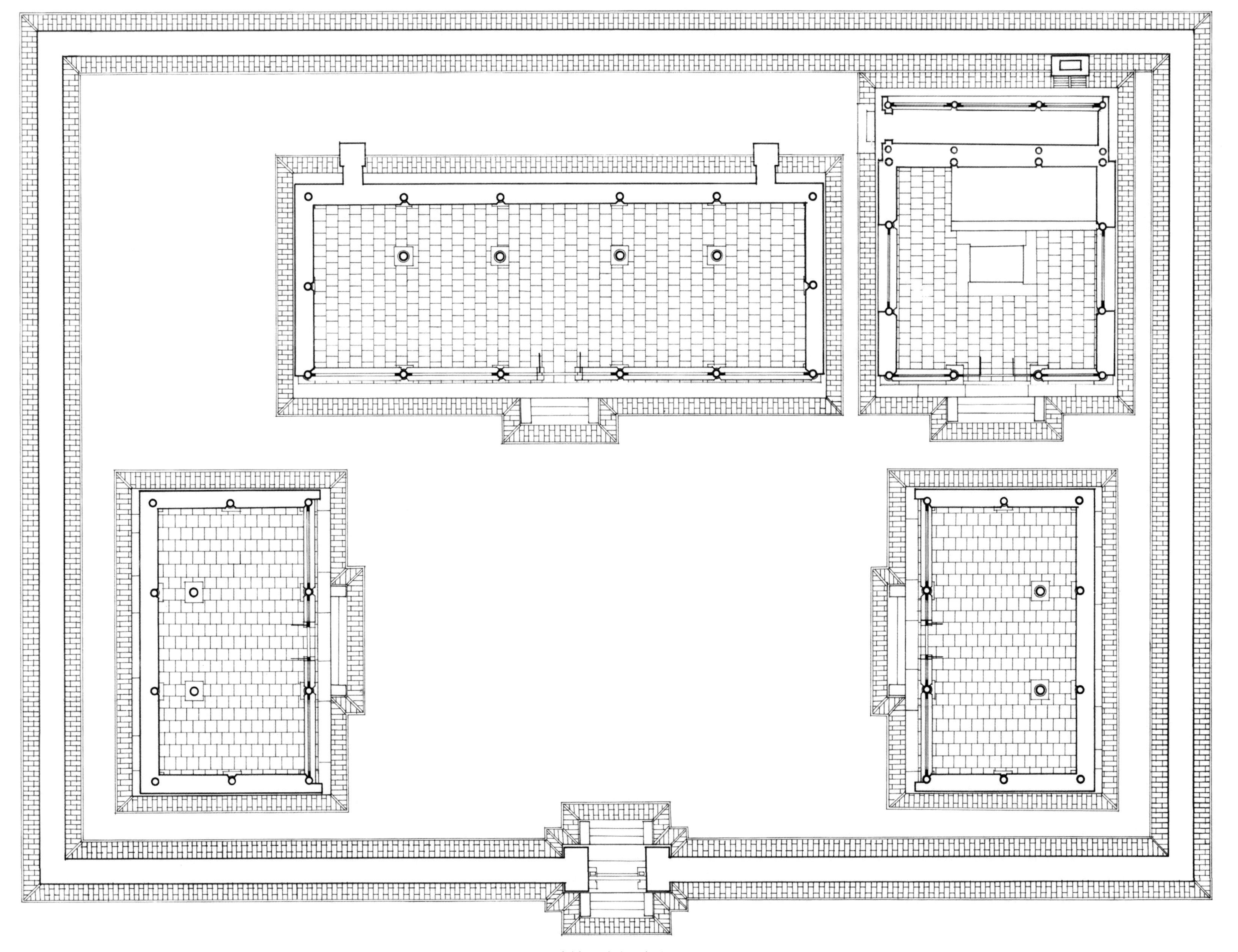

景陵神厨库组群平面图
Plan of the building complex of the Culinary Courtyard for Sacrifices, Tomb of Admiration

景陵神厨库宰牲亭正立面图

Front elevation of the Ritual Abattoir in the Culinary Courtyard for Sacrifices, Tomb of Admiration

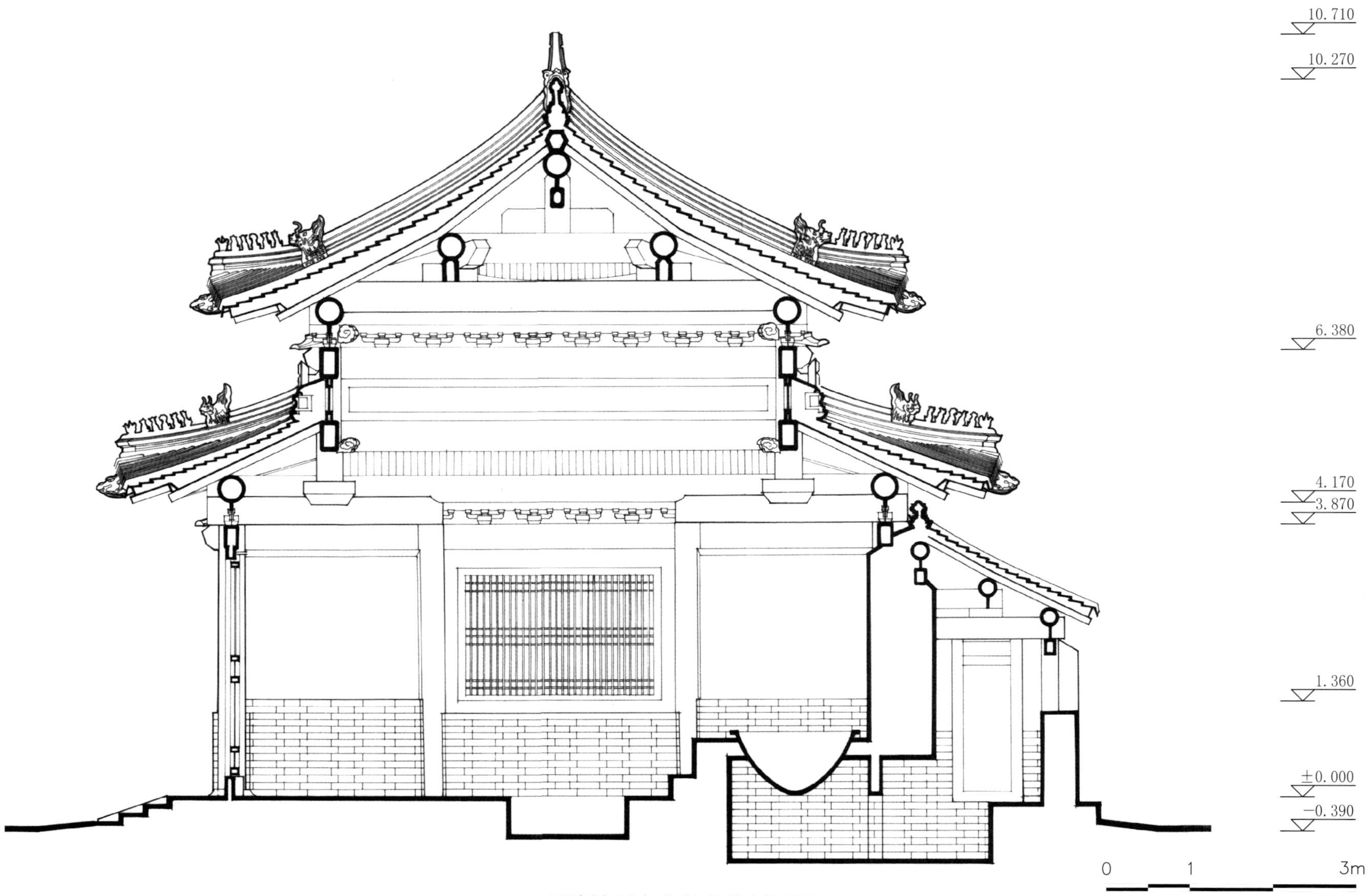

景陵神厨库宰牲亭横剖面图

Cross section of the Ritual Abattoir in the Culinary Courtyard for Sacrifices, Tomb of Admiration

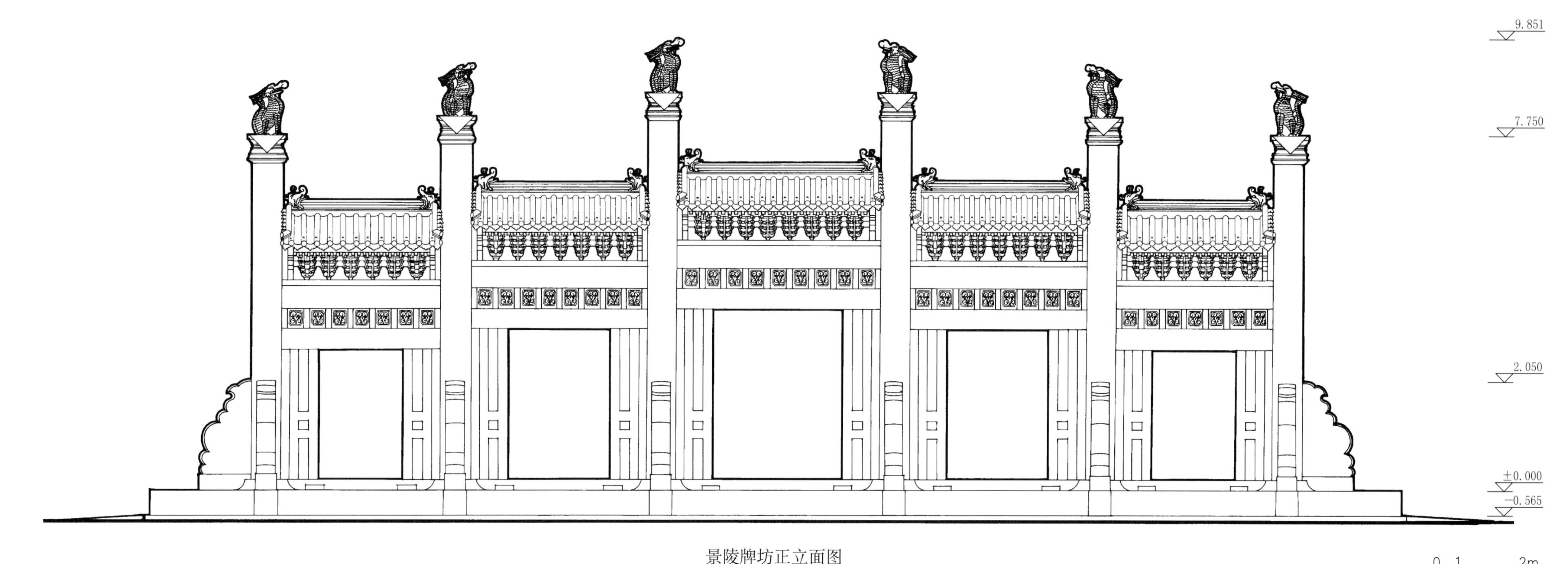

景陵牌坊正立面图

Front elevation of the Memorial Gateway, Tomb of Admiration

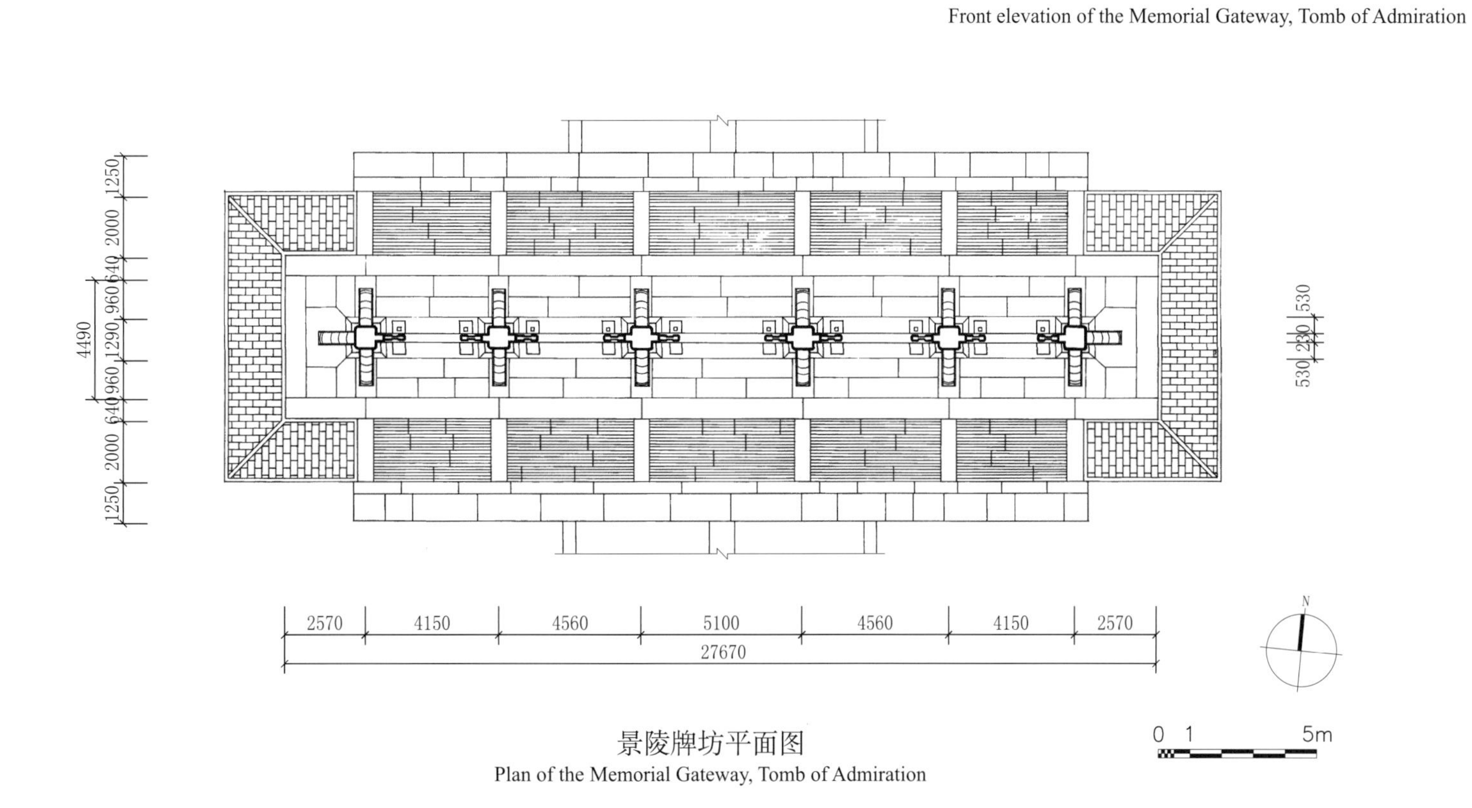

景陵牌坊平面图

Plan of the Memorial Gateway, Tomb of Admiration

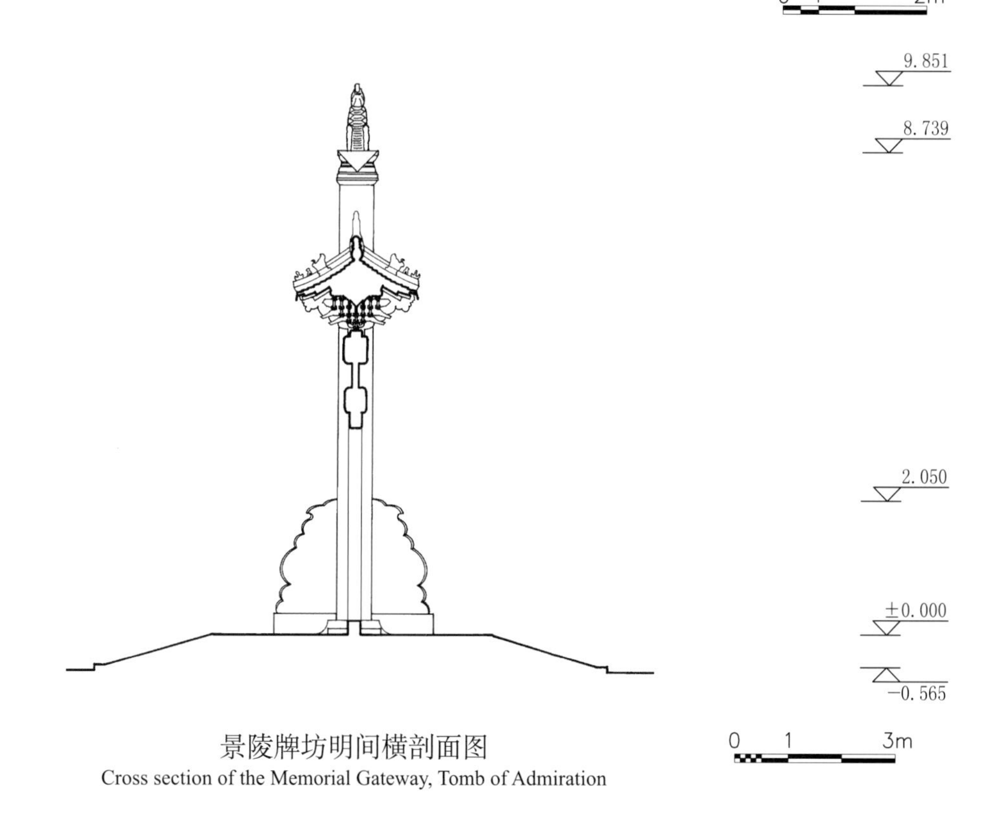

景陵牌坊明间横剖面图

Cross section of the Memorial Gateway, Tomb of Admiration

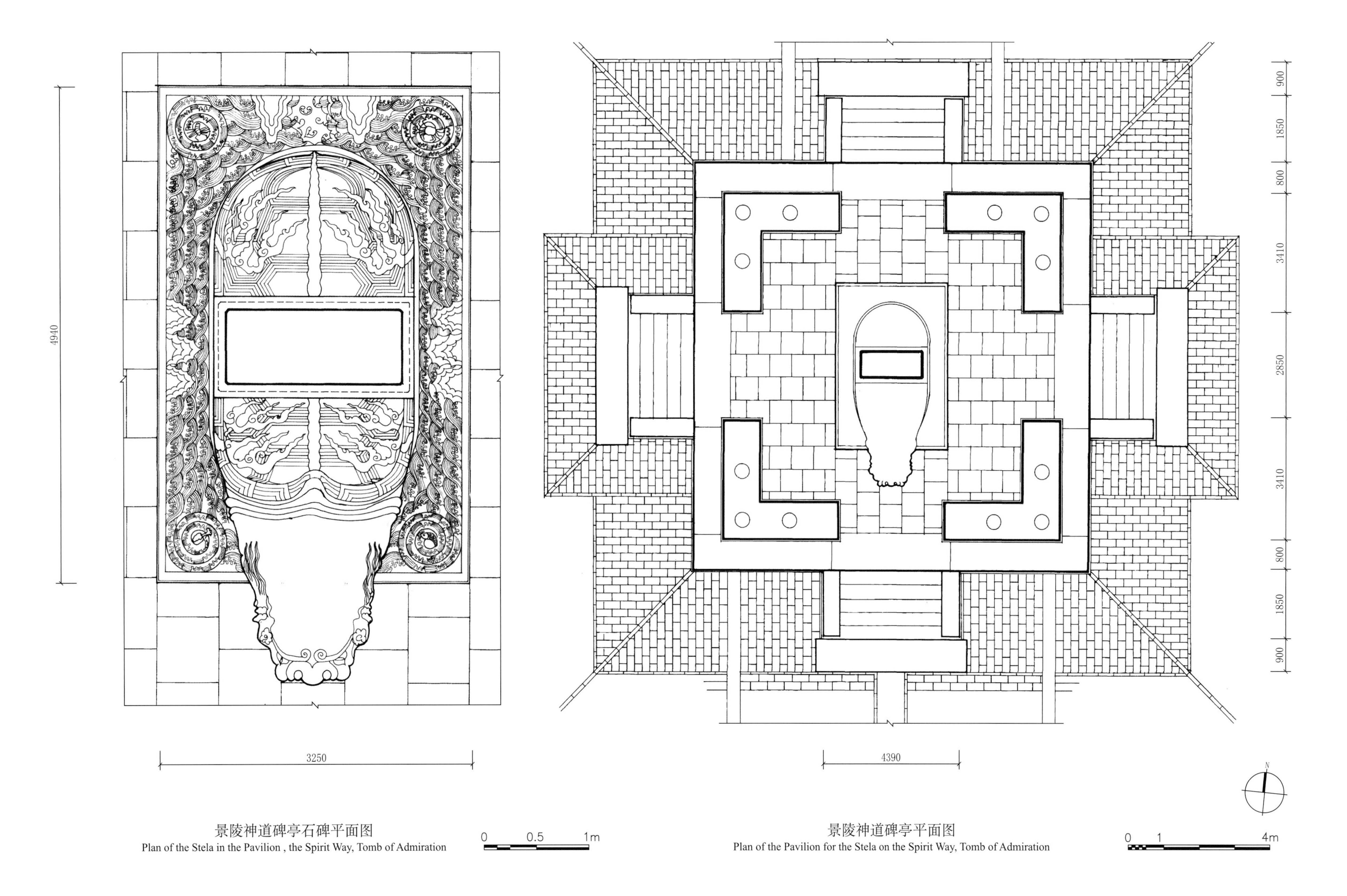

景陵神道碑亭石碑平面图
Plan of the Stela in the Pavilion , the Spirit Way, Tomb of Admiration

景陵神道碑亭平面图
Plan of the Pavilion for the Stela on the Spirit Way, Tomb of Admiration

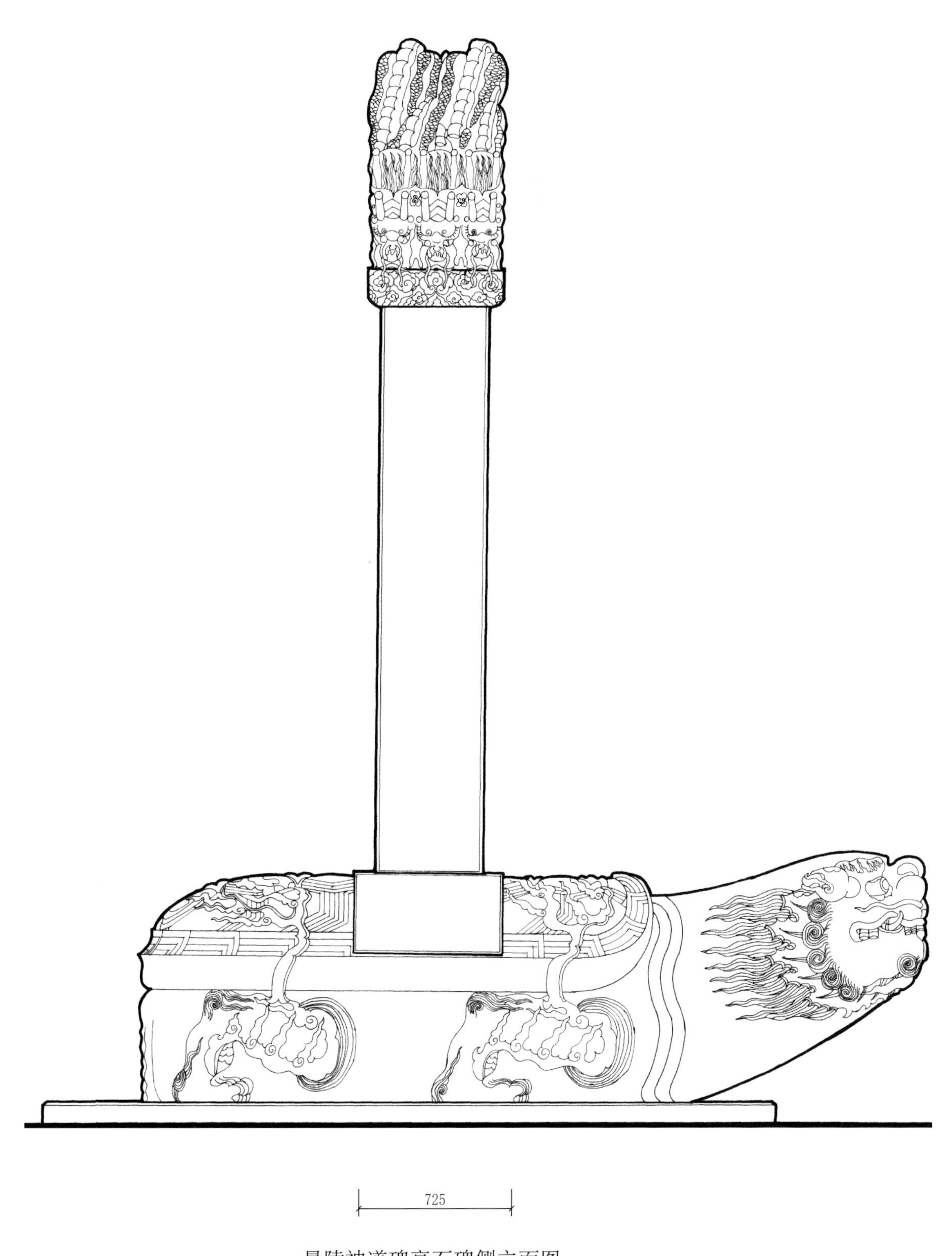

景陵神道碑亭石碑侧立面图

Side elevation of the Stela in the Pavilion on the Spirit Way, Tomb of Admiration

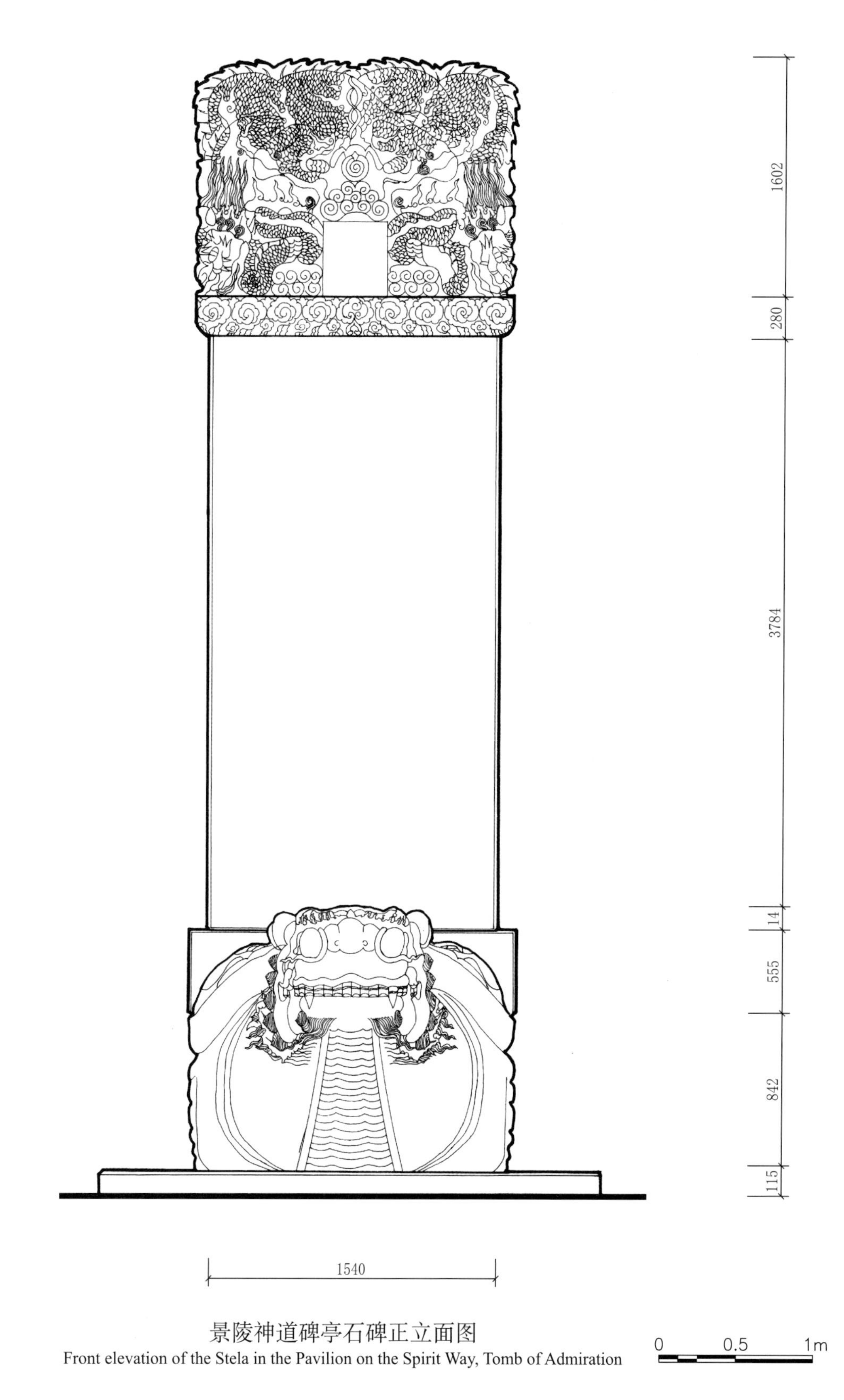

景陵神道碑亭石碑正立面图

Front elevation of the Stela in the Pavilion on the Spirit Way, Tomb of Admiration

景陵神道碑亭正立面图
Front elevation of the Pavilion for the Stela, the Spirit Way, Tomb of Admiration

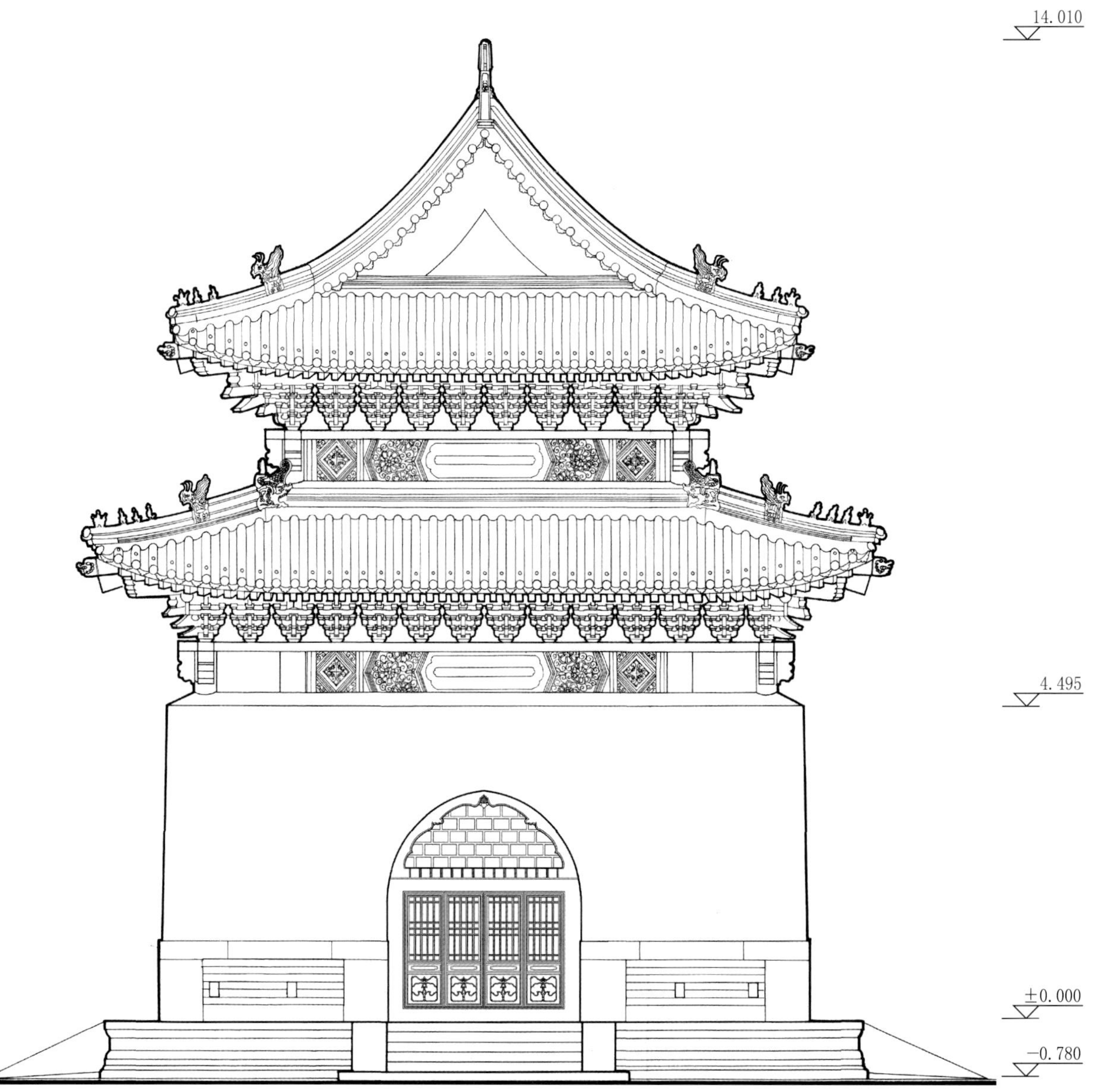

景陵神道碑亭侧立面图
Side elevation of the Pavilion for the Stela on the Spirit Way, Tomb of Admiration

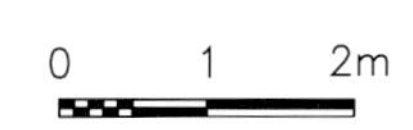

景陵神道碑亭纵剖面图
Longitudinal section of the Pavilion for the Stela on the Spirit Way, Tomb of Admiration

景陵神道碑亭横剖面图
Cross section of the Pavilion for the Stela on the Spirit Way, Tomb of Admiration

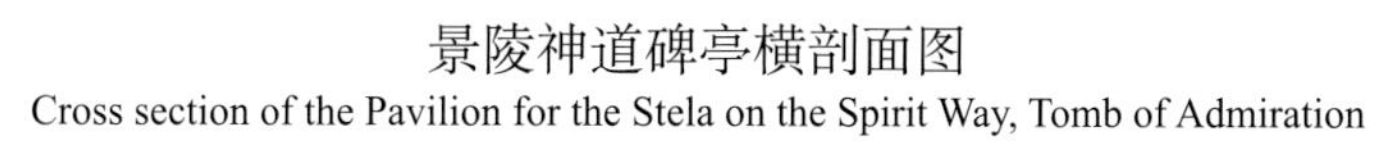

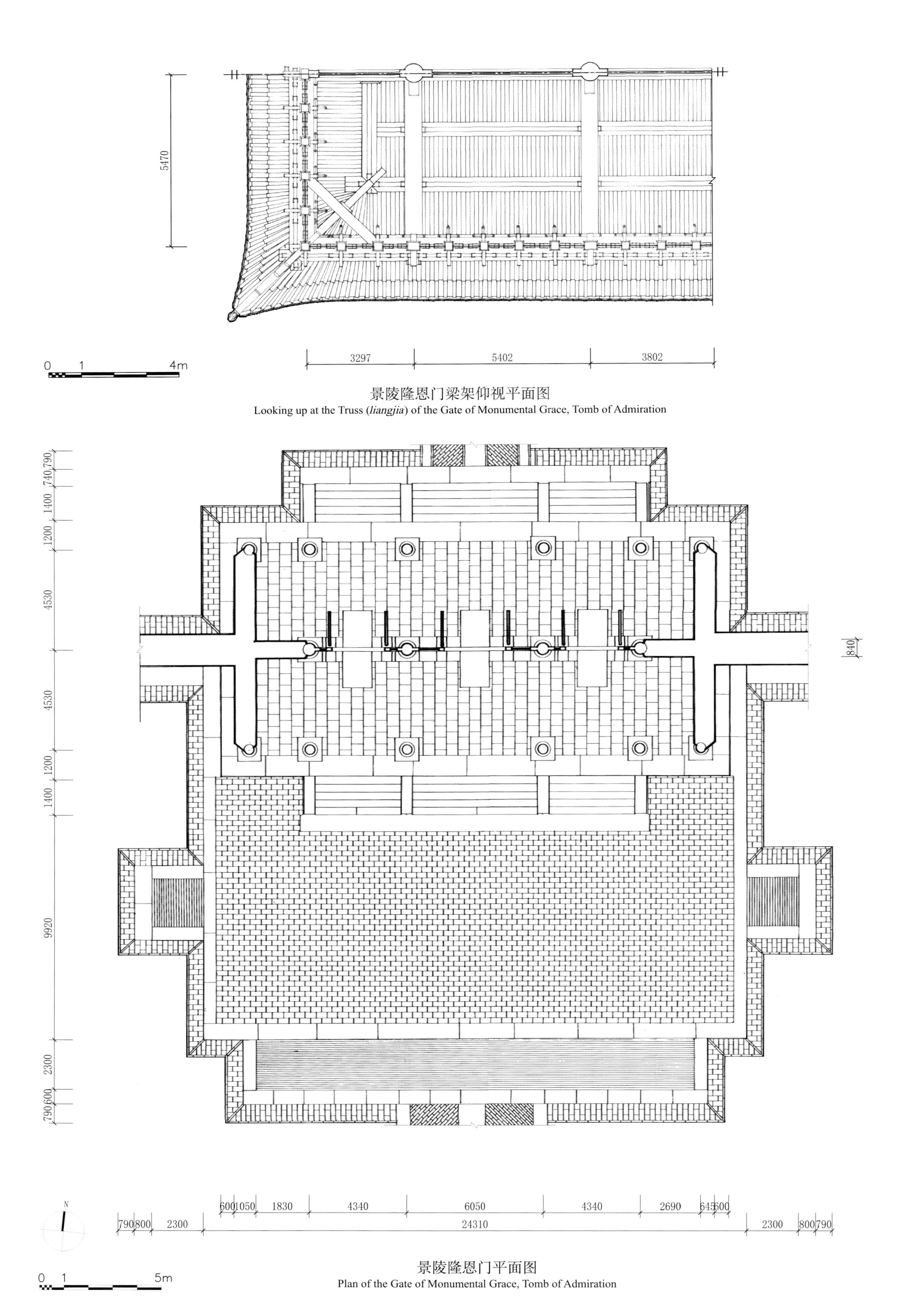

景陵隆恩门梁架仰视平面图
Looking up at the Truss (*liangjia*) of the Gate of Monumental Grace, Tomb of Admiration

景陵隆恩门平面图
Plan of the Gate of Monumental Grace, Tomb of Admiration

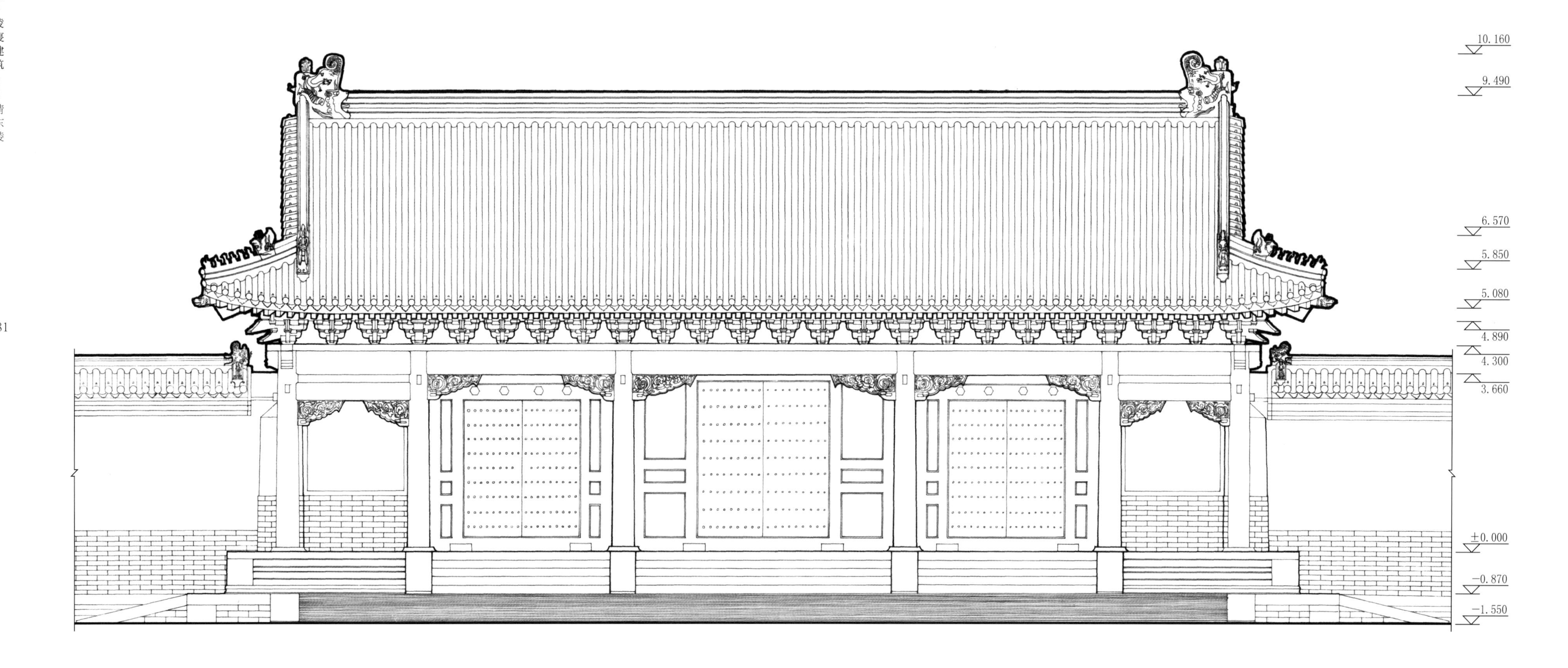

景陵隆恩门正立面图

Front elevation of the Gate of Monumental Grace, Tomb of Admiration

10.160
8.360
6.884
6.000
5.060
4.300
±0.000

景陵隆恩门明间横剖面图
Cross section of the Central Chamber within the Gate of Monumental Grace, Tomb of Admiration

10.160
8.360
6.884
6.000
4.890
4.300
±0.000

景陵隆恩门梢间横剖面图
Cross section of the Chamber at the end within the Gate of Monumental Grace, Tomb of Admiration

0 1 2m

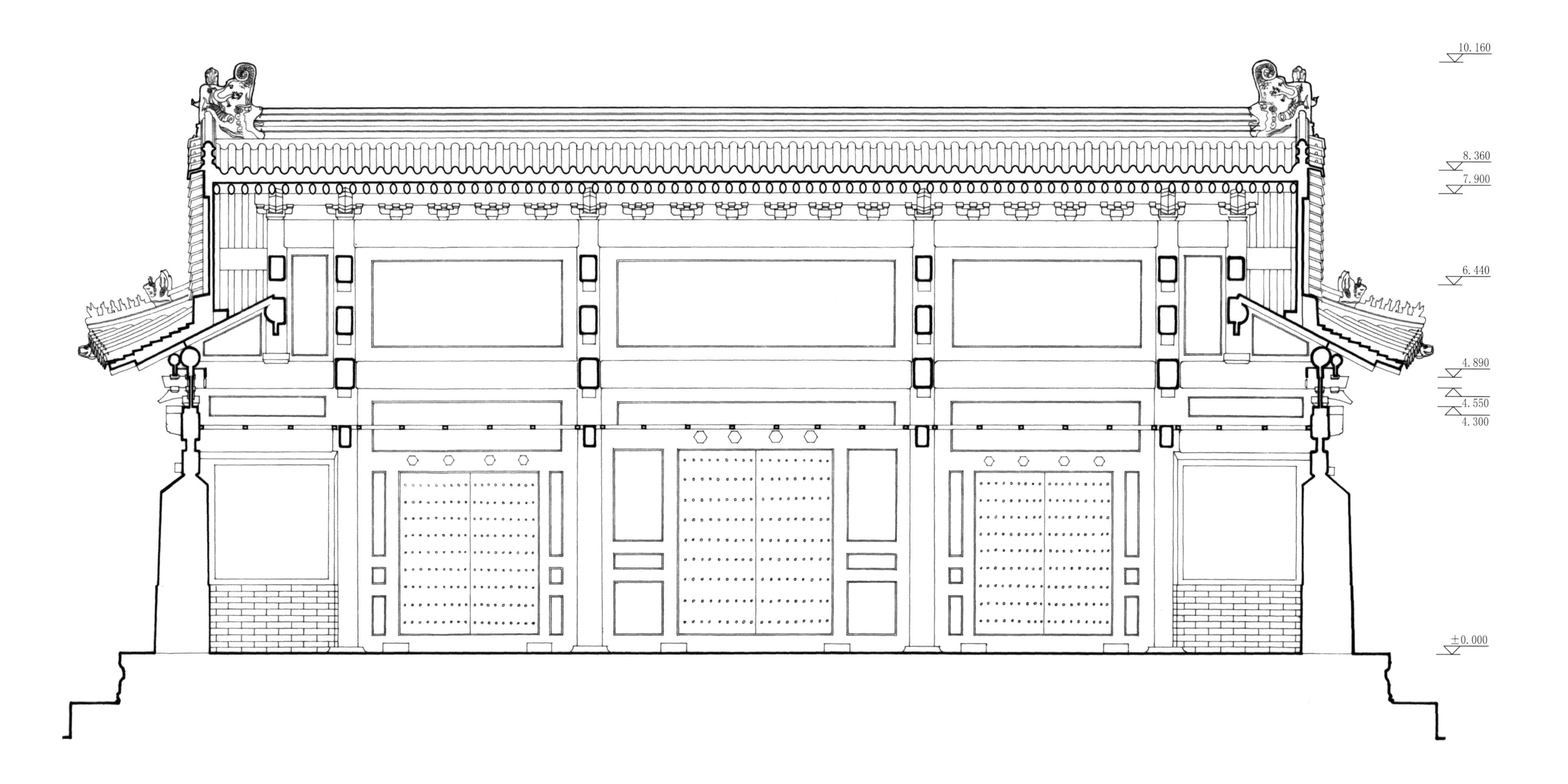

景陵隆恩门纵剖面图

Longitudinal section of the Gate of Monumental Grace, Tomb of Admiration

景陵东配殿正立面图
Front elevation of the Eastern Side Hall, Tomb of Admiration

景陵东配殿平面图
Plan of the Eastern Side Hall, Tomb of Admiration

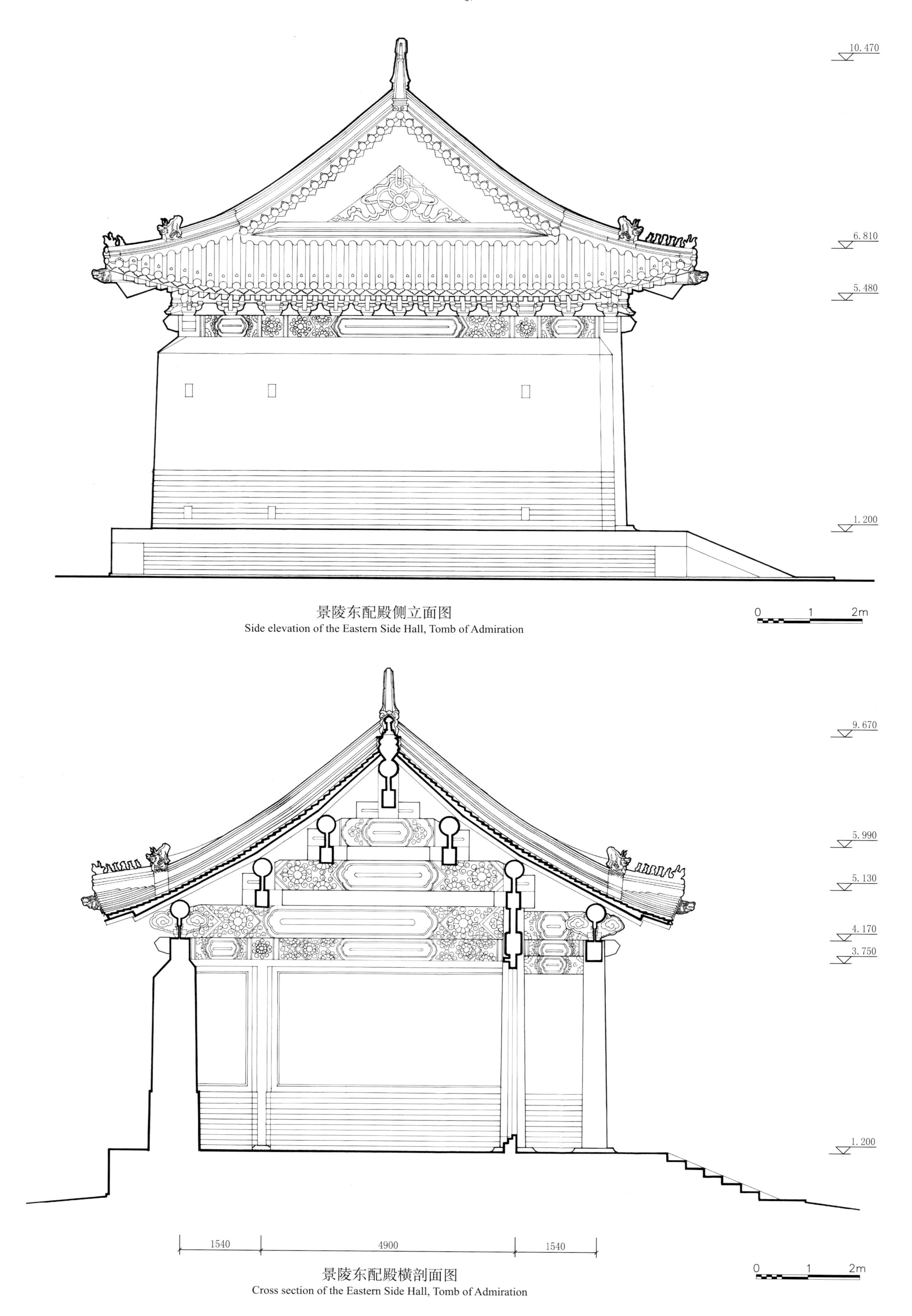

景陵东配殿侧立面图
Side elevation of the Eastern Side Hall, Tomb of Admiration

景陵东配殿横剖面图
Cross section of the Eastern Side Hall, Tomb of Admiration

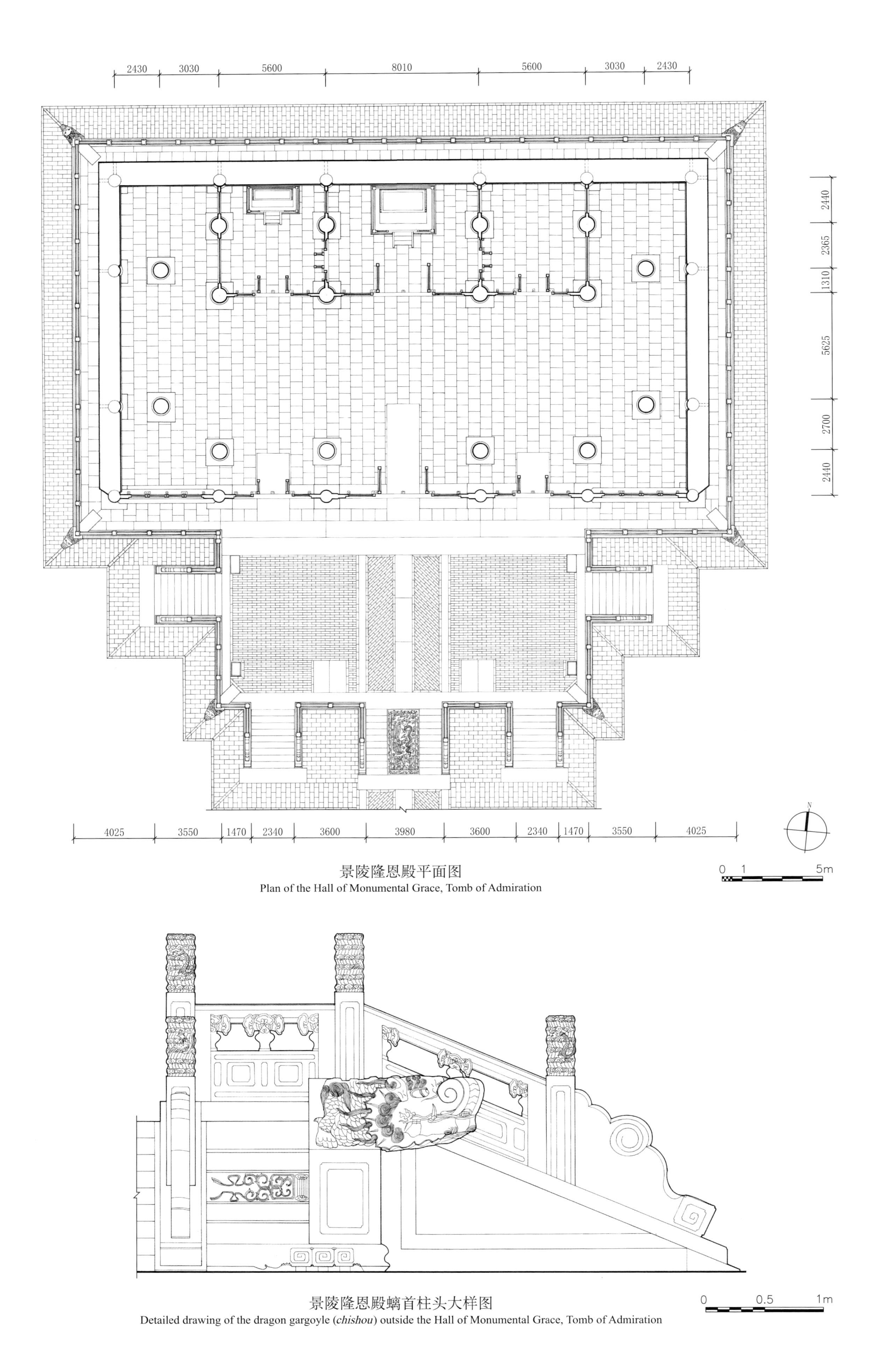

景陵隆恩殿平面图
Plan of the Hall of Monumental Grace, Tomb of Admiration

景陵隆恩殿螭首柱头大样图
Detailed drawing of the dragon gargoyle (*chishou*) outside the Hall of Monumental Grace, Tomb of Admiration

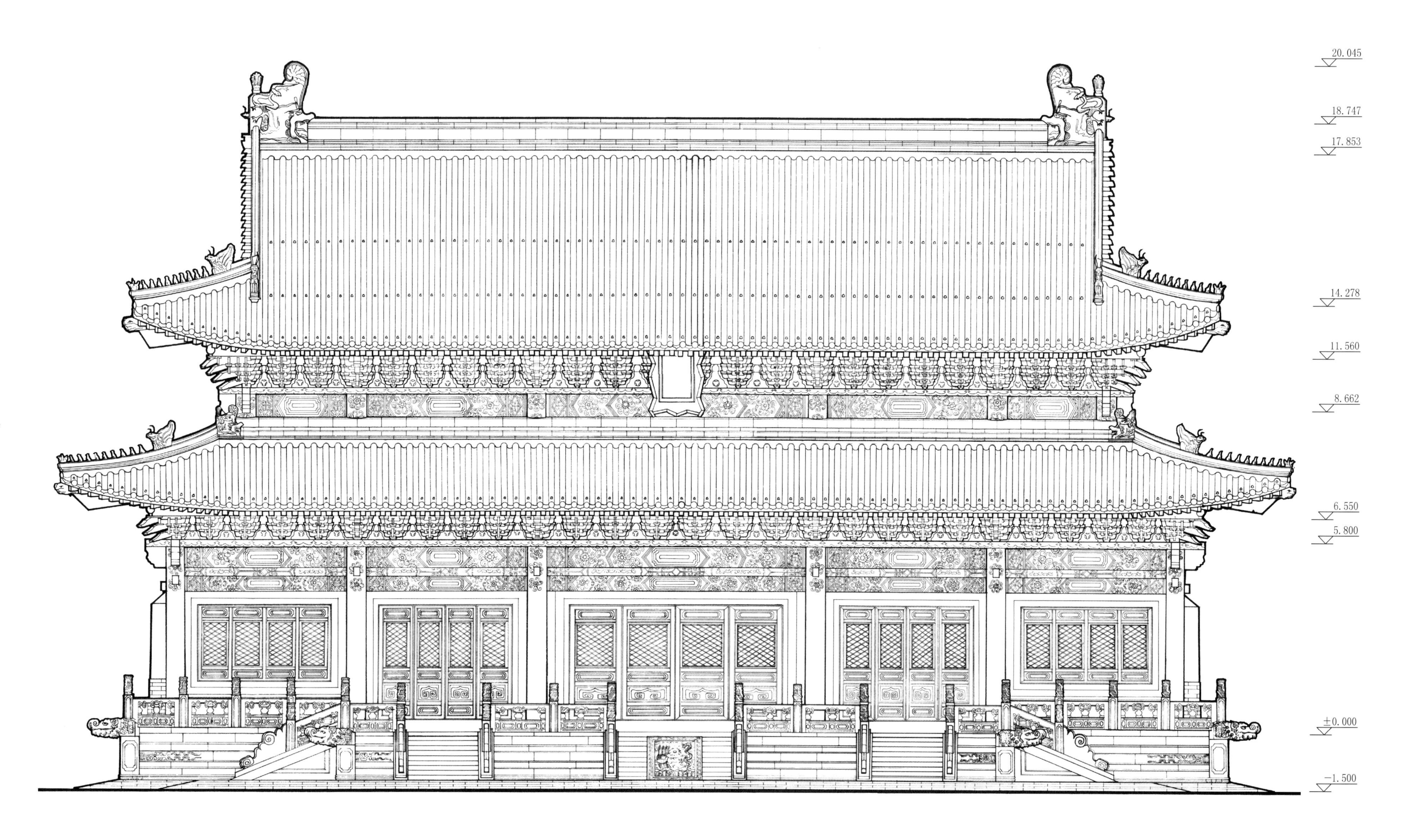

景陵隆恩殿正立面图

Front elevation of the Hall of Monumental Grace, Tomb of Admiration

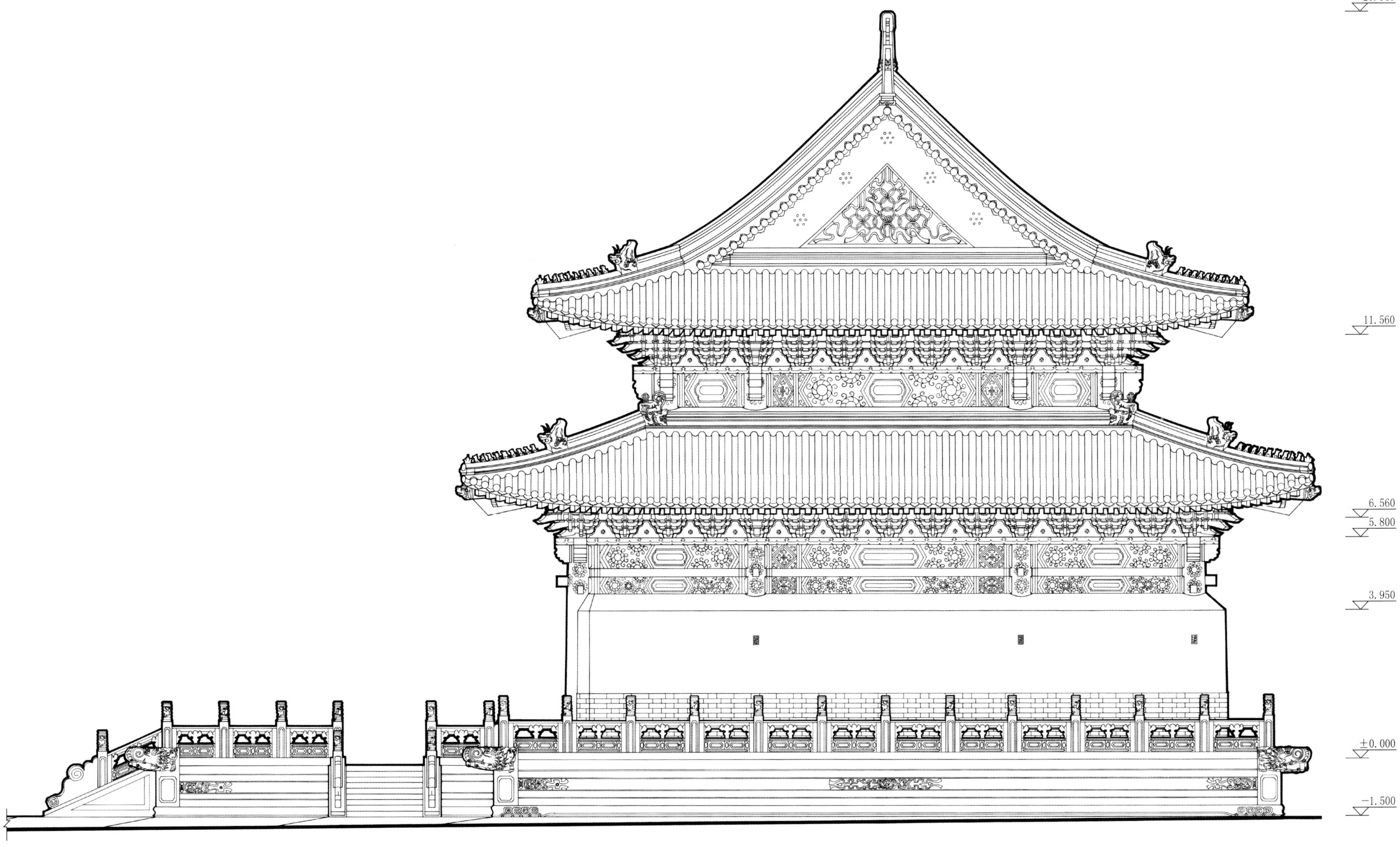

20.045

11.560

6.560

5.800

3.950

±0.000

−1.500

景陵隆恩殿侧立面图

Side elevation of the Hall of Monumental Grace, Tomb of Admiration

0 1 3m

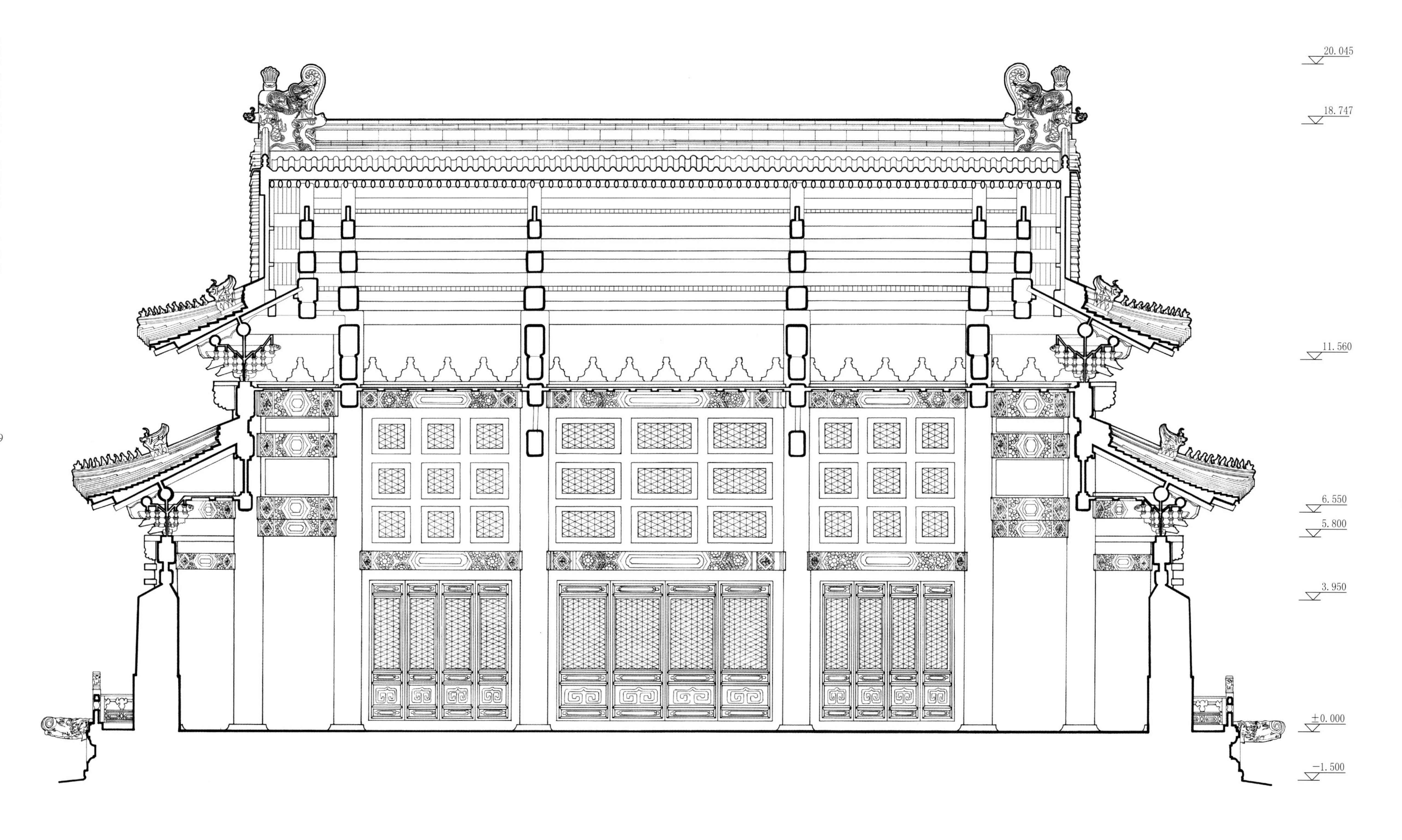

景陵隆恩殿纵剖面图

Longitudinal section of the Hall of Monumental Grace, Tomb of Admiration

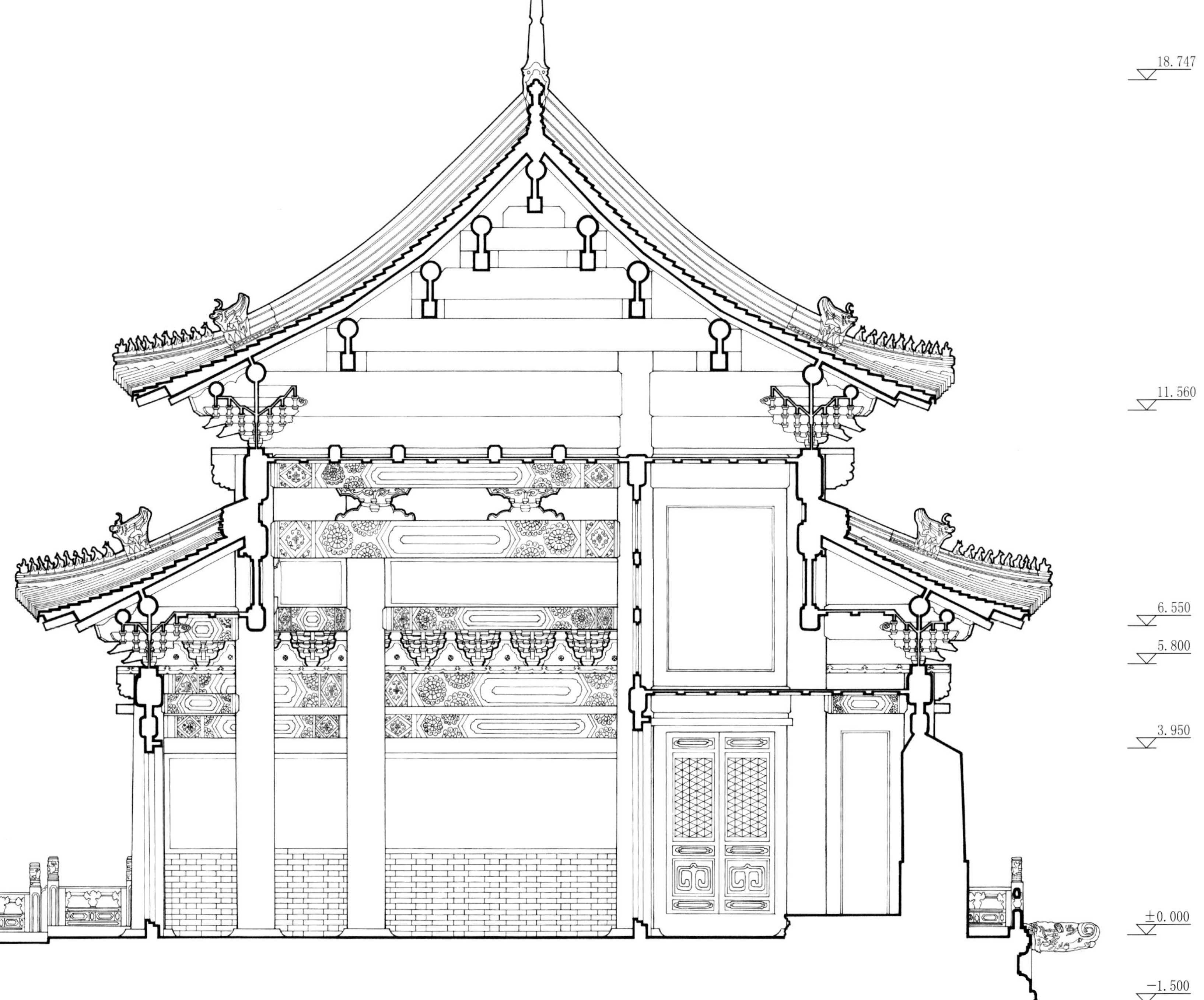

景陵隆恩殿横剖面图
Cross section of the Hall of Monumental Grace, Tomb of Admiration

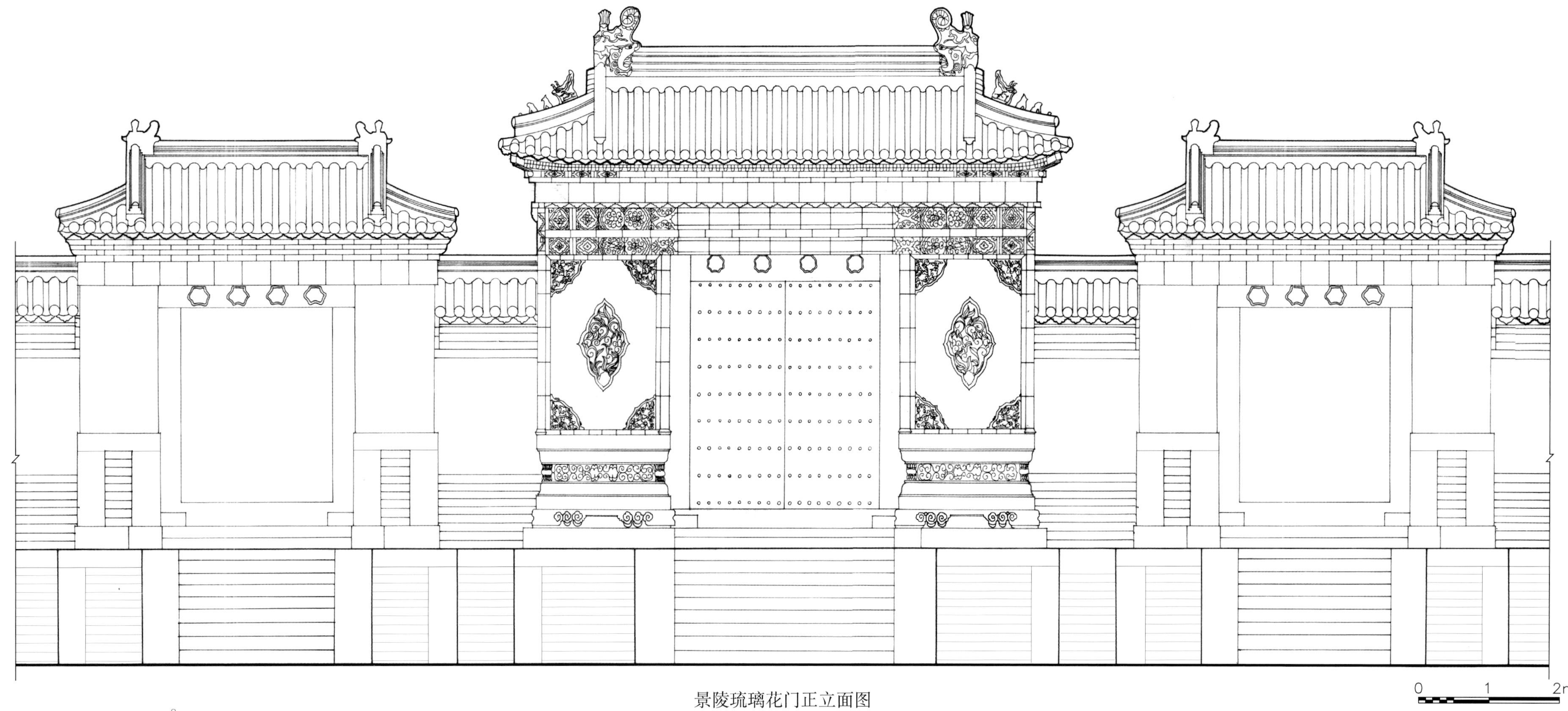

景陵琉璃花门正立面图
Front elevation of the Gate with Glazed Roof Tiles, Tomb of Admiration

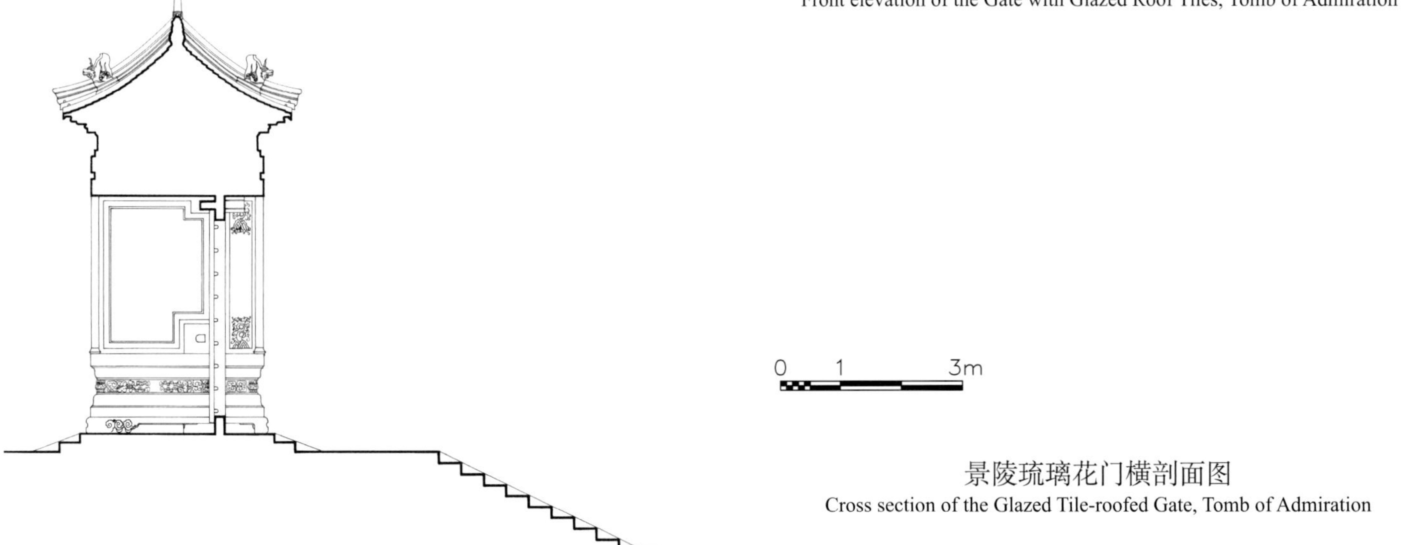

景陵琉璃花门横剖面图
Cross section of the Glazed Tile-roofed Gate, Tomb of Admiration

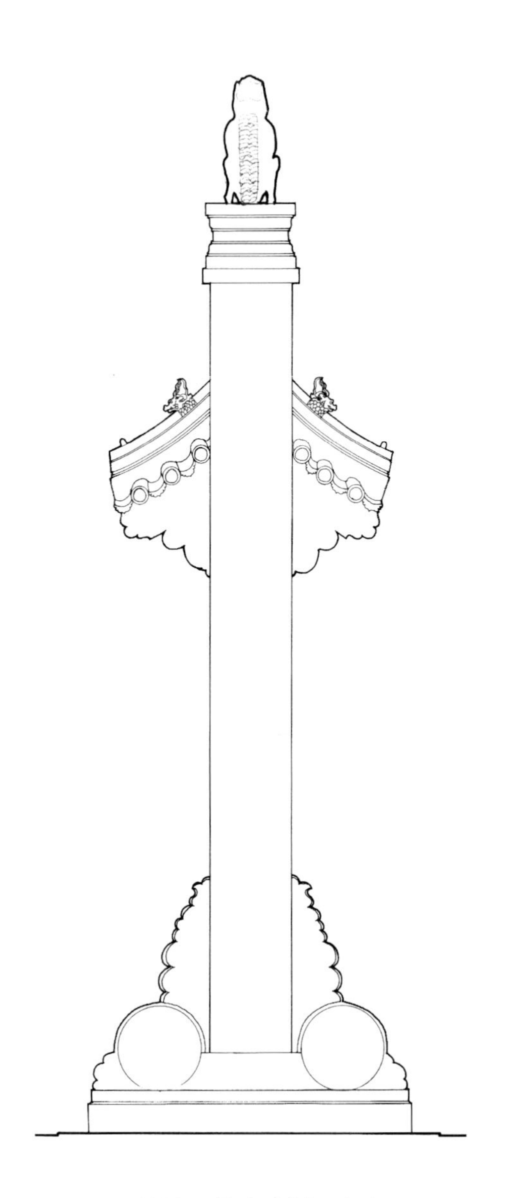

景陵二柱门侧立面图
Side elevation of the Gate with Two Columns, Tomb of Admiration

景陵二柱门正立面图
Front elevation of the Gate with Two Columns, Tomb of Admiration

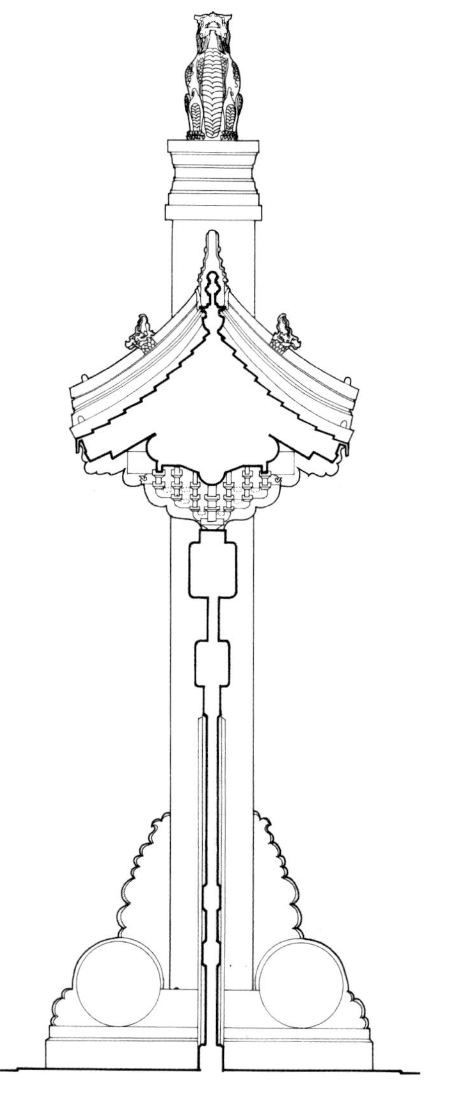

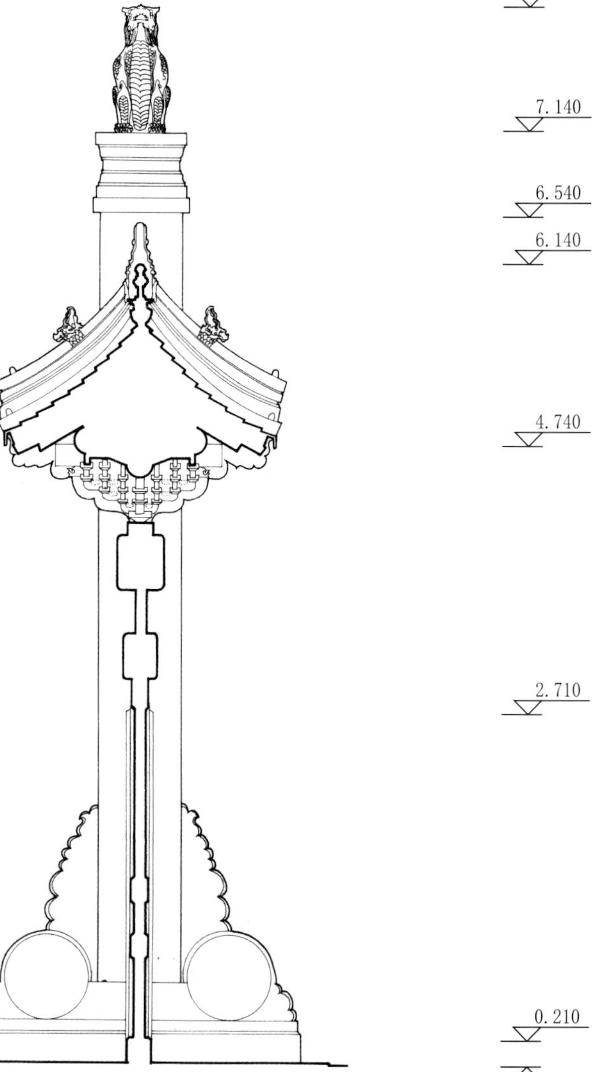

景陵二柱门剖面图
Cross section of the Gate with Two Columns, Tomb of Admiration

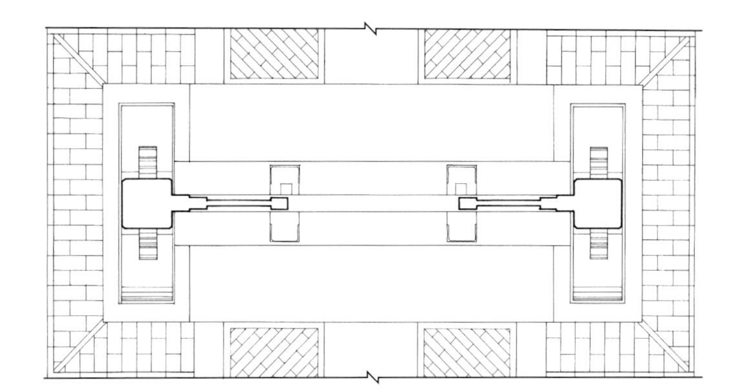

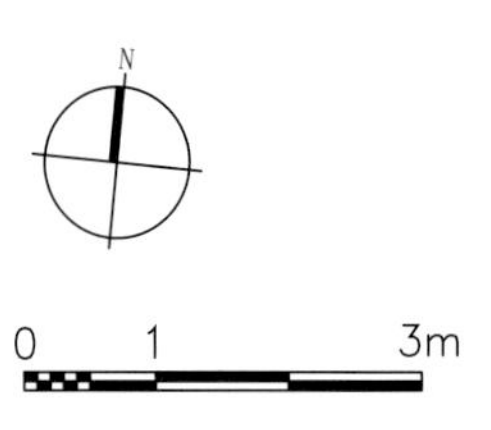

景陵二柱门平面图
Plan of the Gate with Two Columns, Tomb of Admiration

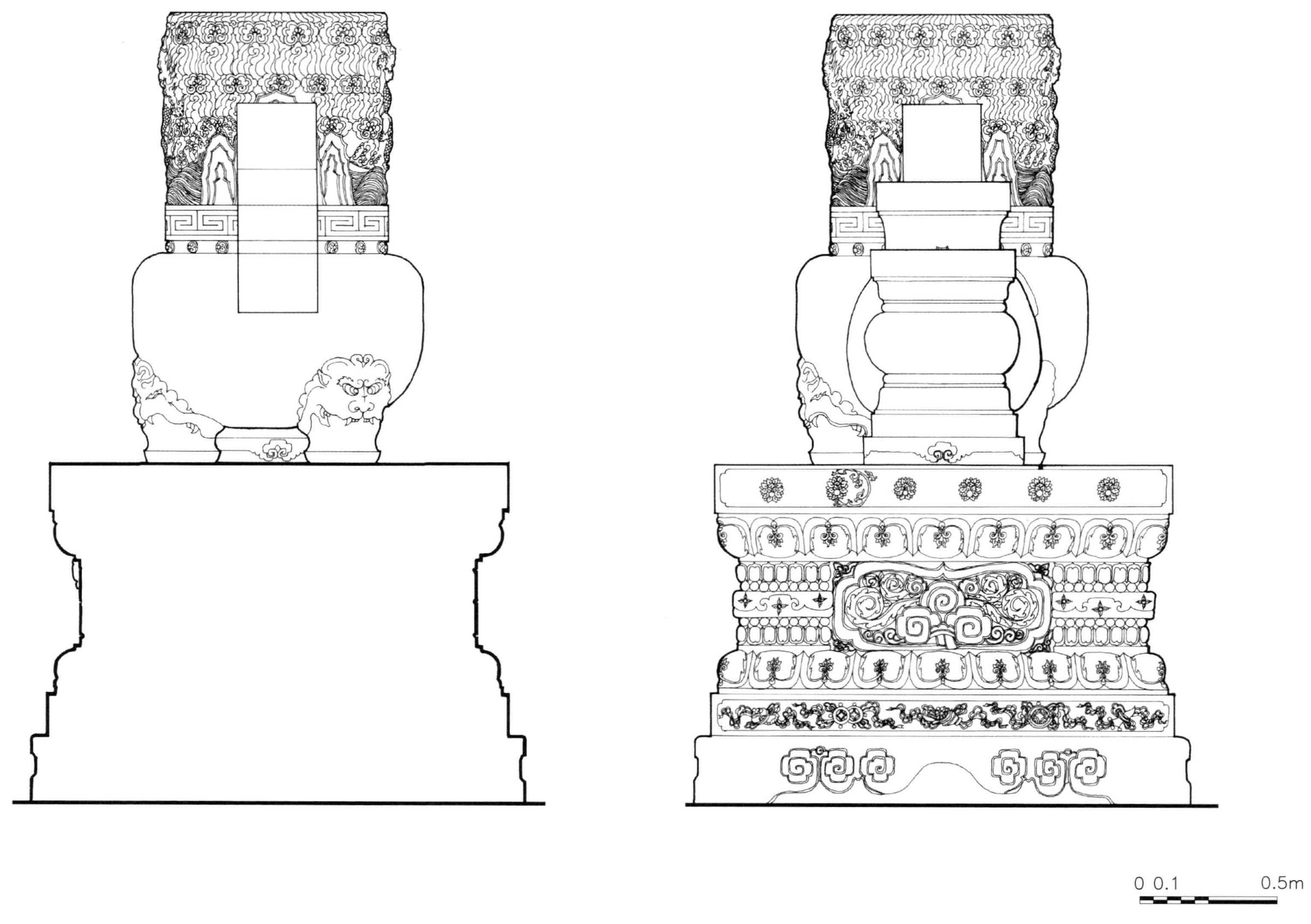

景陵石五供剖面图
Cross section of the Five Stone Ritual Vessels, Tomb of Admiration

景陵石五供侧立面图
Side elevation of the Five Stone Ritual Vessels, Tomb of Admiration

景陵石五供正立面图
Front elevation of the Five Stone Ritual Vessels, Tomb of Admiration

23.680
22.500
17.430
13.480
9.100
8.400
4.250
±0.000
−3.750
−4.460

0 1 5m

景陵方城明楼正立面图
Front elevation of the Square Walled Terrace and the Memorial Tower, Tomb of Admiration

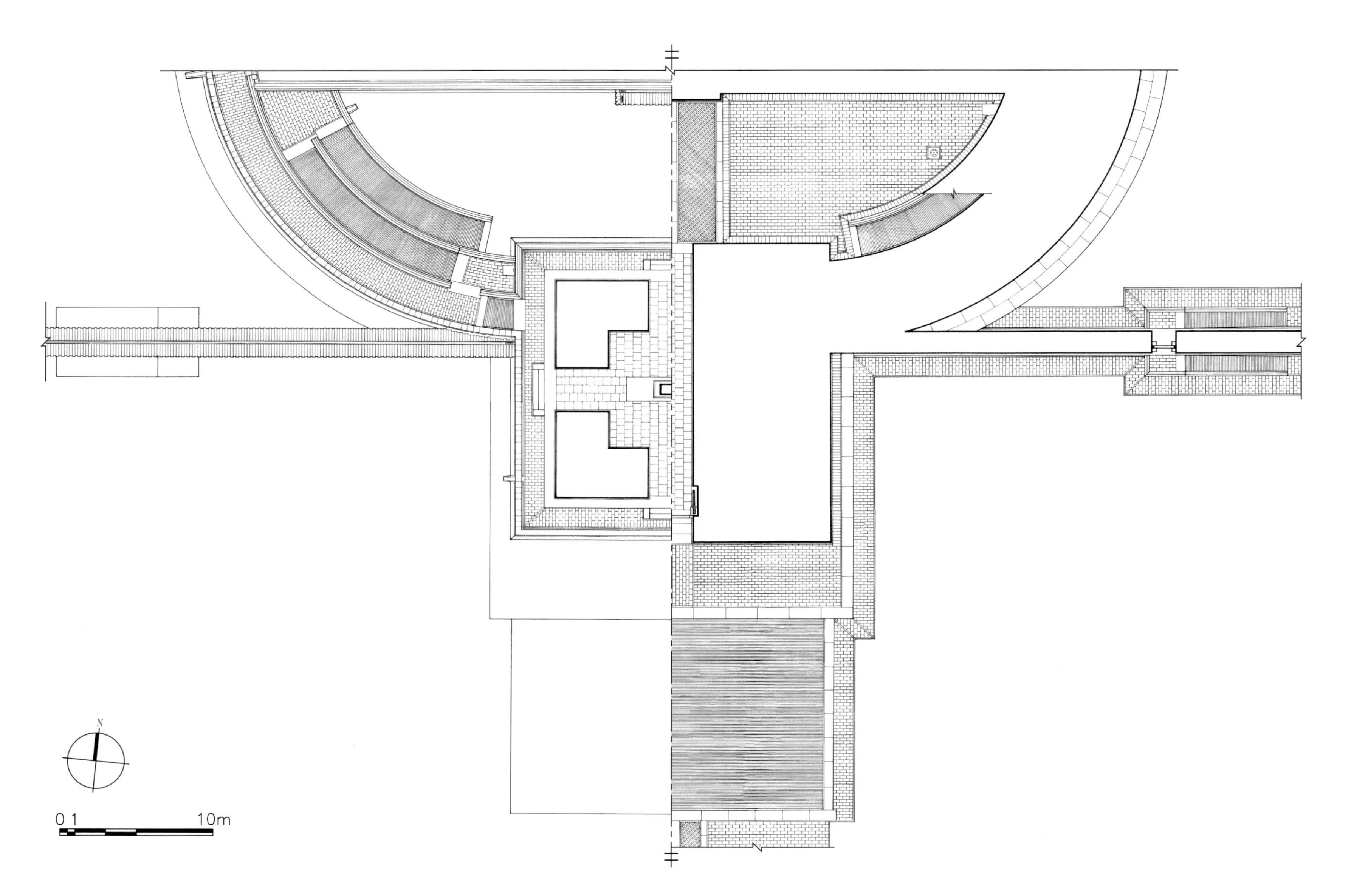

景陵方城明楼平面图
Plan of the Square Walled Terrace and the Memorial Tower, Tomb of Admiration

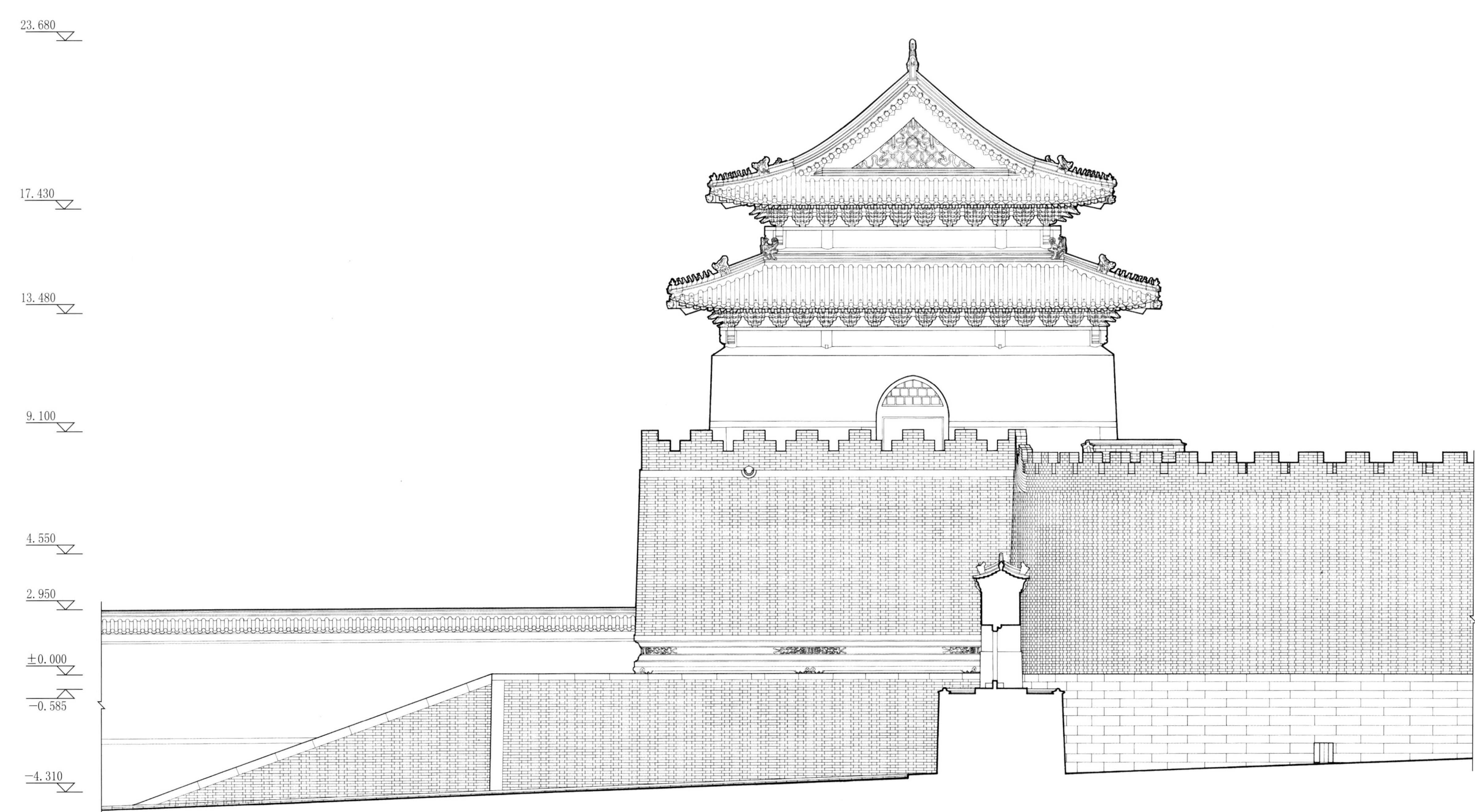

景陵方城明楼侧立面图

Side elevation of the Square Walled Terrace and the Memorial Tower, Tomb of Admiration

景陵方城明楼横剖面图

Cross section of the Square Walled Terrace and the Memorial Tower, Tomb of Admiration

景陵妃园寝
Imperial Consorts' Tombs affiliated with the Tomb of Admiration (Jingling Feiyuanqin)

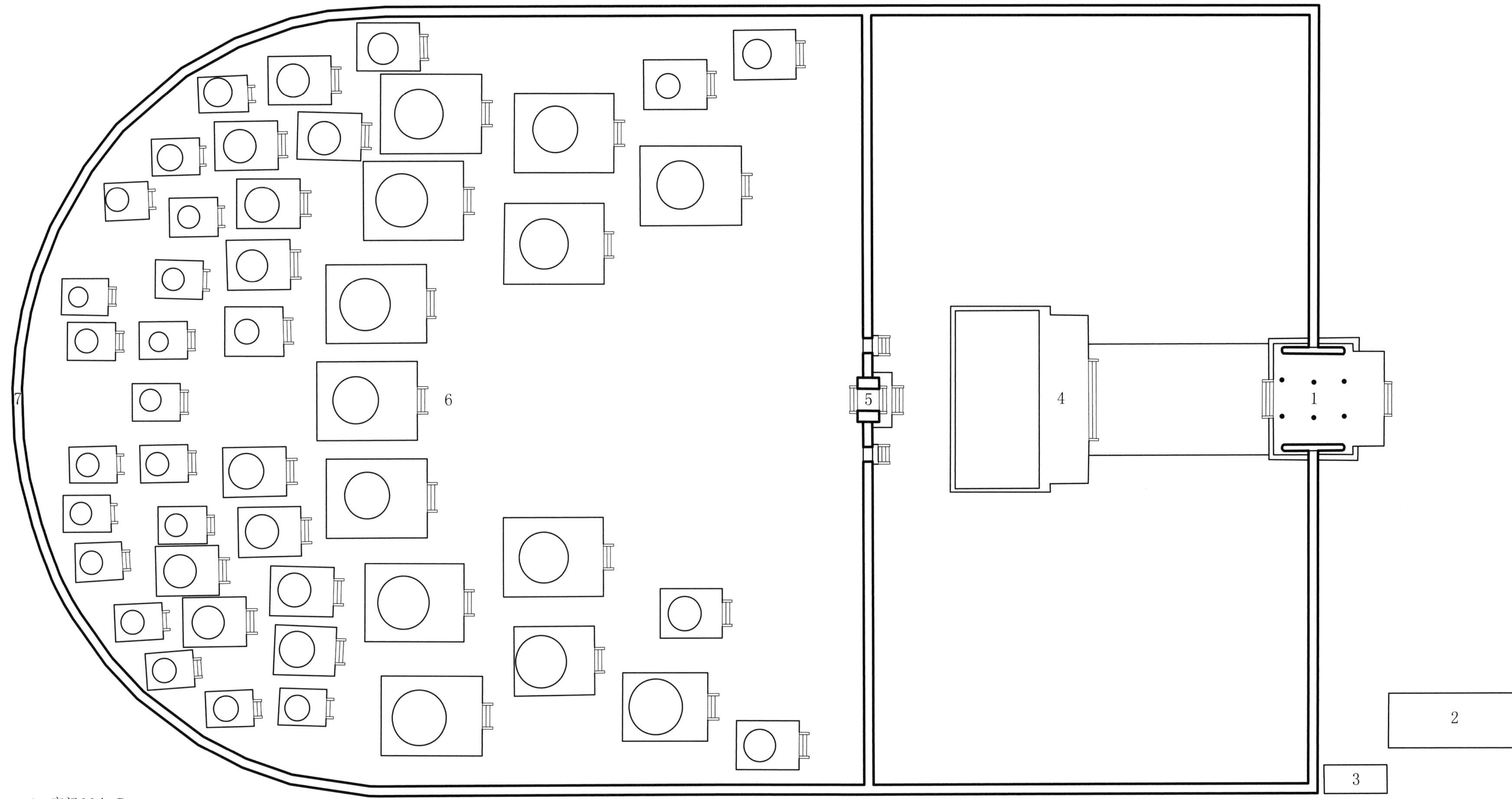

1 宫门 Main Gate
2 朝房 Reception Halls for Court Officials
3 班房 Duty Houses for the Guards
4 享殿 Main Sacrificial Hall
5 琉璃花门 Gate with Glazed Roof Tiles
6 妃嫔宝顶 Concubines' Tumulus
7 罗圈墙 Wall Surrounding the Tomb

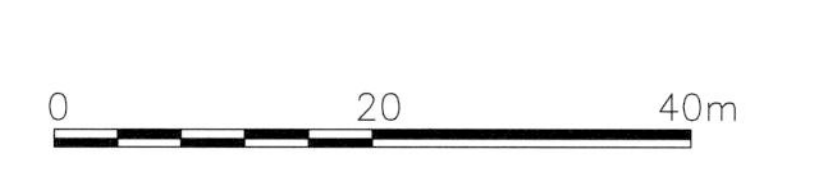

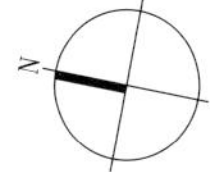

景陵妃园寝总平面图
Site plan of the Imperial Consorts' Tombs affiliated with the Tomb of Admiration

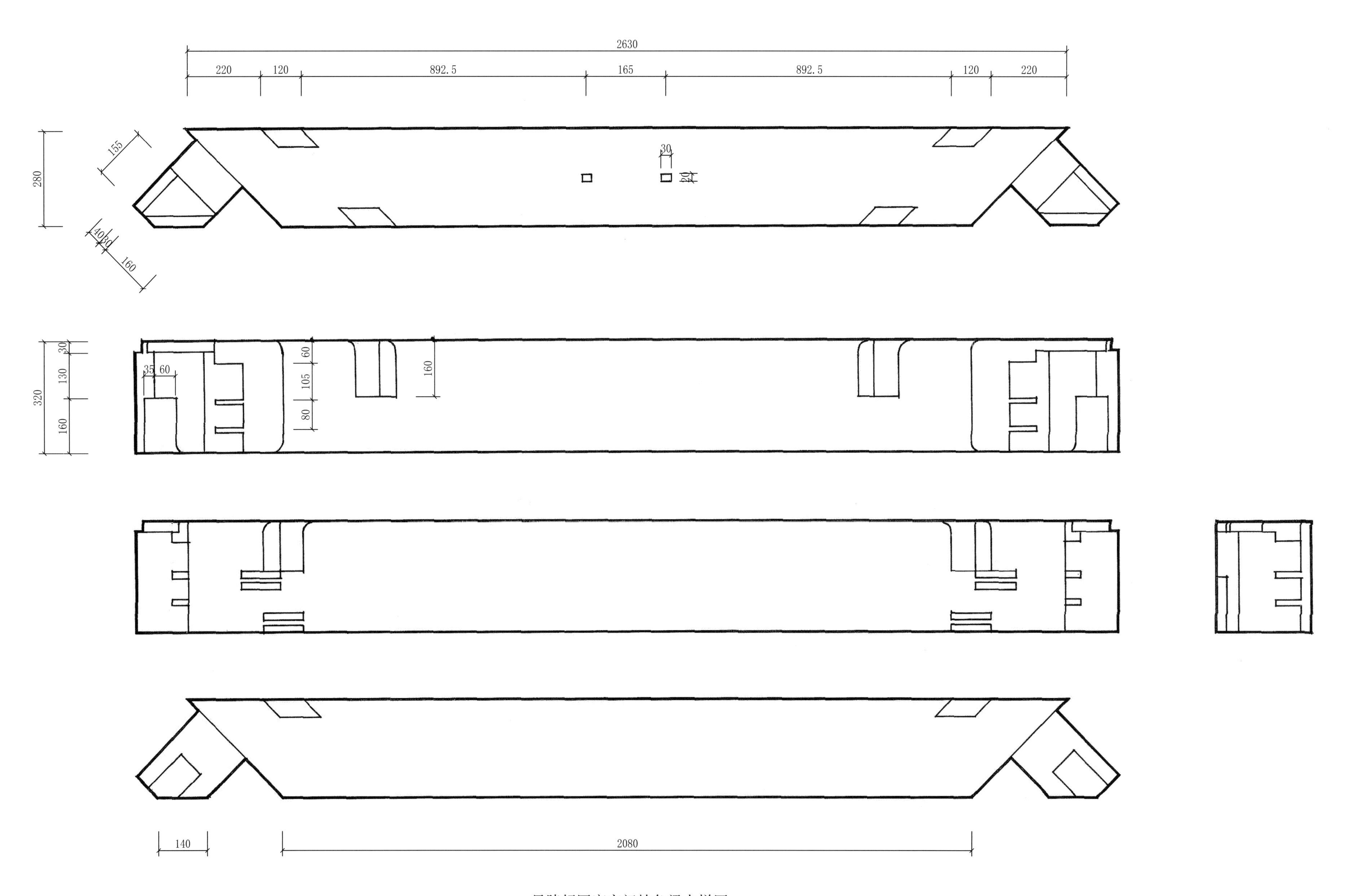

景陵妃园寝宫门抹角梁大样图

Detailed drawing of the Overlapping Corner Beam (*mojiaoliang*) of the Main Gate, the Imperial Consorts' Tombs affiliated with the Tomb of Admiration

0 0.1 0.4m

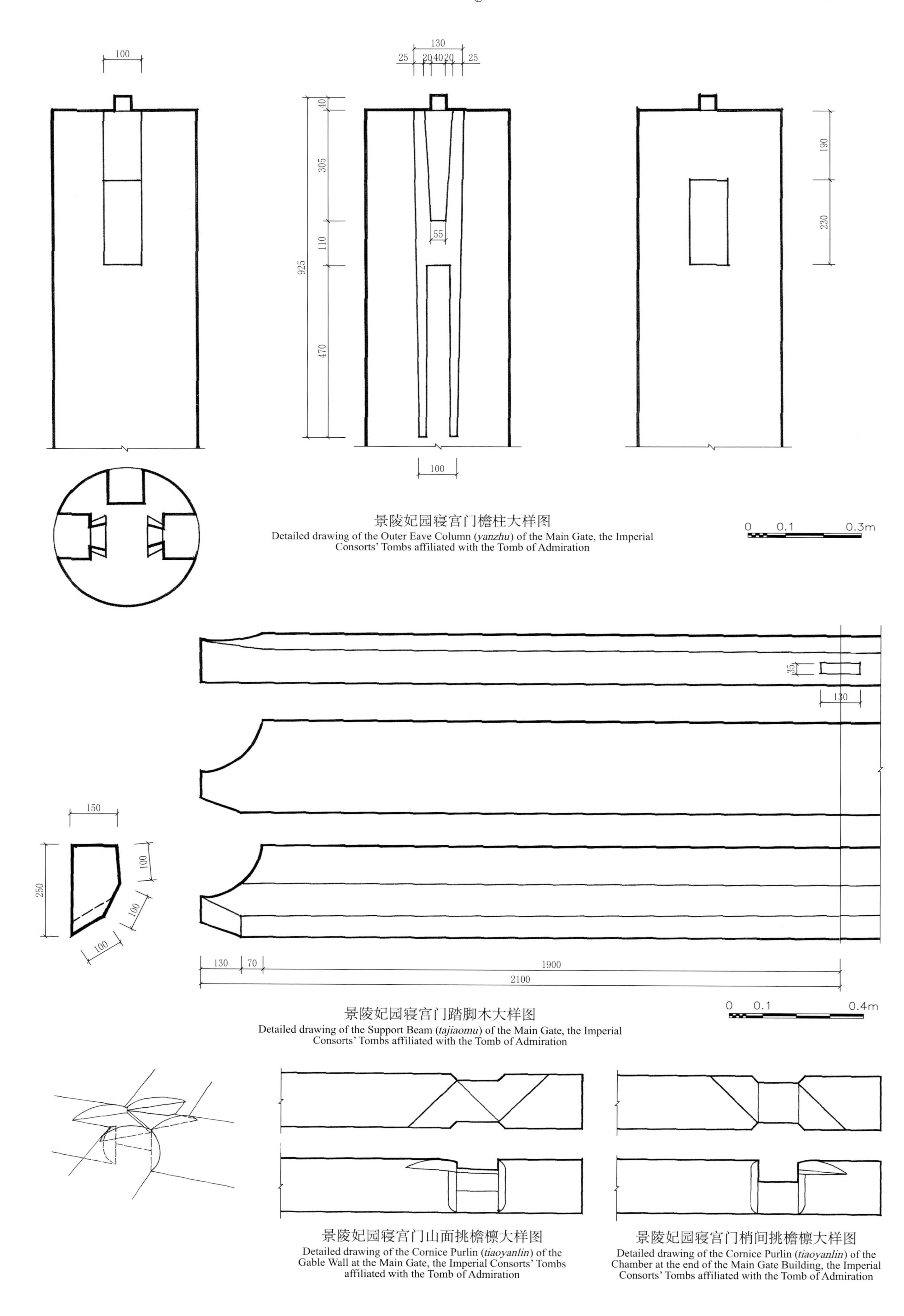

景陵妃园寝宫门檐柱大样图
Detailed drawing of the Outer Eave Column (*yanzhu*) of the Main Gate, the Imperial Consorts' Tombs affiliated with the Tomb of Admiration

景陵妃园寝宫门踏脚木大样图
Detailed drawing of the Support Beam (*tajiaomu*) of the Main Gate, the Imperial Consorts' Tombs affiliated with the Tomb of Admiration

景陵妃园寝宫门山面挑檐檩大样图
Detailed drawing of the Cornice Purlin (*tiaoyanlin*) of the Gable Wall at the Main Gate, the Imperial Consorts' Tombs affiliated with the Tomb of Admiration

景陵妃园寝宫门梢间挑檐檩大样图
Detailed drawing of the Cornice Purlin (*tiaoyanlin*) of the Chamber at the end of the Main Gate Building, the Imperial Consorts' Tombs affiliated with the Tomb of Admiration

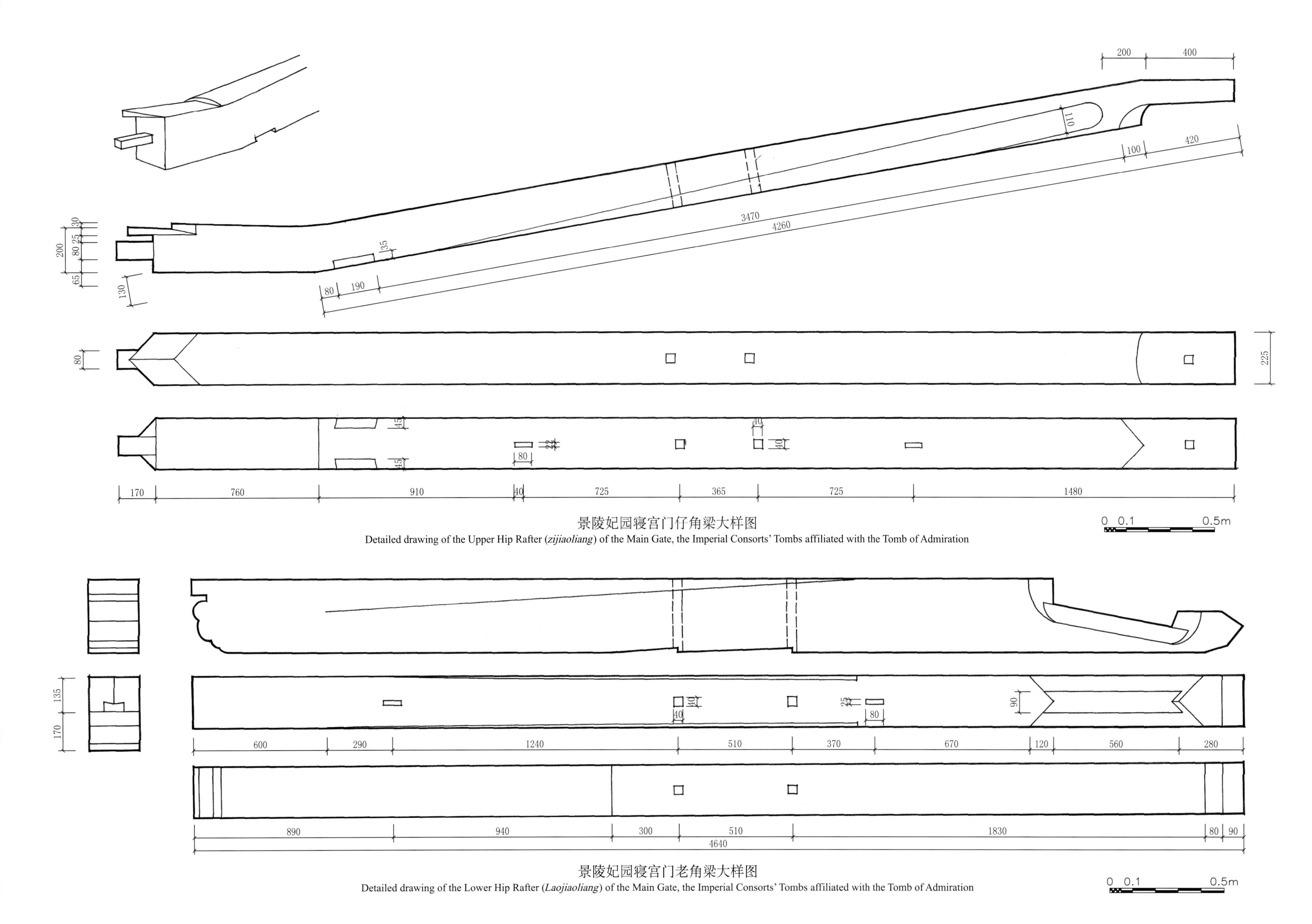

景陵妃园寝宫门仔角梁大样图

Detailed drawing of the Upper Hip Rafter (*zijiaoliang*) of the Main Gate, the Imperial Consorts' Tombs affiliated with the Tomb of Admiration

景陵妃园寝宫门老角梁大样图

Detailed drawing of the Lower Hip Rafter (*Laojiaoliang*) of the Main Gate, the Imperial Consorts' Tombs affiliated with the Tomb of Admiration

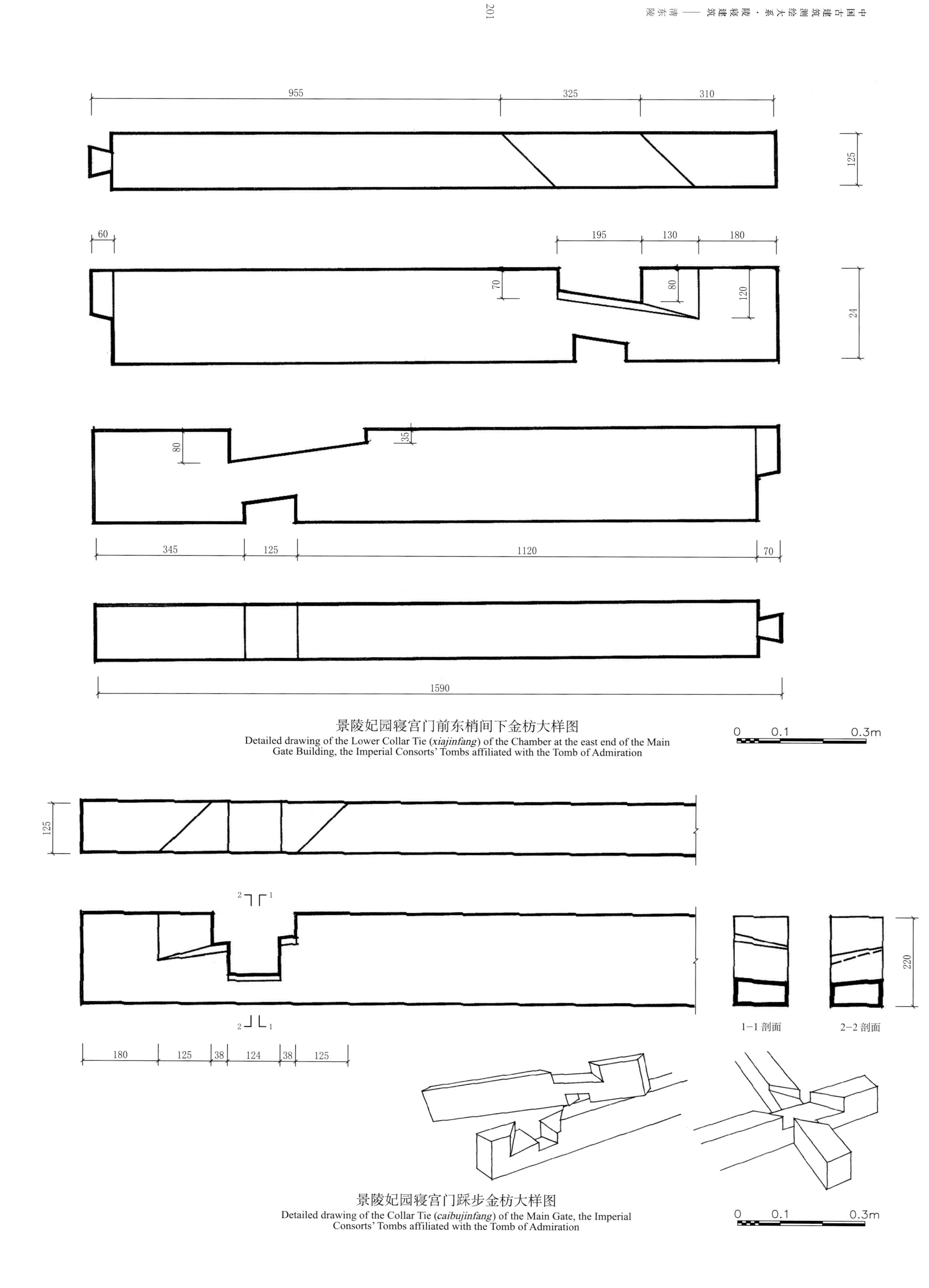

景陵妃园寝宫门前东梢间下金枋大样图

Detailed drawing of the Lower Collar Tie (*xiajinfang*) of the Chamber at the east end of the Main Gate Building, the Imperial Consorts' Tombs affiliated with the Tomb of Admiration

景陵妃园寝宫门踩步金枋大样图

Detailed drawing of the Collar Tie (*caibujinfang*) of the Main Gate, the Imperial Consorts' Tombs affiliated with the Tomb of Admiration

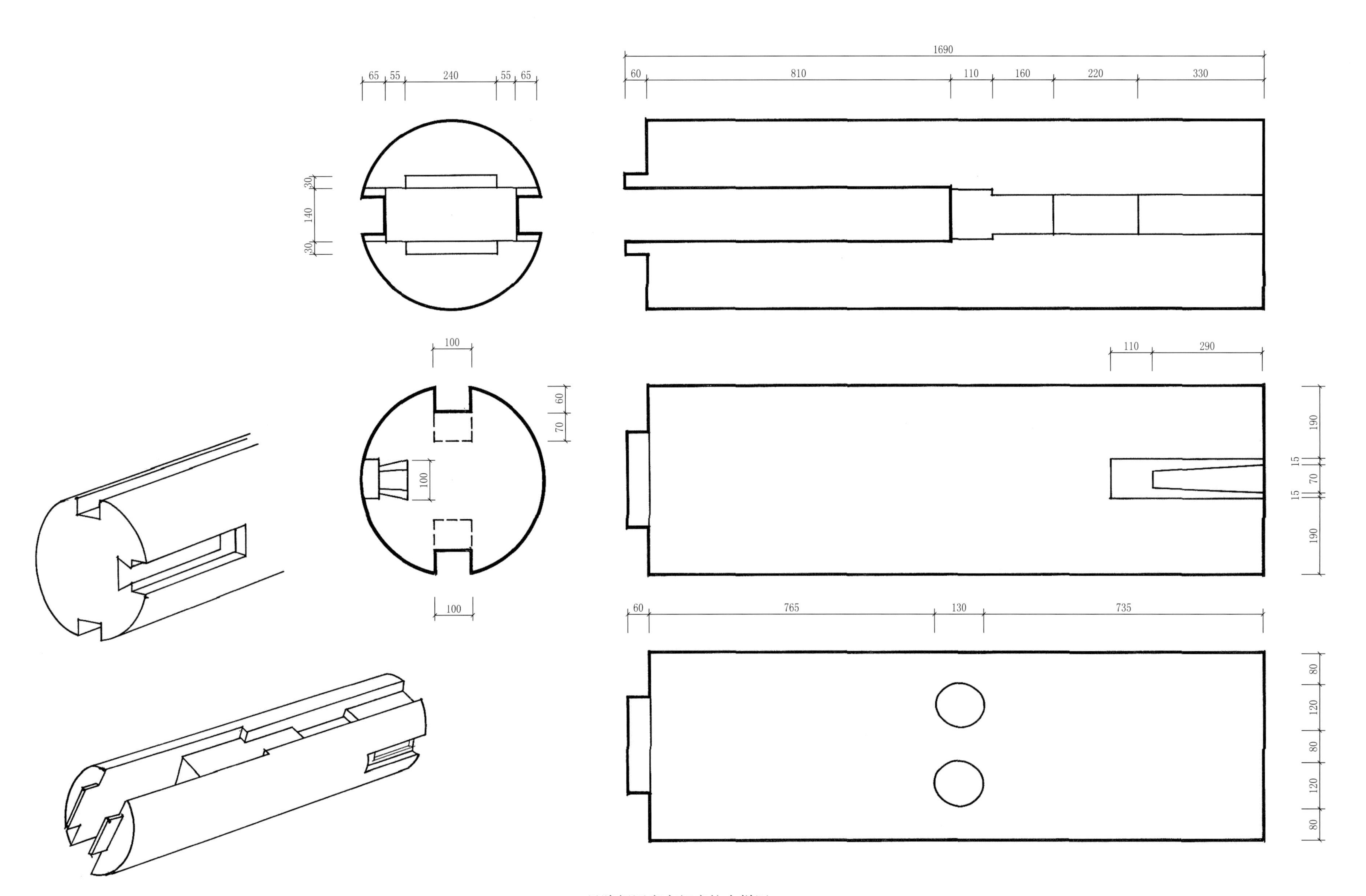

景陵妃园寝宫门童柱大样图

Detailed drawing of the Short Post (*tongzhu*) of the Main Gate, the Imperial Consorts' Tombs affiliated with the Tomb of Admiration

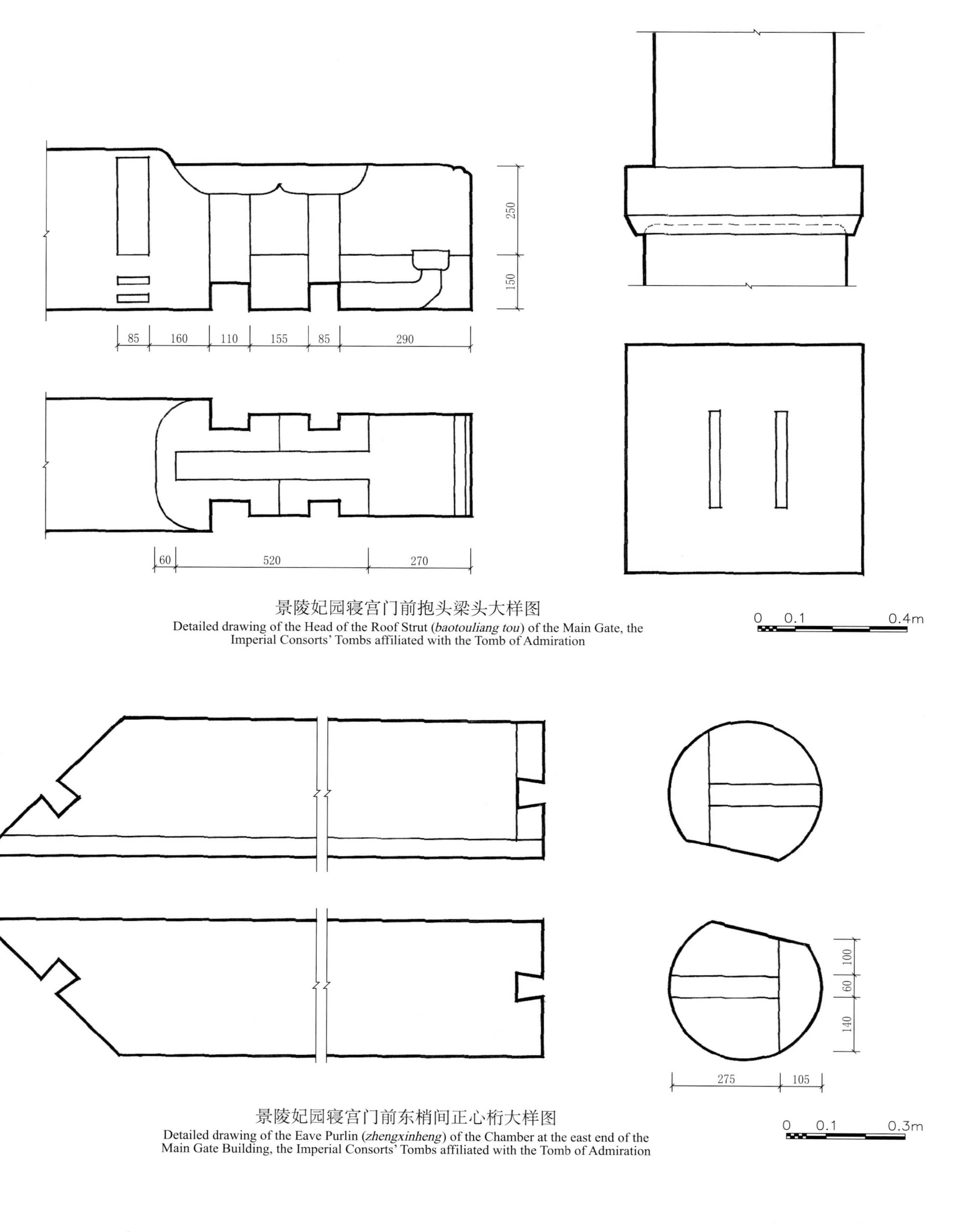

景陵妃园寝宫门前抱头梁头大样图

Detailed drawing of the Head of the Roof Strut (*baotouliang tou*) of the Main Gate, the Imperial Consorts' Tombs affiliated with the Tomb of Admiration

景陵妃园寝宫门前东梢间正心桁大样图

Detailed drawing of the Eave Purlin (*zhengxinheng*) of the Chamber at the east end of the Main Gate Building, the Imperial Consorts' Tombs affiliated with the Tomb of Admiration

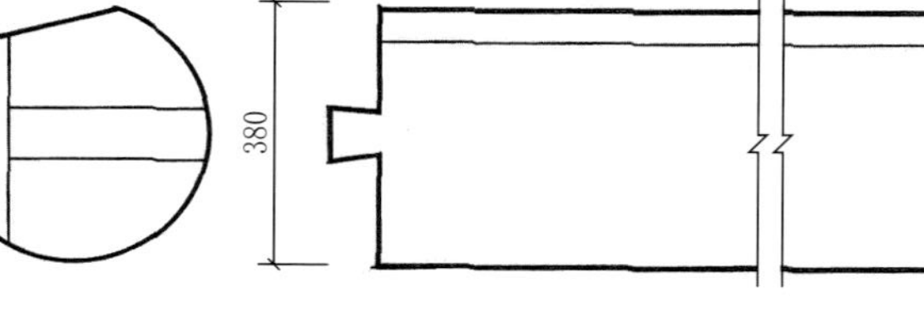

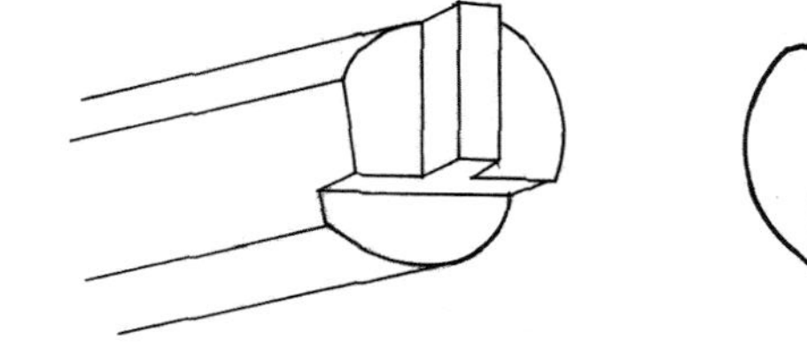

景陵妃园寝宫门东山南面正心梁大样图

Detailed drawing of the south facing Beam (*zhengxinliang*) along the east Gable Wall of the Main Gate, the Imperial Consorts' Tombs affiliated with the Tomb of Admiration

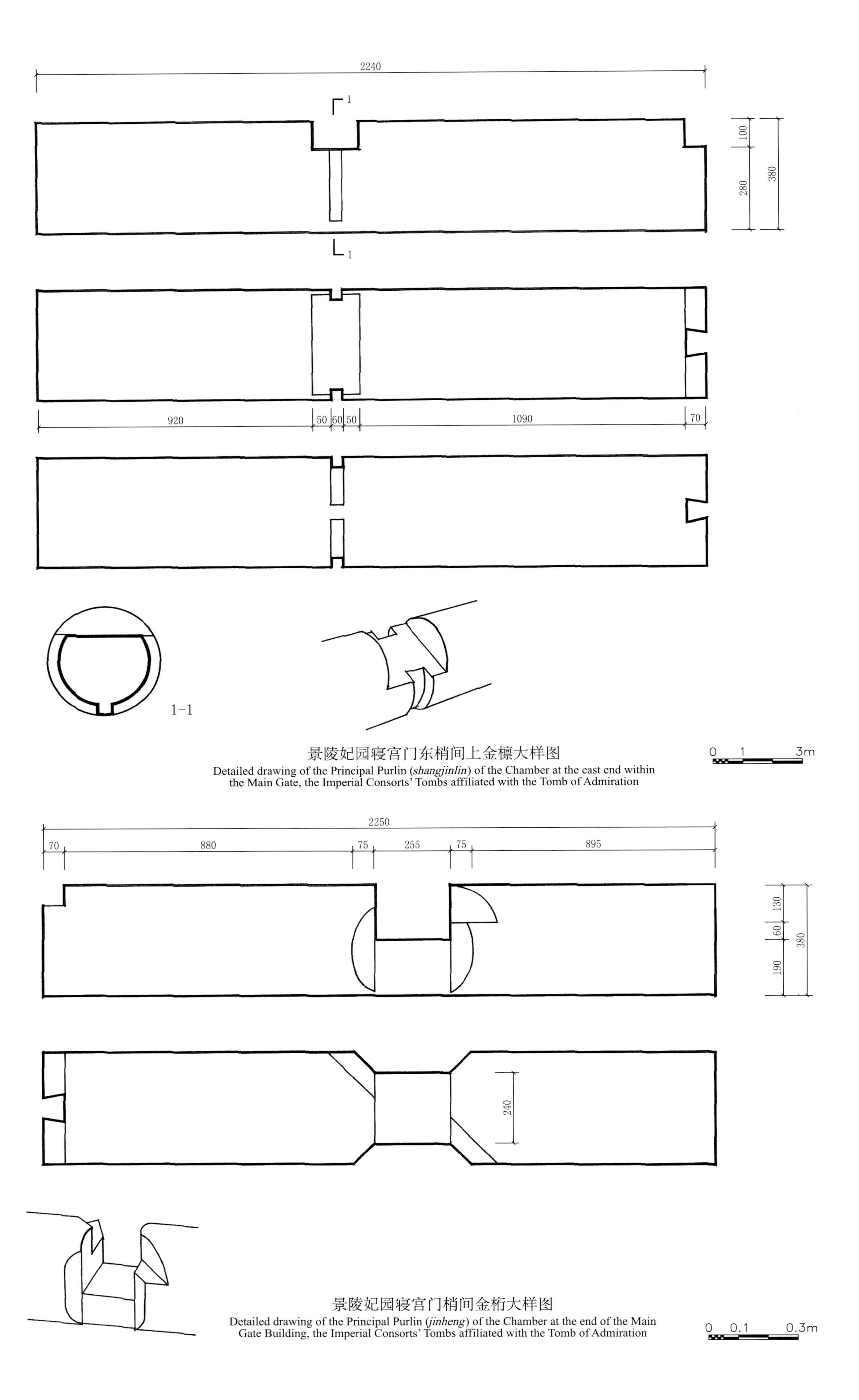

景陵妃园寝宫门东梢间上金檩大样图

Detailed drawing of the Principal Purlin (*shangjinlin*) of the Chamber at the east end within the Main Gate, the Imperial Consorts' Tombs affiliated with the Tomb of Admiration

景陵妃园寝宫门梢间金桁大样图

Detailed drawing of the Principal Purlin (*jinheng*) of the Chamber at the end of the Main Gate Building, the Imperial Consorts' Tombs affiliated with the Tomb of Admiration

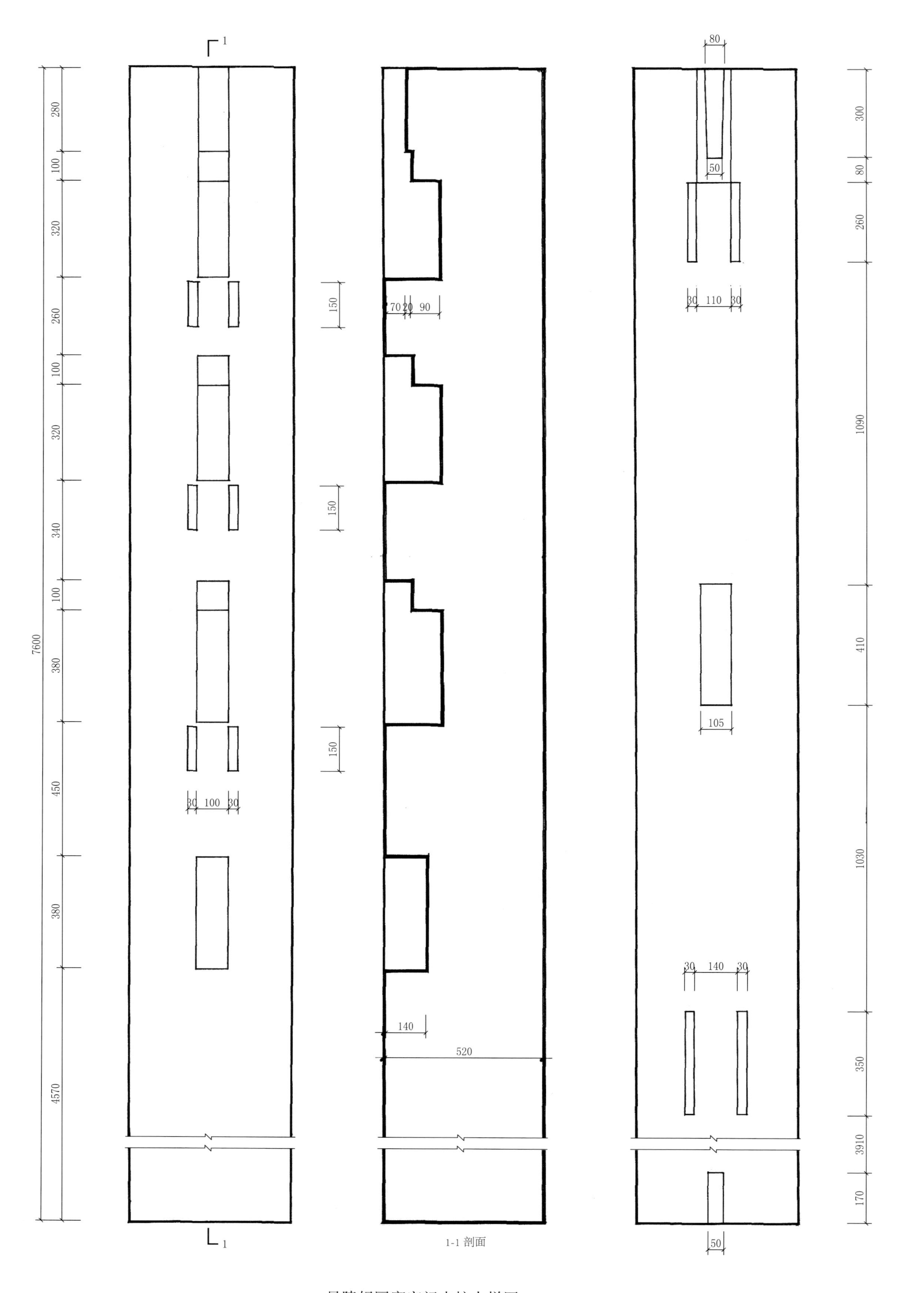

景陵妃园寝宫门中柱大样图

Detailed drawing of the Column for the Roof-ridge Strut (*zhongzhu*) of the Main Gate, the Imperial Consorts' Tombs affiliated with the Tomb of Admiration

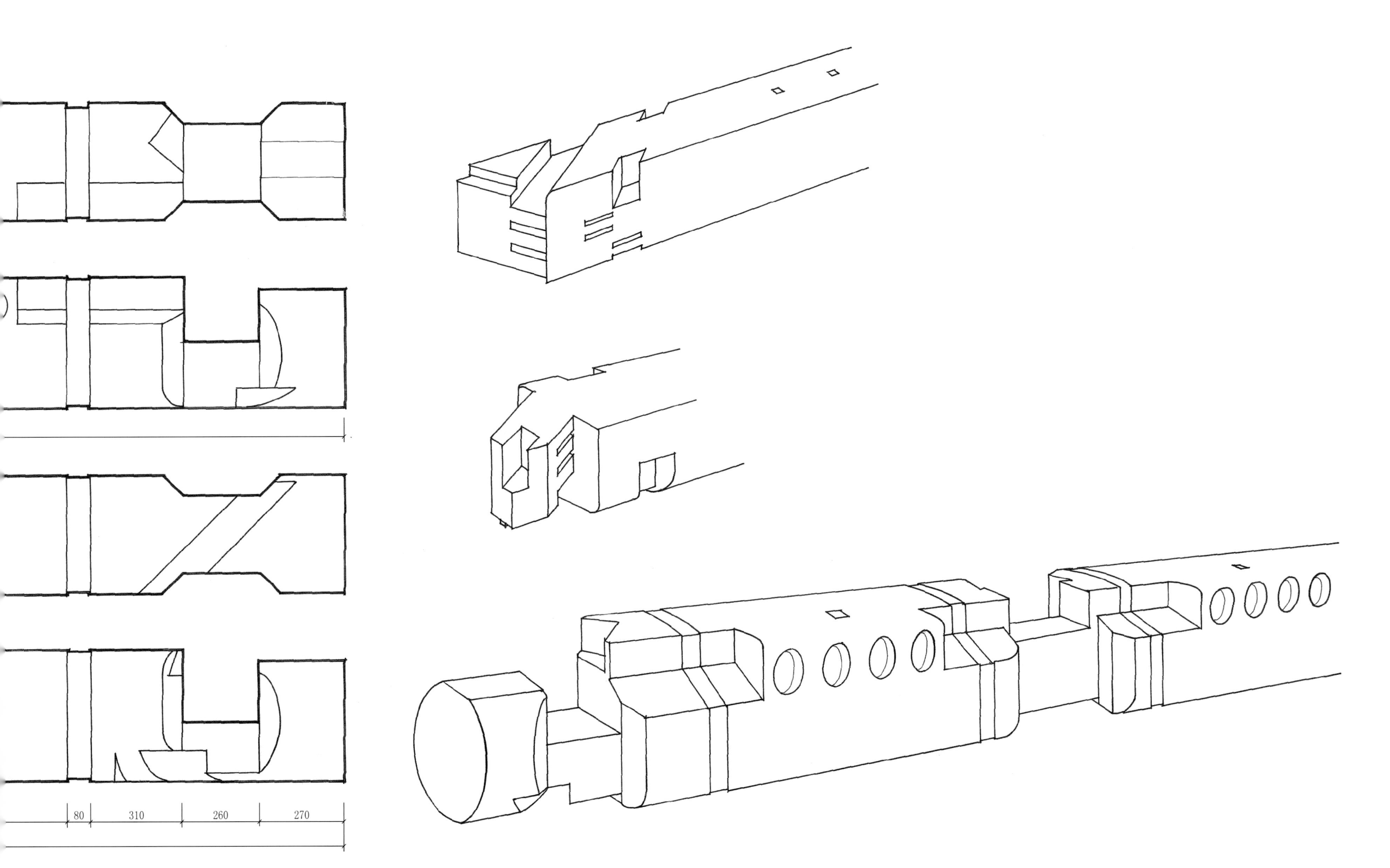
80
310
260
270
0 0.1 0.3m

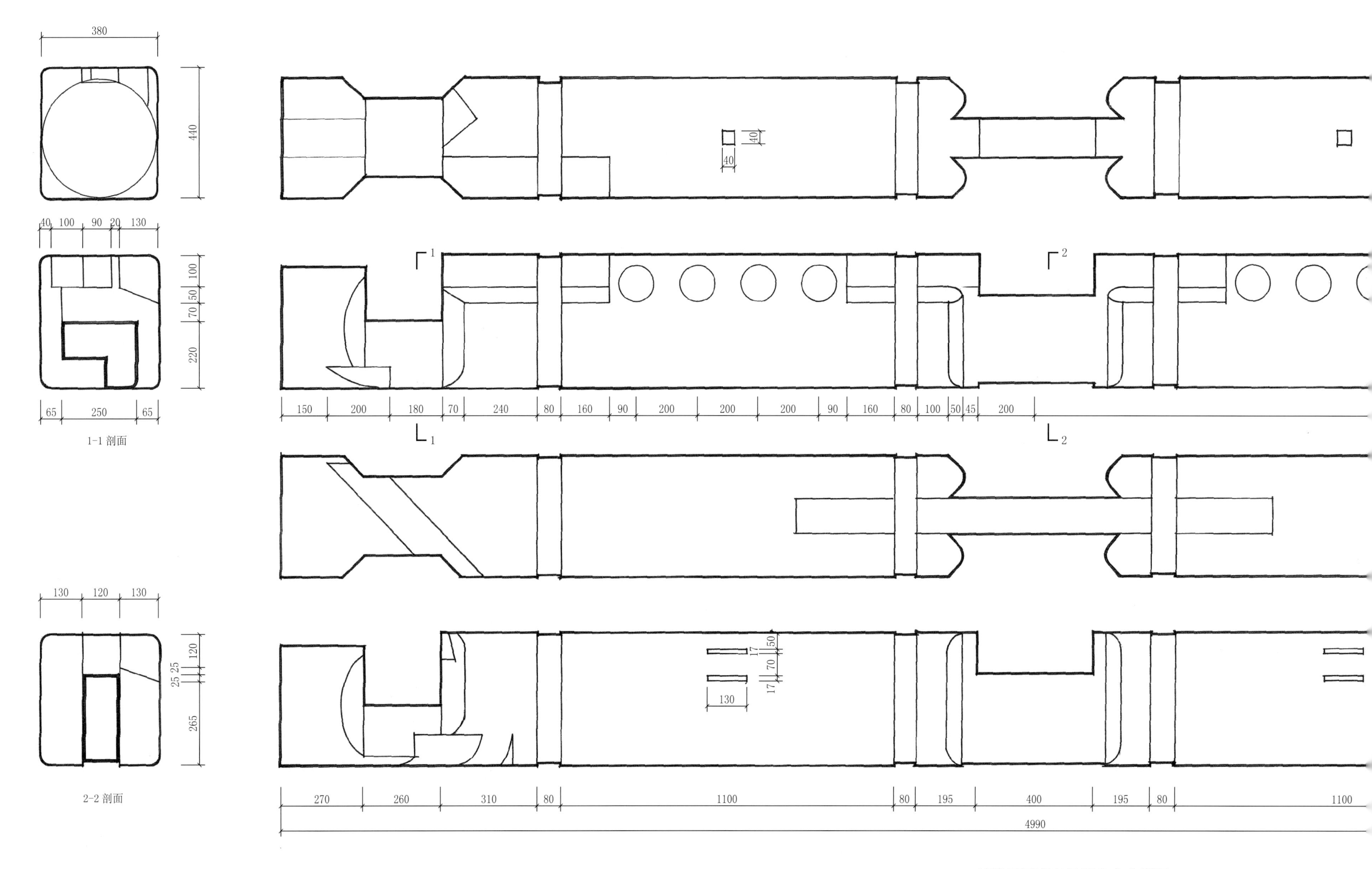

景陵妃园寝宫门踩步金大样图

Detailed drawing of the Collar Tie (*caibujin*) of the Main Gate, Imperial Consorts' Tombs affiliated with the Tomb of Admiration

景陵双妃园寝

Two Imperial Consorts' Tombs affiliated with the Tomb of Admiration (Jingling Shuangfeiyuanqin)

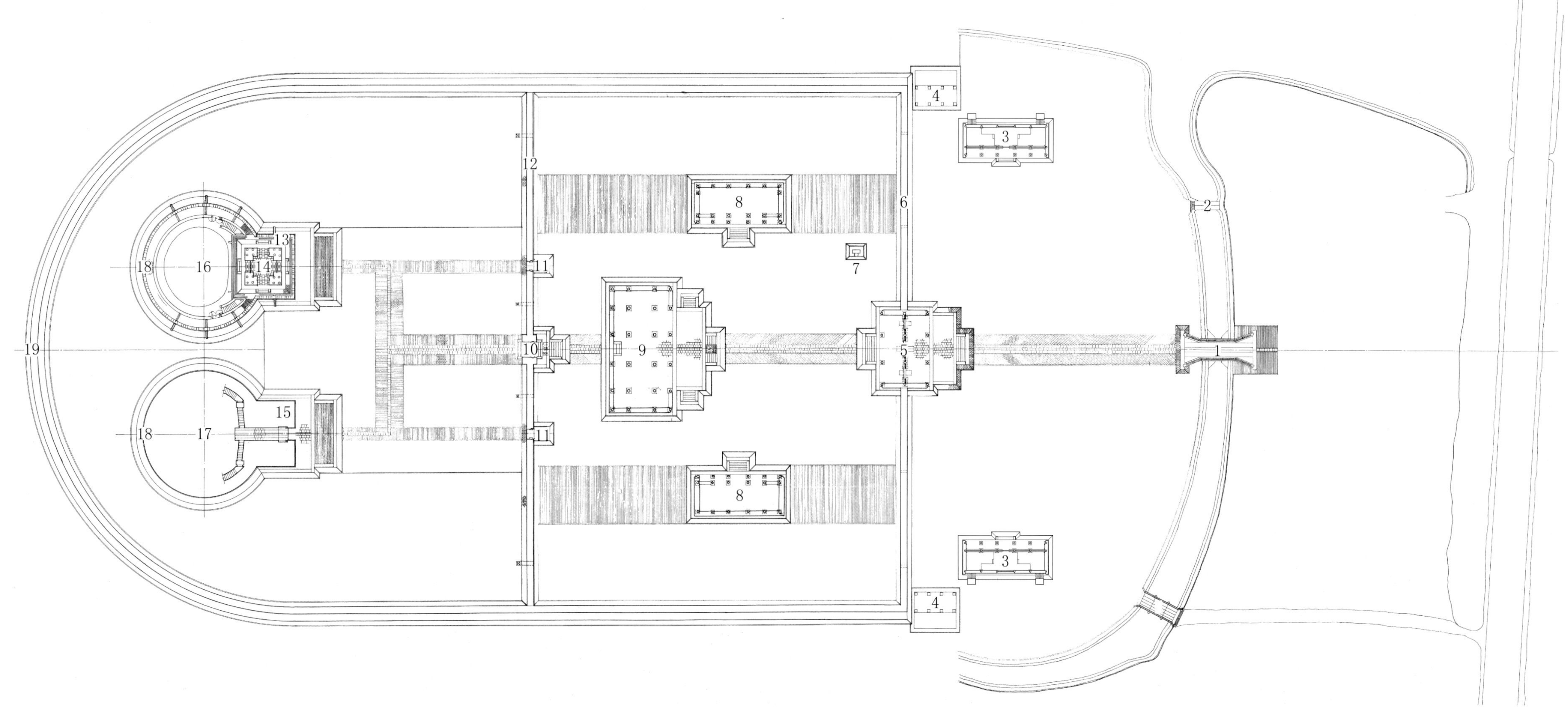

1 单孔桥 Single-Arch Bridge
2 石平桥 Flat Stone Bridge
3 朝房 Reception Halls for Court Officials
4 班房 Duty Houses for the Guards
5 宫门 Main Gate
6 陵墙 Tomb Wall
7 焚帛炉 Sacrificial Burners
8 配殿 Side Hall
9 享殿 Main Sacrificial Hall
10 琉璃花门 Gate with Glazed Roof Tiles
11 随墙角门 Corner Doors on Each Side Along the Wall
12 卡子墙 Spacer Wall
13 方城 Square Walled Terrace
14 明楼 Memorial Tower
15 方城（首层）Square Walled Terrace (First Floor)
16 宝顶 Tumulus
17 宝顶（首层）Tumulus (First Floor)
18 宝城 Encircled Realm of Treasure
19 罗圈墙 Wall Surrounding the Tomb

景陵双妃园寝总平面图

Site plan of the Tombs of the Two Consorts' buried side by side affiliated with the Tomb of Admiration

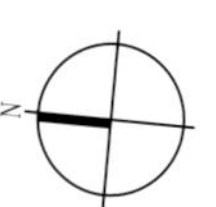

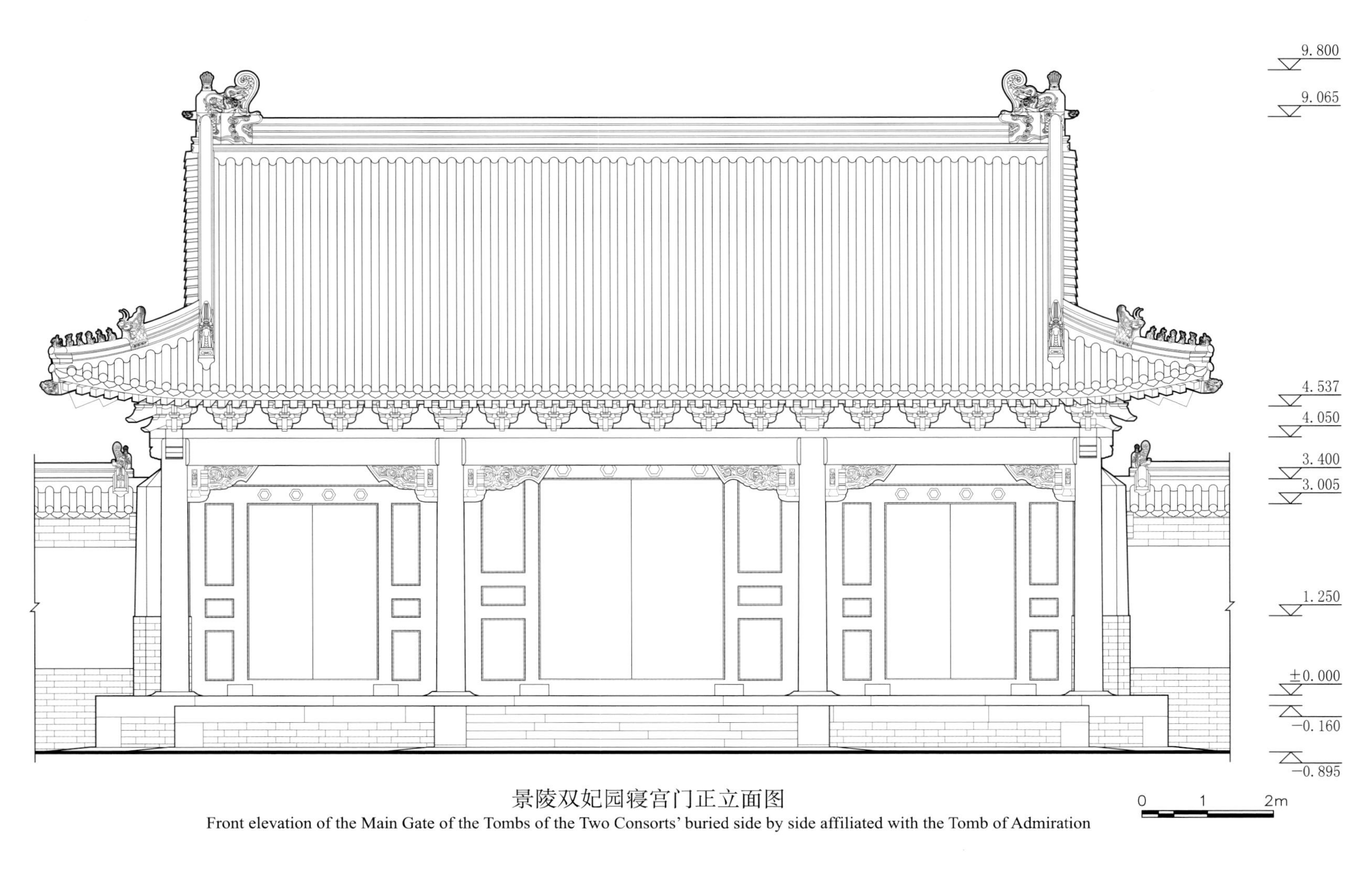

景陵双妃园寝宫门正立面图
Front elevation of the Main Gate of the Tombs of the Two Consorts' buried side by side affiliated with the Tomb of Admiration

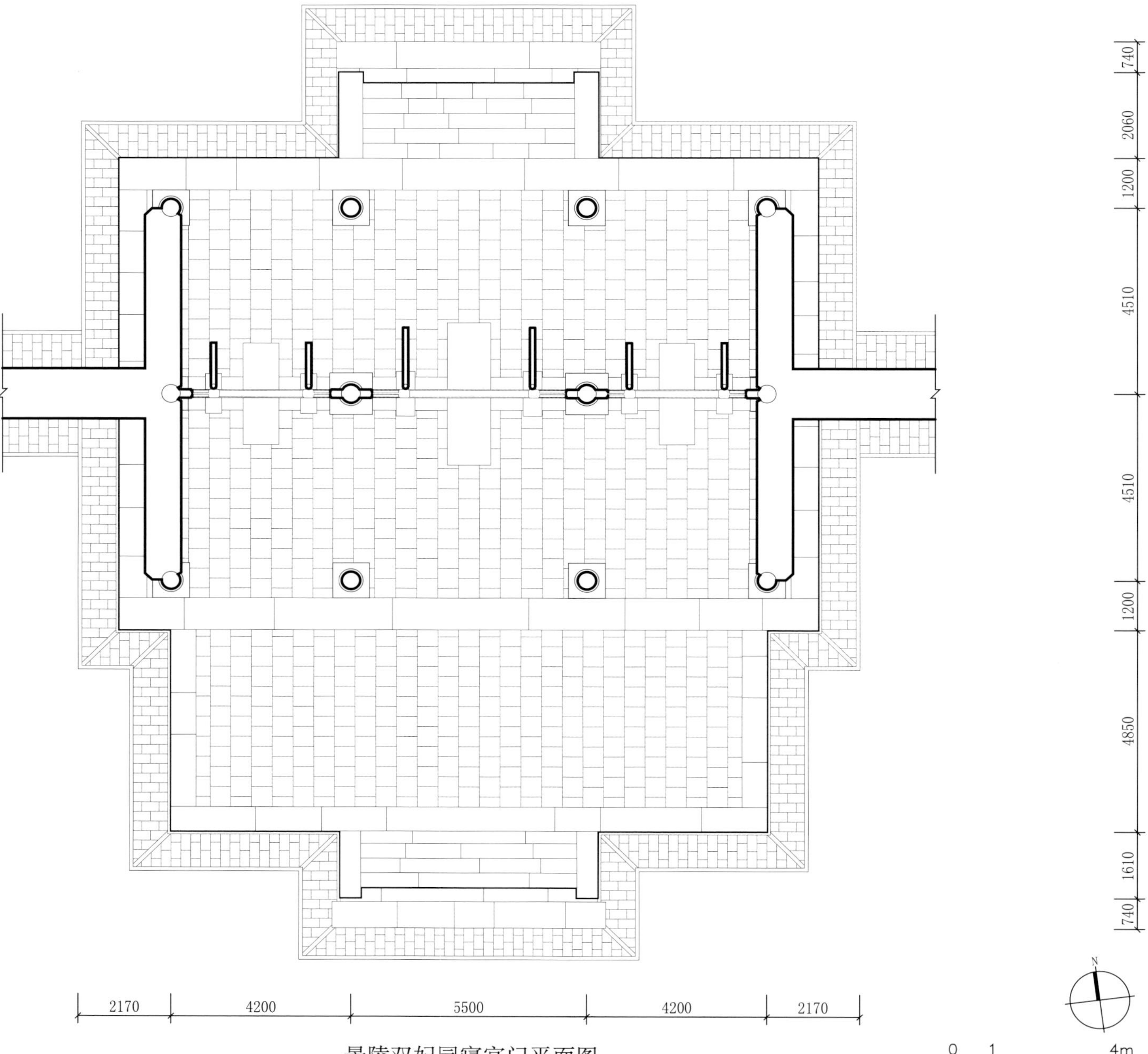

景陵双妃园寝宫门平面图
Plan of the Main Gate of the Tombs of the Two Consorts' buried side by side affiliated with the Tomb of Admiration

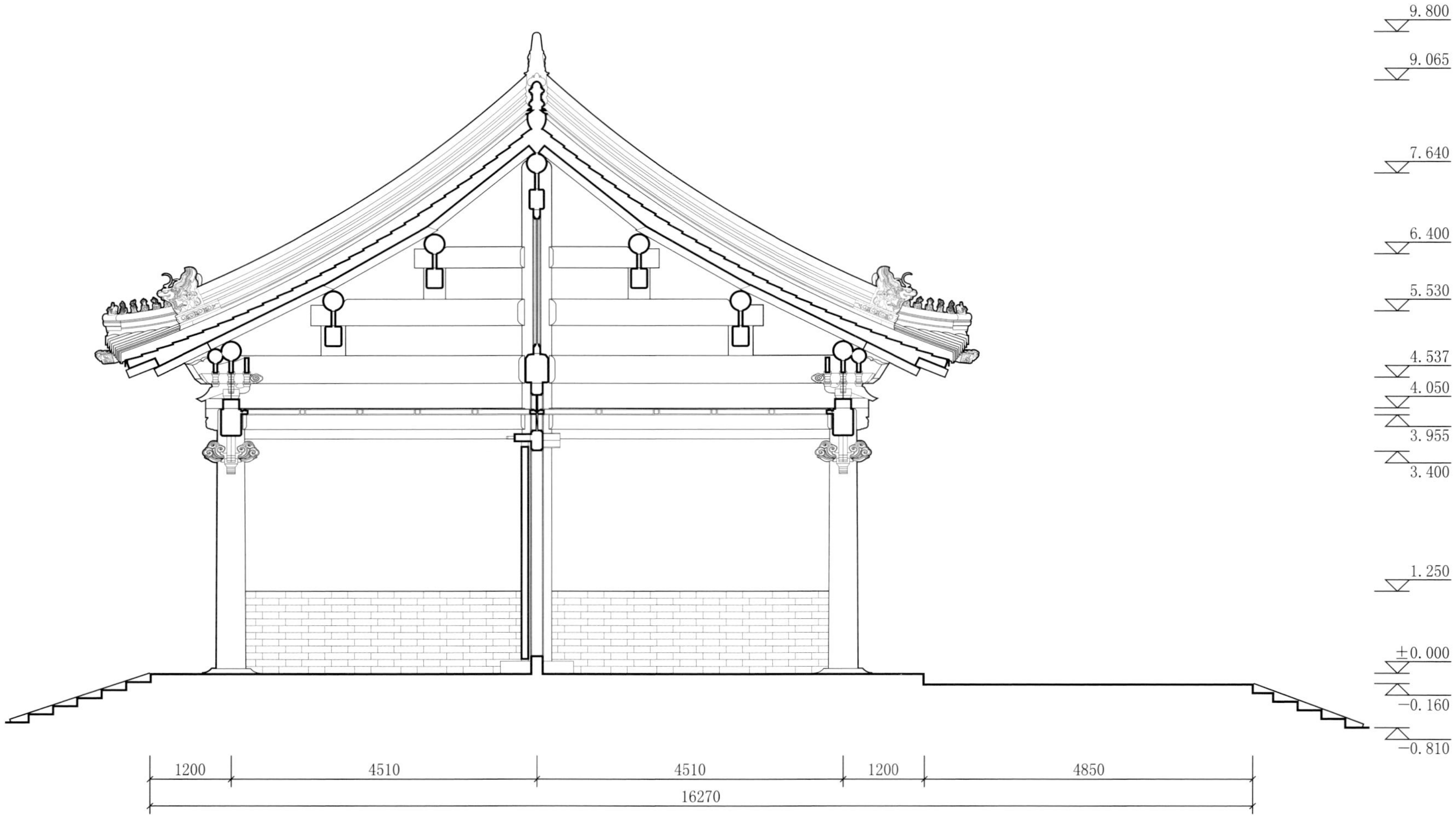

景陵双妃园寝宫门横剖面图
Cross section of the Main Gate of the Tombs of the Two Consorts' buried side by side affiliated with the Tomb of Admiration

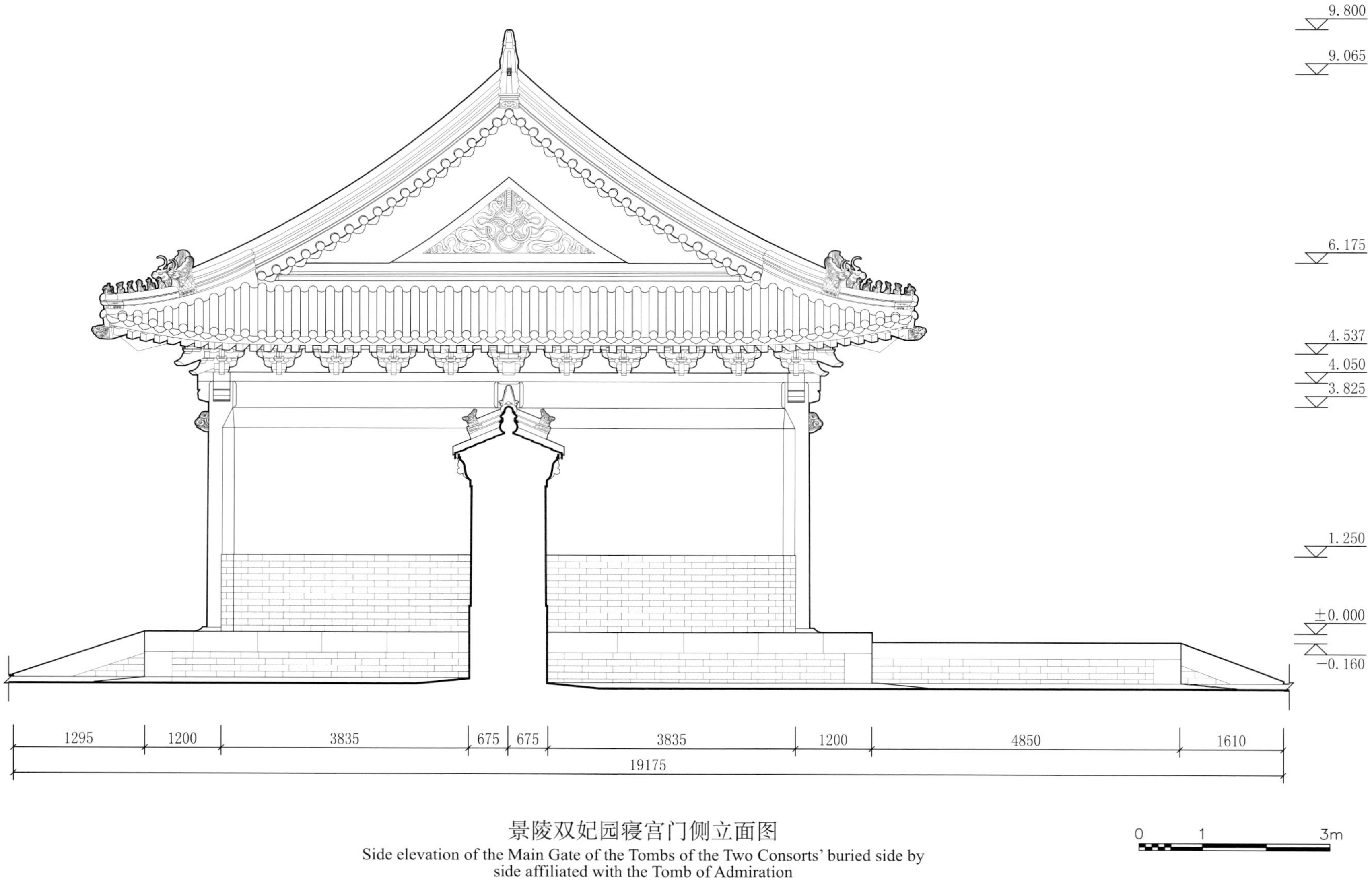

景陵双妃园寝宫门侧立面图
Side elevation of the Main Gate of the Tombs of the Two Consorts' buried side by side affiliated with the Tomb of Admiration

景陵双妃园寝享殿丹陛大样图
Detailed drawing of the marble carving on the steps to the Hall of Ritual Sacrifice, the Tombs of the Two Consorts' buried side by side affiliated with the Tomb of Admiration

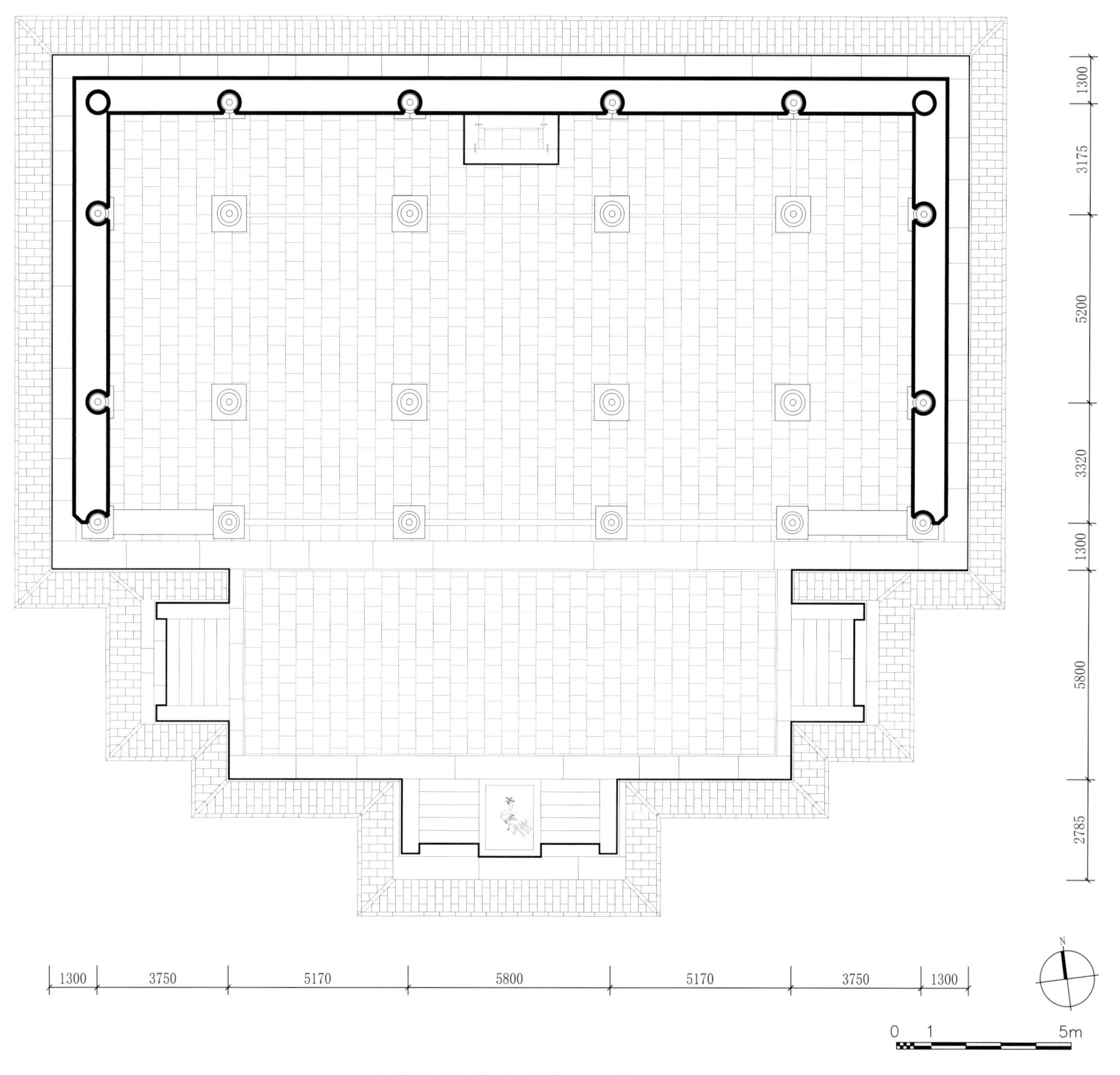

景陵双妃园寝享殿遗址平面图
Plan of the Relics of the Hall of Ritual Sacrifice, the Tombs of the Two Consorts' buried side by side affiliated with the Tomb of Admiration

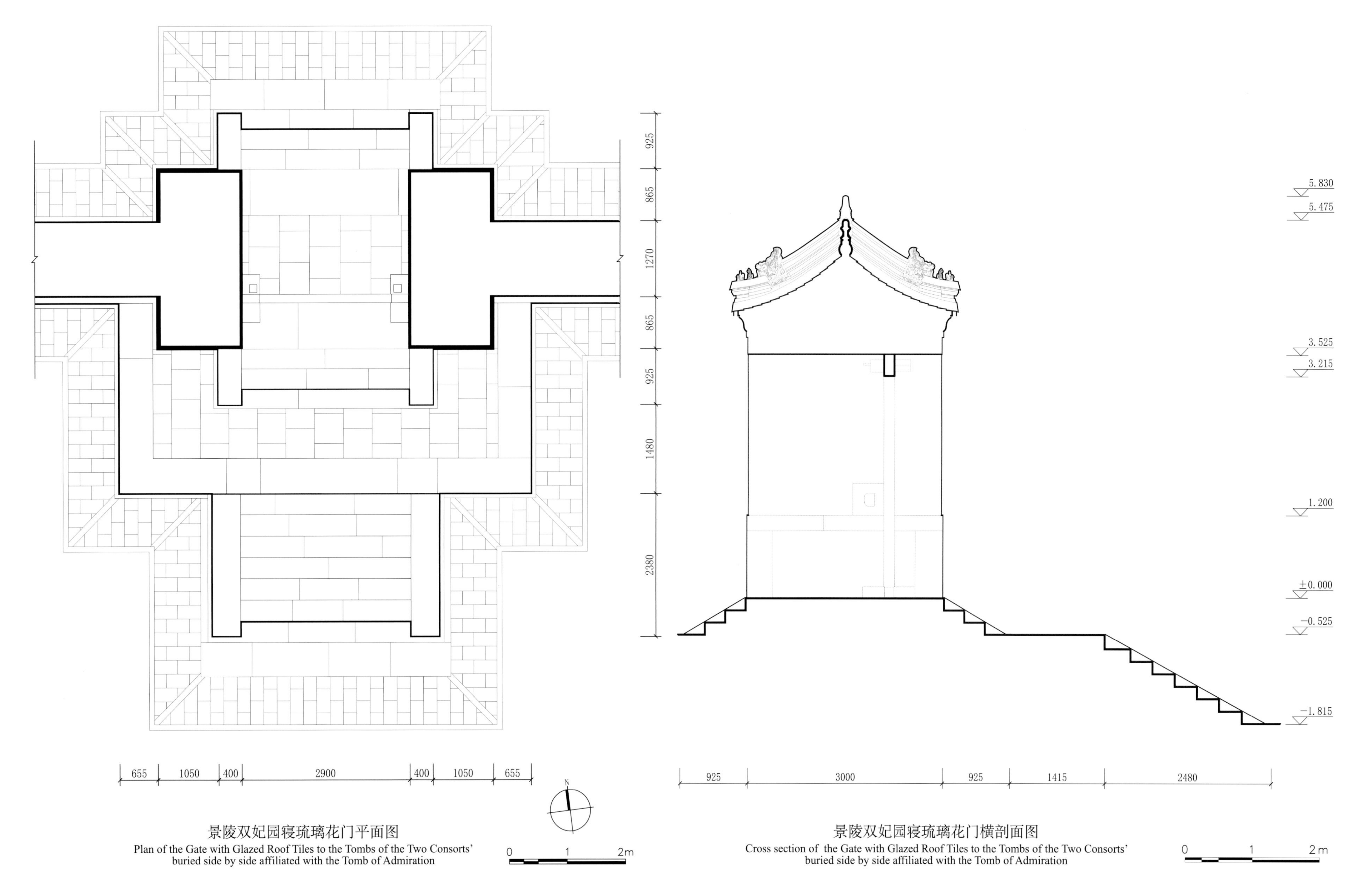

景陵双妃园寝琉璃花门平面图

Plan of the Gate with Glazed Roof Tiles to the Tombs of the Two Consorts' buried side by side affiliated with the Tomb of Admiration

景陵双妃园寝琉璃花门横剖面图

Cross section of the Gate with Glazed Roof Tiles to the Tombs of the Two Consorts' buried side by side affiliated with the Tomb of Admiration

925 865 1370 765 925 1480 2380

景陵双妃园寝琉璃花门侧立面图

Side elevation of the Gate with Glazed Roof Tiles to the Tombs of the Two Consorts' buried side by side affiliated with the Tomb of Admiration

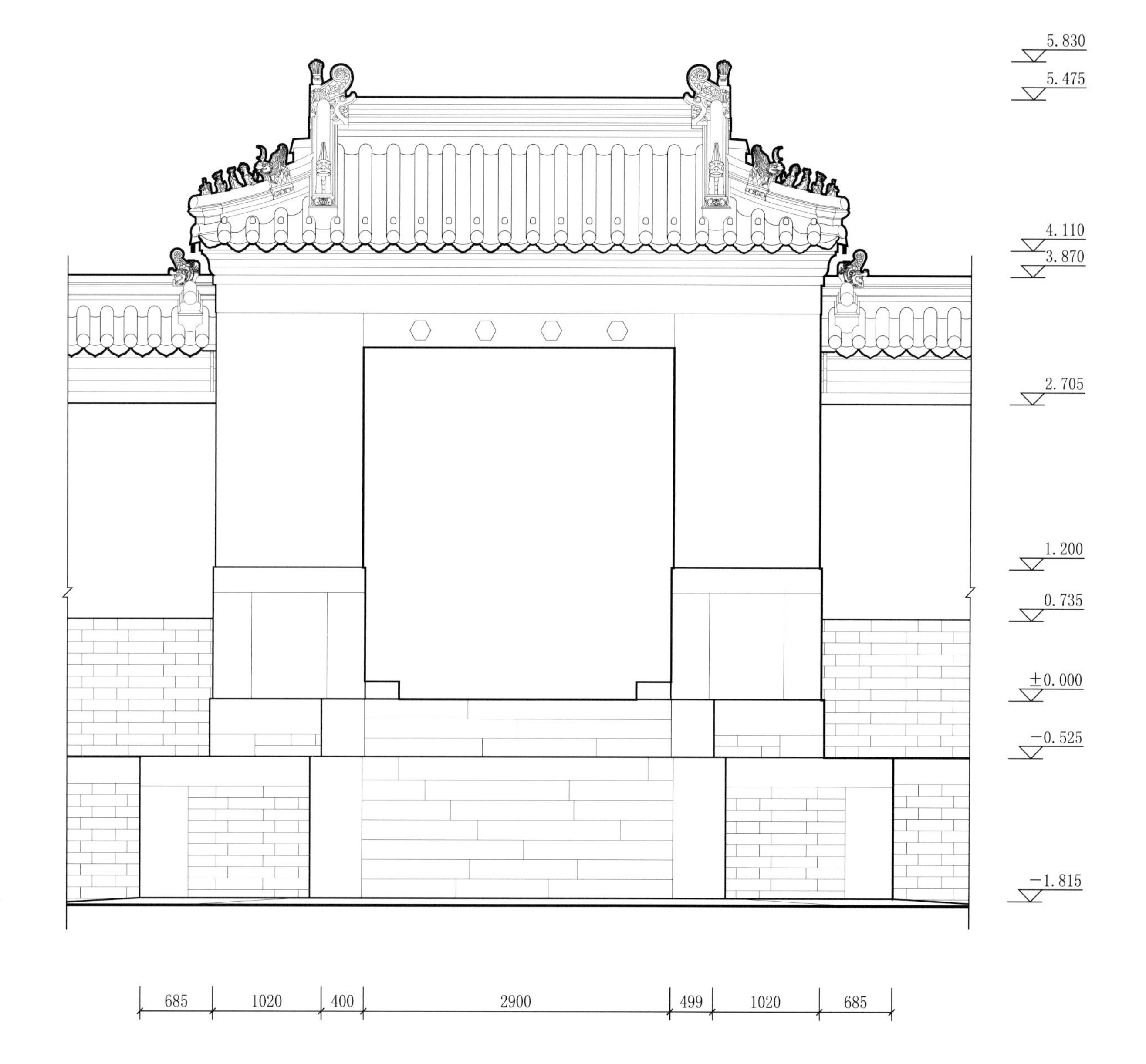

景陵双妃园寝琉璃花门正立面图

Front elevation of the Gate with Glazed Roof Tiles to the Tombs of the Two Consorts' buried side by side affiliated with the Tomb of Admiration

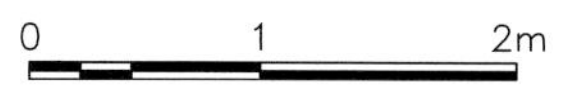

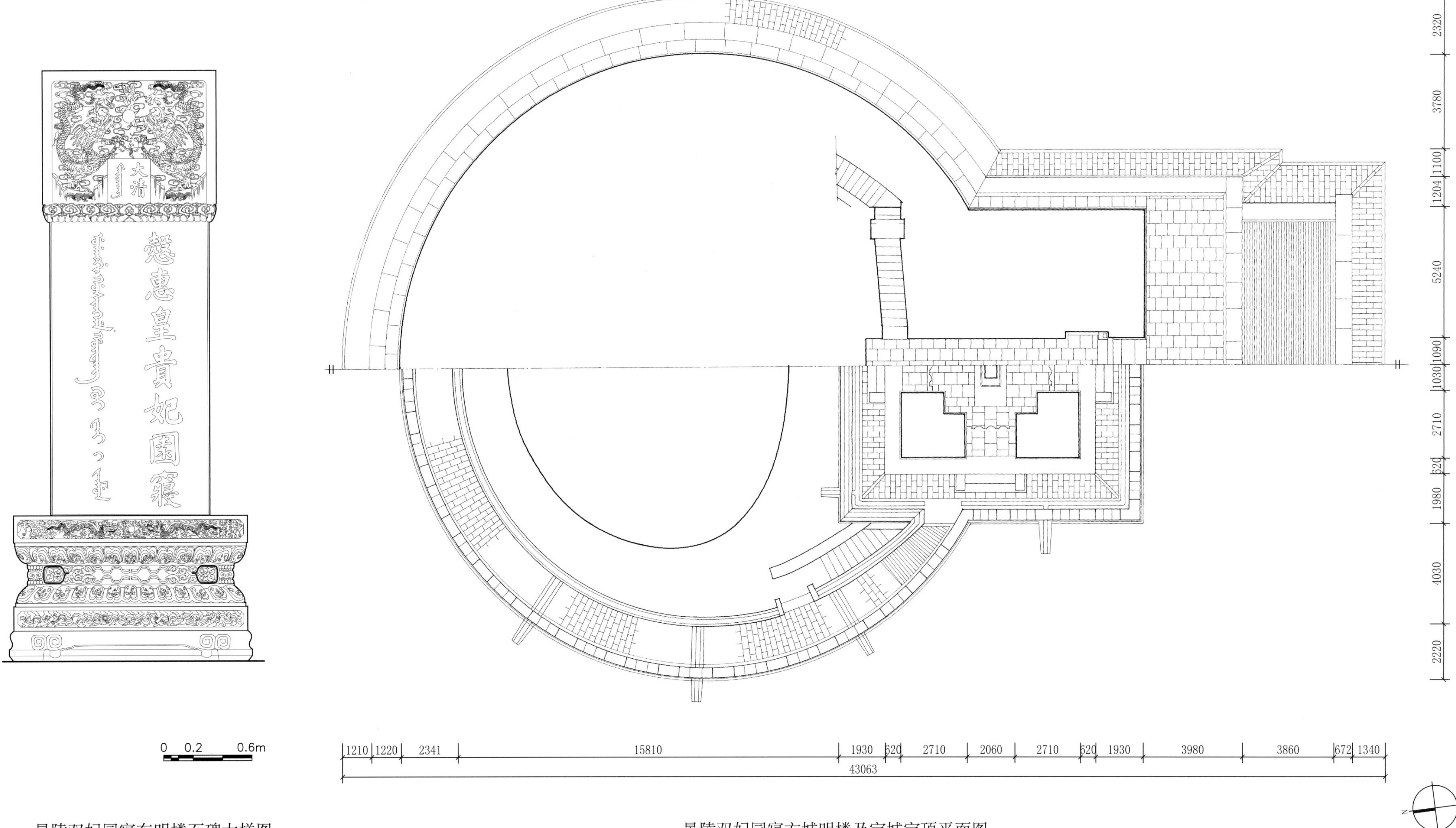

景陵双妃园寝东明楼石碑大样图

Detailed drawing of the Marble Stela in the Eastern Memorial Tower, the Tombs of the Two Consorts' buried side by side affiliated with the Tomb of Admiration

景陵双妃园寝方城明楼及宝城宝顶平面图

Plan of the Square Walled Terrace and the Memorial Tower as well as the Encircled Realm of Treasure and the Tumulus, the Tombs of the Two Consorts' buried side by side affiliated with the Tomb of Admiration

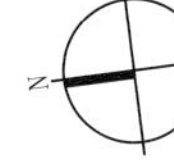

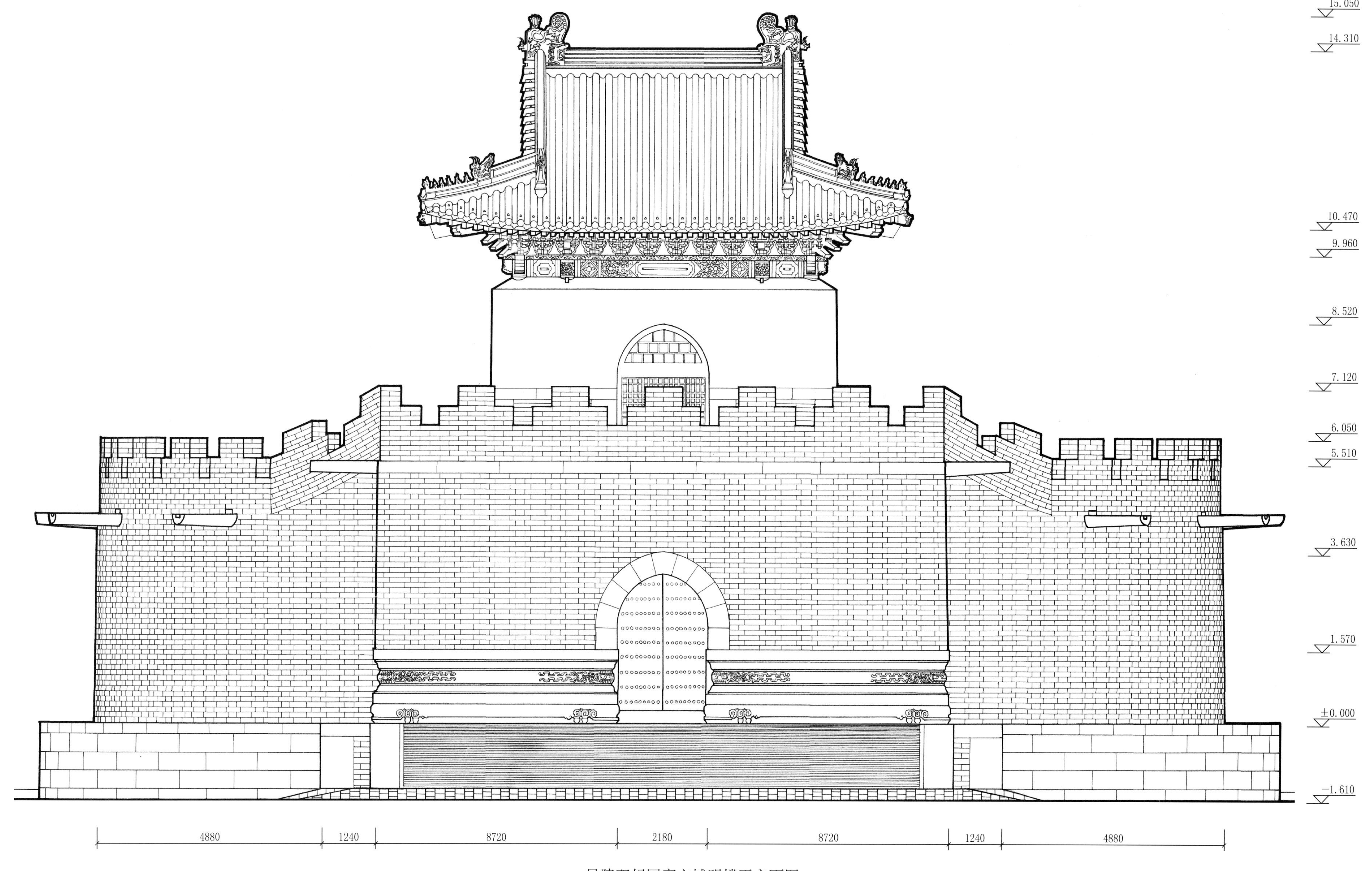

景陵双妃园寝方城明楼正立面图

Front elevation of the Square Walled Terrace and the Memorial Tower, the Tombs of the Two Consorts' buried side by side affiliated with the Tomb of Admiration

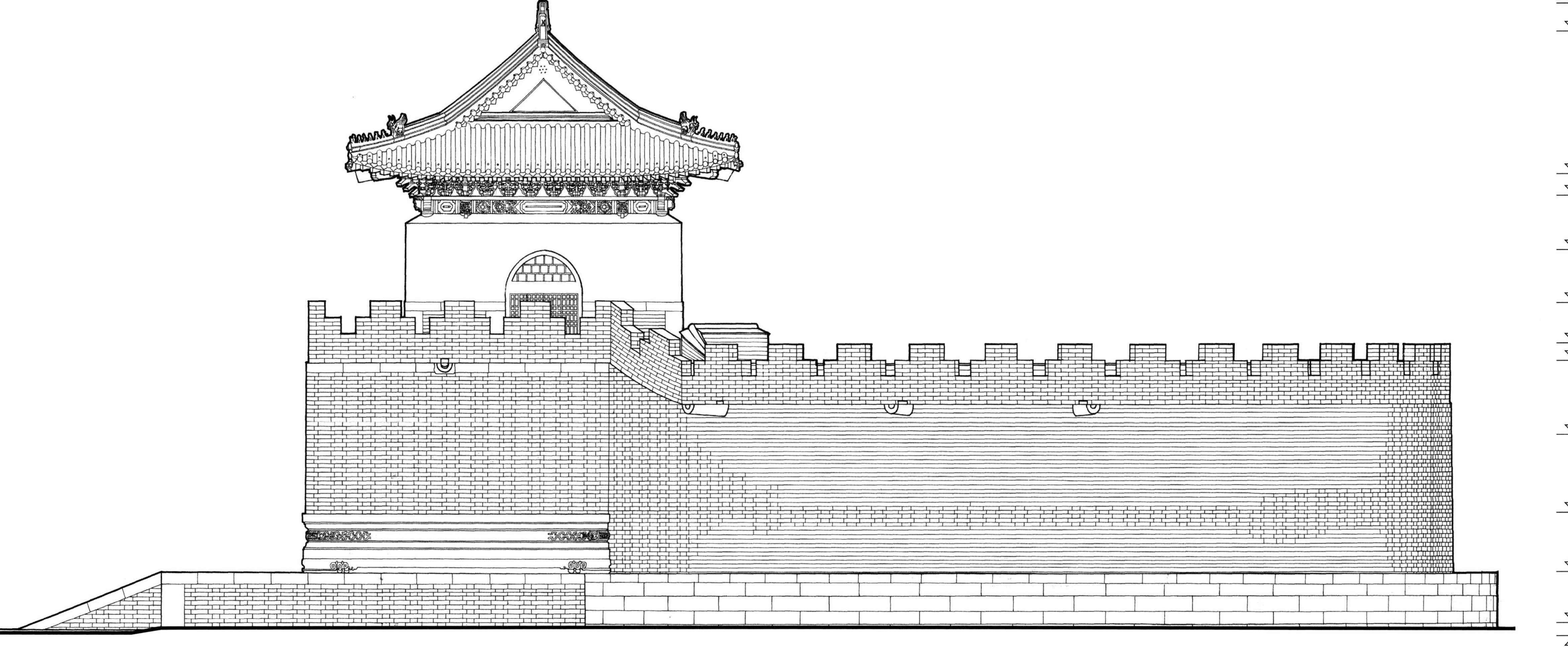

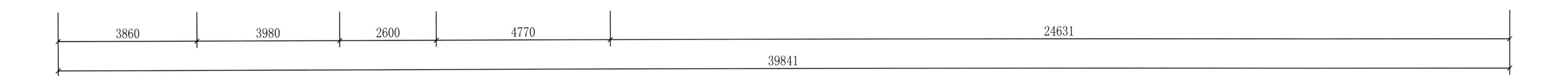

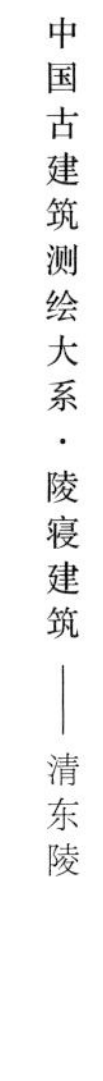

景陵双妃园寝方城明楼及宝城宝顶侧立面图

Side elevation of the Square Walled Terrace and the Memorial Tower as well as the Encircled Realm of Treasure and the Tumulus, the Tombs of the Two Consorts' buried side by side affiliated with the Tomb of Admiration

0 1 3m

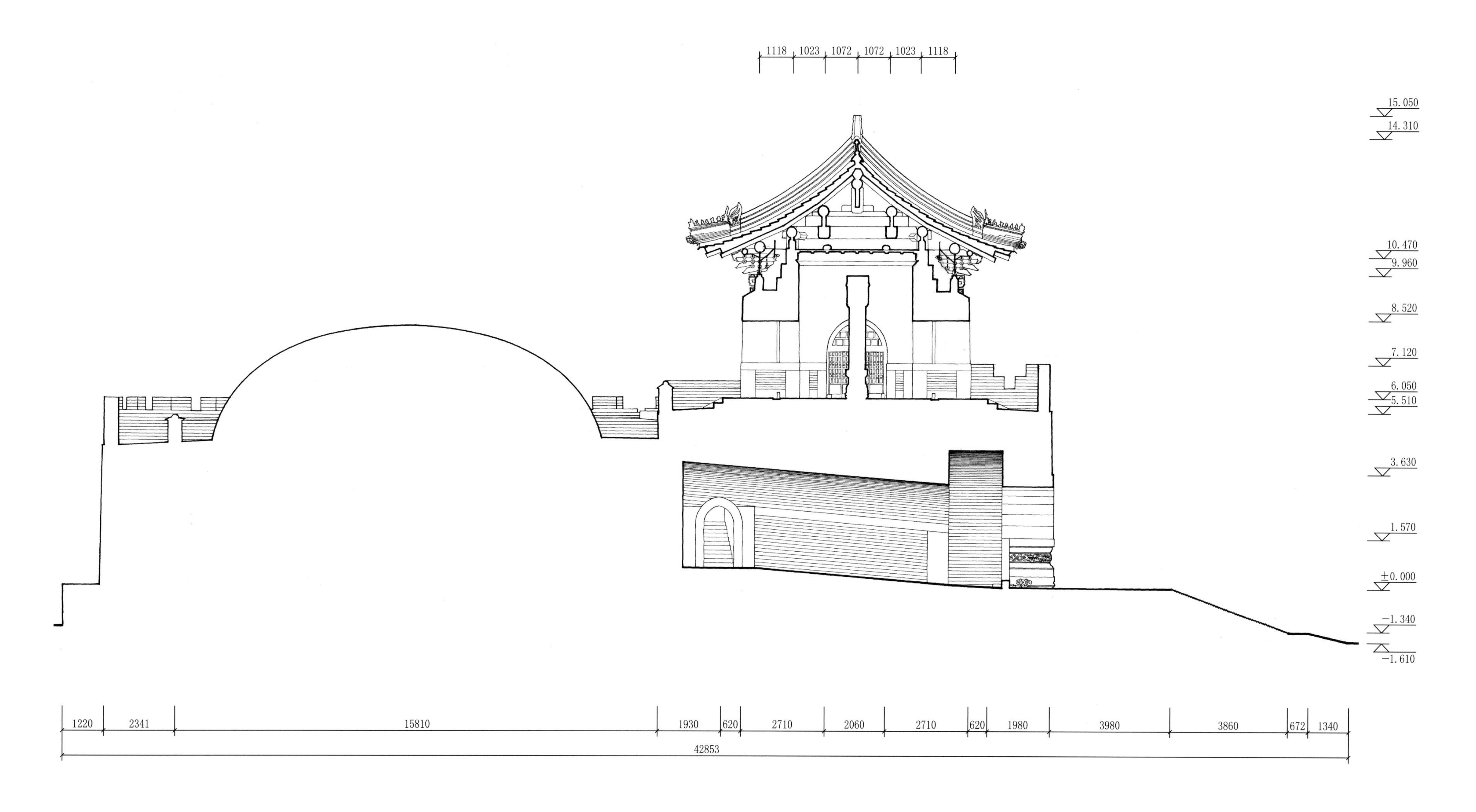

景陵双妃园寝方城明楼横剖面图

Cross section of the Square Walled Terrace and the Memorial Tower, the Tombs of the Two Consorts' buried side by side affiliated with the Tomb of Admiration

裕陵

Tomb of Prosperity (Yuling)

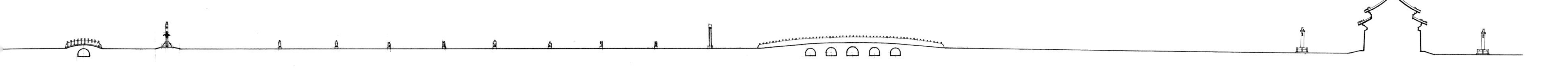

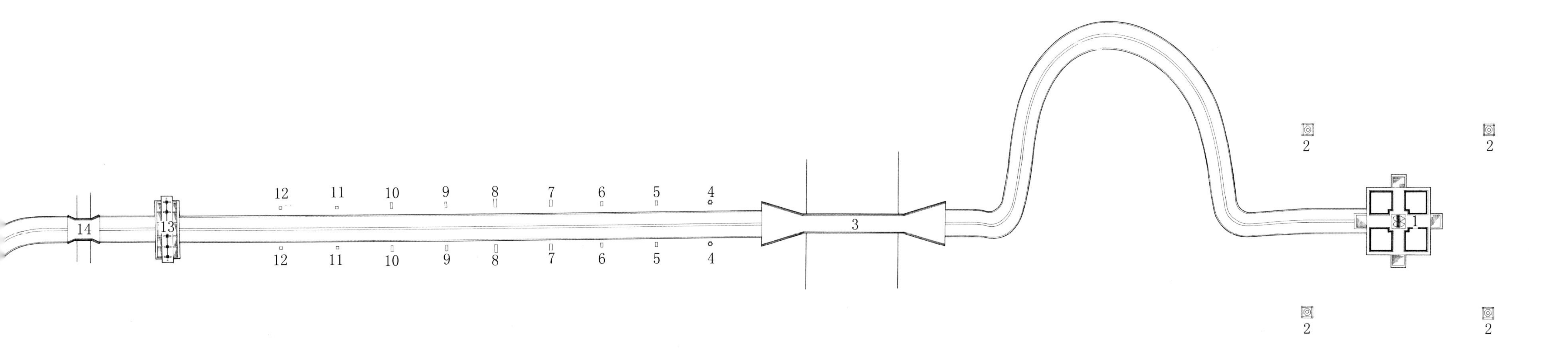

1 圣德神功碑亭 Pavilion for the Stela of Sagely Virtue and Divine Merit
2 华表 Ceremonial Columns
3 五孔石券桥 Five-Arch Stone Bridge
4 望柱 Ornamental Column
5 狮子 Lion
6 獬豸 Haechi
7 骆驼 Camel
8 大象 Elephant
9 麒麟 Kylin
10 马 Horse
11 武将 Martial Official
12 文臣 Literary Official
13 牌坊 Marble Memorial Gateway
14 单孔桥 Single-Arch Bridge
15 神道碑亭 Pavilion for the Stela on the Spirit Way
16 神厨库院落 Culinary Courtyard for Sacrifices
17 朝房 Reception Halls for Court Officials
18 三路三孔桥 Three-Way Three-Arch Bridges
19 石平桥 Flat Stone Bridge
20 班房 Duty Houses for the Guards
21 隆恩门 Gate of Monumental Grace
22 焚帛炉 Sacrificial Burners
23 配殿 Side Hall
24 隆恩殿 Hall of Monumental Grace
25 三路单孔桥 Three-Way Single-Arch Bridges
26 玉带河 Jade Ribbon River
27 琉璃花门 Gate with Glazed Roof Tiles
28 二柱门 Gate with Two Columns
29 石五供 Five Stone Ritual Vessels
30 石平桥 Flat Stone Bridge
31 月牙河 Crescent River
32 方城 Square Walled Terrace
33 明楼 Memorial Tower
34 哑巴院 Courtyard of the Mute
35 月牙城 Crescent Wall
36 琉璃影壁 Screen Wall of Glazed Tiles
37 宝顶 Tumulus
38 宝城 Encircled Realm of Treasure
39 罗圈墙 Wall Surrounding the Tomb
40 卡子墙 Spacer Wall
41 神厨库院门 Culinary Courtyard Gate for Sacrifices
42 神厨 Sacrificial Kitchen
43 神库 Sacrificial Storehouse
44 宰牲亭 Ritual Abattoir
45 井亭 Well Pavilion
46 下马牌 Stela Marking the Place for Dismounting from one's Horse
47 神道 Spirit Way

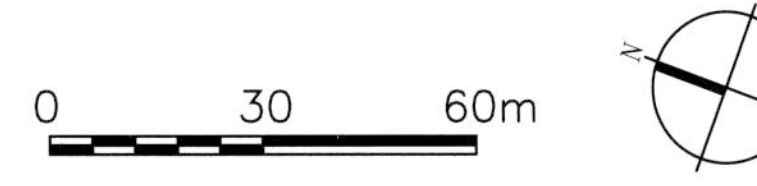

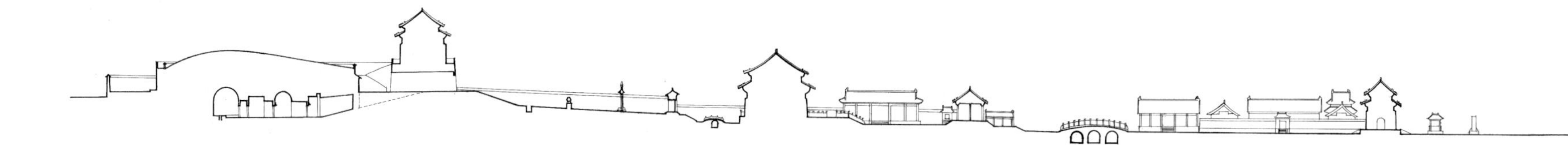

裕陵总剖面图

Site section of the Tomb of Prosperity

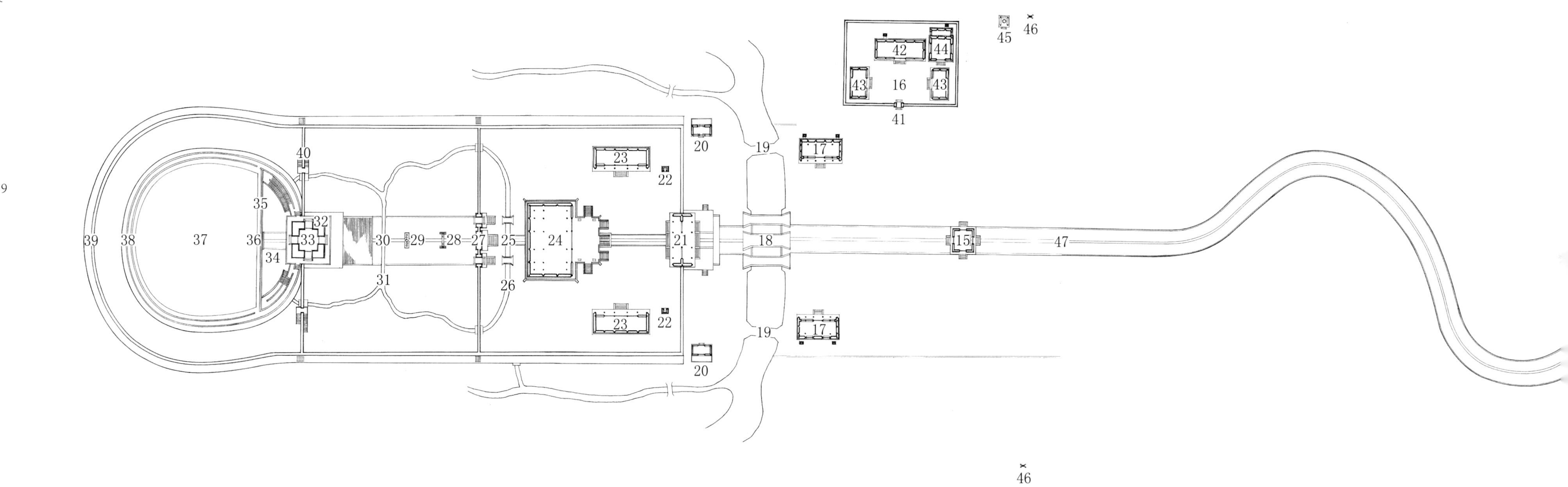

裕陵总平面图

Site plan of the Tomb of Prosperity

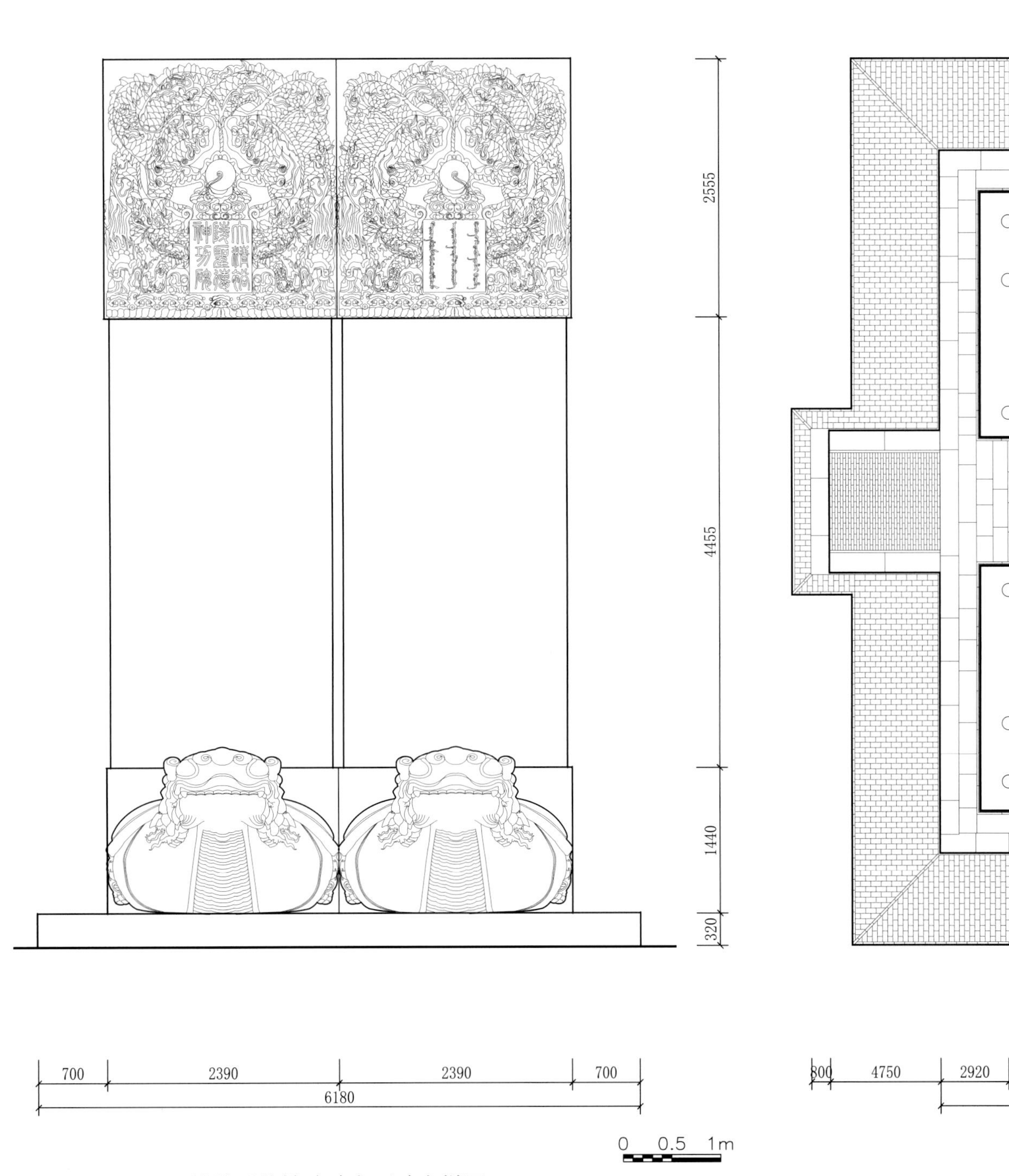

裕陵圣德神功碑亭石碑大样图

Detailed drawing of Stelae of Sage Virtue and Divine Merit in the Pavilion, Tomb of Prosperity

785
4830
1670
8146
2049
5200
25590
2049
8146
1670
4830
785

800 4750 2920 2410 2305 3180 7300 3180 2305 2410 2920 4750 800
28930

0 1 5m

裕陵圣德神功碑亭平面图

Plan of the Pavilion for the Stelae of Sage Virtue and Divine Merit, Tomb of Prosperity

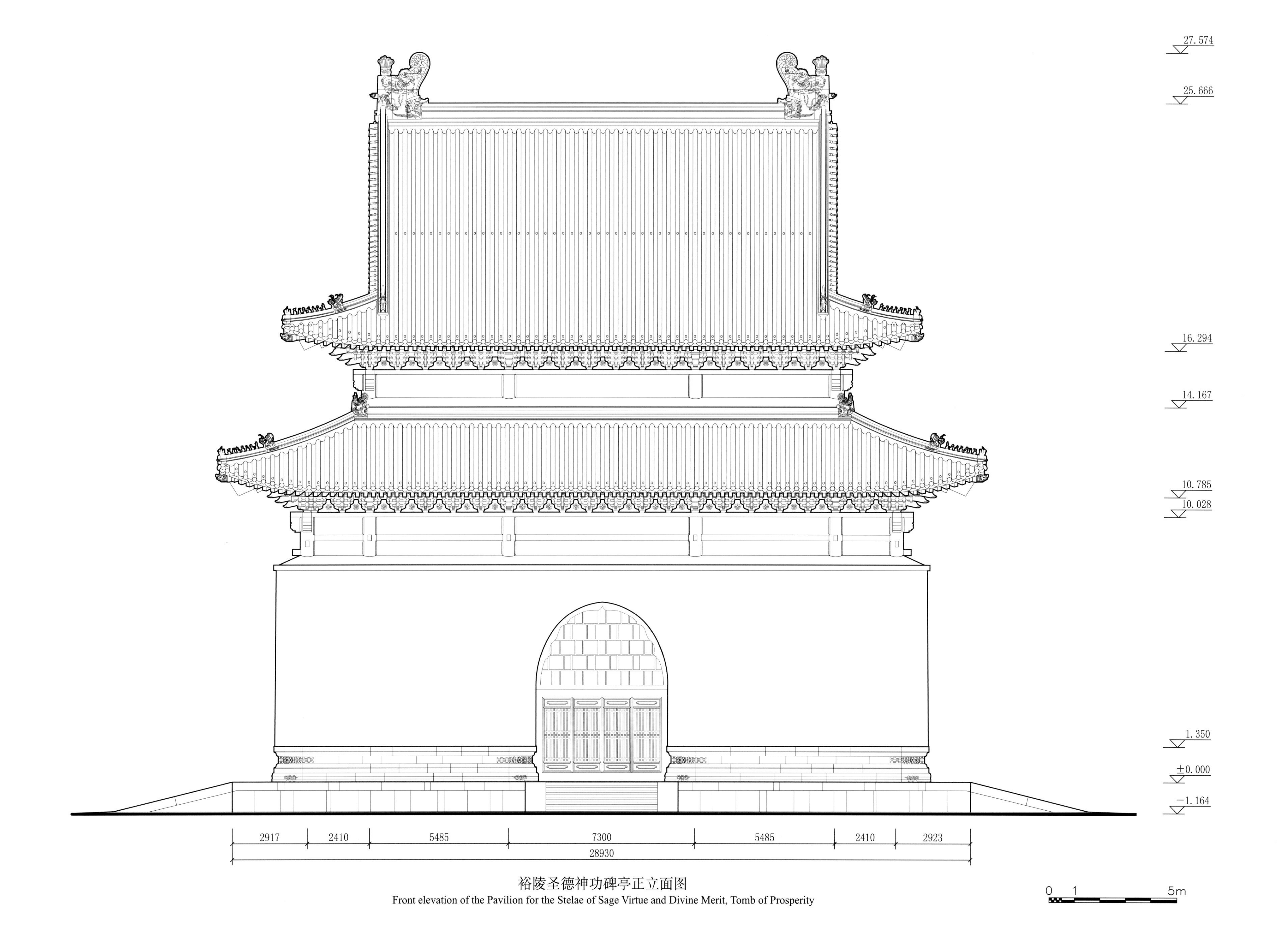

裕陵圣德神功碑亭正立面图

Front elevation of the Pavilion for the Stelae of Sage Virtue and Divine Merit, Tomb of Prosperity

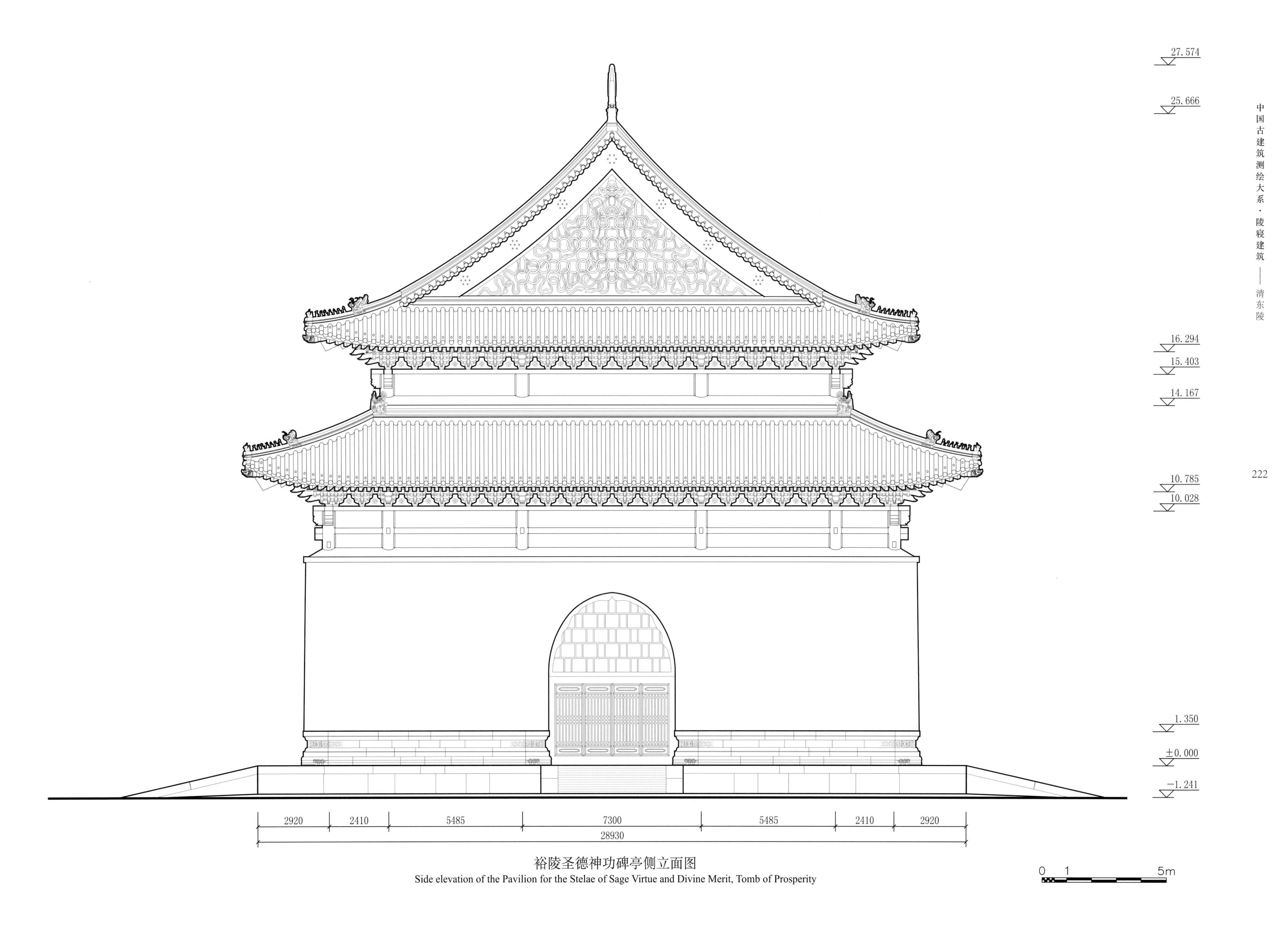

裕陵圣德神功碑亭侧立面图
Side elevation of the Pavilion for the Stelae of Sage Virtue and Divine Merit, Tomb of Prosperity

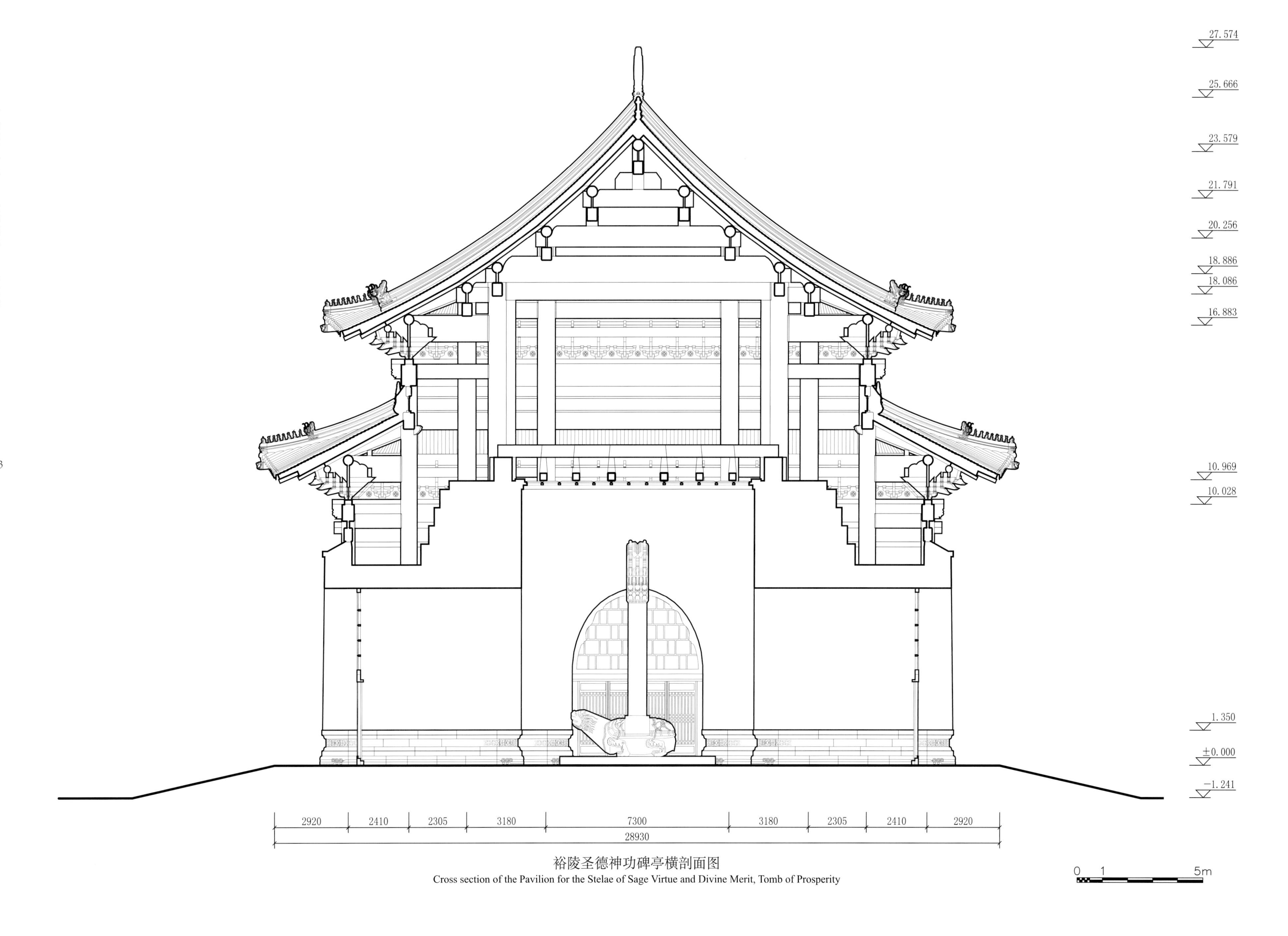

裕陵圣德神功碑亭横剖面图

Cross section of the Pavilion for the Stelae of Sage Virtue and Divine Merit, Tomb of Prosperity

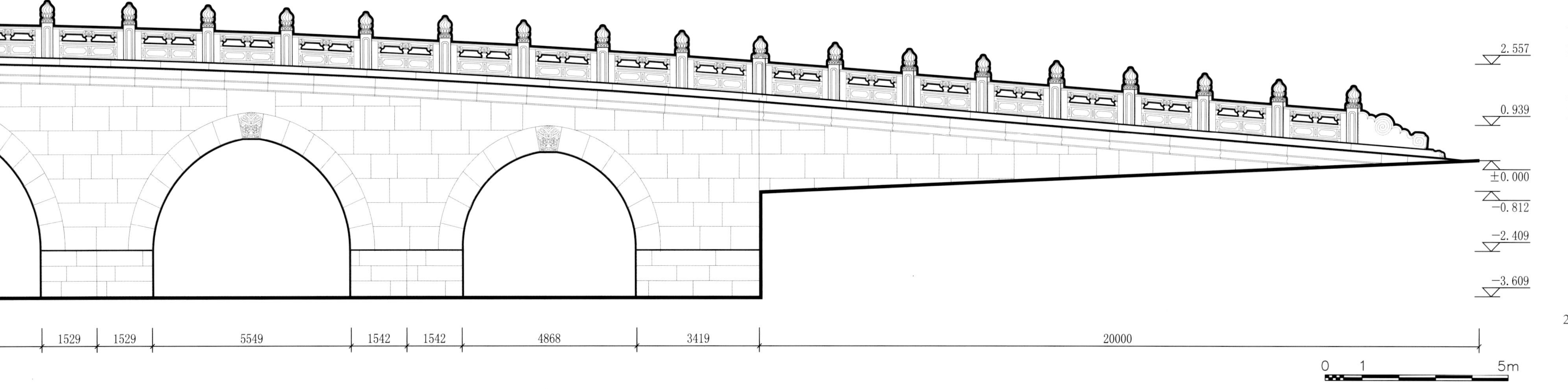

2.557
0.939
±0.000
−0.812
−2.409
−3.609
1529
1529
5549
1542
1542
4868
3419
20000
0
1
5m

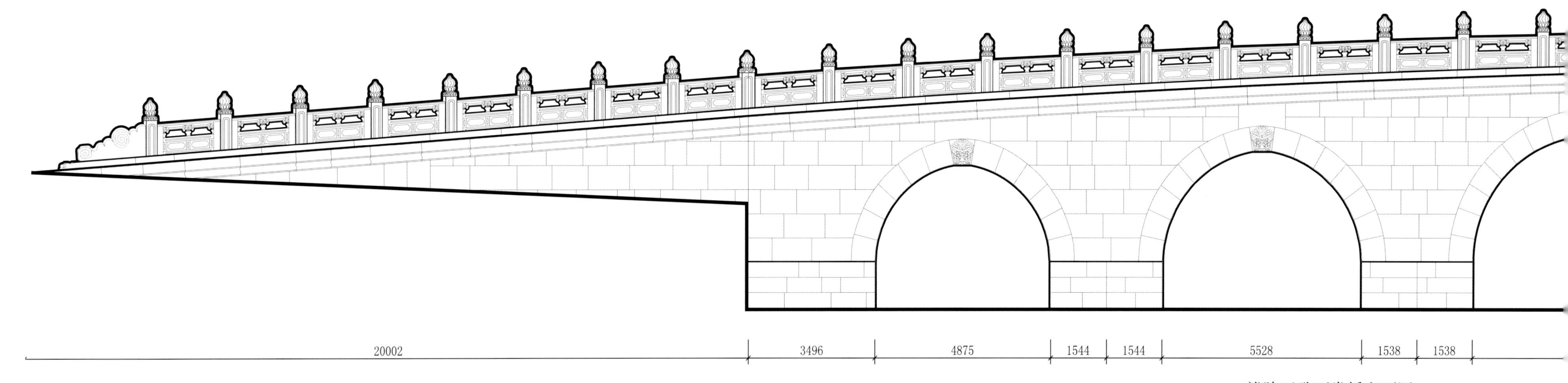

裕陵五孔石券桥立面图
Elevation of the Five-arch Stone Bridge, Tomb of Prosperity

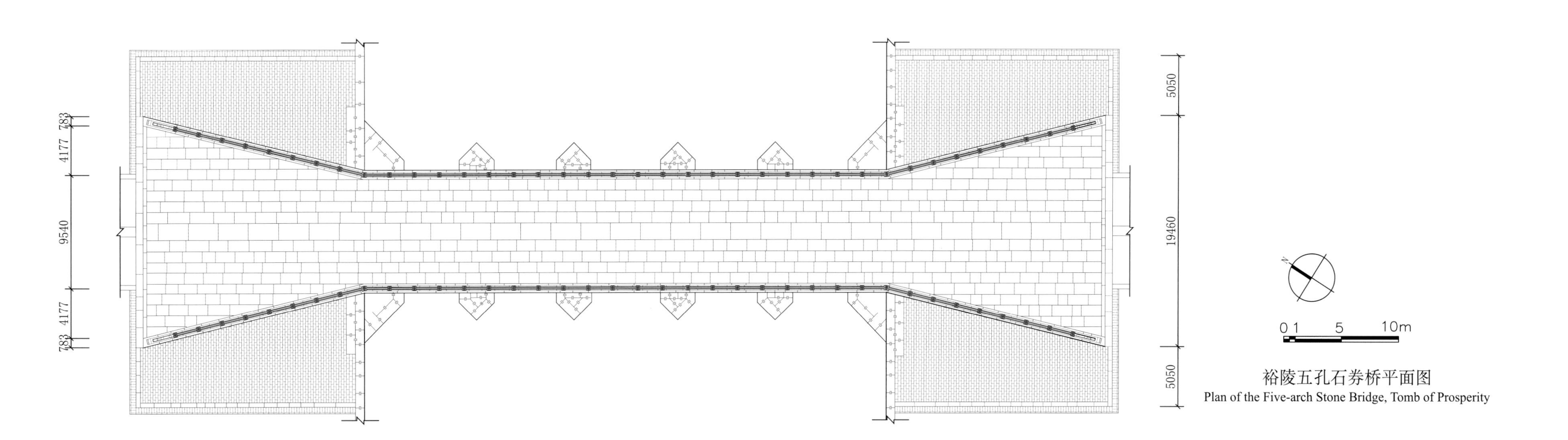

裕陵五孔石券桥平面图
Plan of the Five-arch Stone Bridge, Tomb of Prosperity

裕陵神道望柱平面图
Plan of the Ornamental Column of the Spirit Way, Tomb of Prosperity

裕陵神道望柱立面图
Elevation of the Ornamental Column of the Spirit Way, Tomb of Prosperity

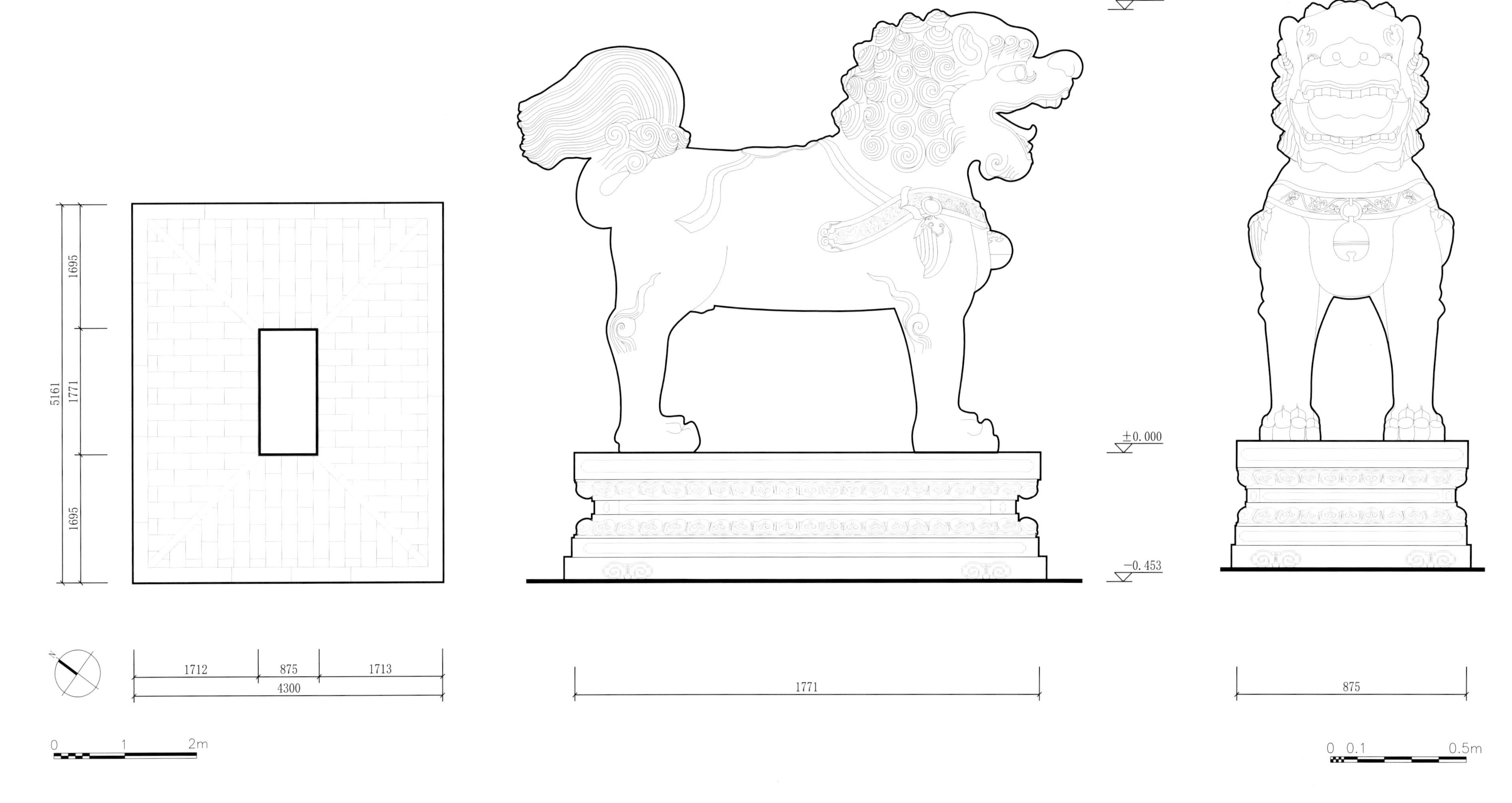

裕陵神道石像生狮子大样图

Detailed drawing of one of the Lions among the Stone Statues along the Spirit Way, Tomb of Prosperity

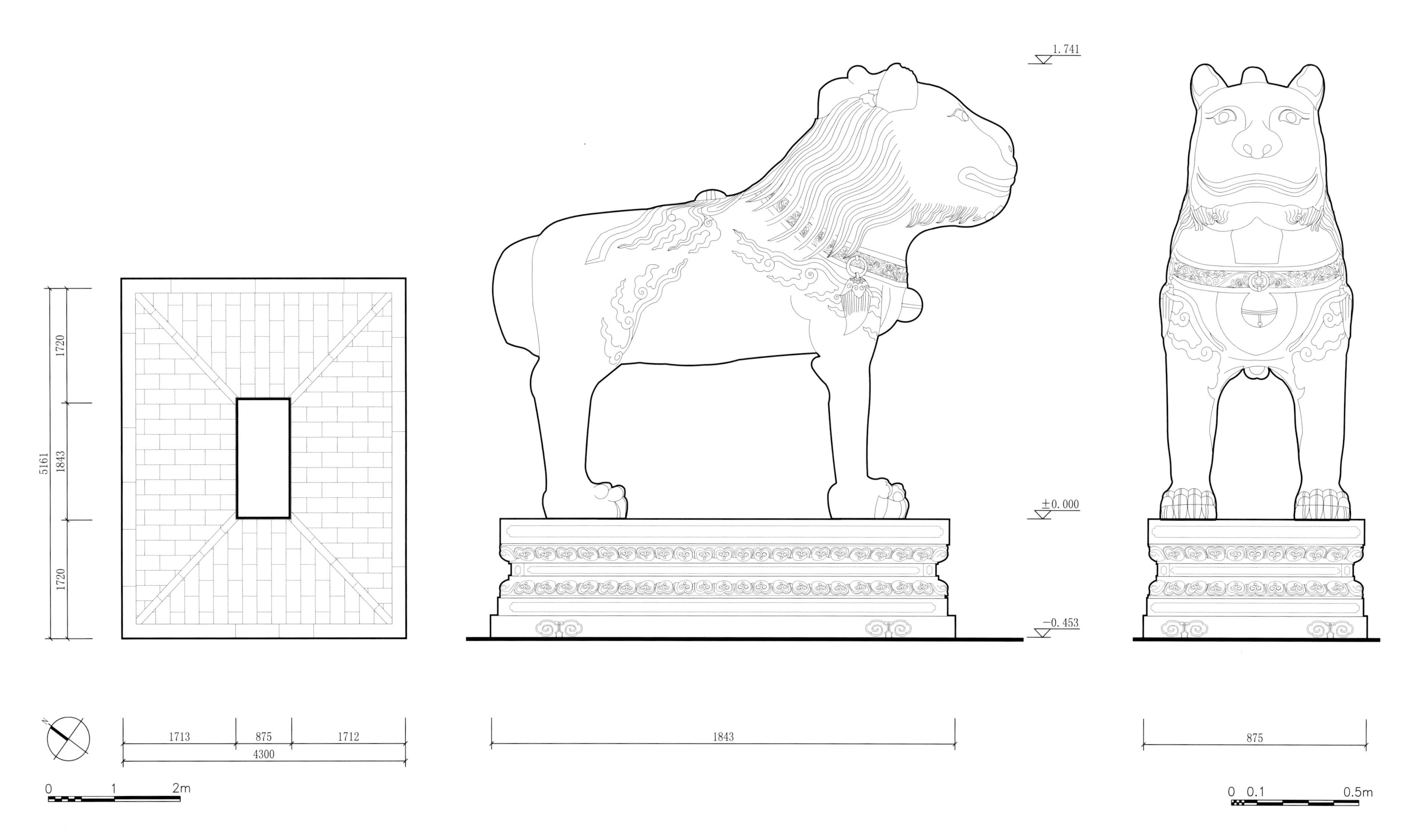

裕陵神道石像生獬豸大样图
Detailed drawing of one of the Xizhis among the Stone Statues along the Spirit Way, Tomb of Prosperity

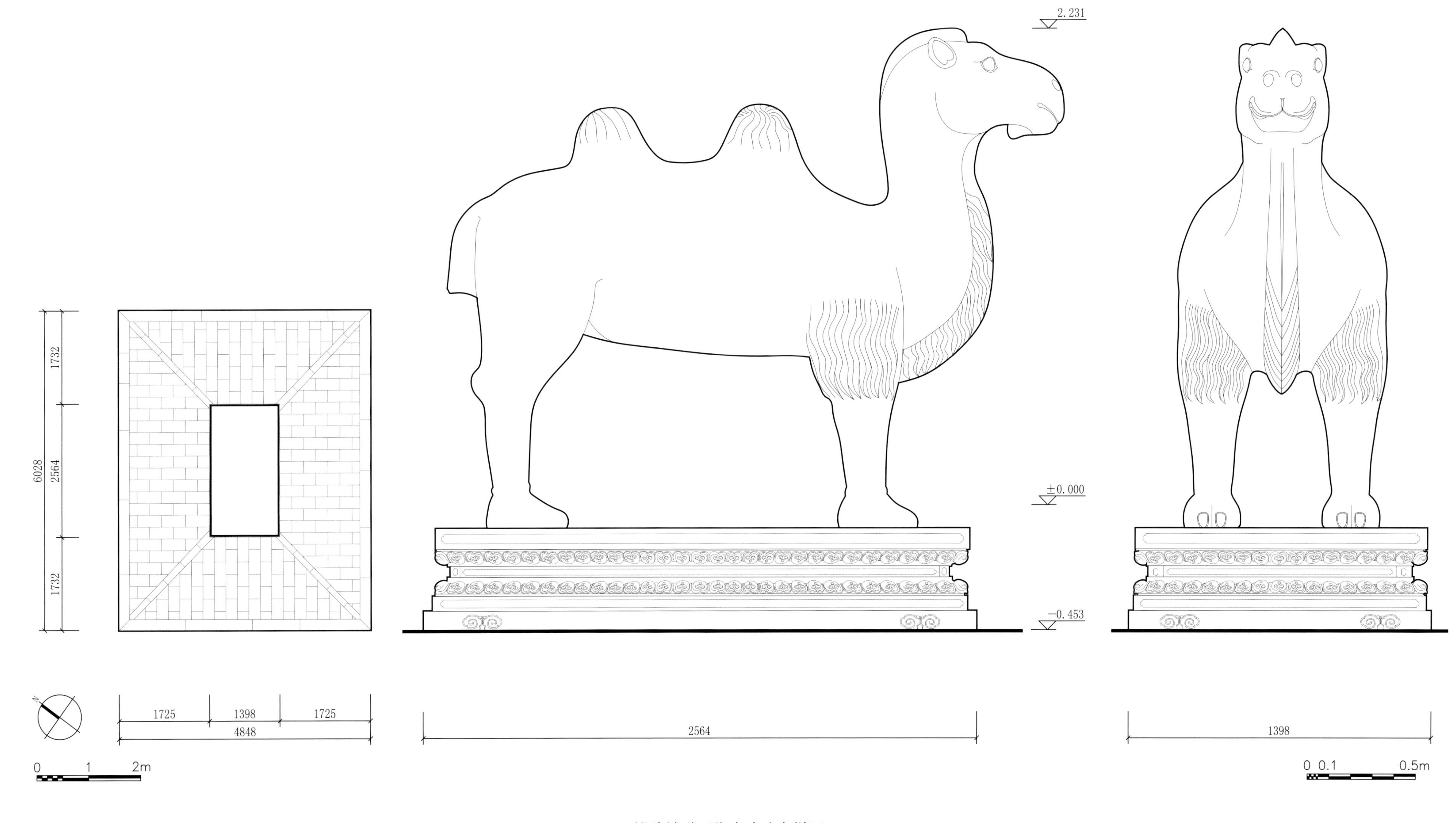

裕陵神道石像生骆驼大样图
Detailed drawing of one of the Camels among the Stone Statues along the Spirit Way, Tomb of Prosperity

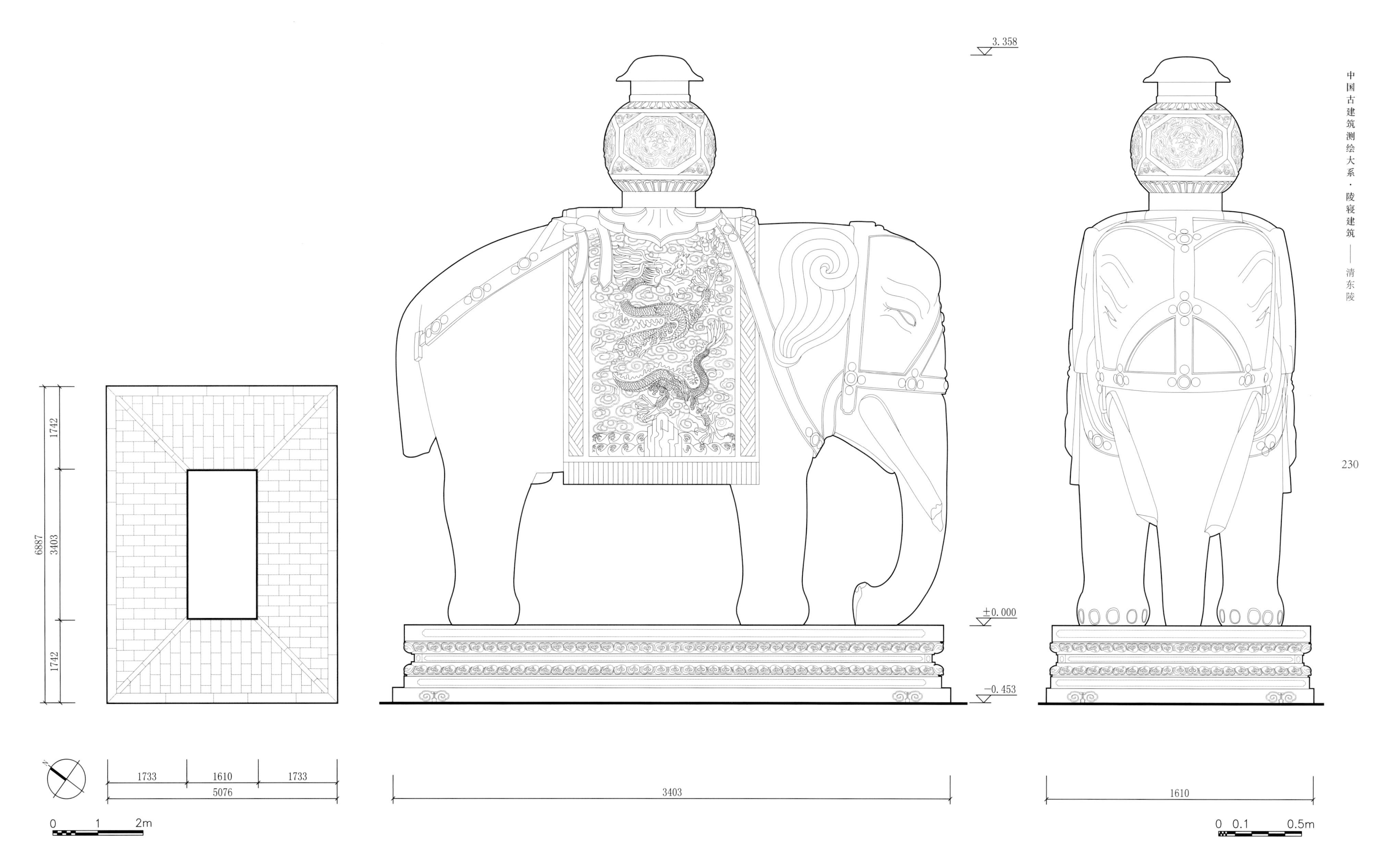

裕陵神道石像生大象大样图

Detailed drawing of one of the Elephants among the Stone Statues along the Spirit Way, Tomb of Prosperity

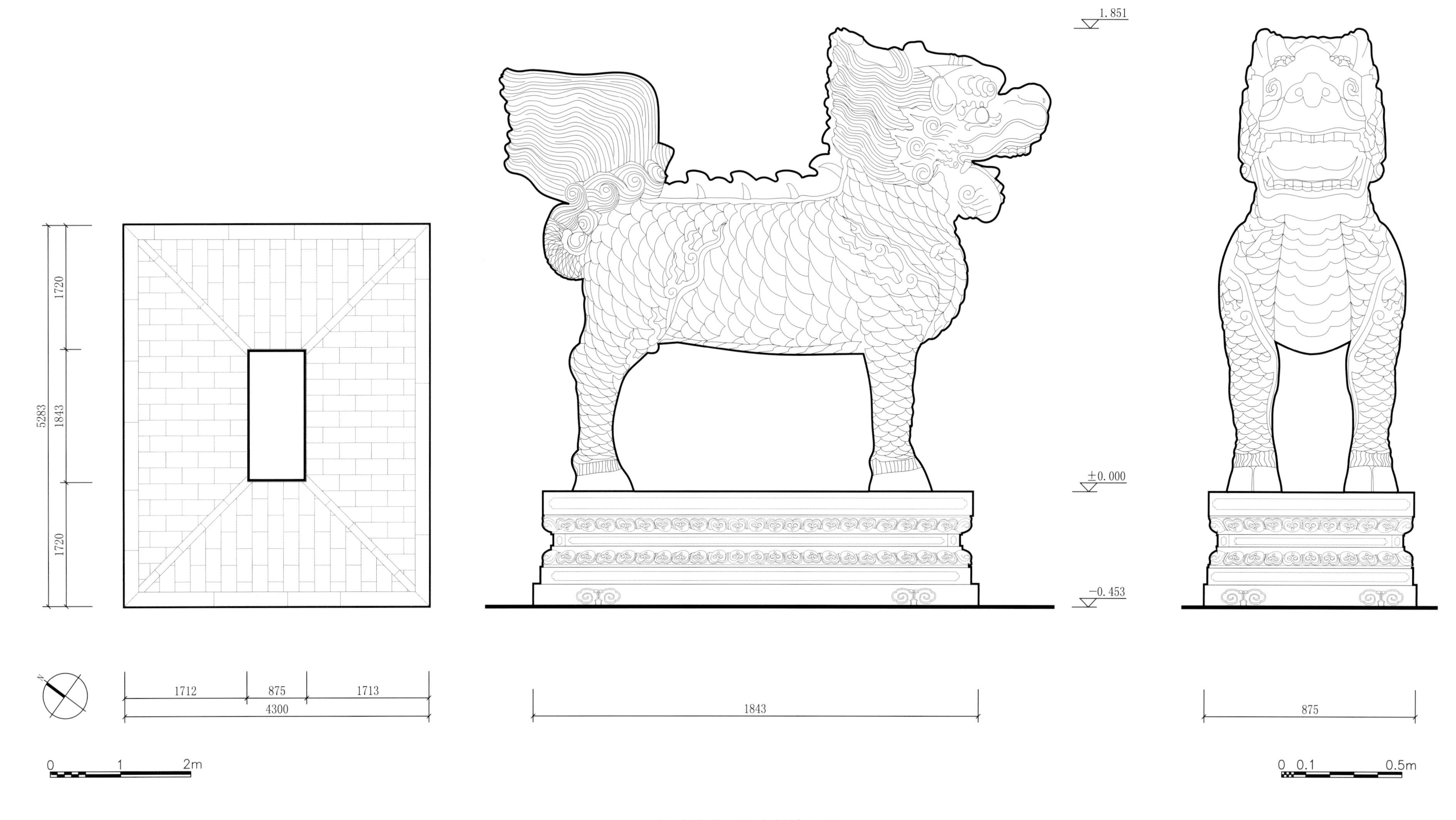

裕陵神道石像生麒麟大样图

Detailed drawing of one of the Qilins among the Stone Statues along the Spirit Way, Tomb of Prosperity

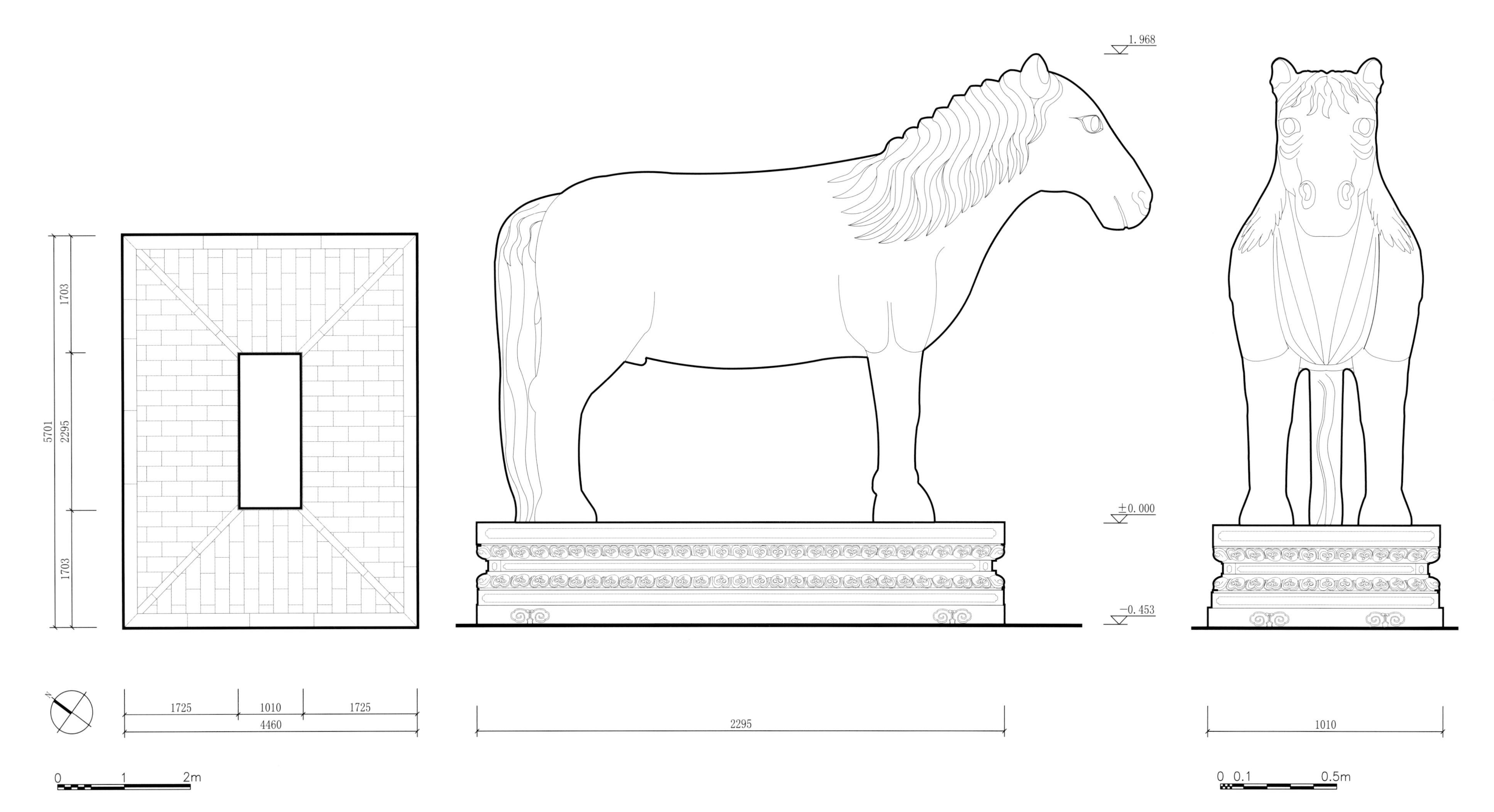

裕陵神道石像生马大样图

Detailed drawing of one of the Horses among the Stone Statues along the Spirit Way, Tomb of Prosperity

1740
844
1740
1725
1098
1725
4548
0 1 2m

裕陵神道石像生武臣大样图

Detailed drawing of one of the Martial Officials among the Stone Statues along the Spirit Way, Tomb of Prosperity

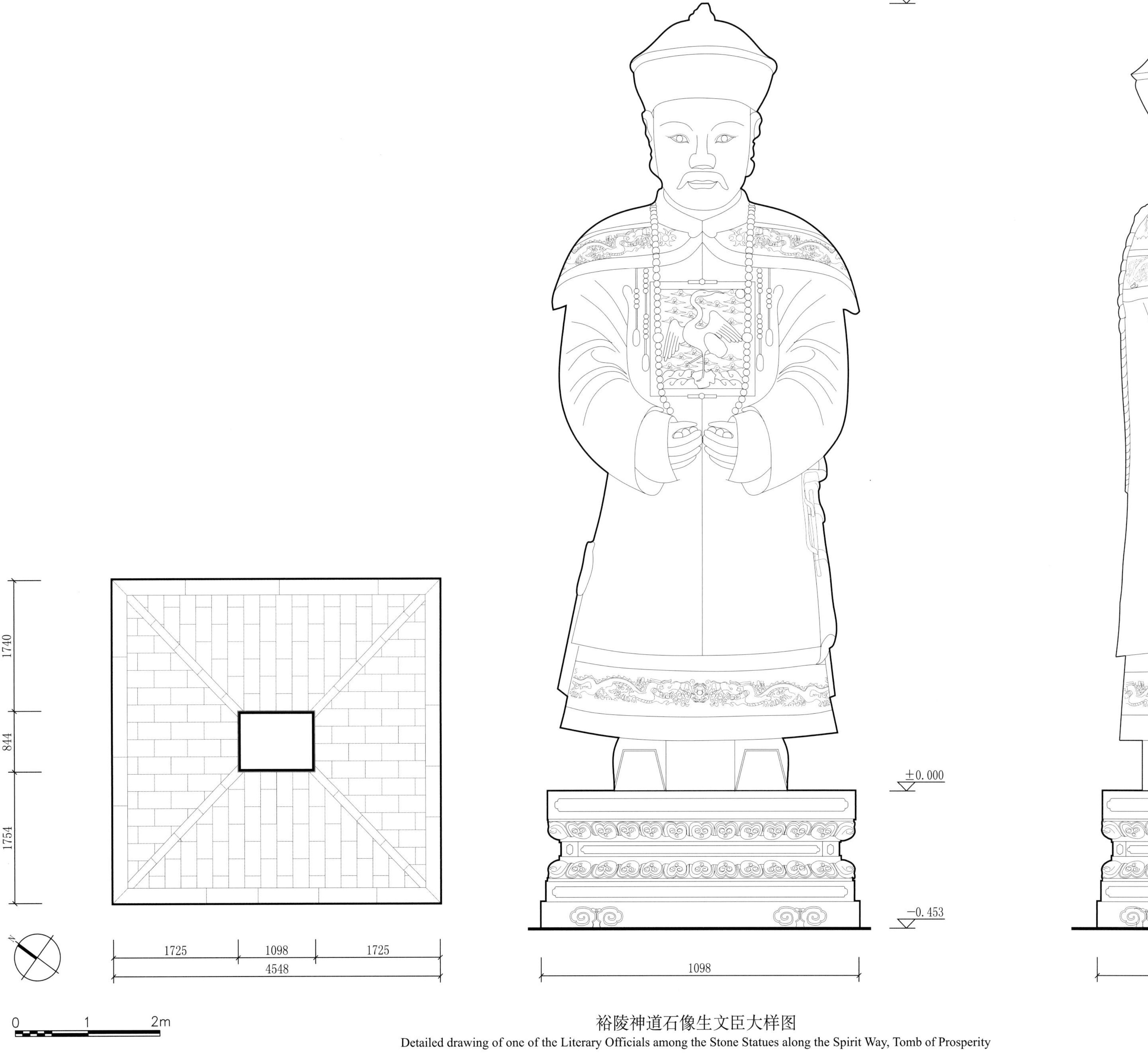

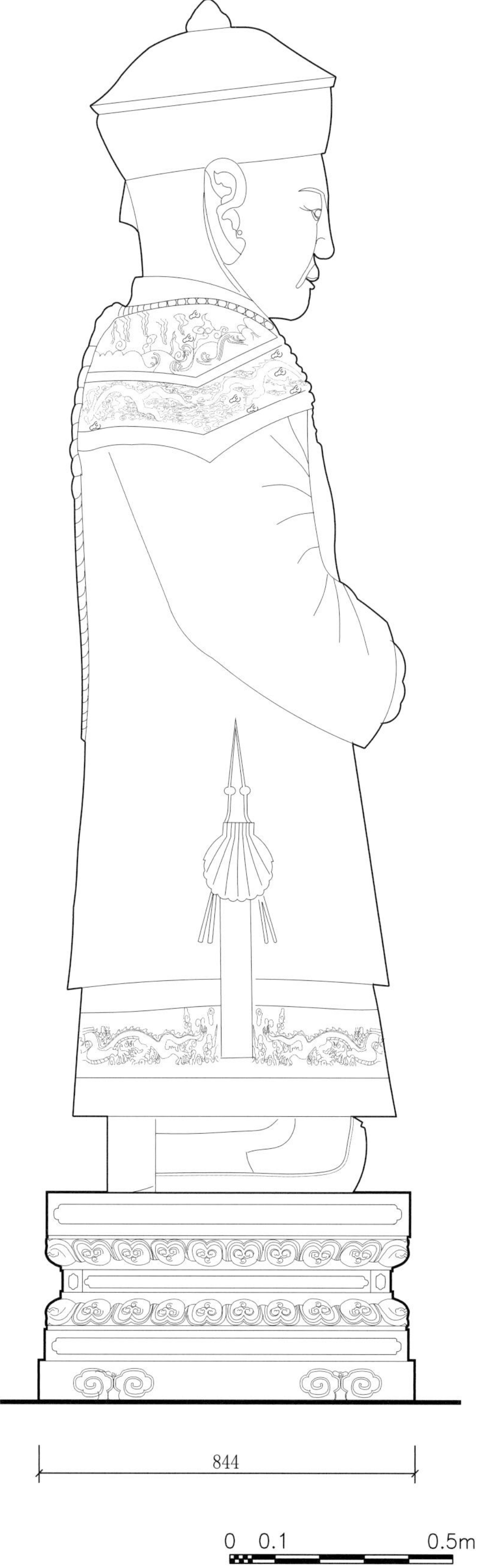

裕陵神道石像生文臣大样图

Detailed drawing of one of the Literary Officials among the Stone Statues along the Spirit Way, Tomb of Prosperity

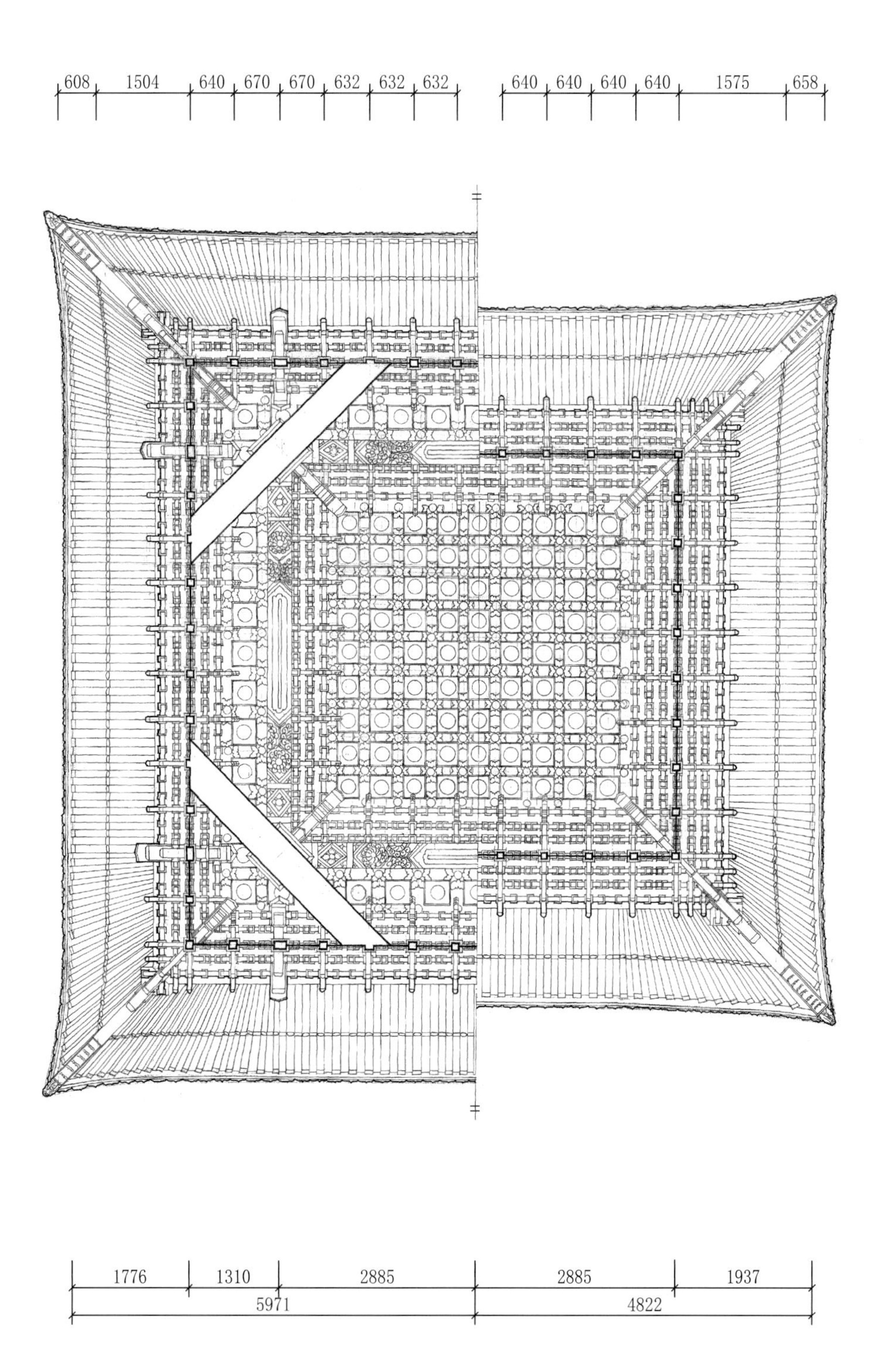

裕陵神道碑亭梁架仰视平面图

Looking up at the Truss (*liangjia*) of the Pavilion for the Stela on the Spirit Way, Tomb of Prosperity

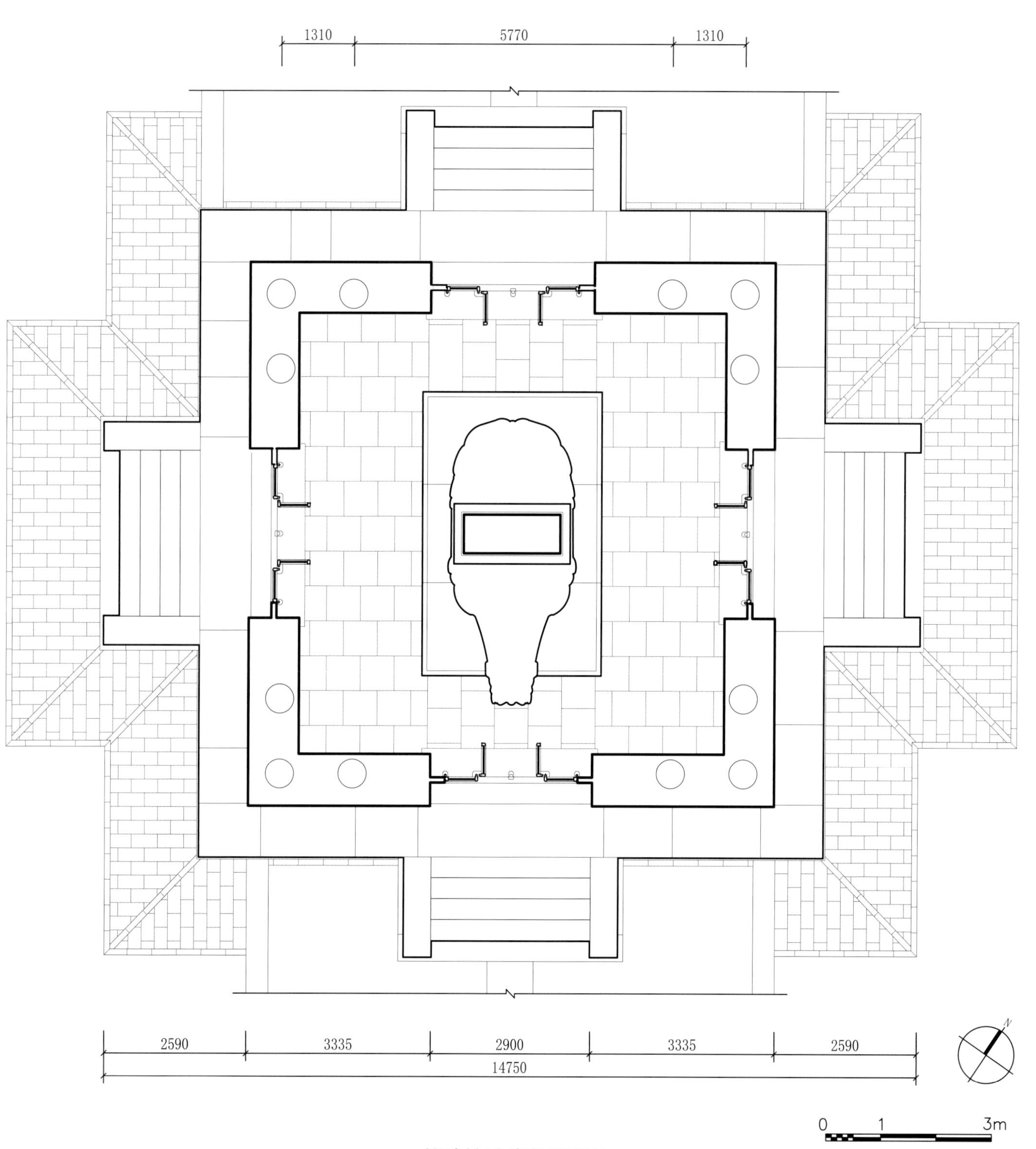

裕陵神道碑亭平面图

Plan of the Pavilion for the Stela on the Spirit Way, Tomb of Prosperity

13.610
12.890
9.238
8.695
8.535
6.128
5.695
5.530
1.355
±0.000
−0.926

1455 1310 5770 1310 1455
11300

0 1 2m

裕陵神道碑亭正立面图
Front elevation of the Pavilion for the Stela on the Spirit Way, Tomb of Prosperity

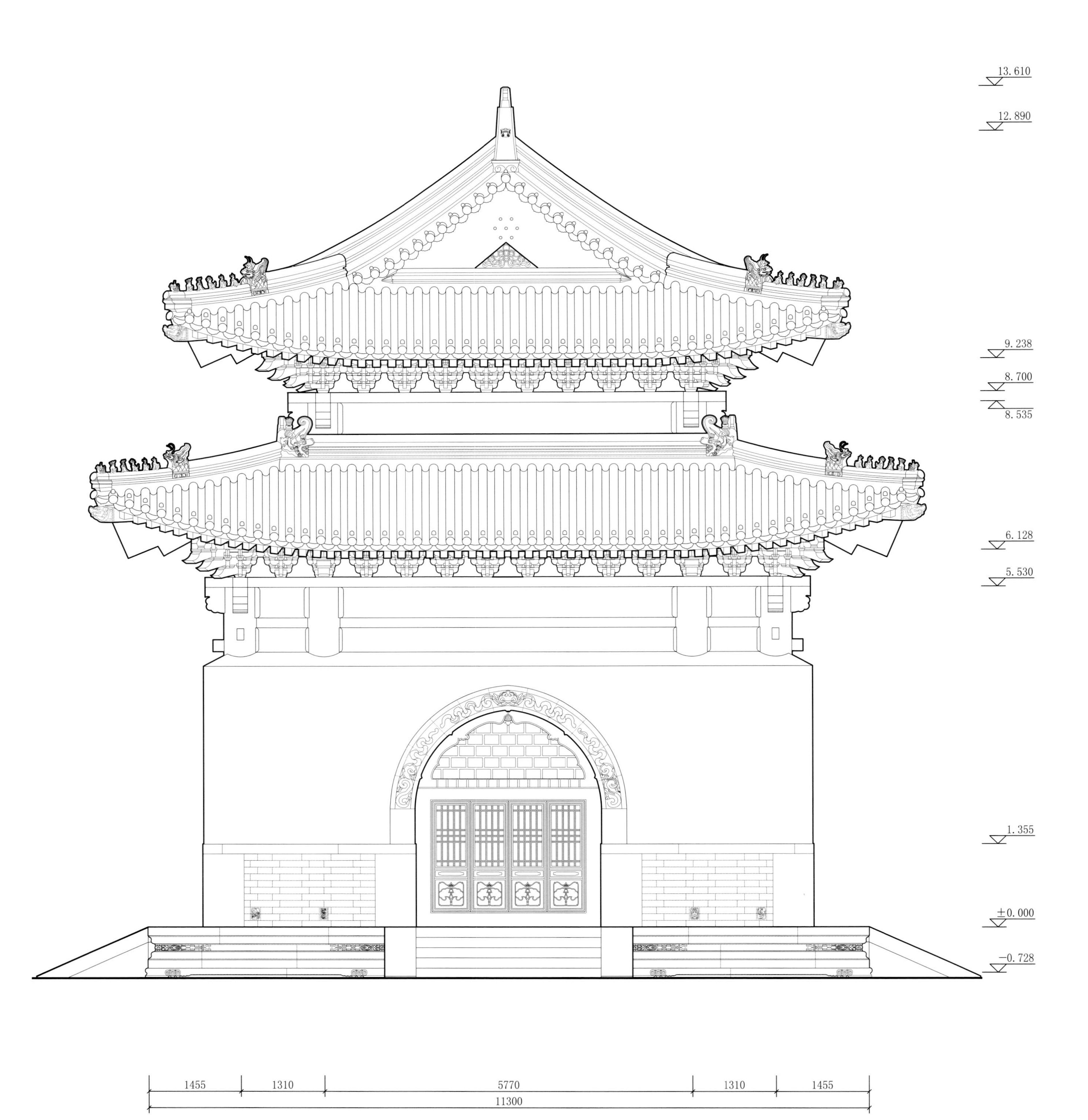

裕陵神道碑亭侧立面图
Side elevation of the Pavilion for the Stela on the Spirit Way, Tomb of Prosperity

裕陵神道碑亭横剖面图

Cross section of the Pavilion for the Stela on the Spirit Way, Tomb of Prosperity

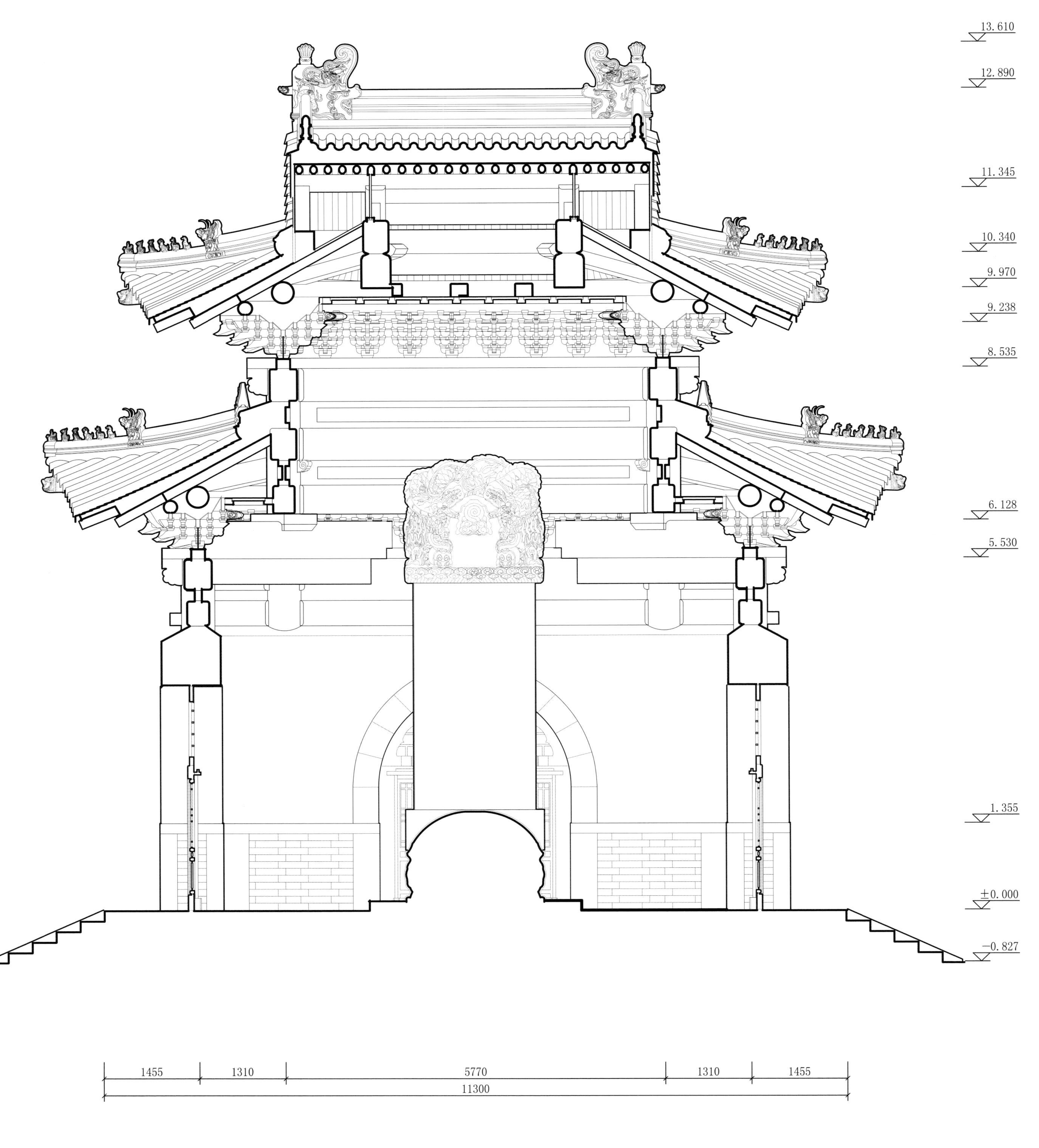

裕陵神道碑亭纵剖面图

Longitudinal section of the Pavilion for the Stela on the Spirit Way, Tomb of Prosperity

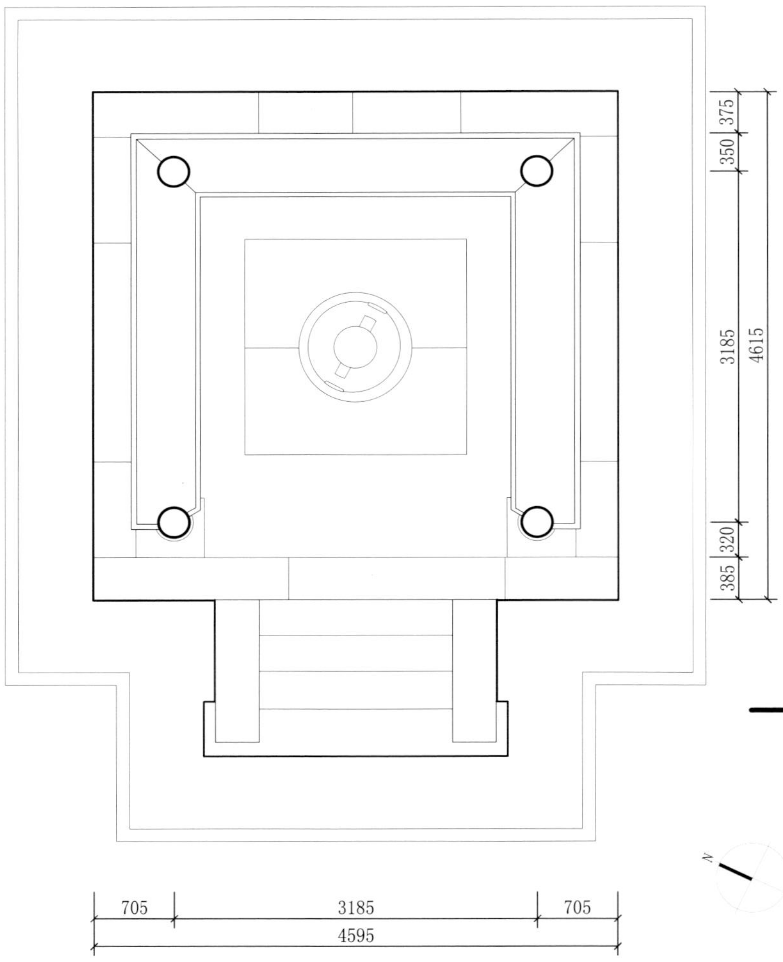

裕陵井亭平面图

Plan of the Pavilion for the Well, Tomb of Prosperity

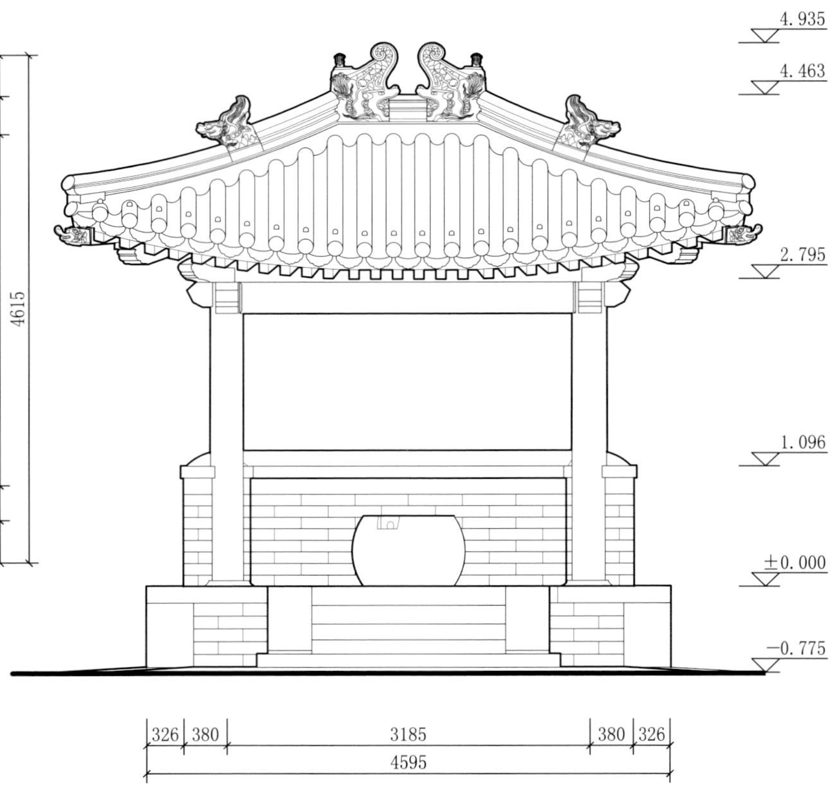

裕陵井亭正立面图

Front elevation of the Pavilion for the Well, Tomb of Prosperity

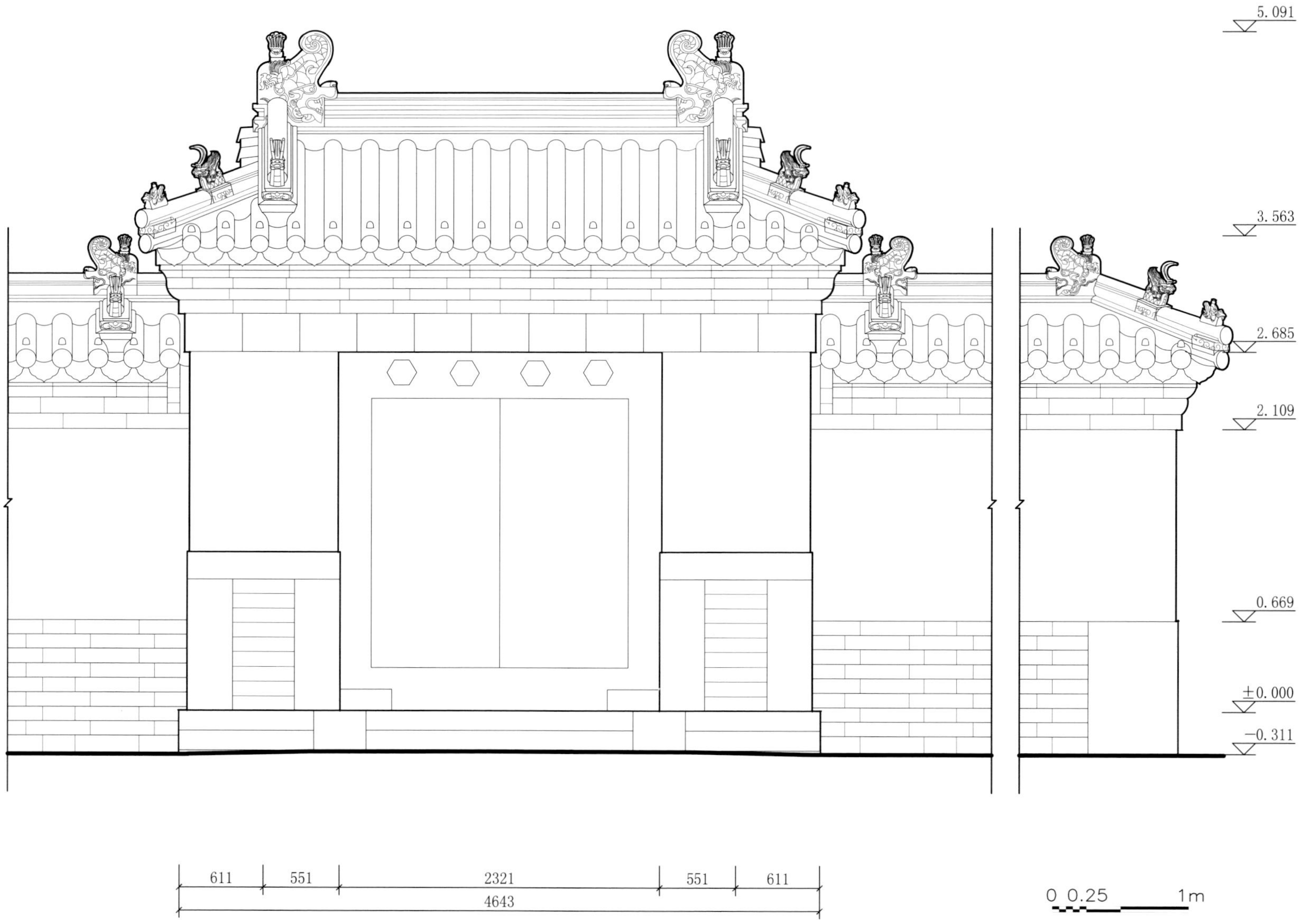

裕陵神厨库院门正立面图

Front elevation of the Gate in the Culinary Courtyard for Sacrifices, Tomb of Prosperity

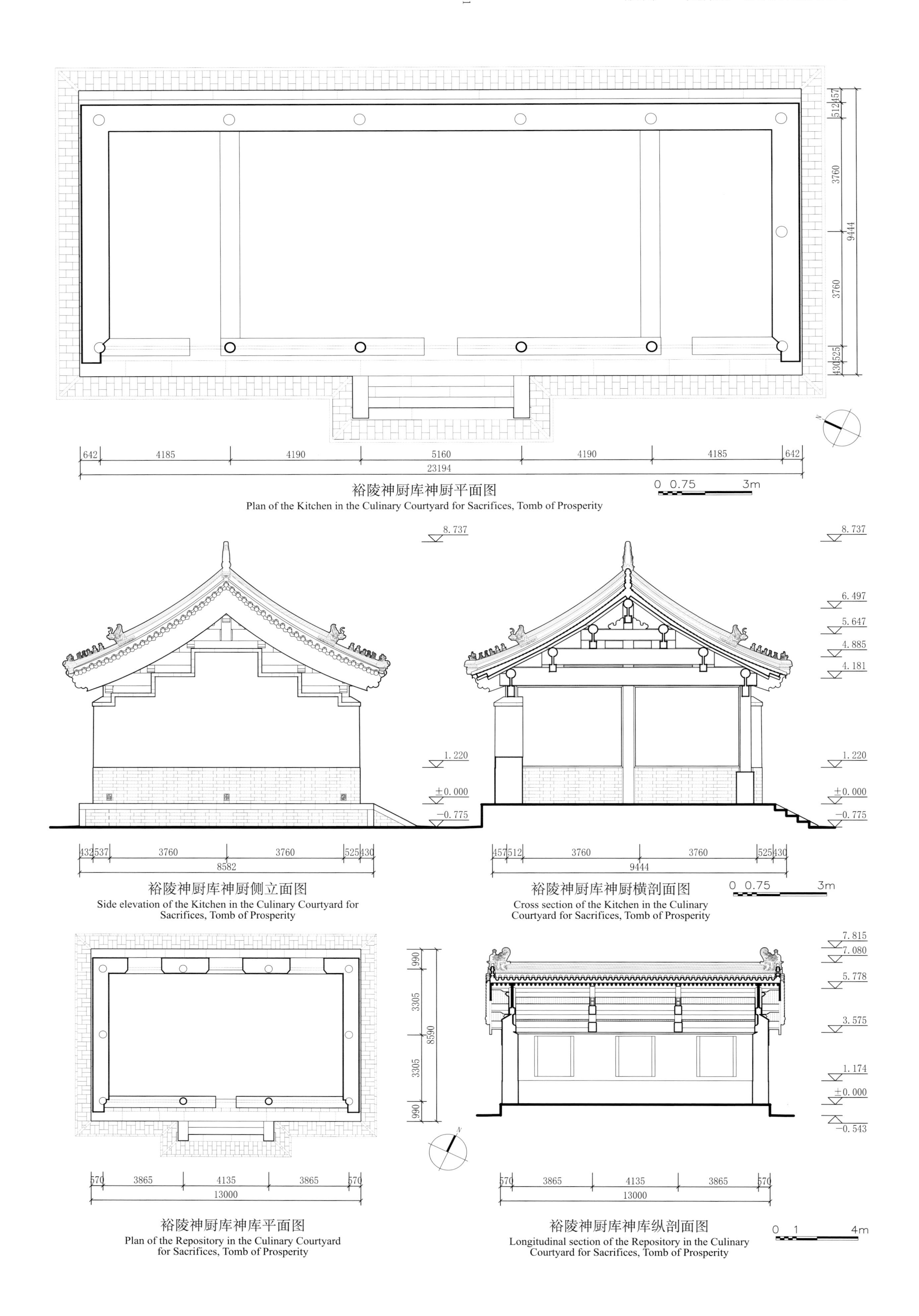

裕陵神厨库神厨平面图
Plan of the Kitchen in the Culinary Courtyard for Sacrifices, Tomb of Prosperity

裕陵神厨库神厨侧立面图
Side elevation of the Kitchen in the Culinary Courtyard for Sacrifices, Tomb of Prosperity

裕陵神厨库神厨横剖面图
Cross section of the Kitchen in the Culinary Courtyard for Sacrifices, Tomb of Prosperity

裕陵神厨库神库平面图
Plan of the Repository in the Culinary Courtyard for Sacrifices, Tomb of Prosperity

裕陵神厨库神库纵剖面图
Longitudinal section of the Repository in the Culinary Courtyard for Sacrifices, Tomb of Prosperity

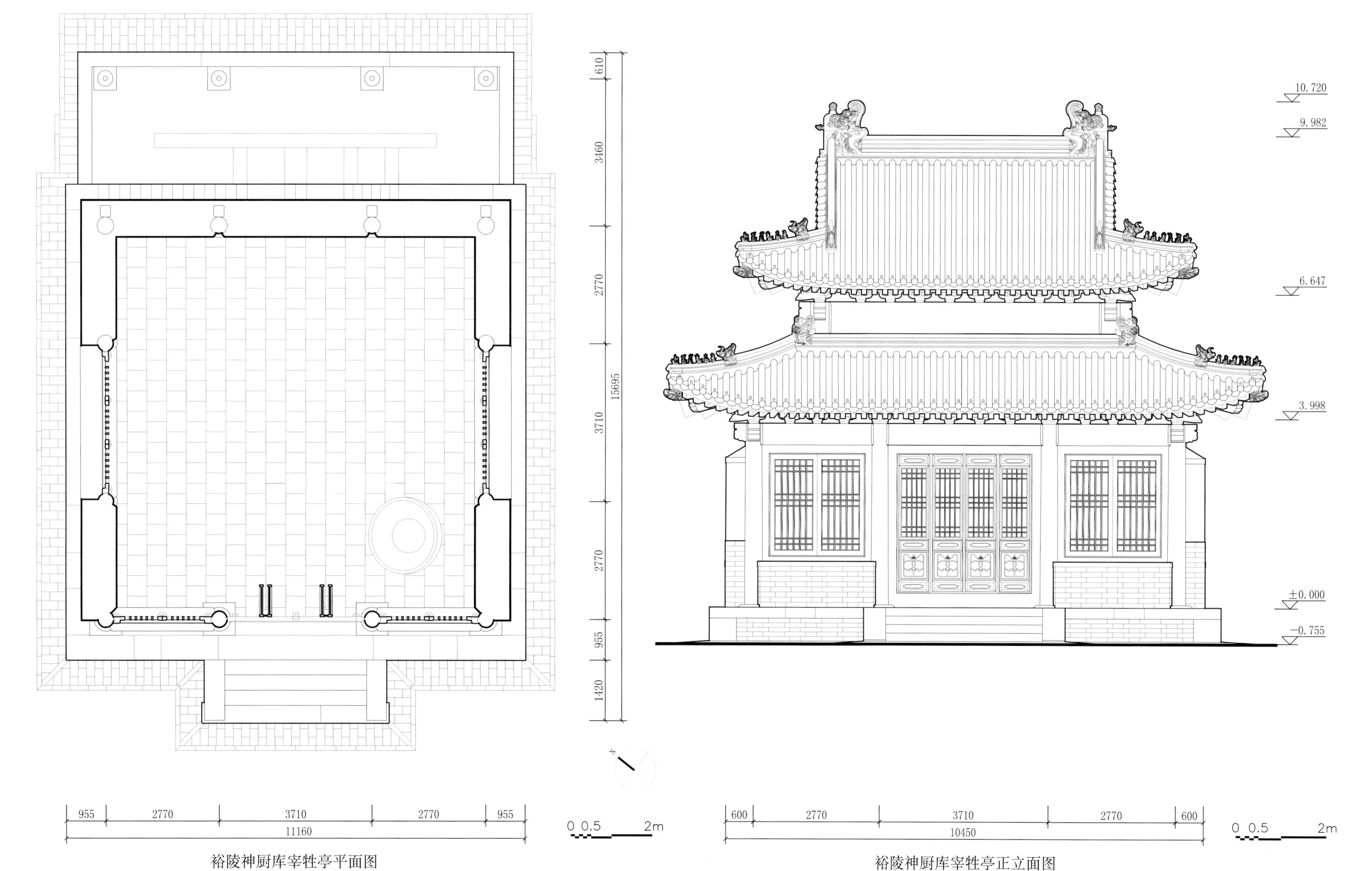

裕陵神厨库宰牲亭平面图

Plan of the Ritual Abattoir in the Culinary Courtyard for Sacrifices, Tomb of Prosperity

裕陵神厨库宰牲亭正立面图

Front elevation of the Ritual Abattoir in the Culinary Courtyard for Sacrifices, Tomb of Prosperity

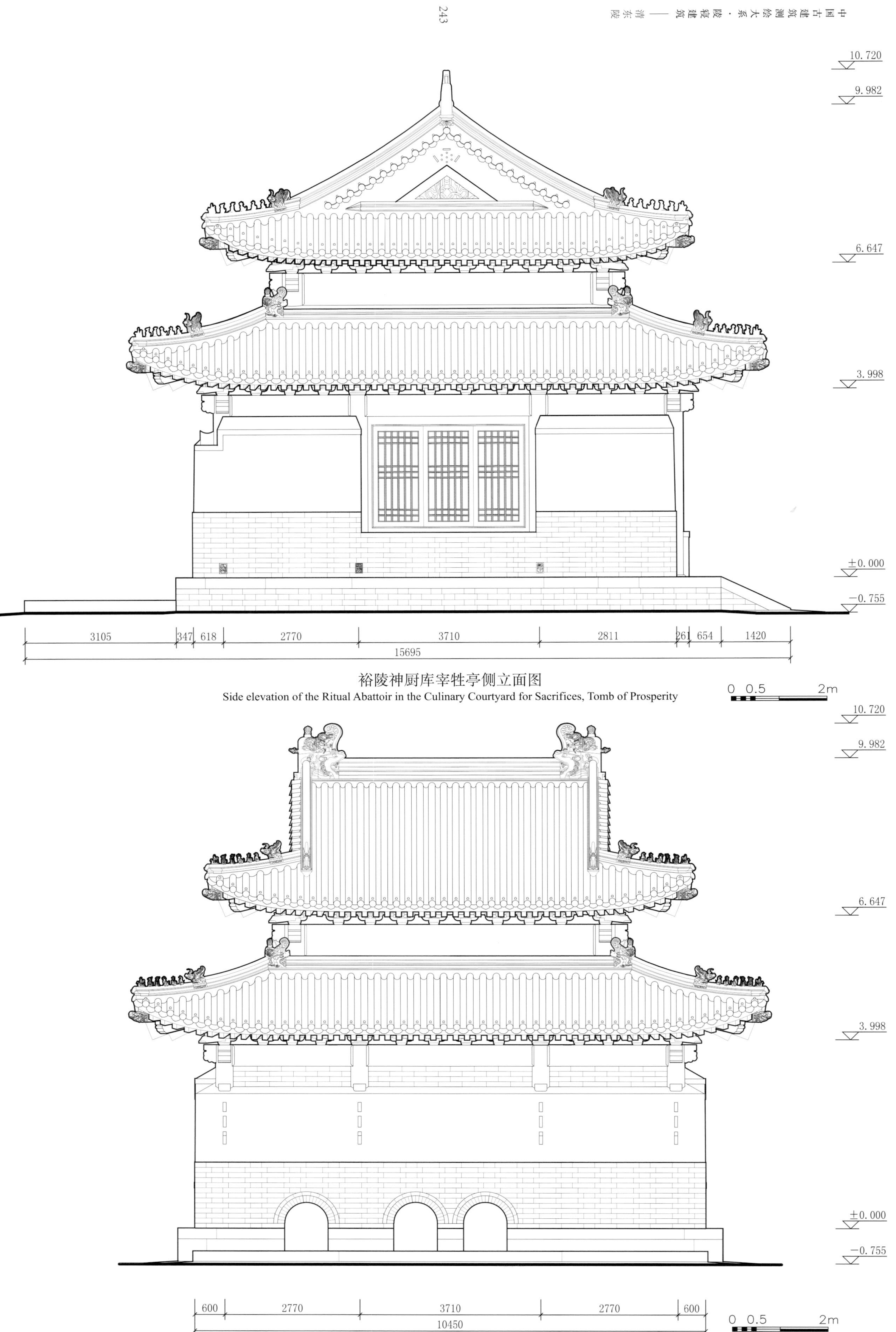

裕陵神厨库宰牲亭侧立面图

Side elevation of the Ritual Abattoir in the Culinary Courtyard for Sacrifices, Tomb of Prosperity

裕陵神厨库宰牲亭背立面图

Back elevation of the Ritual Abattoir in the Culinary Courtyard for Sacrifices, Tomb of Prosperity

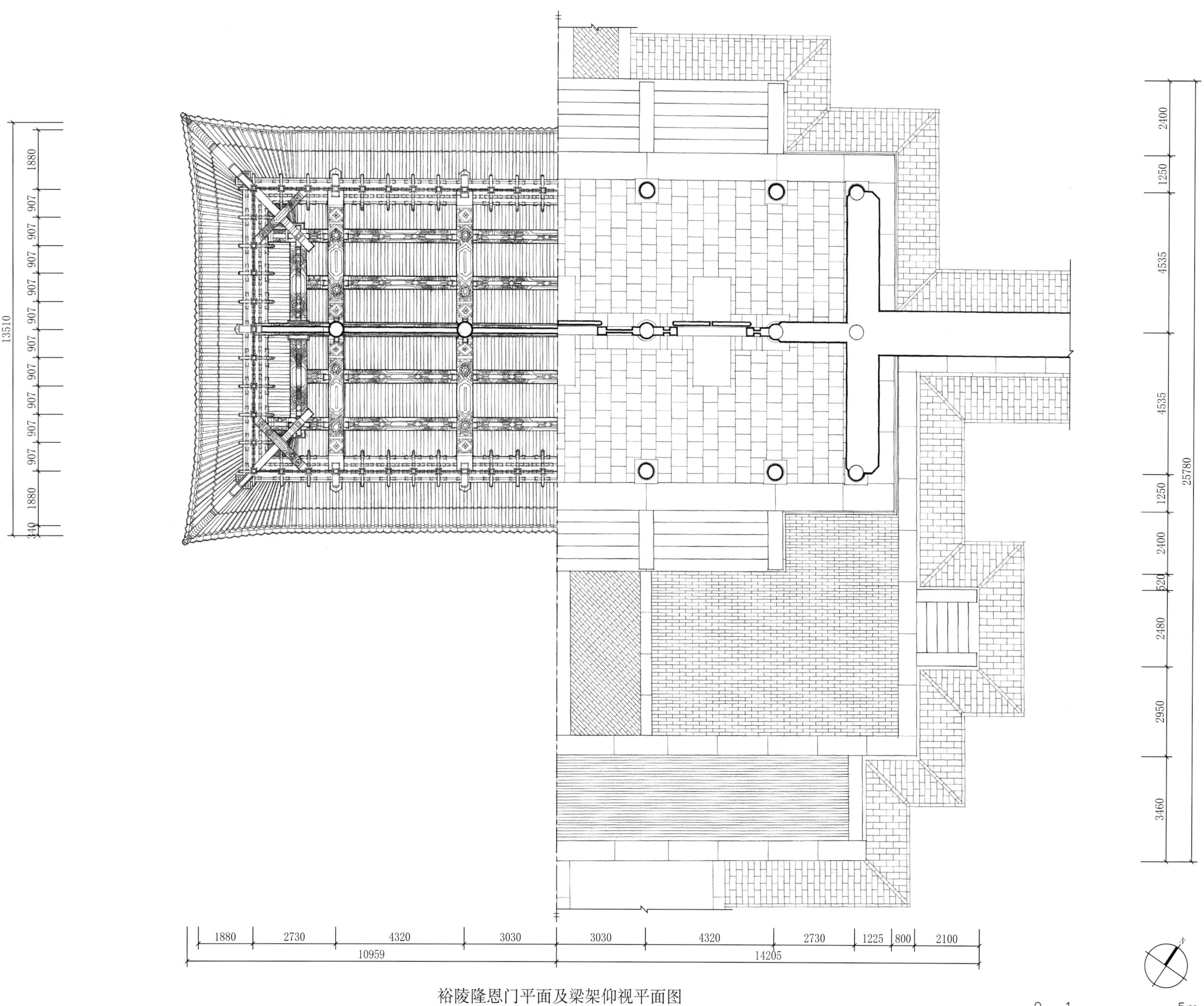

裕陵隆恩门平面及梁架仰视平面图

Ground plan and looking up at the Truss of the Gate of Monumental Grace, Tomb of Prosperity

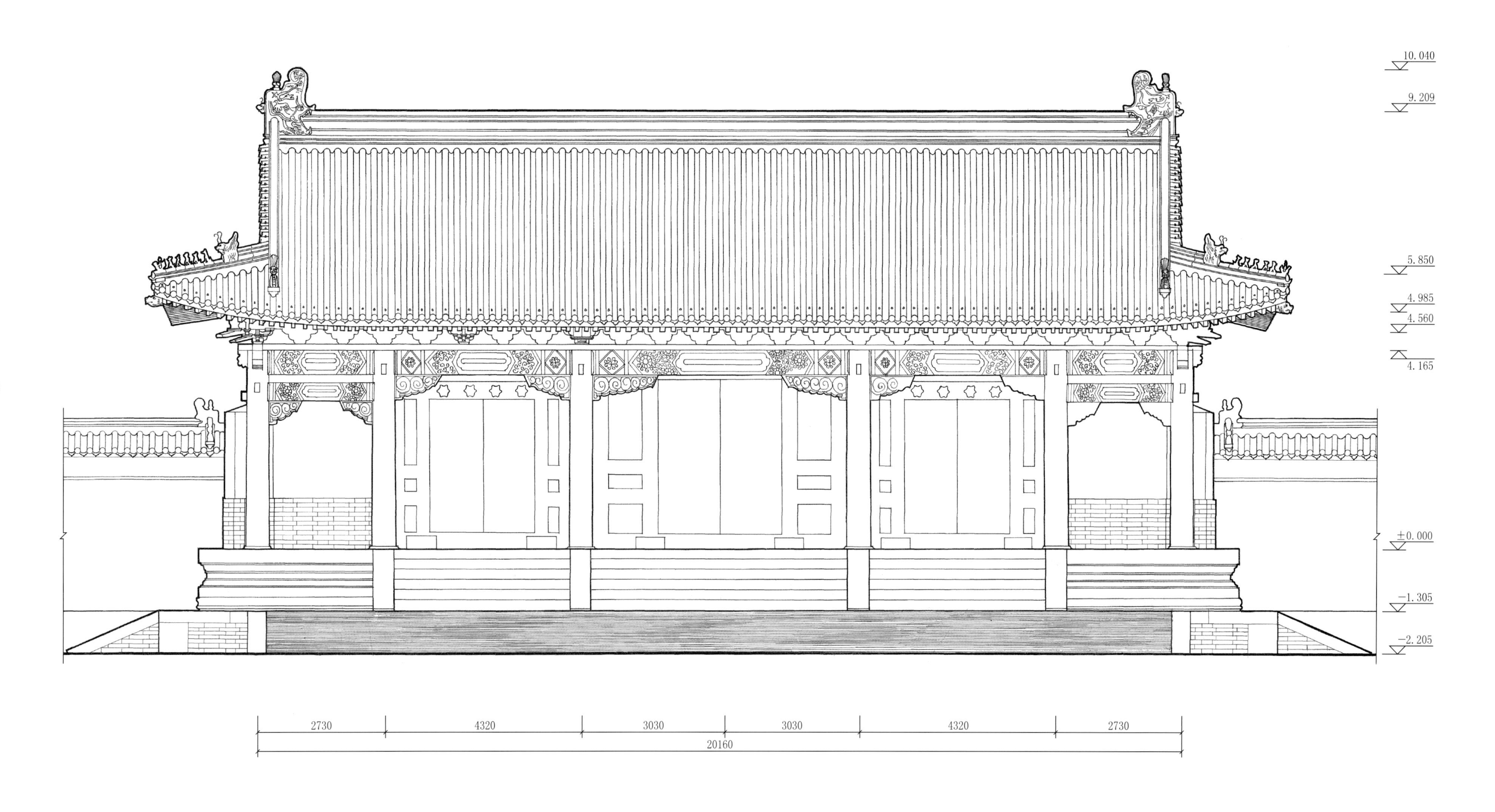

裕陵隆恩门正立面图

Front elevation of the Gate of Monumental Grace, Tomb of Prosperity

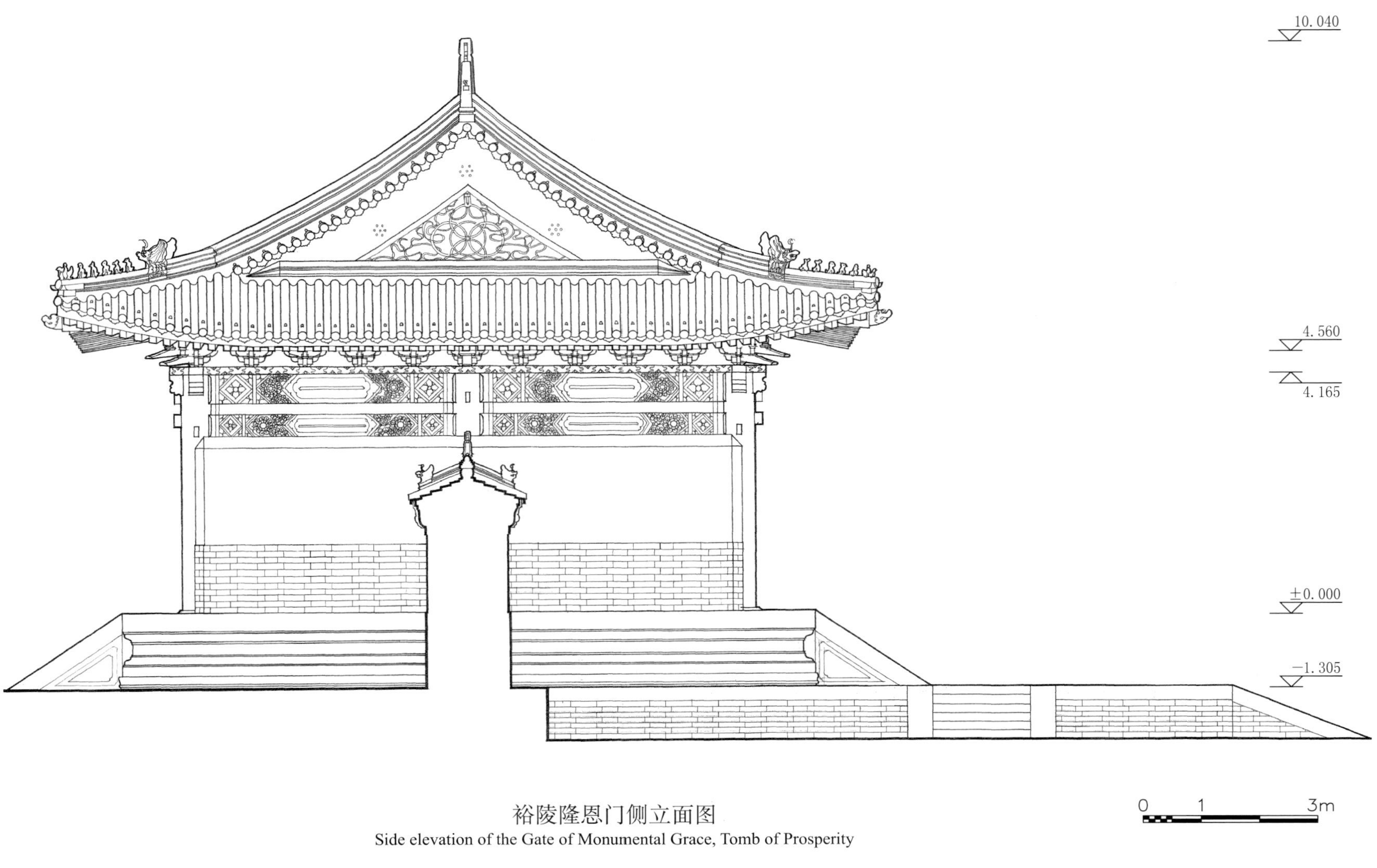

裕陵隆恩门侧立面图
Side elevation of the Gate of Monumental Grace, Tomb of Prosperity

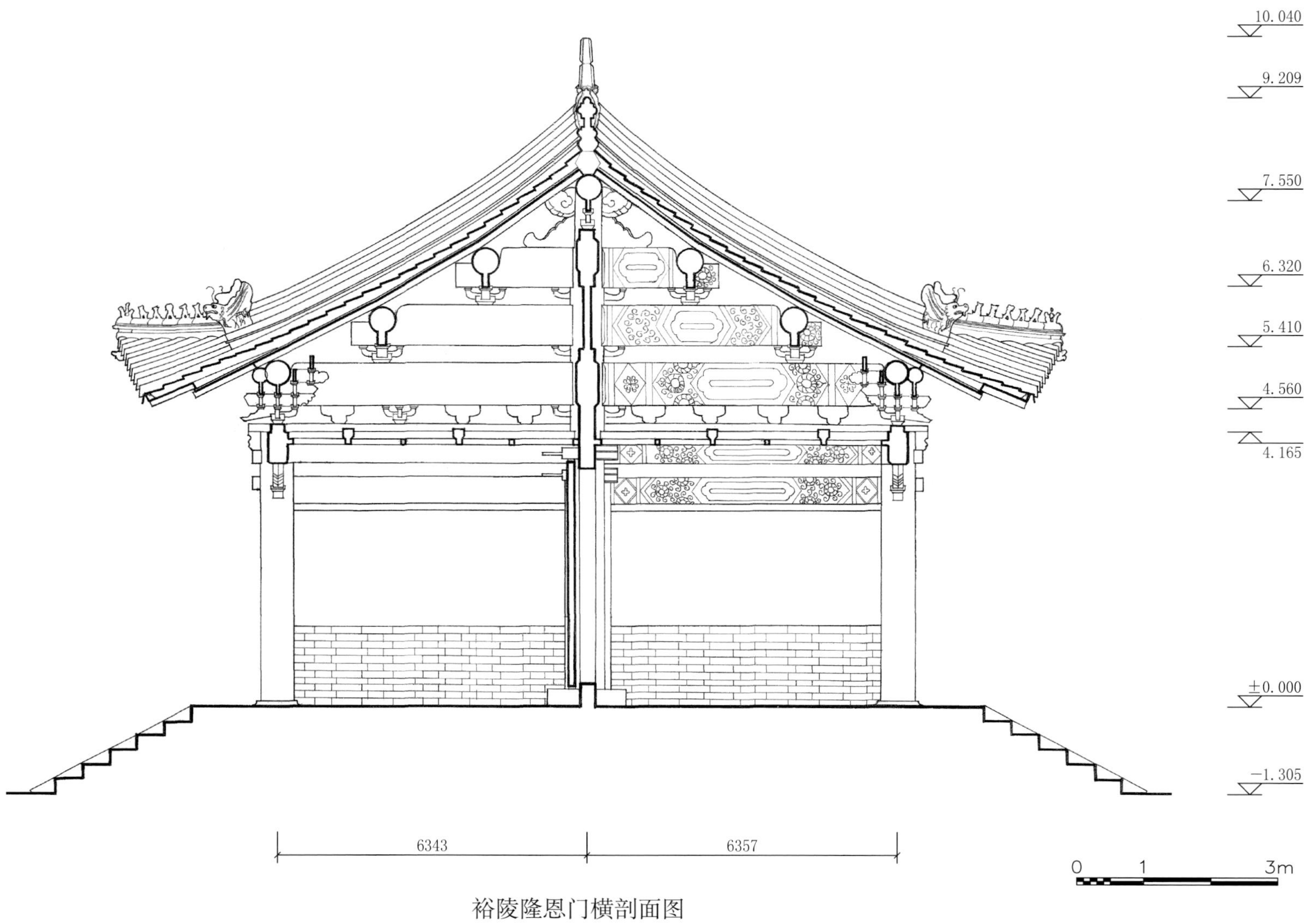

裕陵隆恩门横剖面图
Cross section of the Gate of Monumental Grace, Tomb of Prosperity

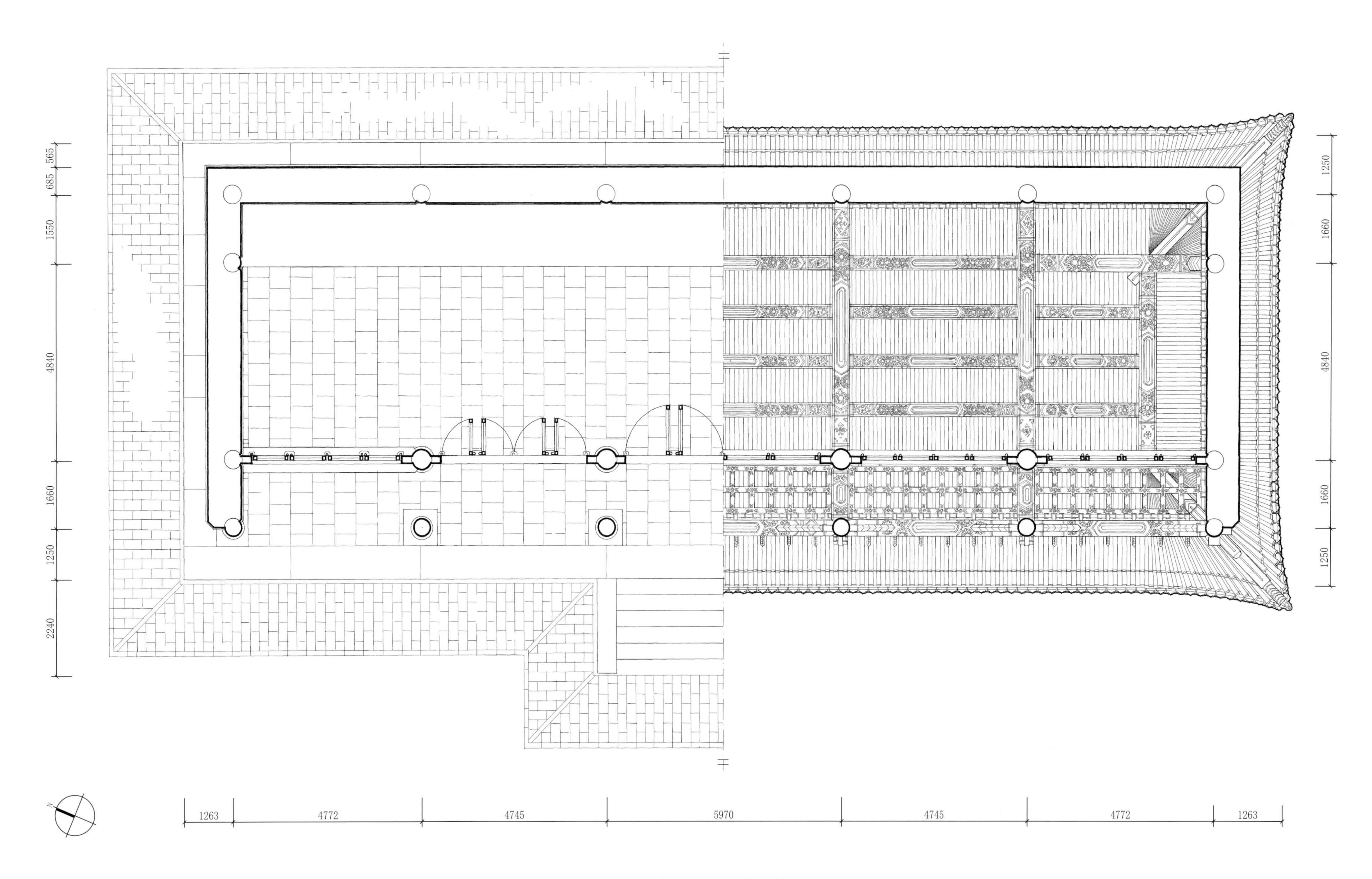

裕陵东配殿平面及梁架仰视平面图

Ground plan and looking up at the Truss of the Eastern Side Hall, Tomb of Prosperity

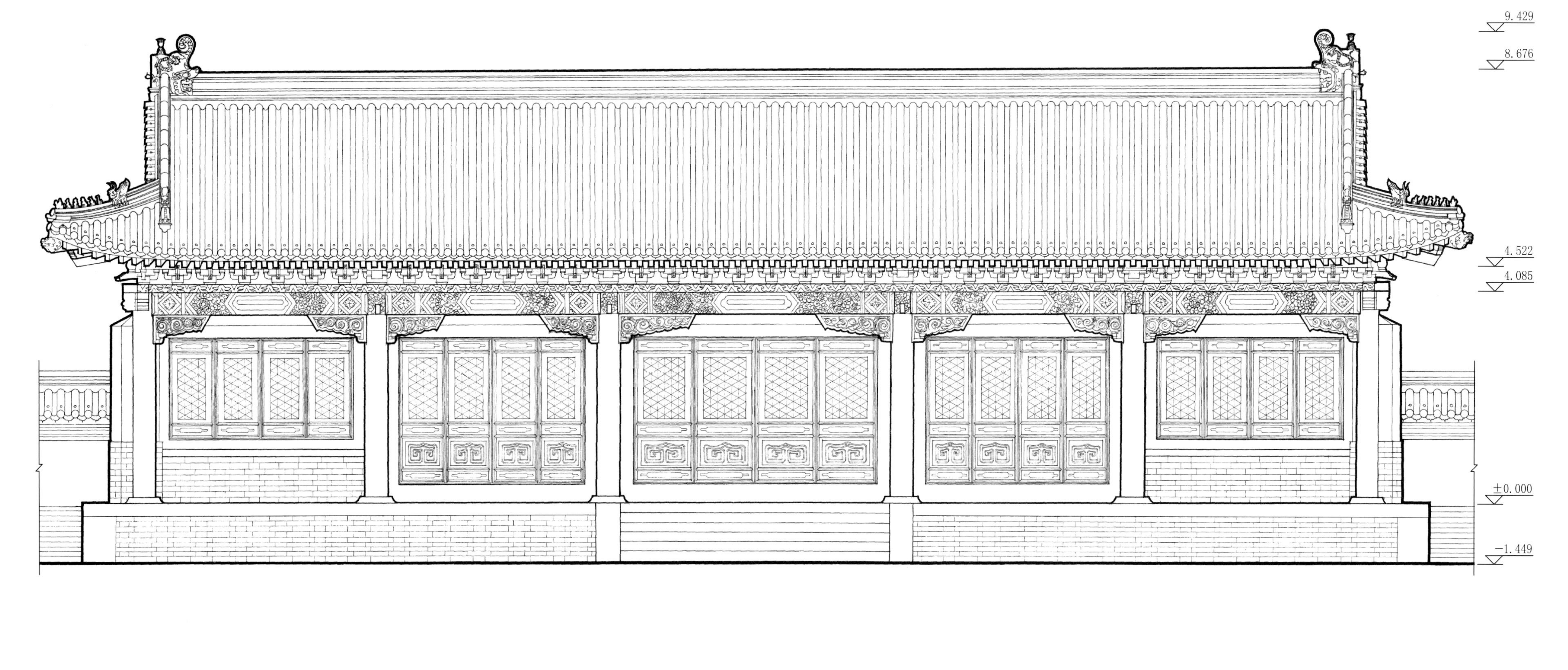

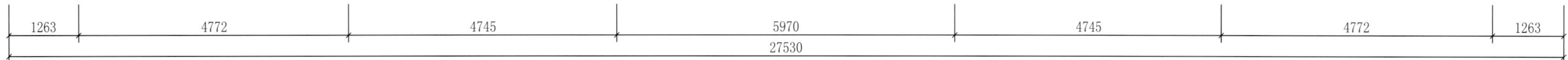

裕陵东配殿正立面图

Front elevation of the Eastern Side Hall, Tomb of Prosperity

0 1 3m

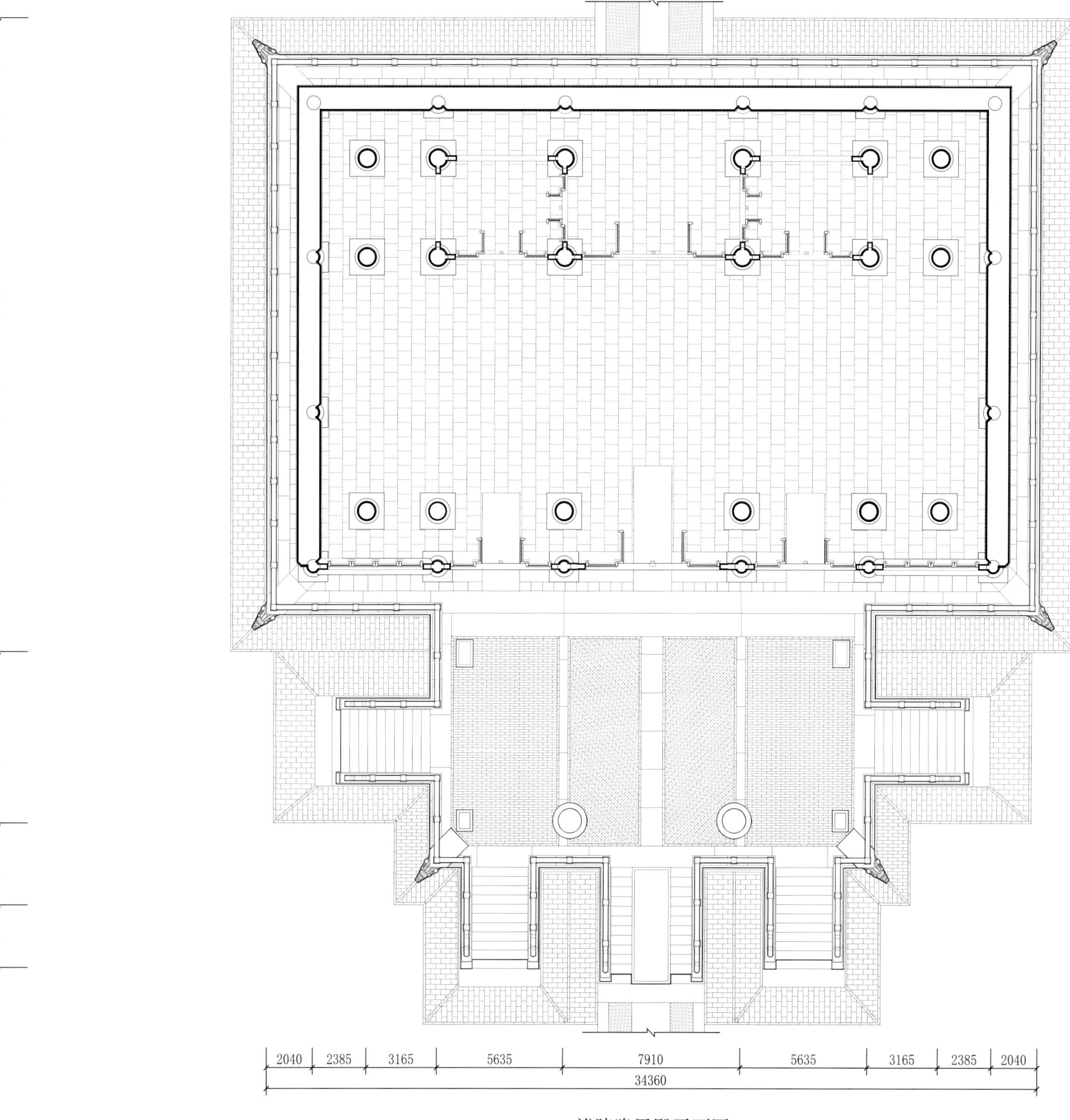

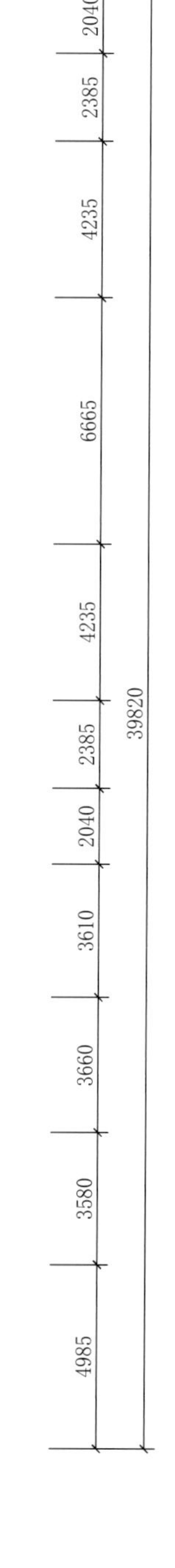

裕陵隆恩殿平面图

Plan of the Hall of Monumental Grace, Tomb of Prosperity

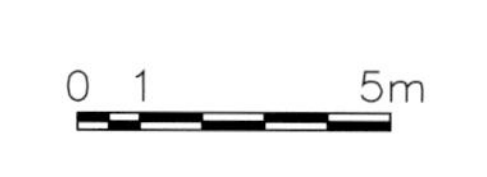

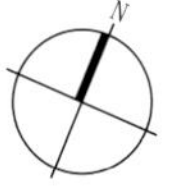

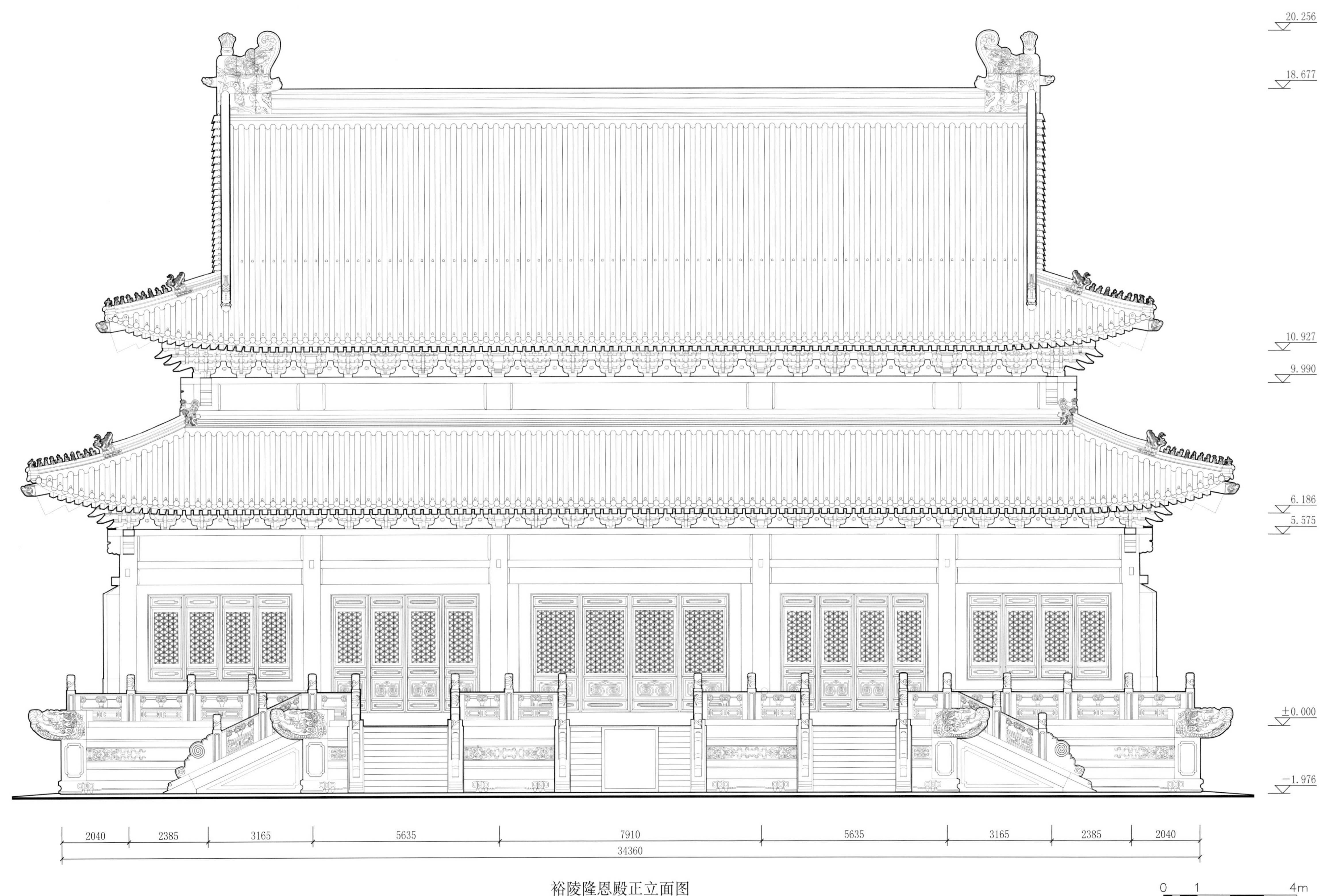

裕陵隆恩殿正立面图

Front elevation of the Hall of Monumental Grace, Tomb of Prosperity

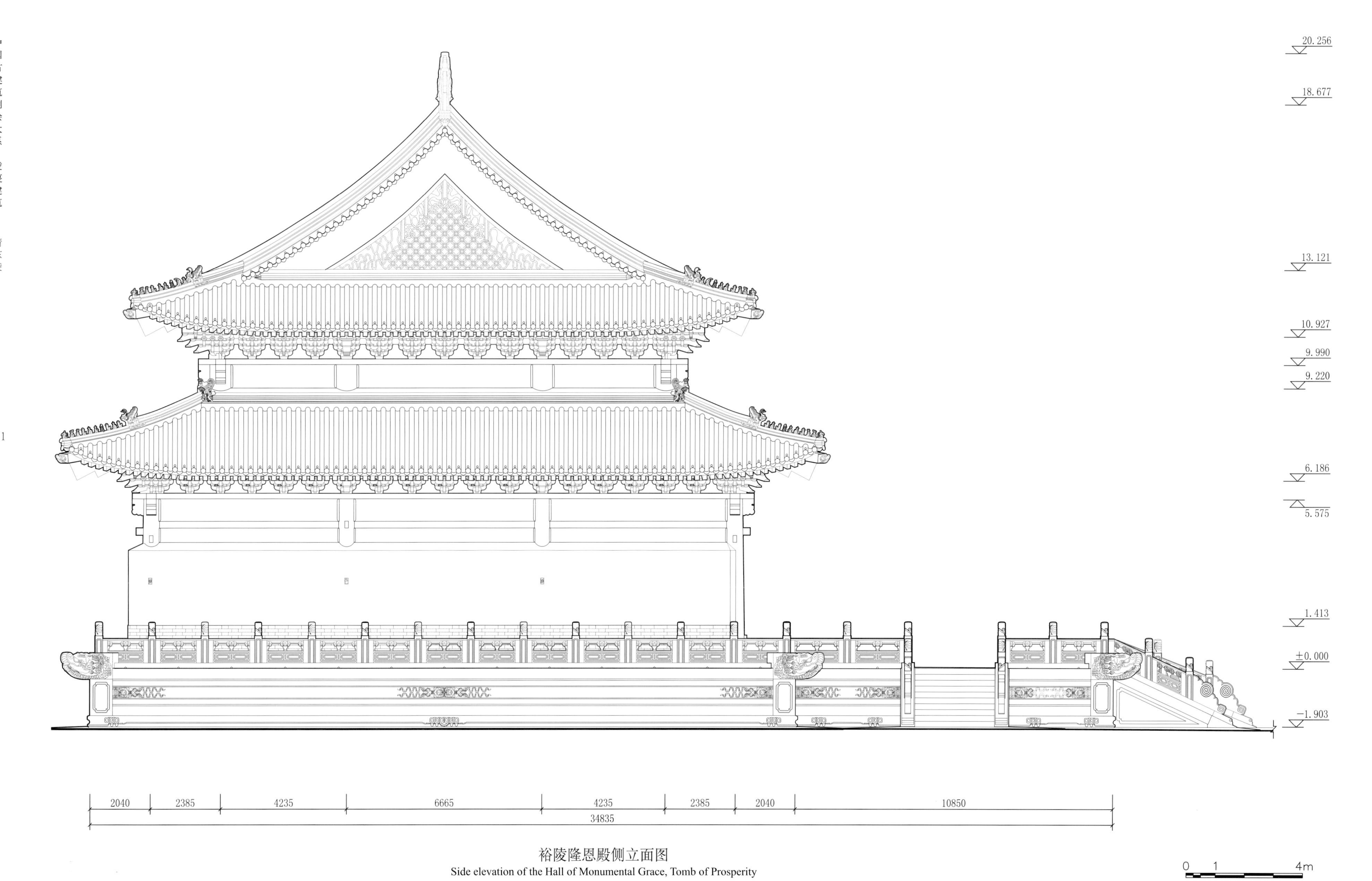

裕陵隆恩殿侧立面图

Side elevation of the Hall of Monumental Grace, Tomb of Prosperity

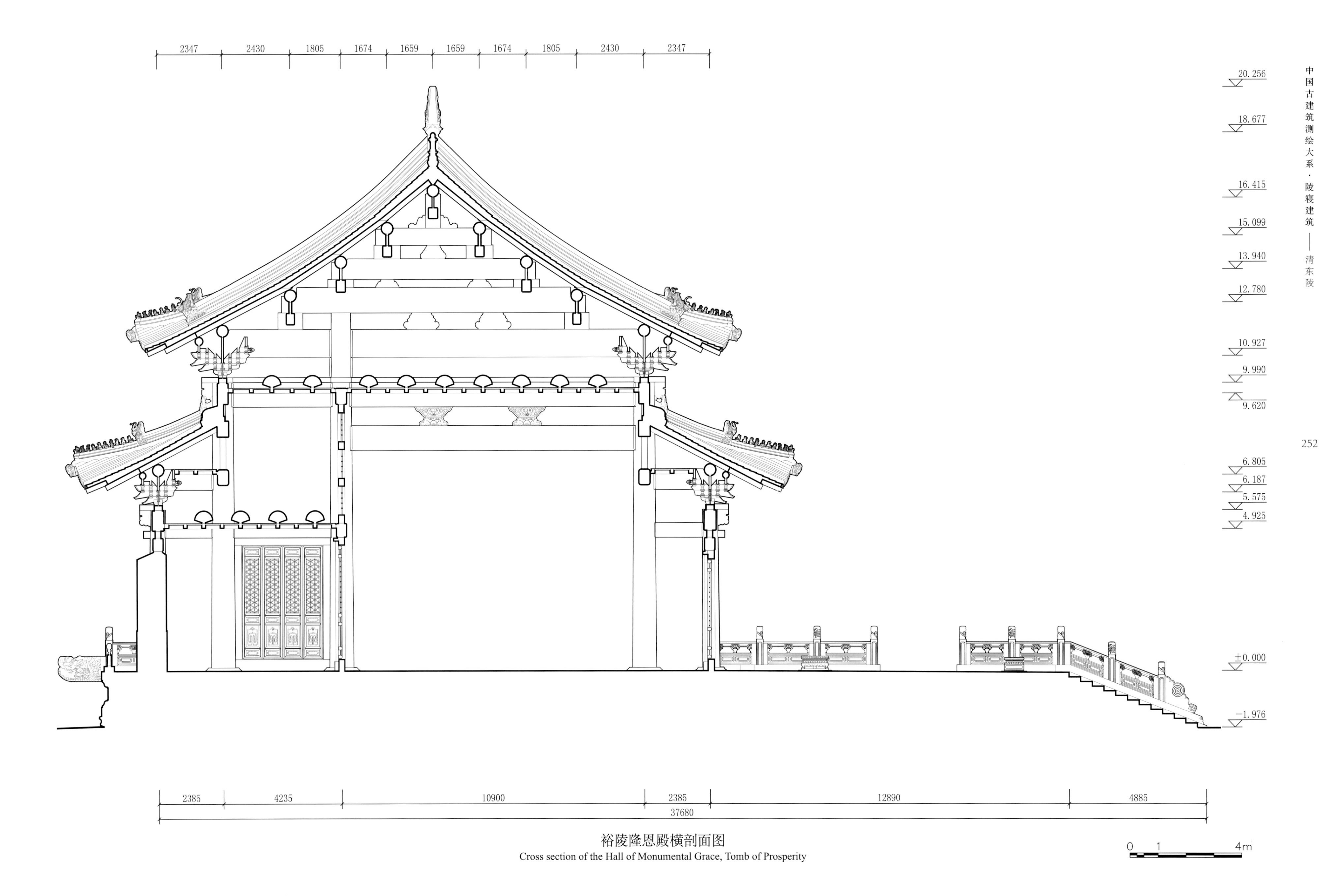

裕陵隆恩殿横剖面图

Cross section of the Hall of Monumental Grace, Tomb of Prosperity

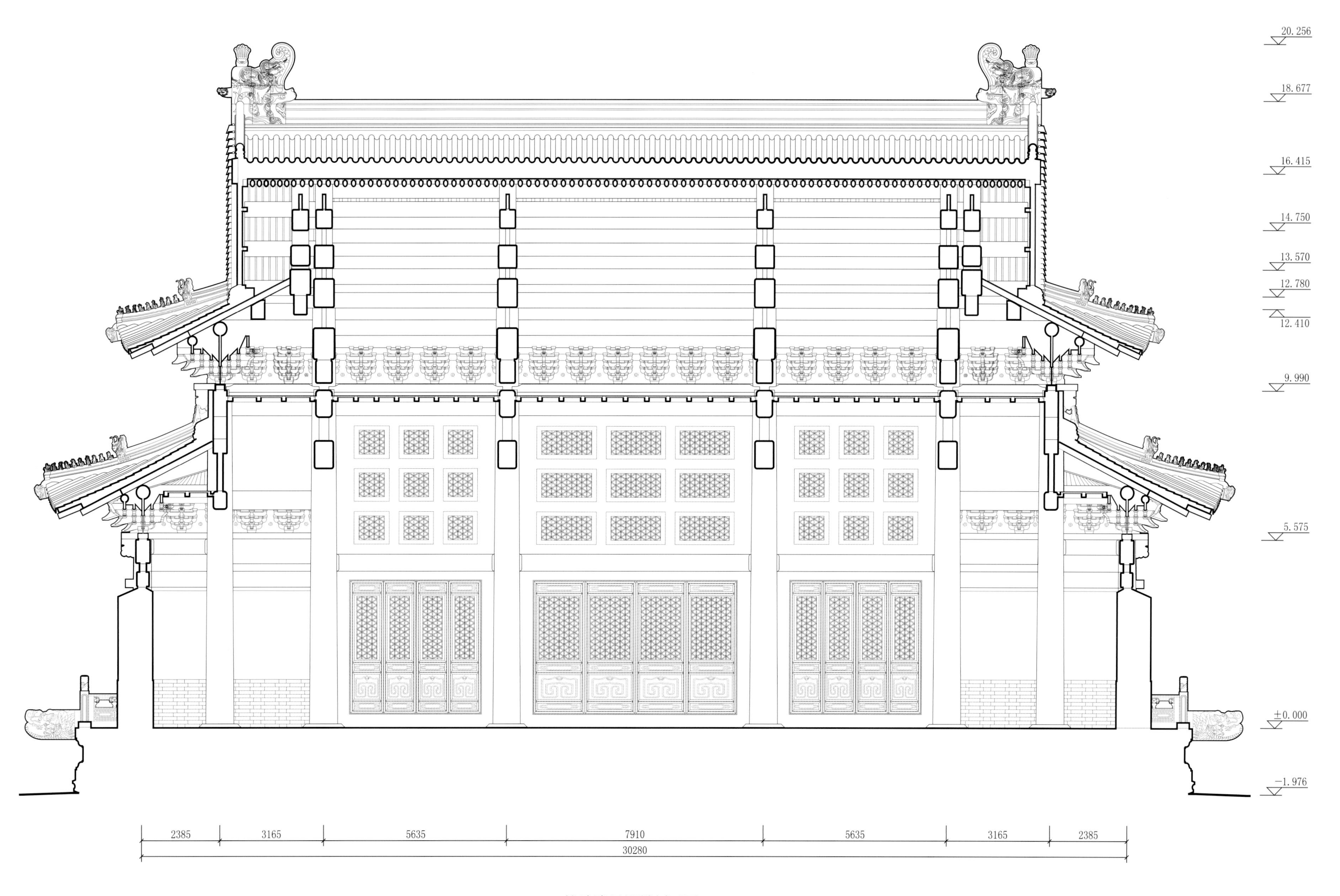

裕陵隆恩殿纵剖面图

Longitudinal section of the Hall of Monumental Grace, Tomb of Prosperity

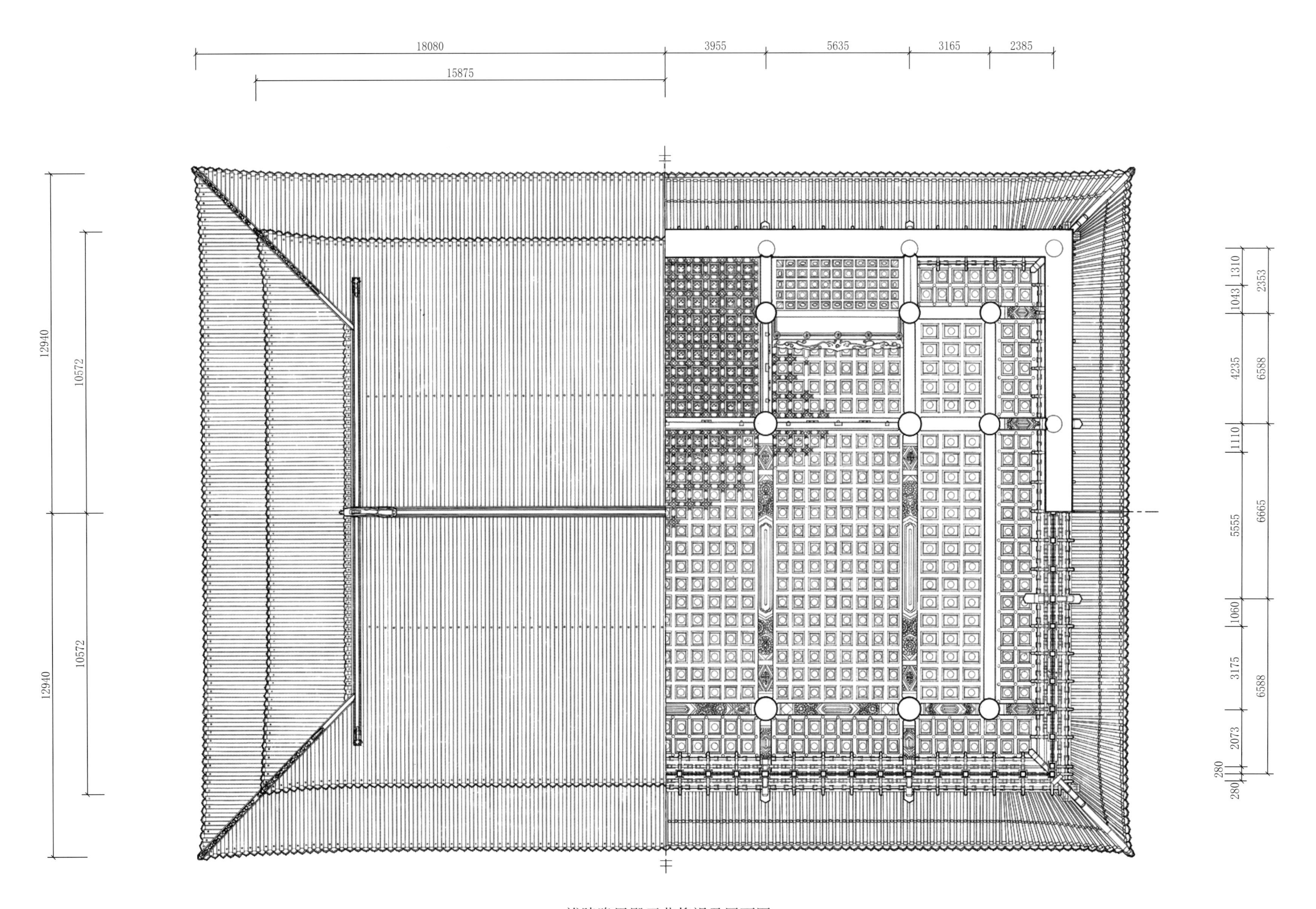

裕陵隆恩殿天花仰视及屋面图

Roof plan and looking up at the ceiling of the Hall of Monumental Grace, Tomb of Prosperity

裕陵隆恩殿丹陛大样图

Detailed drawing of the marble carving on the steps to the Hall of Monumental Grace, Tomb of Prosperity

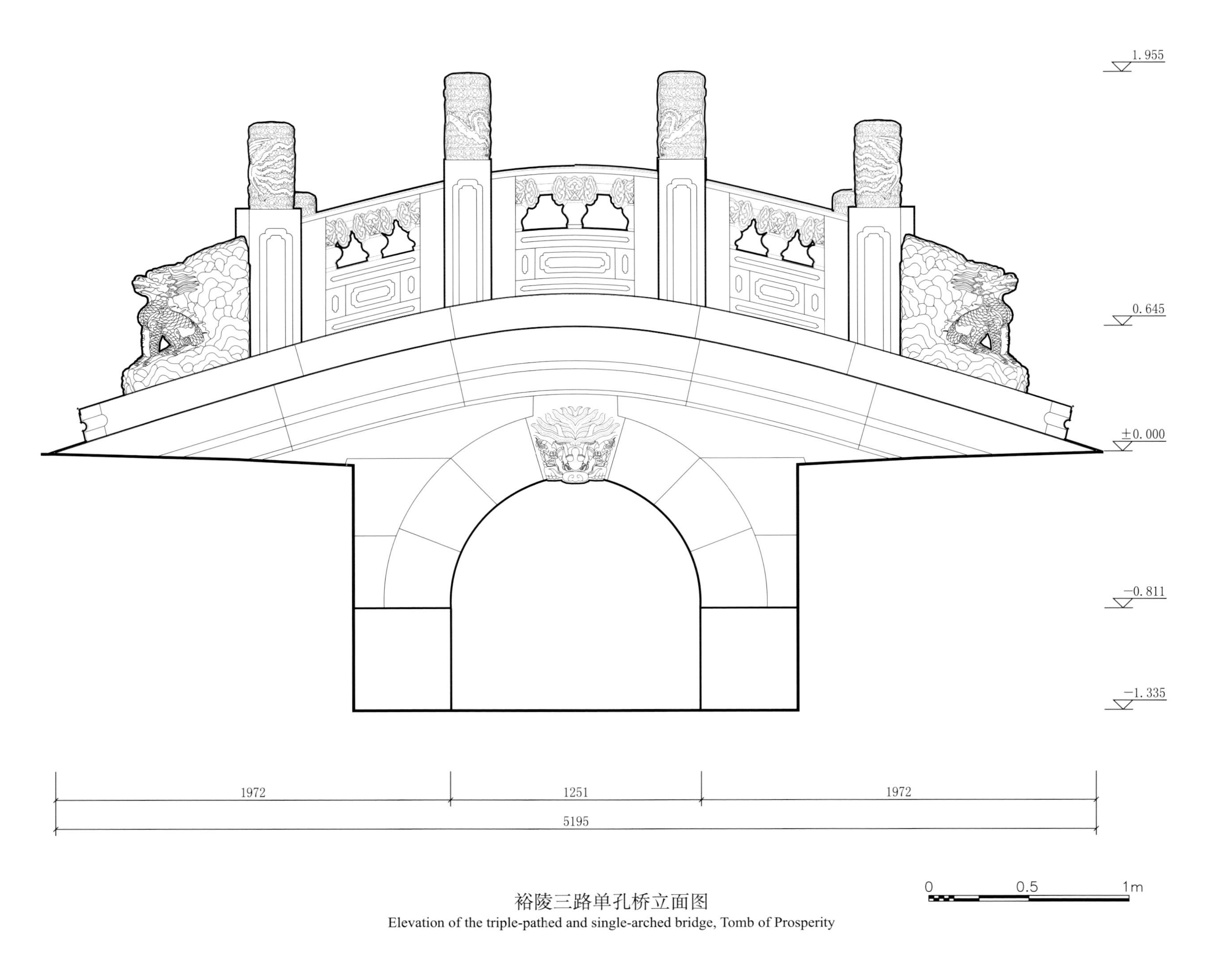

裕陵三路单孔桥立面图
Elevation of the triple-pathed and single-arched bridge, Tomb of Prosperity

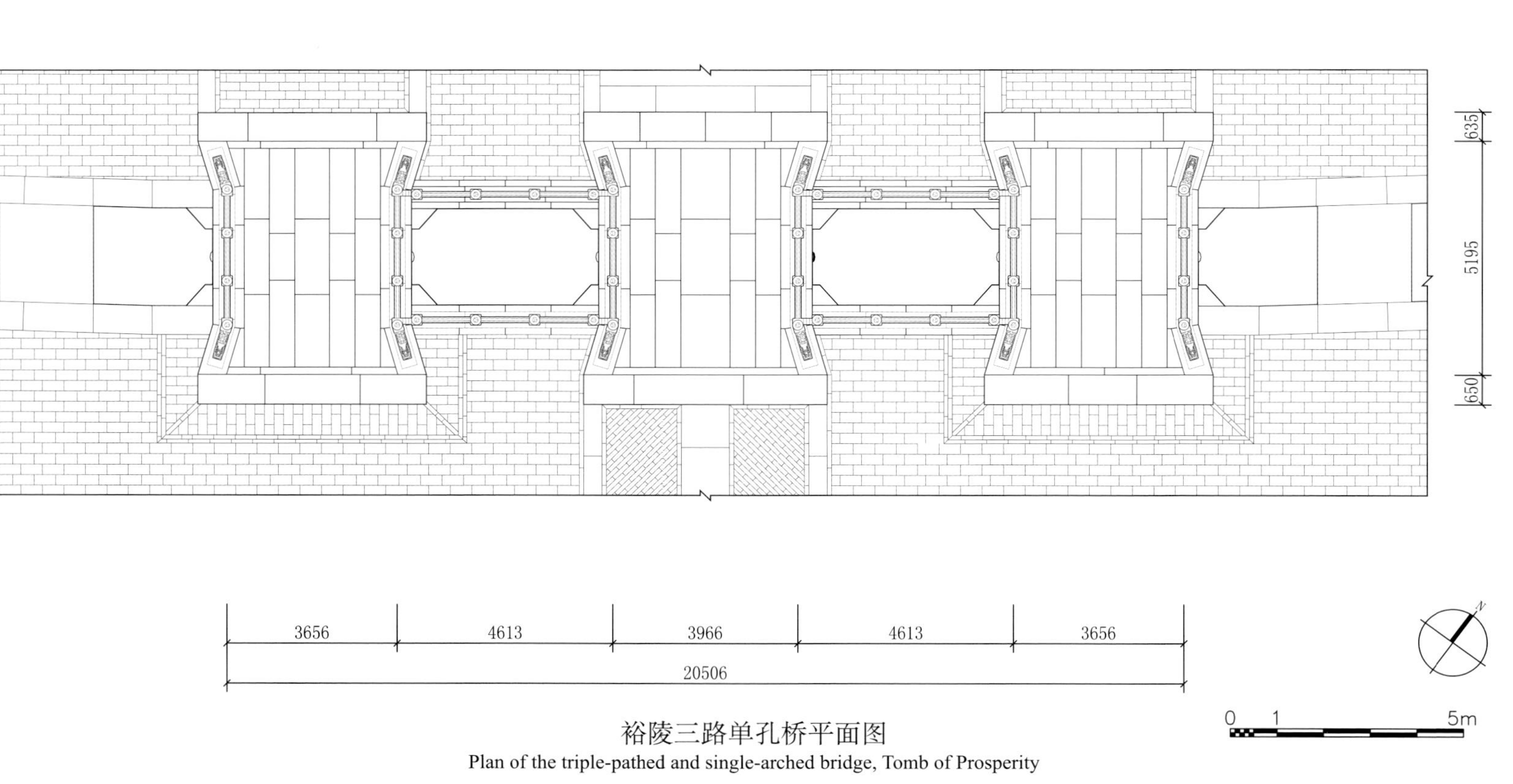

裕陵三路单孔桥平面图
Plan of the triple-pathed and single-arched bridge, Tomb of Prosperity

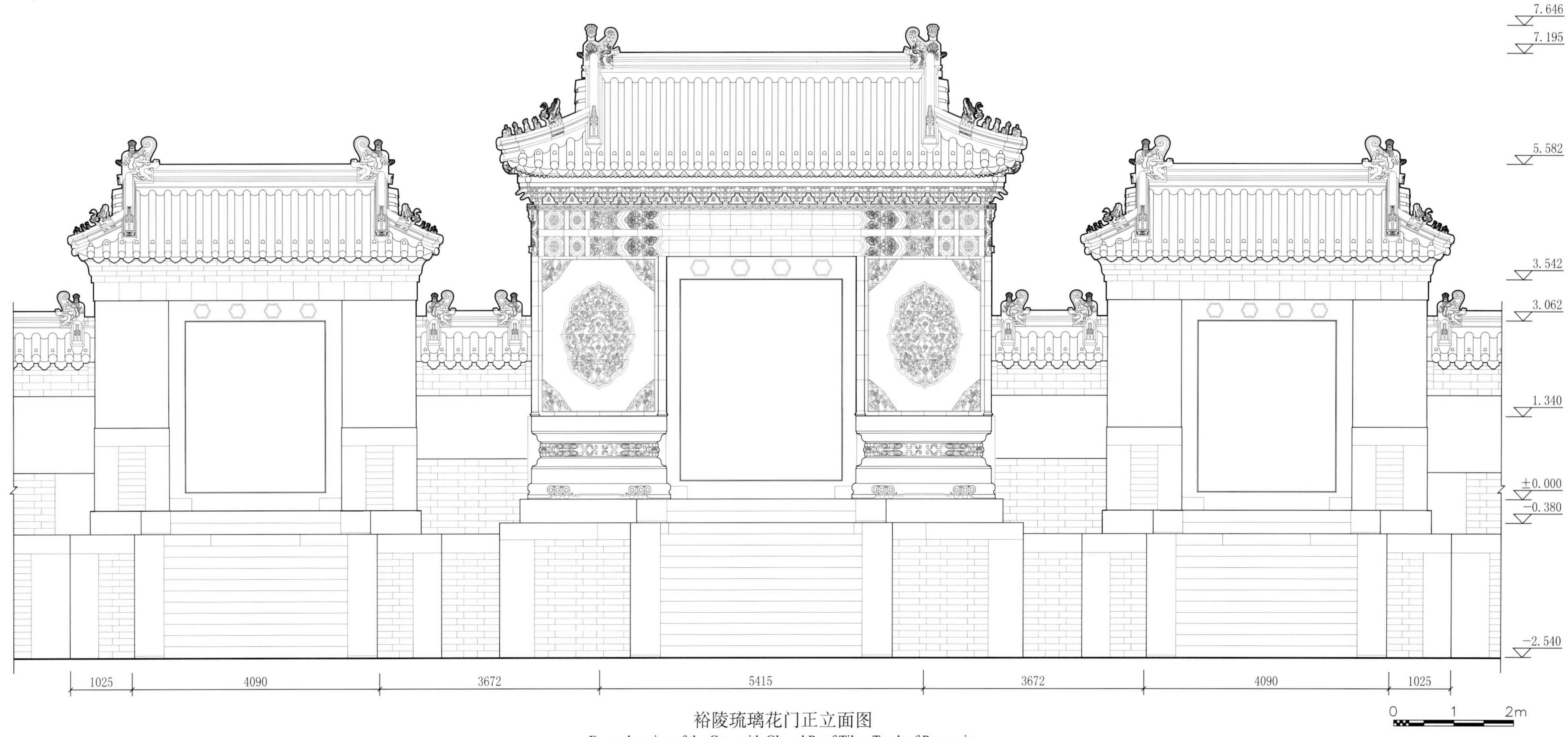

裕陵琉璃花门正立面图
Front elevation of the Gate with Glazed Roof Tiles, Tomb of Prosperity

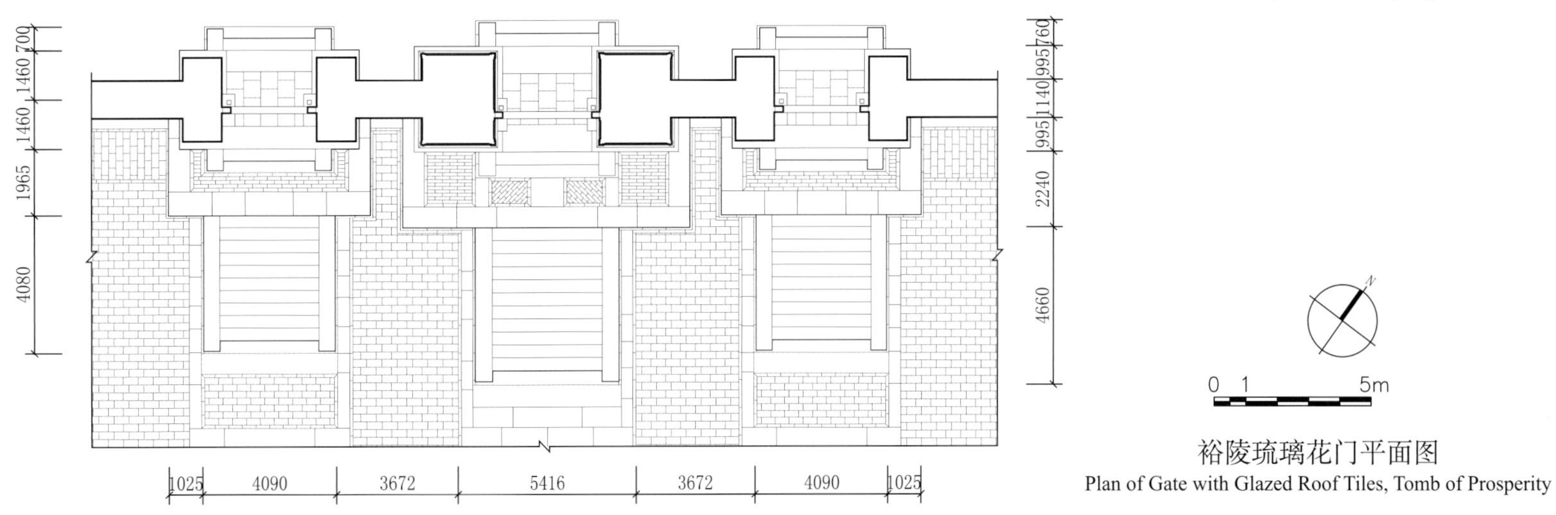

裕陵琉璃花门平面图
Plan of Gate with Glazed Roof Tiles, Tomb of Prosperity

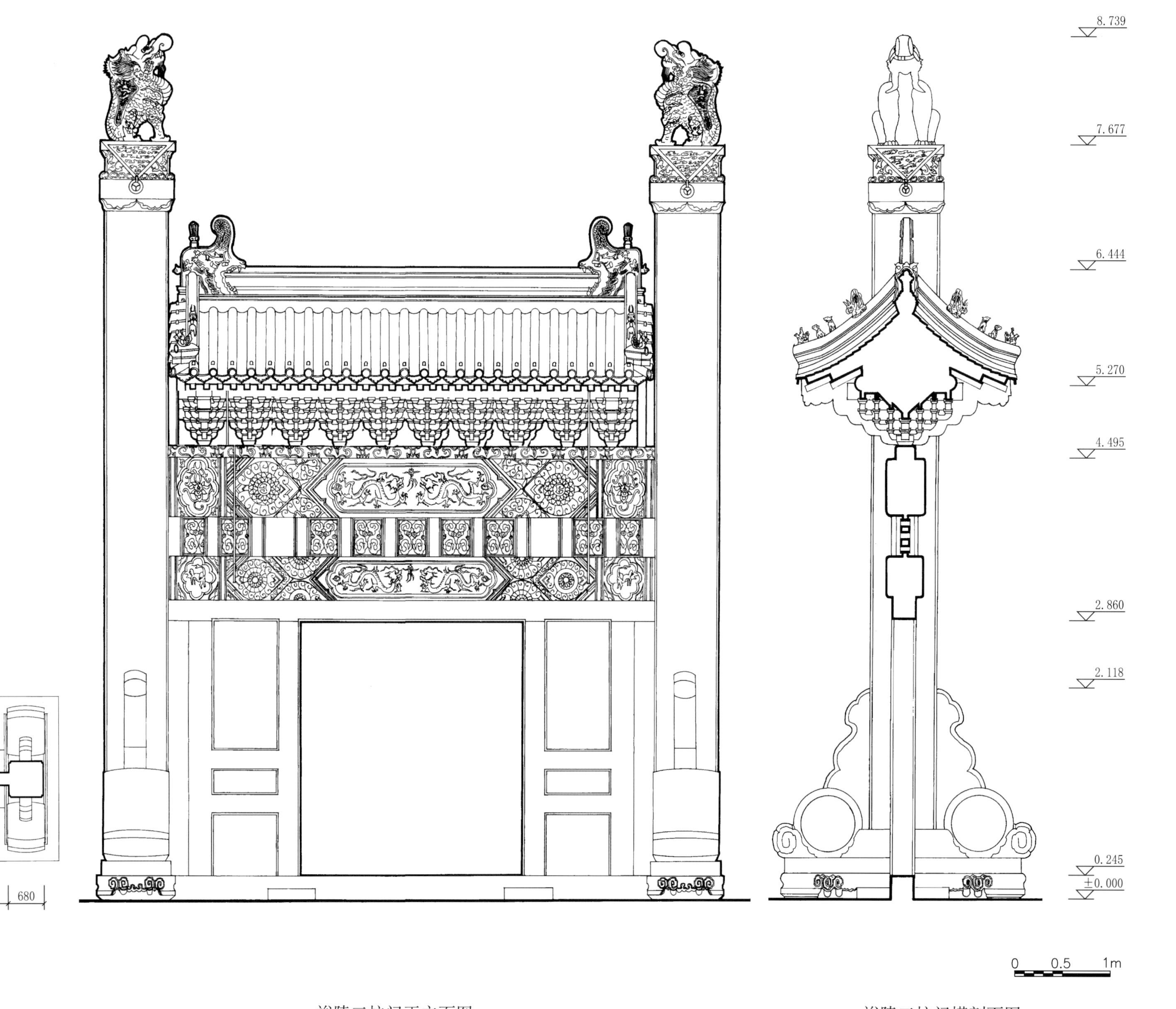

裕陵二柱门平面图
Plan of the Gate with Two Columns, Tomb of Prosperity

裕陵二柱门正立面图
Front elevation of the Gate with Two Columns, Tomb of Prosperity

裕陵二柱门横剖面图
Cross section of the Gate with Two Columns, Tomb of Prosperity

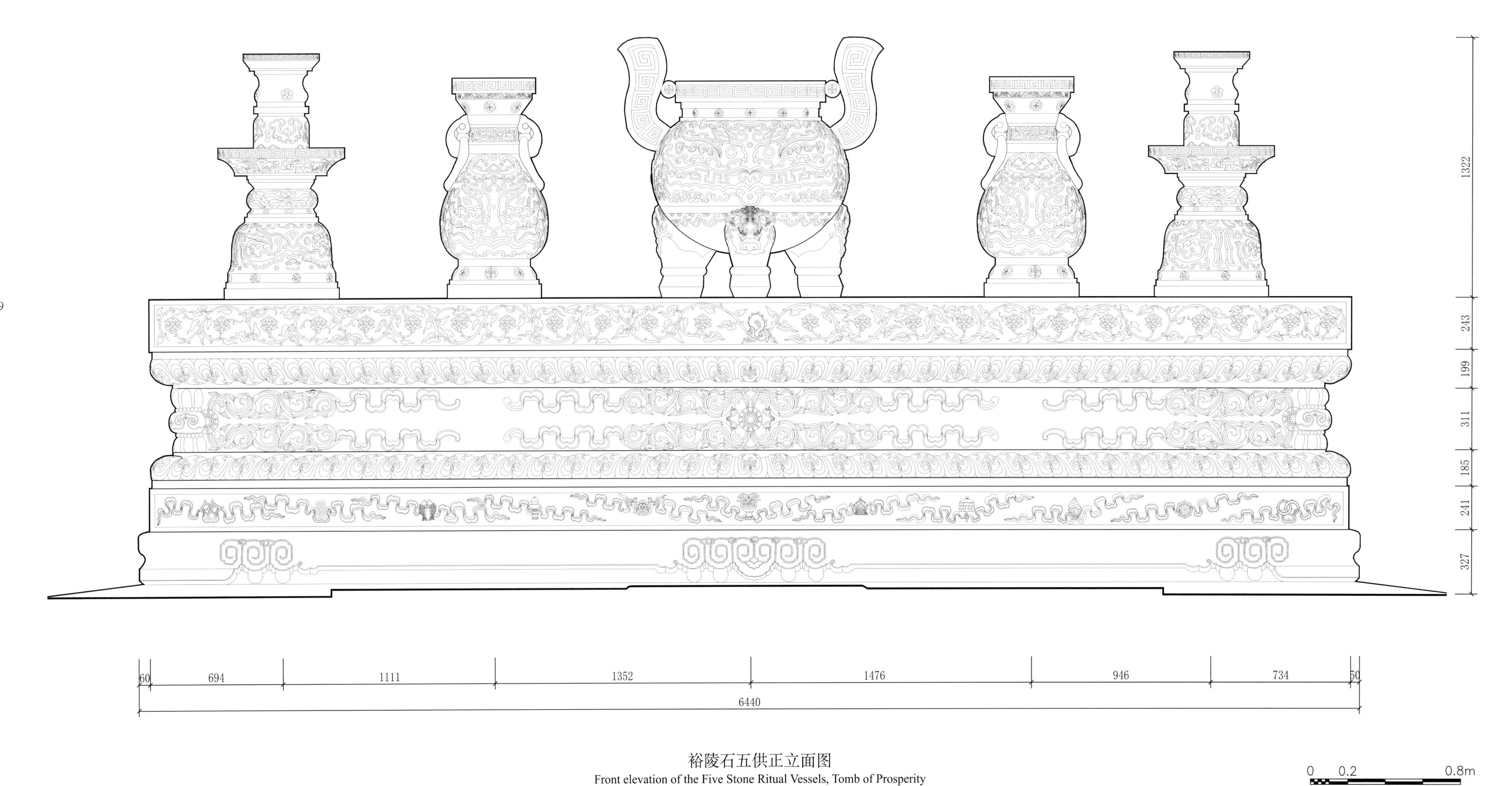

裕陵石五供正立面图
Front elevation of the Five Stone Ritual Vessels, Tomb of Prosperity

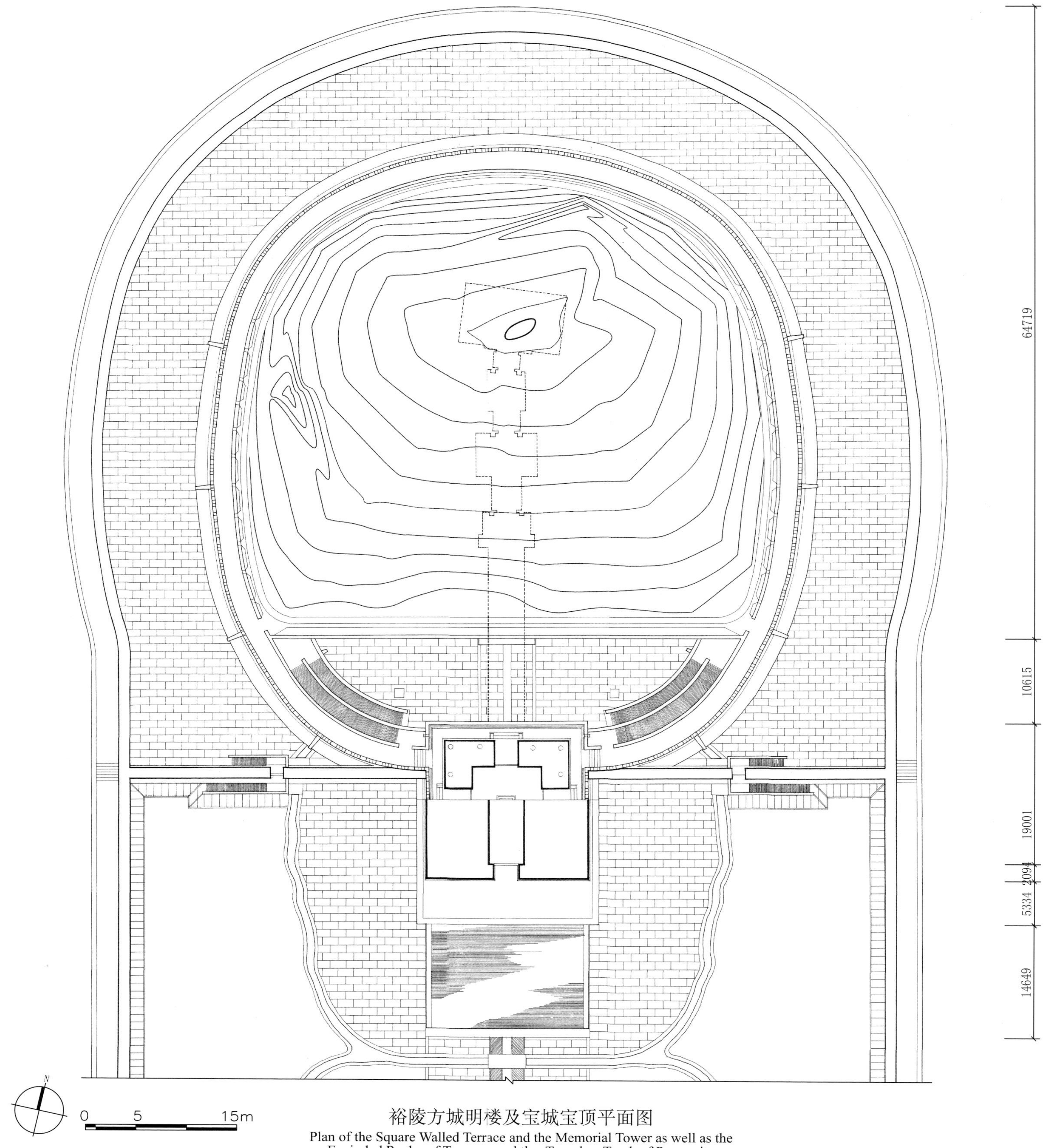

裕陵方城明楼及宝城宝顶平面图

Plan of the Square Walled Terrace and the Memorial Tower as well as the Encircled Realm of Treasure and the Tumulus, Tomb of Prosperity

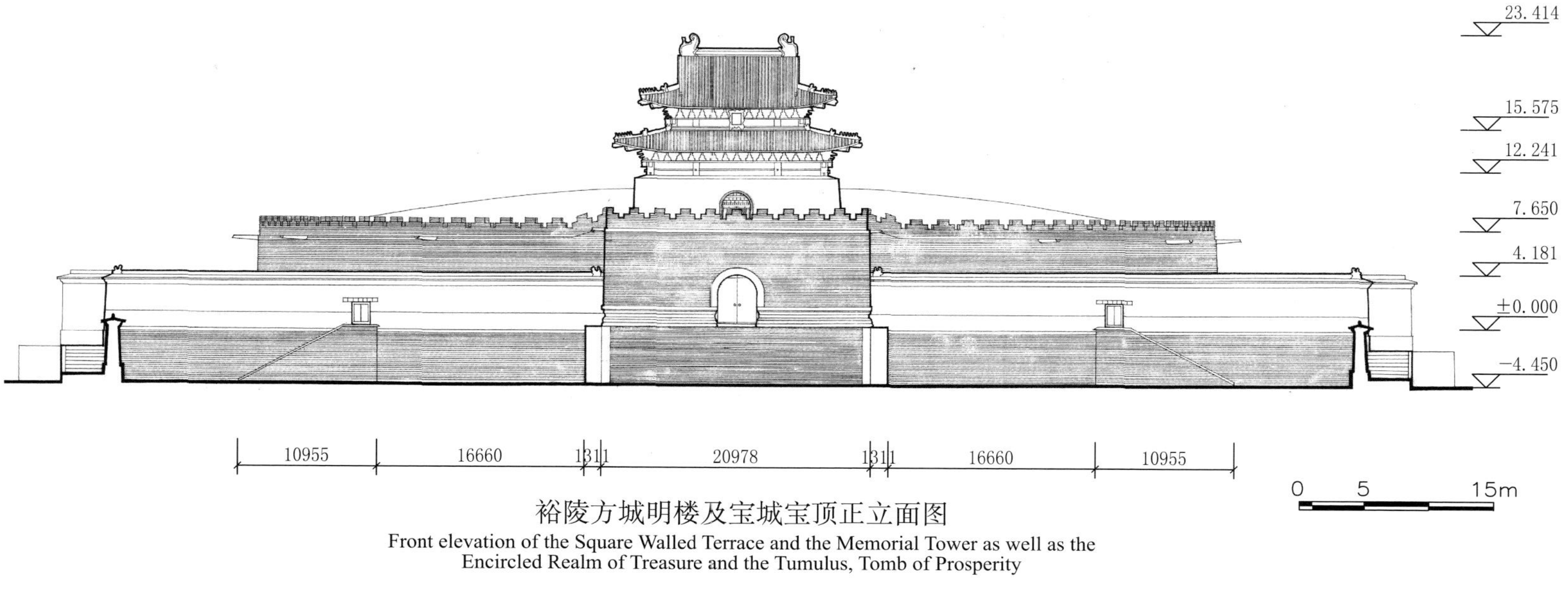

裕陵方城明楼及宝城宝顶正立面图

Front elevation of the Square Walled Terrace and the Memorial Tower as well as the Encircled Realm of Treasure and the Tumulus, Tomb of Prosperity

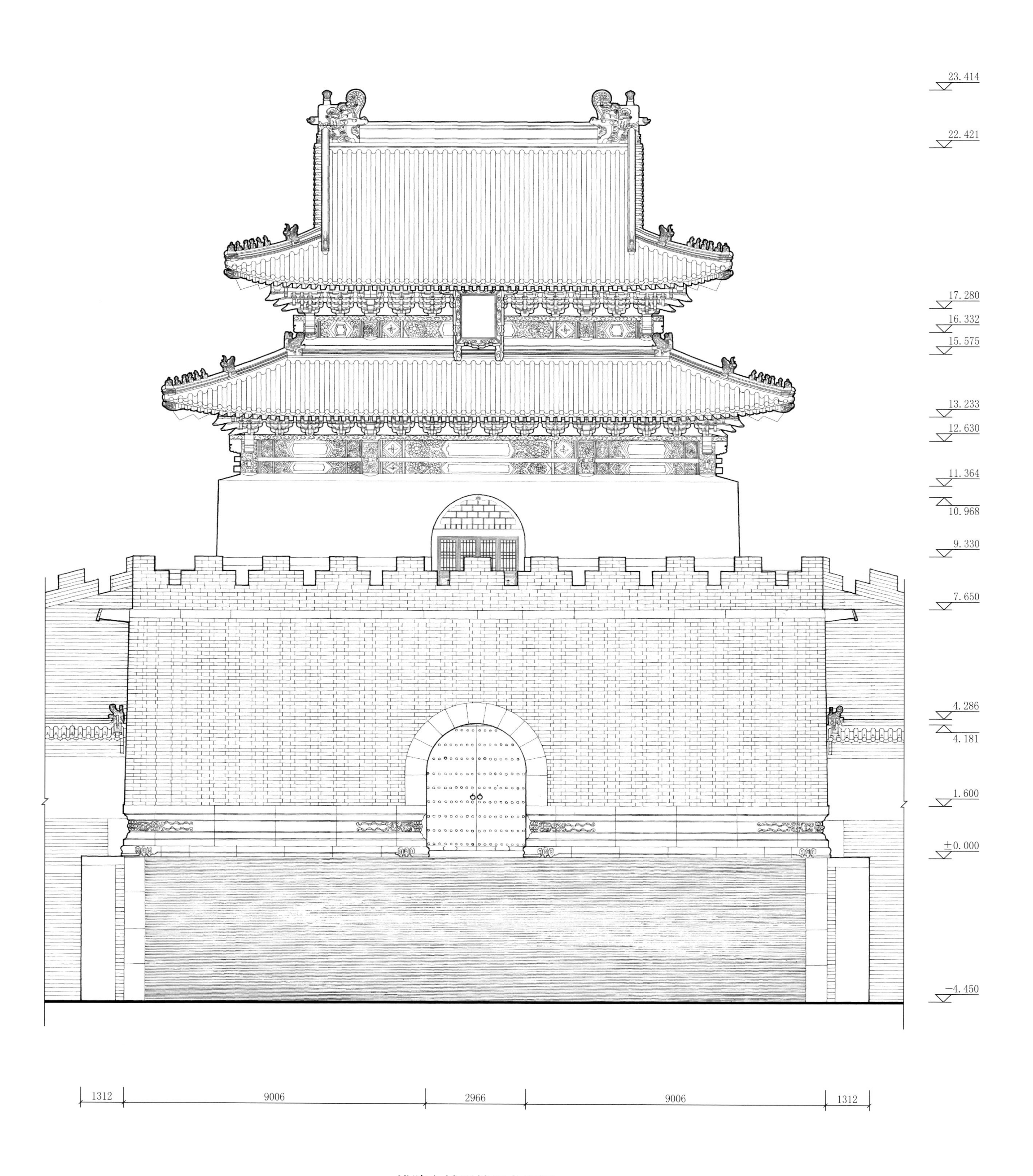

裕陵方城明楼正立面图
Front elevation of the Square Walled Terrace and the Memorial Tower, Tomb of Prosperity

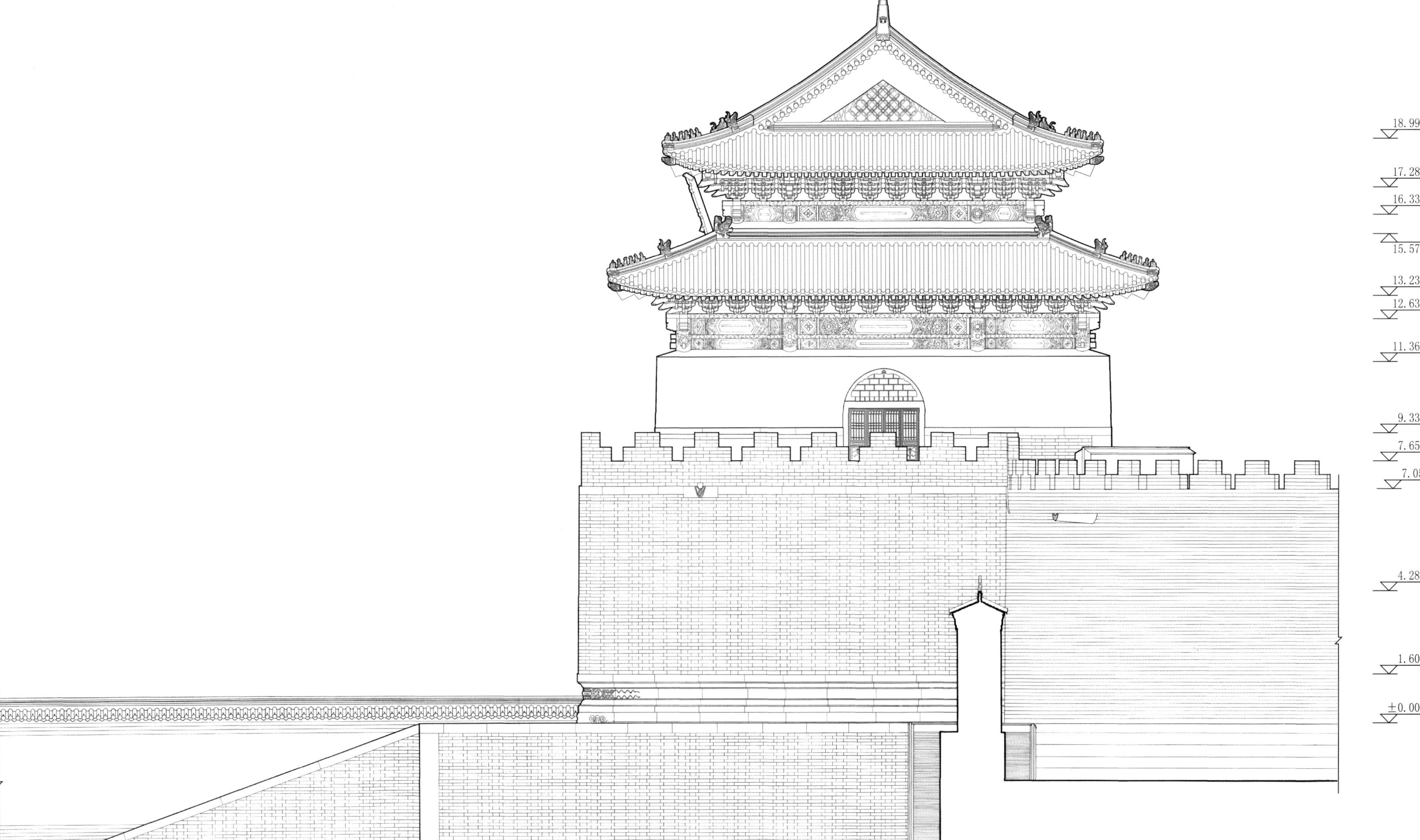

裕陵方城明楼侧立面图

Side elevation of the Square Walled Terrace and the Memorial Tower, Tomb of Prosperity

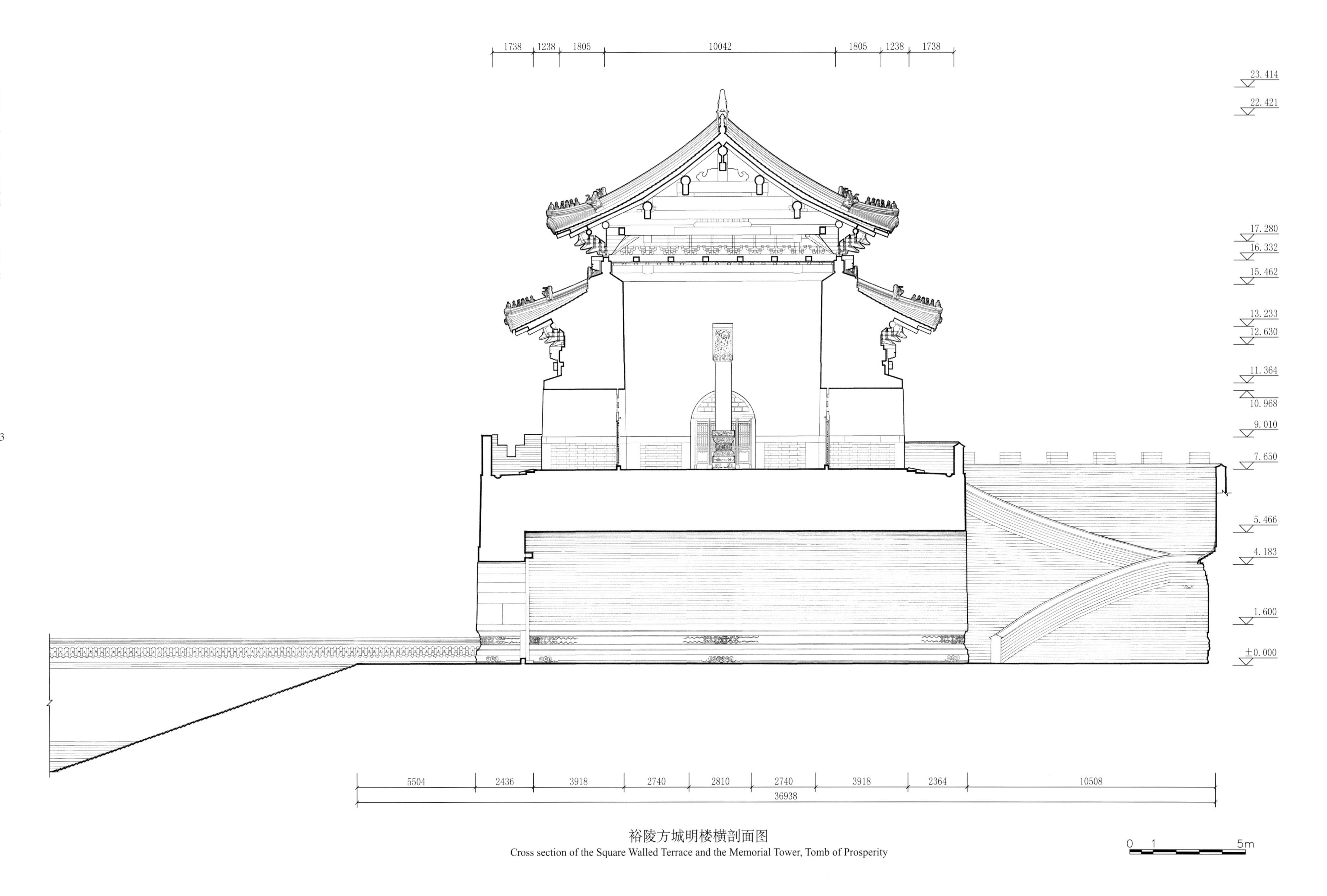

裕陵方城明楼横剖面图

Cross section of the Square Walled Terrace and the Memorial Tower, Tomb of Prosperity

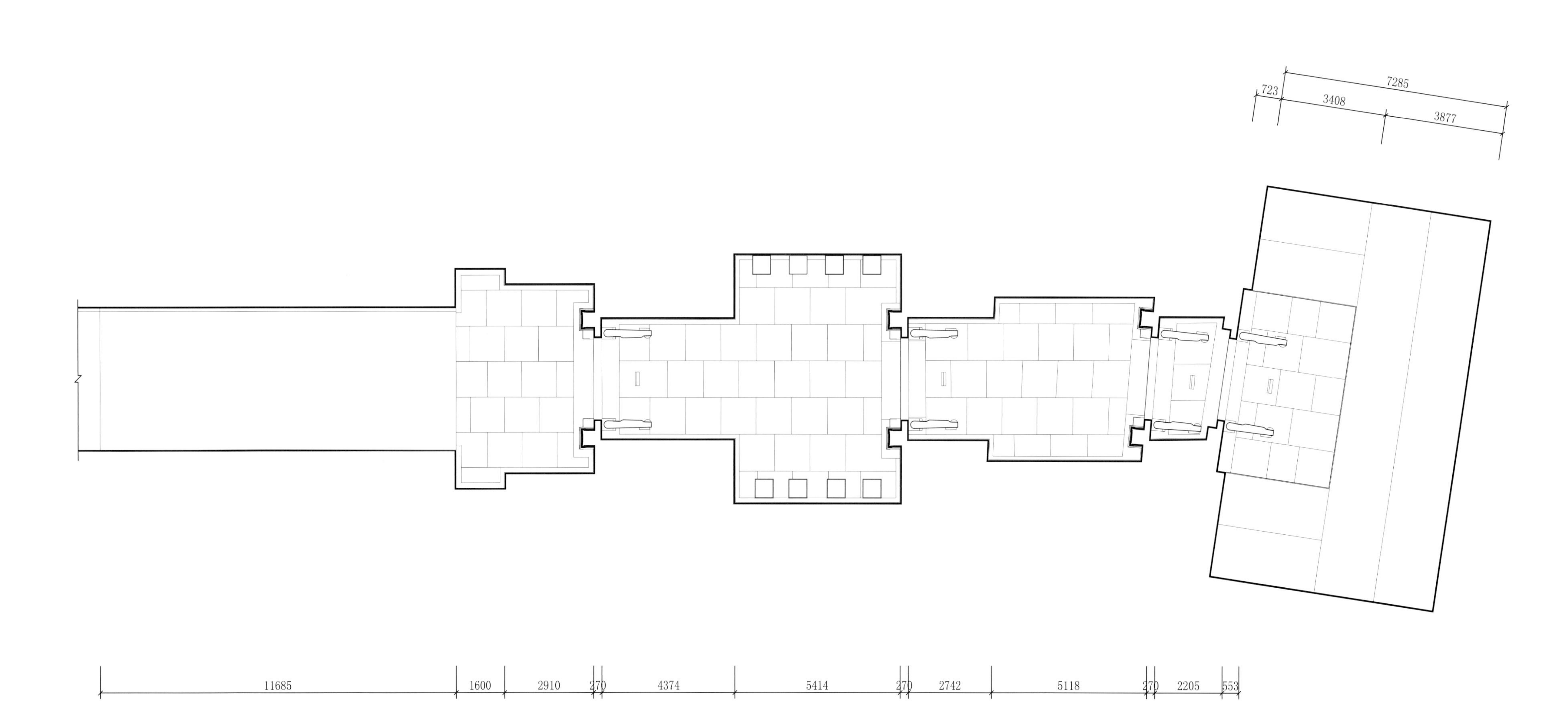

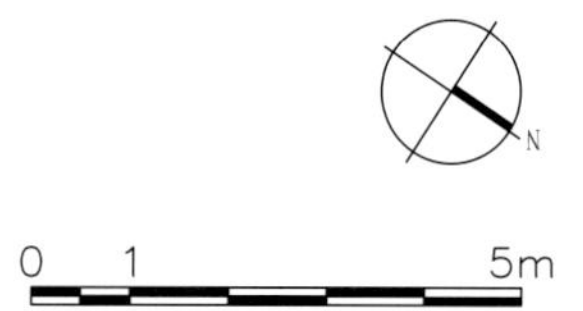

裕陵地宫平面图

Plan of the Underground Palace, Tomb of Prosperity

裕陵地宫头道石门大样图

Detailed drawing of the First Marble Gate in the Underground Palace, Tomb of Prosperity

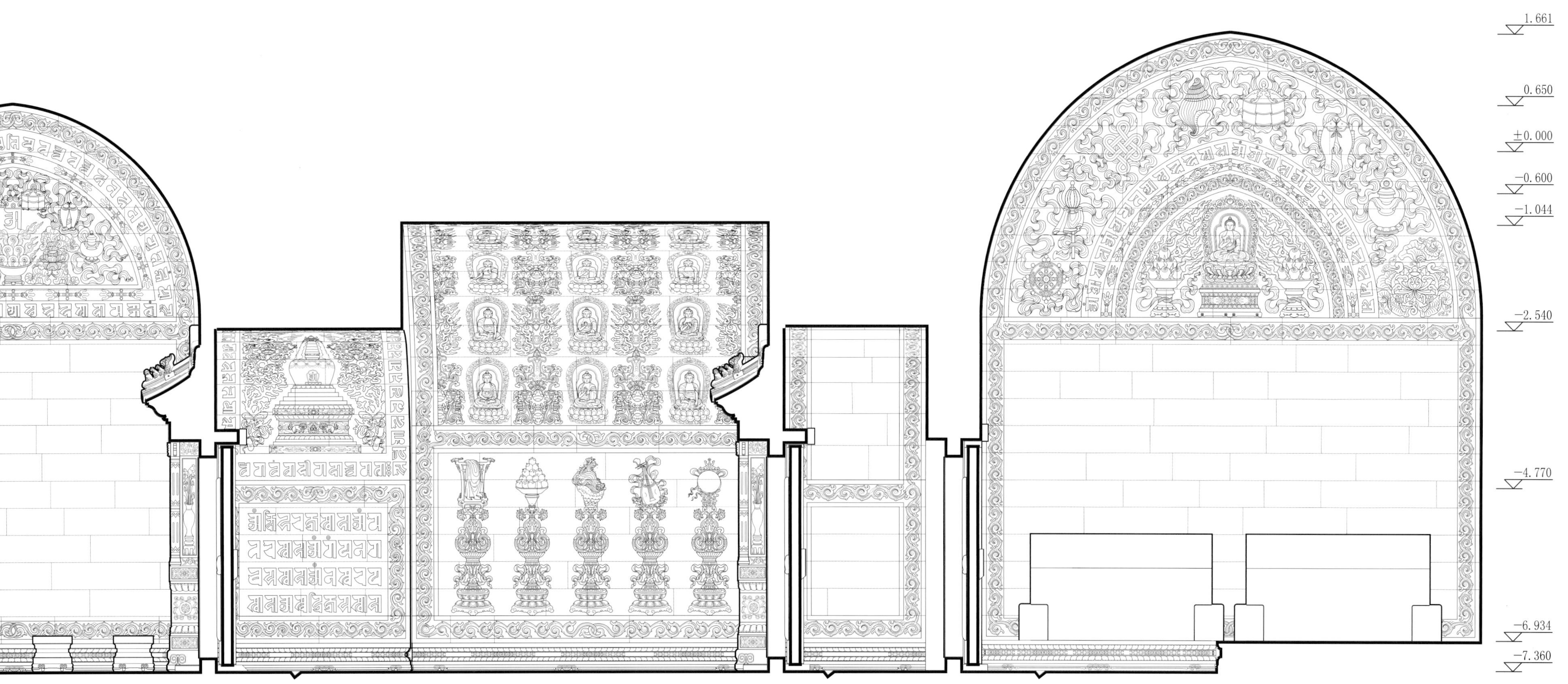
1.661
0.650
±0.000
−0.600
−1.044
−2.540
−4.770
−6.934
−7.360
0 1 2m

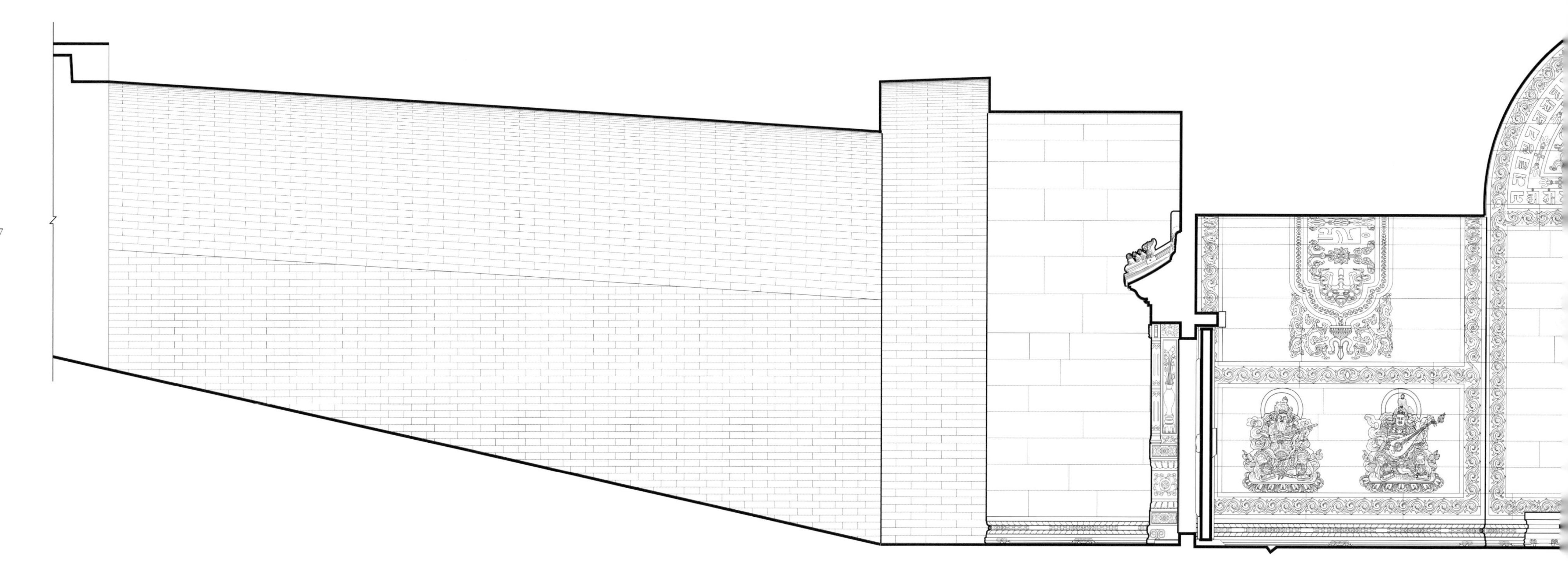

裕陵地宫横剖面图

Cross section of the Underground Palace, Tomb of Prosperity

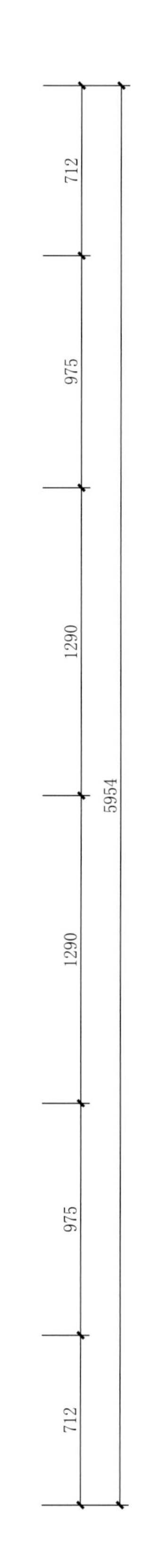

裕陵地宫罩门券纵剖面图

Longitudinal section of the Arched Vault covering the Main Gate of the Underground Palace, Tomb of Prosperity

裕陵地宫头层门洞券纵剖面图
Longitudinal section of the Vault of the First Doorway of the Underground Palace, Tomb of Prosperity

裕陵地宫明堂券纵剖面图
Longitudinal section of the Vault of the Ceremonial Hall (*mingtang*) of the Underground Palace, Tomb of Prosperity

裕陵地宫穿堂券纵剖面图

Longitudinal section of the Vault of the Hall of Passage in the Underground Palace, Tomb of Prosperity

裕陵地宫三层门洞券纵剖面图
Longitudinal section of the Vault of the Third Doorway of the Underground Palace, Tomb of Prosperity

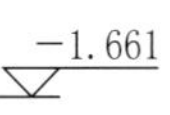

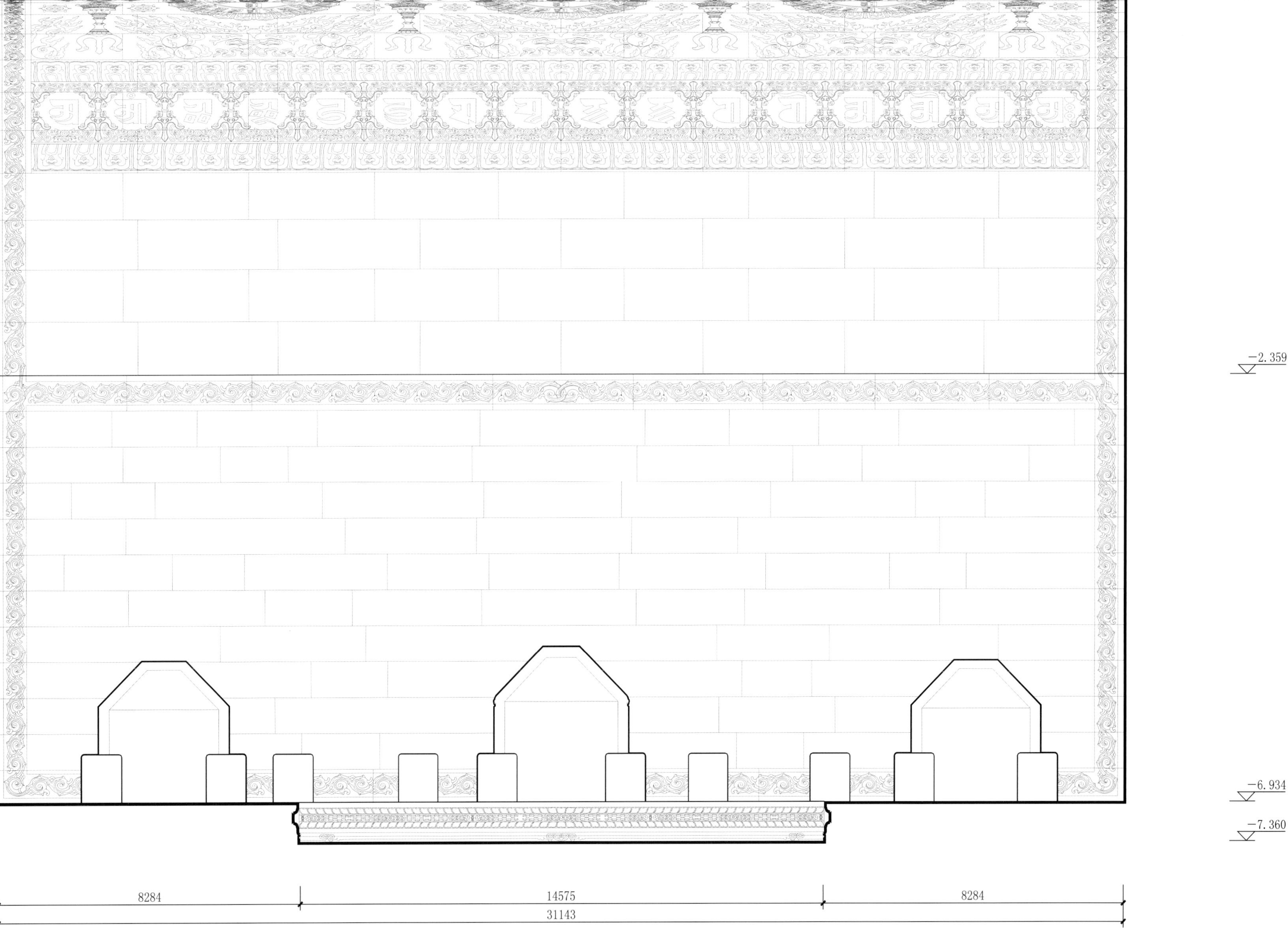

裕陵地宫金券纵剖面图

Longitudinal section of the Golden Vault at the Main Chamber of the Underground Palace, Tomb of Prosperity

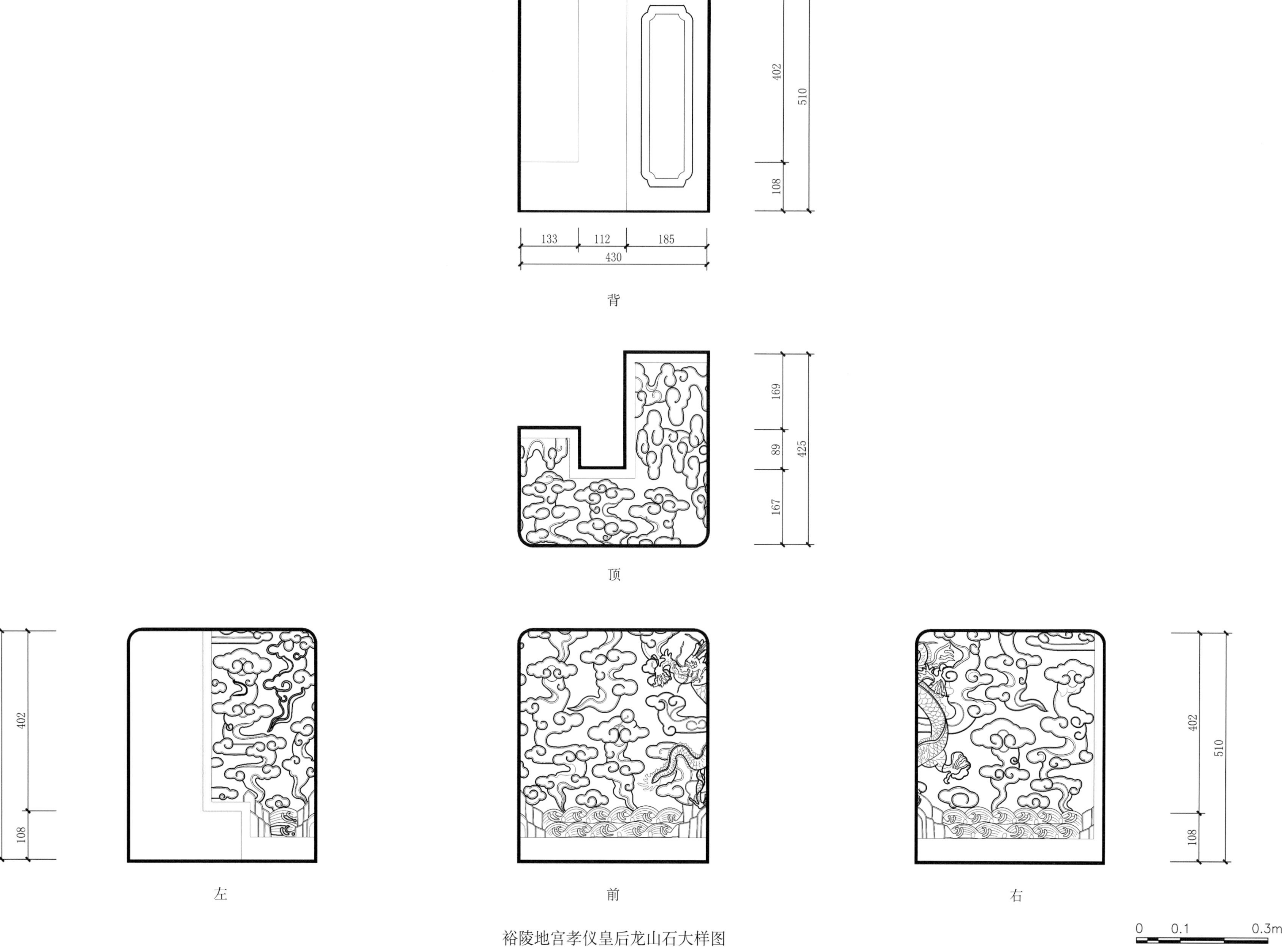

裕陵地宫孝仪皇后龙山石大样图

Detailed drawing of the Reinforcement Stone for Empress Xiaoyi's coffin in the Underground Palace, Tomb of Prosperity

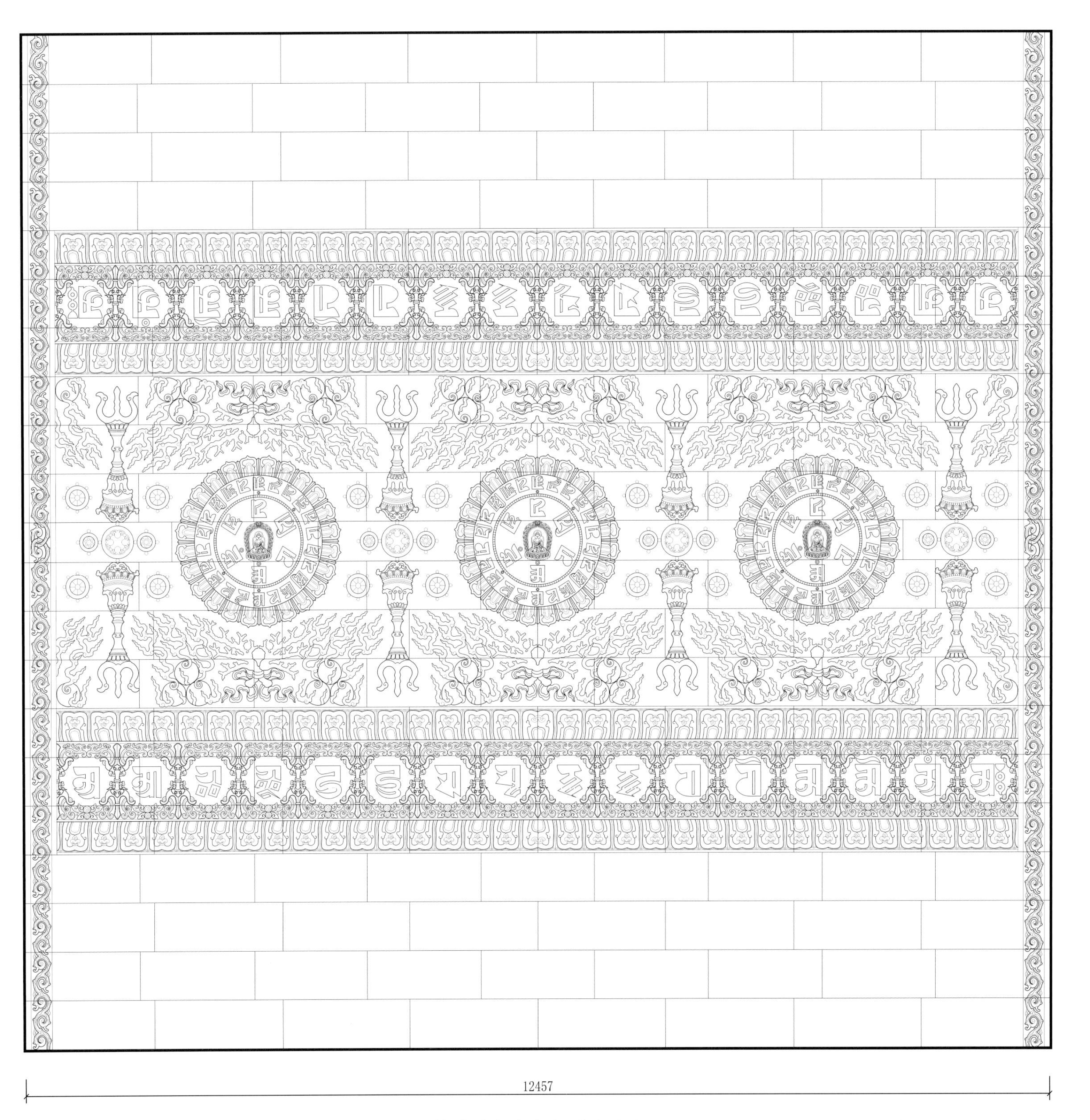

裕陵地宫金券券顶展开图

Detailed plan of the Golden Vault at the Main Chamber in the Underground Palace, Tomb of Prosperity

160
386
386
386
386
386
386
386
386
386
386

7285

裕陵地宫金券东壁月光墙大样图

Detailed drawing of the Crescent Arch with Lustrous Stones on the Eastern Wall at the Main Chamber in the Underground Palace, Tomb of Prosperity

0 0.1 0.5m

5026

500 501 500 501 500 501 500 501 396 501 500 501 500 501 500 501 500

5210

裕陵地宫穿堂券券顶展开图

Detailed plan of the Vault of the Hall of Passage in the Underground Palace, Tomb of Prosperity

0 0.5 1m

裕陵地宫穿堂券西壁五欲供大样图

Detailed drawing of the Western Wall with Sacrificial Offerings for the Five Desires in the Hall of Passage in the Underground Palace, Tomb of Prosperity

0 0.1 0.5m

裕陵地宫明堂券券顶展开图
Detailed plan of the Vault of the Ceremonial Hall in the Underground Palace, Tomb of Prosperity

裕陵地宫明堂券东壁月光墙大样图

Detailed drawing of the Crescent Arch with Lustrous Stones on the Eastern Wall at the Ceremonial Hall in the Underground Palace, Tomb of Prosperity

487 487 487 487 487 487 396 487 487 487 487 487 487

4347

0 0.2 1m

裕陵地宫头层门洞券顶展开图

Detailed plan of the Vault of the First Doorway in the Underground Palace, Tomb of Prosperity

增长天王

持国天王

裕陵地宫头层门洞券西壁天王大样图

Detailed drawing of the Guardian of the World on the Western Wall of the Vault of the First Doorway in the Underground Palace, Tomb of Prosperity

财宝天王

广目天王

裕陵地宫头层门洞券东壁天王大样图

Detailed drawing of the Guardian of the World on the Eastern Wall of the Vault of the First Doorway in the Underground Palace, Tomb of Prosperity

裕陵地宫头道石门菩萨大样图
Detailed drawing of the Figure of Bodhisattva on the First Stone Gate in the Underground Palace, Tomb of Prosperity

裕陵地宫二道石门菩萨大样图
Detailed drawing of the Figure of Bodhisattva on the Second Stone Gate in the Underground Palace, Tomb of Prosperity

裕陵地宫三道石门菩萨大样图

Detailed drawing of the Figure of Bodhisattva on the Third Stone Gate in the Underground Palace, Tomb of Prosperity

普贤菩萨

弥勒菩萨

裕陵地宫四道石门菩萨大样图

Detailed drawing of the Figure of Bodhisattva on the Fourth Stone Gate in the Underground Palace, Tomb of Prosperity

0　0.1　0.3m

裕陵妃园寝
Imperial Consorts' Tombs affiliated with the Tomb of Prosperity (Yuling Feiyuanqin)

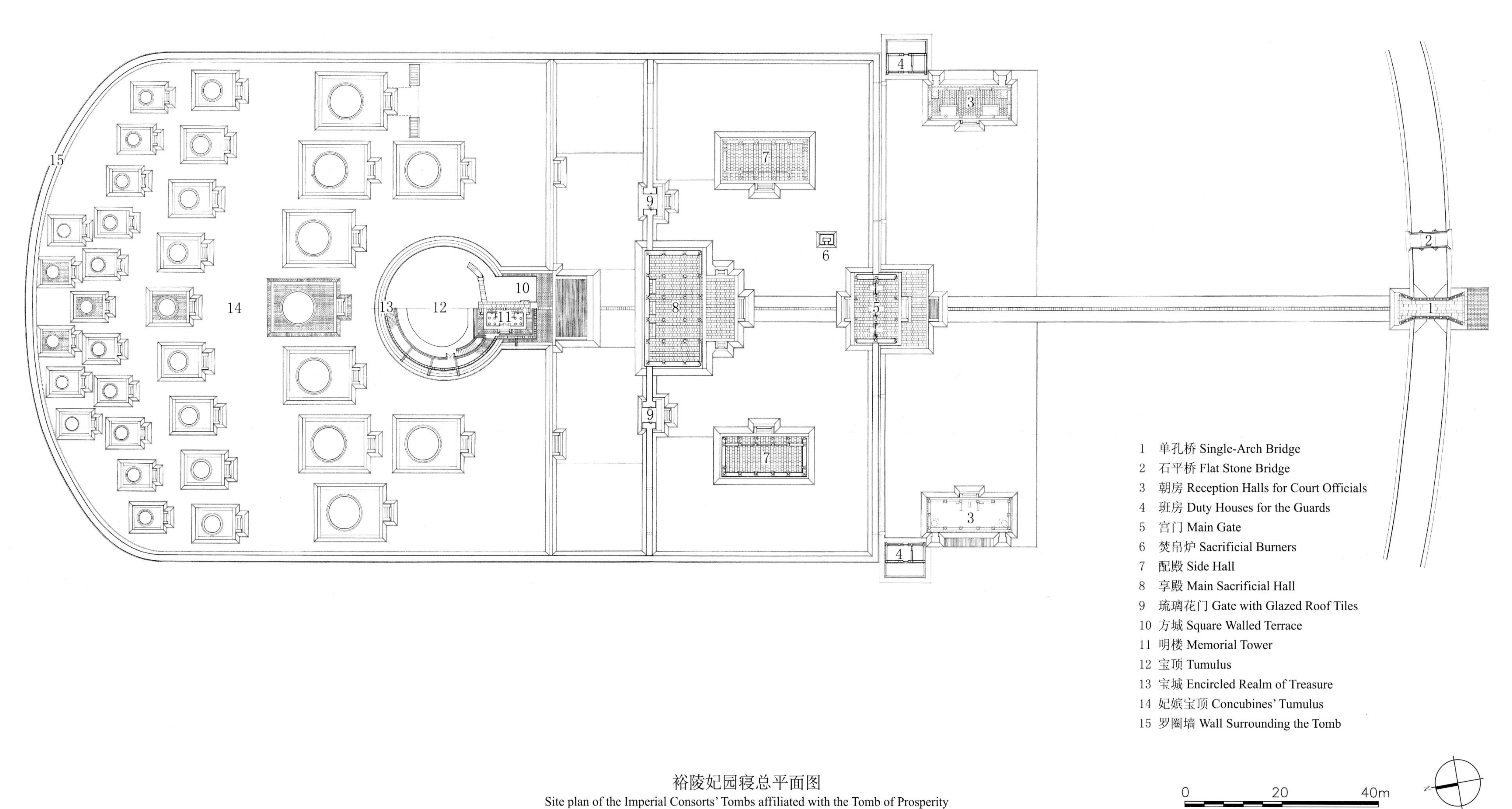

裕陵妃园寝总平面图
Site plan of the Imperial Consorts' Tombs affiliated with the Tomb of Prosperity

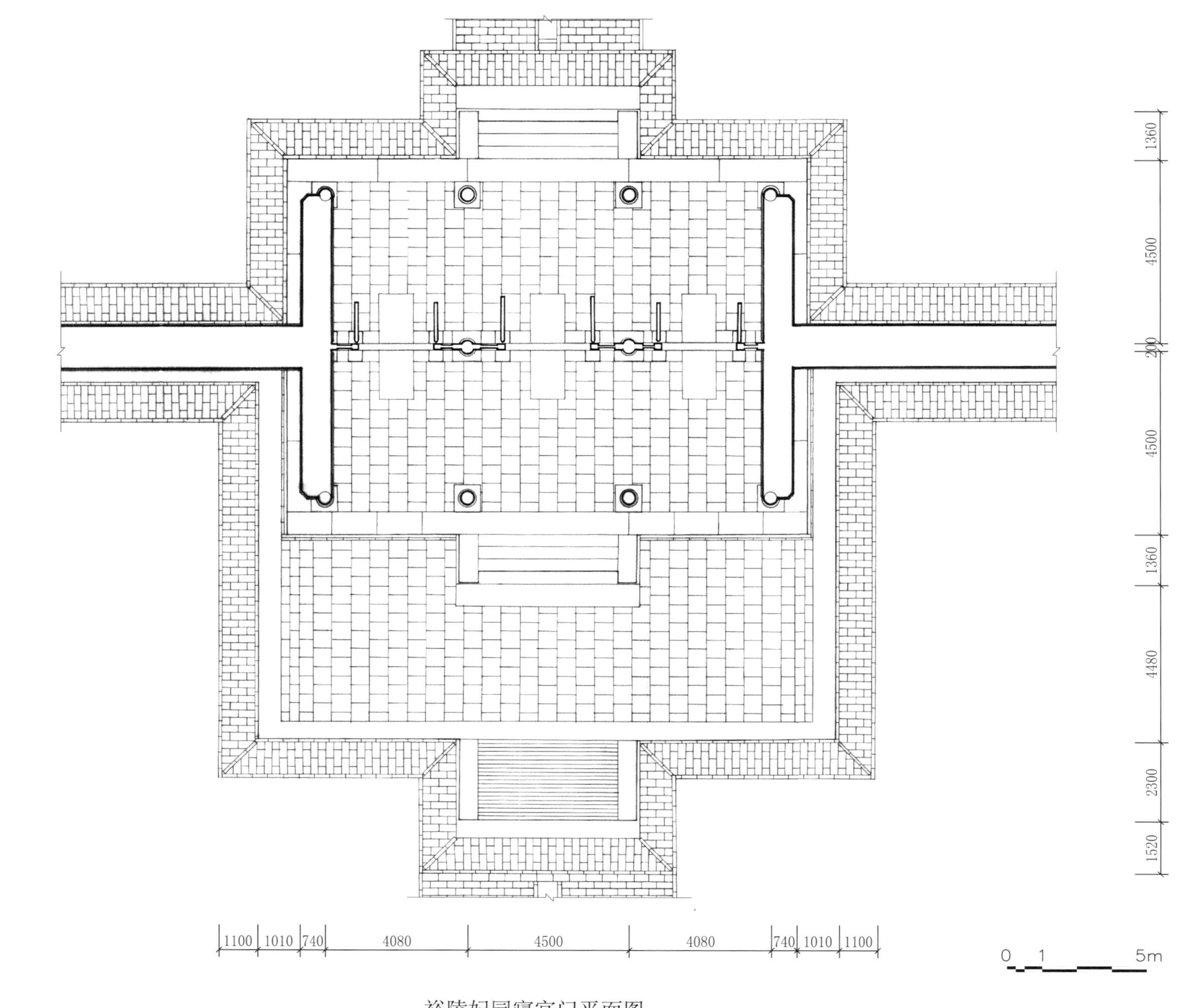

裕陵妃园寝宫门平面图
Plan of the Main Gate, Imperial Consorts' Tombs affiliated with the Tomb of Prosperity

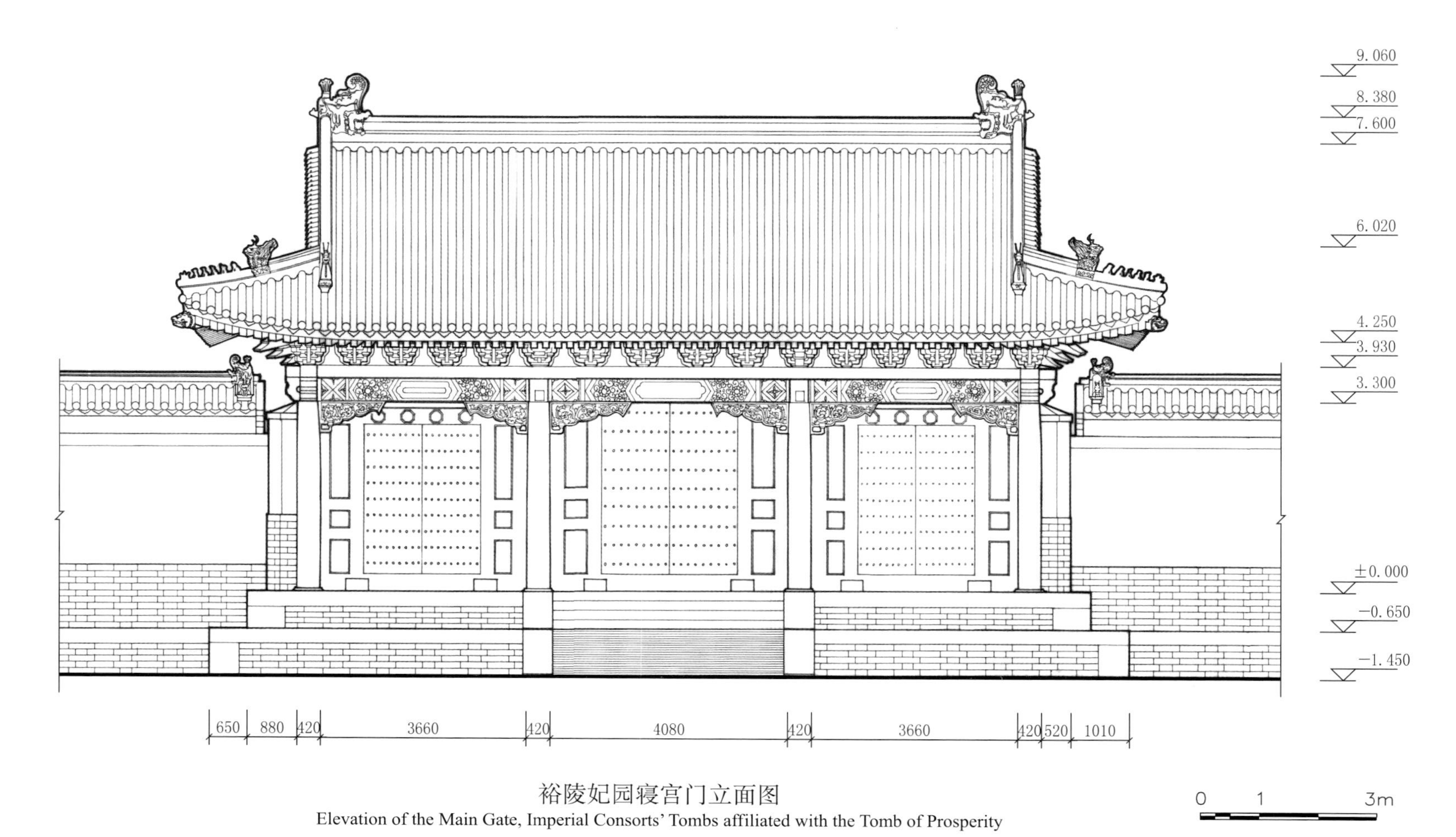

裕陵妃园寝宫门立面图
Elevation of the Main Gate, Imperial Consorts' Tombs affiliated with the Tomb of Prosperity

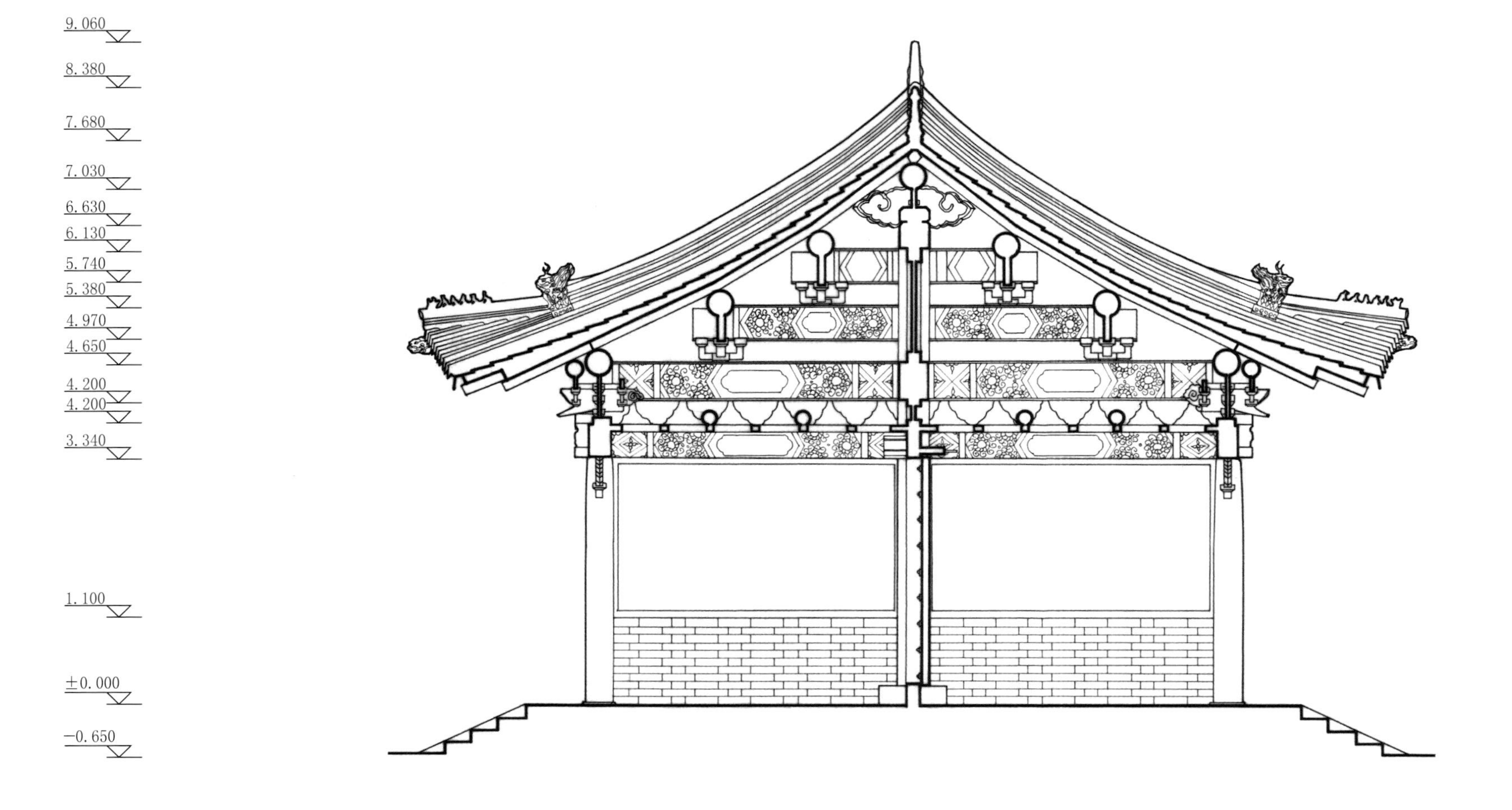

裕陵妃园寝宫门横剖面图
Cross Section of the Main Gate, Imperial Consorts' Tombs affiliated with the Tomb of Prosperity

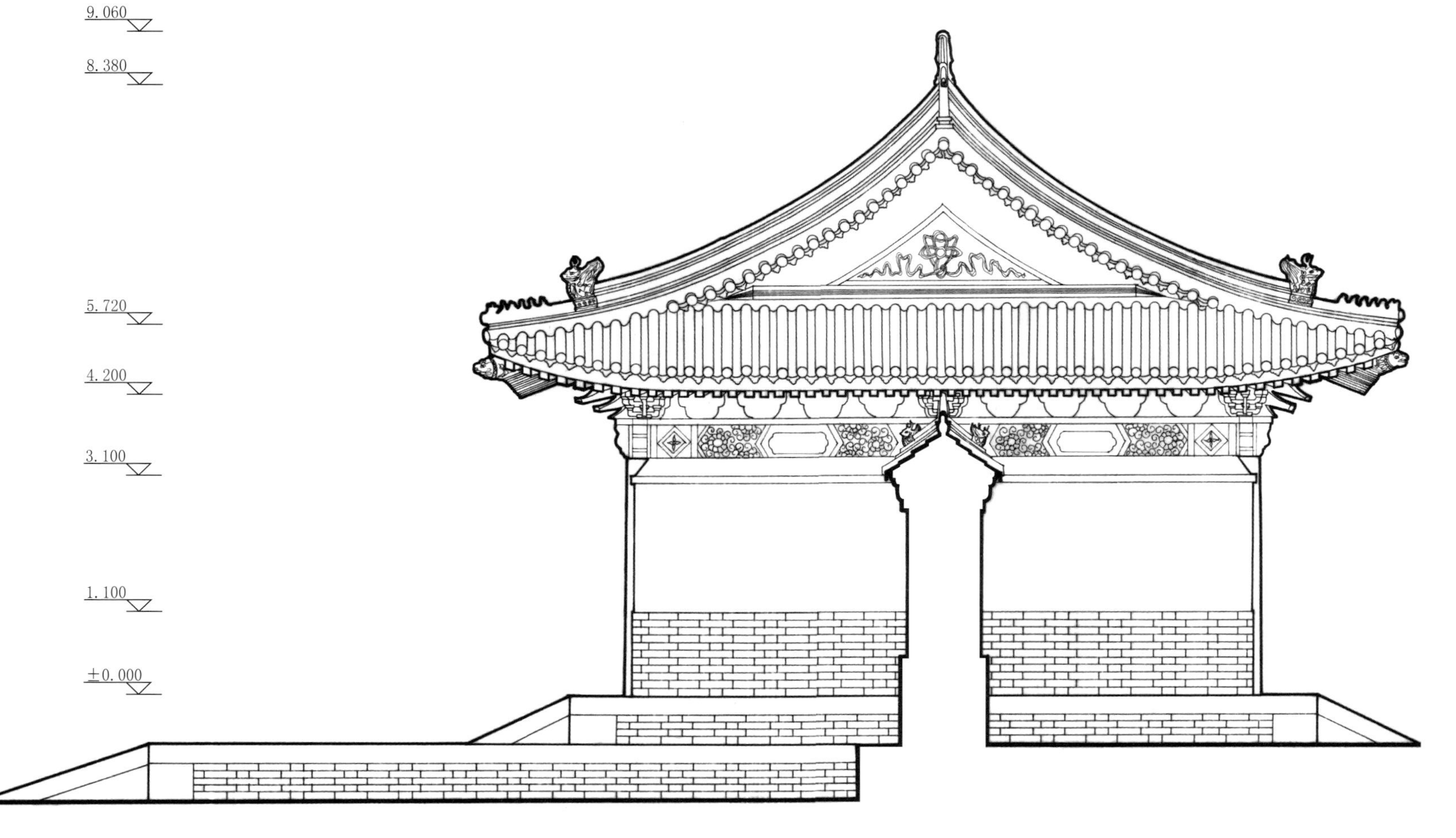

裕陵妃园寝宫门侧立面图
Side Elevation of the Main Gate, Imperial Consorts' Tombs affiliated with the Tomb of Prosperity

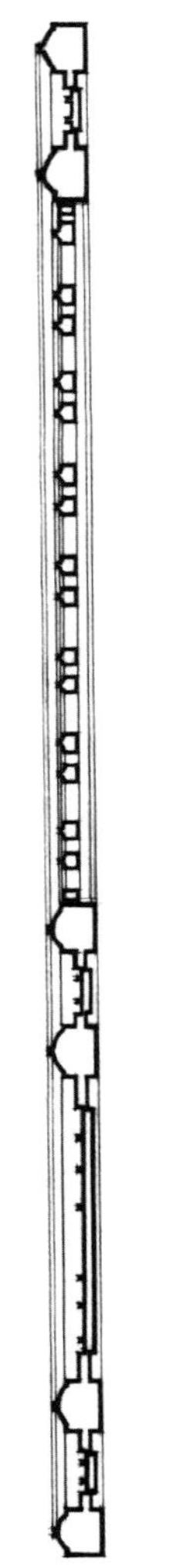

裕陵妃园寝西配殿门窗大样图
Detailed drawing of windows and doors of the Western Side Hall, Imperial Consorts' Tombs affiliated with the Tomb of Prosperity

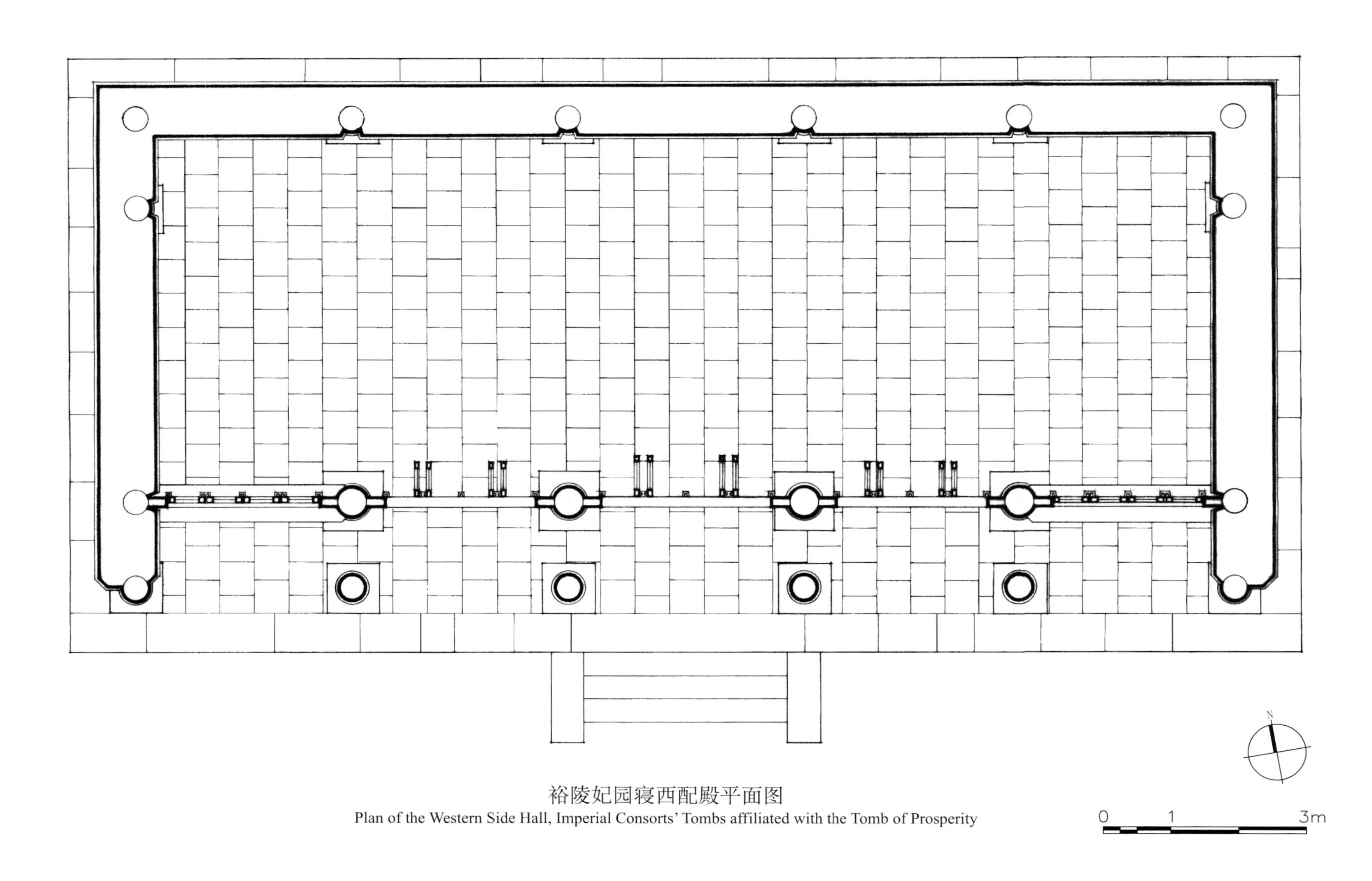

裕陵妃园寝西配殿平面图
Plan of the Western Side Hall, Imperial Consorts' Tombs affiliated with the Tomb of Prosperity

裕陵妃园寝西配殿正立面图

Front elevation of the Western Side Hall, Imperial Consorts' Tombs affiliated with the Tomb of Prosperity

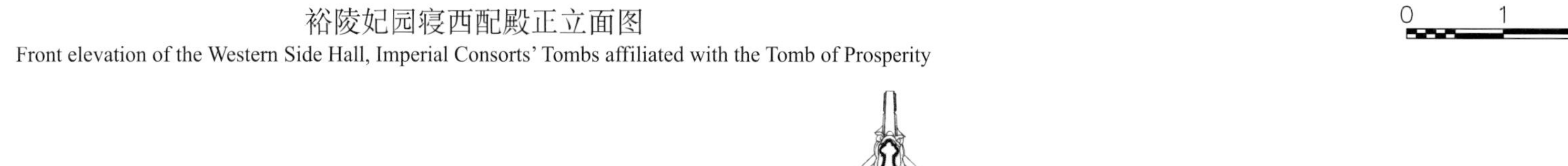

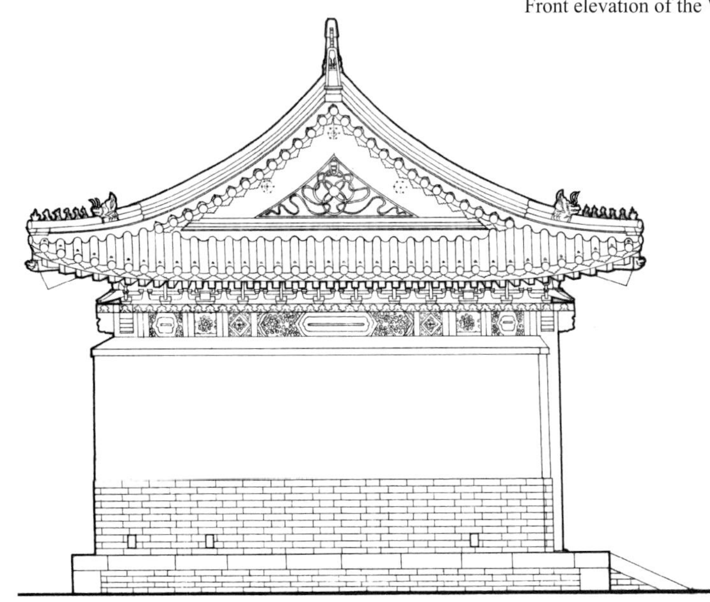

裕陵妃园寝西配殿侧立面图

Side elevation of the Western Side Hall, Imperial Consorts' Tombs affiliated with the Tomb of Prosperity

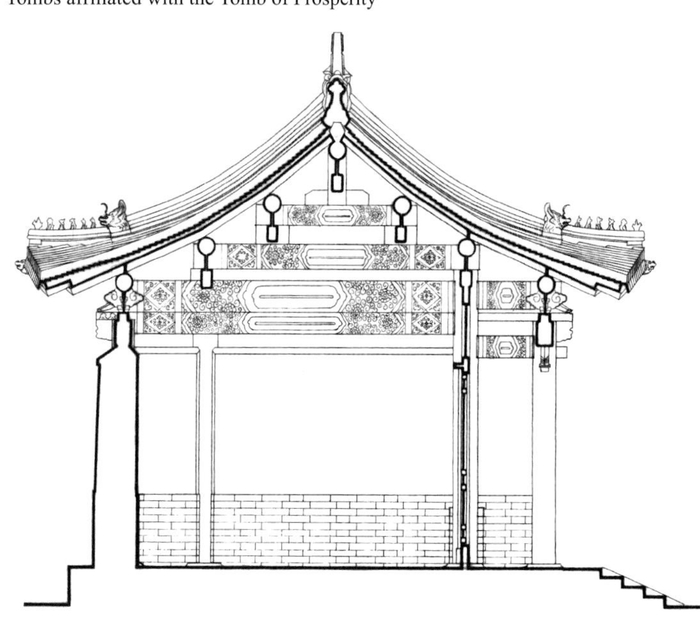

裕陵妃园寝西配殿横剖面图

Cross section of the Western Side Hall, Imperial Consorts' Tombs affiliated with the Tomb of Prosperity

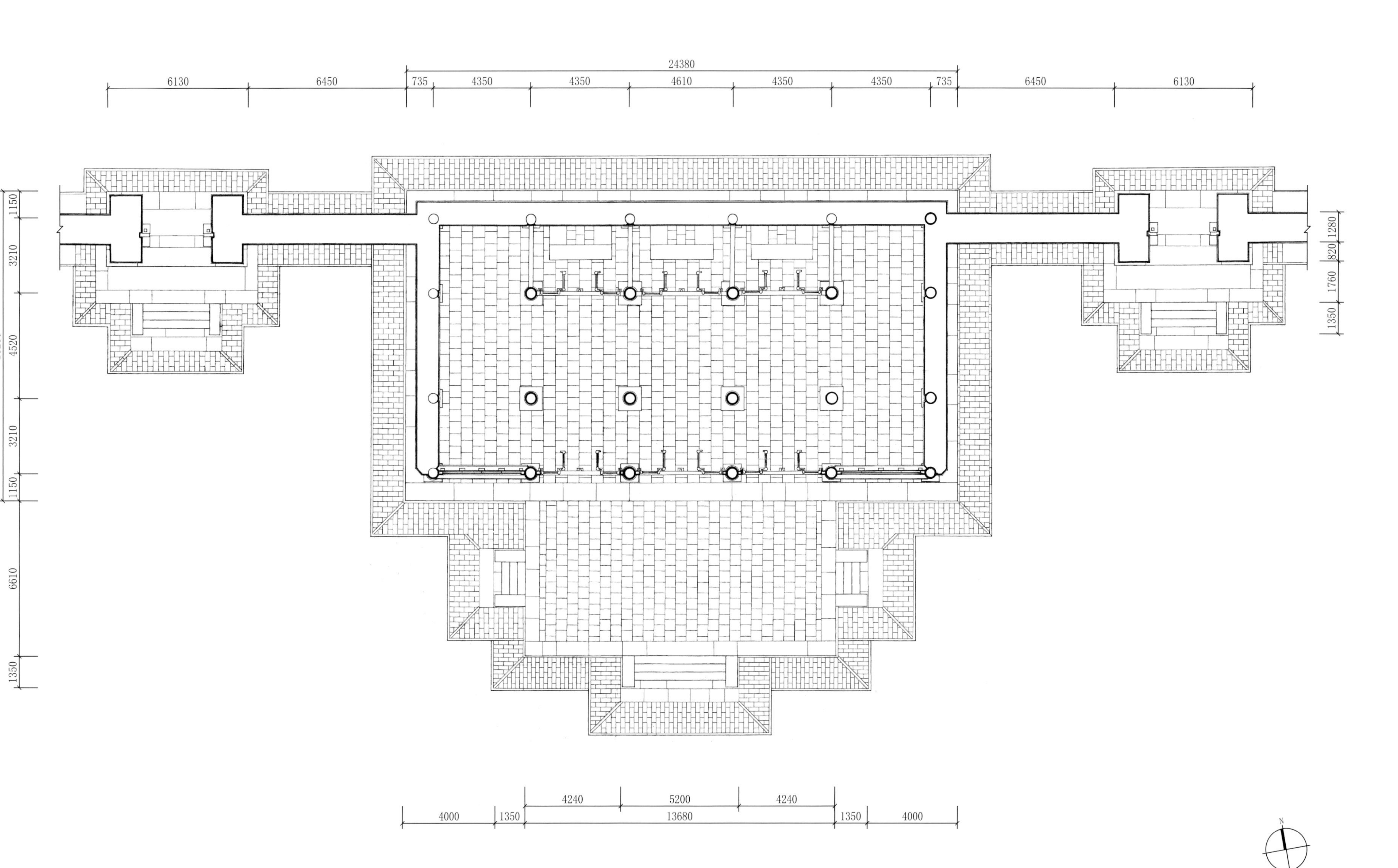

裕陵妃园寝享殿平面图

Plan of the Hall of Ritual Sacrifice, Imperial Consorts' Tombs affiliated with the Tomb of Prosperity

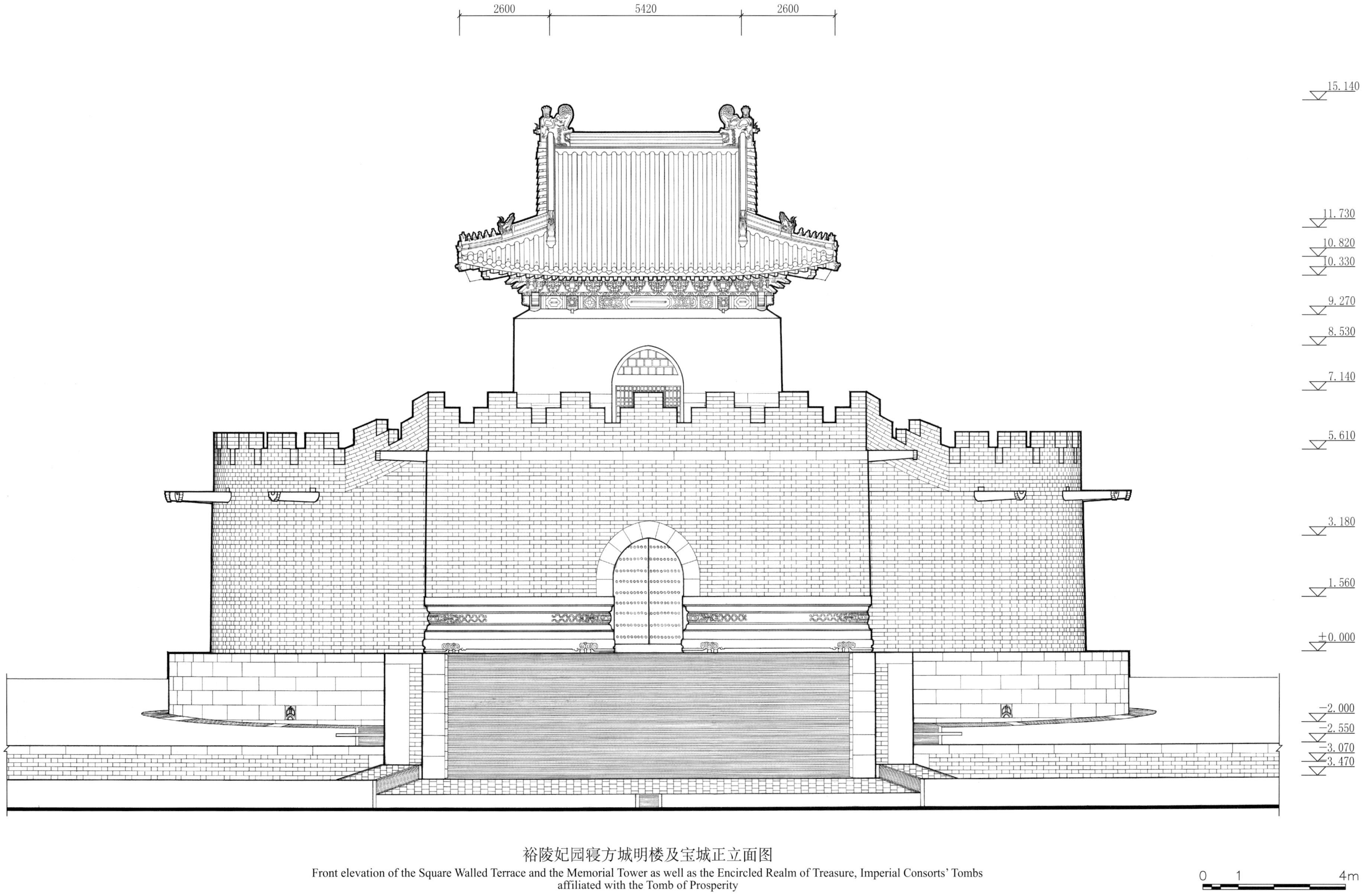

2600 5420 2600

15.140
11.730
10.820
10.330
9.270
8.530
7.140
5.610
3.180
1.560
±0.000
−2.000
−2.550
−3.070
−3.470

裕陵妃园寝方城明楼及宝城正立面图

Front elevation of the Square Walled Terrace and the Memorial Tower as well as the Encircled Realm of Treasure, Imperial Consorts' Tombs affiliated with the Tomb of Prosperity

0 1 4m

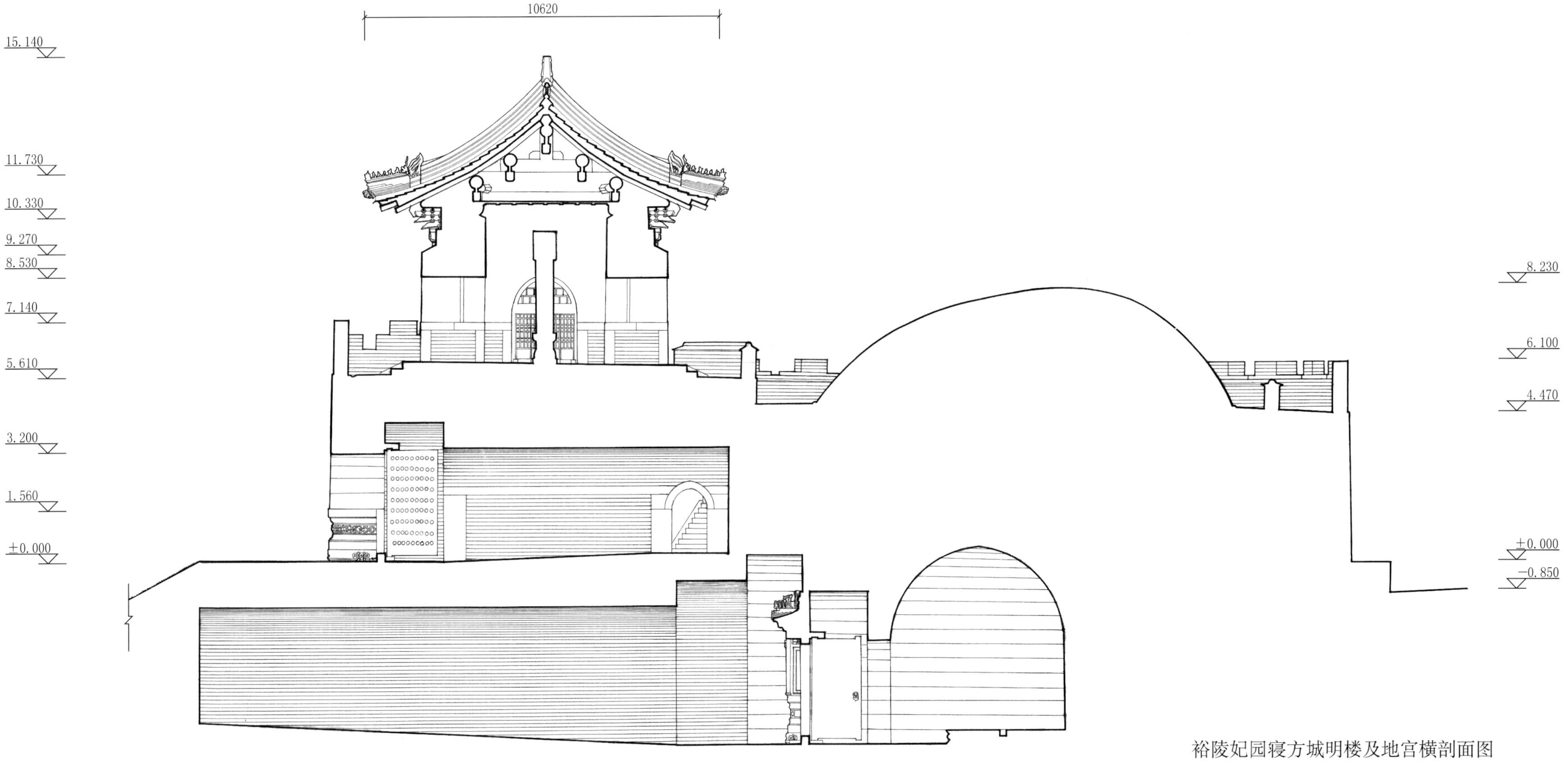

10620

15.140
11.730
10.330
9.270
8.530
7.140
5.610
3.200
1.560
±0.000

8.230
6.100
4.470
±0.000
-0.850

裕陵妃园寝方城明楼及地宫横剖面图

Cross section of the Square Walled Terrace and the Memorial Tower and the Underground Palace, Imperial Consorts' Tombs affiliated with the Tomb of Prosperity

744 752 3000 752 744

14550 2250 1653 2752 5300

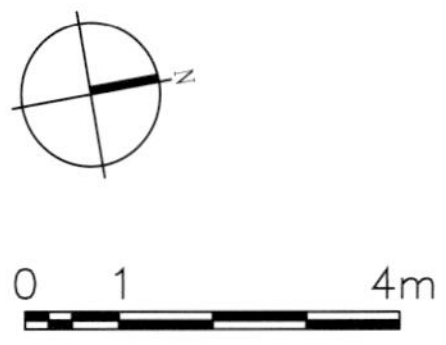

0 1 4m

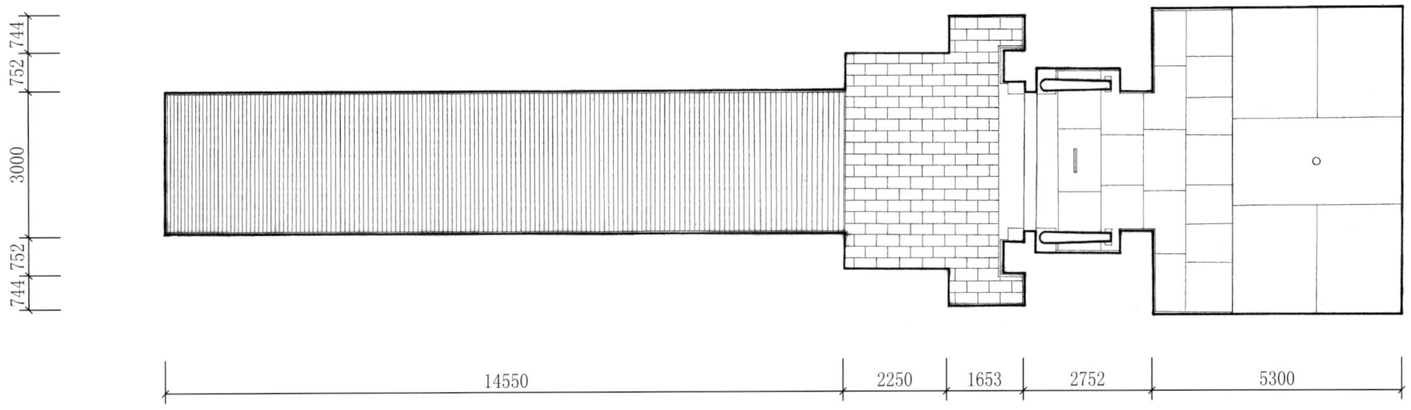

裕陵妃园寝地宫平面图

Plan of the Underground Palace, Imperial Consorts' Tombs affiliated with the Tomb of Prosperity

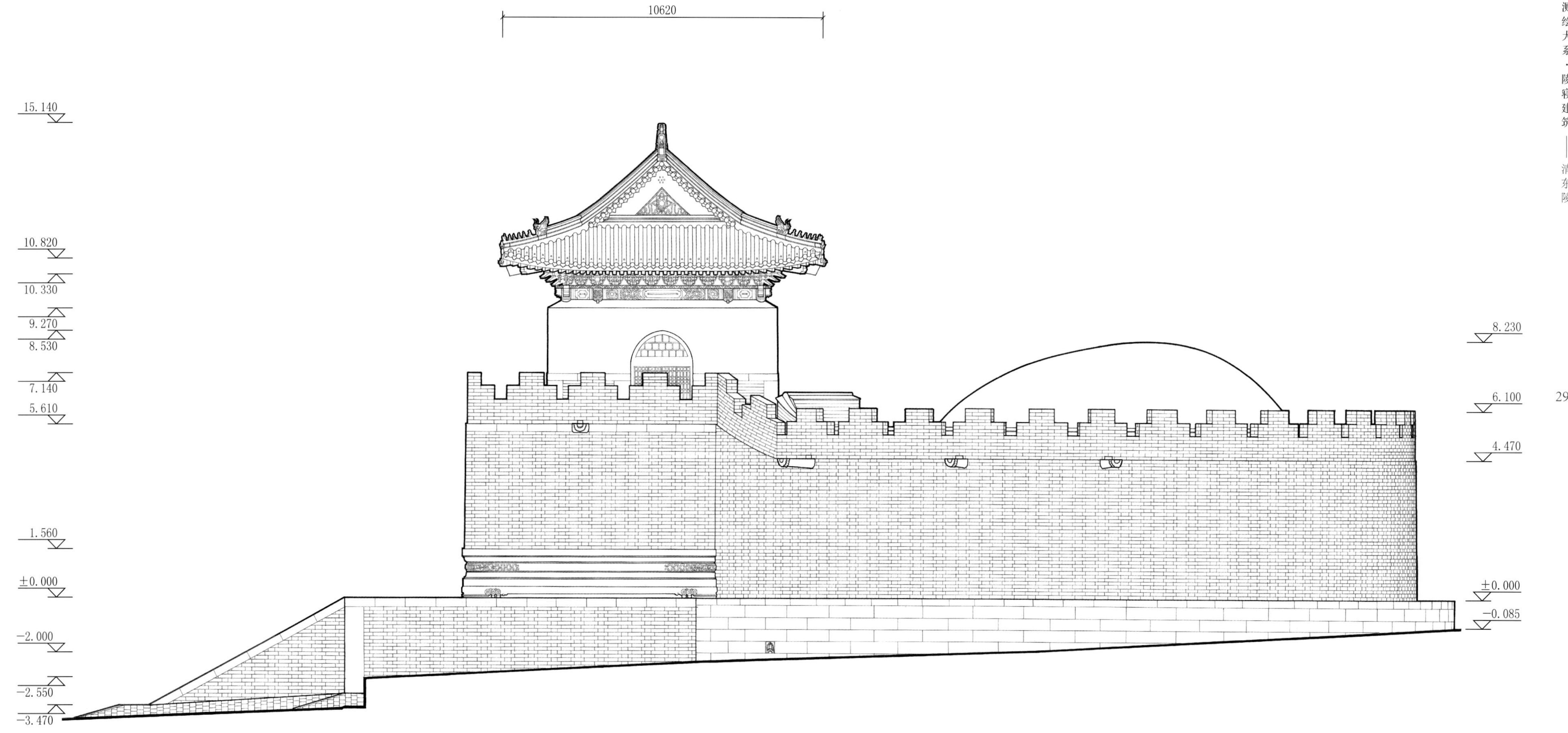

裕陵妃园寝方城明楼及宝城宝顶侧立面图

Side elevation of the Square Walled Terrace and the Memorial Tower as well as the City of Treasure(the superstructure of the tomb chamber) and the Tumulus, Imperial Consorts' Tombs affiliated with the Tomb of Prosperity

3.650
±0.000
−1.420
−1.980
−7.140

裕陵妃园寝容妃地宫剖面图
Section of the Underground Palace, the Tomb of Imperial Consort Rong at the Imperial Consorts' Tombs affiliated with the Tomb of Prosperity

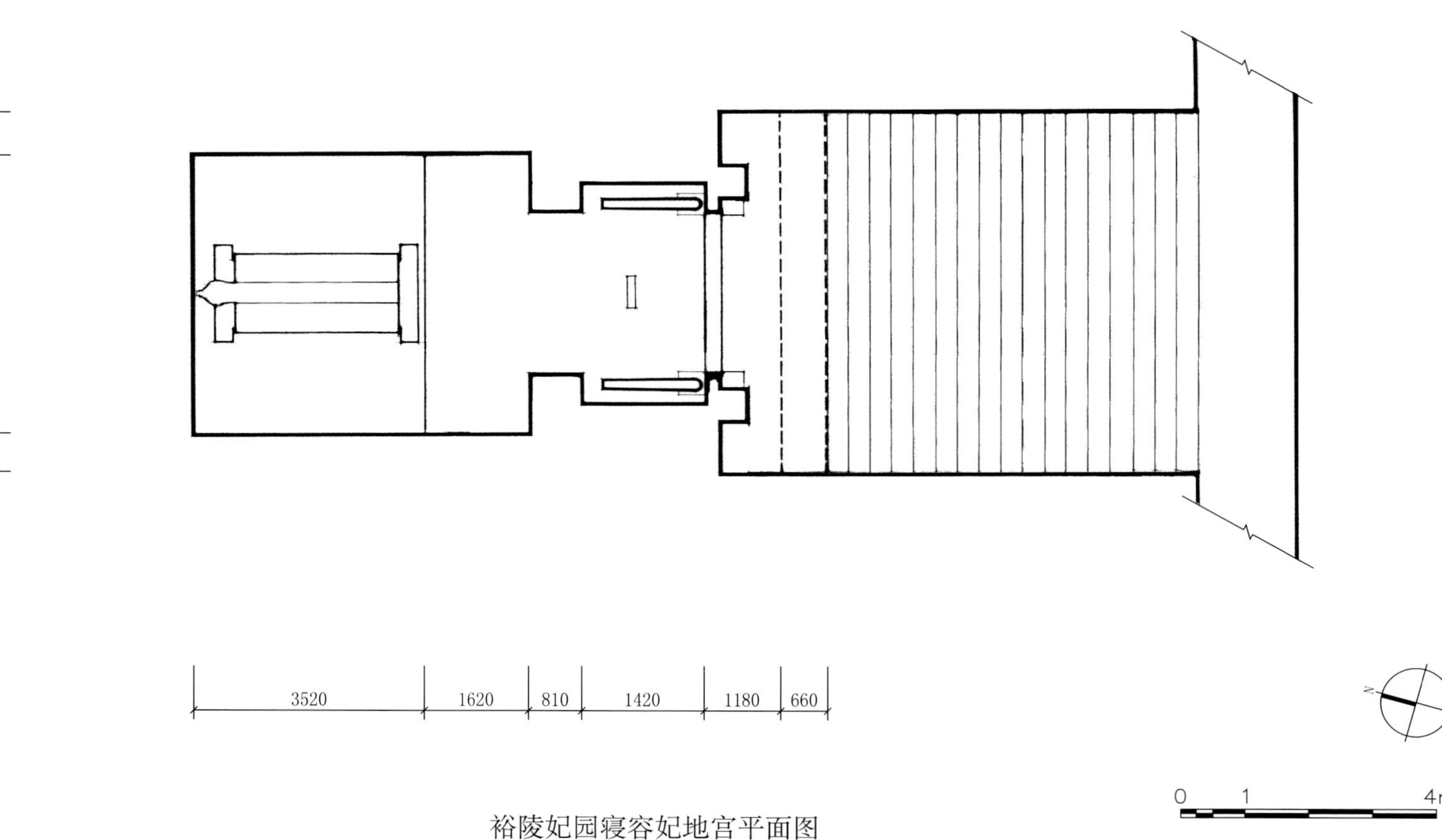

裕陵妃园寝容妃地宫平面图
Plan of the Underground Palace, the Tomb of Imperial Consort Rong at the Imperial Consorts' Tombs affiliated with the Tomb of Prosperity

定陵
Tomb of Stability (Dingling)

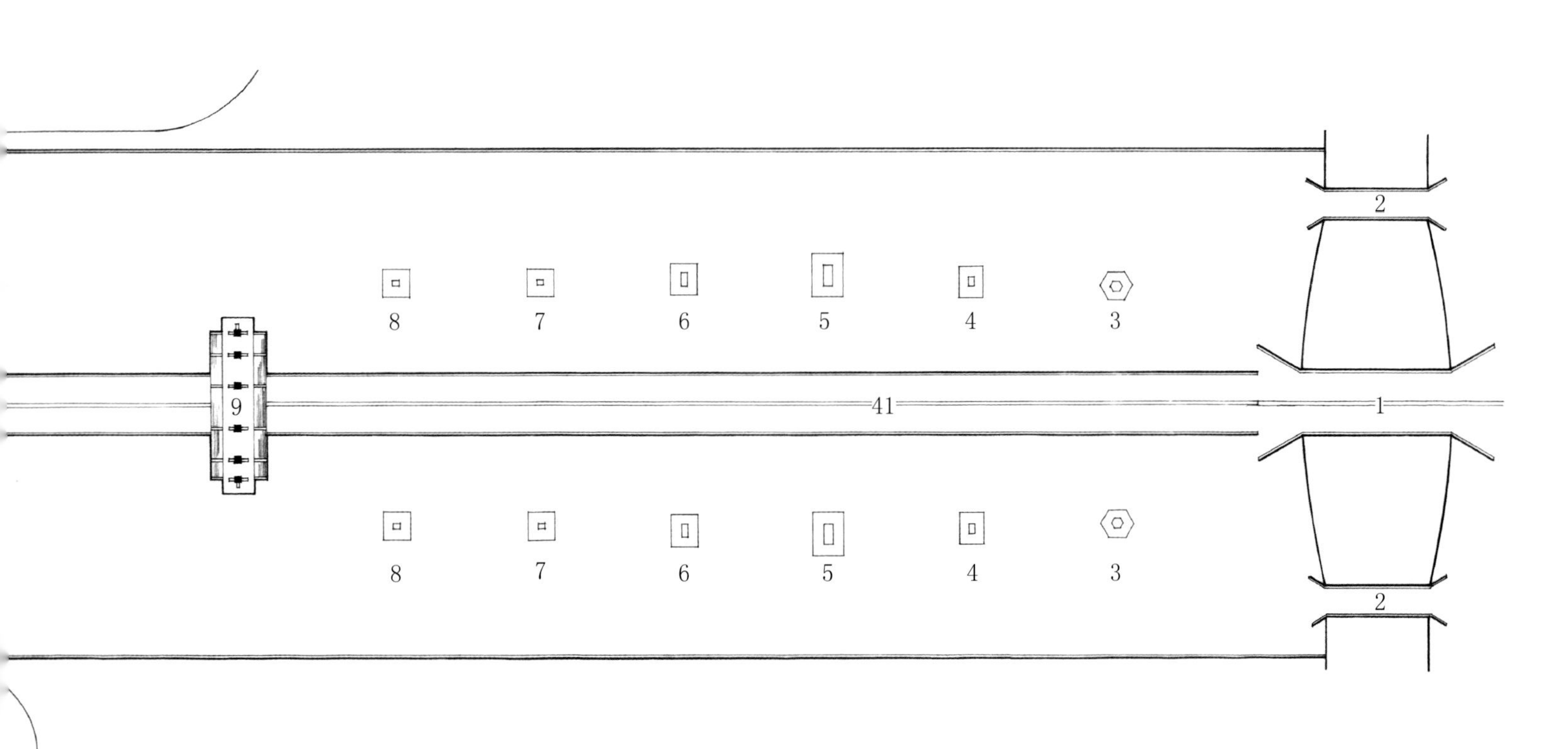

1 五孔石券桥 Five-Arch Stone Bridge
2 石平桥 Flat Stone Bridge
3 望柱 Ornamental Column
4 狮子 Lion
5 大象 Elephant
6 马 Horse
7 武将 Martial Official
8 文臣 Literary Official
9 牌坊 Marble Memorial Gateway
10 礓磜坡道 Ramp
11 神道碑亭 Pavilion for the Stela on the Spirit Way
12 神厨库院落 Culinary Courtyard for Sacrifices
13 三路三孔桥 Three-Way Three-Arch Bridges
14 便桥 Convenient Bridge
15 朝房 Reception Halls for Court Officials
16 班房 Duty Houses for the Guards
17 隆恩门 Gate of Monumental Grace
18 焚帛炉 Sacrificial Burners
19 配殿 Side Hall
20 隆恩殿 Hall of Monumental Grace
21 石平桥 Flat Stone Bridge
22 玉带河 Jade Ribbon River
23 琉璃花门 Gate with Glazed Roof Tiles
24 石五供 Five Stone Ritual Vessels
25 石平桥 Flat Stone Bridge
26 月牙河 Crescent River
27 方城 Square Walled Jerrace
28 明楼 Memorial Tower
29 哑巴院 Courtyard of the Mute
30 月牙城 Crescent Wall
31 琉璃影壁 Screen Wall of Glazed Tiles
32 宝顶 Tumulus
33 宝城 Encircled Realm of Treasure
34 罗圈墙 Wall Surrounding the Tomb
35 卡子墙 Spacer Wall
36 院门 Yard Gate
37 神厨 Sacrificial Kitchen
38 神库 Sacrificial Storehouse
39 宰牲亭 Ritual Abattoir
40 下马牌 Stela Marking the Place for Dismounting from one's Horse
41 神道 Spirit Way

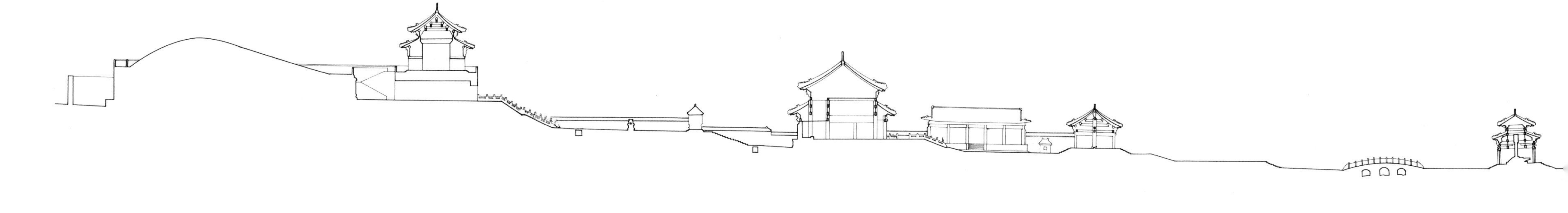

定陵总剖面图
Site section of the Tomb of Stability

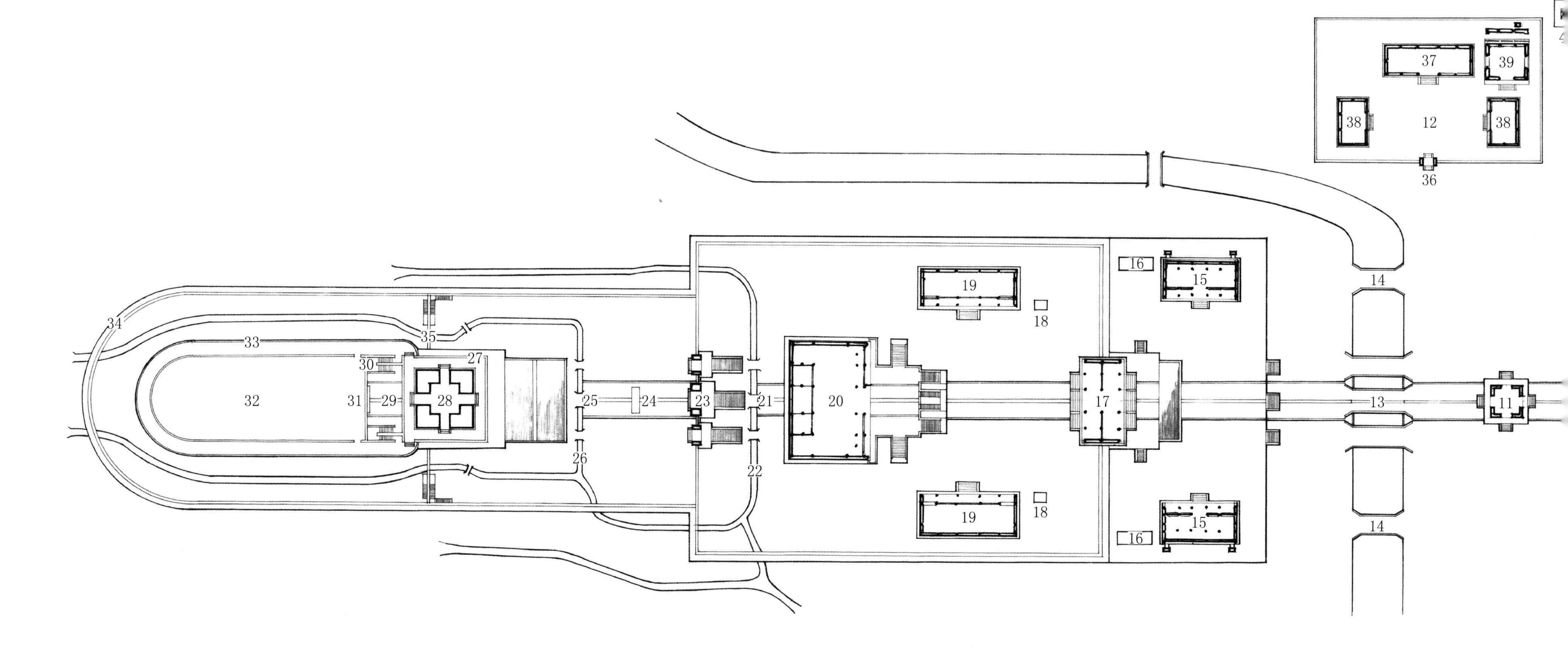

定陵总平面图
Site plan of the Tomb of Stability

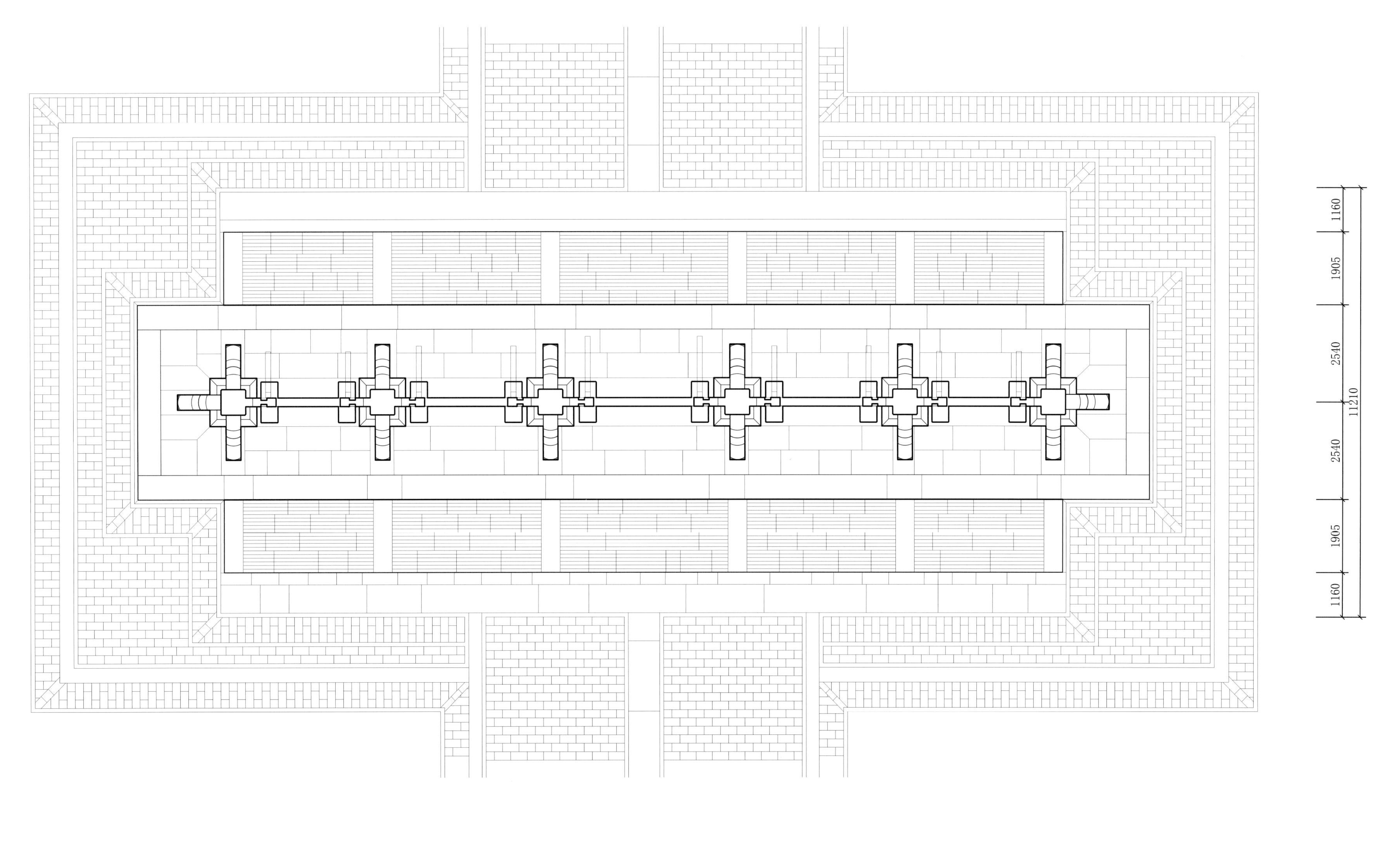

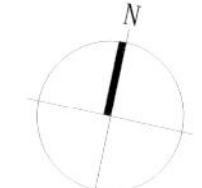

0 1.5 3m

定陵牌坊平面图
Plan of the Gateway, Tomb of Stability

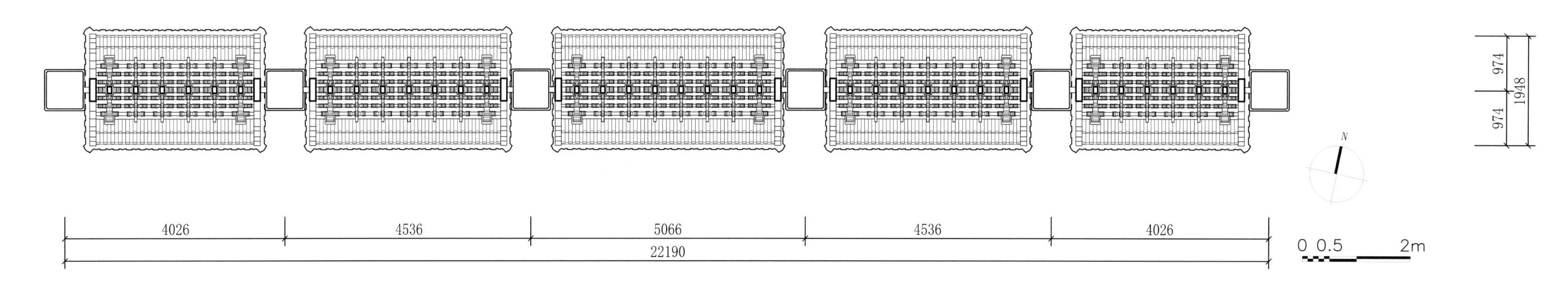

定陵牌坊梁架仰视图

Looking up at the Truss (*liangjia*) of the Gateway, Tomb of Stability

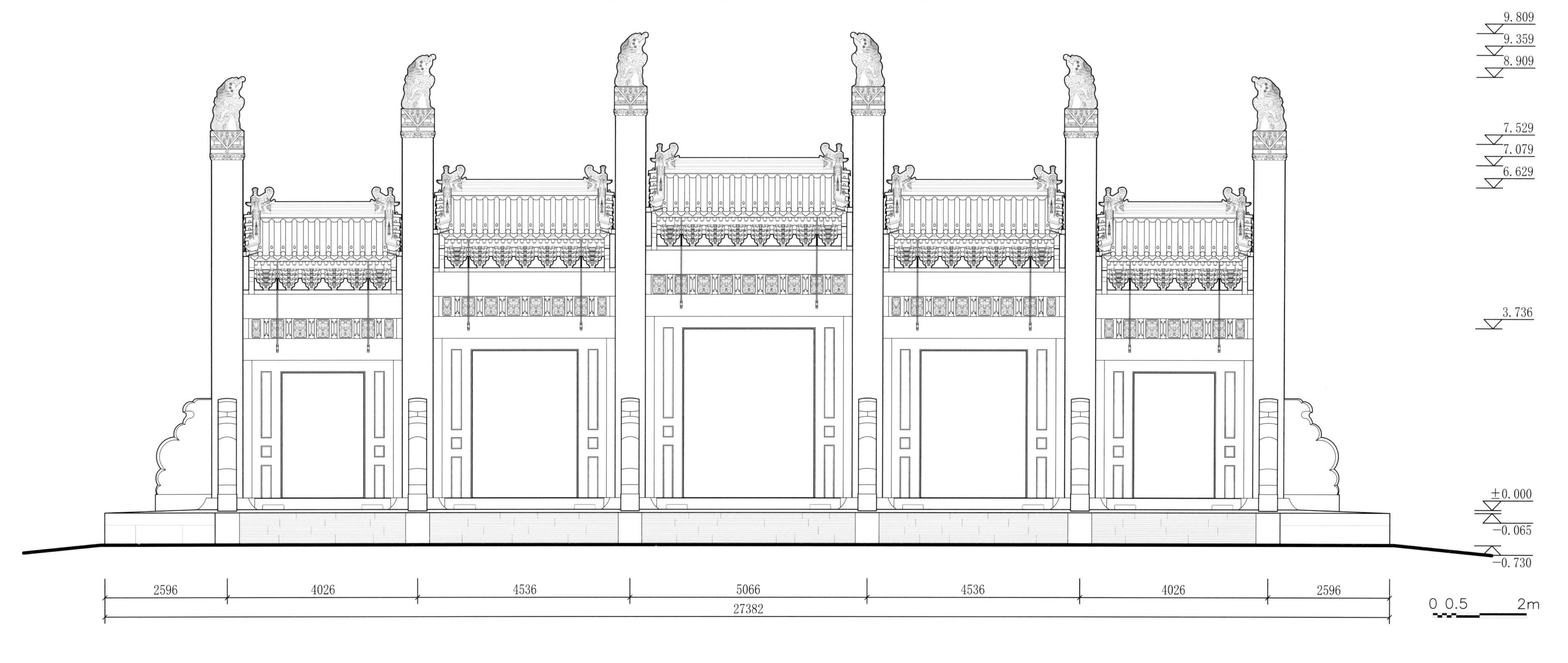

定陵牌坊正立面图

Front elevation of the Gateway, Tomb of Stability

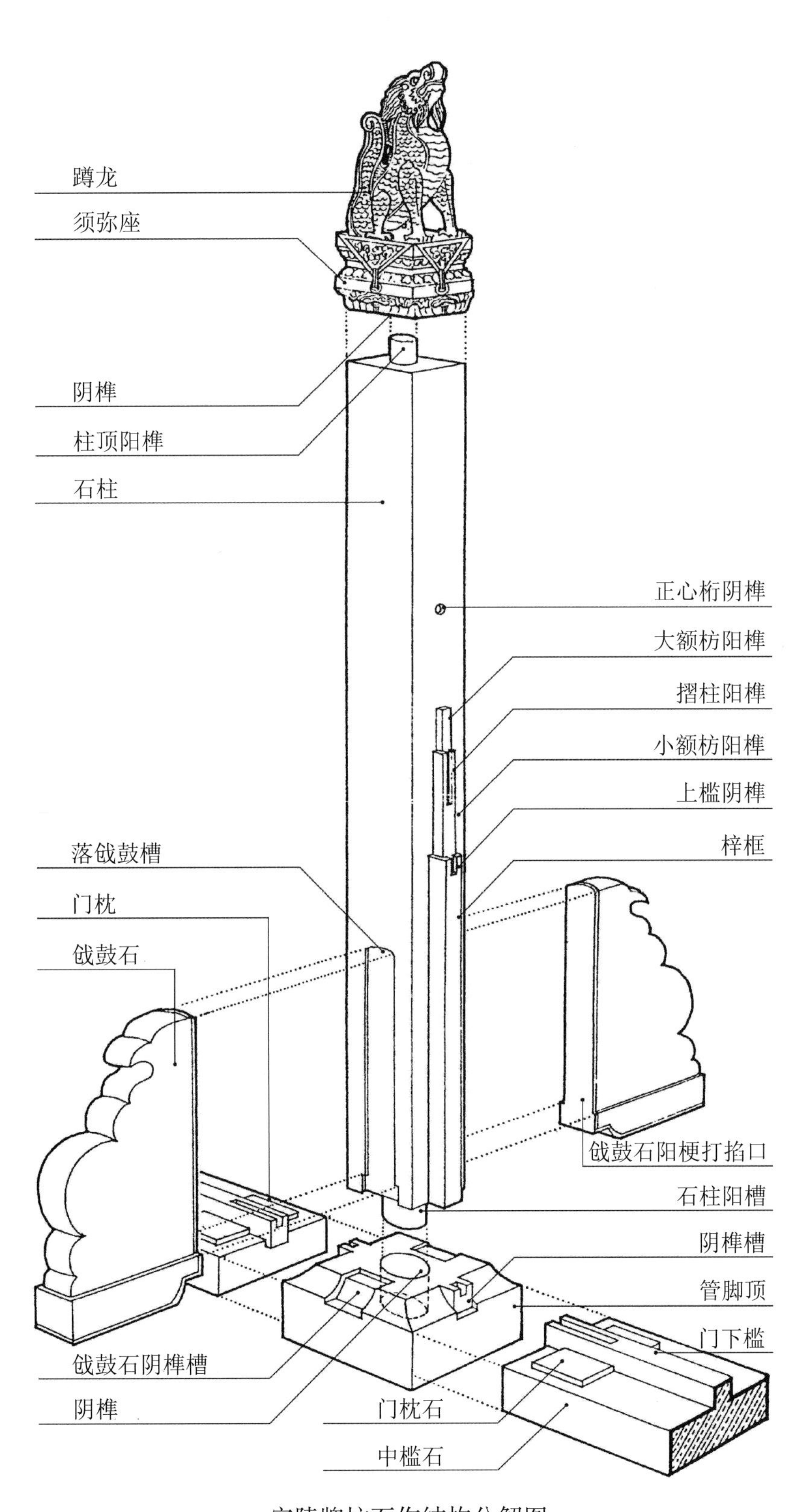

定陵牌坊石作结构分解图
Stone structural decomposition plan of the Gateway, Tomb of Stability

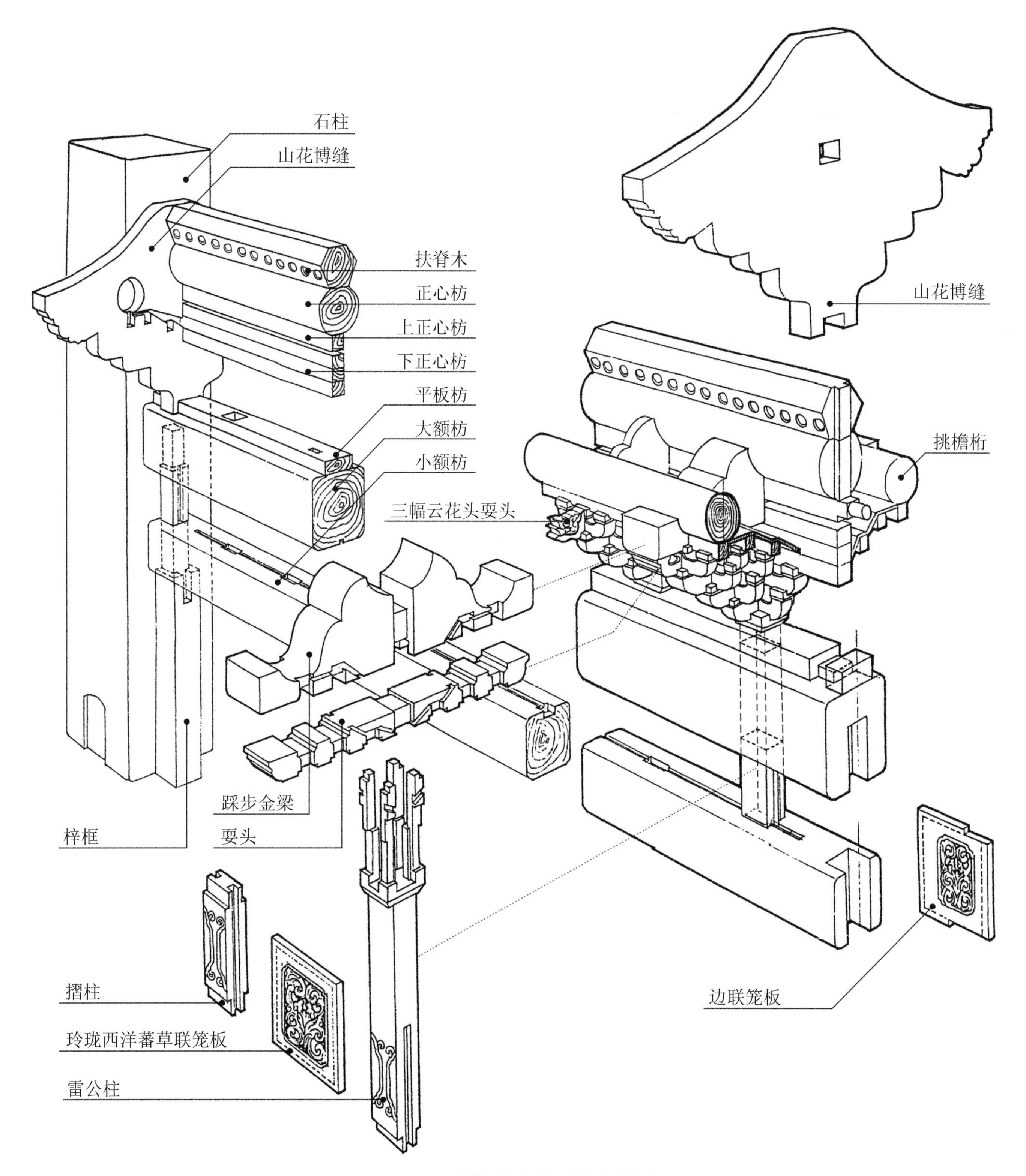

定陵牌坊木作结构分解图
Wooden structural decomposition plan of the Gateway, Tomb of Stability

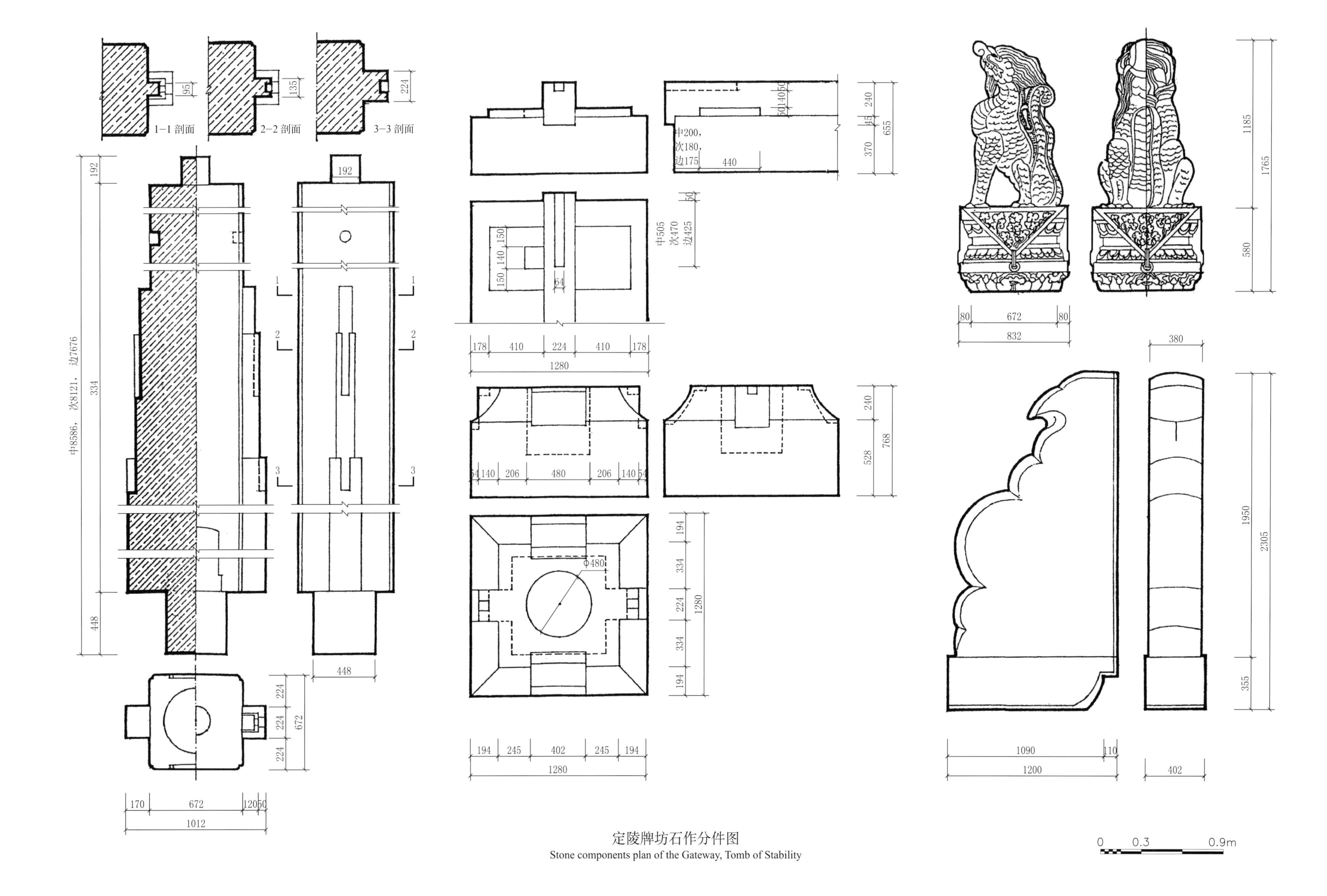

定陵牌坊石作分件图
Stone components plan of the Gateway, Tomb of Stability

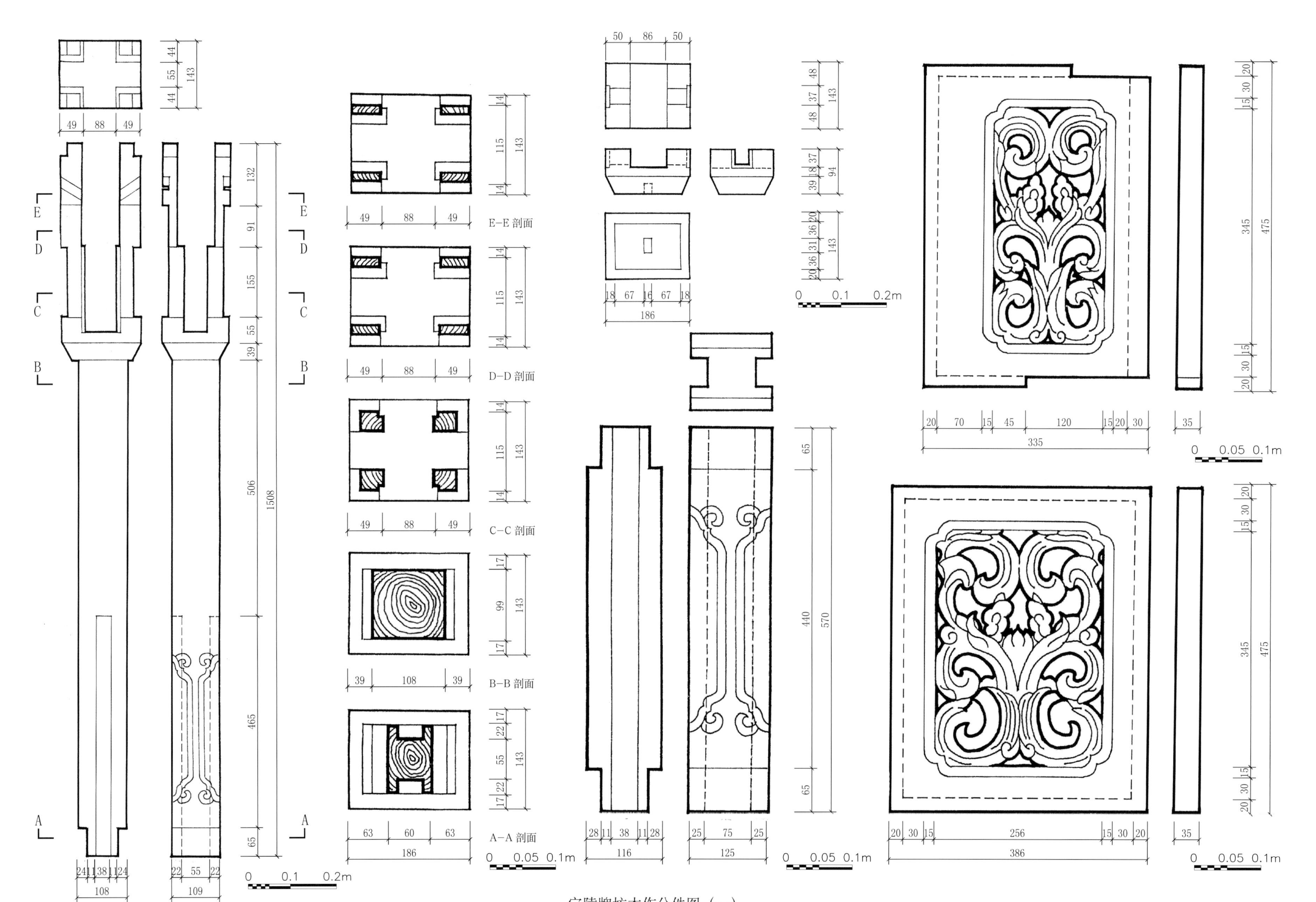

定陵牌坊木作分件图（一）

Wooden components plan of the Gateway, Tomb of Stability (1)

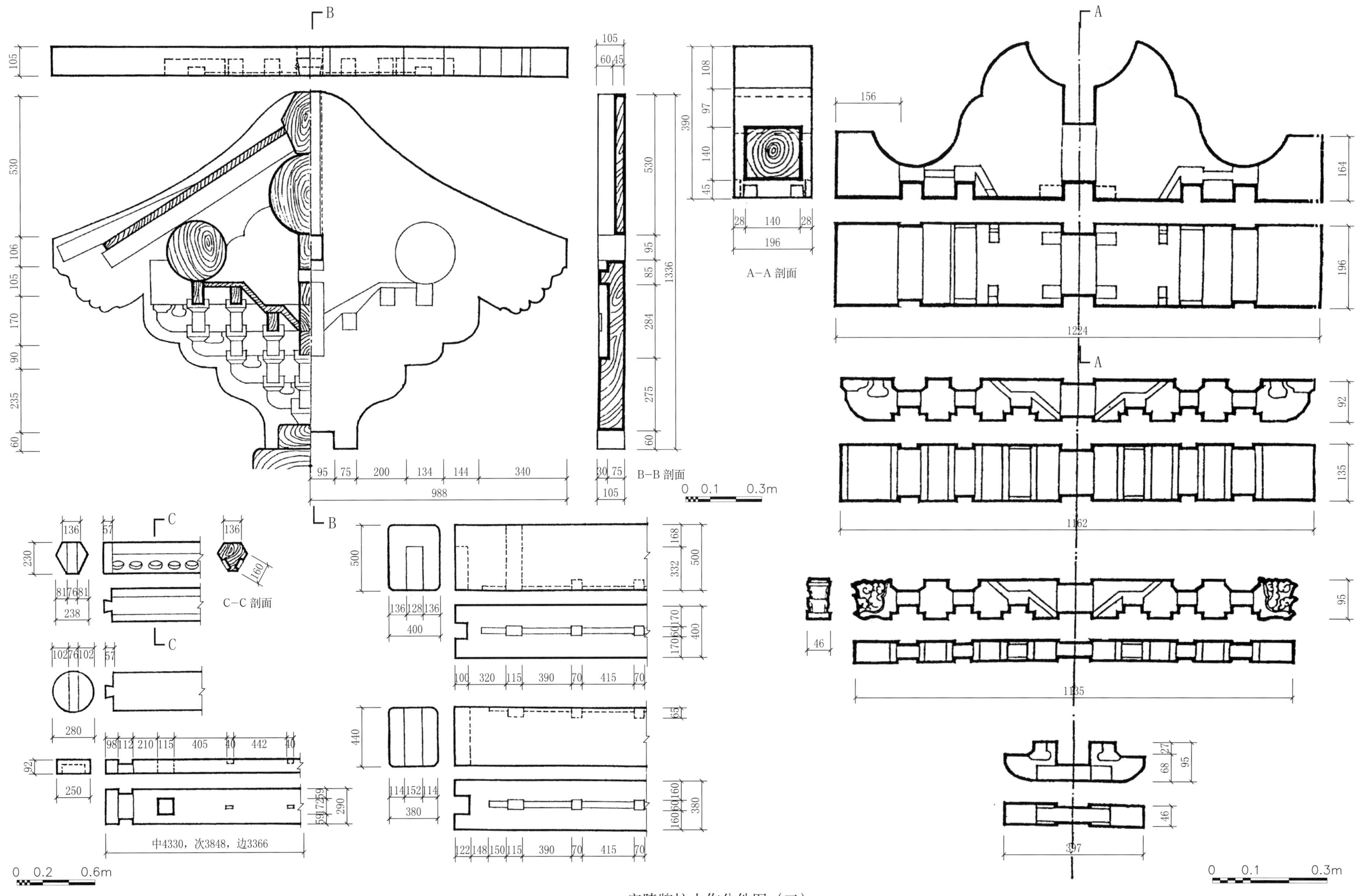

定陵牌坊木作分件图（二）

Wooden components plan of the Gateway, Tomb of Stability (2)

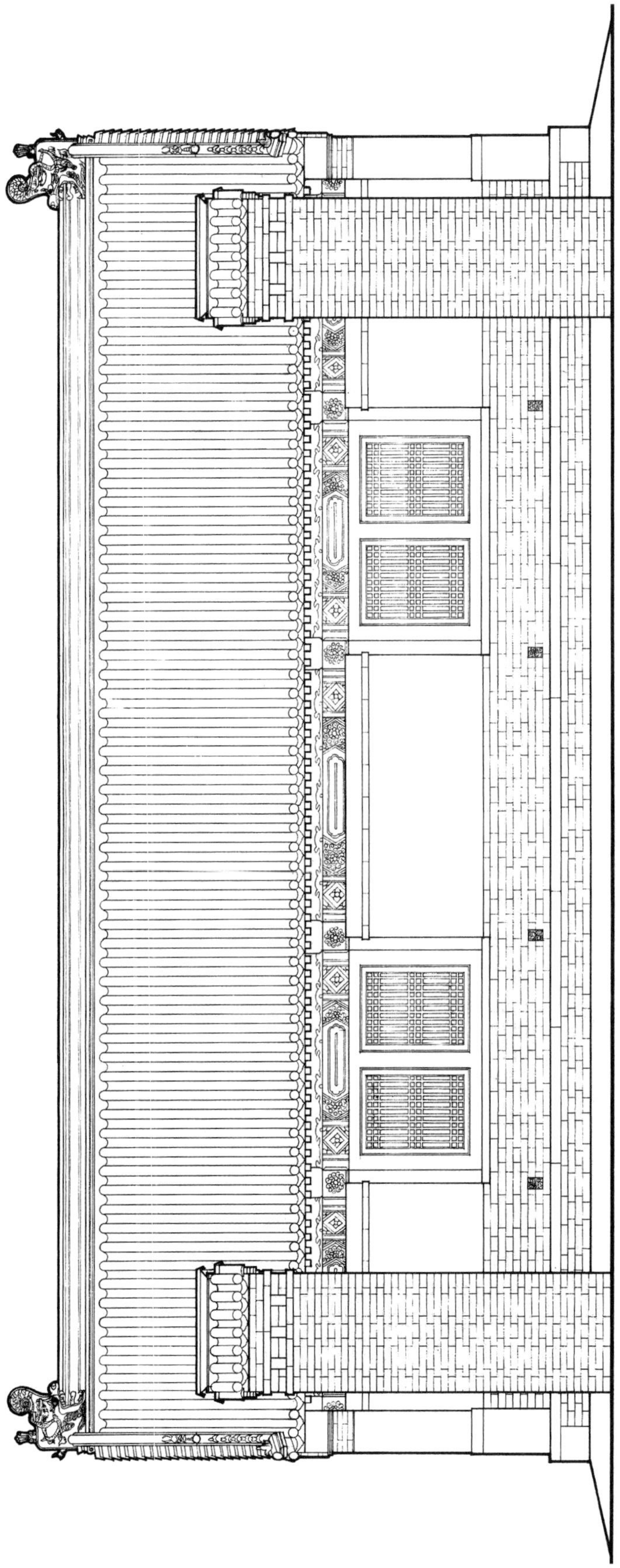

0 1 3m

定陵东朝房背立面图
Back elevation of the Eastern Reception Hall for Court Officials, Tomb of Stability

0 1 3m

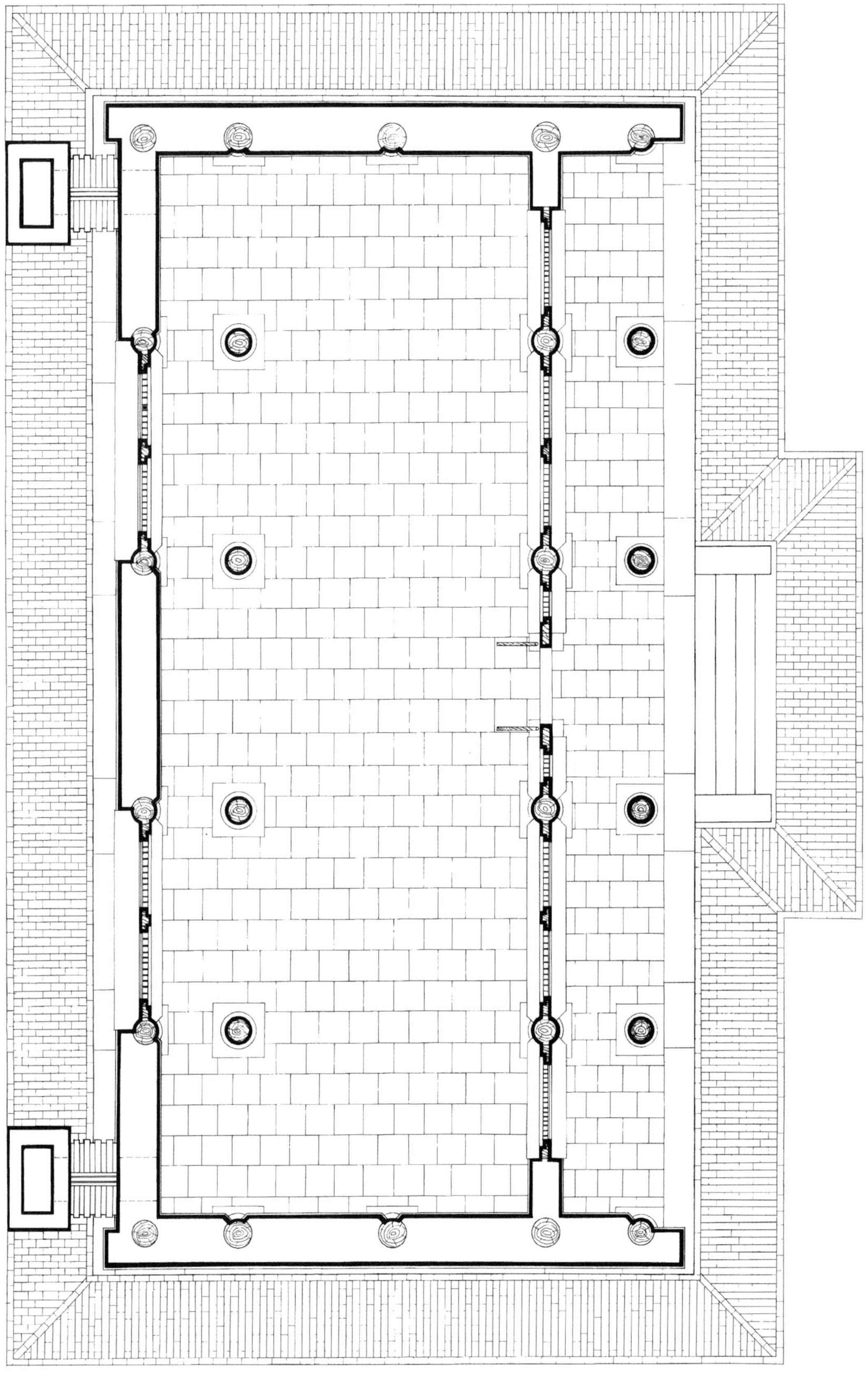

定陵东朝房平面图
Plan of the Eastern Reception Hall for Court Officials, Tomb of Stability

0 1 2m

定陵东朝房正立面图

Front elevation of the Eastern Reception Hall for Court Officials, Tomb of Stability

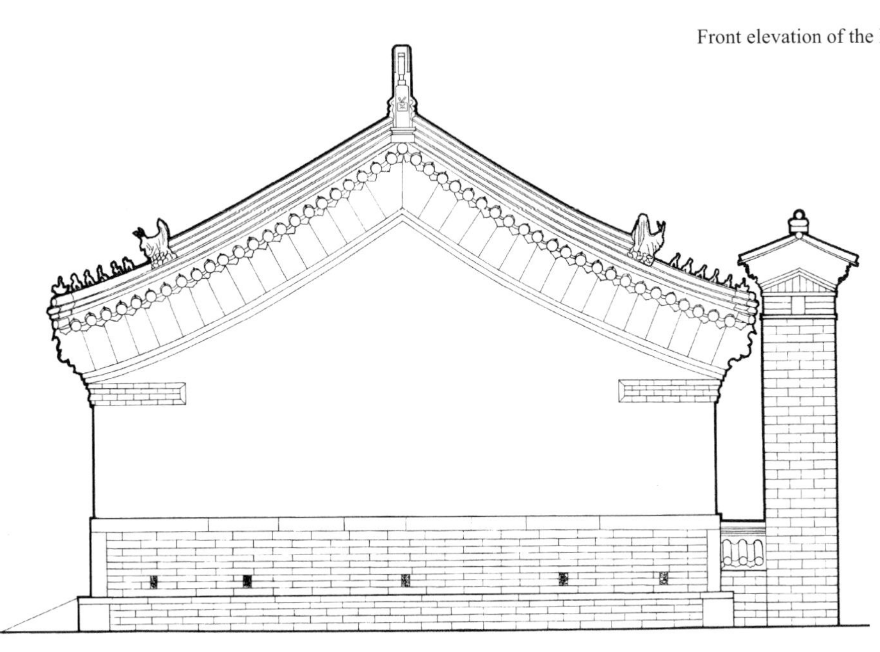

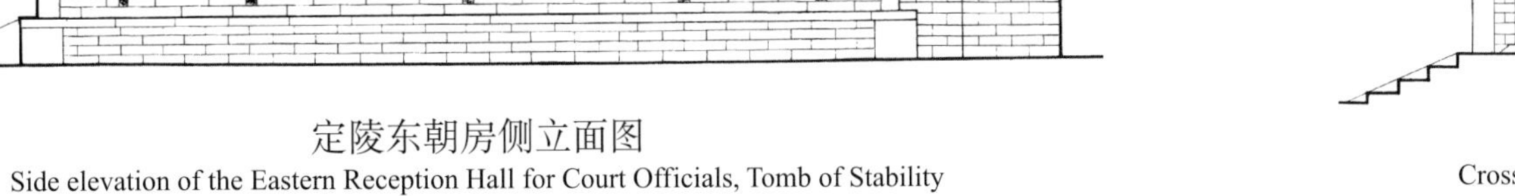

定陵东朝房侧立面图

Side elevation of the Eastern Reception Hall for Court Officials, Tomb of Stability

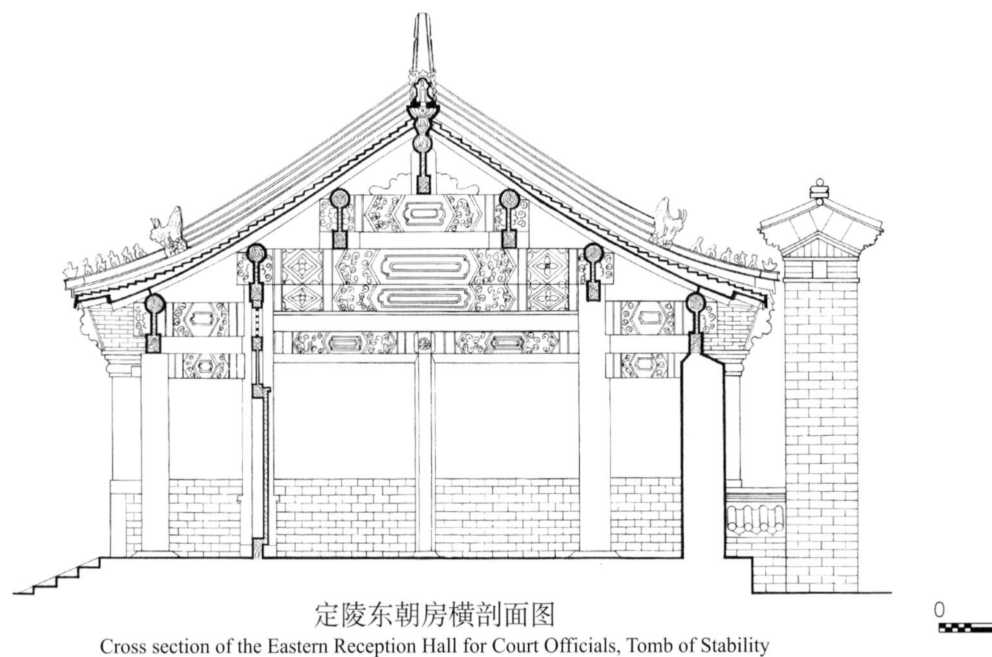

0 1 3m

定陵东朝房横剖面图

Cross section of the Eastern Reception Hall for Court Officials, Tomb of Stability

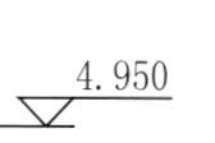
4.950
3.645
3.315
3.015
0.350
210
1090
26450
0
1
4m

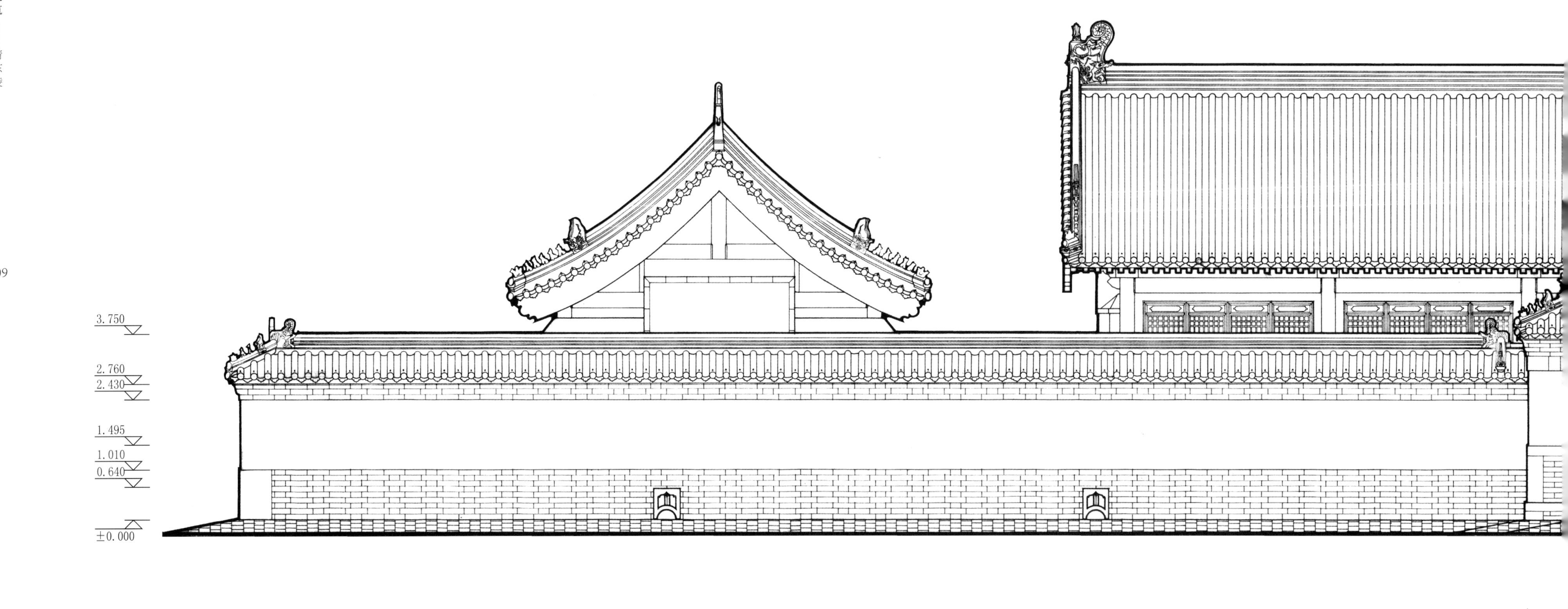

定陵神厨库组群立面图

Elevation of the building complex of the Culinary Courtyard for Sacrifices, Tomb of Stability

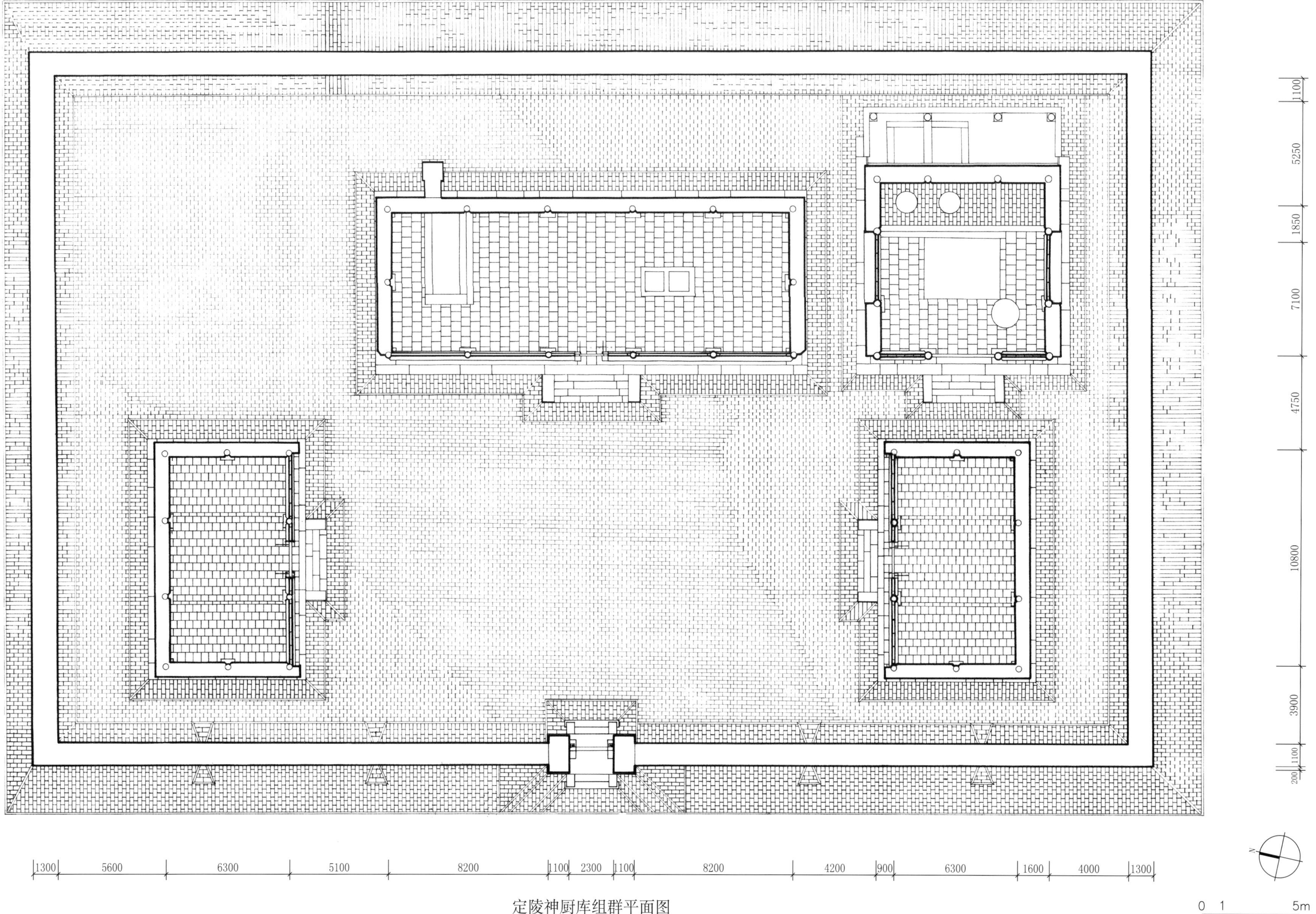

定陵神厨库组群平面图

Site plan of the building complex of the Culinary Courtyard for Sacrifices, Tomb of Stability

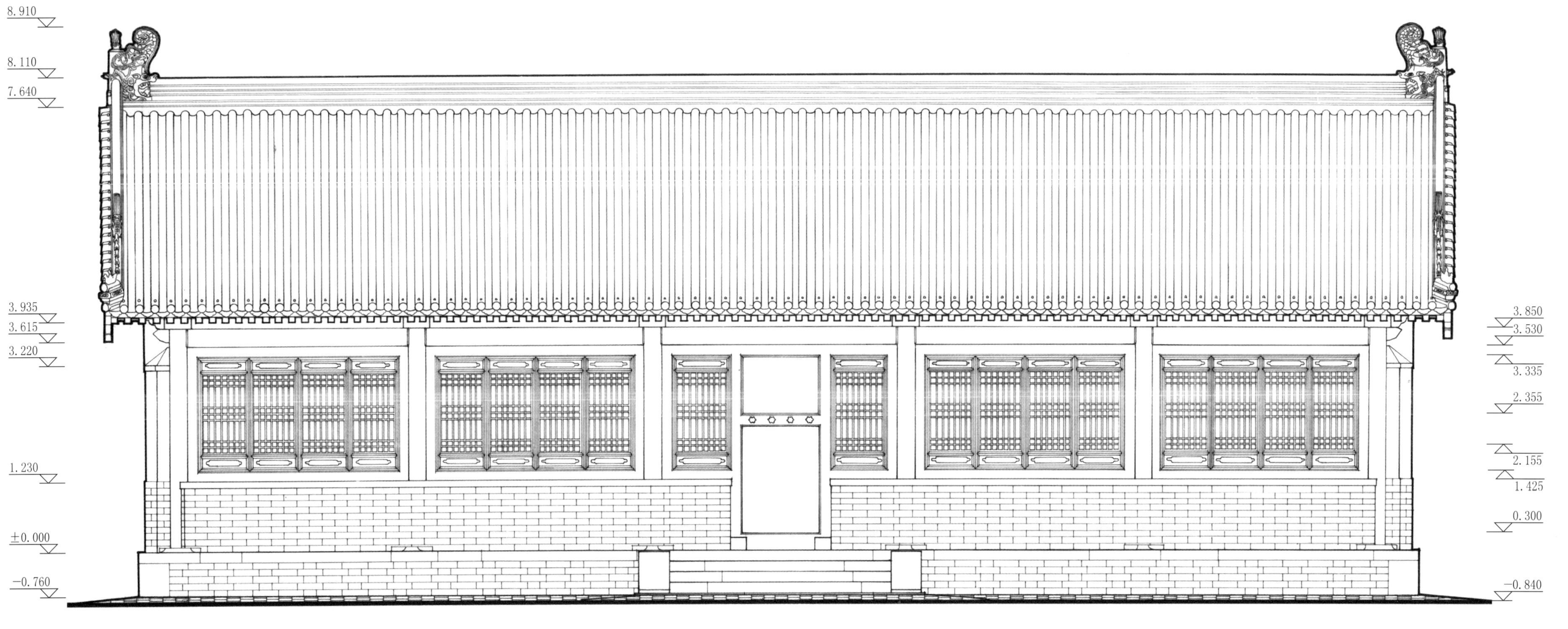

定陵神厨库神厨立面图

Elevation of the Kitchen in the Culinary Courtyard for Sacrifices, Tomb of Stability

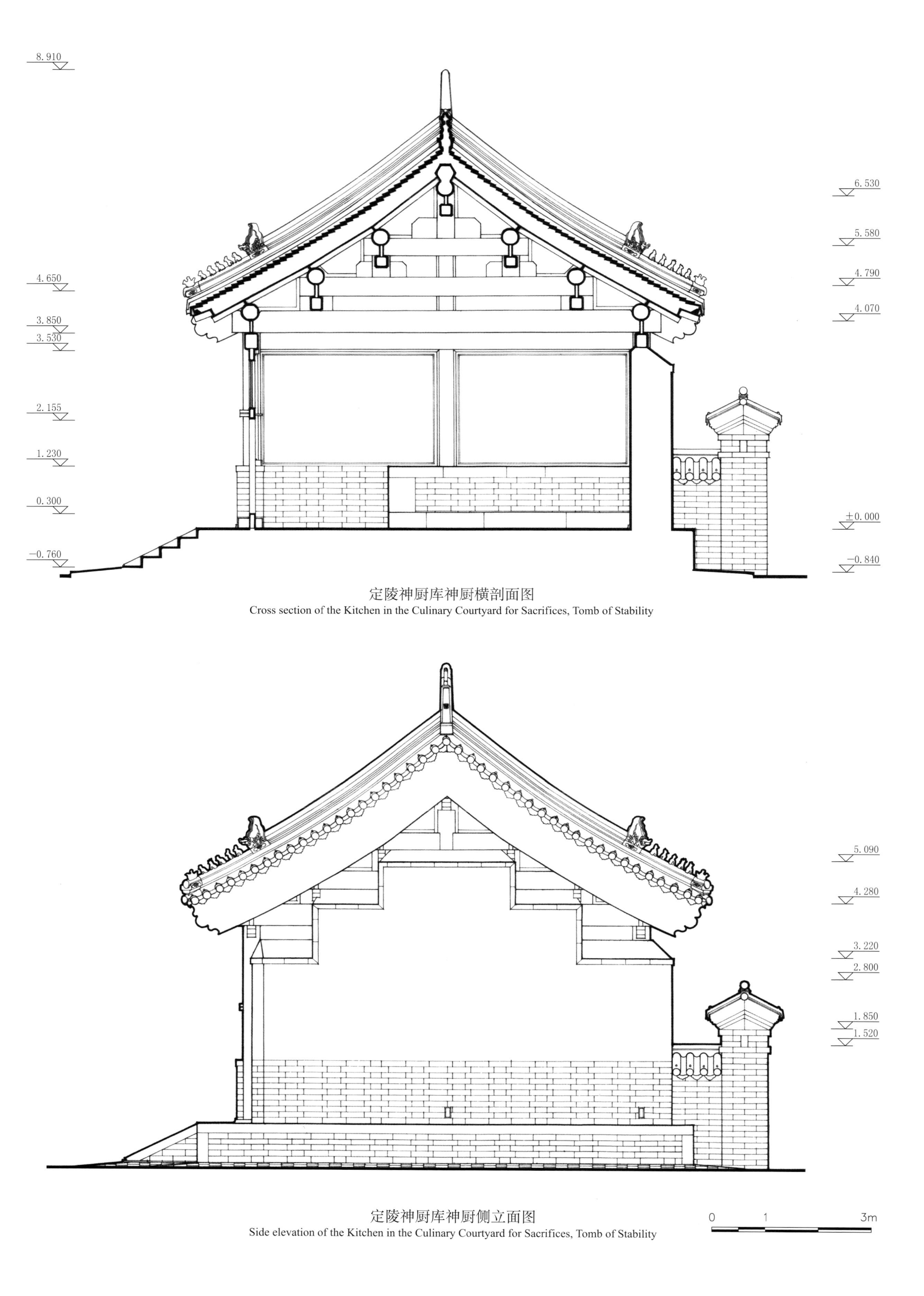

定陵神厨库神厨横剖面图
Cross section of the Kitchen in the Culinary Courtyard for Sacrifices, Tomb of Stability

定陵神厨库神厨侧立面图
Side elevation of the Kitchen in the Culinary Courtyard for Sacrifices, Tomb of Stability

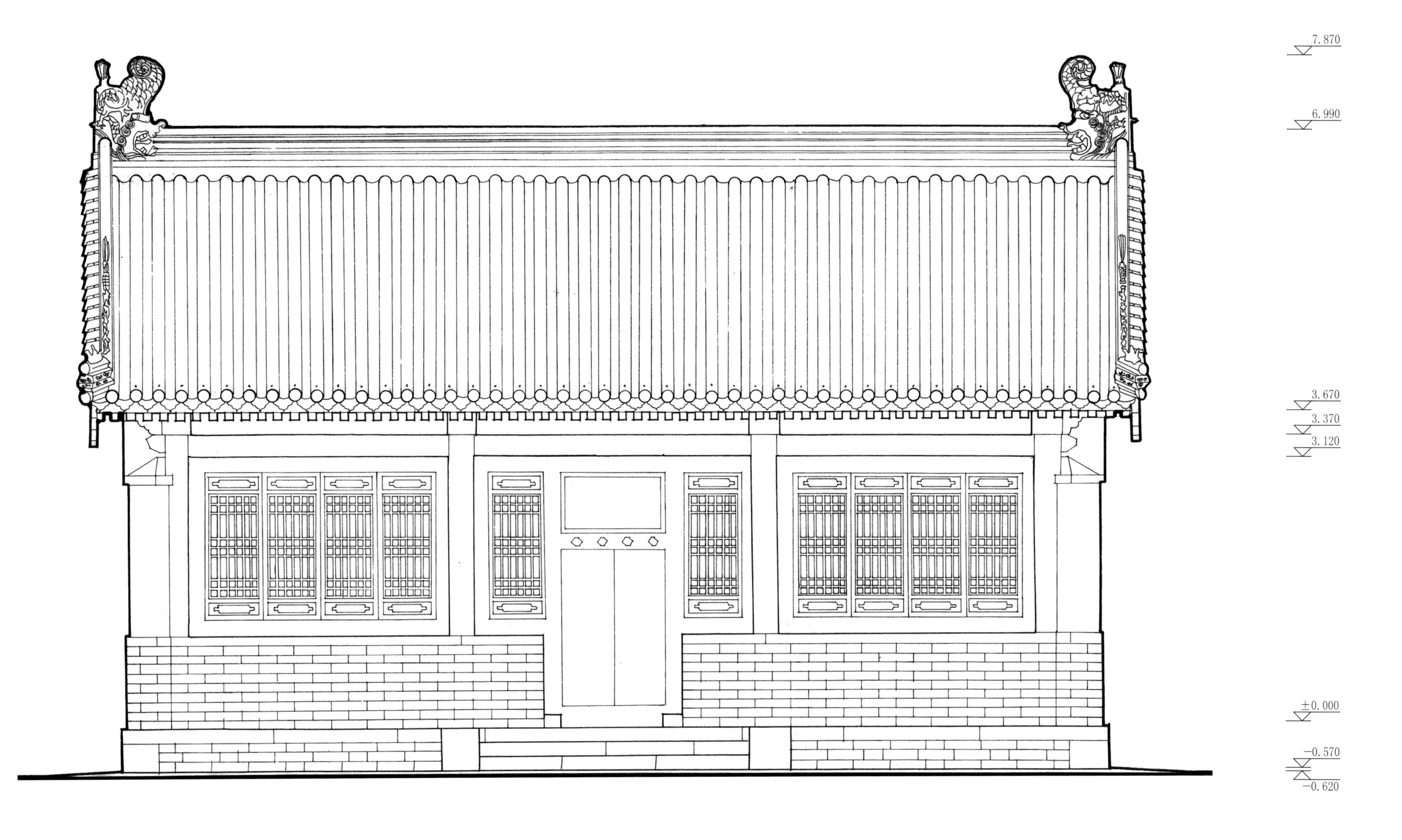

定陵神厨库神库立面图

Elevation of the Storehouse in the Culinary Courtyard for Sacrifices, Tomb of Stability

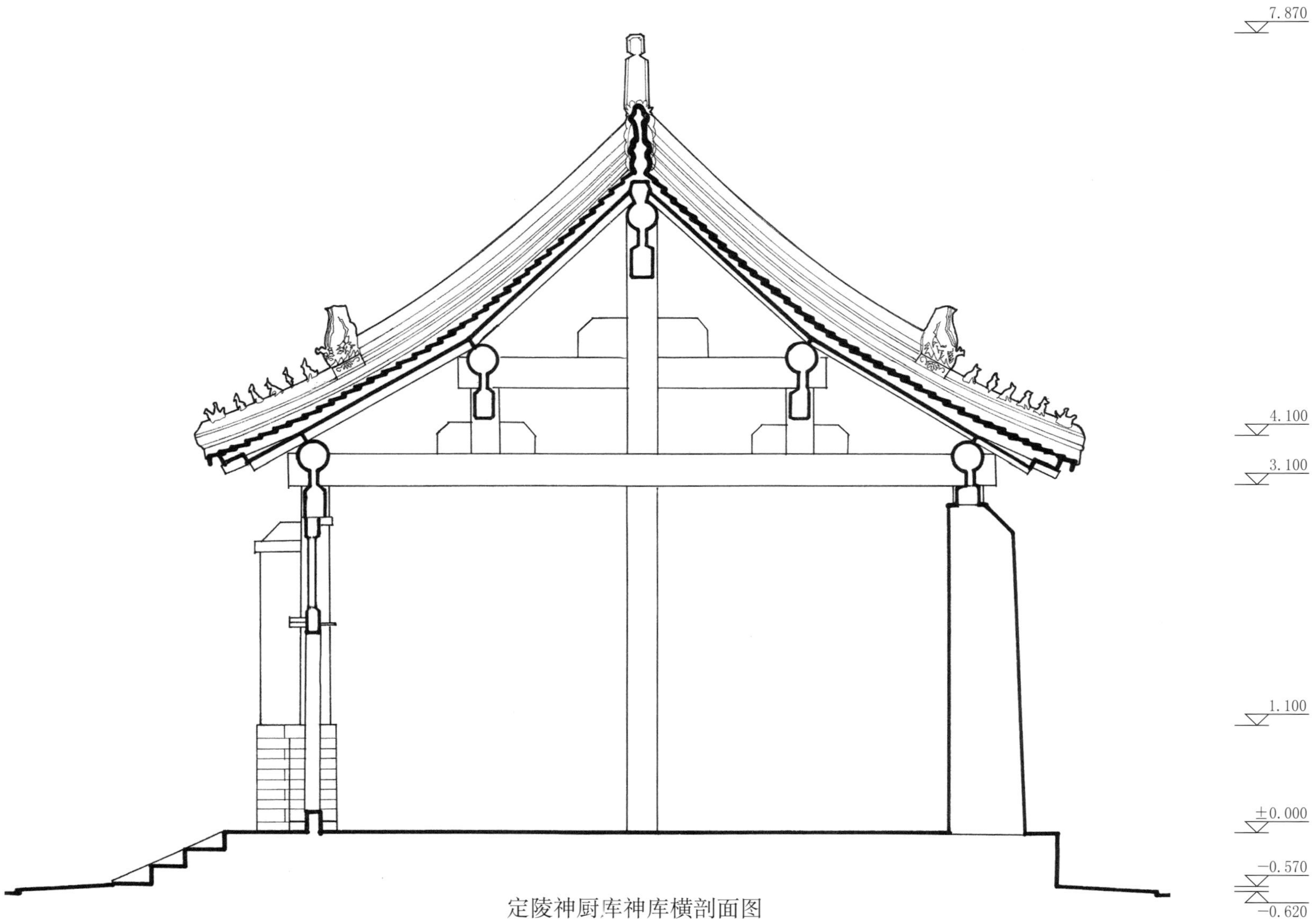

7.870
4.100
3.100
1.100
±0.000
−0.570
−0.620

定陵神厨库神库横剖面图
Cross section of the Storehouse in the Culinary Courtyard for Sacrifices, Tomb of Stability

7.870
7.160
3.300
1.100
±0.000
−0.570
−0.620

0 1 2m

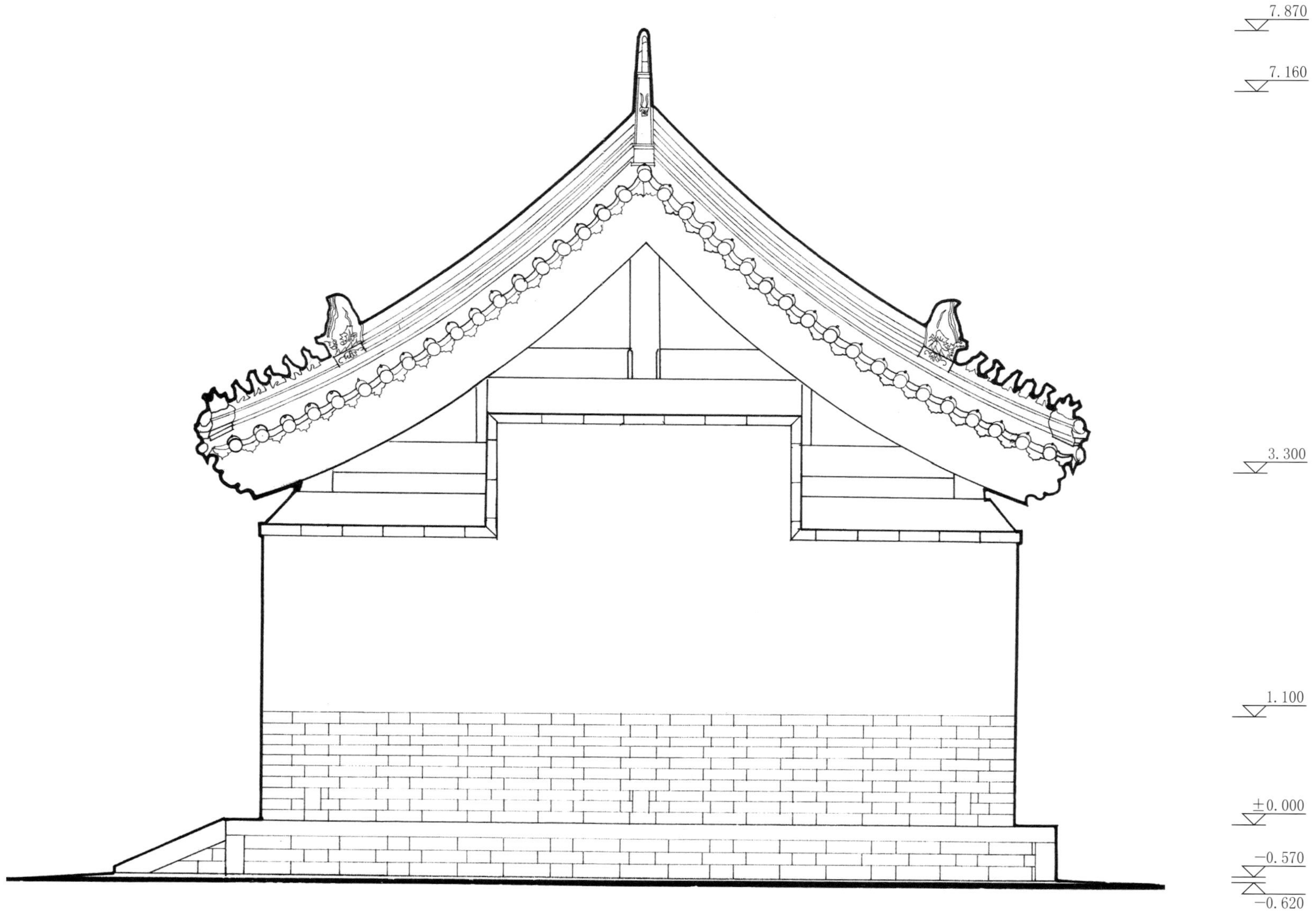

定陵神厨库神库侧立面图
Side elevation of the Storehouse in the Culinary Courtyard for Sacrifices, Tomb of Stability

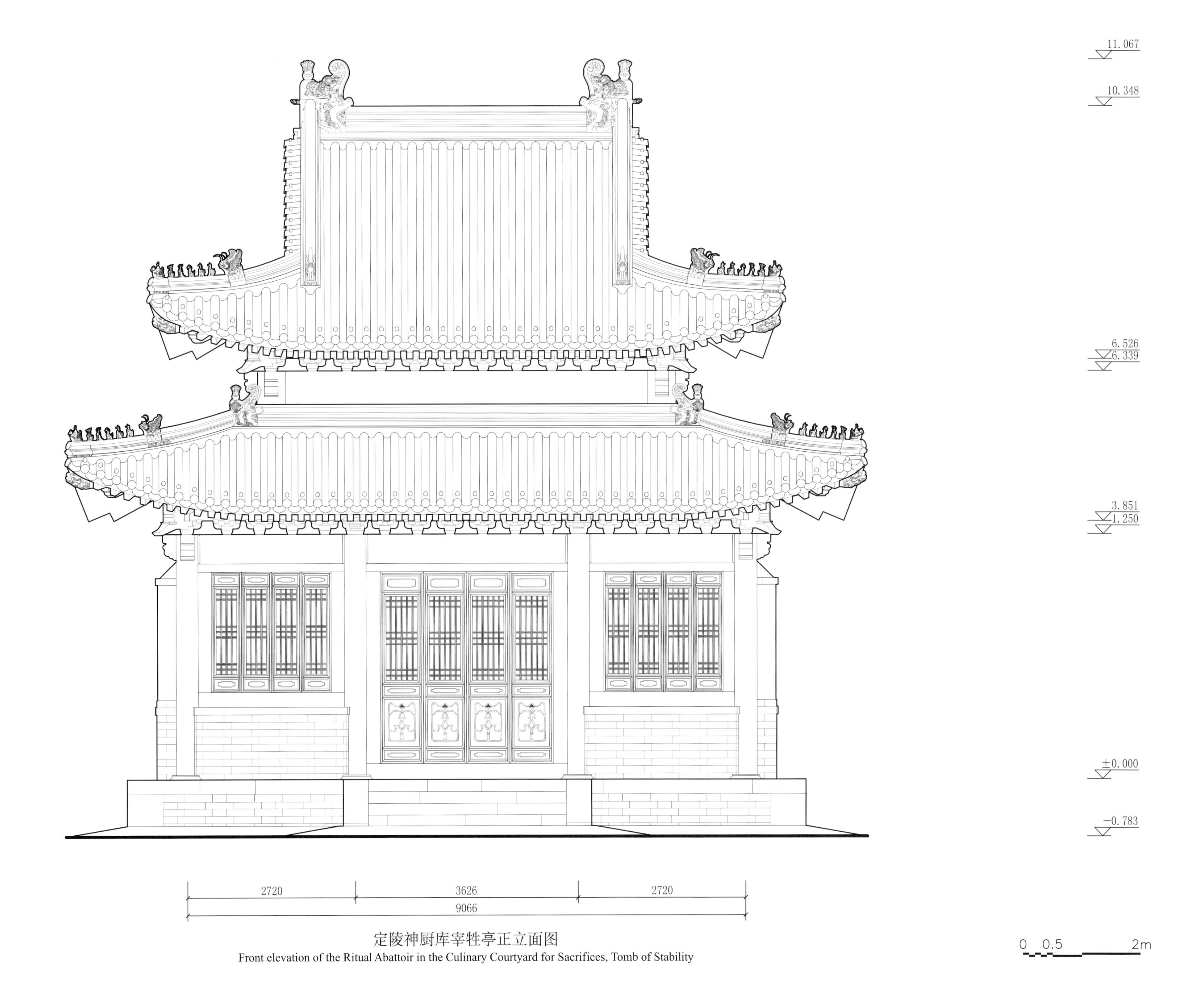

定陵神厨库宰牲亭正立面图

Front elevation of the Ritual Abattoir in the Culinary Courtyard for Sacrifices, Tomb of Stability

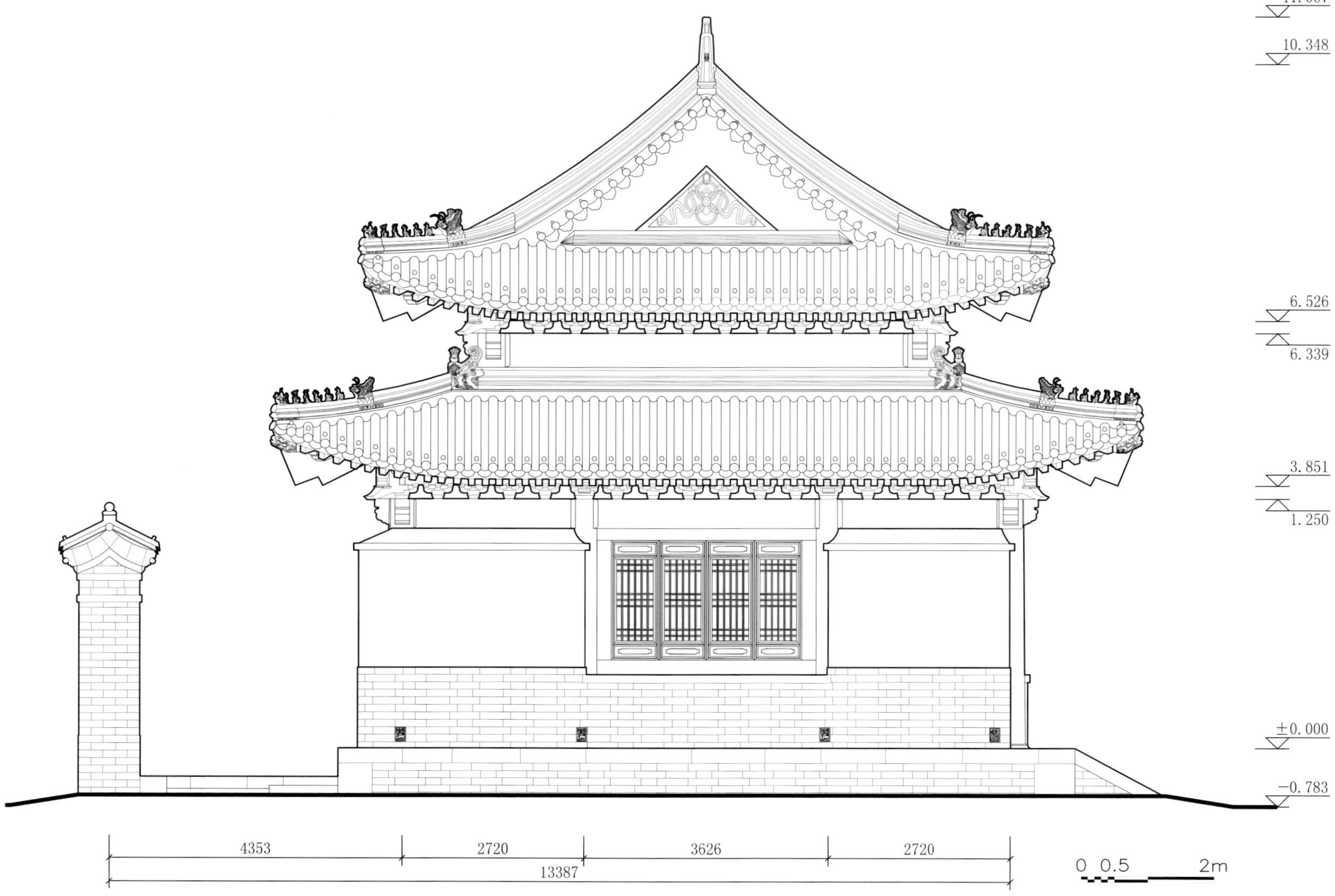

11.067
10.348
6.526
6.339
3.851
1.250
±0.000
−0.783

4353 | 2720 | 3626 | 2720
13387

0 0.5 2m

定陵神厨库宰牲亭侧立面图

Side elevation of the Ritual Abattoir in the Culinary Courtyard for Sacrifices, Tomb of Stability

11.067
9.004
7.604
6.695
6.339
4.160
1.250
±0.000
−0.783

2720 | 3626 | 2720
9066

0 0.5 2m

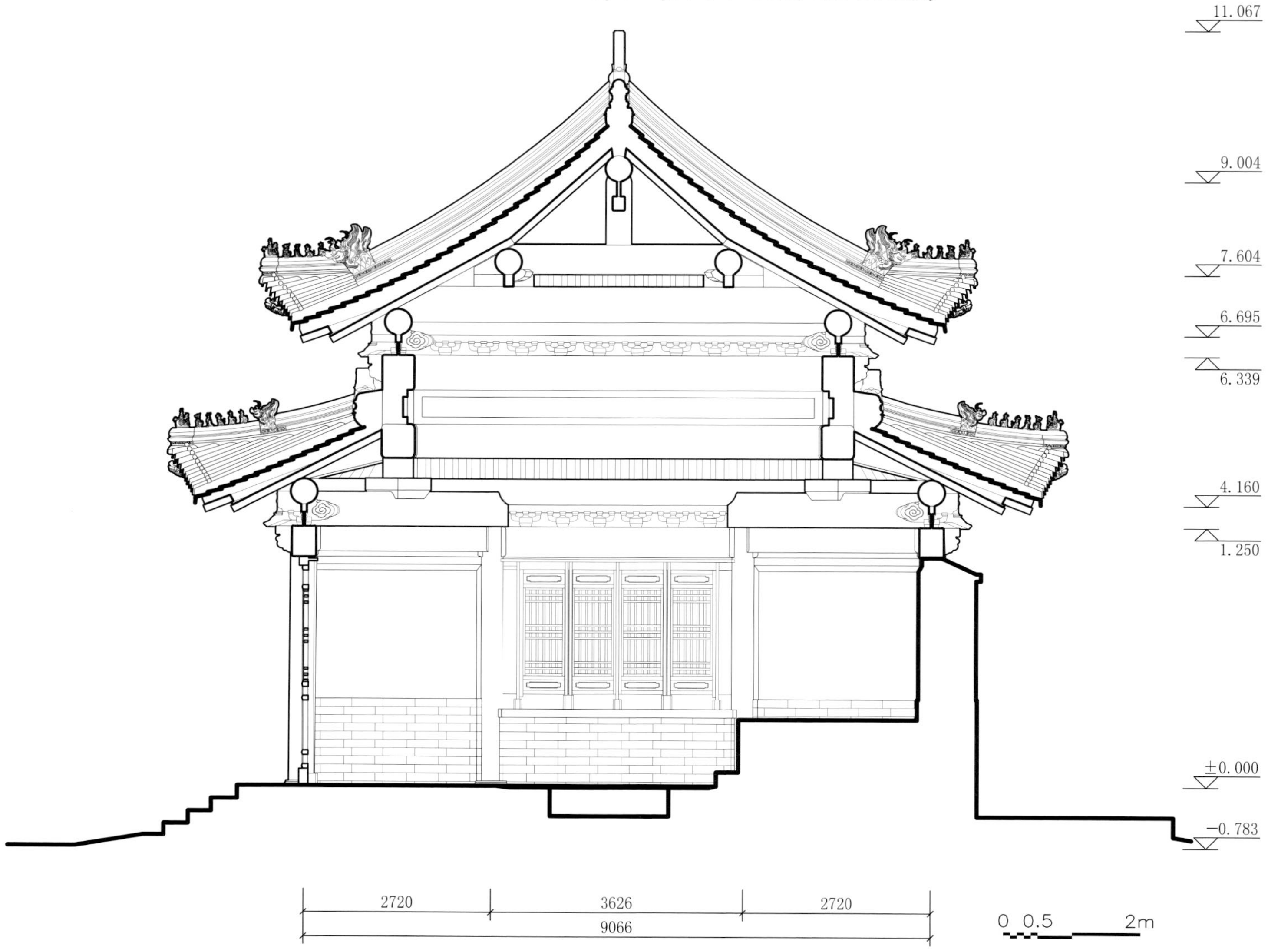

定陵神厨库宰牲亭横剖面图

Cross section of the Ritual Abattoir in the Culinary Courtyard for Sacrifices, Tomb of Stability

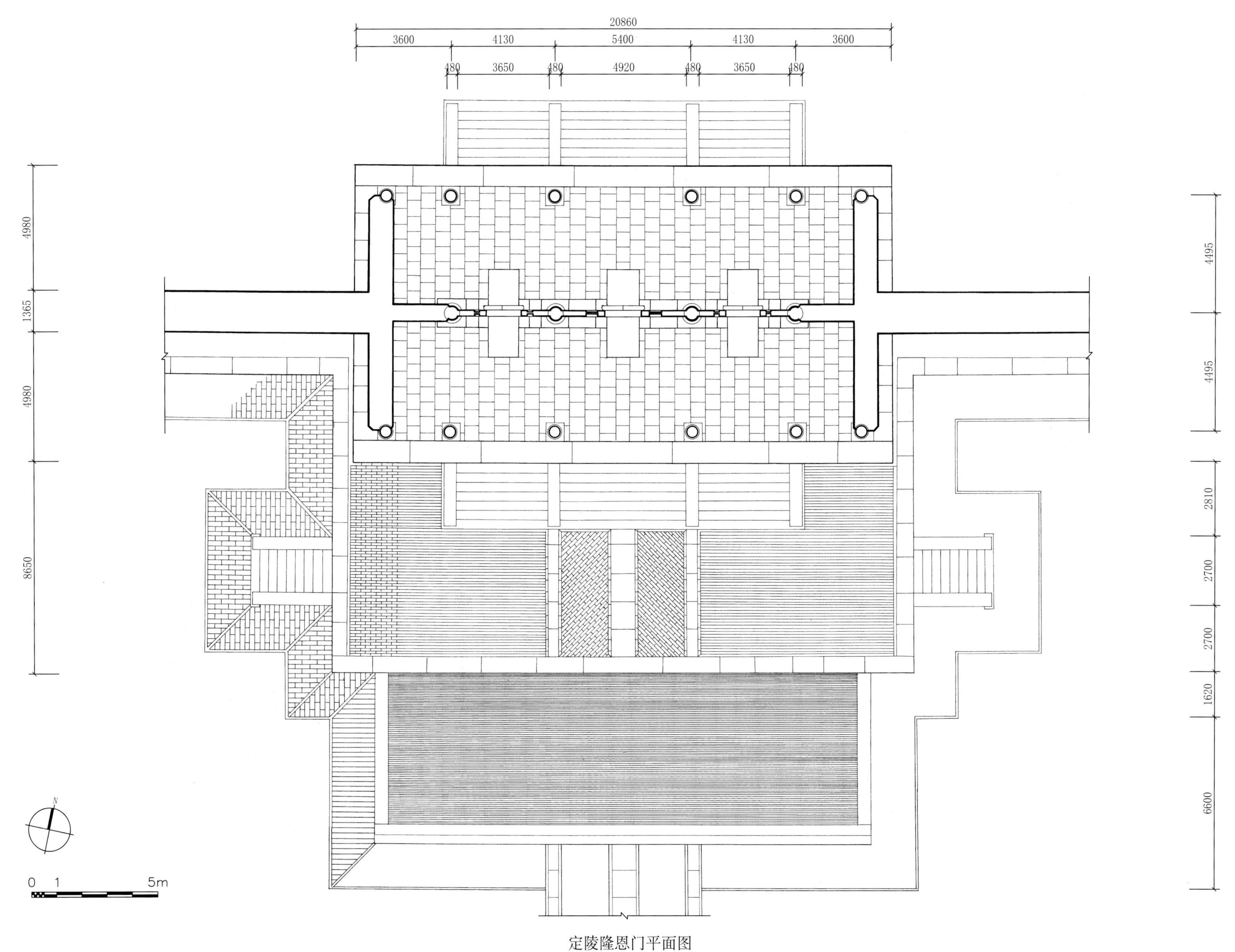

定陵隆恩门平面图

Plan of the Gate of Monumental Grace, Tomb of Stability

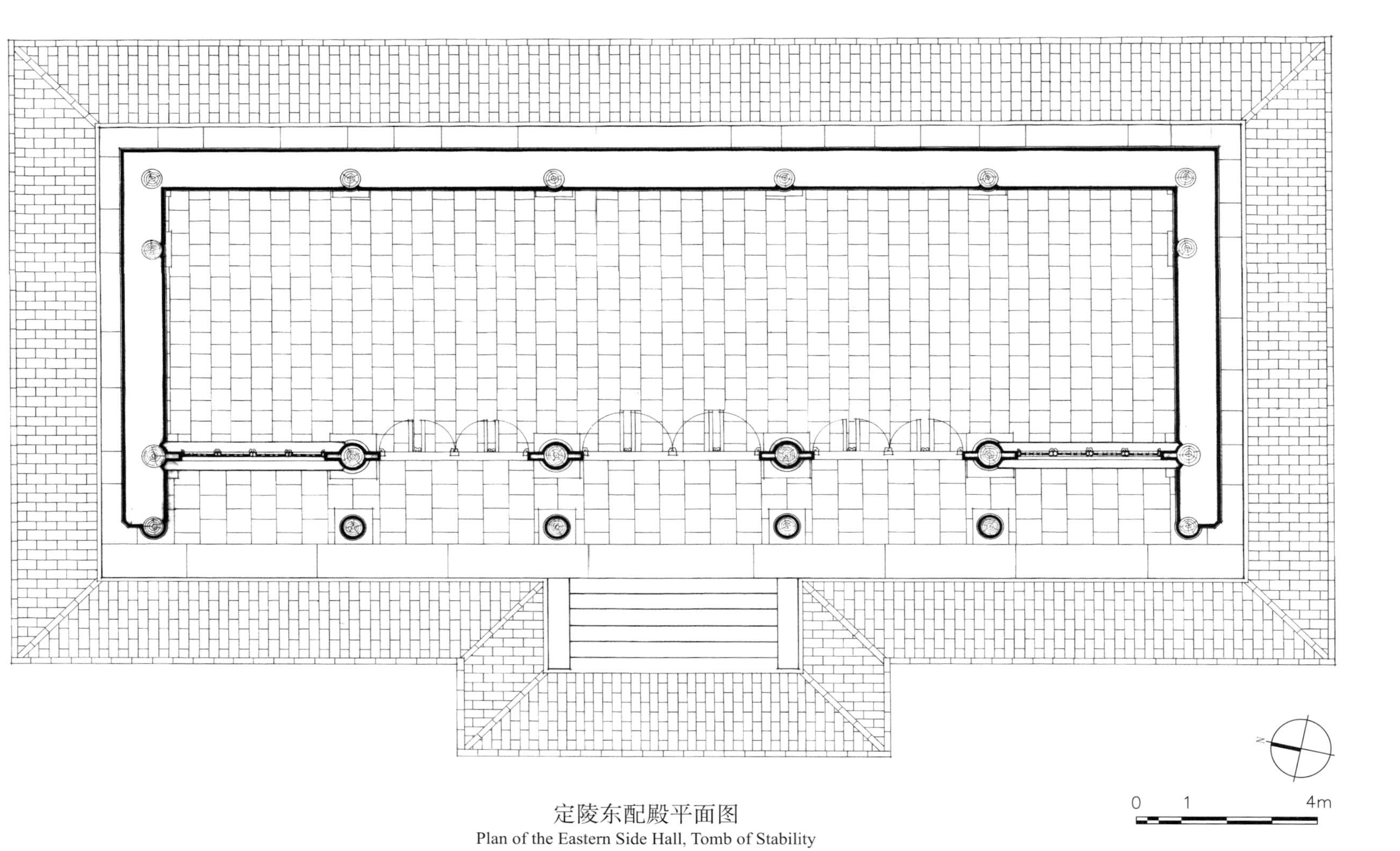

定陵东配殿平面图
Plan of the Eastern Side Hall, Tomb of Stability

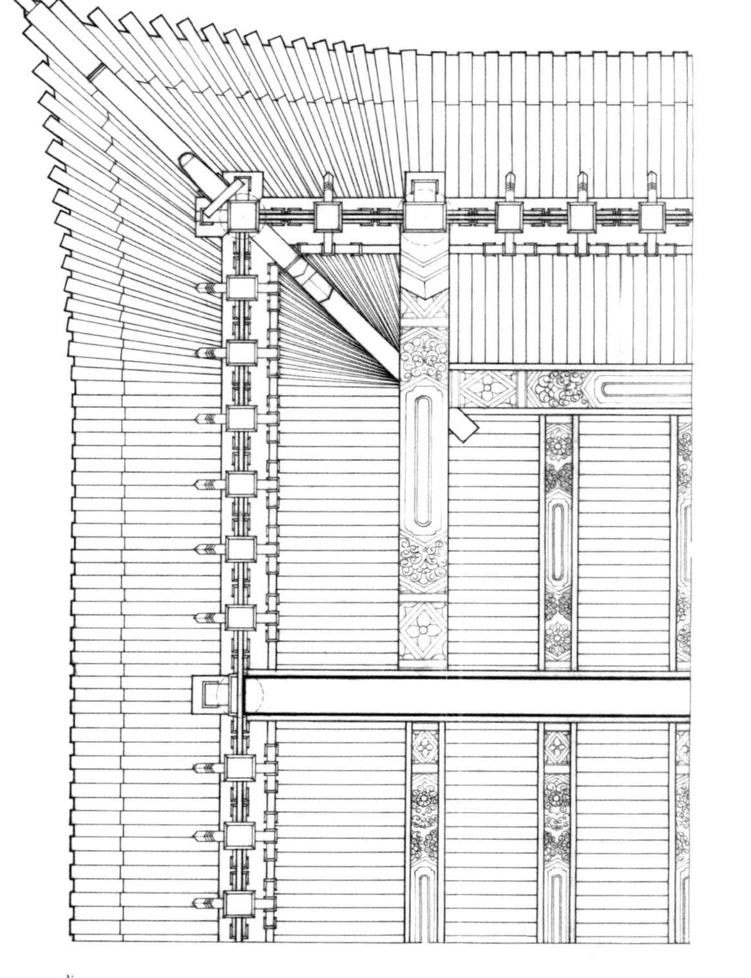

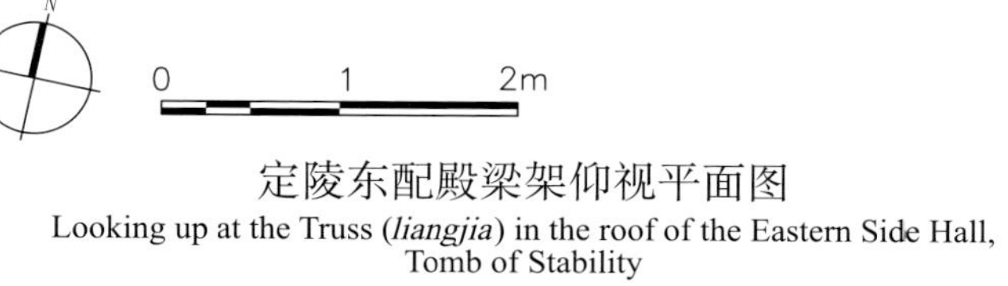

定陵东配殿梁架仰视平面图
Looking up at the Truss (*liangjia*) in the roof of the Eastern Side Hall, Tomb of Stability

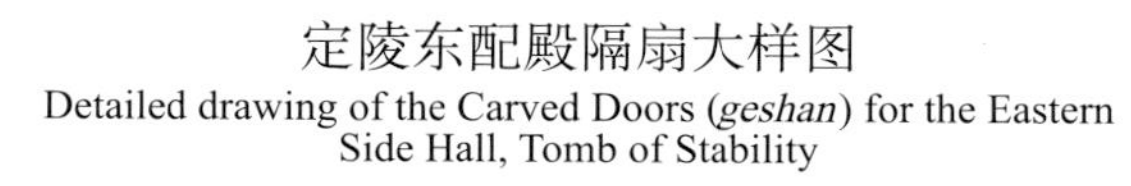

定陵东配殿隔扇大样图
Detailed drawing of the Carved Doors (*geshan*) for the Eastern Side Hall, Tomb of Stability

定陵东配殿正立面图
Front elevation of the Eastern Side Hall, Tomb of Stability

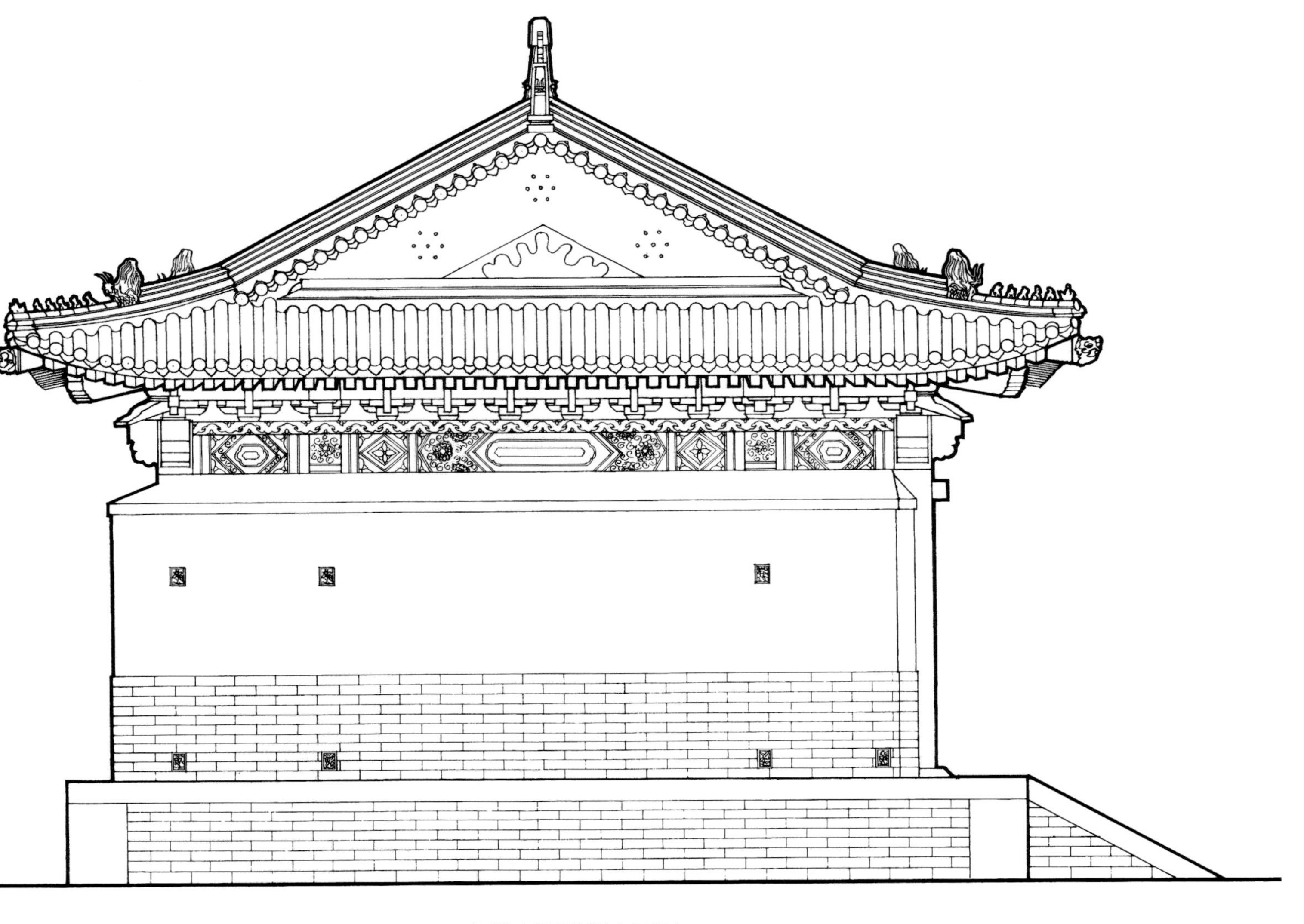

定陵东配殿侧立面图
Side elevation of the Eastern Side Hall, Tomb of Stability

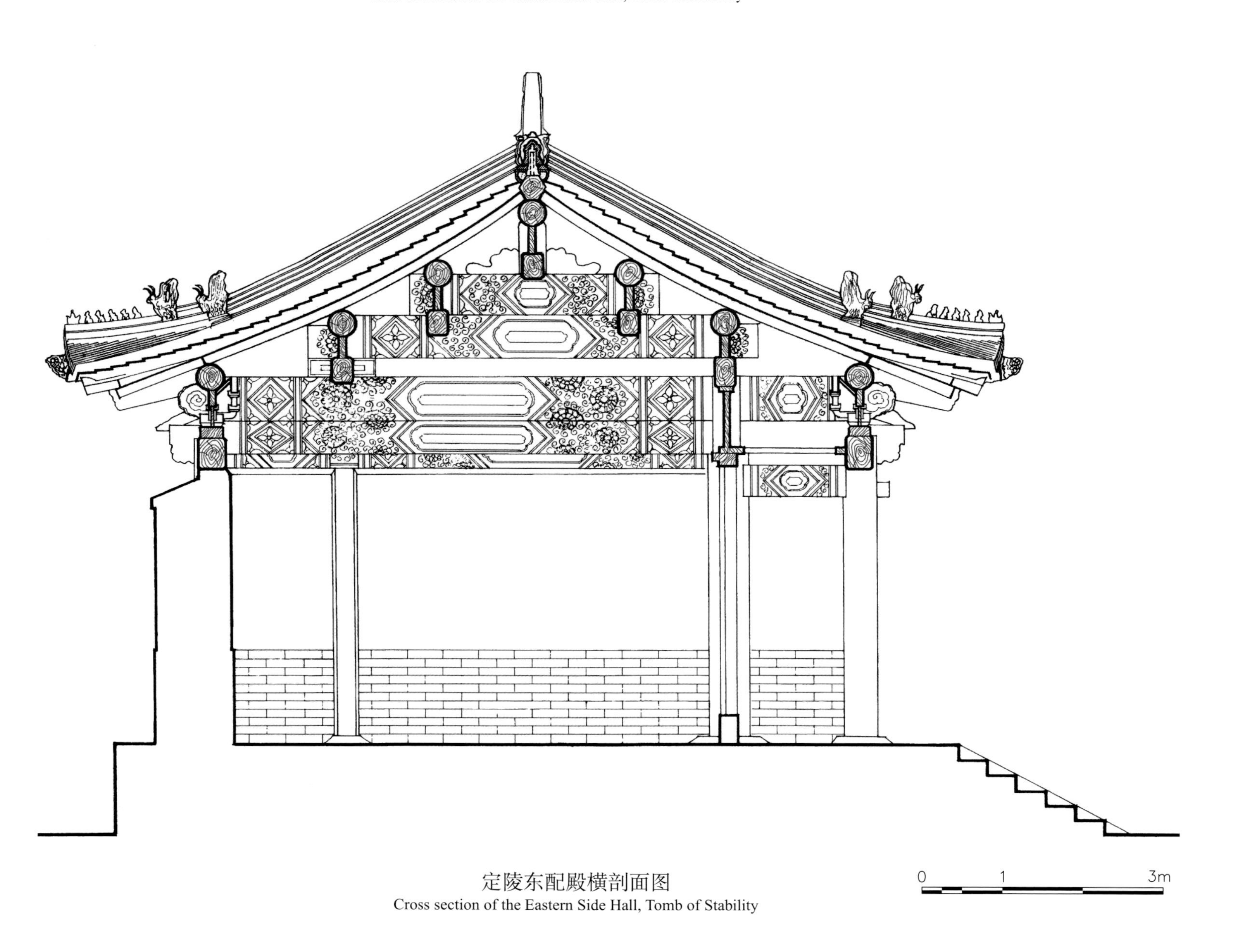

定陵东配殿横剖面图
Cross section of the Eastern Side Hall, Tomb of Stability

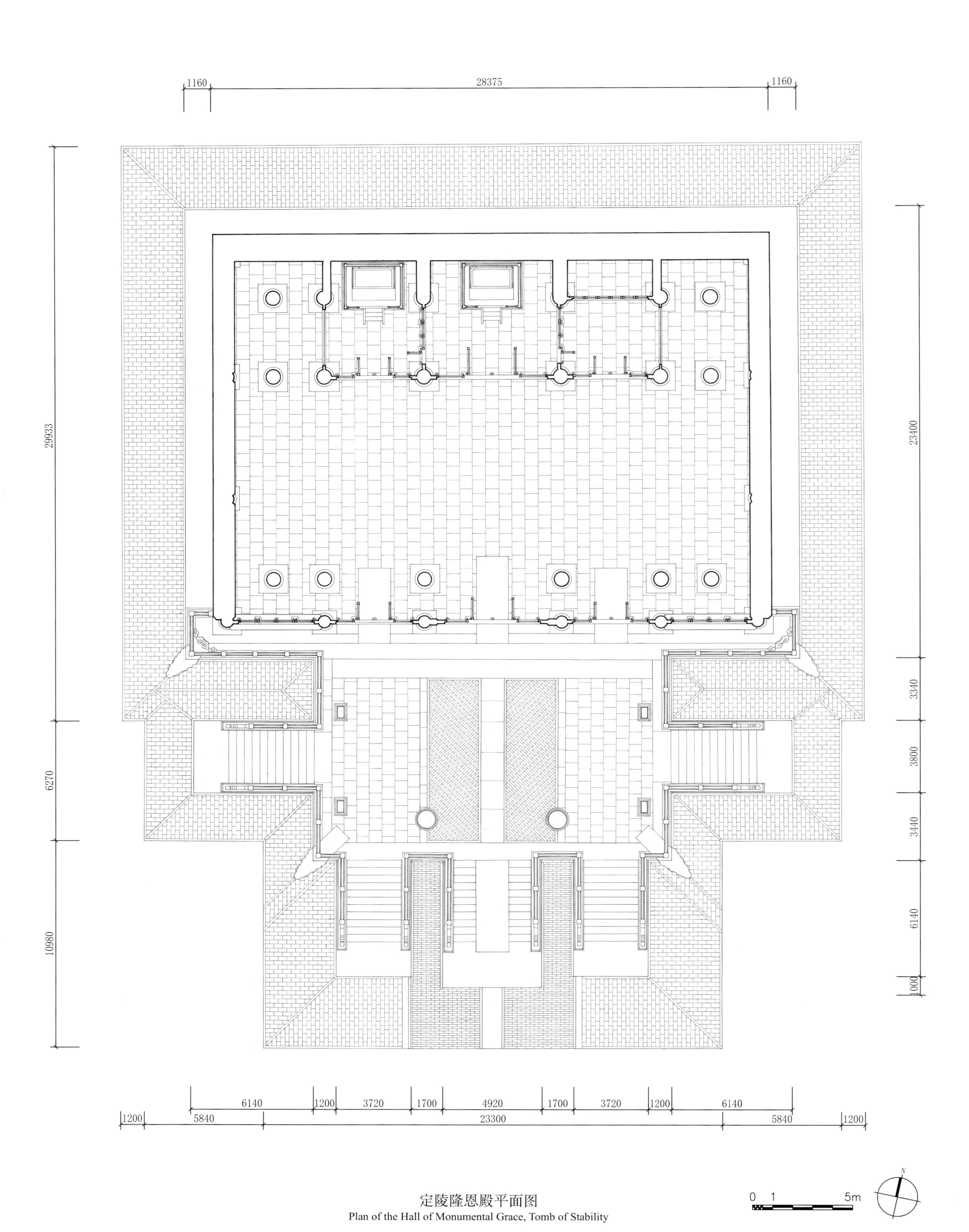

定陵隆恩殿平面图
Plan of the Hall of Monumental Grace, Tomb of Stability

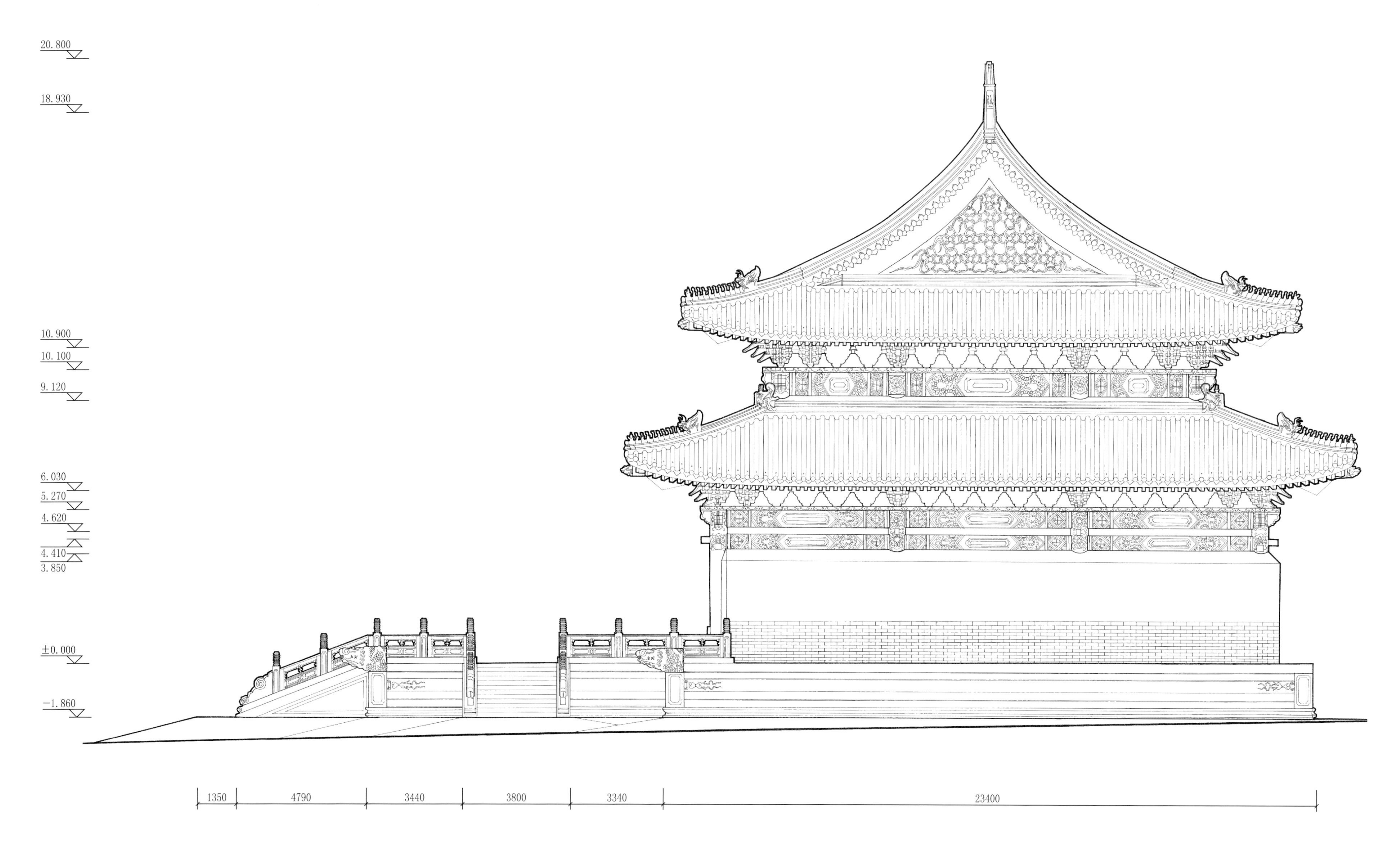

定陵隆恩殿侧立面图

Side elevation of the Hall of Monumental Grace, Tomb of Stability

定陵隆恩殿横剖面图

Cross section of the Hall of Monumental Grace, Tomb of Stability

0 1 3m

140

4950

240

1730

0 0.1 0.5m

定陵隆恩殿丹陛石大样图

Detailed drawing of the Marble Carving on the steps to the Hall of Monumental Grace, Tomb of Stability

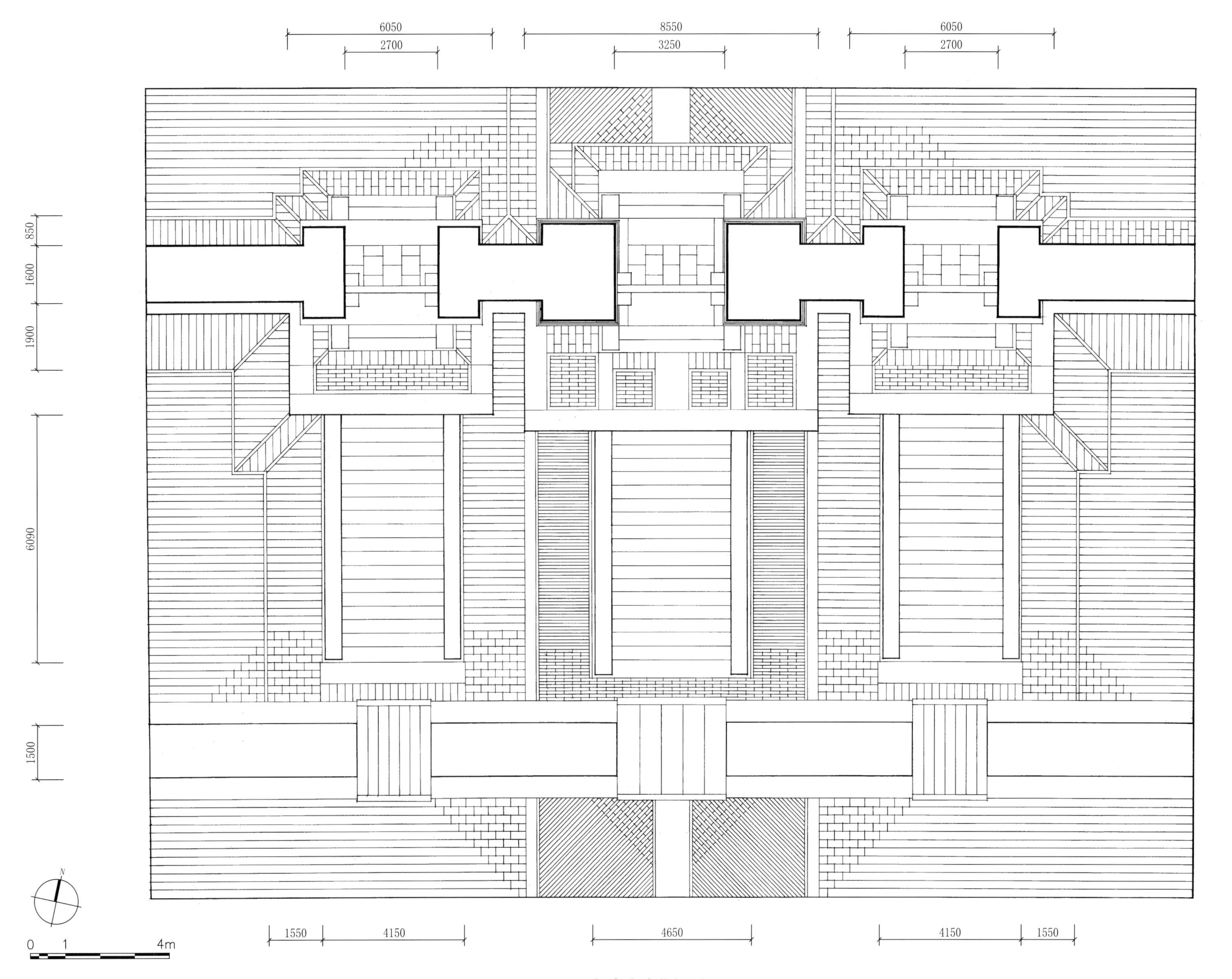

定陵琉璃花门平面图
Plan of the Gate with Glazed Roof Tiles, Tomb of Stability

定陵琉璃花门正立面图
Front elevation of the Gate with Glazed Roof Tiles, Tomb of Stability

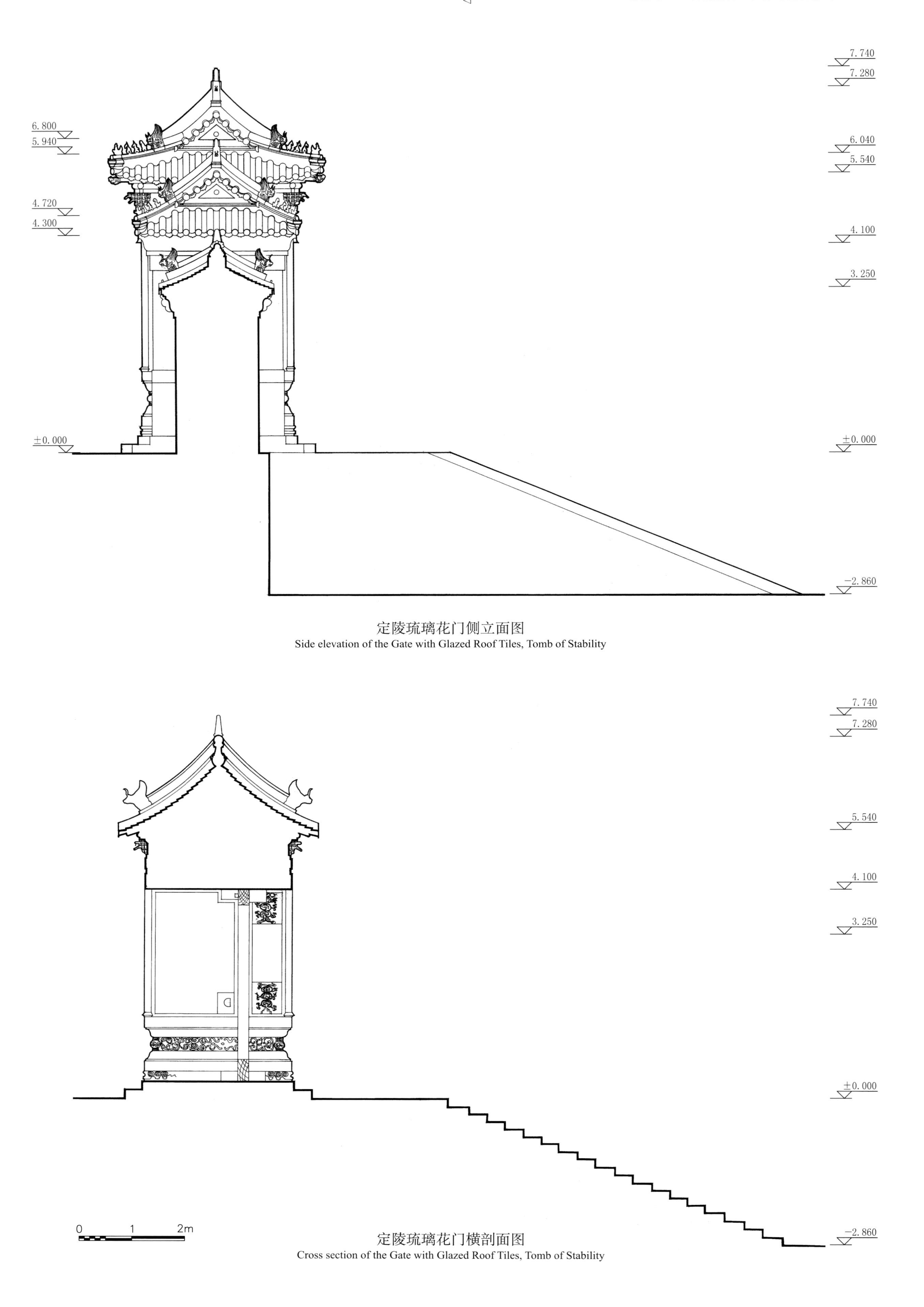

定陵琉璃花门侧立面图
Side elevation of the Gate with Glazed Roof Tiles, Tomb of Stability

定陵琉璃花门横剖面图
Cross section of the Gate with Glazed Roof Tiles, Tomb of Stability

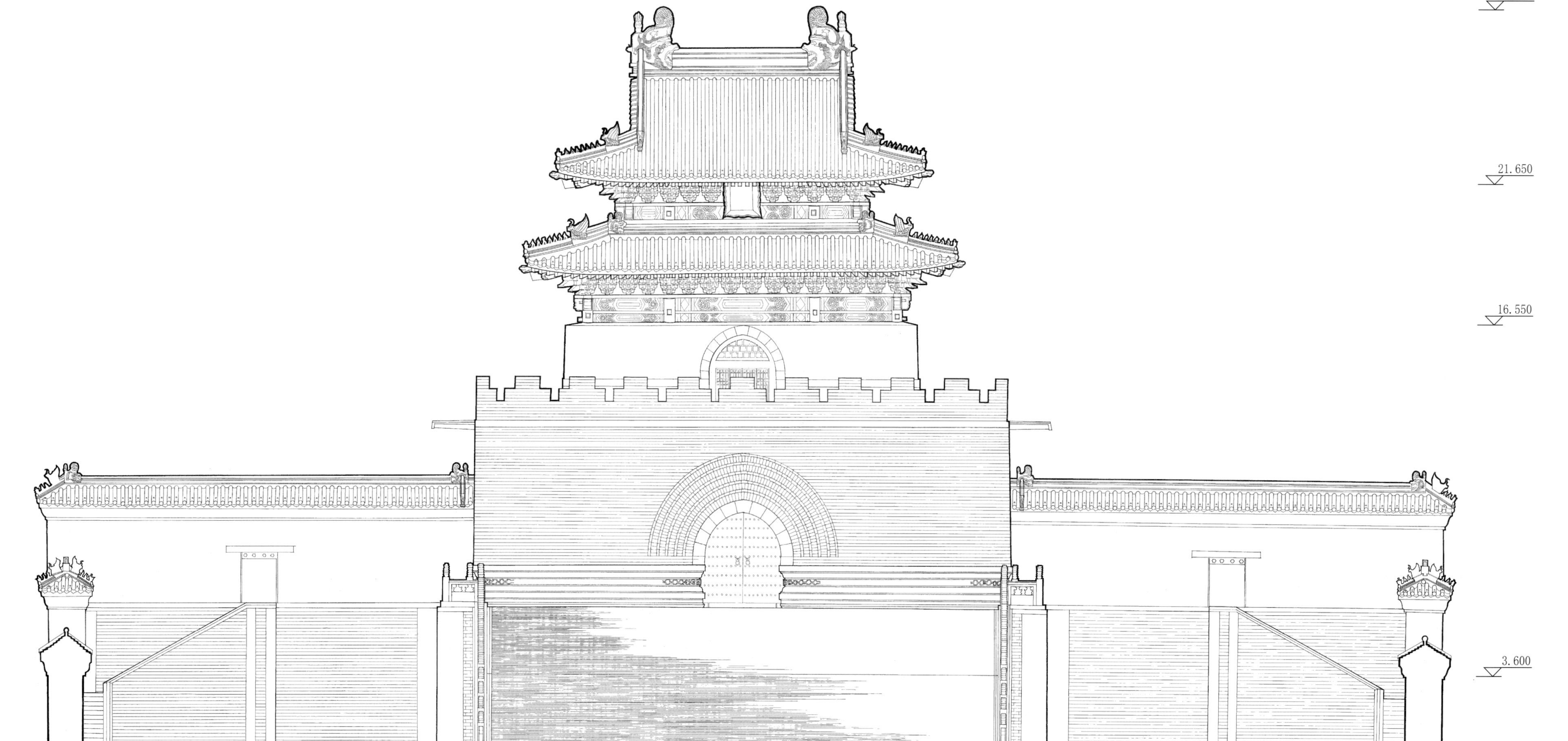

定陵方城明楼正立面图
Front elevation of the Square Walled Terrace and the Memorial Tower, Tomb of Stability

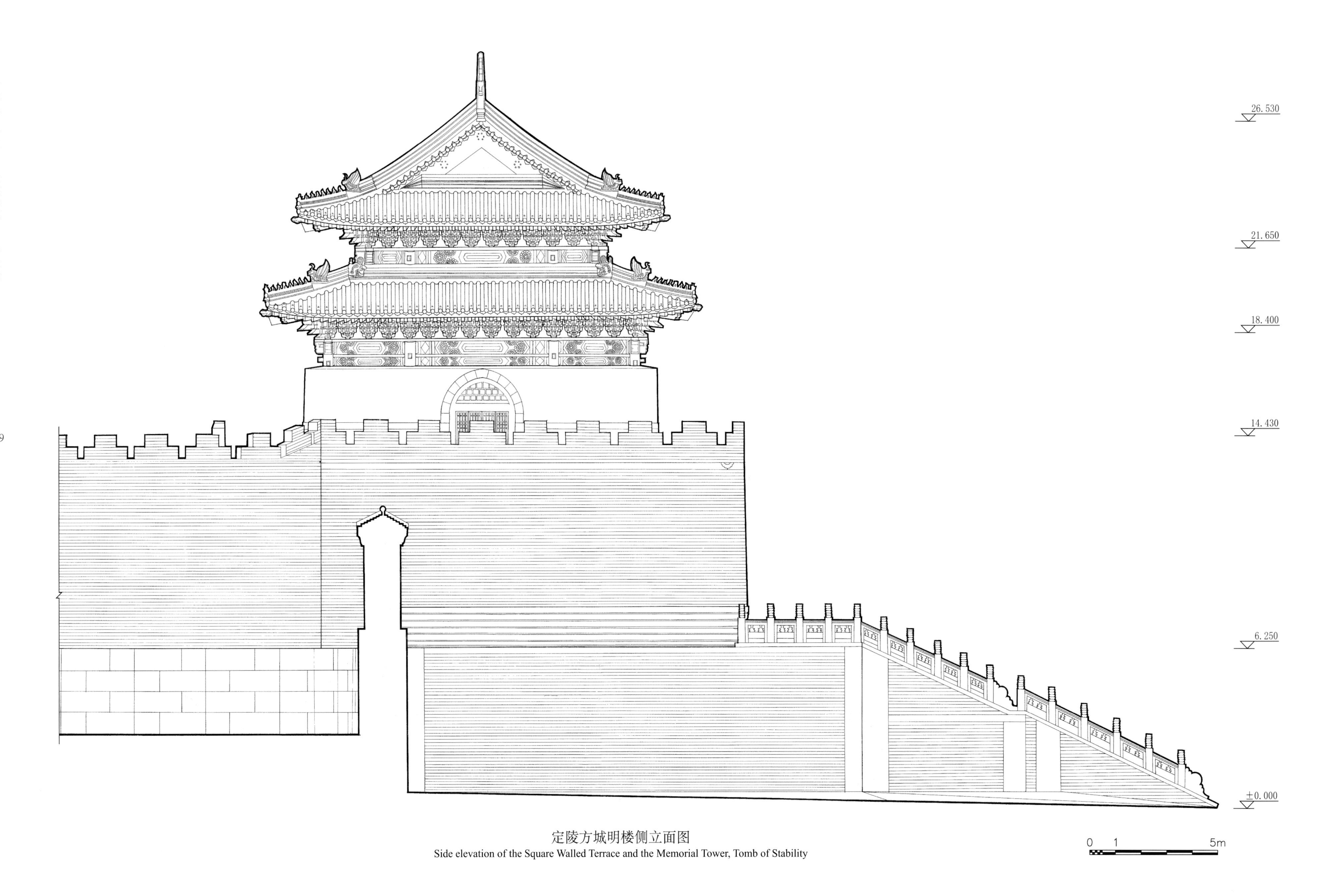

定陵方城明楼侧立面图

Side elevation of the Square Walled Terrace and the Memorial Tower, Tomb of Stability

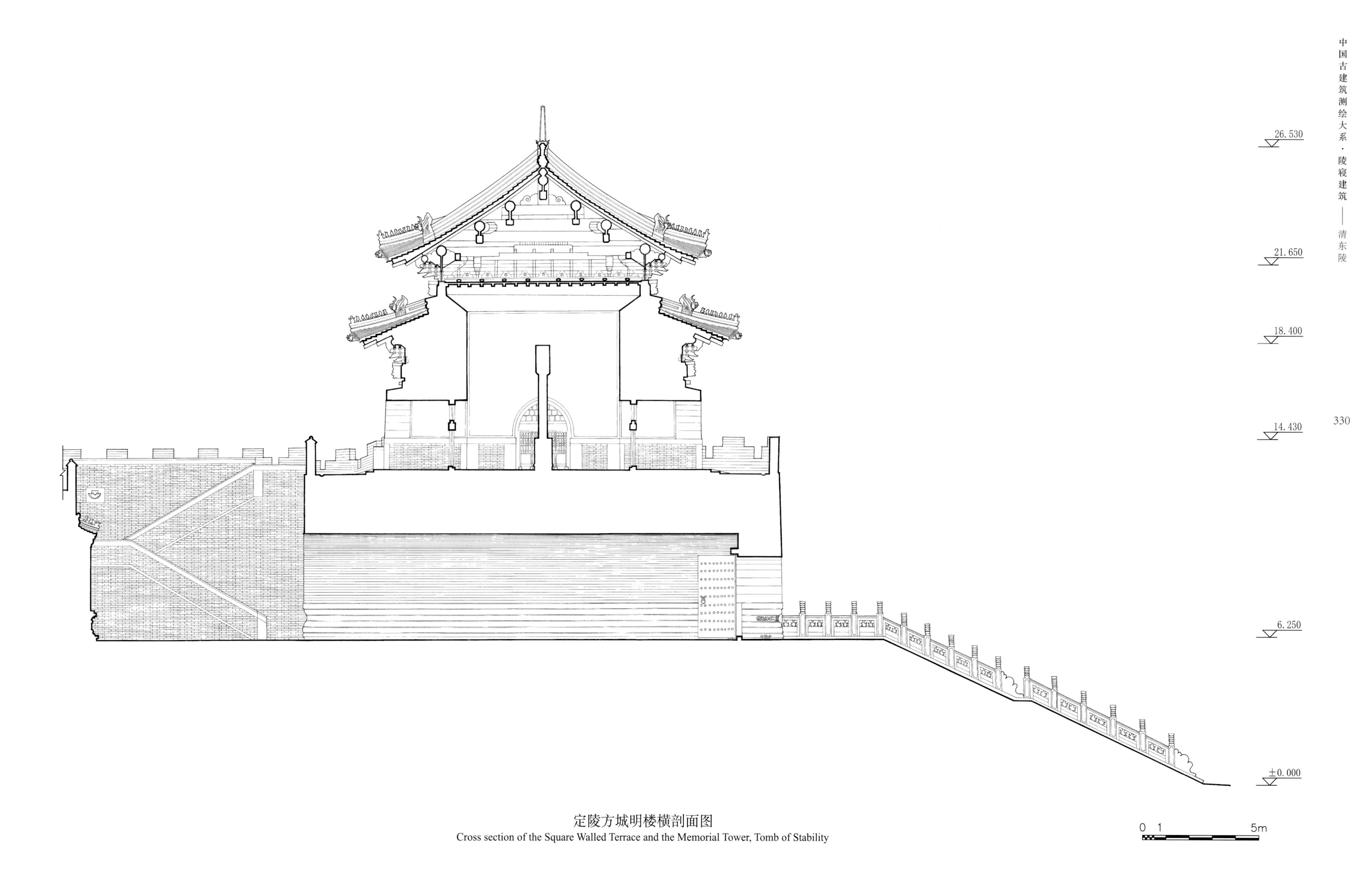

定陵方城明楼横剖面图

Cross section of the Square Walled Terrace and the Memorial Tower, Tomb of Stability

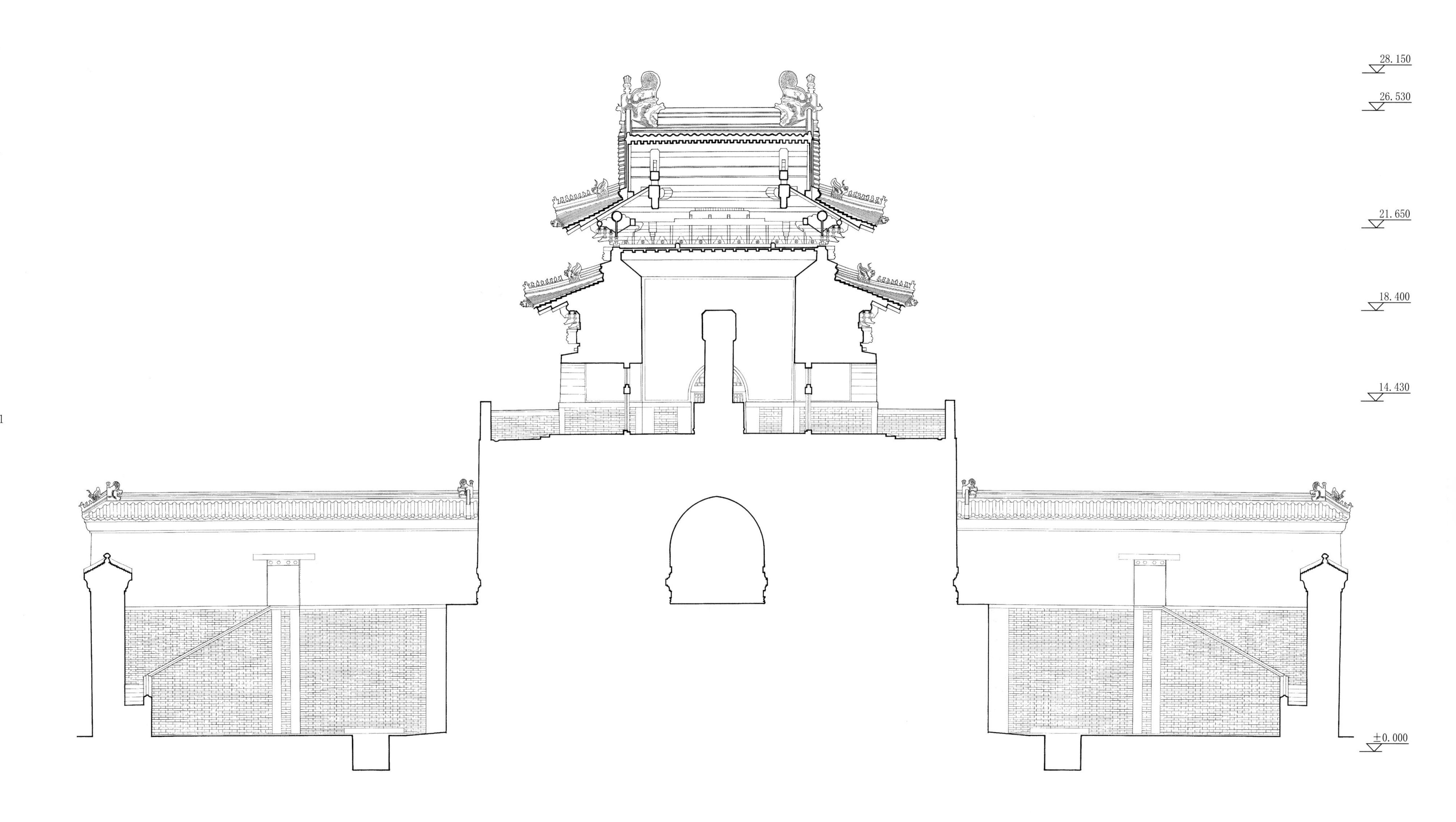

定陵方城明楼纵剖面图

Longitudinal section of the Square Walled Terrace and the Memorial Tower, Tomb of Stability

定东陵

East Tomb of Stability (Ding Dongling)

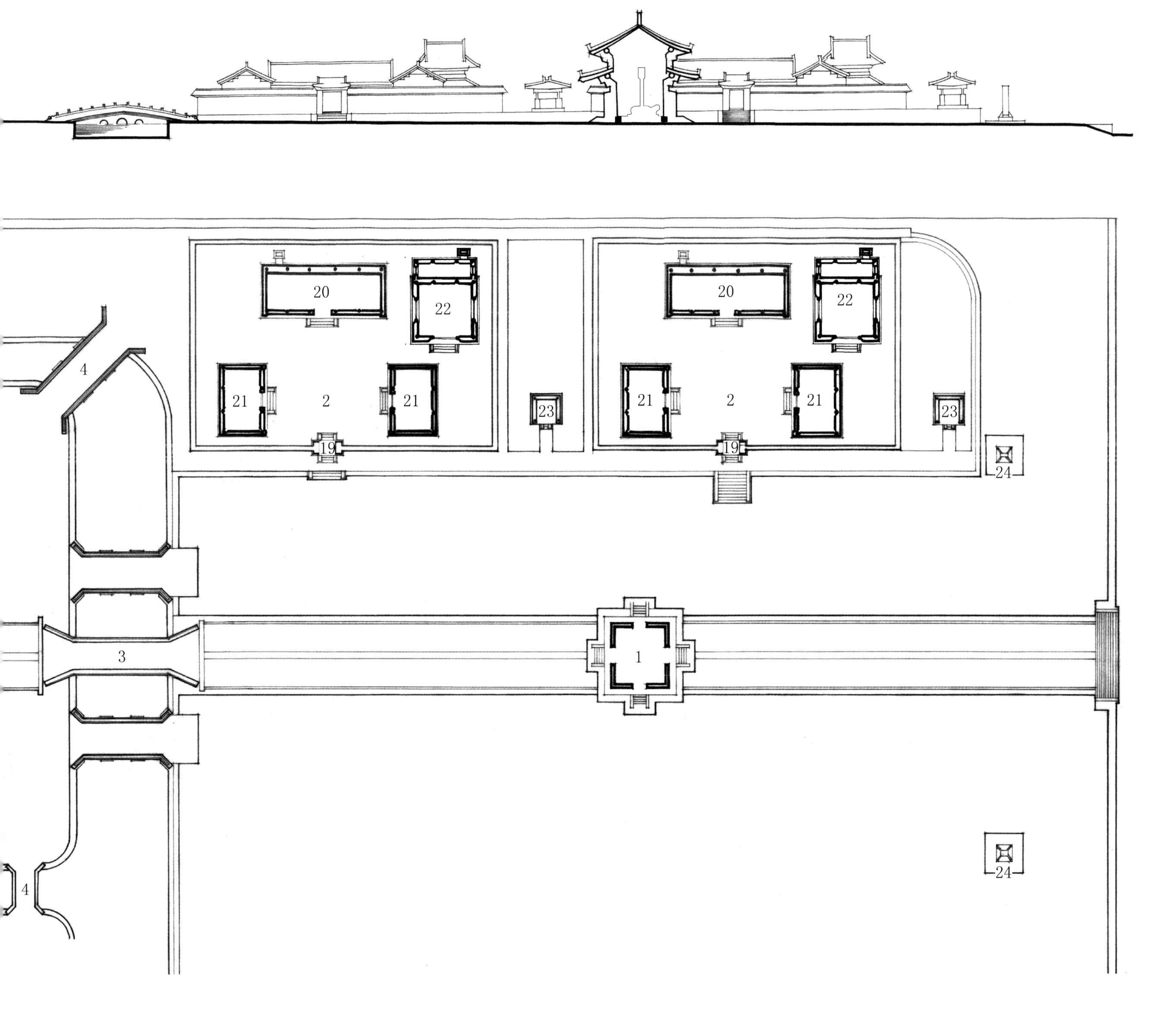

1 神道碑亭 Pavilion for the Stela on the Spirit Way
2 神厨库院落 Culinary Courtyard for Sacrifices
3 三路三孔桥 Three-Way Three-Arch Bridges
4 石平桥 Flat Stone Bridge
5 朝房 Reception Halls for Court Officials
6 班房 Duty Houses for the Guards
7 隆恩门 Gate of Monumental Grace
8 焚帛炉 Sacrificial Burners
9 配殿 Side Hall
10 隆恩殿 Hall of Monumental Grace
11 琉璃花门 Gate with Glazed Roof Tiles
12 石五供 Five Stone Ritual Vessels
13 方城 Square Walled Terrace
14 明楼 Memorial Tower
15 宝顶 Tumulus
16 宝城 Encircled Realm of Treasure
17 罗圈墙 Luoquan Wall
18 宝城 Encircled Realm of Treasure
19 院门 Yard Gate
20 神厨 Sacrificial Kitchen
21 神库 Sacrificial Storehouse
22 宰牲亭 Ritual Abattoir
23 井亭 Well Pavilion
24 下马牌 Stela Marking the Place for Dismounting from one's Horse

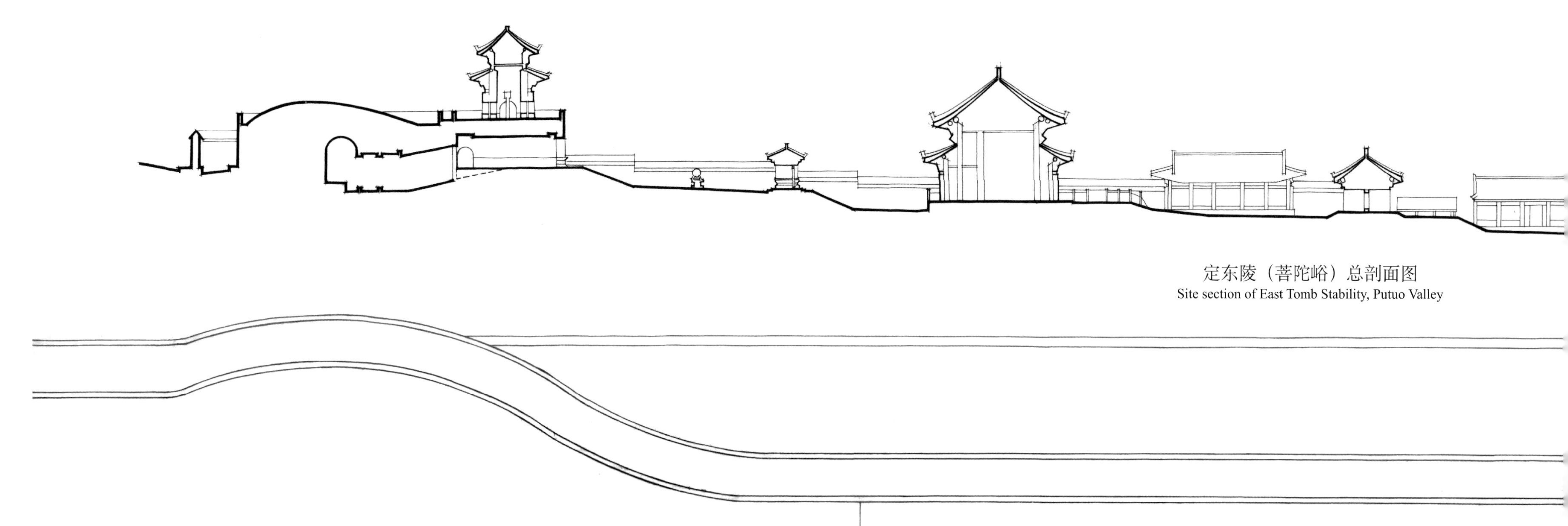

定东陵（菩陀峪）总剖面图
Site section of East Tomb Stability, Putuo Valley

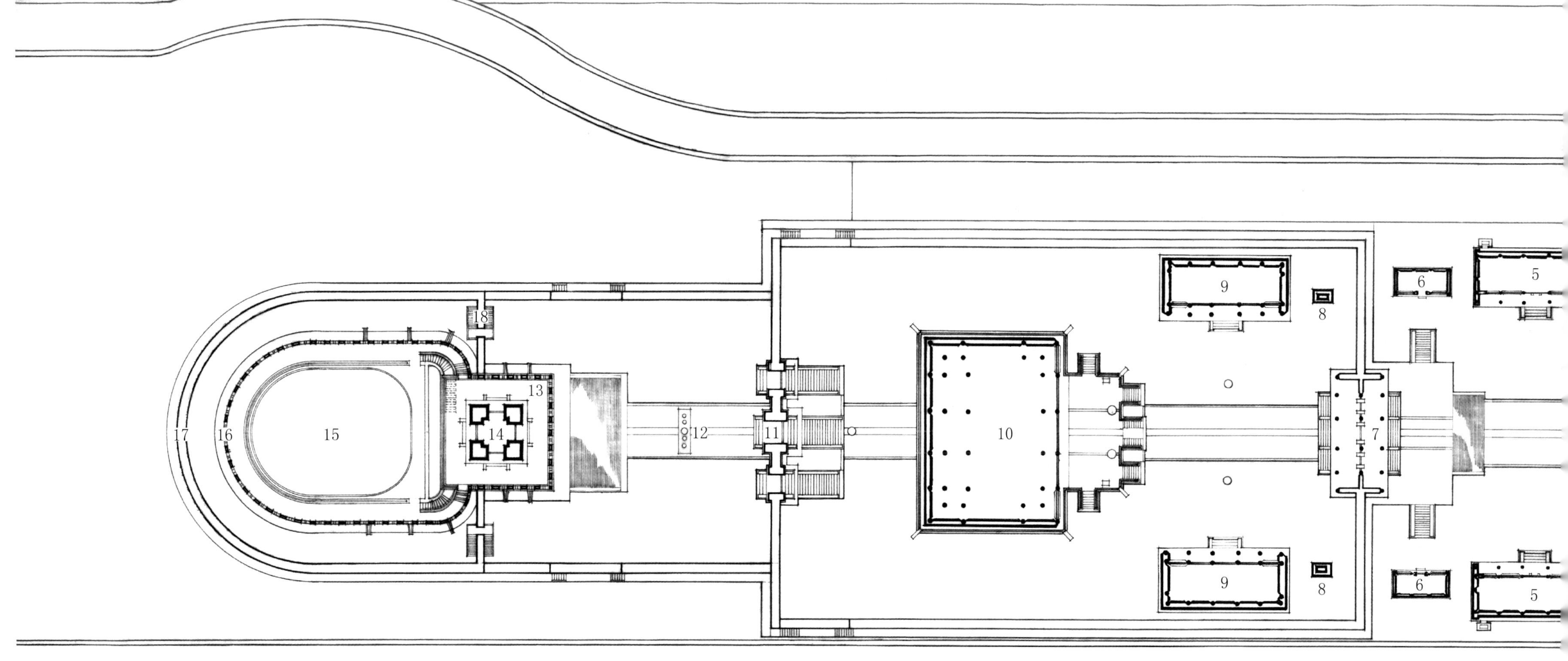

定东陵（菩陀峪）总平面图
Site plan of East Tomb of Stability, Putuo Valley

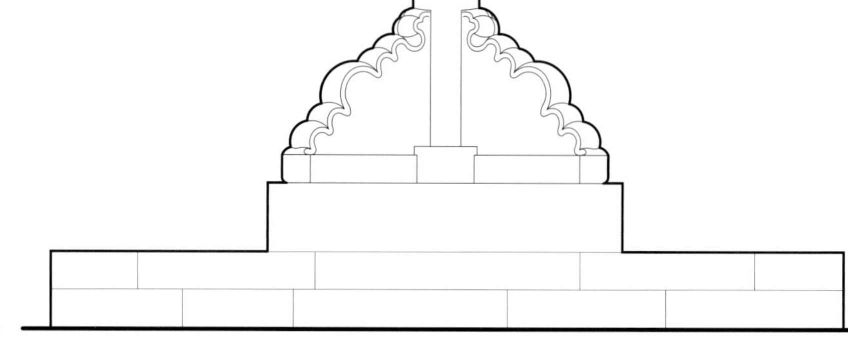

0 1 2m

0 0.5 1m

定东陵（菩陀峪）下马牌大样图

Detailed drawing of the Stela marking the place for dismounting from one's horse, East Tomb of Stability, Putuo Valley

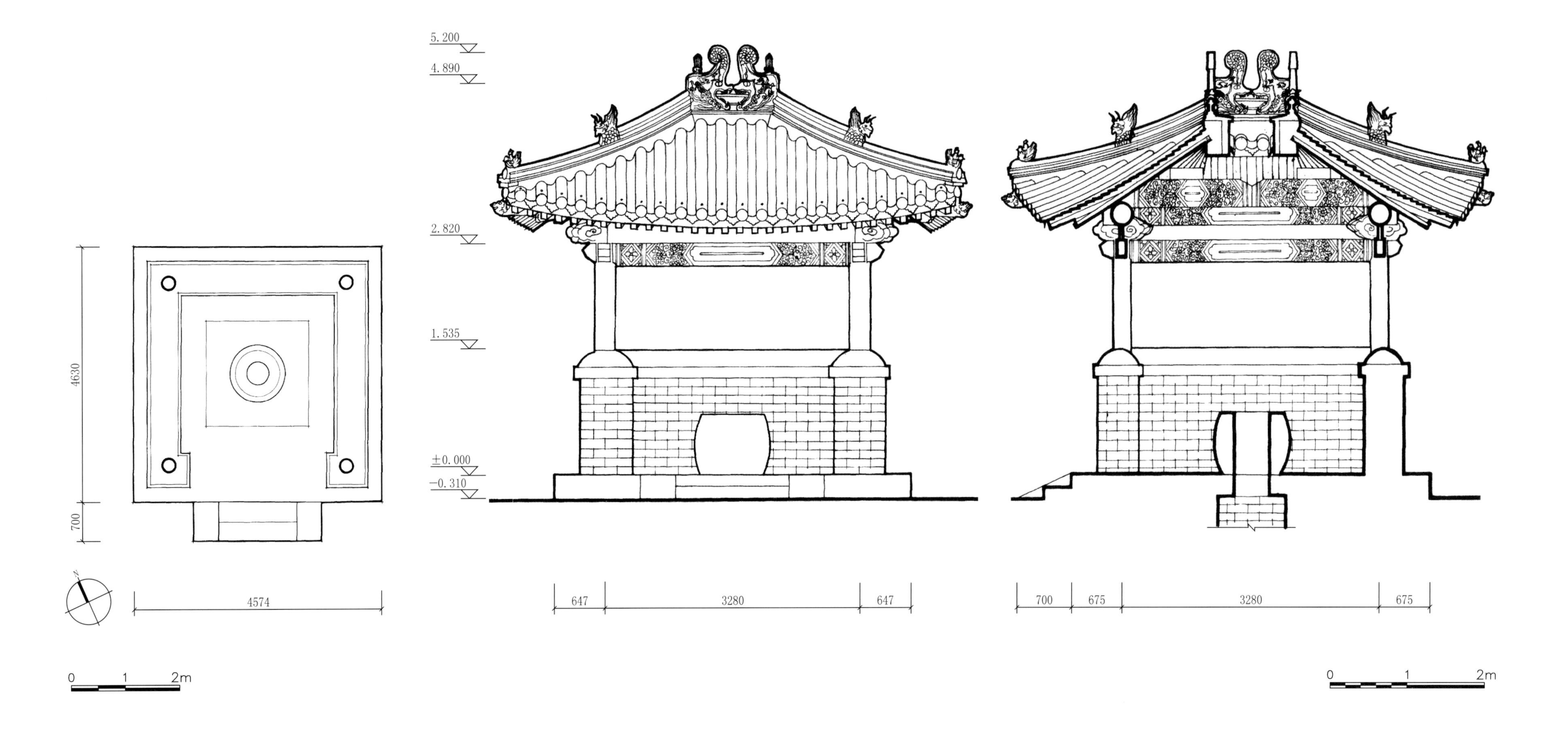

定东陵（菩陀峪）井亭平面图
Plan of the Pavilion for the Well, East Tomb of Stability, Putuo Valley

定东陵（菩陀峪）井亭正立面图
Front elevation of the Pavilion for the Well, East Tomb of Stability, Putuo Valley

定东陵（菩陀峪）井亭横剖面图
Cross section of the Pavilion for the Well, East Tomb of Stability, Putuo Valley

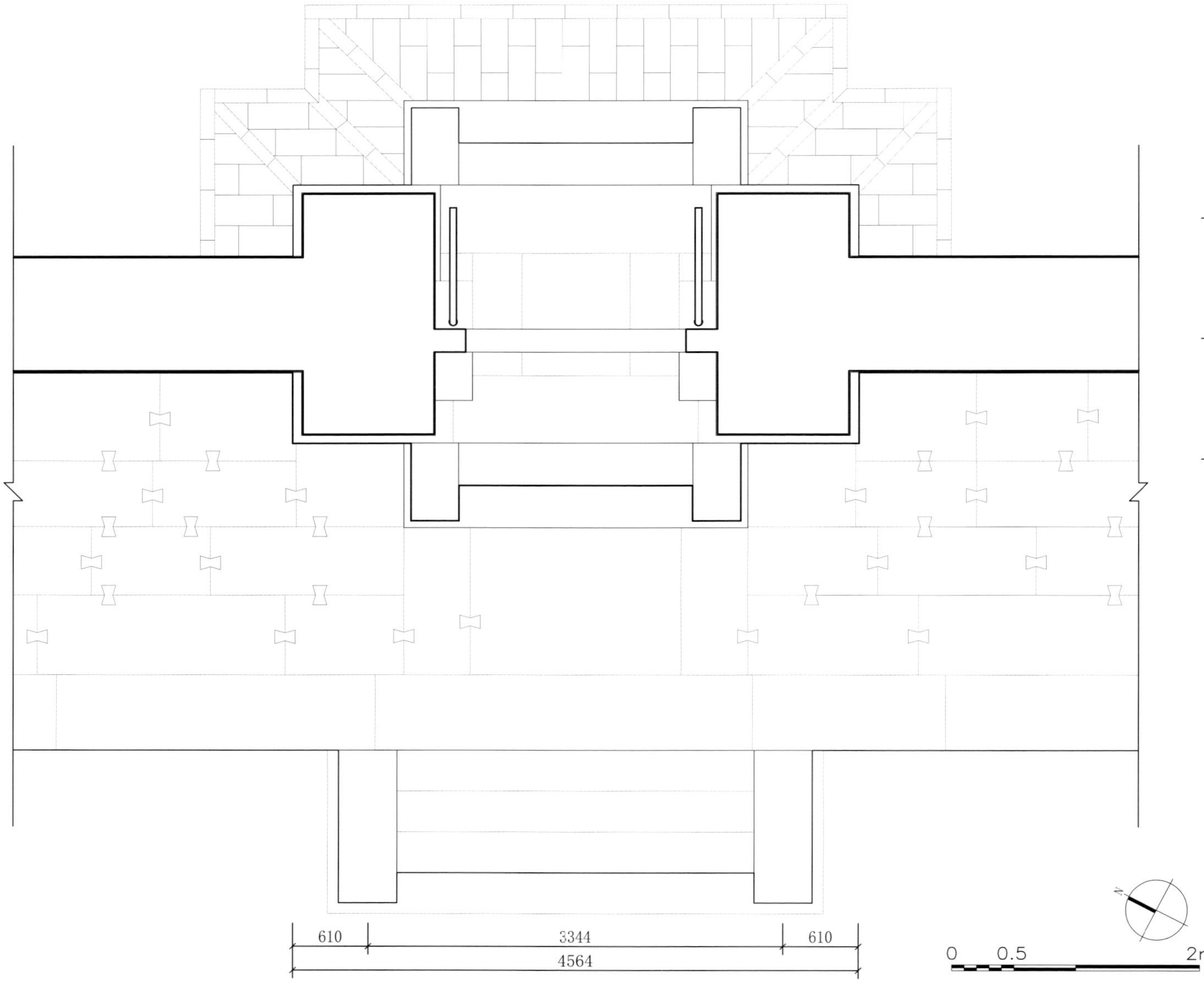

1008
1008
2016
610
3344
610
4564
0
0.5
2m
N

定东陵（菩陀峪）神厨库院门平面图

Plan of the Gate of the Culinary Courtyard for Sacrifices, East Tomb of Stability, Putuo Valley

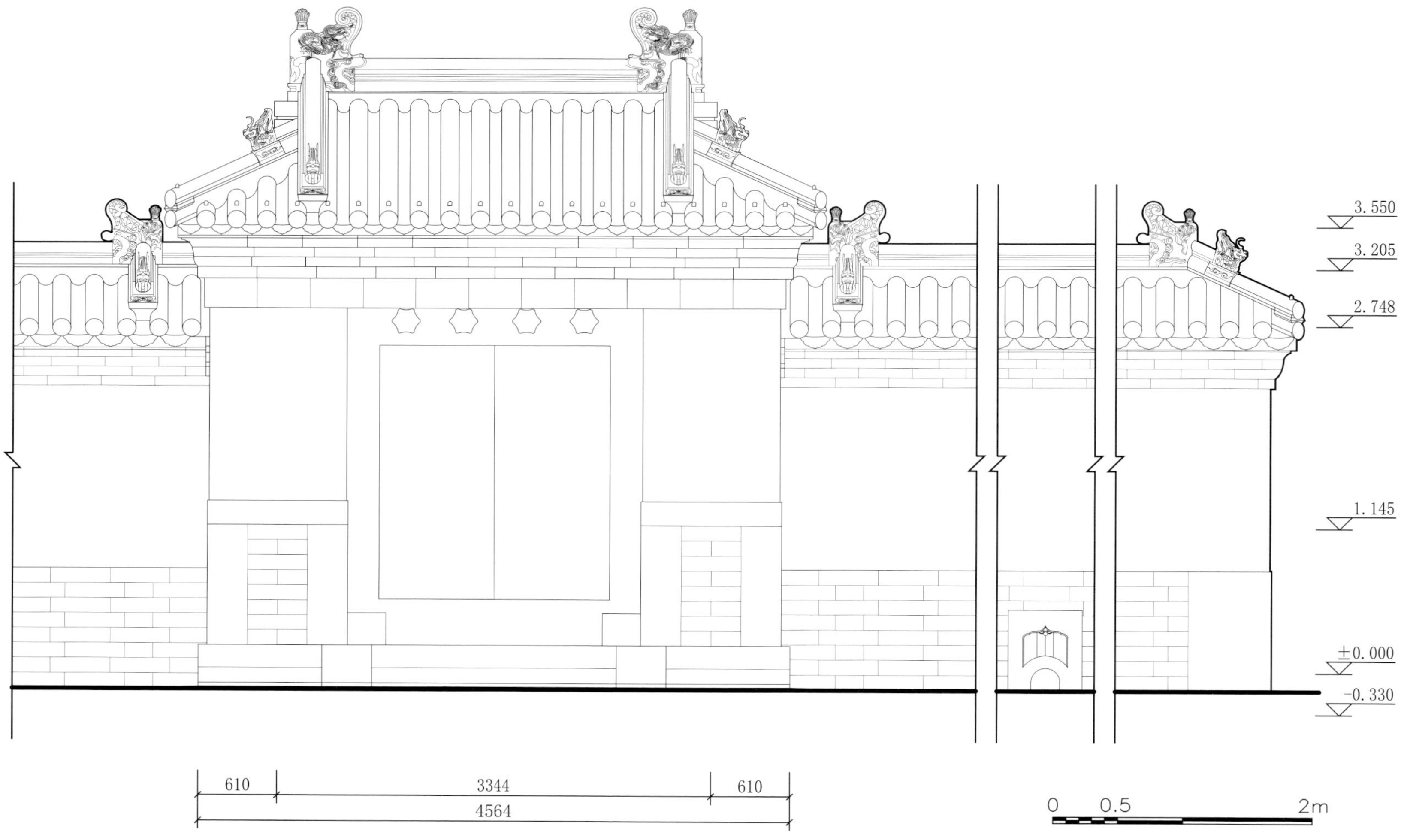

3.550
3.205
2.748
1.145
±0.000
-0.330
610
3344
610
4564
0
0.5
2m

定东陵（菩陀峪）神厨库院门正立面图

Front elevation of the Gate of the Culinary Courtyard for Sacrifices, East Tomb of Stability, Putuo Valley

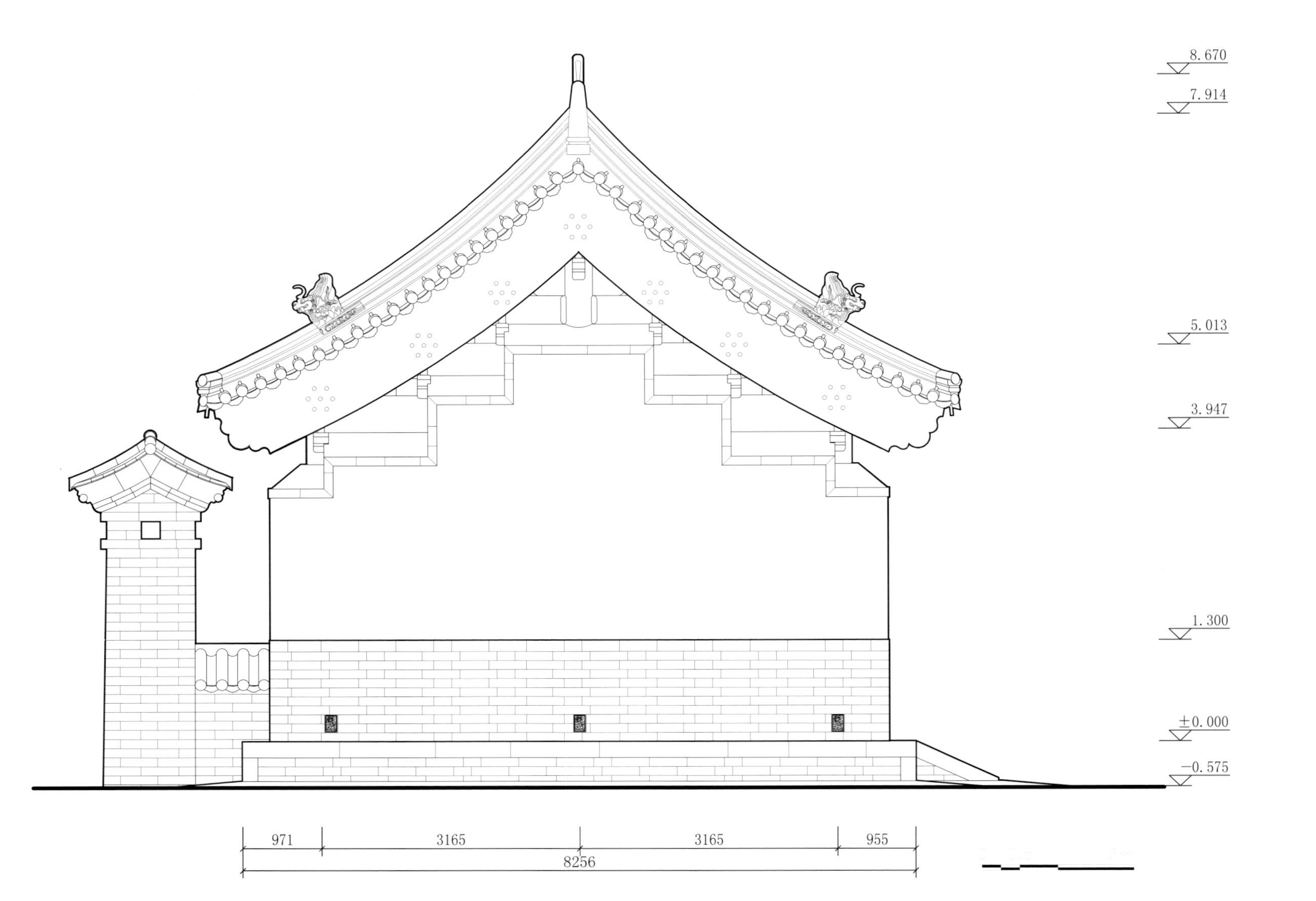

定东陵（菩陀峪）神厨库神厨立面图
Front elevation of the Kitchen in the Culinary Courtyard for Sacrifices, East Tomb of Stability, Putuo Valley

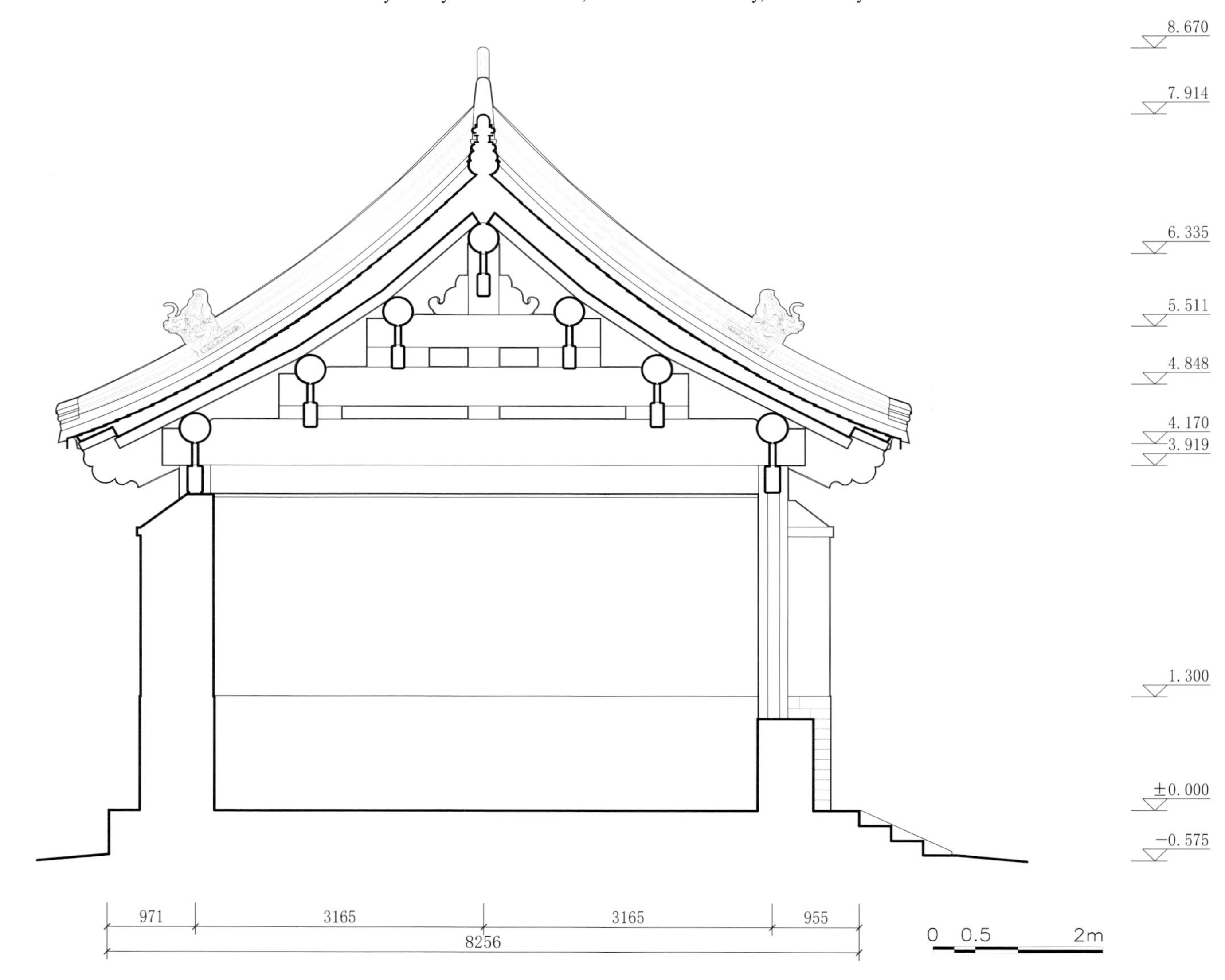

定东陵（菩陀峪）神厨库神厨剖面图
Cross section of the Kitchen in the Culinary Courtyard for Sacrifices, East Tomb of Stability, Putuo Valley

定东陵（菩陀峪）神厨库神库平面图
Plan of the Storehouse in the Culinary Courtyard for Sacrifices, East Tomb of Stability, Putuo Valley

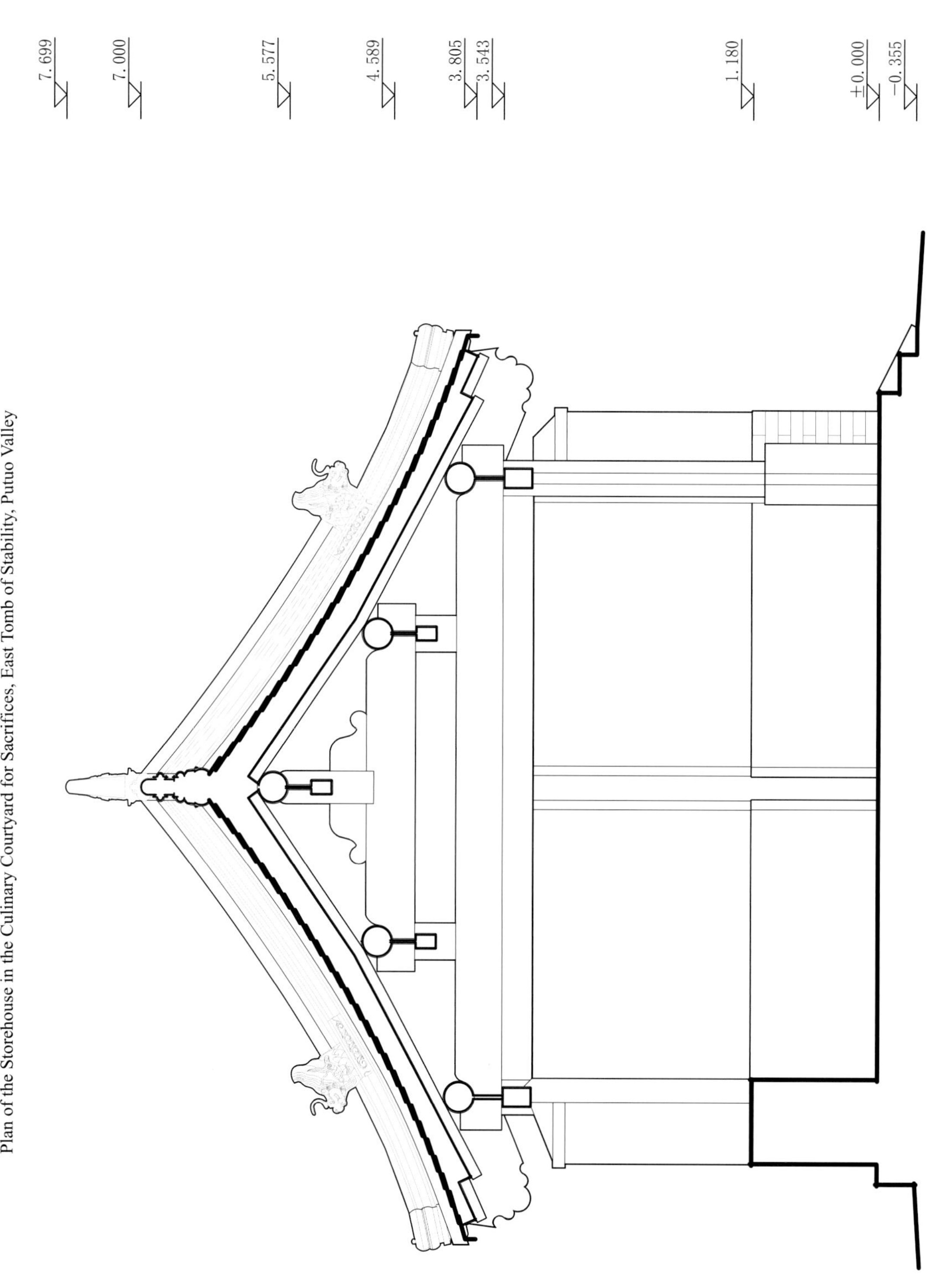

定东陵（菩陀峪）神厨库神库剖面图
Plan of the Storehouse in the Culinary Courtyard for Sacrifices, East Tomb of Stability, Putuo Valley

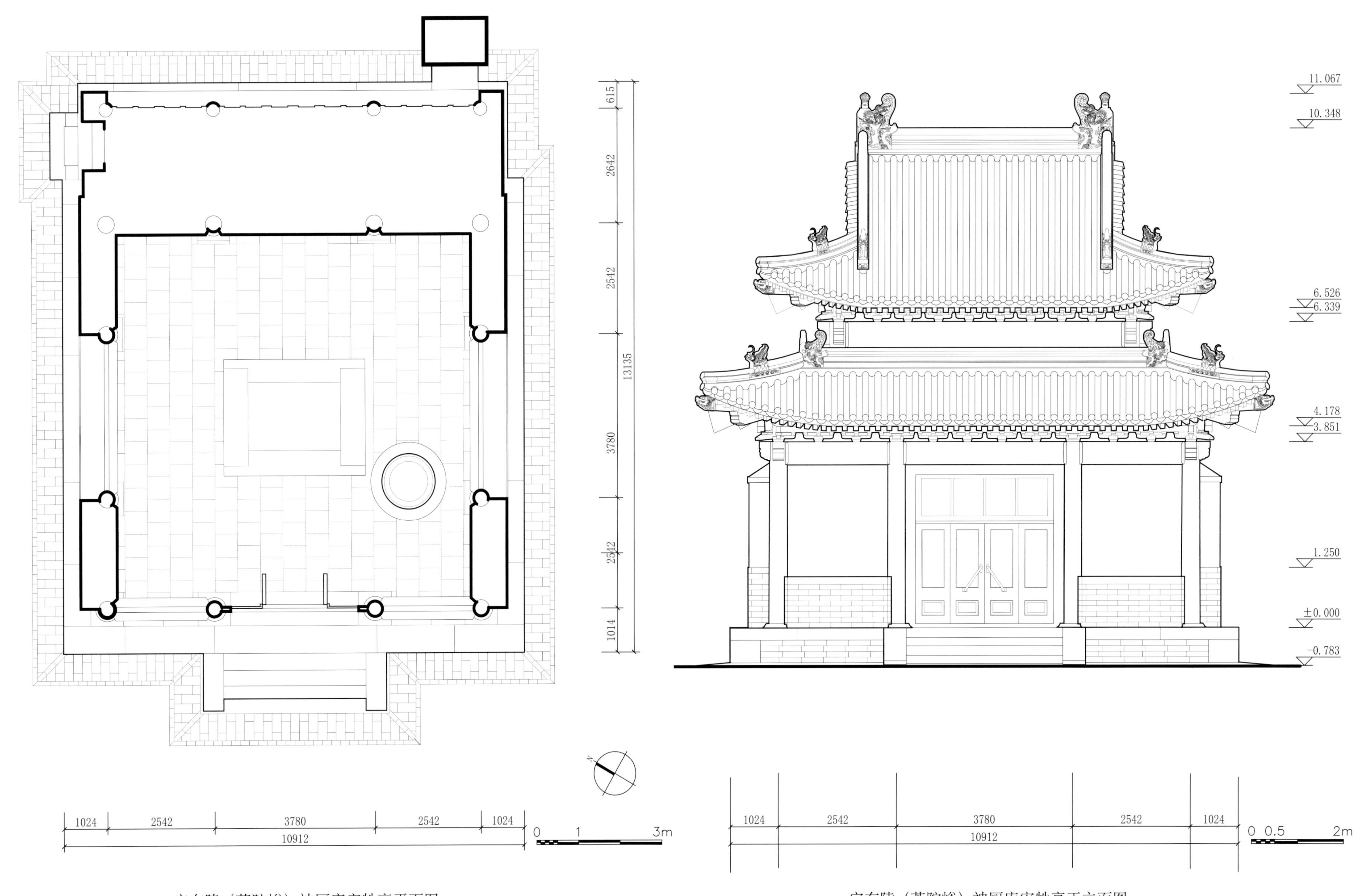

定东陵（菩陀峪）神厨库宰牲亭平面图

Plan of the Ritual Abattoir in the Culinary Courtyard for Sacrifices, East Tomb of Stability, Putuo Valley

定东陵（菩陀峪）神厨库宰牲亭正立面图

Front elevation of the Ritual Abattoir in the Culinary Courtyard for Sacrifices, East Tomb of Stability, Putuo Valley

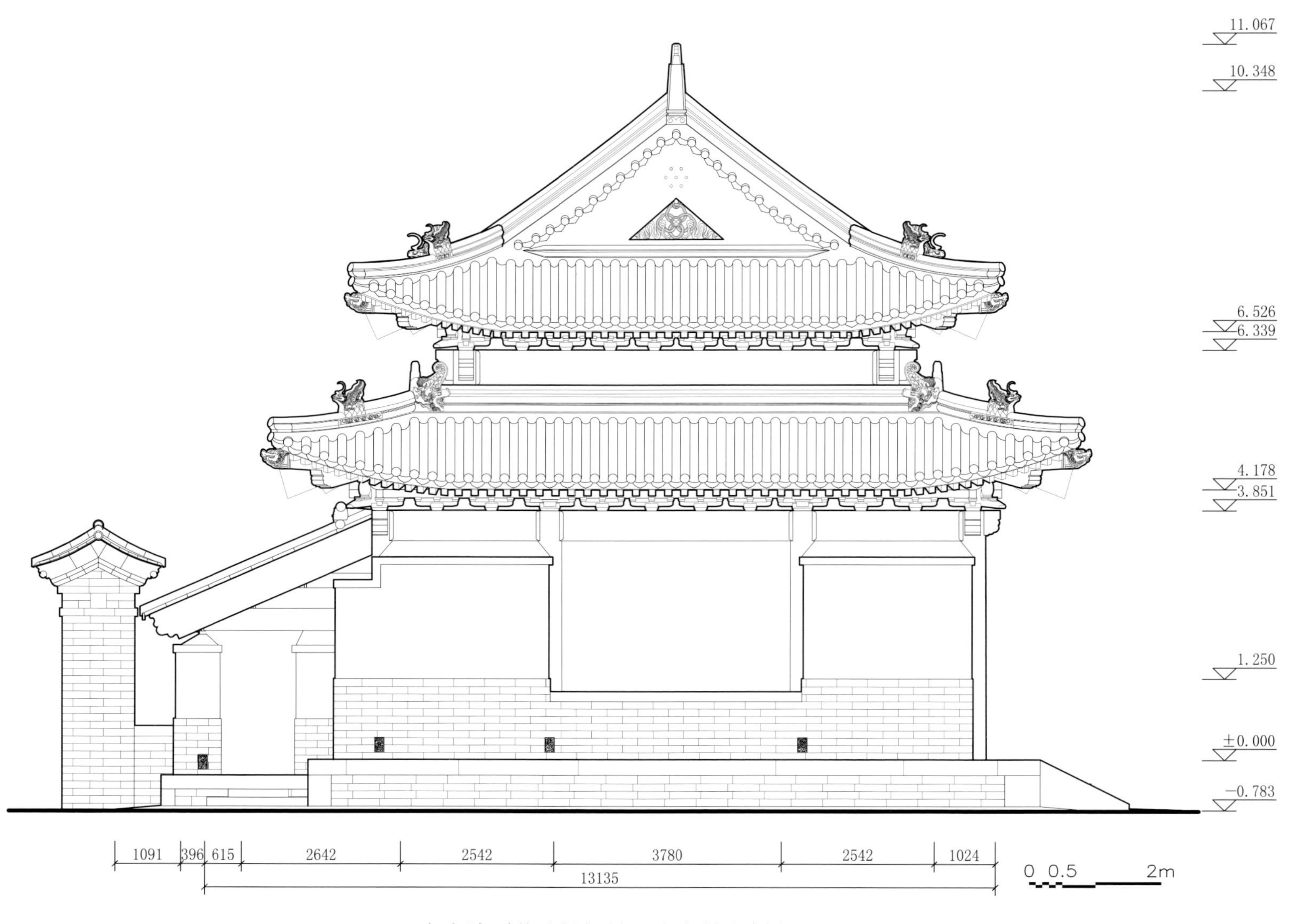

定东陵（菩陀峪）神厨库宰牲亭侧立面图
Side elevation of the Ritual Abattoir in the Culinary Courtyard for Sacrifices, East Tomb of Stability, Putuo Valley, East Tomb of Stability, Putuo Valley

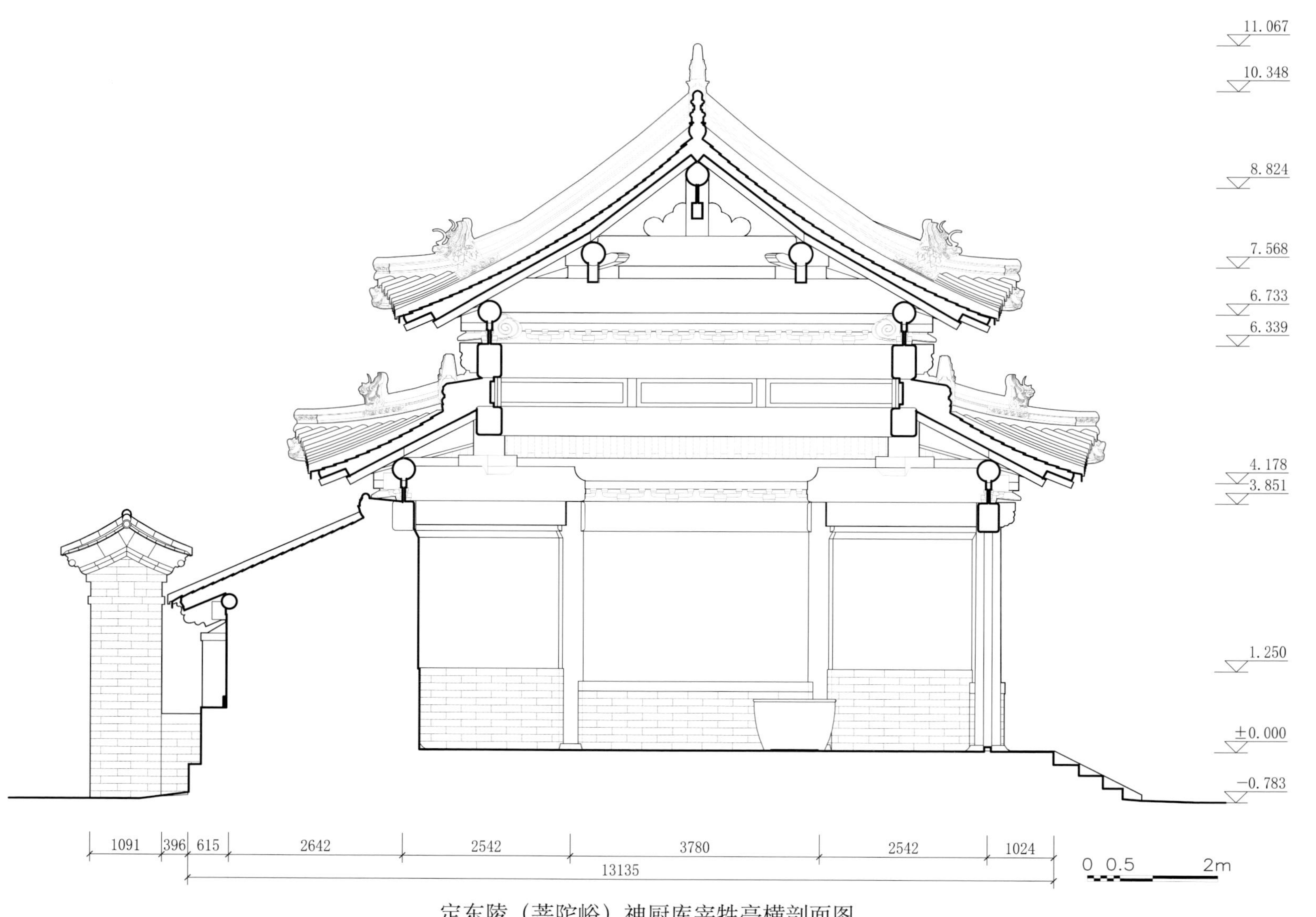

定东陵（菩陀峪）神厨库宰牲亭横剖面图
Cross section of the Ritual Abattoir in the Culinary Courtyard for Sacrifices, East Tomb of Stability, Putuo Valley.

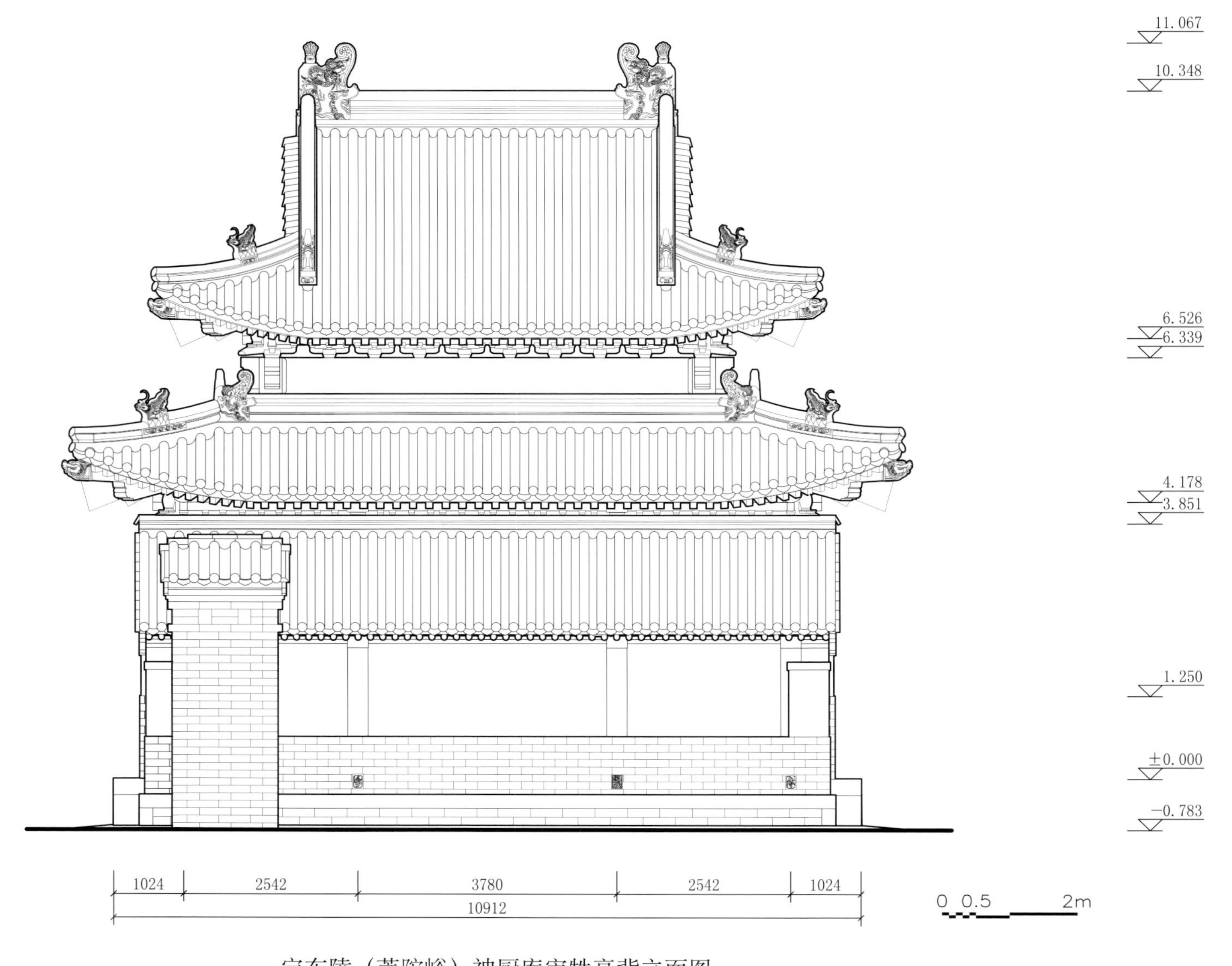

定东陵（菩陀峪）神厨库宰牲亭背立面图
Back elevation of the Ritual Abattoir in the Culinary Courtyard for Sacrifices, East Tomb of Stability, Putuo Valley

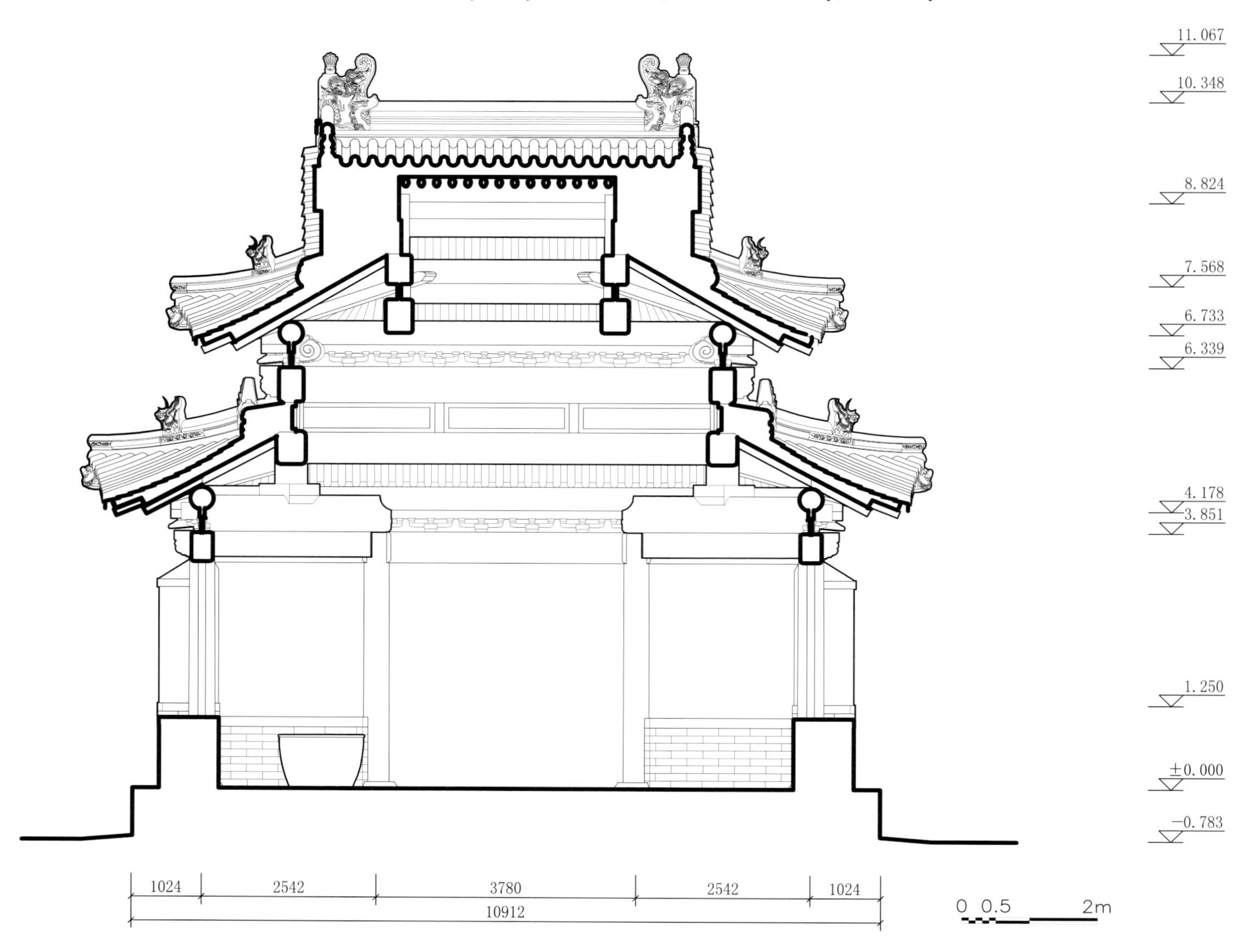

定东陵（菩陀峪）神厨库宰牲亭纵剖面图
Longitudinal section of the Ritual Abattoir in the Culinary Courtyard for Sacrifices, East Tomb of Stability, Putuo Valley

定东陵（菩陀峪）东朝房正立面图
Front elevation of the Eastern Reception Hall for Court Officials, East Tomb of Stability, Putuo Valley

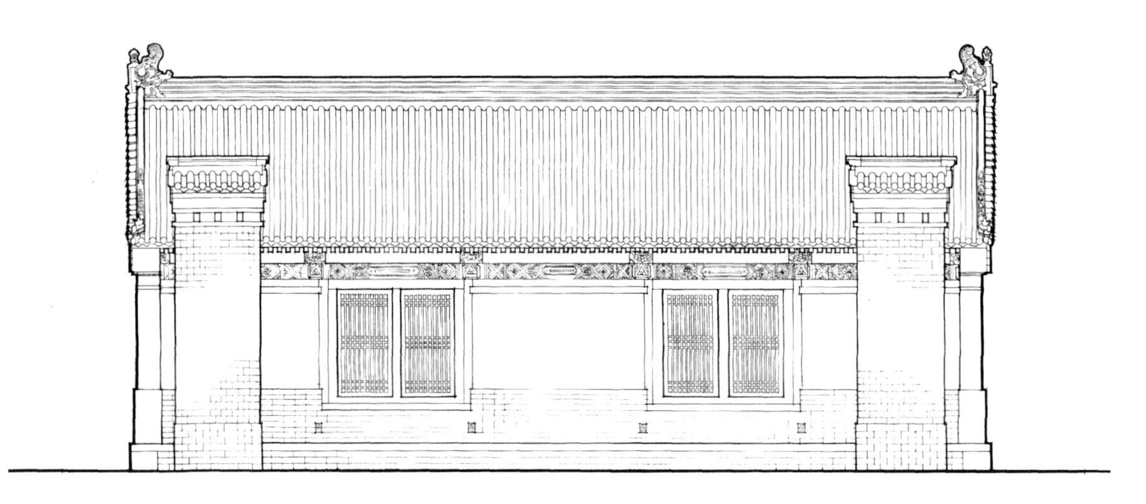

定东陵（菩陀峪）东朝房背立面图
Back elevation of the Reception Hall for Court Officials, East Tomb of Stability, Putuo Valley

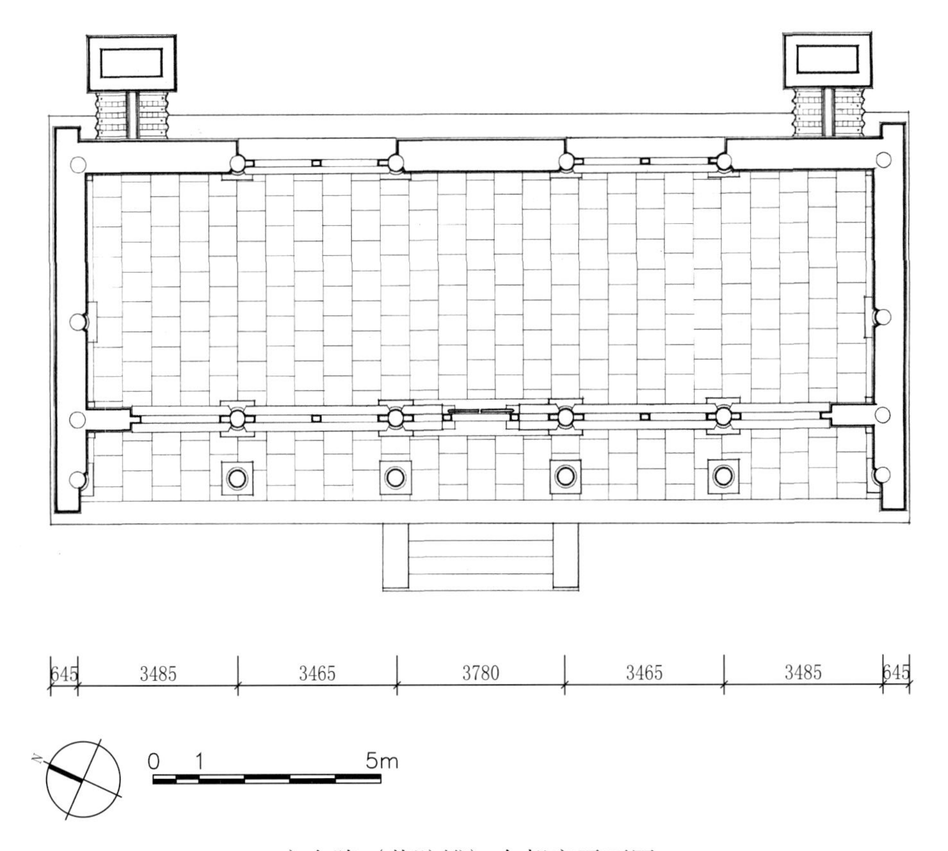

定东陵（菩陀峪）东朝房平面图
Plan of the Eastern Reception Hall for Court Officials, East Tomb of Stability, Putuo Valley

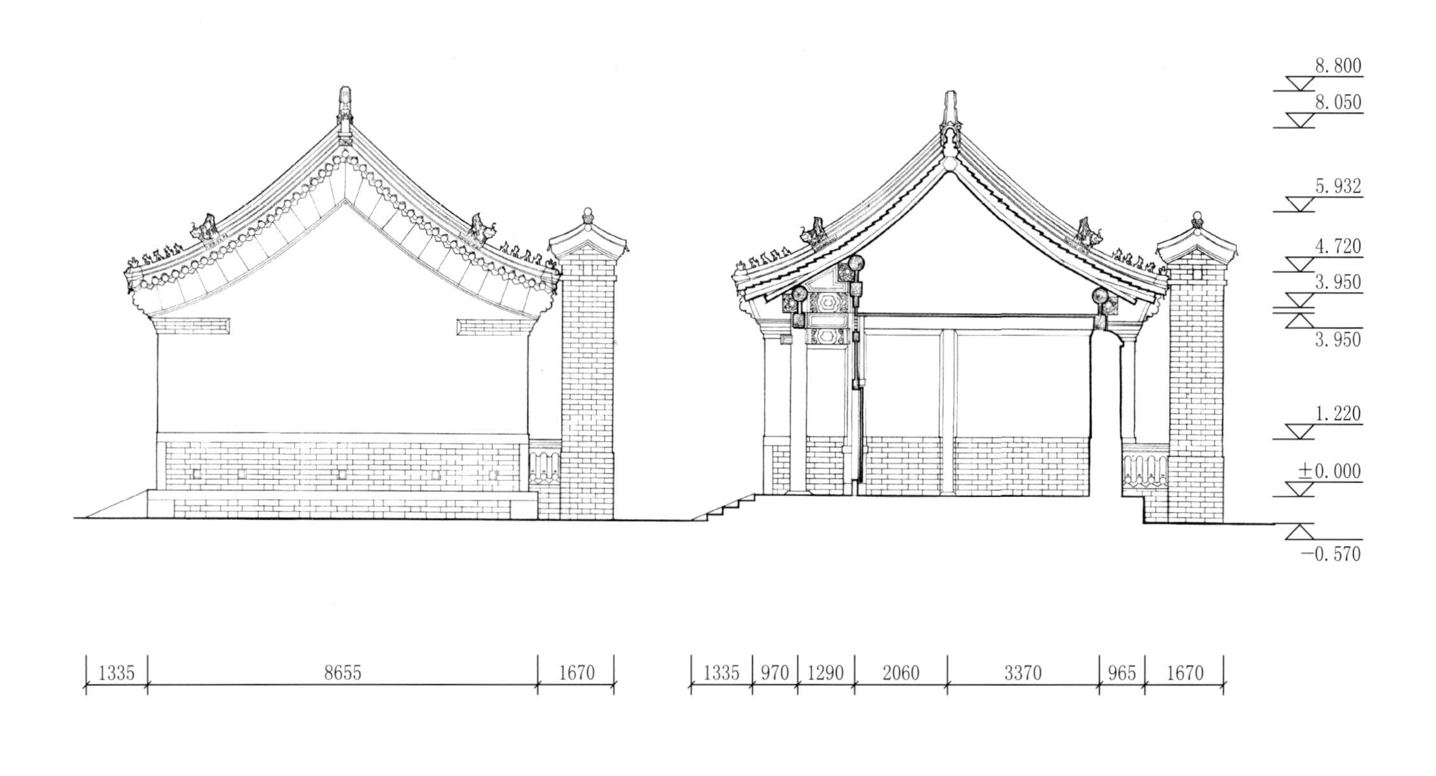

定东陵（菩陀峪）东朝房侧立面图
Side elevation of the Eastern Reception Hall for Court Officials, East Tomb of Stability, Putuo Valley

定东陵（菩陀峪）东朝房横剖面图
Cross section of the Eastern Reception Hall for Court Officials, East Tomb of Stability, Putuo Valley

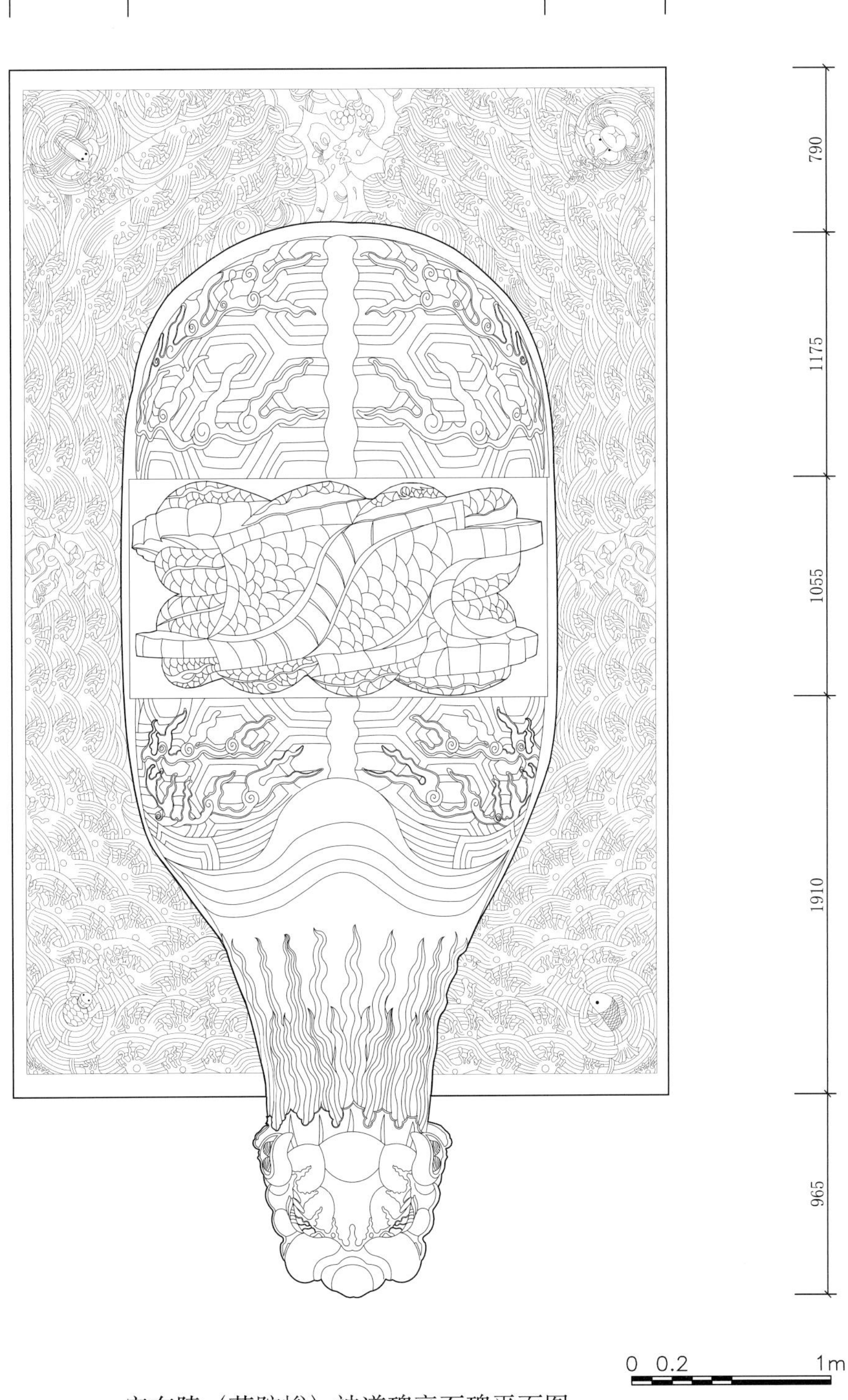

定东陵（菩陀峪）神道碑亭石碑平面图

Plan of the Stela in the Pavilion on the Spirit Way, East Tomb of Stability, Putuo Valley

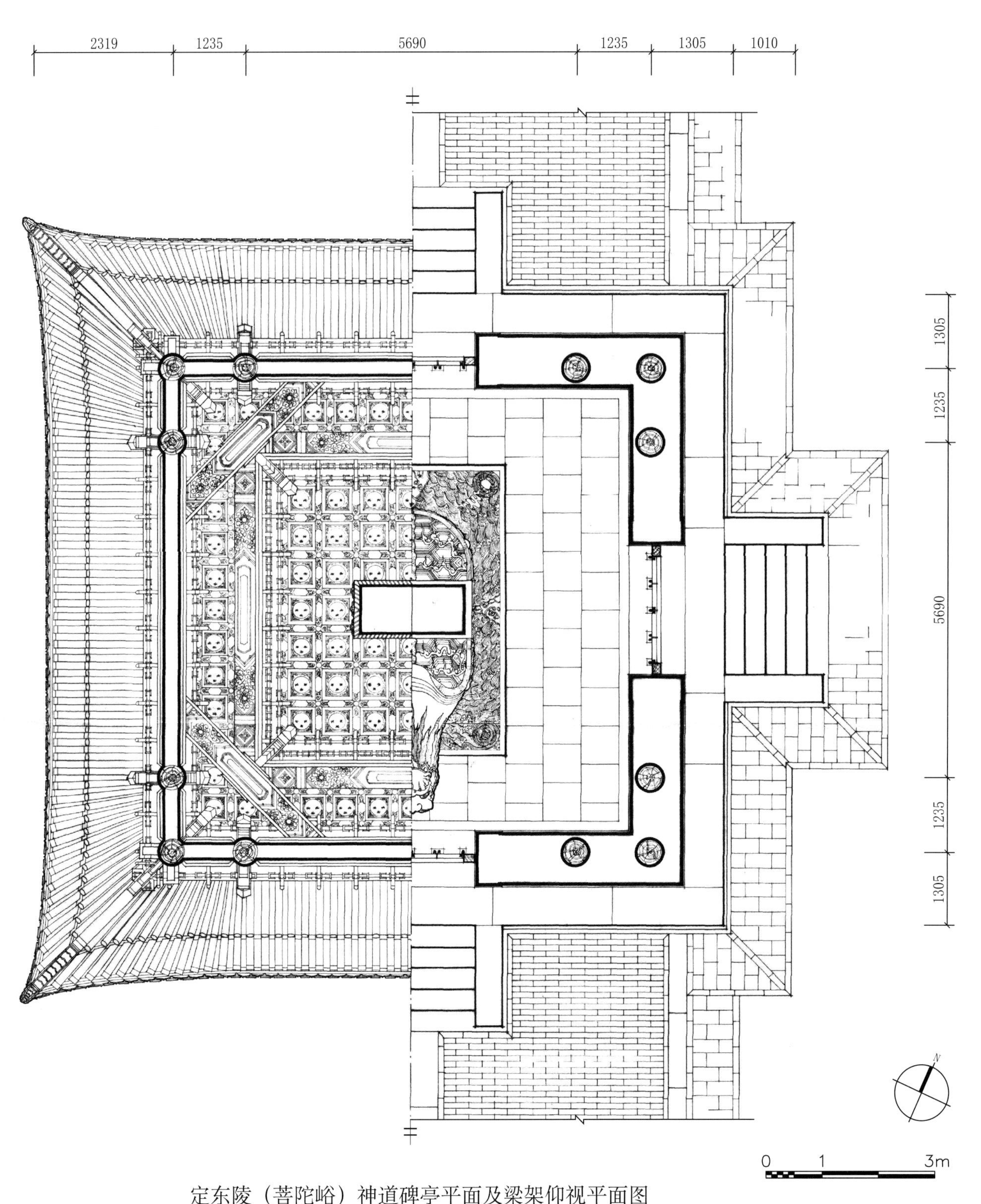

定东陵（菩陀峪）神道碑亭平面及梁架仰视平面图

Ground plan and looking up at the Truss and Ceiling of the Pavilion for the Stela on the Spirit Way, East Tomb of Stability, Putuo Valley

定东陵（菩陀峪）神道碑亭石碑正立面图
Front elevation of the Stela in the Pavilion on the Spirit Way, East Tomb of Stability, Putuo Valley

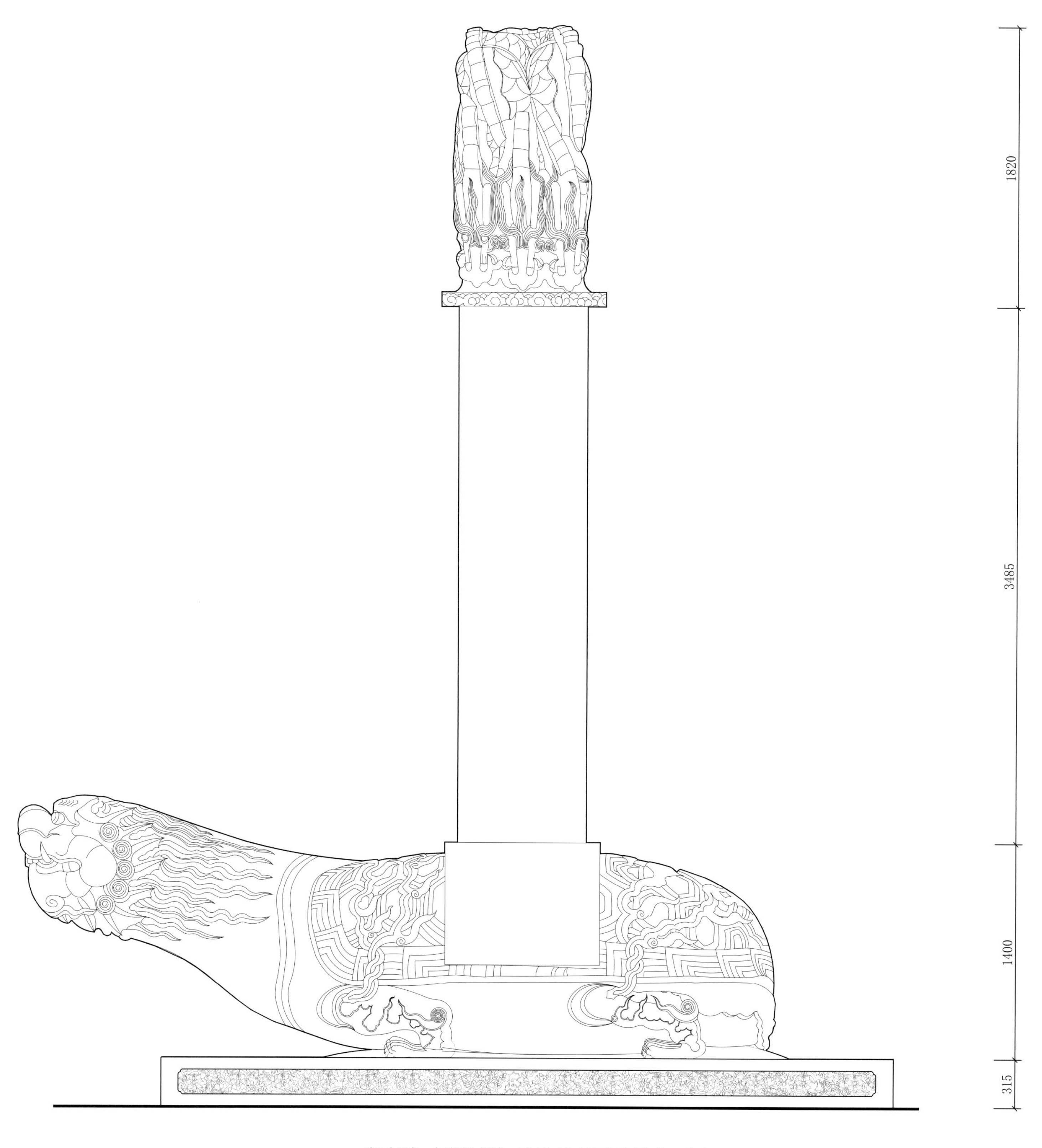

定东陵（菩陀峪）神道碑亭石碑侧立面图
Side elevation of the Stela in the Pavilion on the Spirit Way, East Tomb of Stability, Putuo Valley

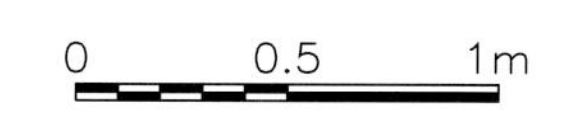

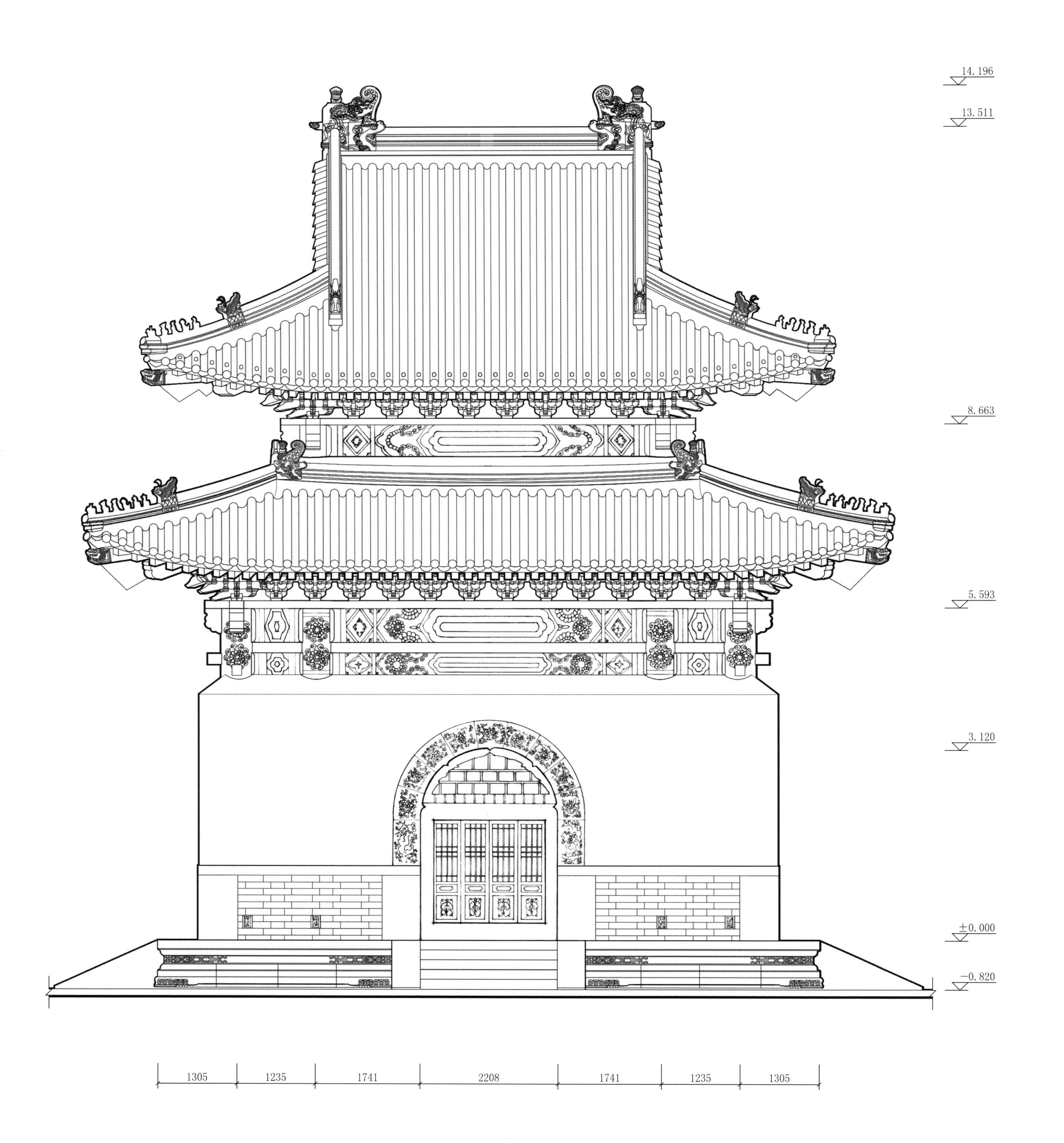

定东陵（菩陀峪）神道碑亭正立面图
Front elevation of the Pavilion for the Stela on the Spirit Way, East Tomb of Stability, Putuo Valley

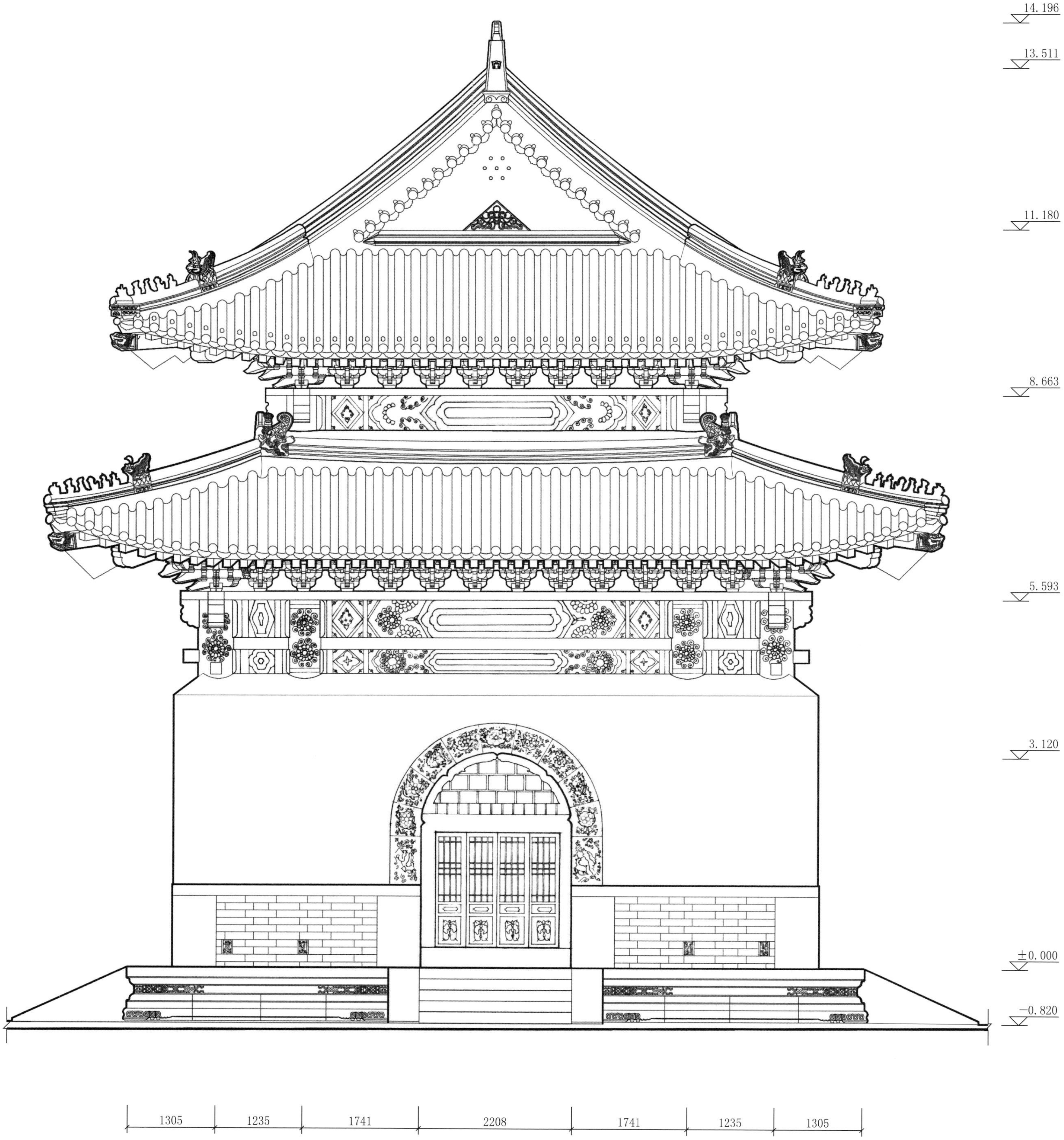

定东陵（菩陀峪）神道碑亭侧立面图
Side elevation of the Pavilion for the Stela on the Spirit Way, East Tomb of Stability, Putuo Valley

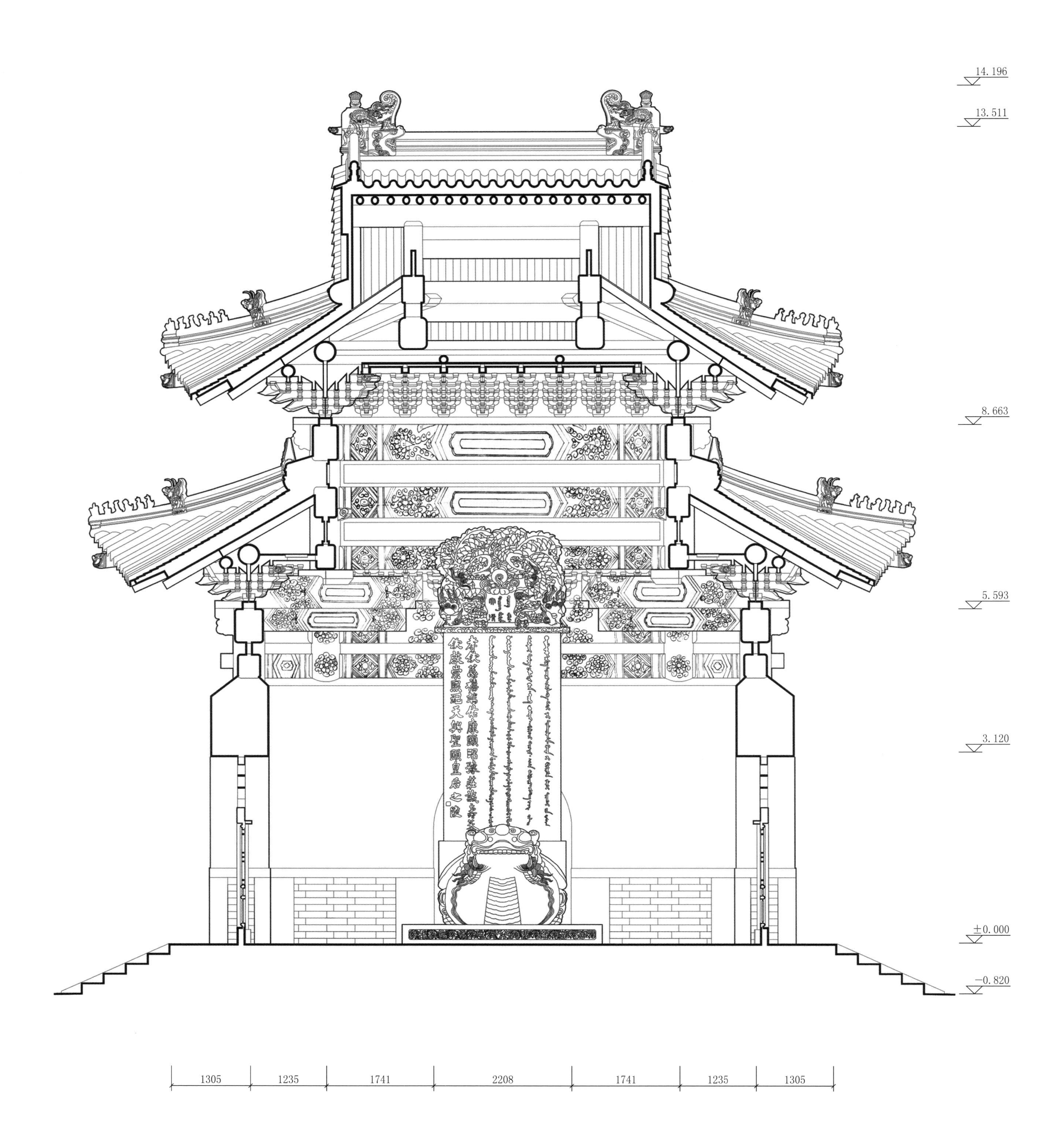

定东陵（菩陀峪）神道碑亭纵剖面图
Longitudinal section of the Pavilion for the Stela on the Spirit Way, East Tomb of Stability, Putuo Valley

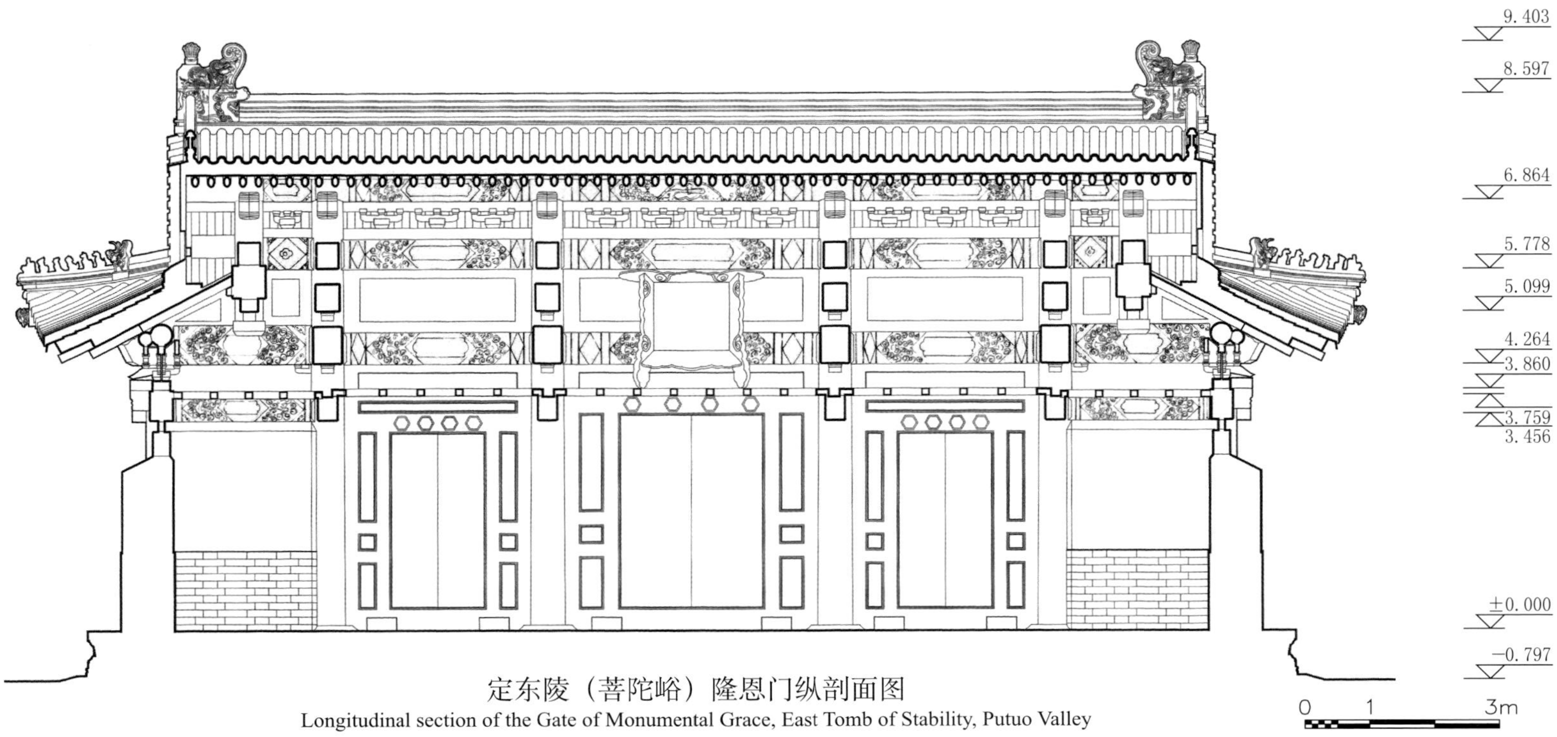

9.403
8.597
6.864
5.778
5.099
4.264
3.860
3.759
3.456
±0.000
−0.797
0 1 3m

定东陵（菩陀峪）隆恩门纵剖面图

Longitudinal section of the Gate of Monumental Grace, East Tomb of Stability, Putuo Valley

1680
1120
3300
3300
1120
1680
7825

2104
2600
3400
4400
3400
2600
1120
1000

N
0 1 3m

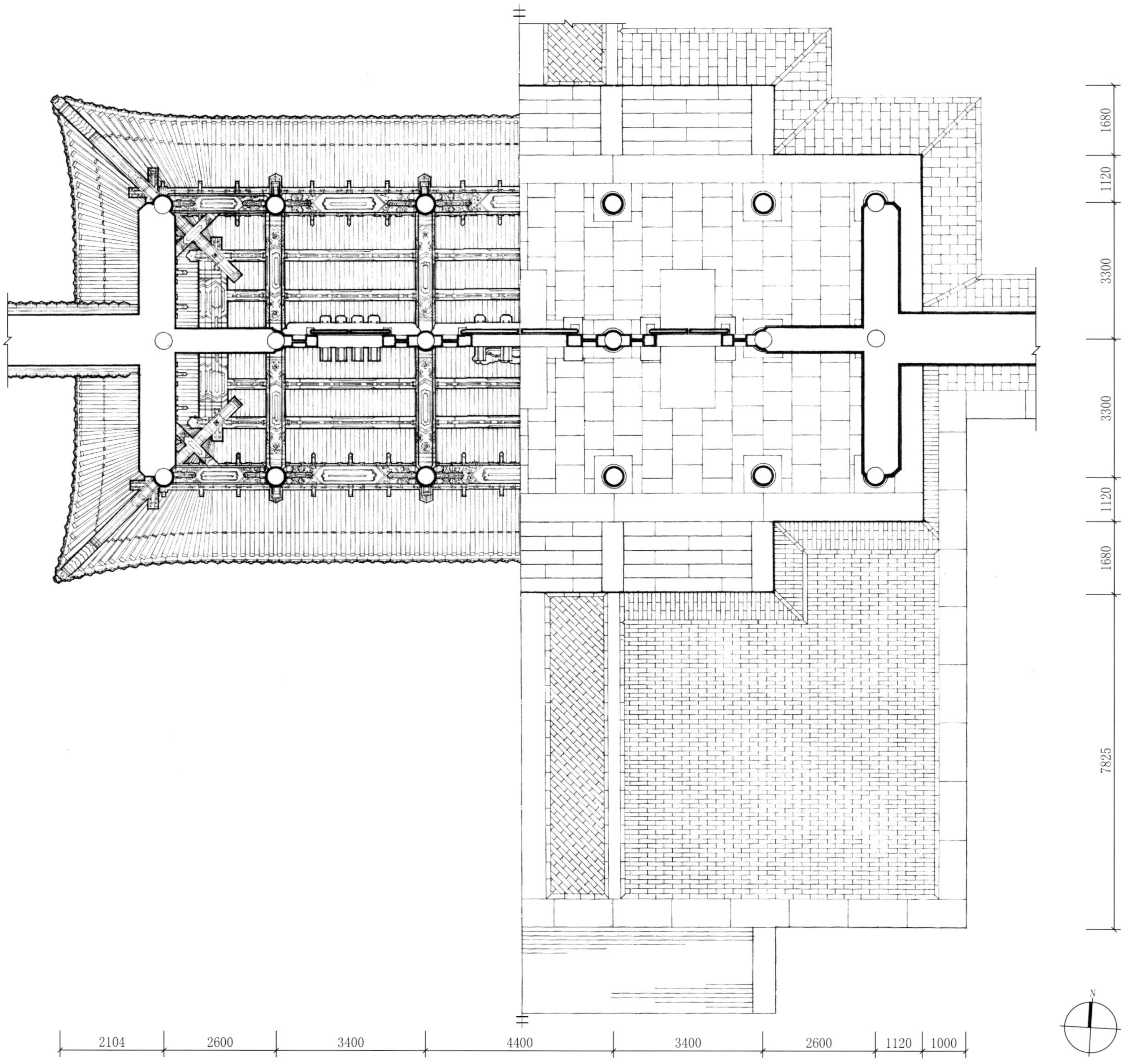

定东陵（菩陀峪）隆恩门平面及梁架仰视平面图

Ground plan and looking up at the Truss of the Gate of Monumental Grace, East Tomb of Stability, Putuo Valley

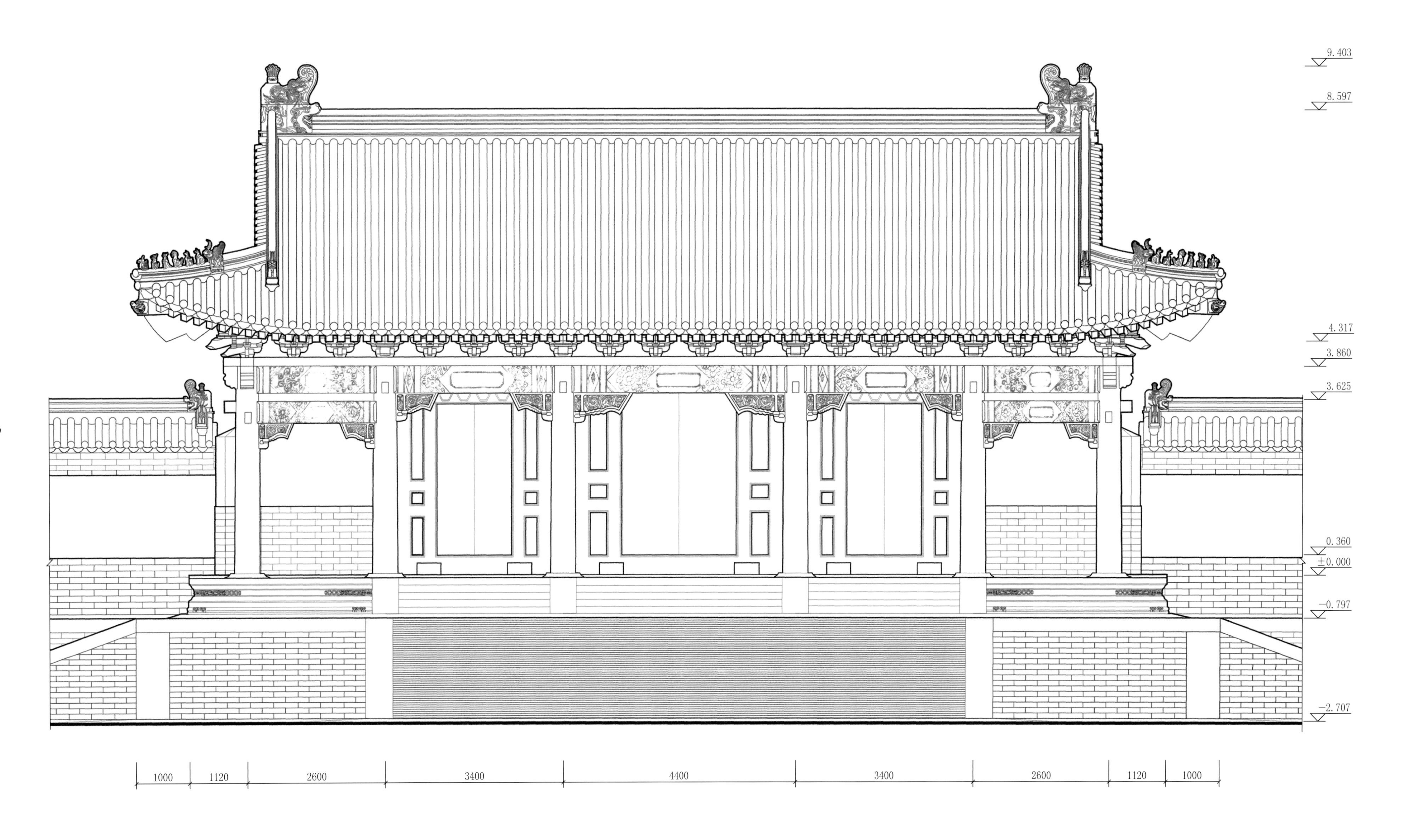

定东陵（菩陀峪）隆恩门正立面图

Front elevation of the Gate of Monumental Grace, East Tomb of Stability, Putuo Valley

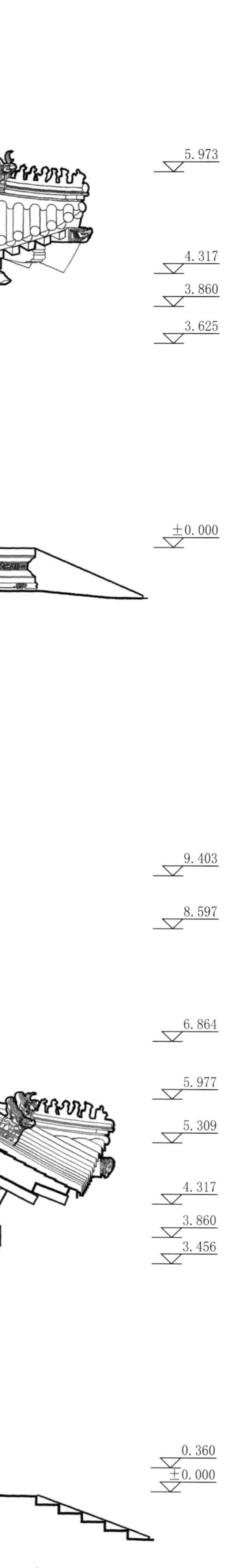

定东陵（菩陀峪）隆恩门侧立面图
Side elevation of the Gate of Monumental Grace, East Tomb of Stability, Putuo Valley

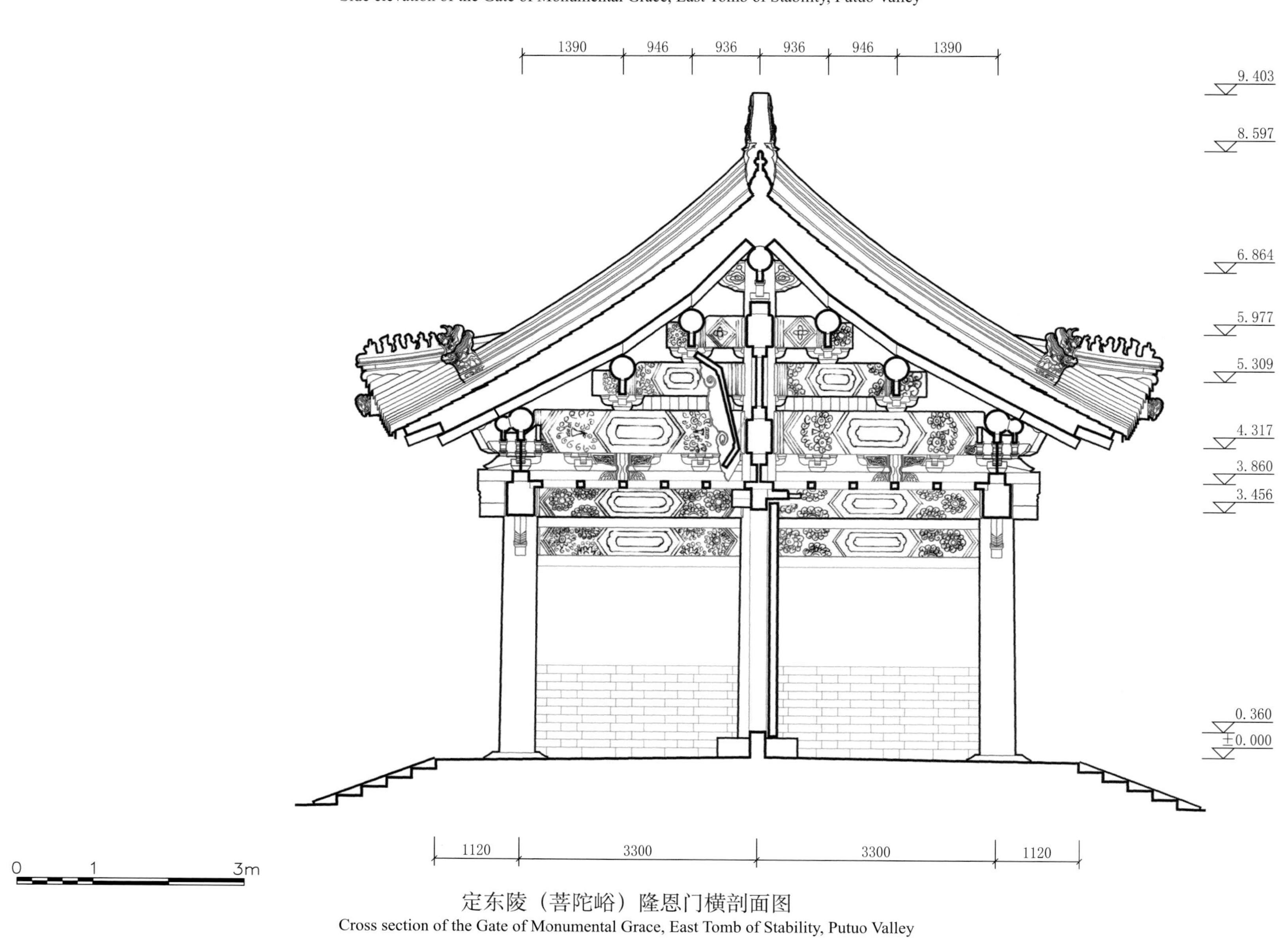

定东陵（菩陀峪）隆恩门横剖面图
Cross section of the Gate of Monumental Grace, East Tomb of Stability, Putuo Valley

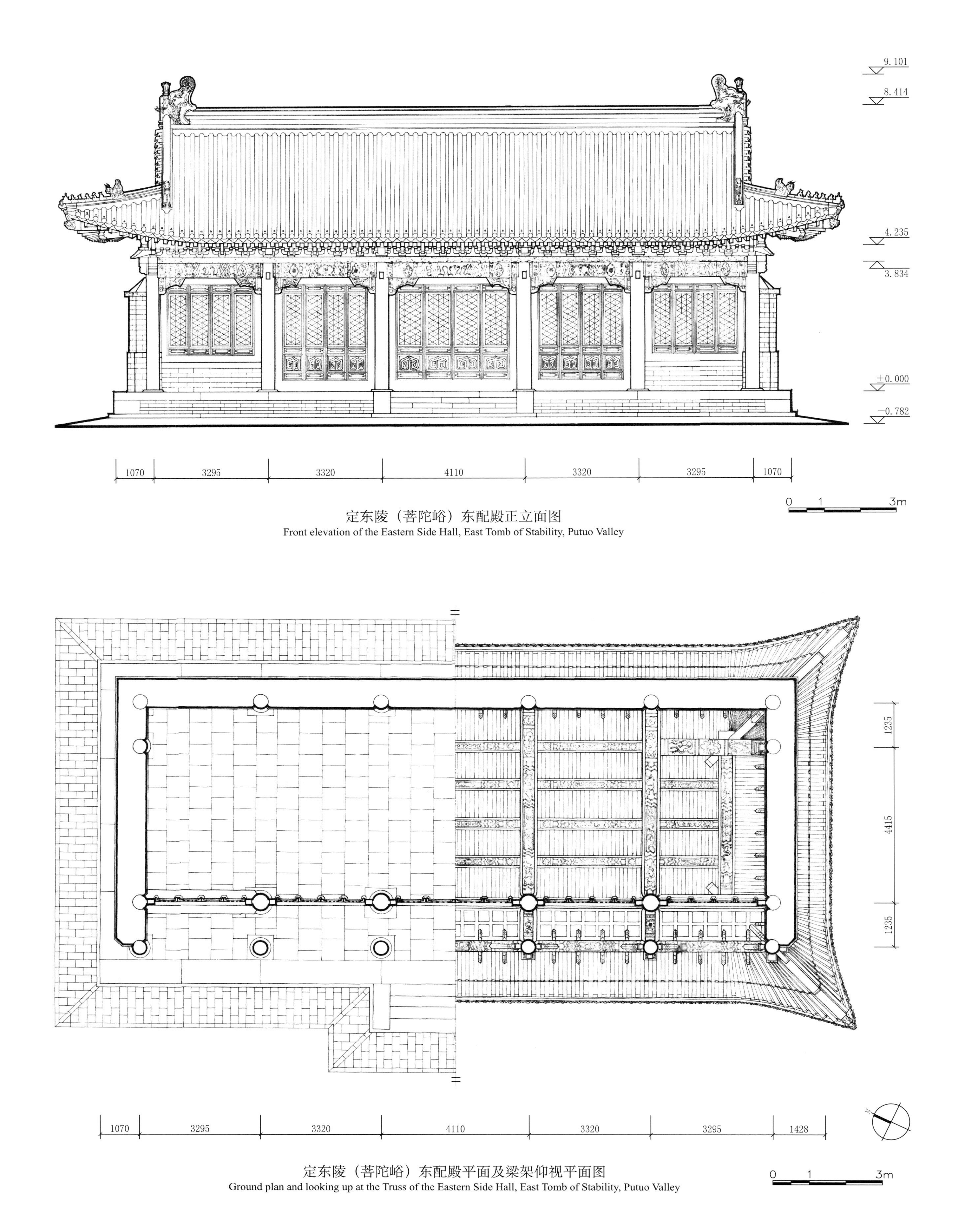

定东陵（菩陀峪）东配殿正立面图
Front elevation of the Eastern Side Hall, East Tomb of Stability, Putuo Valley

定东陵（菩陀峪）东配殿平面及梁架仰视平面图
Ground plan and looking up at the Truss of the Eastern Side Hall, East Tomb of Stability, Putuo Valley

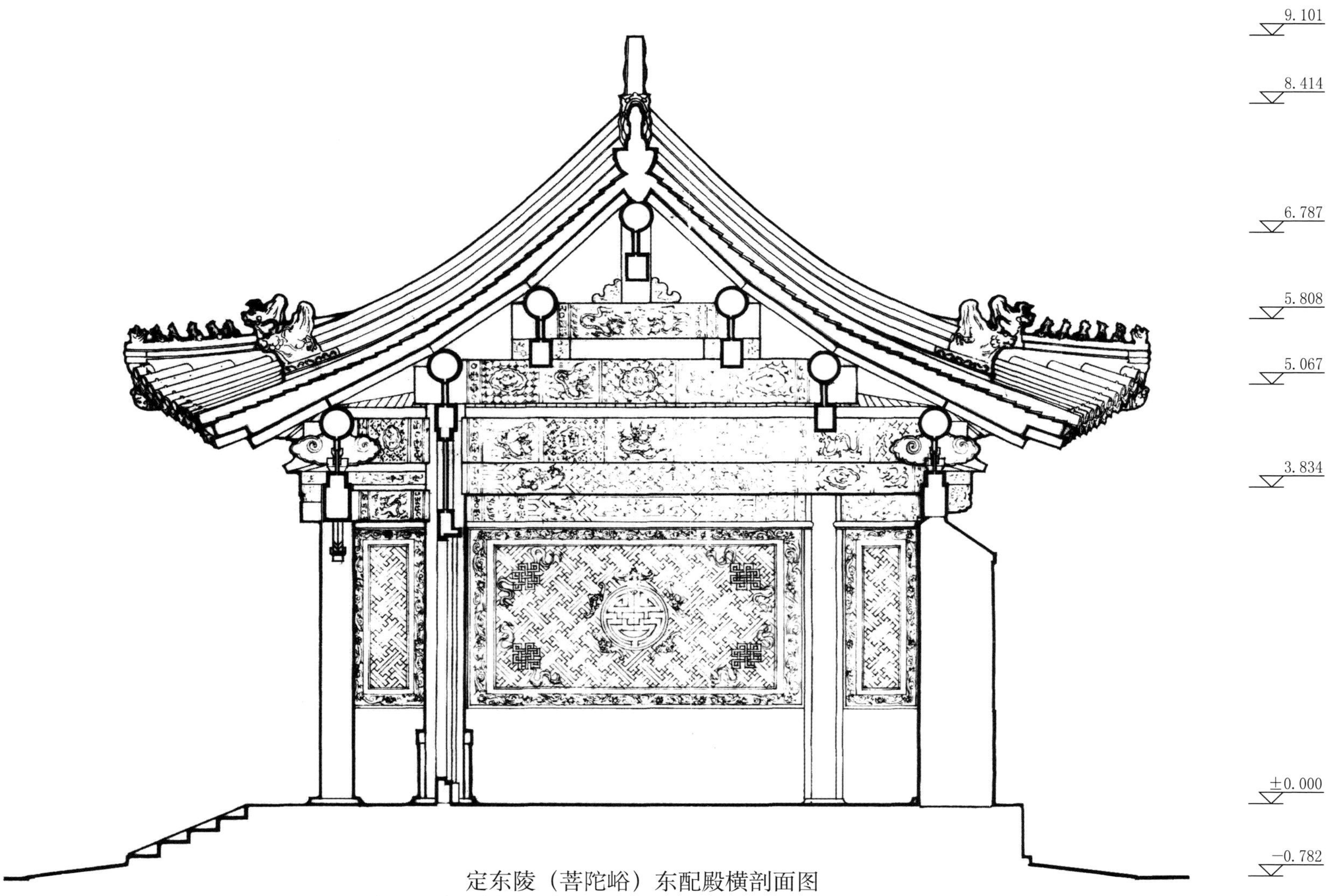

定东陵（菩陀峪）东配殿横剖面图
Cross section of the Eastern Side Hall, East Tomb of Stability, Putuo Valley

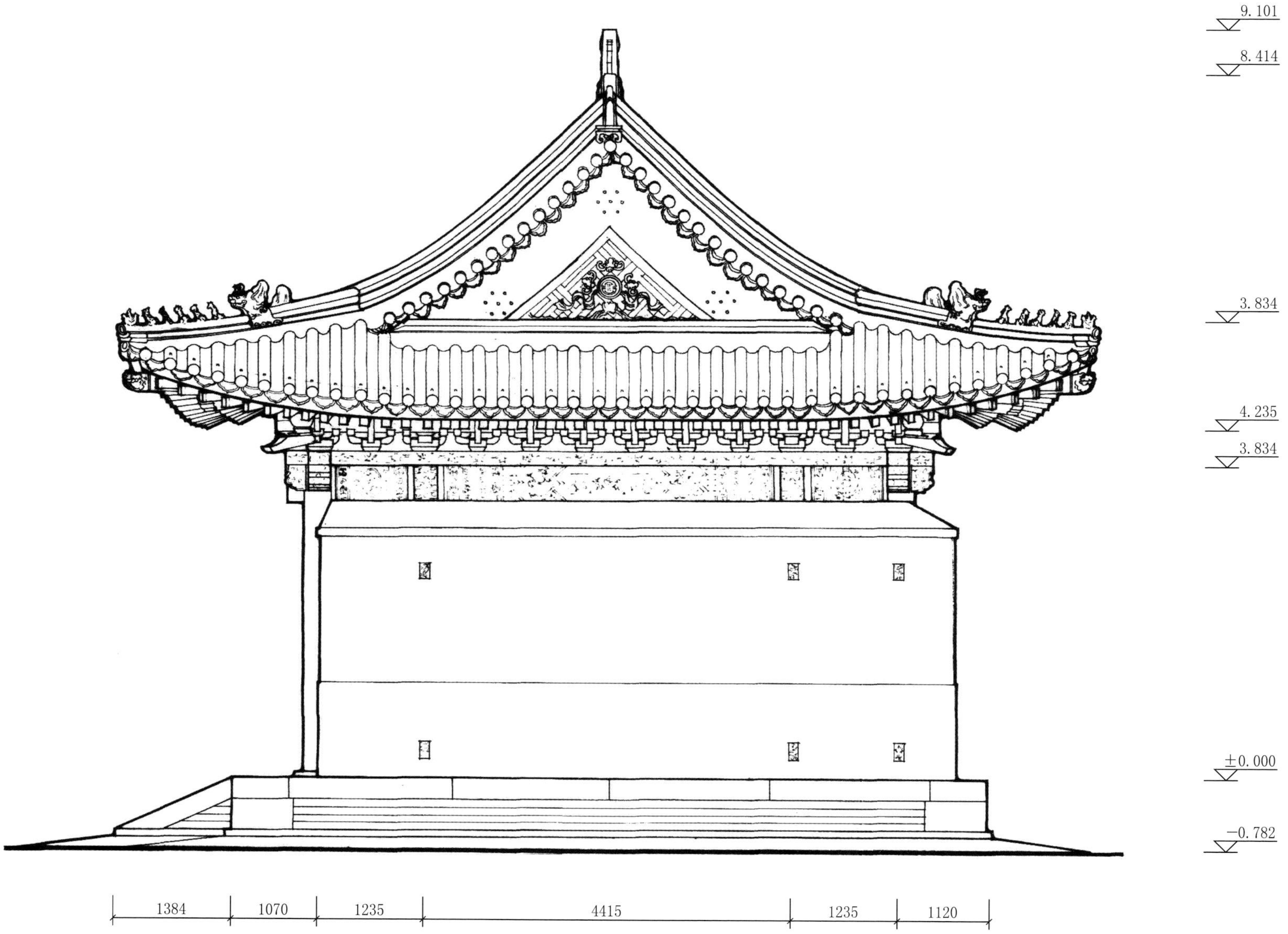

定东陵（菩陀峪）东配殿侧立面图
Side elevation of the Eastern Side Hall, East Tomb of Stability, Putuo Valley

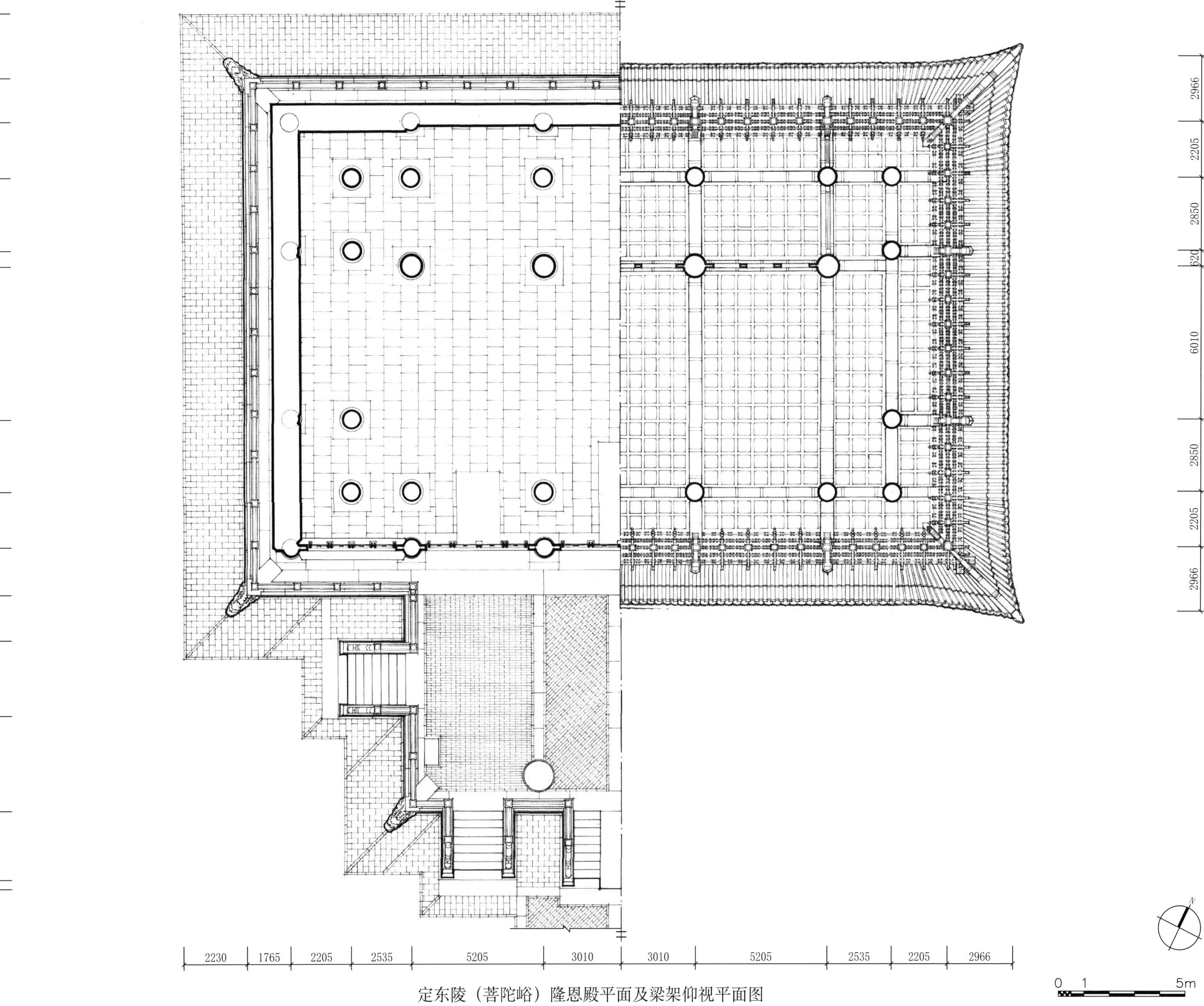

定东陵（菩陀峪）隆恩殿平面及梁架仰视平面图

Ground plan and looking up at the ceiling of the Hall of Monumental Grace, East Tomb of Stability, Putuo Valley

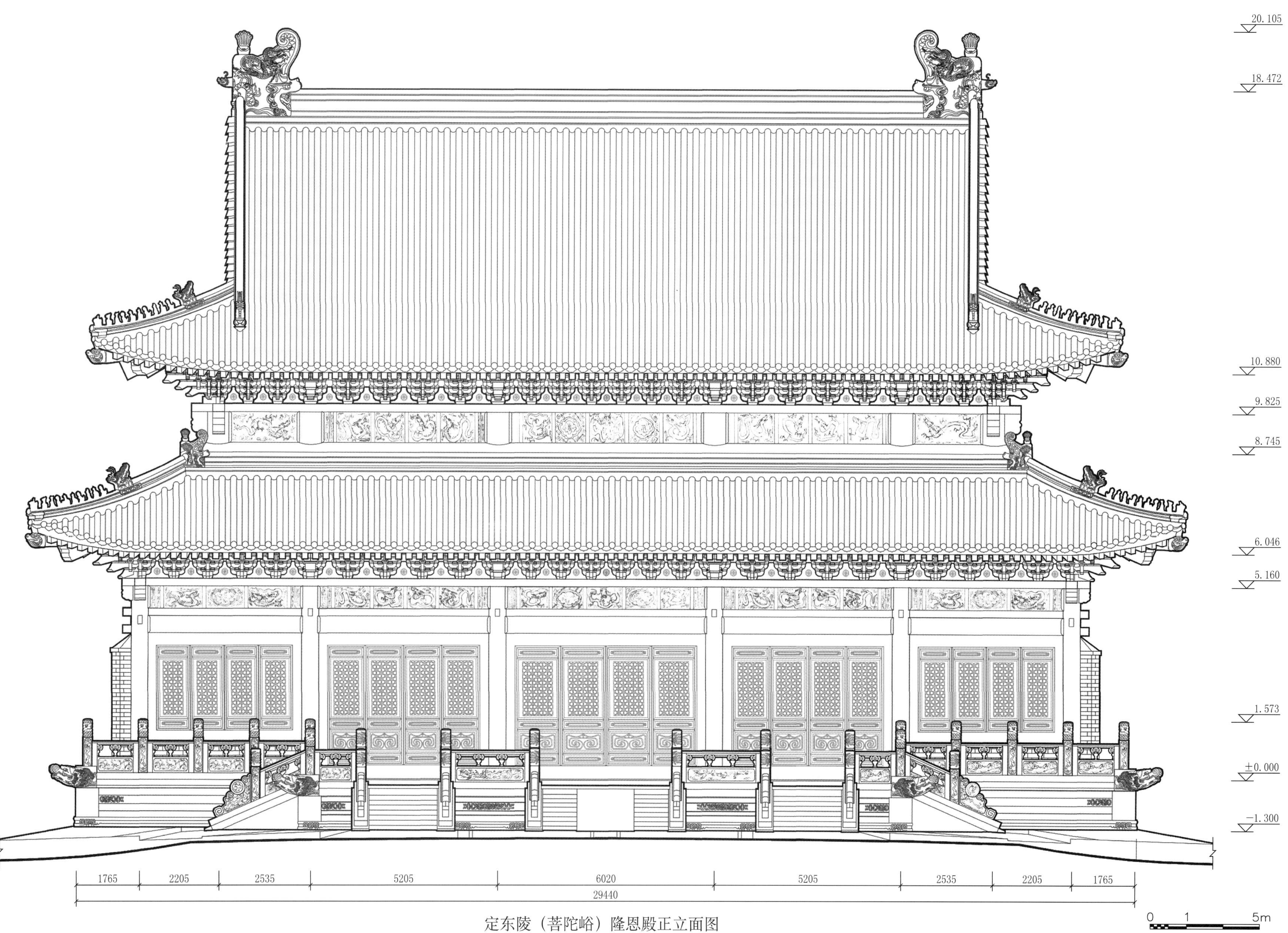

定东陵（菩陀峪）隆恩殿正立面图

Front elevation of the Hall of Monumental Grace, East Tomb of Stability, Putuo Valley

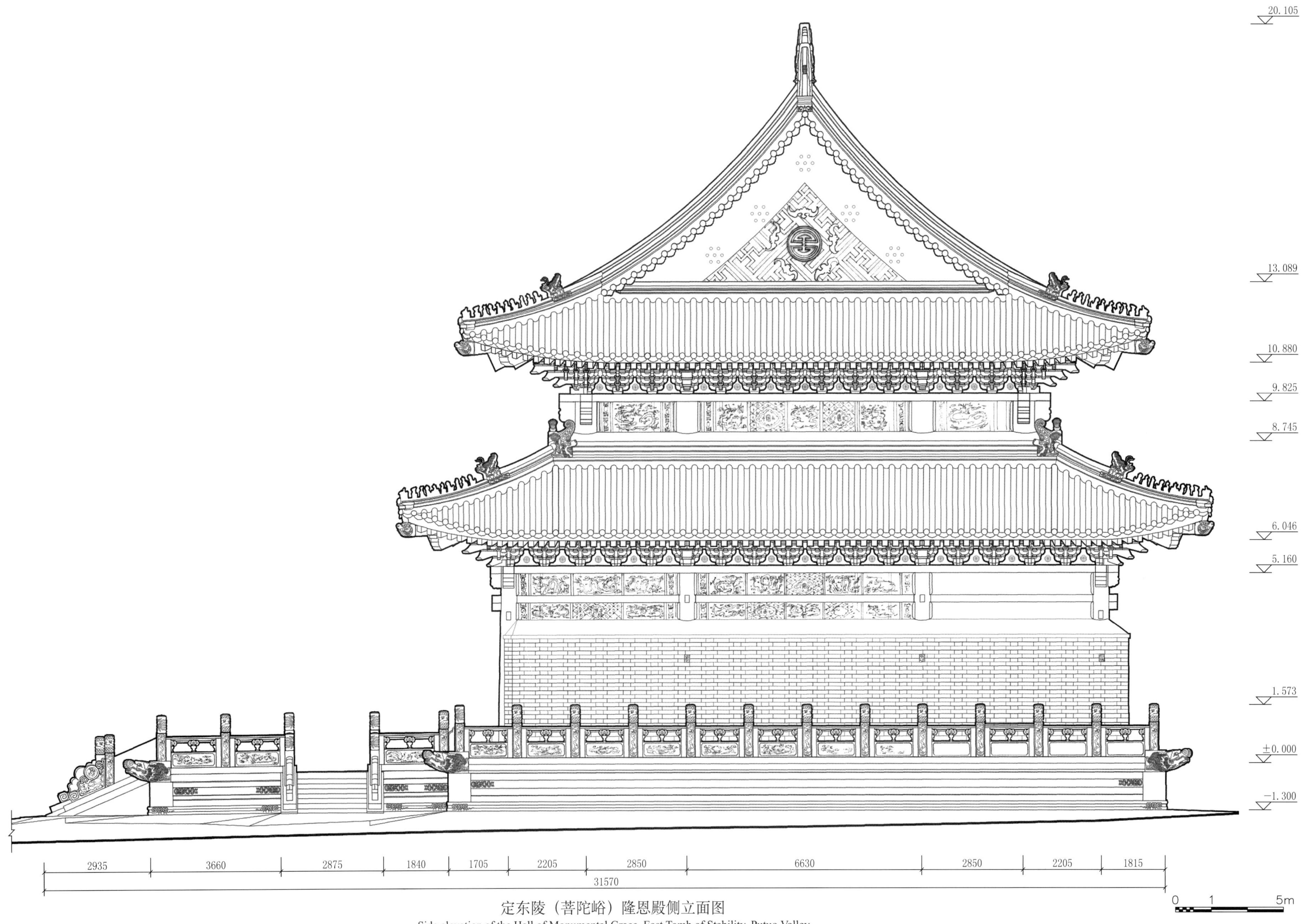

定东陵（菩陀峪）隆恩殿侧立面图

Side elevation of the Hall of Monumental Grace, East Tomb of Stability, Putuo Valley

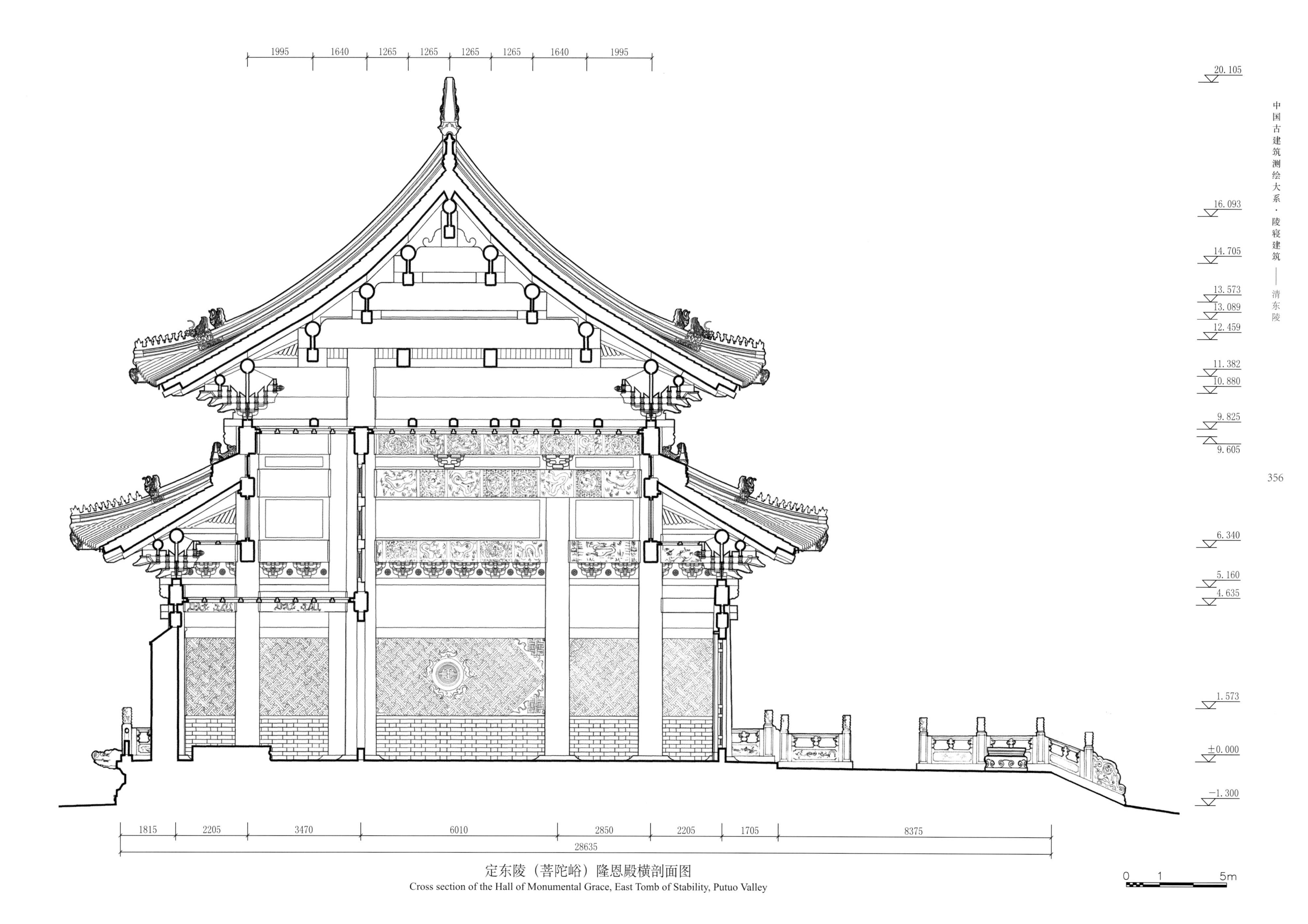

定东陵（菩陀峪）隆恩殿横剖面图

Cross section of the Hall of Monumental Grace, East Tomb of Stability, Putuo Valley

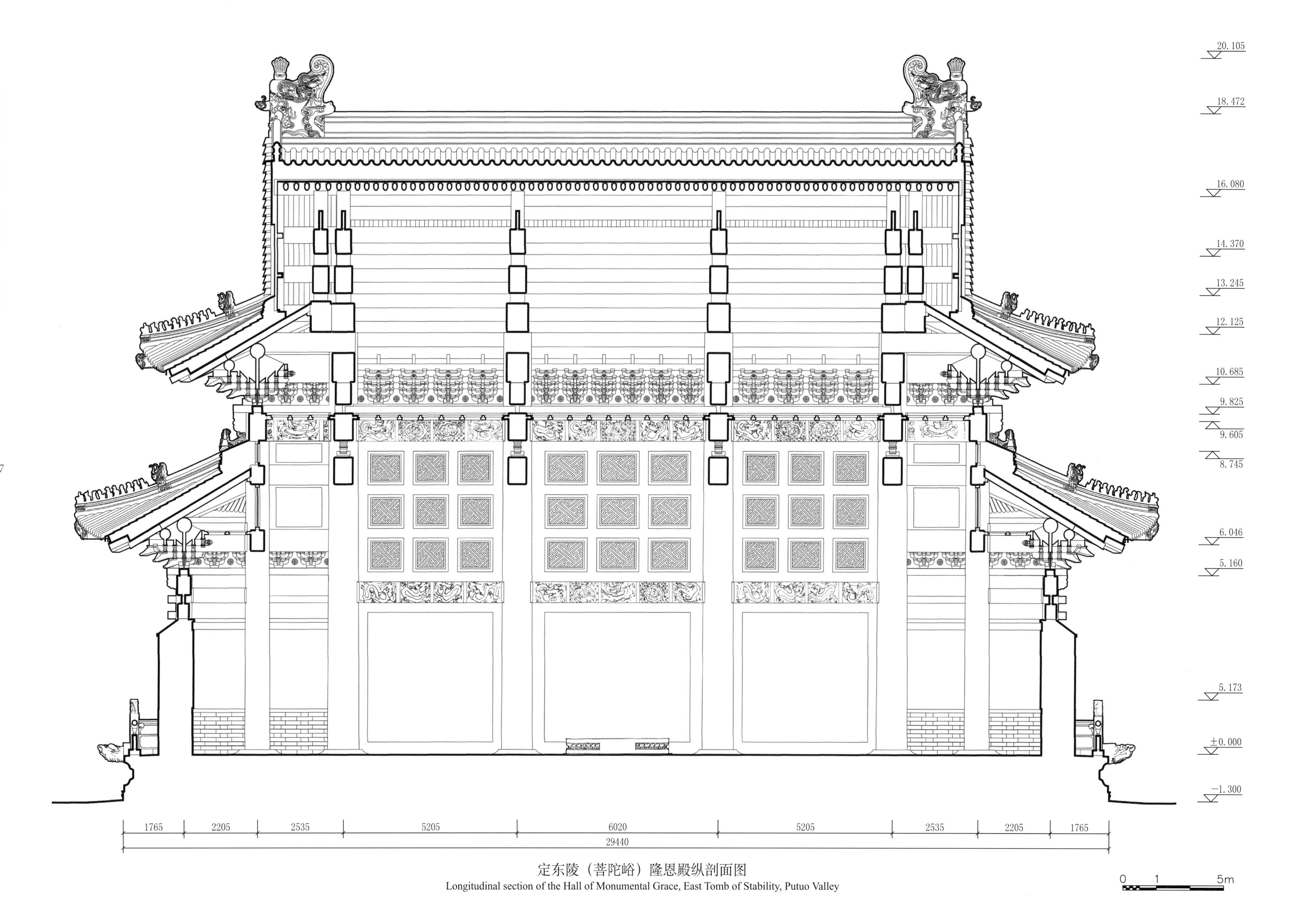

定东陵（菩陀峪）隆恩殿纵剖面图

Longitudinal section of the Hall of Monumental Grace, East Tomb of Stability, Putuo Valley

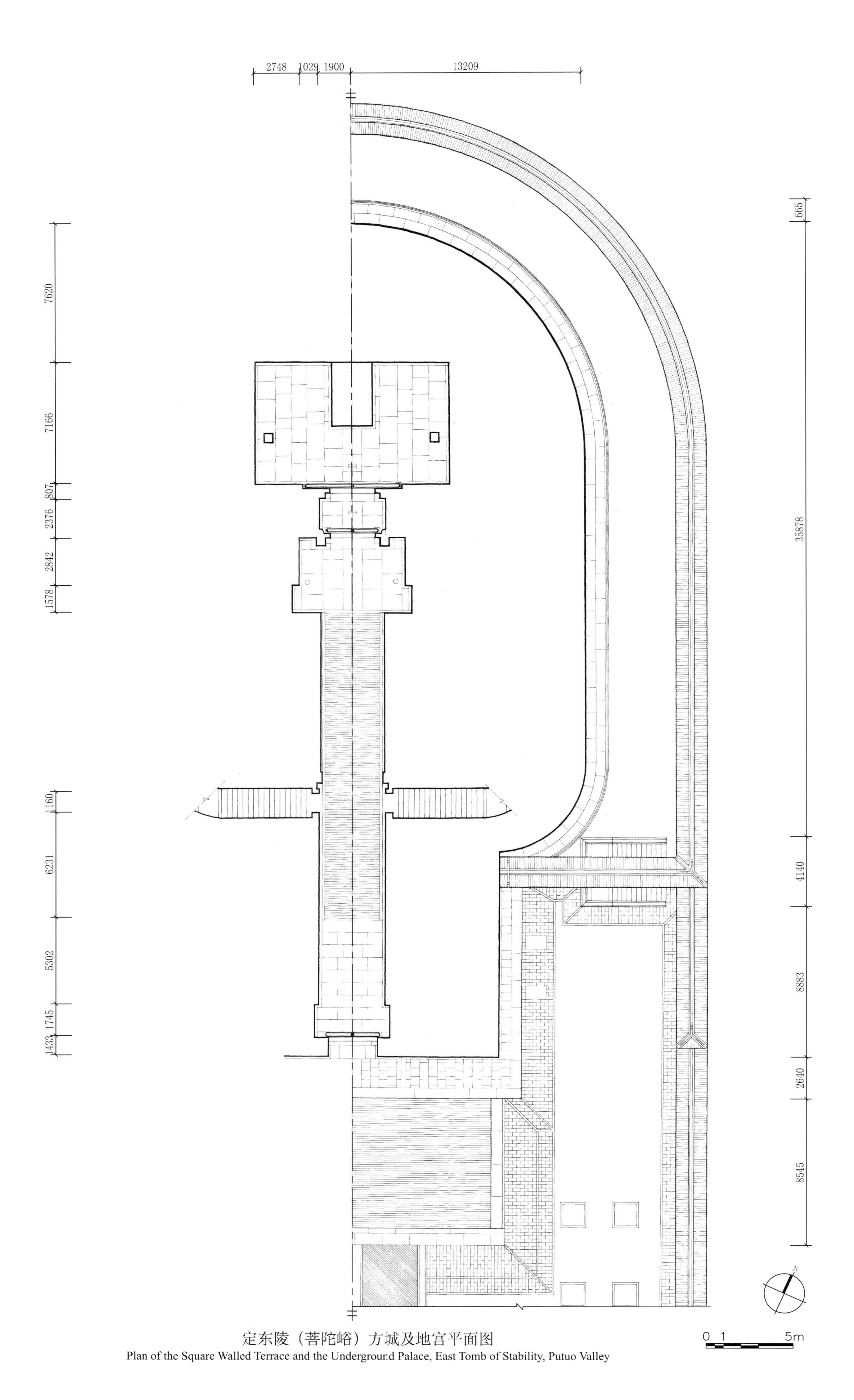

定东陵（菩陀峪）方城及地宫平面图
Plan of the Square Walled Terrace and the Undergrourd Palace, East Tomb of Stability, Putuo Valley

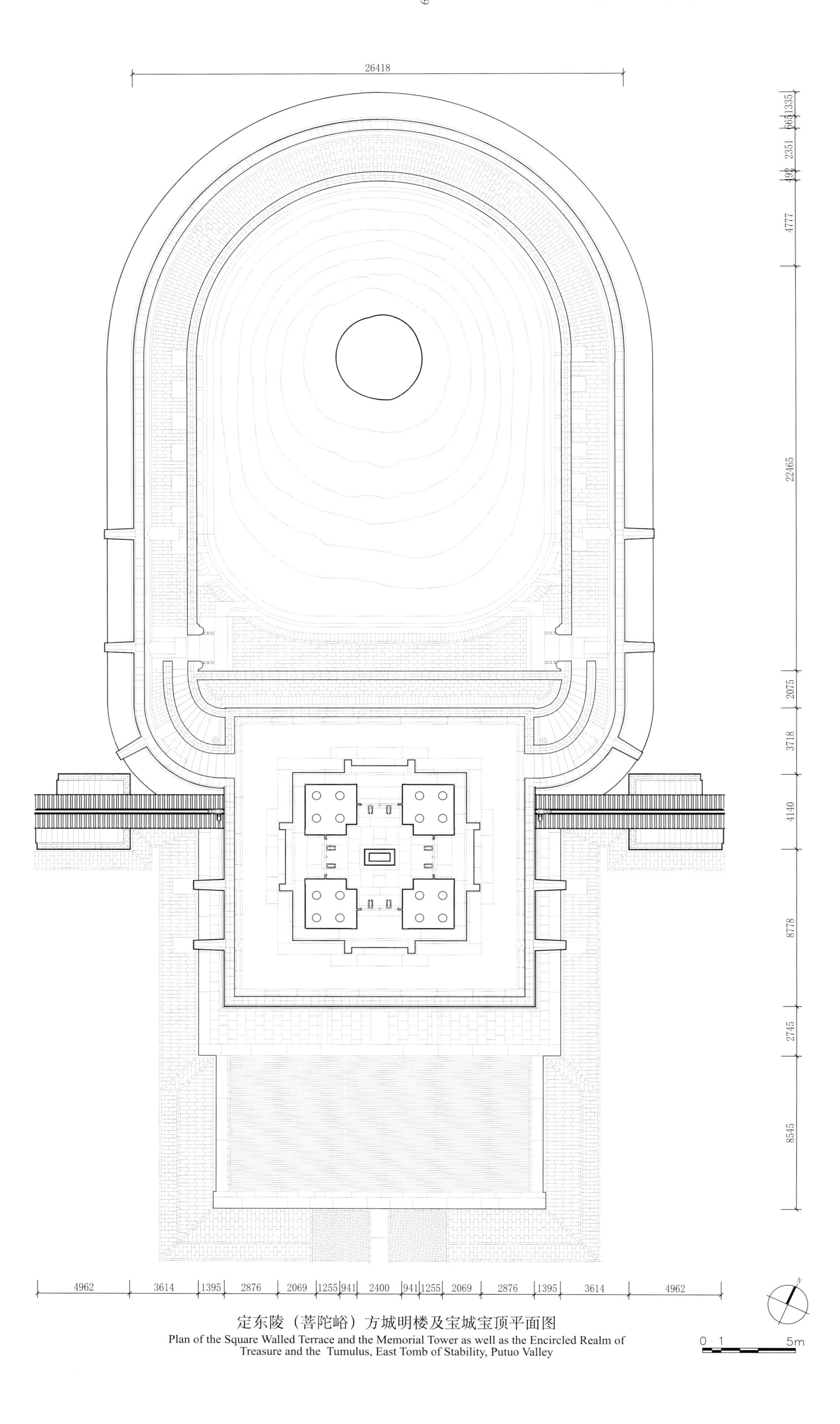

定东陵（菩陀峪）方城明楼及宝城宝顶平面图
Plan of the Square Walled Terrace and the Memorial Tower as well as the Encircled Realm of Treasure and the Tumulus, East Tomb of Stability, Putuo Valley

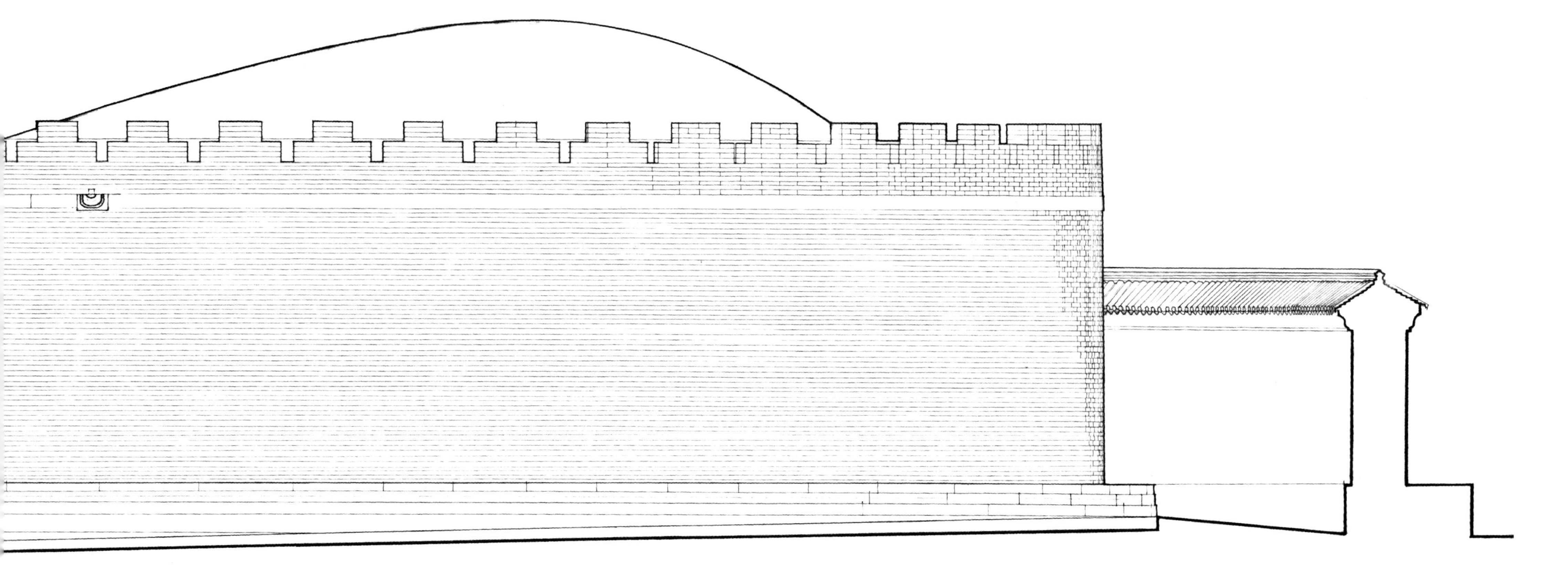

0 1 3m

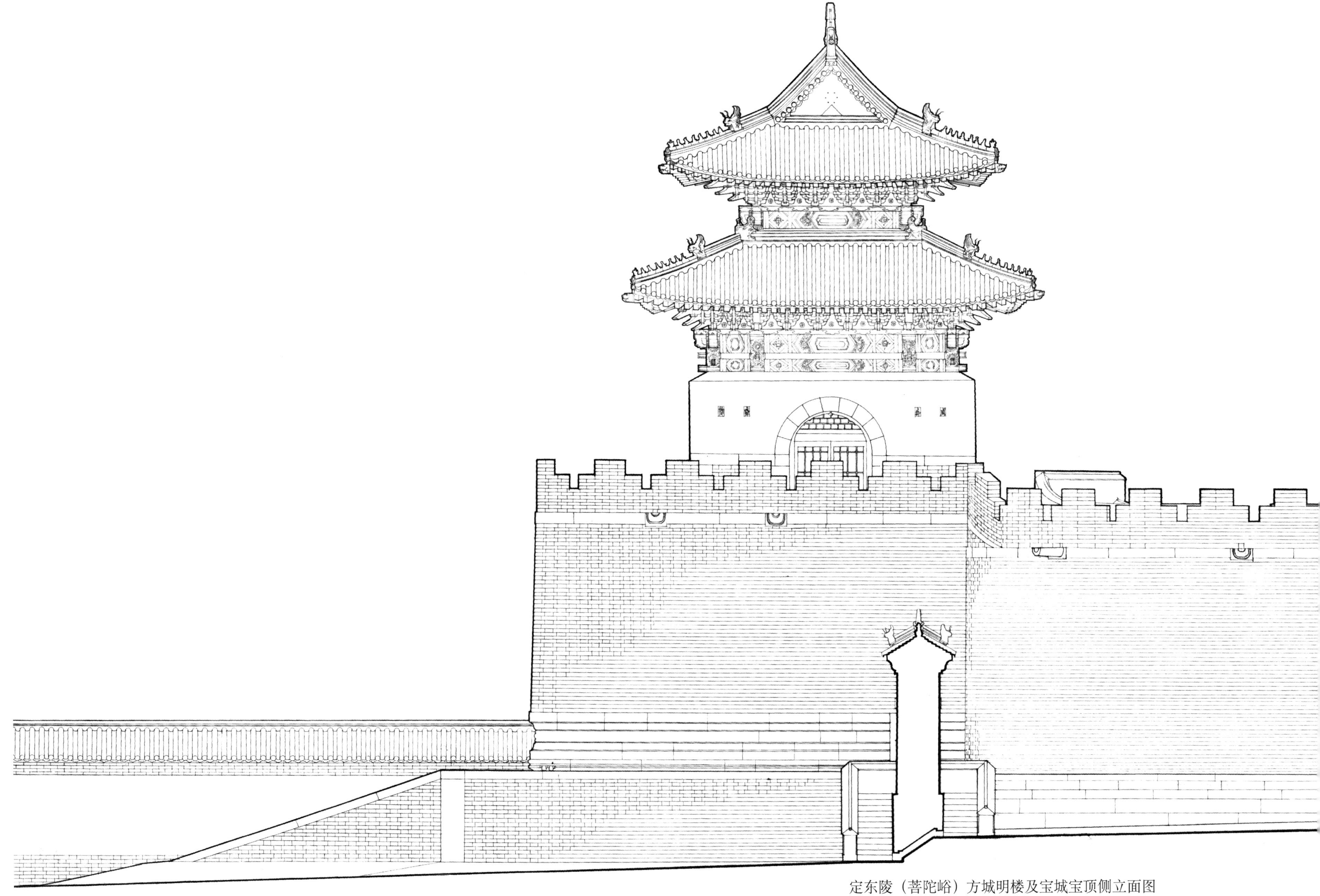

定东陵（菩陀峪）方城明楼及宝城宝顶侧立面图

Side elevation of the Square Walled Terrace and the Memorial Tower as well as the Encircled Realm of Treasure and the Tumulus, East Tomb of Stability, Putuo Valley

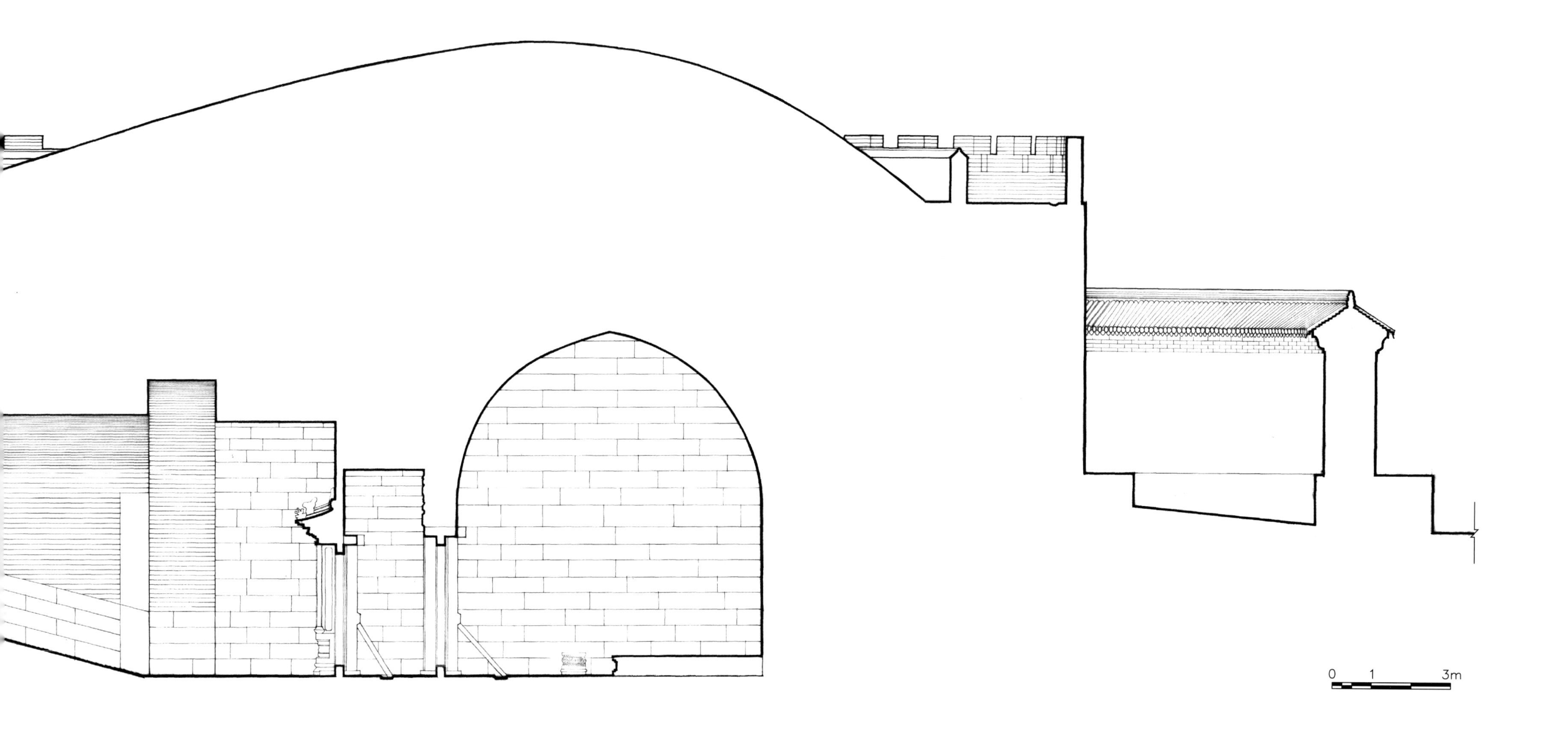
0
1
3m

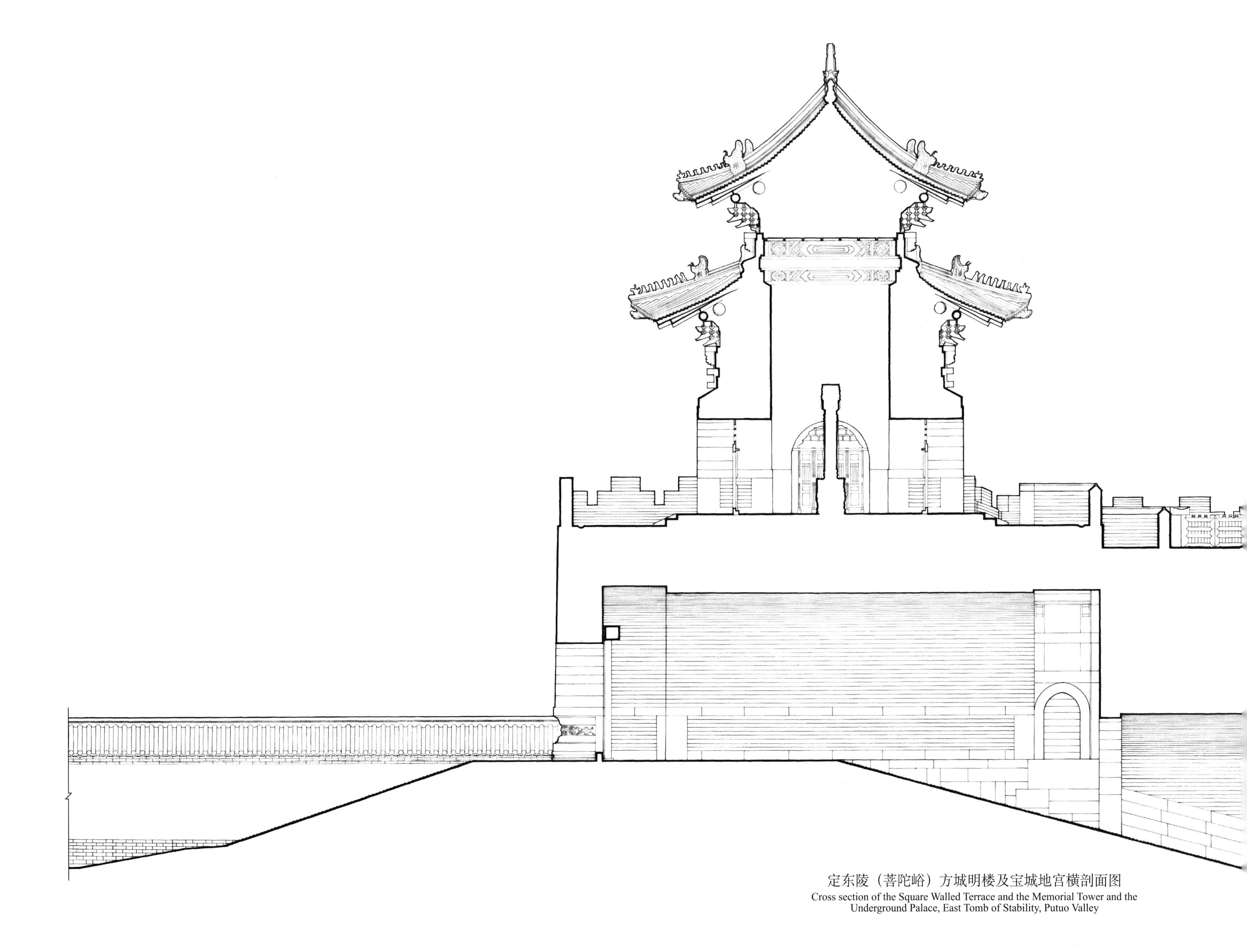

定东陵（菩陀峪）方城明楼及宝城地宫横剖面图

Cross section of the Square Walled Terrace and the Memorial Tower and the Underground Palace, East Tomb of Stability, Putuo Valley

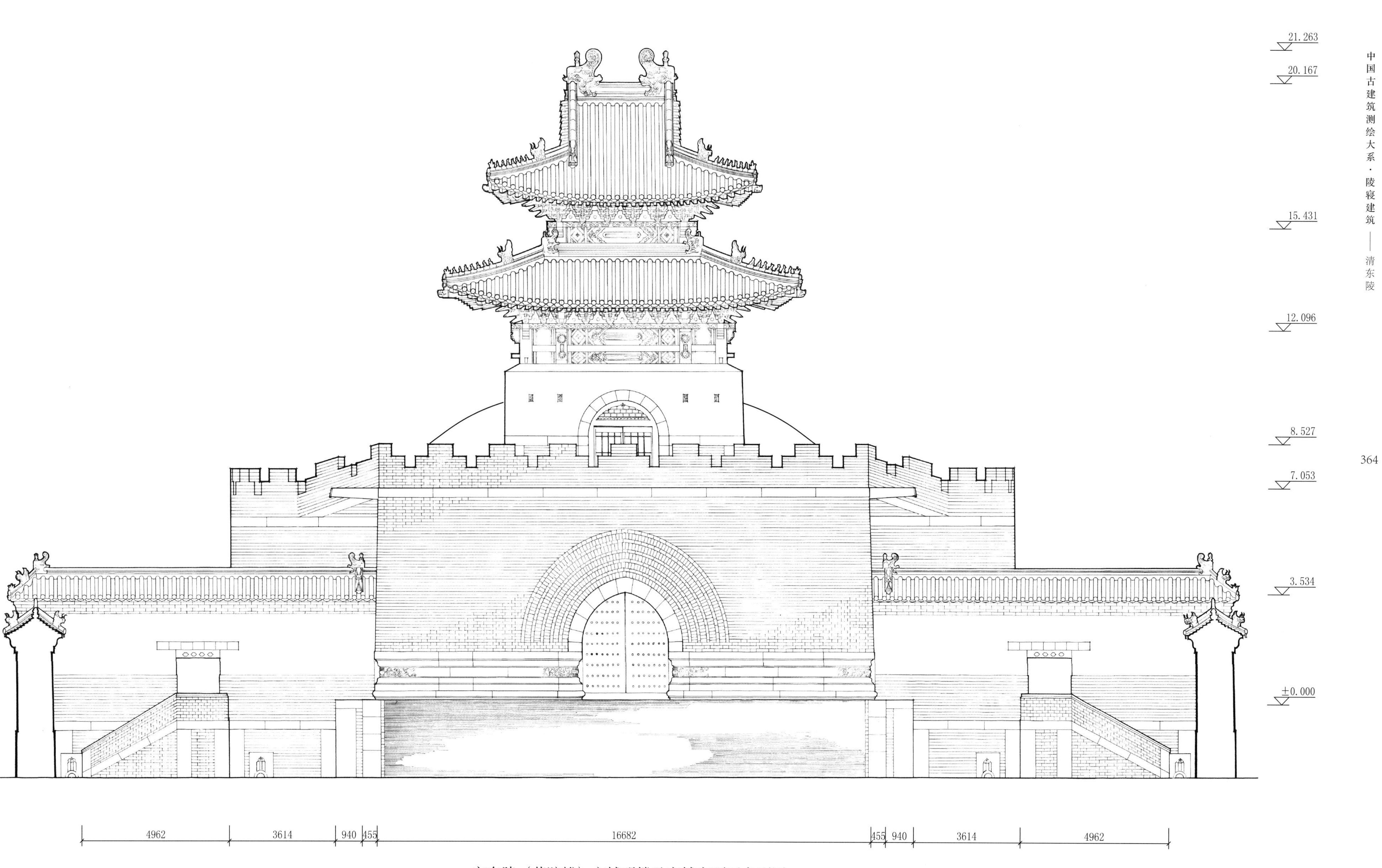

定东陵（菩陀峪）方城明楼及宝城宝顶正立面图

Front elevation of the Square Walled Terrace and the Memorial Tower as well as the Encircled Realm of Treasure and the Tumulus, East Tomb of Stability, Putuo Valley

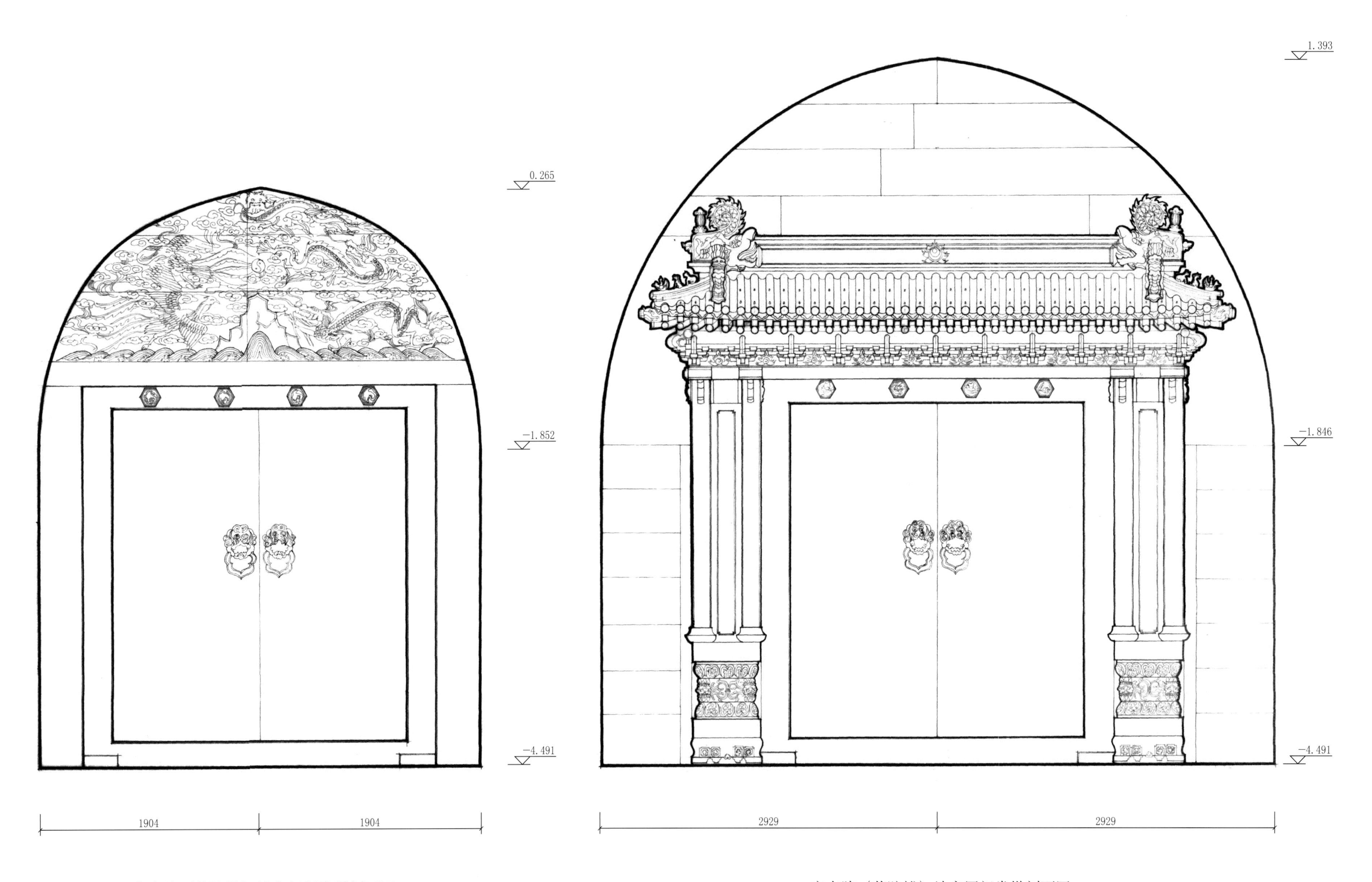

定东陵（菩陀峪）地宫门洞券纵剖面图

Longitudinal section of the Vaulted Passage of the Underground Palace, East Tomb of Stability, Putuo Valley

定东陵（菩陀峪）地宫罩门券纵剖面图

Longitudinal section of the Vaulted Arch at the Main Gate of the Underground Palace, East Tomb of Stability, Putuo Valley

定陵妃园寝

Imperial Consorts' Tombs affiliated with the Tomb of Stability (Dingling Feiyuanqin)

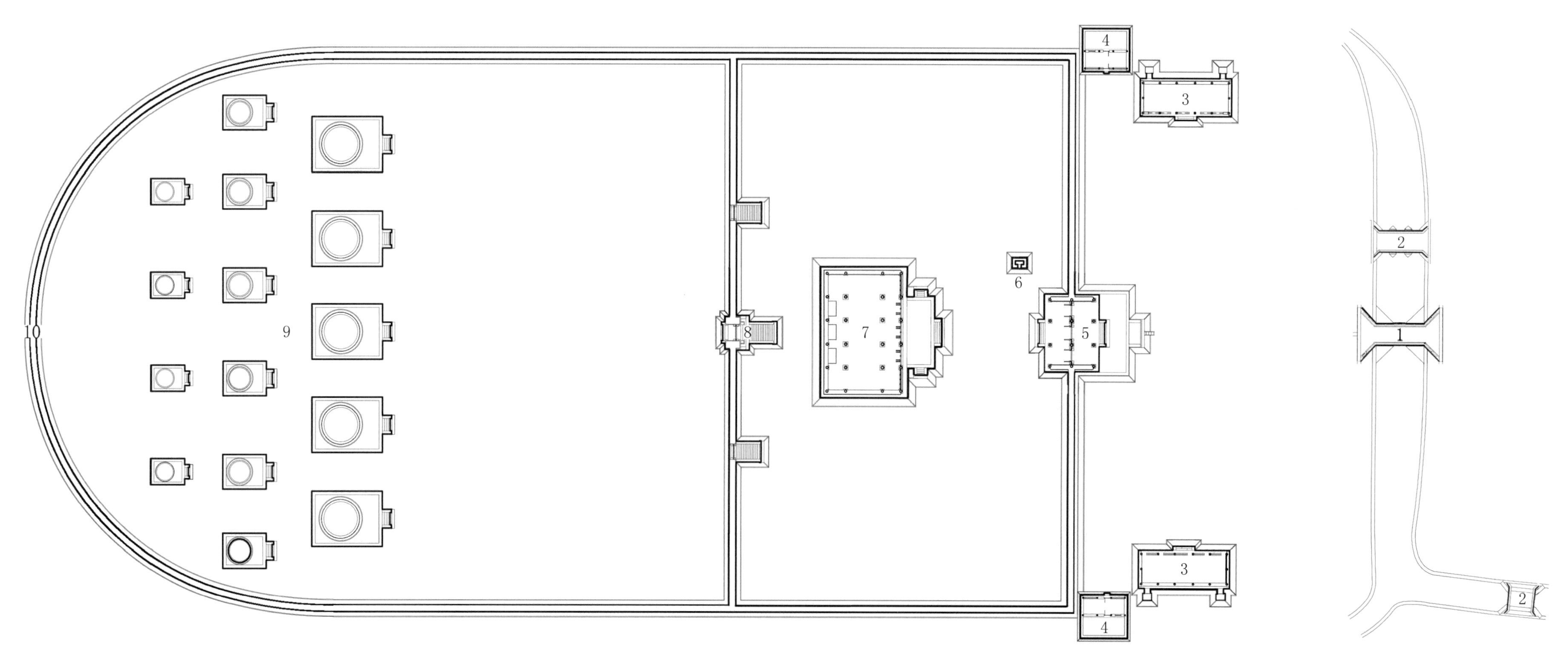

1 单孔桥 Single-Arch Stone Bridge
2 石平桥 Flat Stone Bridge
3 朝房 Reception Halls for Court Officials
4 班房 Duty Houses for the Guards
5 宫门 Main Gate
6 焚帛炉 Sacrificial Burners
7 享殿 Main Sacrificial Hall
8 琉璃花门 Gate with Glazed Roof Tiles
9 妃嫔宝顶 Concubines' Tumulus
10 罗圈墙 Wall Surrounding the Tomb

定陵妃园寝总平面图
Site plan of the Imperial Consorts' Tombs affiliated with the Tomb of Stability

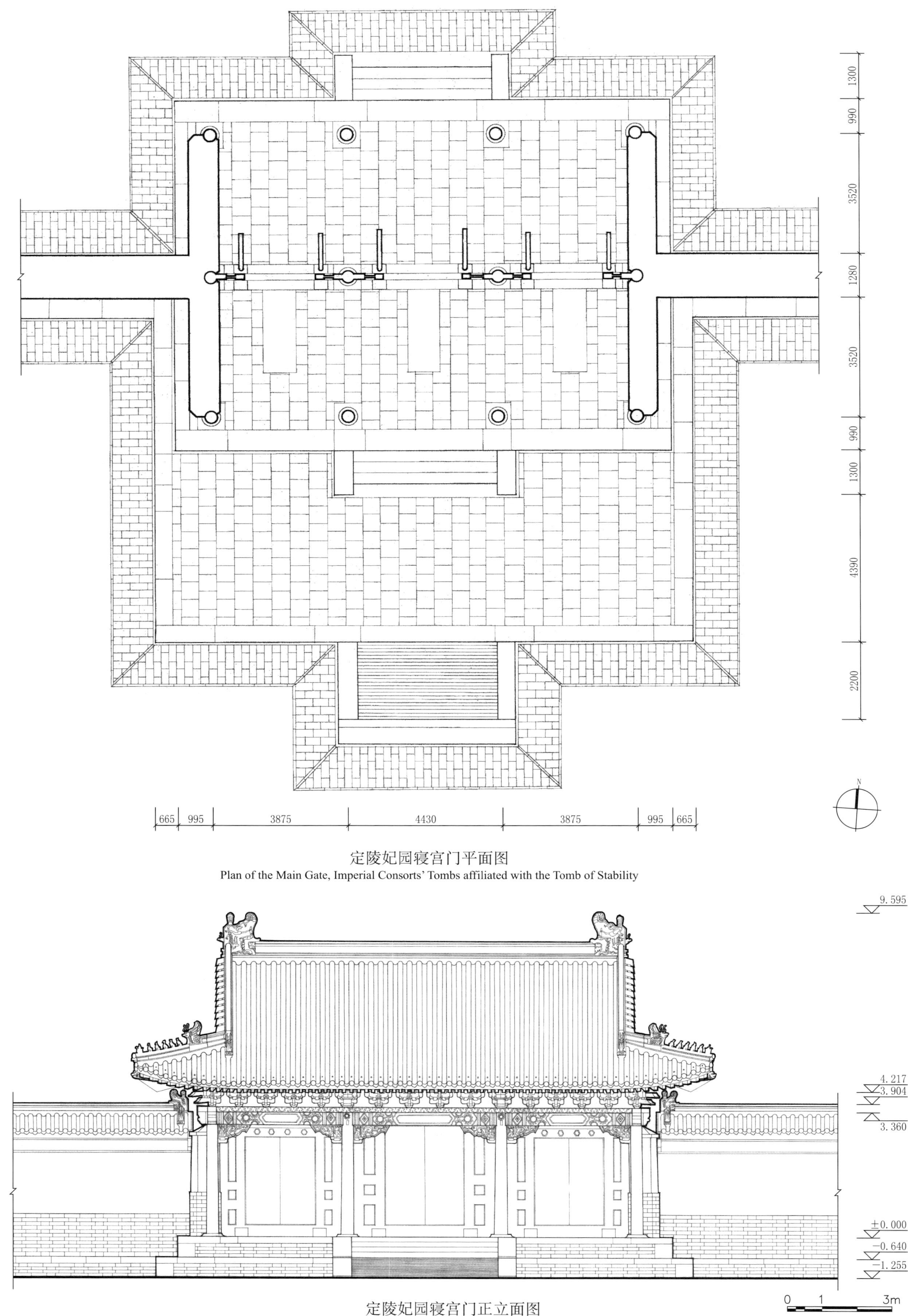

定陵妃园寝宫门平面图
Plan of the Main Gate, Imperial Consorts' Tombs affiliated with the Tomb of Stability

定陵妃园寝宫门正立面图
Front elevation of the Main Gate, Imperial Consorts' Tombs affiliated with the Tomb of Stability

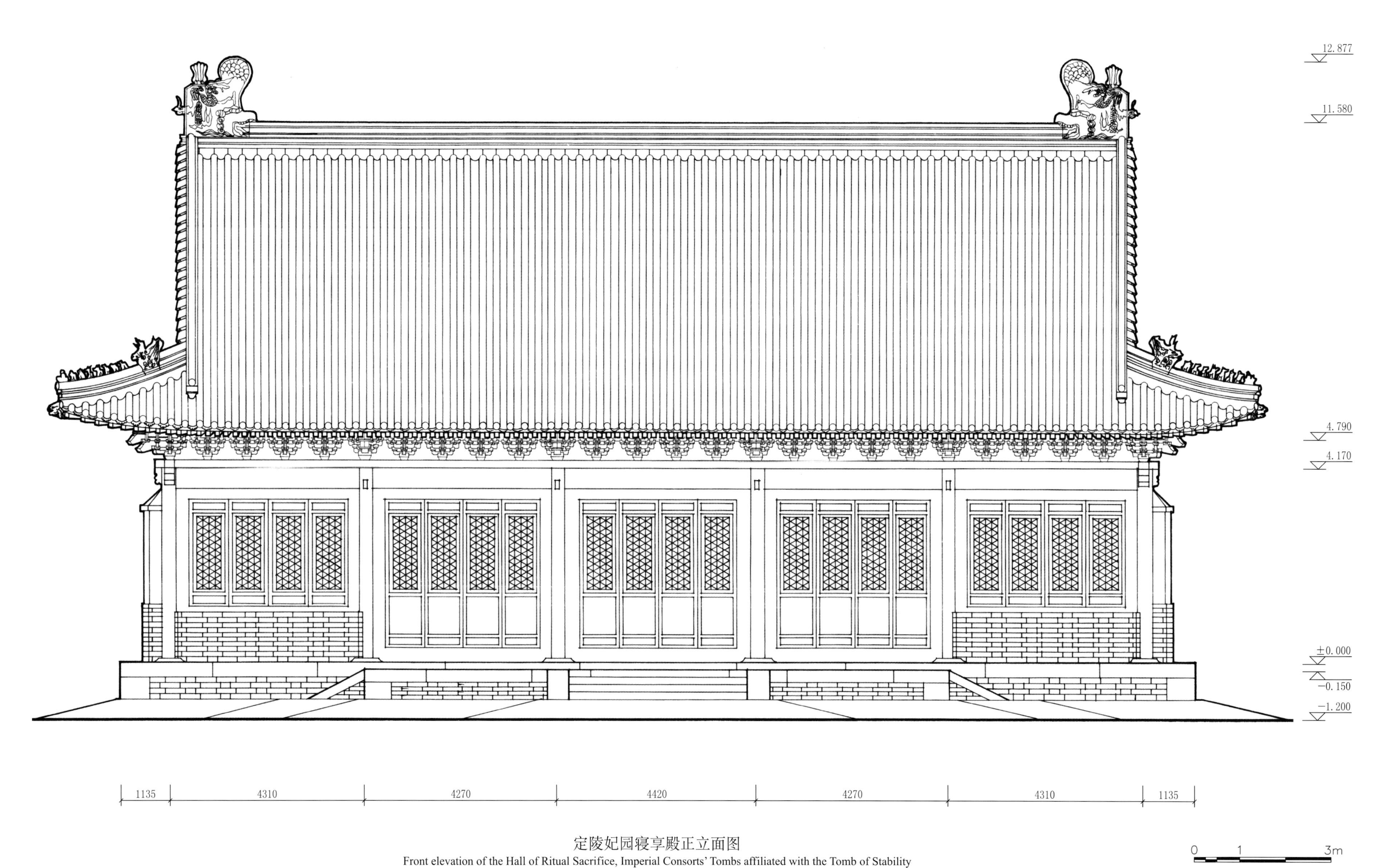

定陵妃园寝享殿正立面图

Front elevation of the Hall of Ritual Sacrifice, Imperial Consorts' Tombs affiliated with the Tomb of Stability

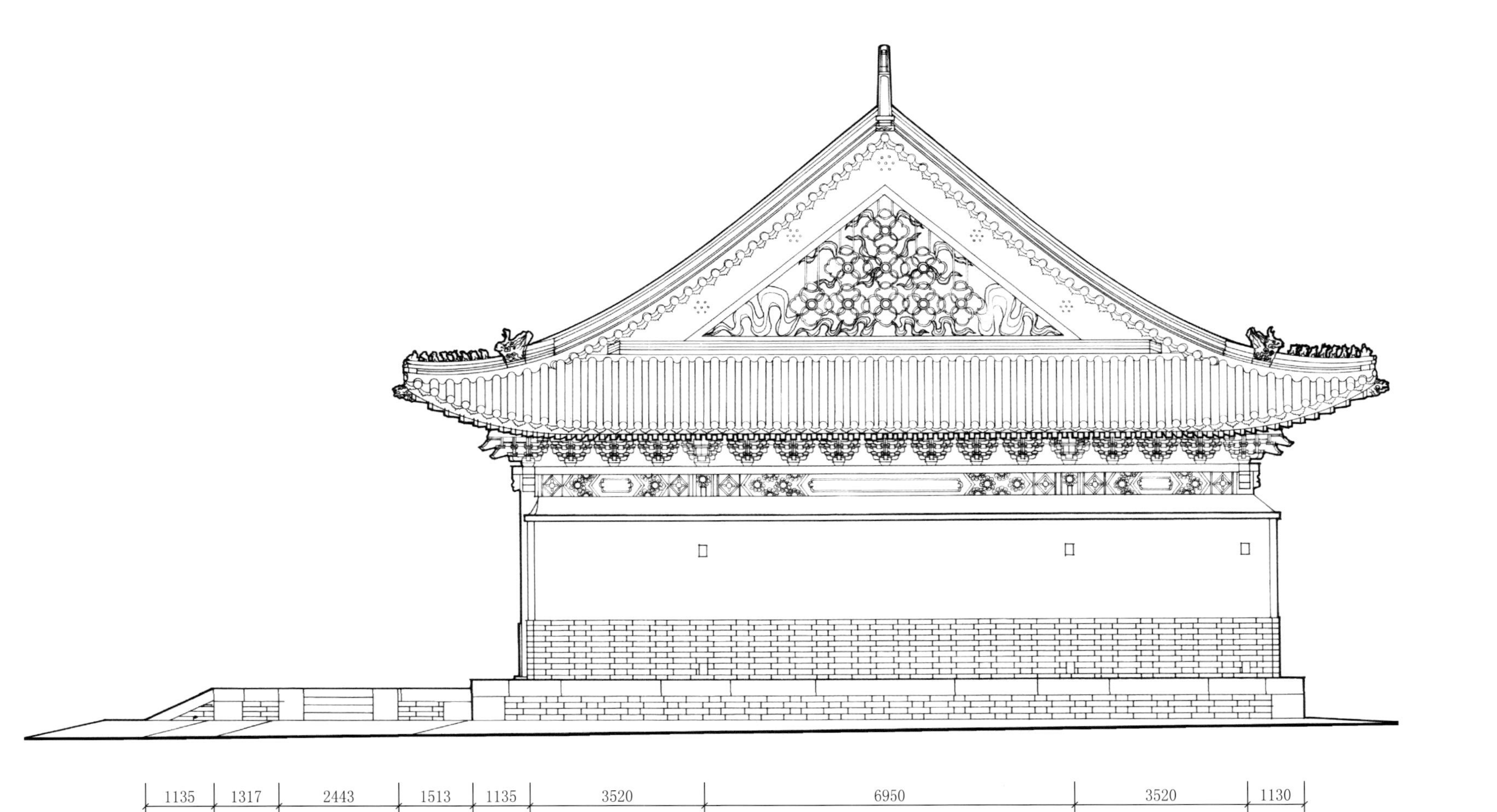

定陵妃园寝享殿侧立面图
Side elevation of the Hall of Ritual Sacrifice, Imperial Consorts' Tombs affiliated with the Tomb of Stability

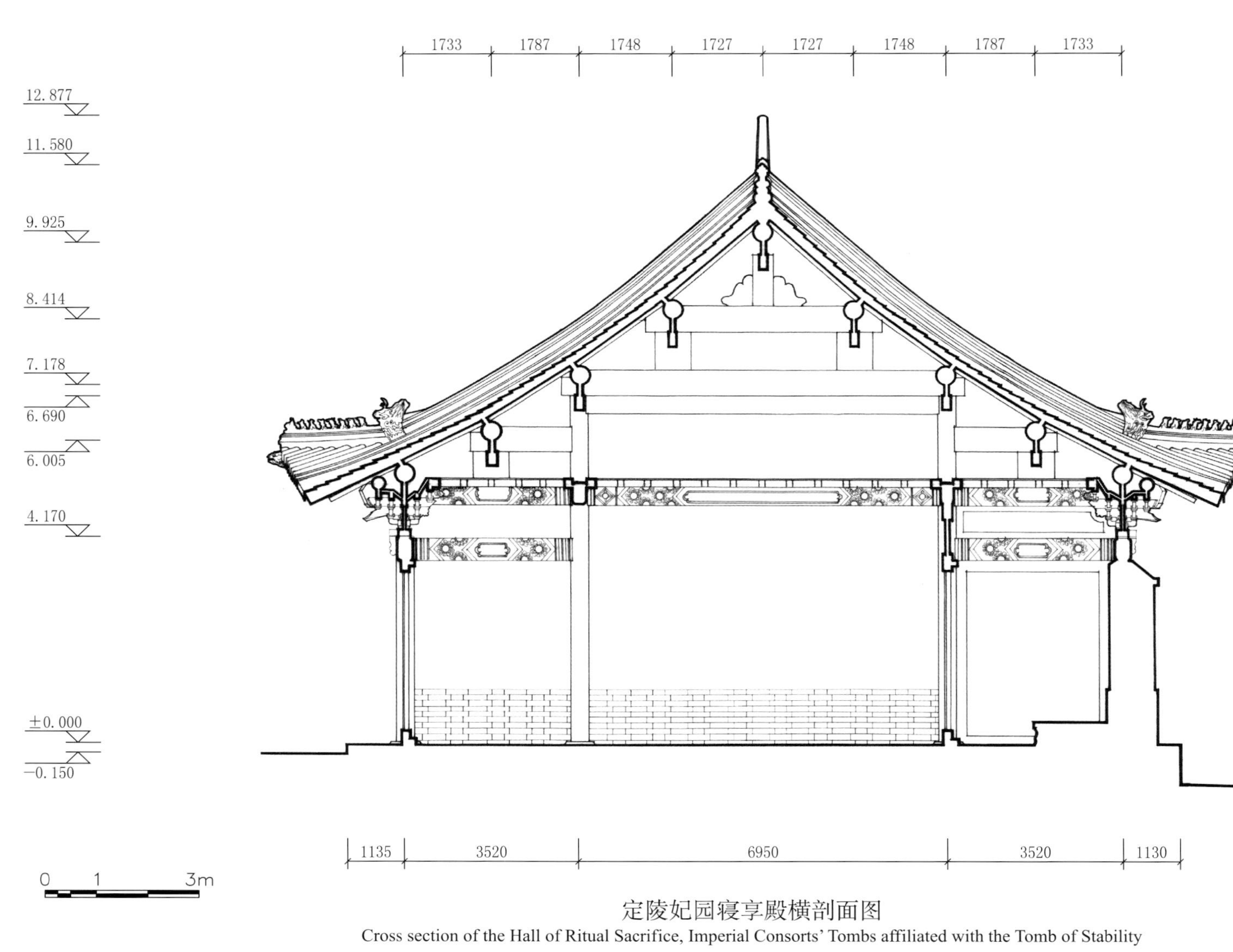

定陵妃园寝享殿横剖面图
Cross section of the Hall of Ritual Sacrifice, Imperial Consorts' Tombs affiliated with the Tomb of Stability

惠陵
Tomb of Benevolence (Huiling)

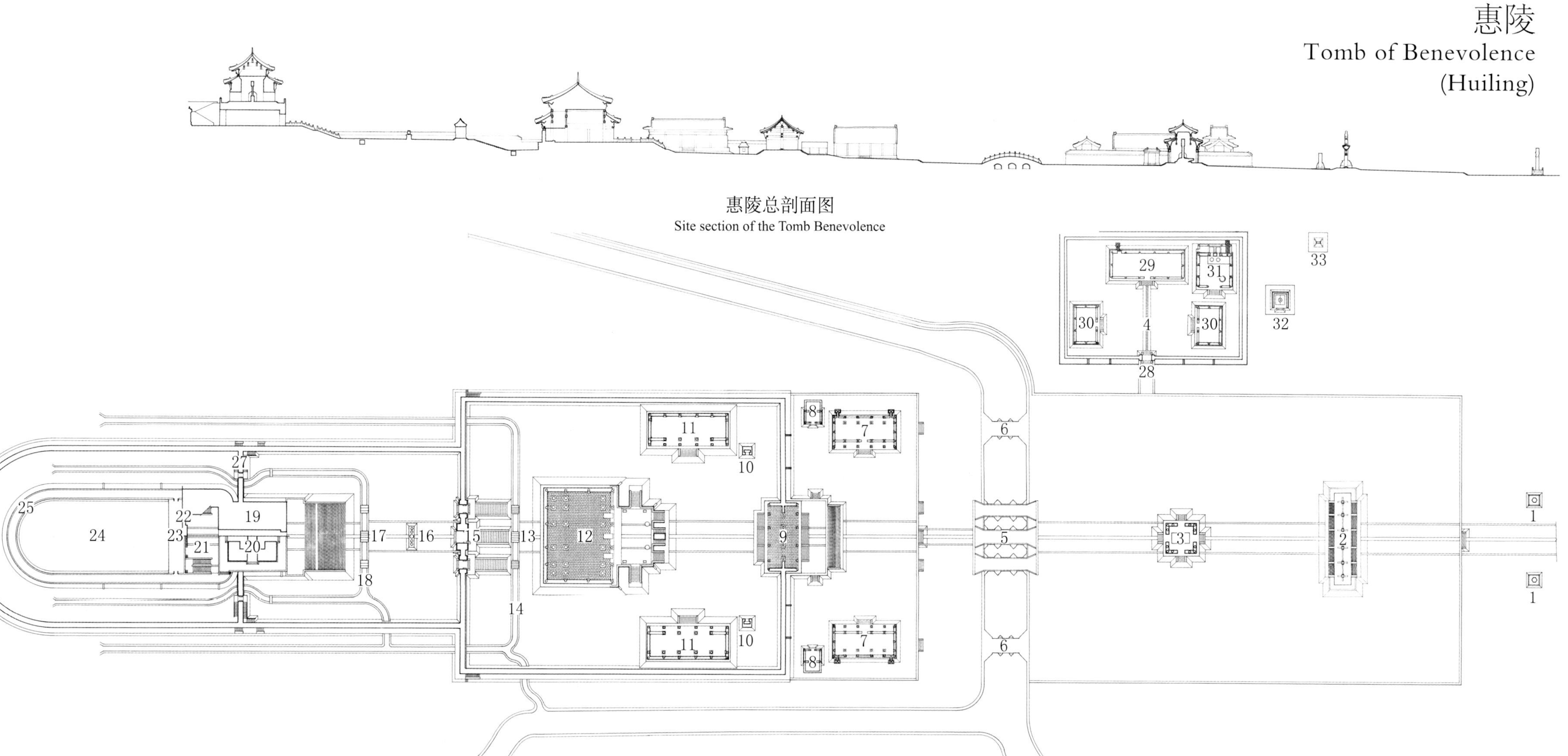

惠陵总剖面图
Site section of the Tomb Benevolence

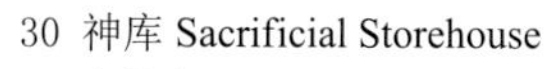

- 1 望柱 Ornamental Column
- 2 牌坊 Marble Memorial Gateway
- 3 神道碑亭 Pavilion for the Stela on the Spirit Way
- 4 神厨库院落 Culinary Courtyard for Sacrifices
- 5 三路三孔桥 Three-Way Three-Arch Bridges
- 6 石平桥 Flat Stone Bridge
- 7 朝房 Reception Halls for Court Officials
- 8 班房 Duty Houses for the Guards
- 9 隆恩门 Gate of Monumental Grace
- 10 焚帛炉 Sacrificial Burners
- 11 配殿 Side Hall
- 12 隆恩殿 Hall of Monumental Grace
- 13 石平桥 Flat Stone Bridge
- 14 玉带河 Jade Ribbon River
- 15 琉璃花门 Gate with Glazed Roof Tiles
- 16 石五供 Five Stone Ritual Vessels
- 17 石平桥 Flat Stone Bridge
- 18 月牙河 Crescent River
- 19 方城 Square Walled Terrace
- 20 明楼 Memorial Tower
- 21 哑巴院 Courtyard of the Mute
- 22 月牙城 Crescent Wall
- 23 琉璃影壁 Screen Wall of Glazed Tiles
- 24 宝顶 Tumulus
- 25 宝城 Encircled Realm of Treasure
- 26 罗圈墙 Luoquan Wall
- 27 卡子墙 Spacer Wall
- 28 院门 Yard Gate
- 29 神厨 Sacrificial Kitchen
- 30 神库 Sacrificial Storehouse
- 31 宰牲亭 Ritual Abattoir
- 32 井亭 Well Pavilion
- 33 下马牌 Stela Marking the Place for Dismounting from one's Horse

惠陵总平面图
Site plan of the Tomb of Benevolence

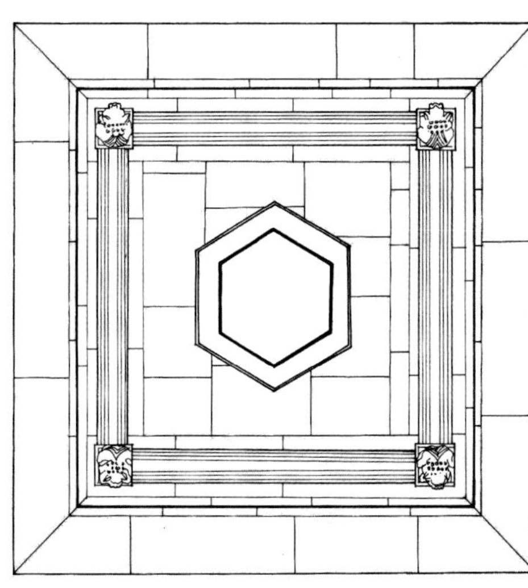

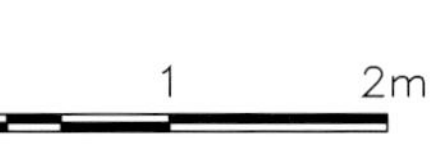

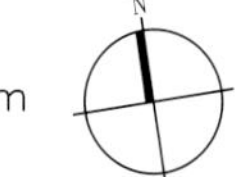

惠陵神道望柱大样图

Detailed drawing of the Column on the Spirit Way, Tomb of Benevolence

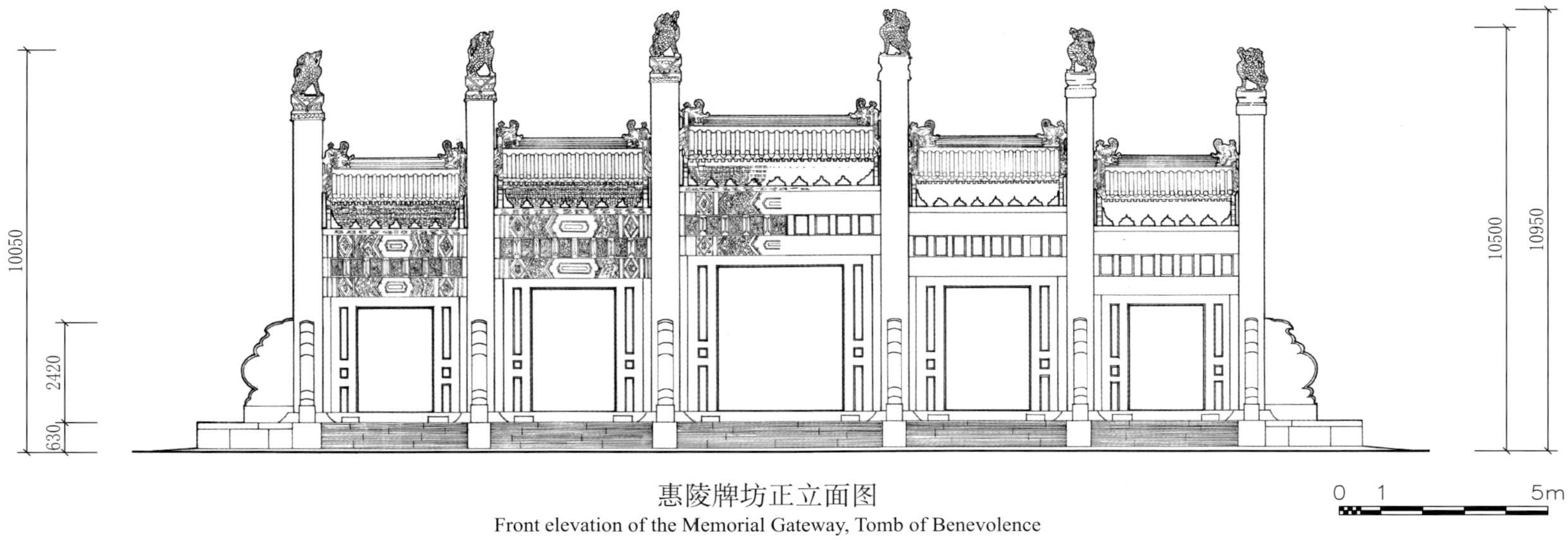

惠陵牌坊正立面图
Front elevation of the Memorial Gateway, Tomb of Benevolence

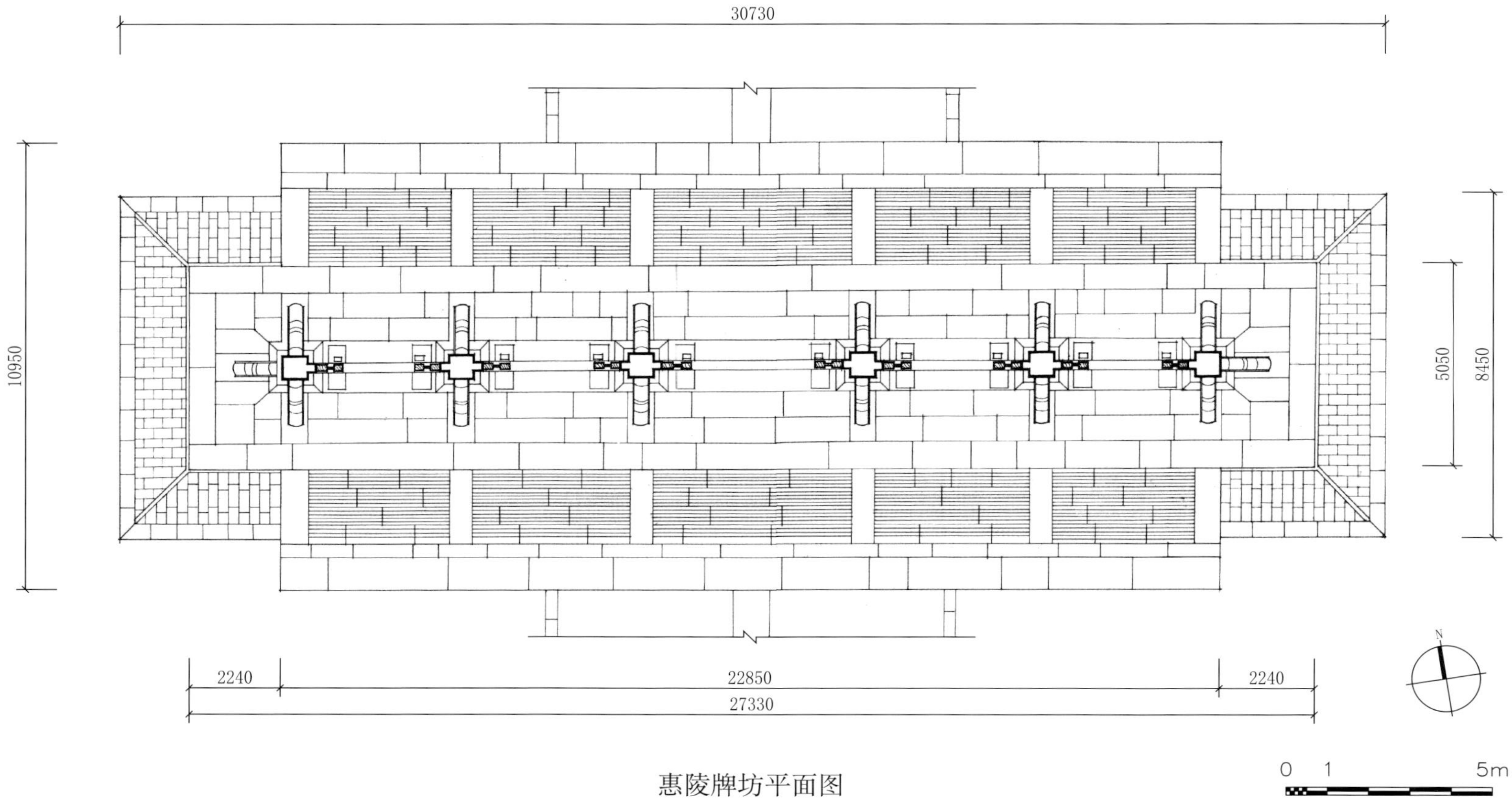

惠陵牌坊平面图
Plan of the Memorial Gateway, Tomb of Benevolence

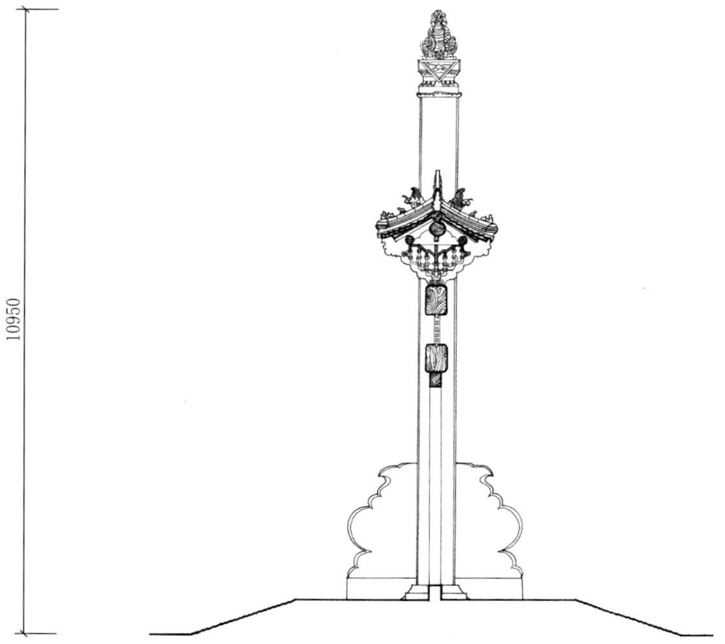

惠陵牌坊横剖面图
Cross section of the Memorial Gateway, Tomb of Benevolence

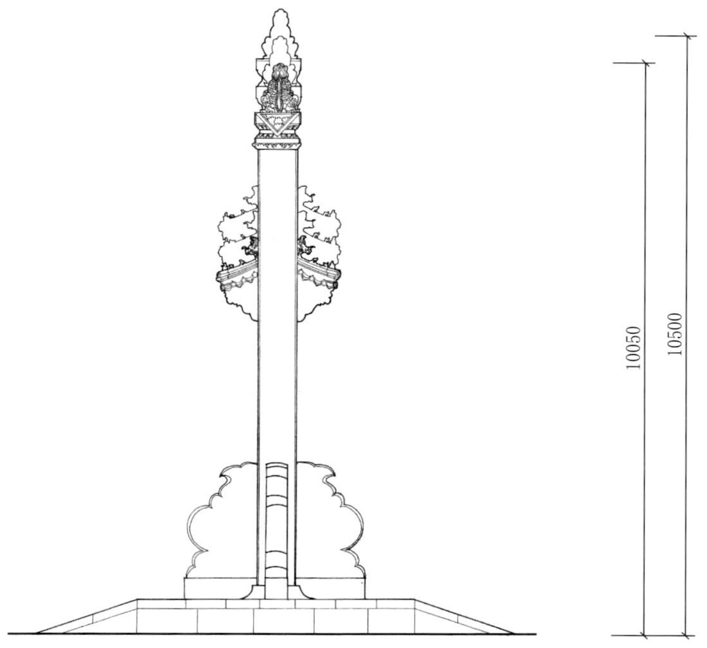

惠陵牌坊侧立面图
Side elevation of the Memorial Gateway, Tomb of Benevolence

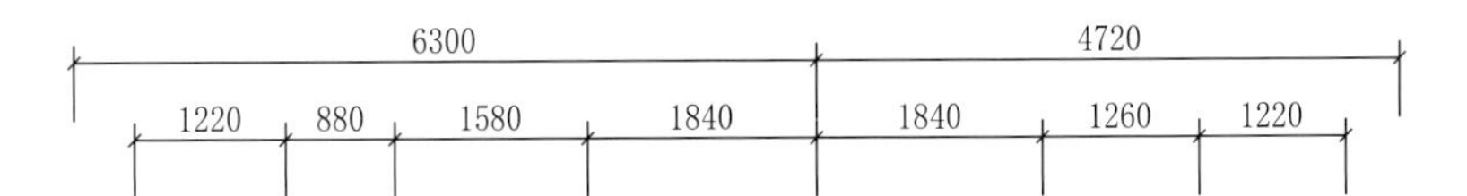

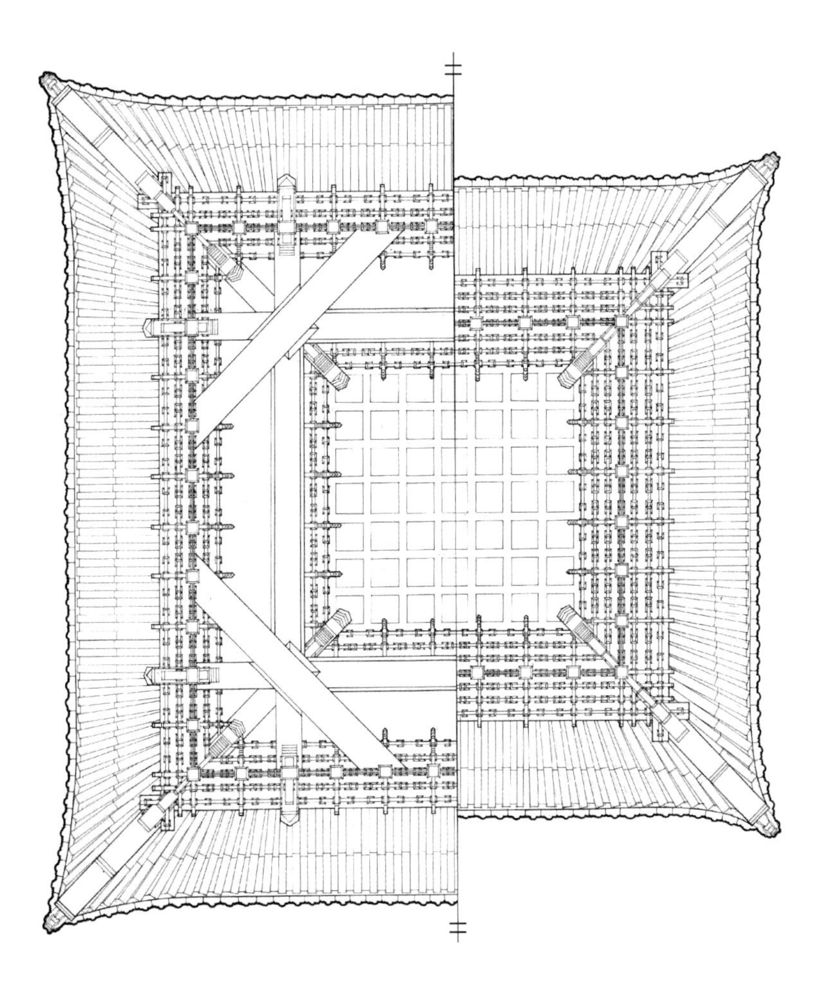

惠陵神道碑亭仰视图
Looking up at the Truss and Ceiling of the Pavilion for the Stela on the Spirit Way, Tomb of Benevolence

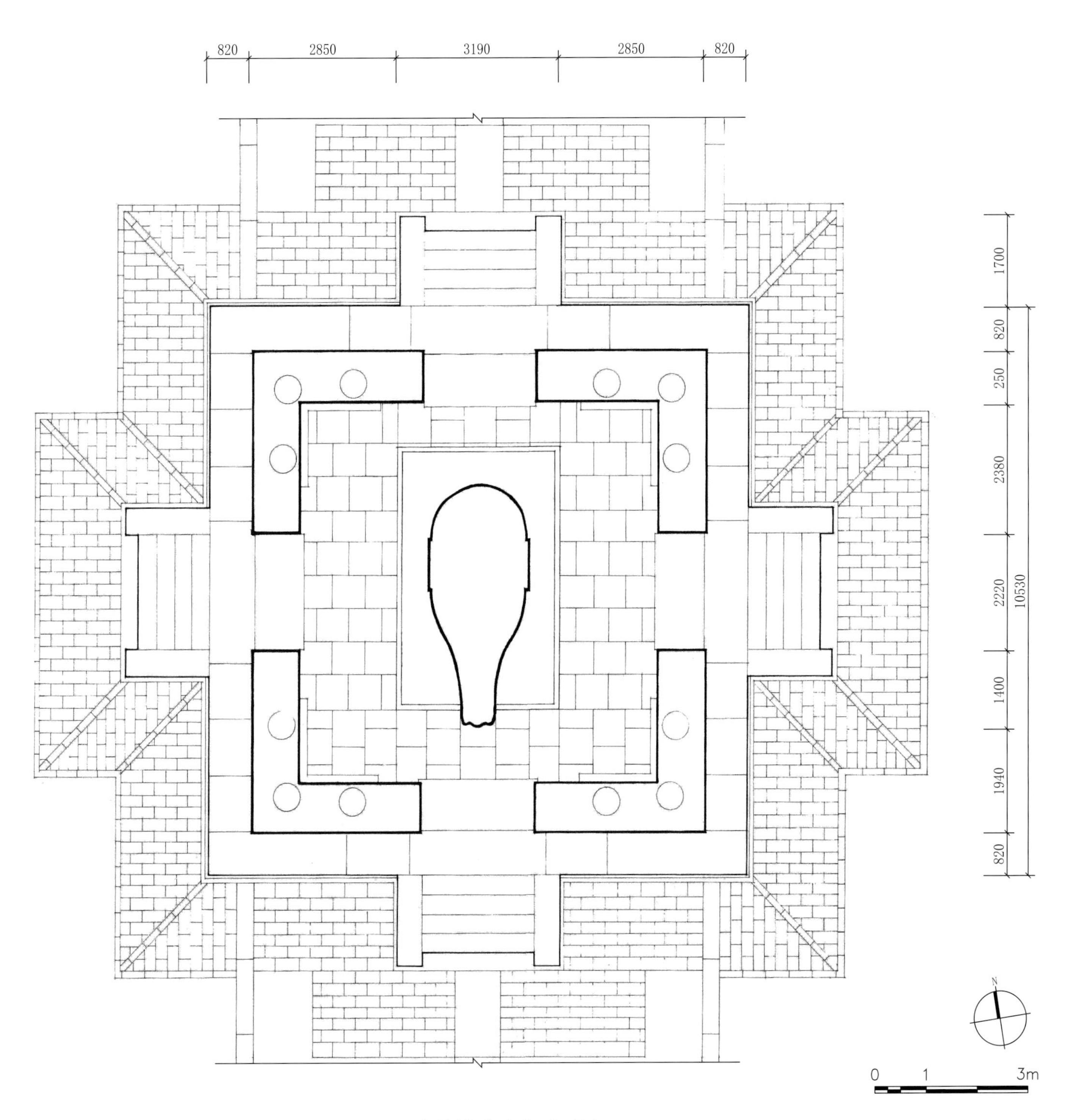

惠陵神道碑亭平面图
Plan of the Pavilion for the Stela on the Spirit Way, Tomb of Benevolence

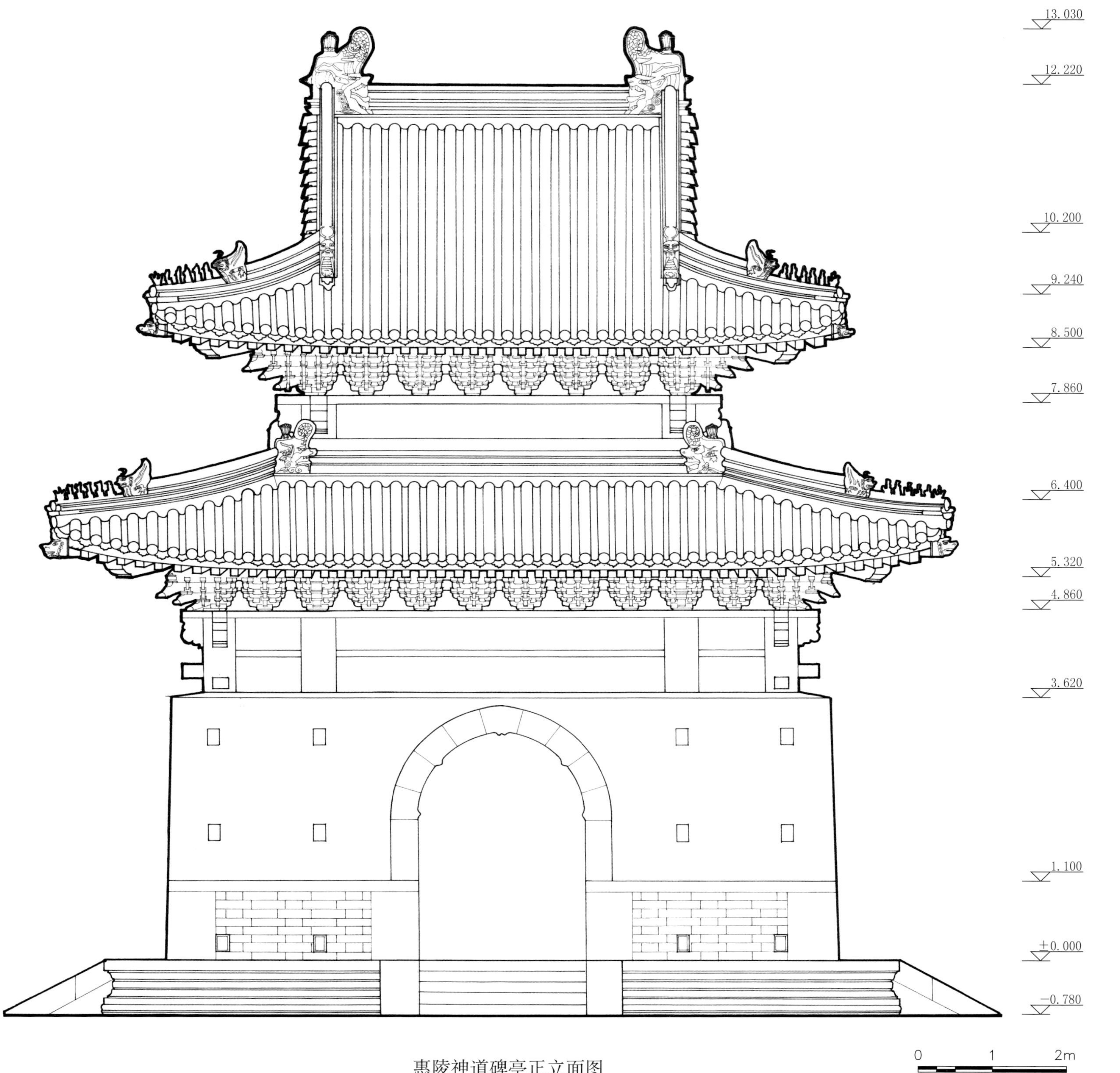

惠陵神道碑亭正立面图

Front elevation of the Pavilion for the Stela on the Spirit Way, Tomb of Benevolence

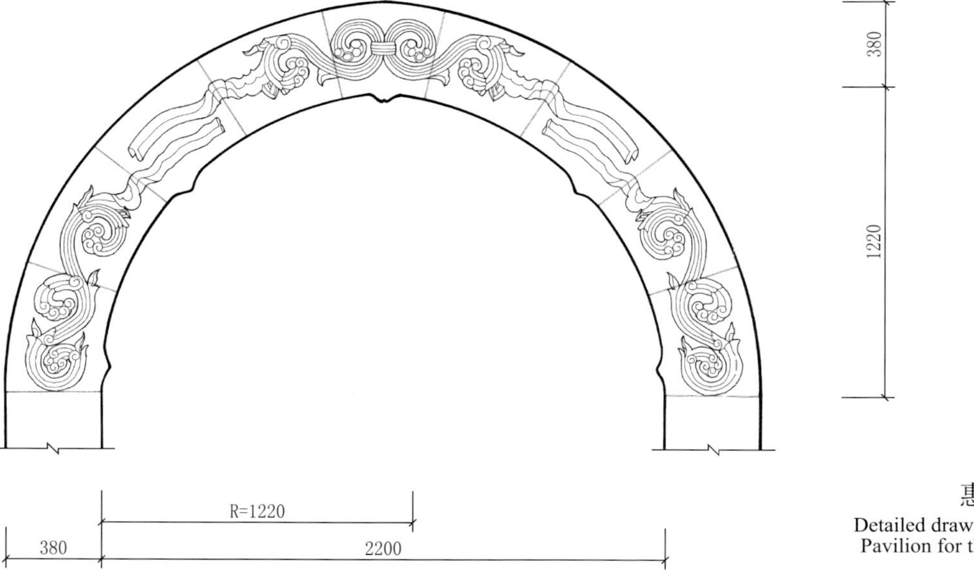

惠陵神道碑亭券脸石大样图

Detailed drawing of the Marble Plate on the Vaulted Arch of the Pavilion for the Stela on the Spirit Way, Tomb of Benevolence

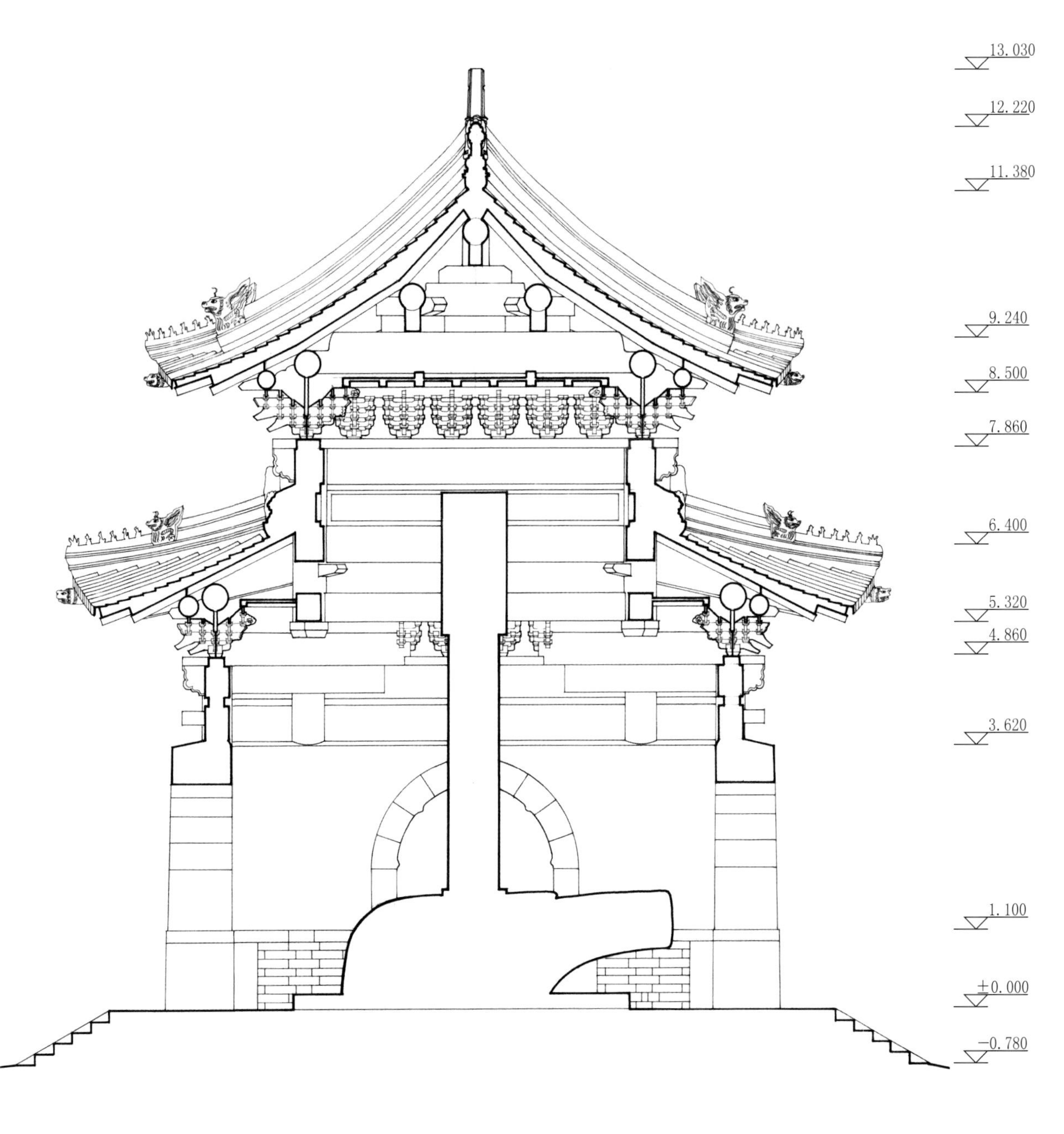

惠陵神道碑亭横剖面图
Cross section of the Pavilion for the Stela on the Spirit Way, Tomb of Benevolence

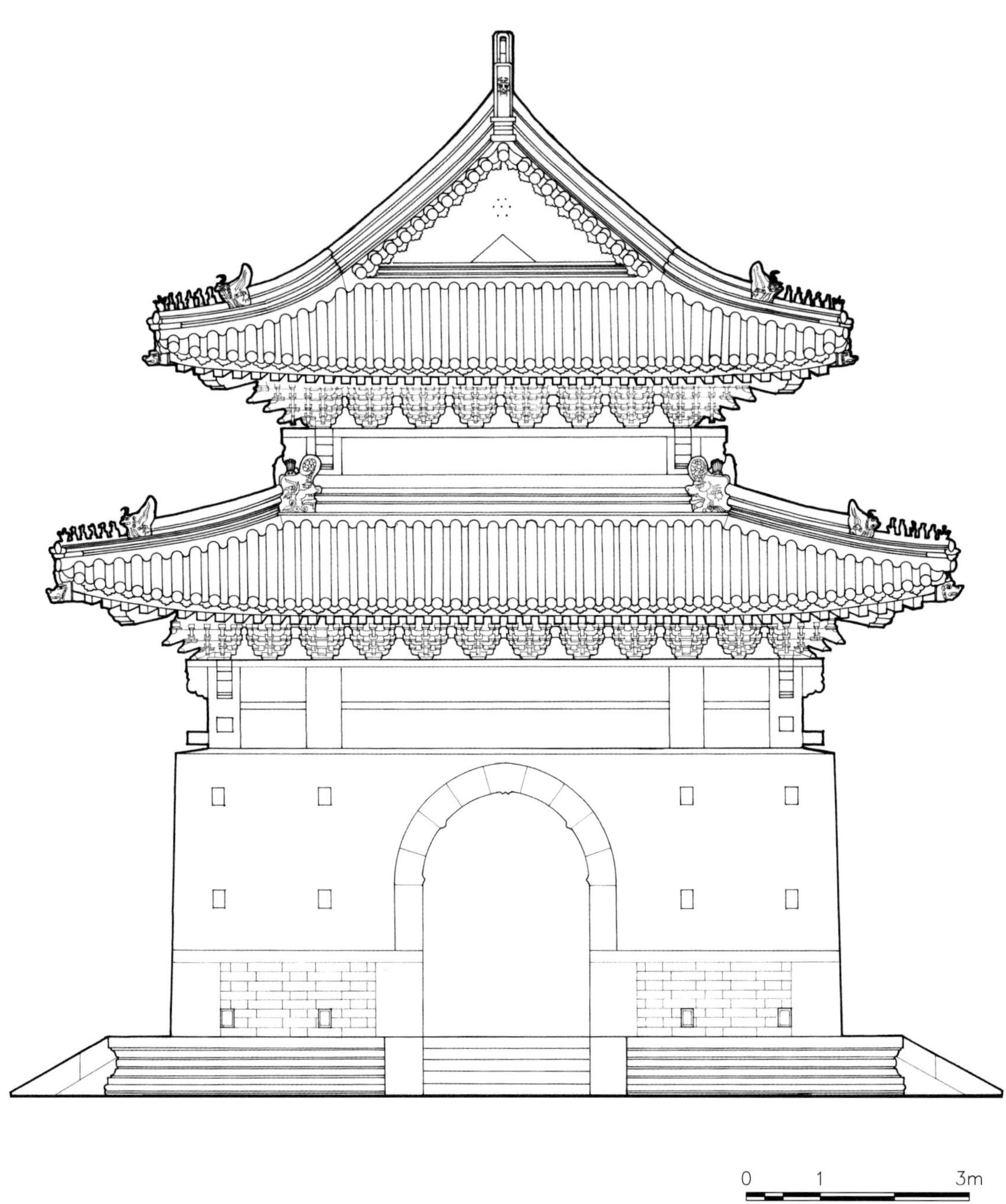

惠陵神道碑亭侧立面图
Side elevation of the Pavilion for the Stela on the Spirit Way, Tomb of Benevolence

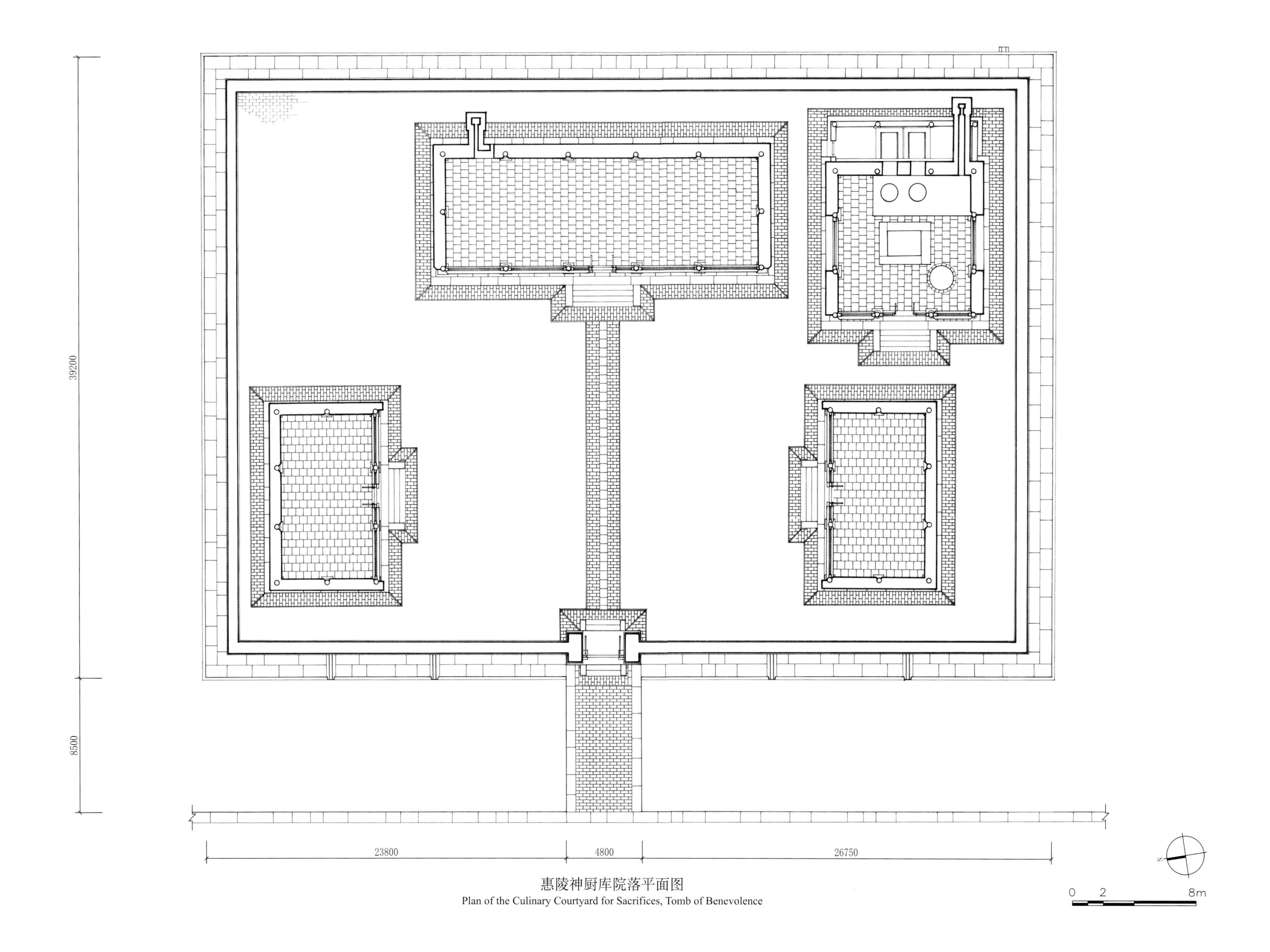

惠陵神厨库院落平面图
Plan of the Culinary Courtyard for Sacrifices, Tomb of Benevolence

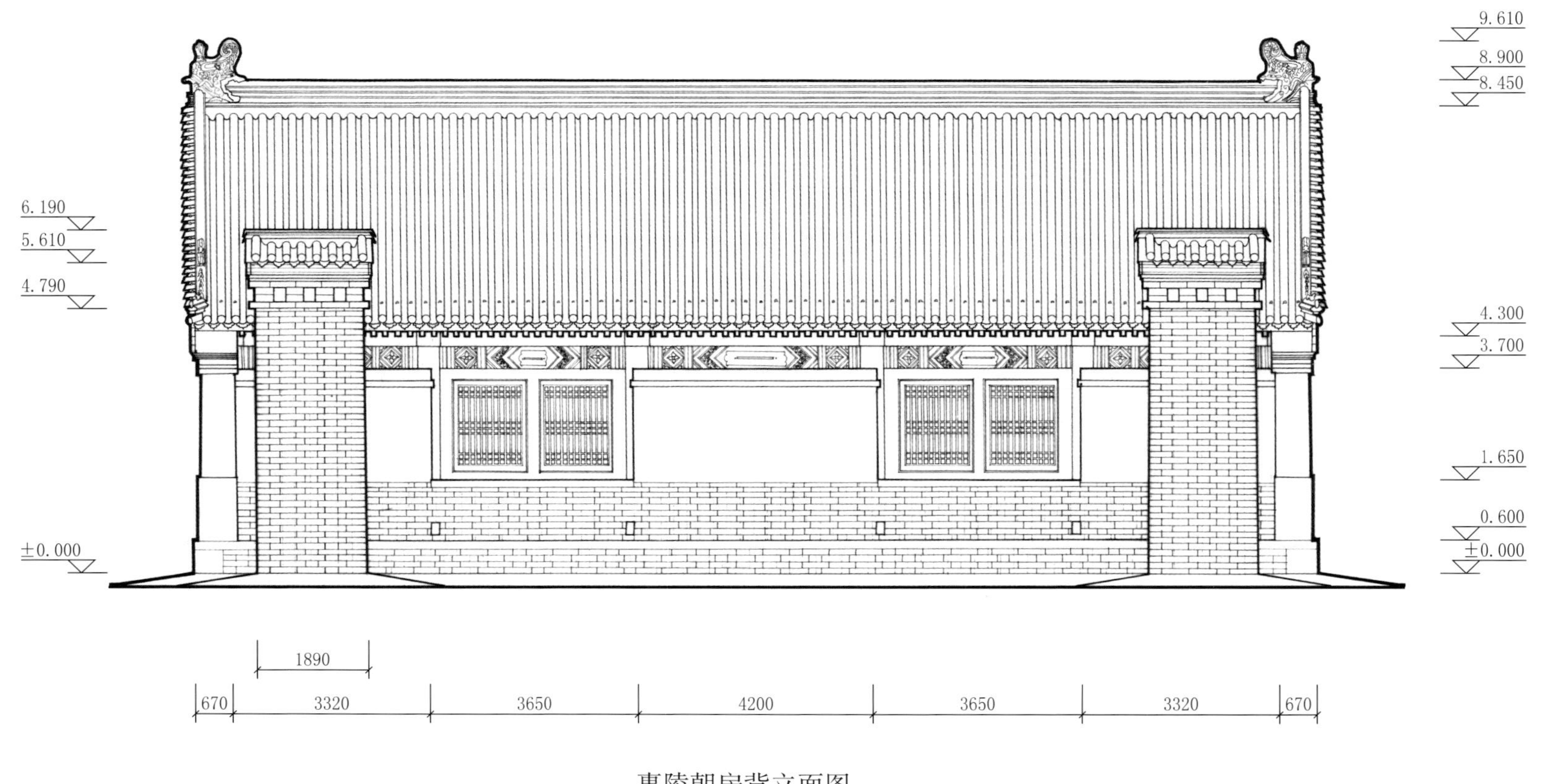

惠陵朝房背立面图
Back elevation of the Reception Hall for Court Officials, Tomb of Benevolence

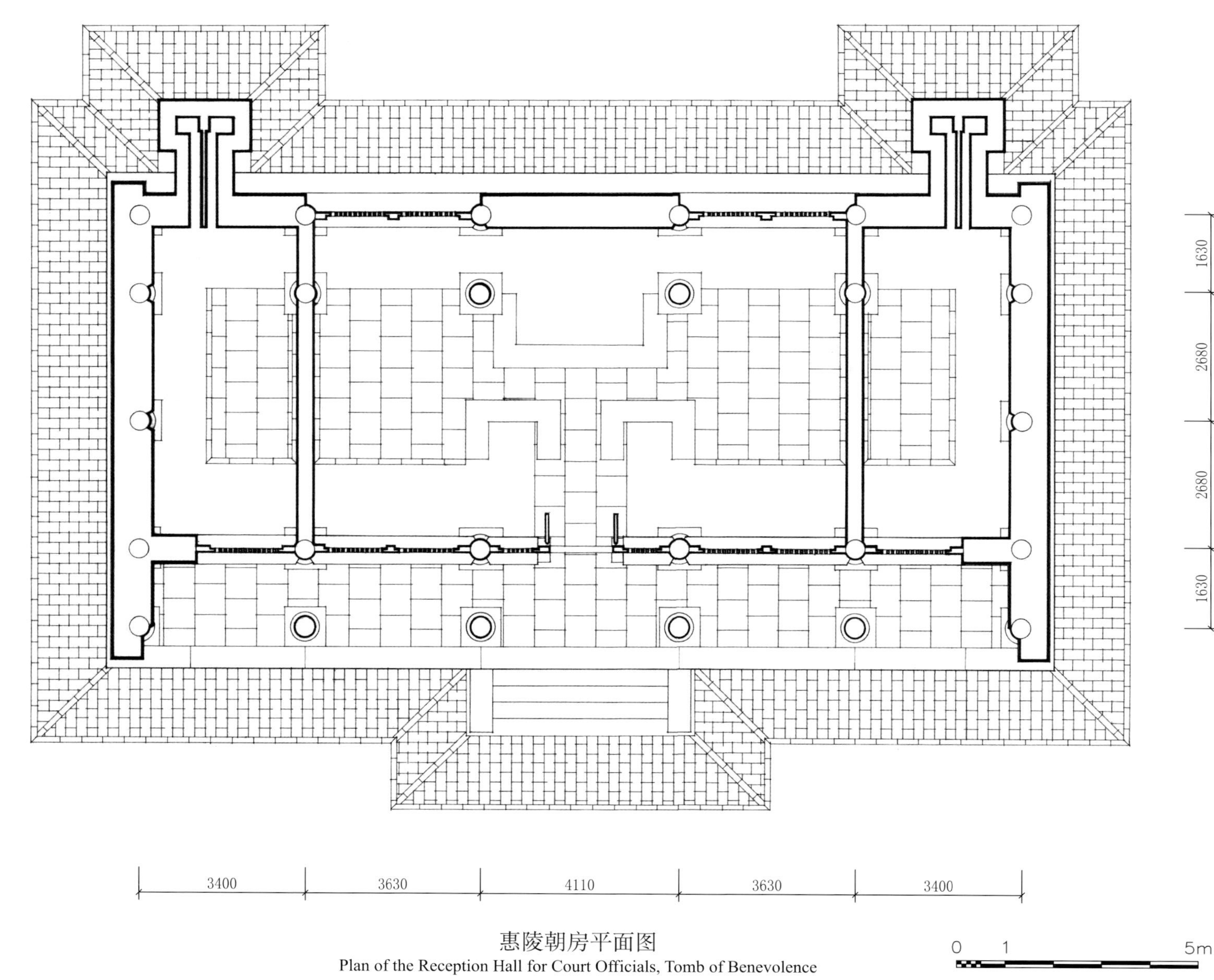

惠陵朝房平面图
Plan of the Reception Hall for Court Officials, Tomb of Benevolence

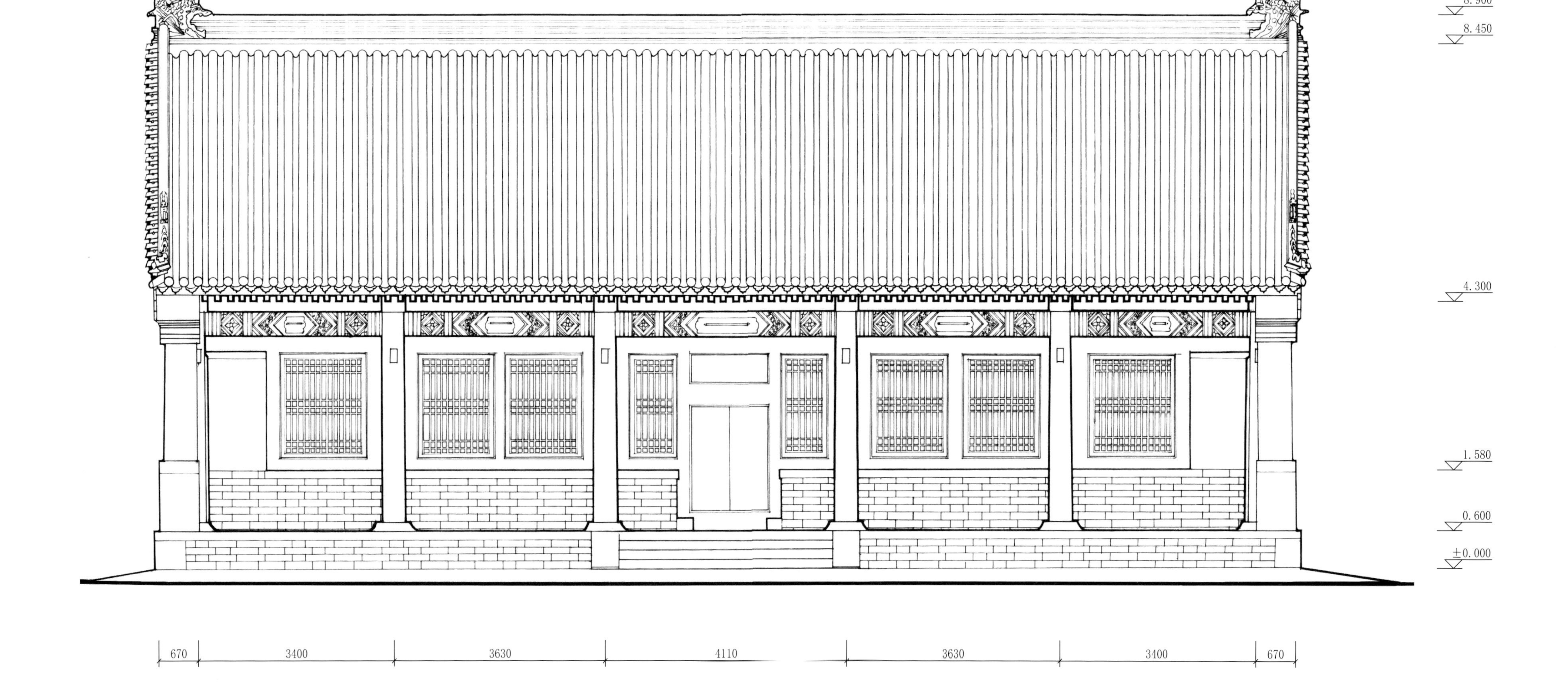

惠陵朝房正立面图

Front elevation of the Reception Hall for Court Officials, Tomb of Benevolence

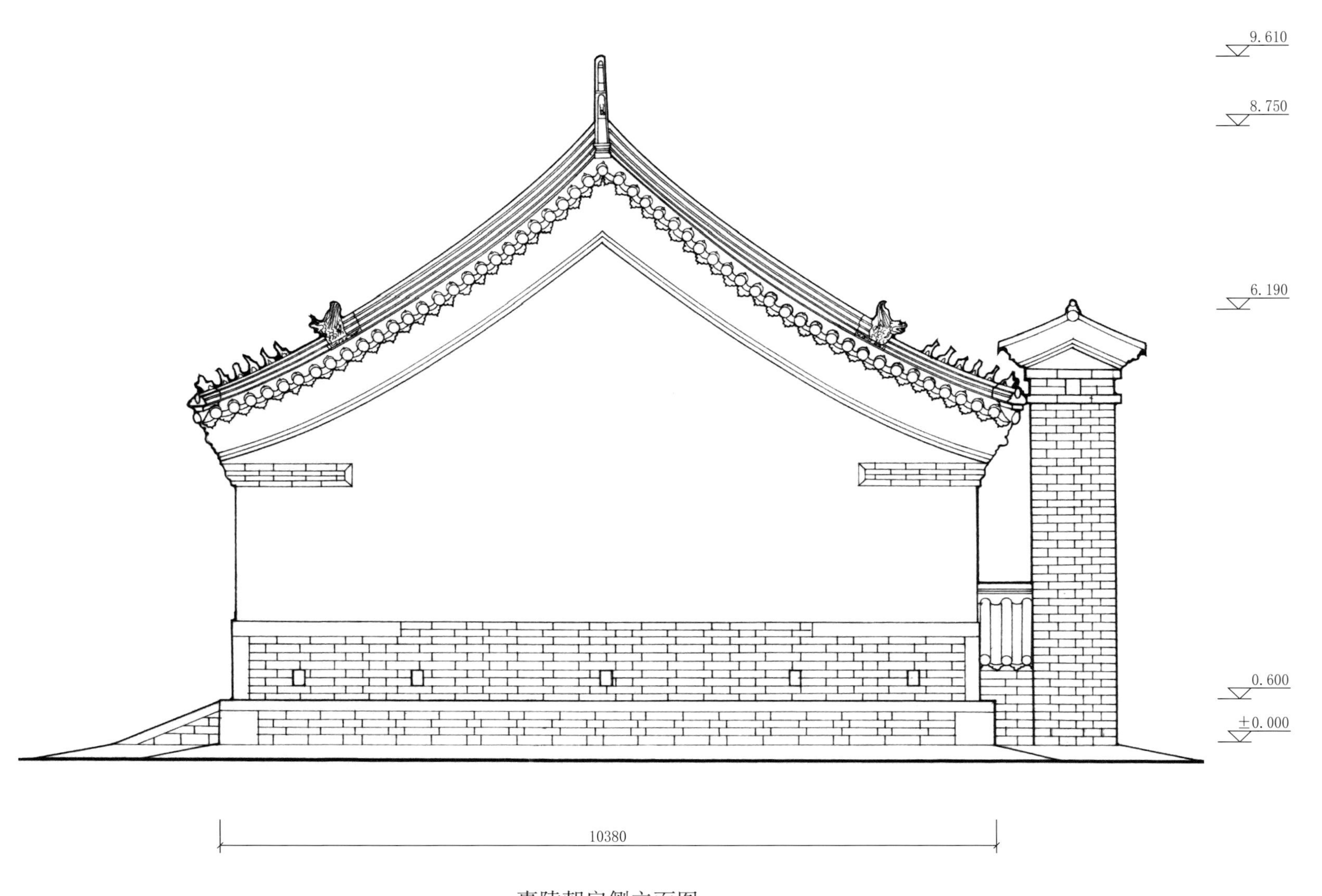

惠陵朝房侧立面图
Side elevation of Reception Hall for Court Officials, Tomb of Benevolence

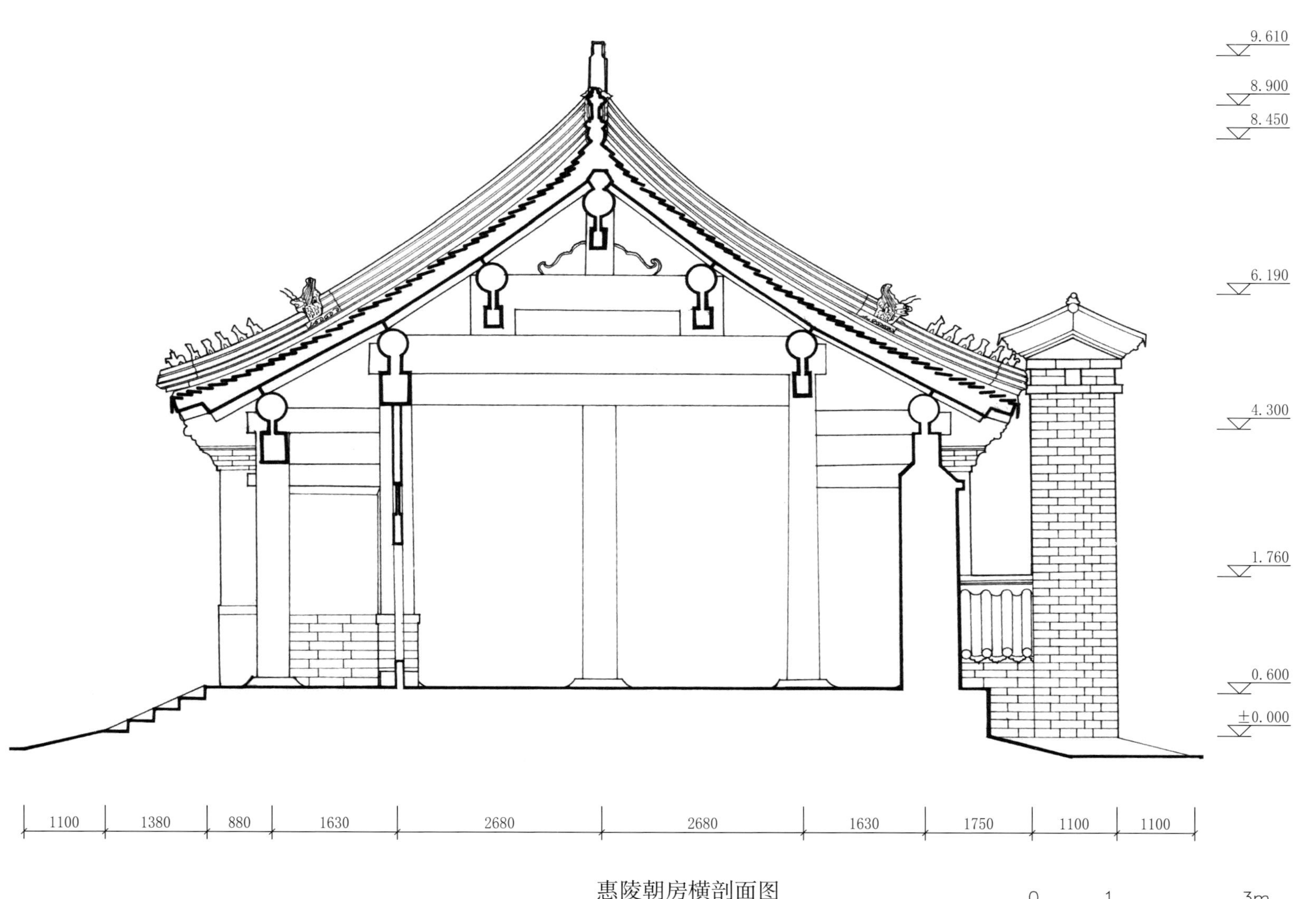

惠陵朝房横剖面图
Cross section of the Reception Hall for Court Officials, Tomb of Benevolence

惠陵隆恩门仰视图
Looking up at the Truss of the Gate of Monumental Grace, Tomb of Benevolence

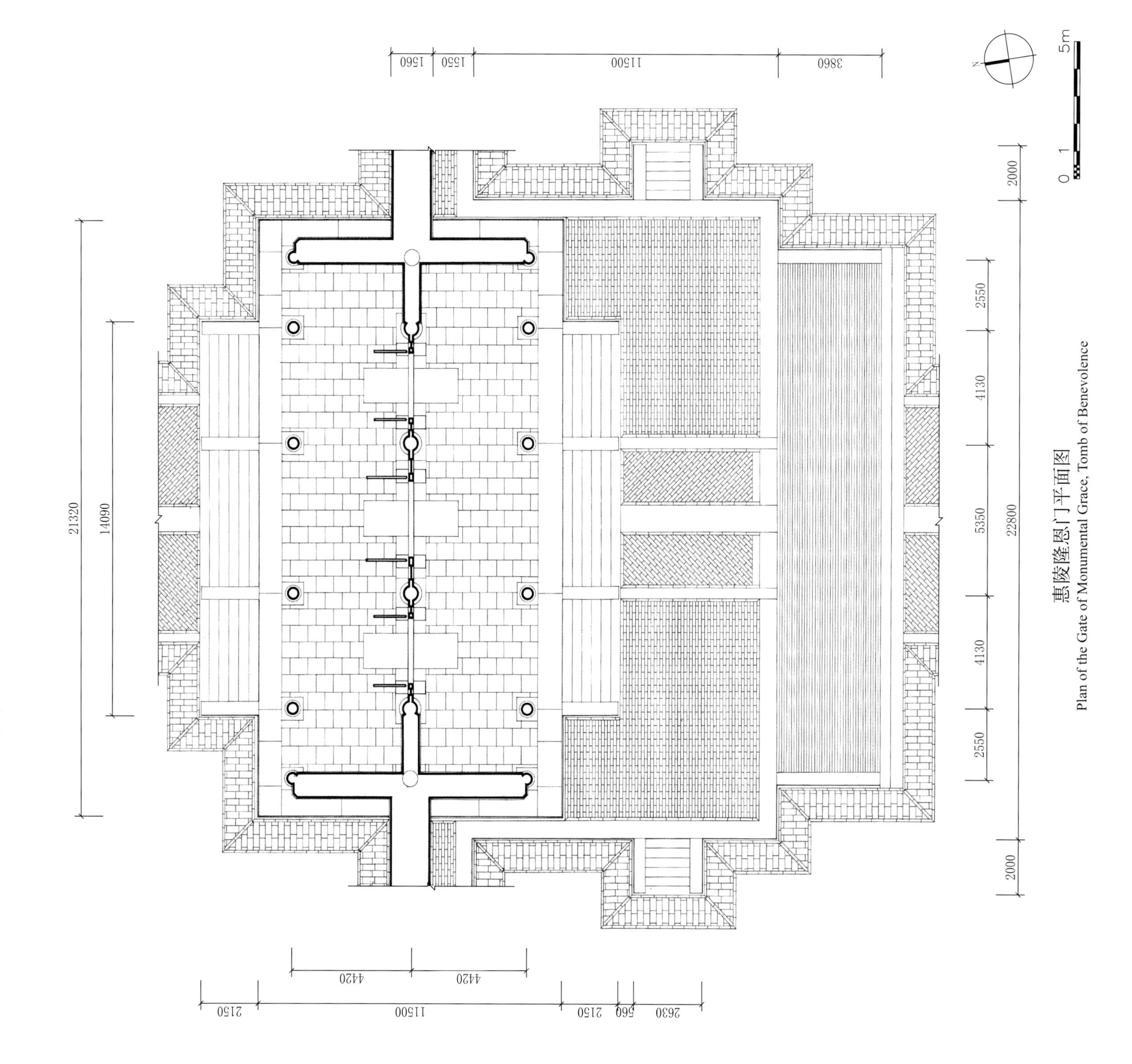

惠陵隆恩门平面图
Plan of the Gate of Monumental Grace, Tomb of Benevolence

惠陵隆恩门正立面图

Front elevation of the Gate of Monumental Grace, Tomb of Benevolence

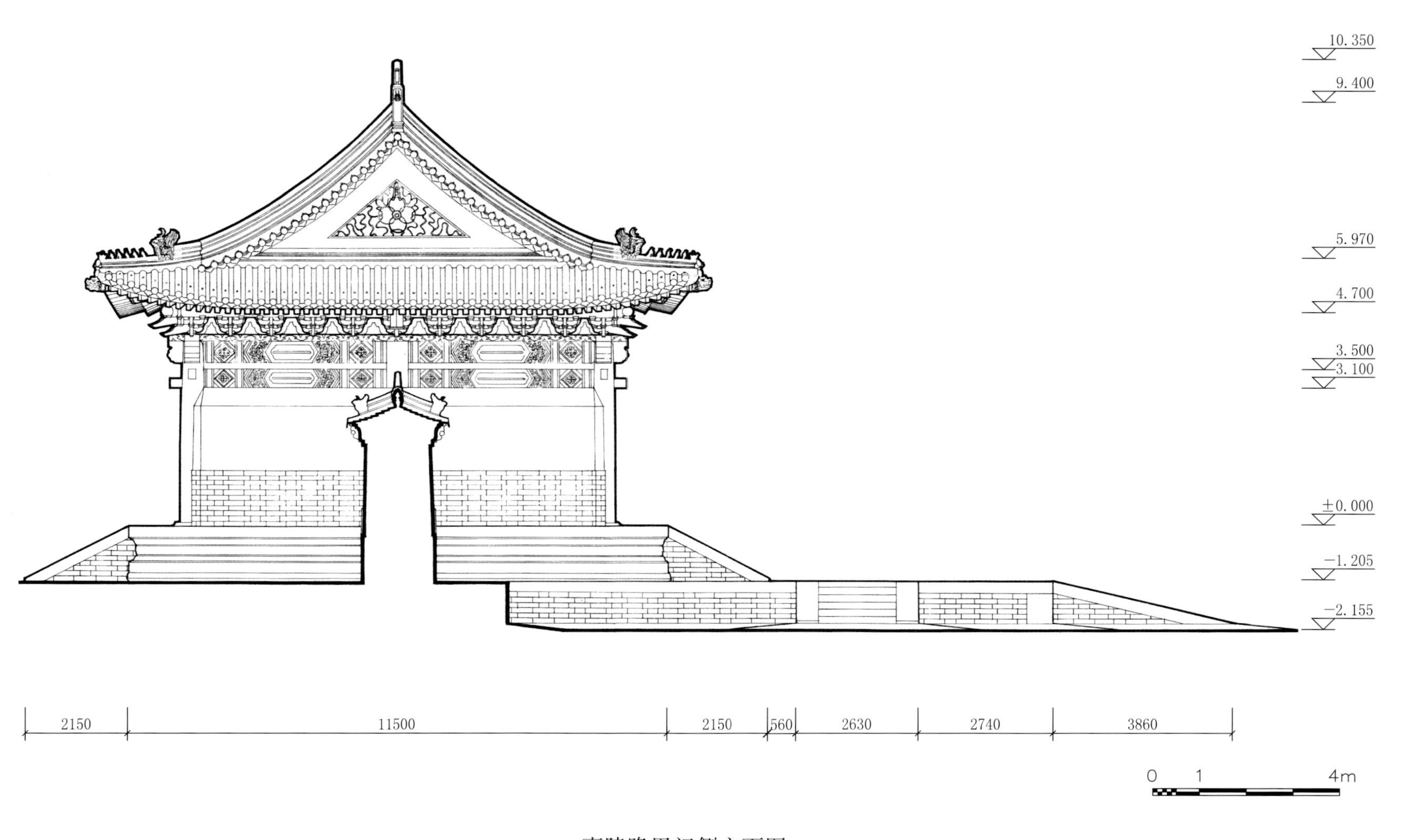

惠陵隆恩门侧立面图
Side elevation of the Gate of Monumental Grace, Tomb of Benevolence

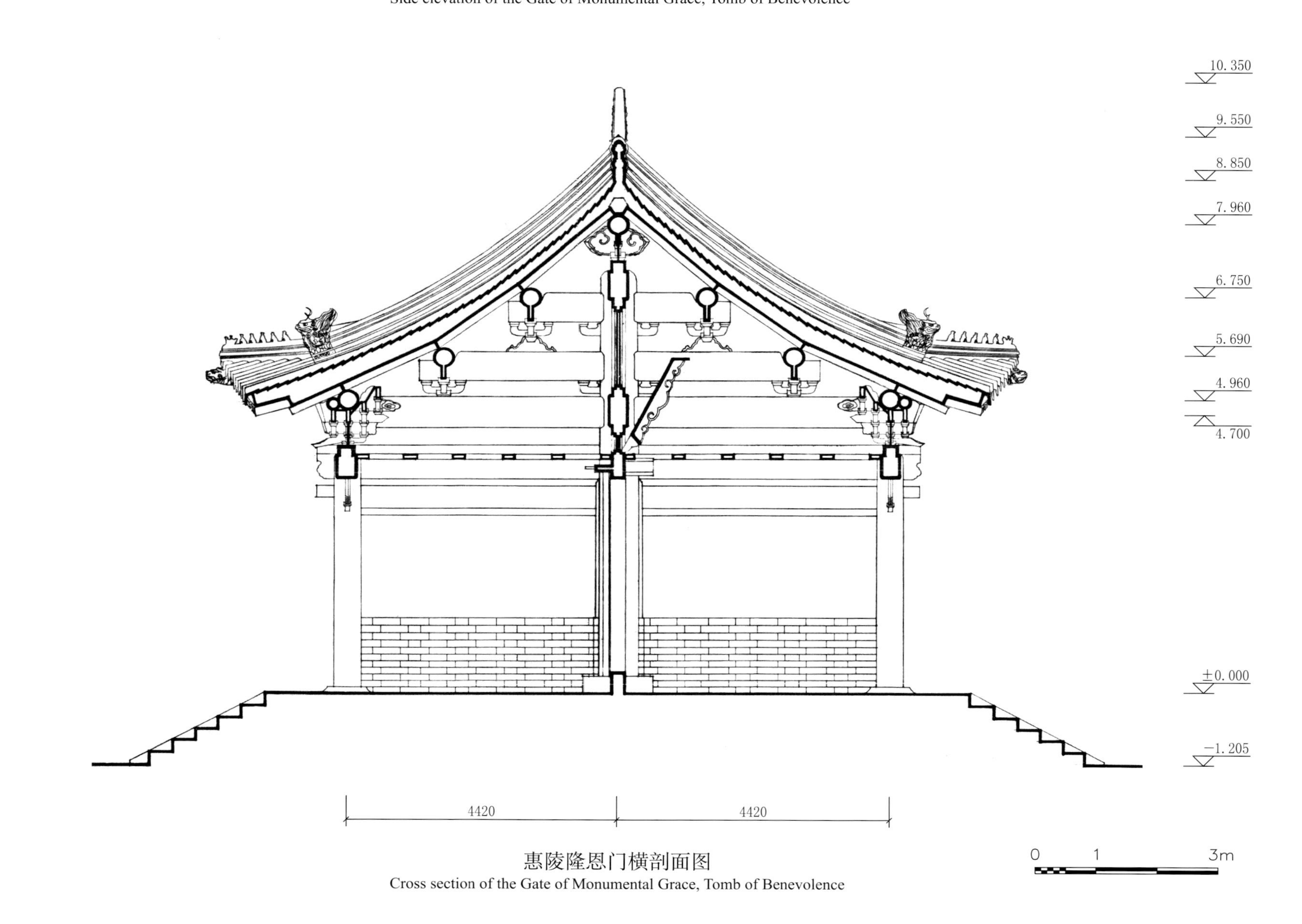

惠陵隆恩门横剖面图
Cross section of the Gate of Monumental Grace, Tomb of Benevolence

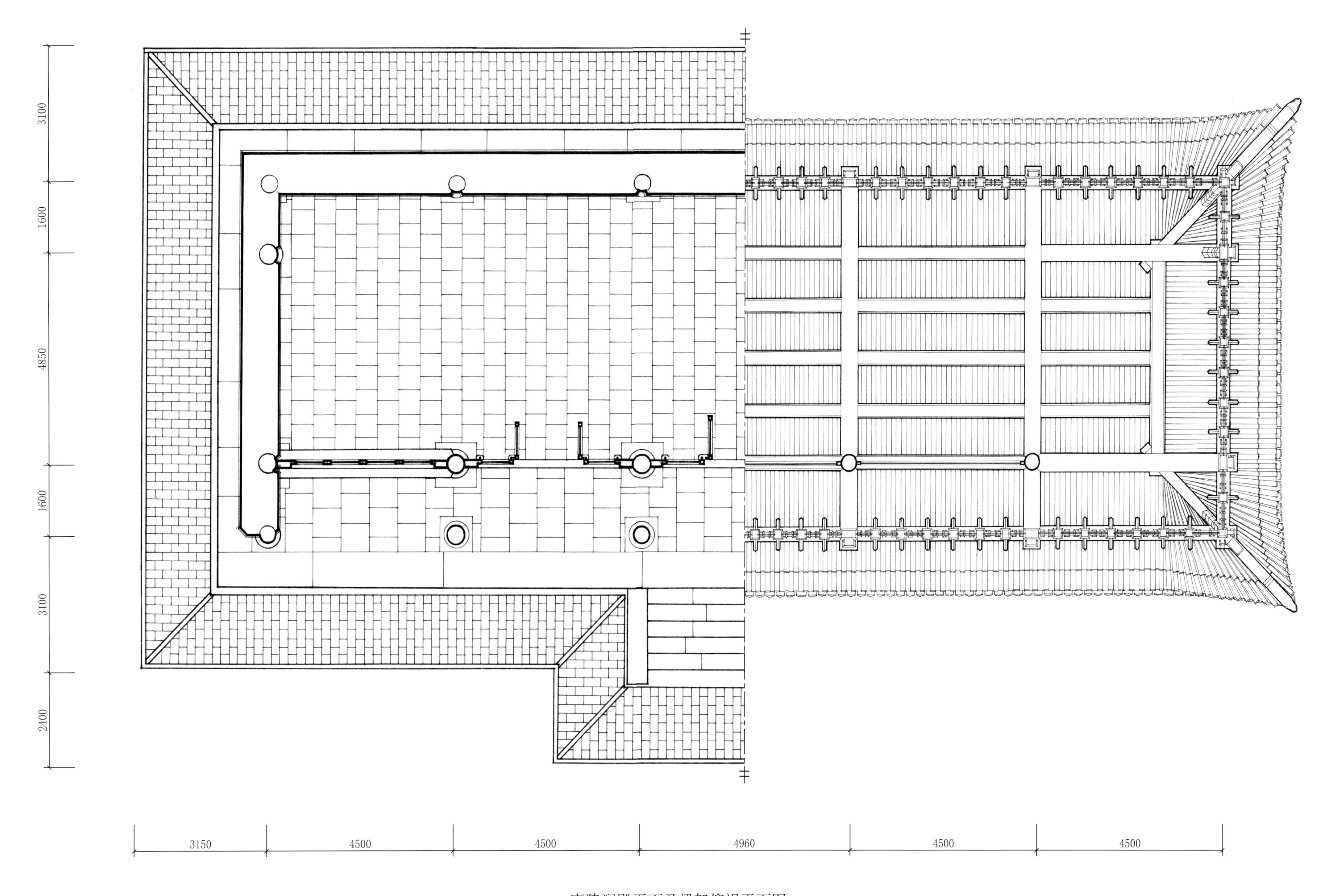

惠陵配殿平面及梁架仰视平面图

Ground plan and looking up at the Truss of the Side Hall, Tomb of Benevolence

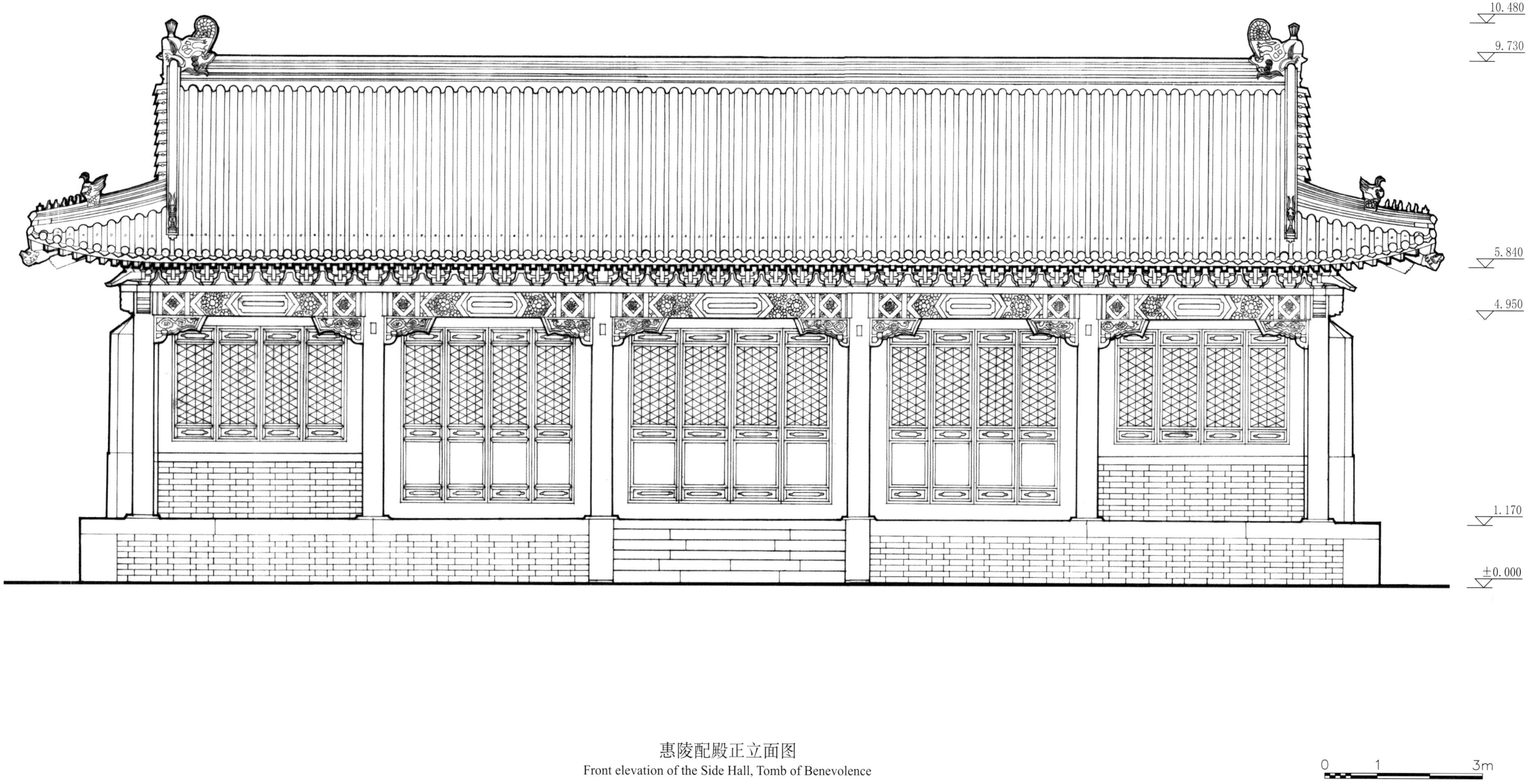

10.480
9.730
5.840
4.950
1.170
±0.000

惠陵配殿正立面图
Front elevation of the Side Hall, Tomb of Benevolence

0 1 3m

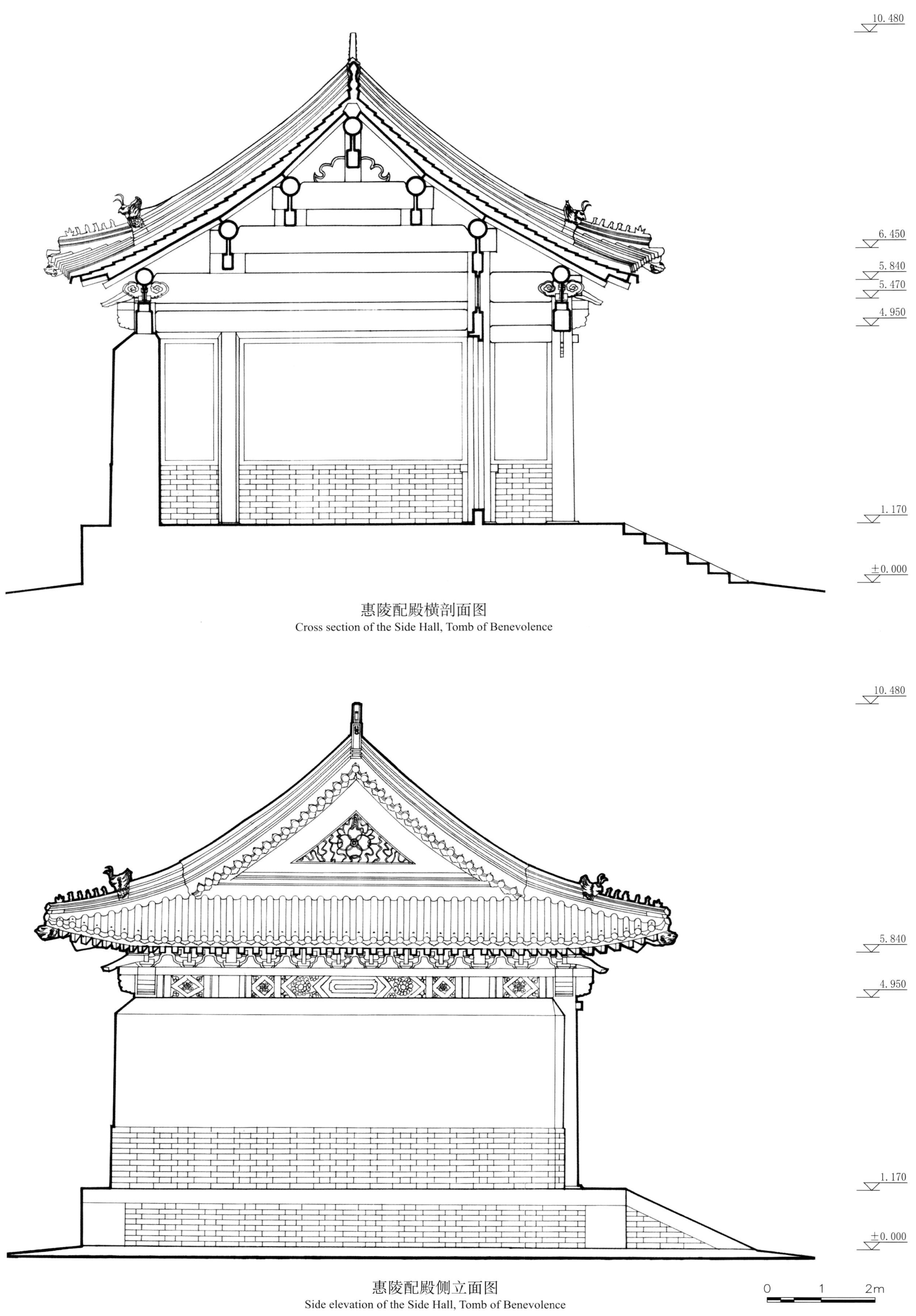

惠陵配殿横剖面图
Cross section of the Side Hall, Tomb of Benevolence

惠陵配殿侧立面图
Side elevation of the Side Hall, Tomb of Benevolence

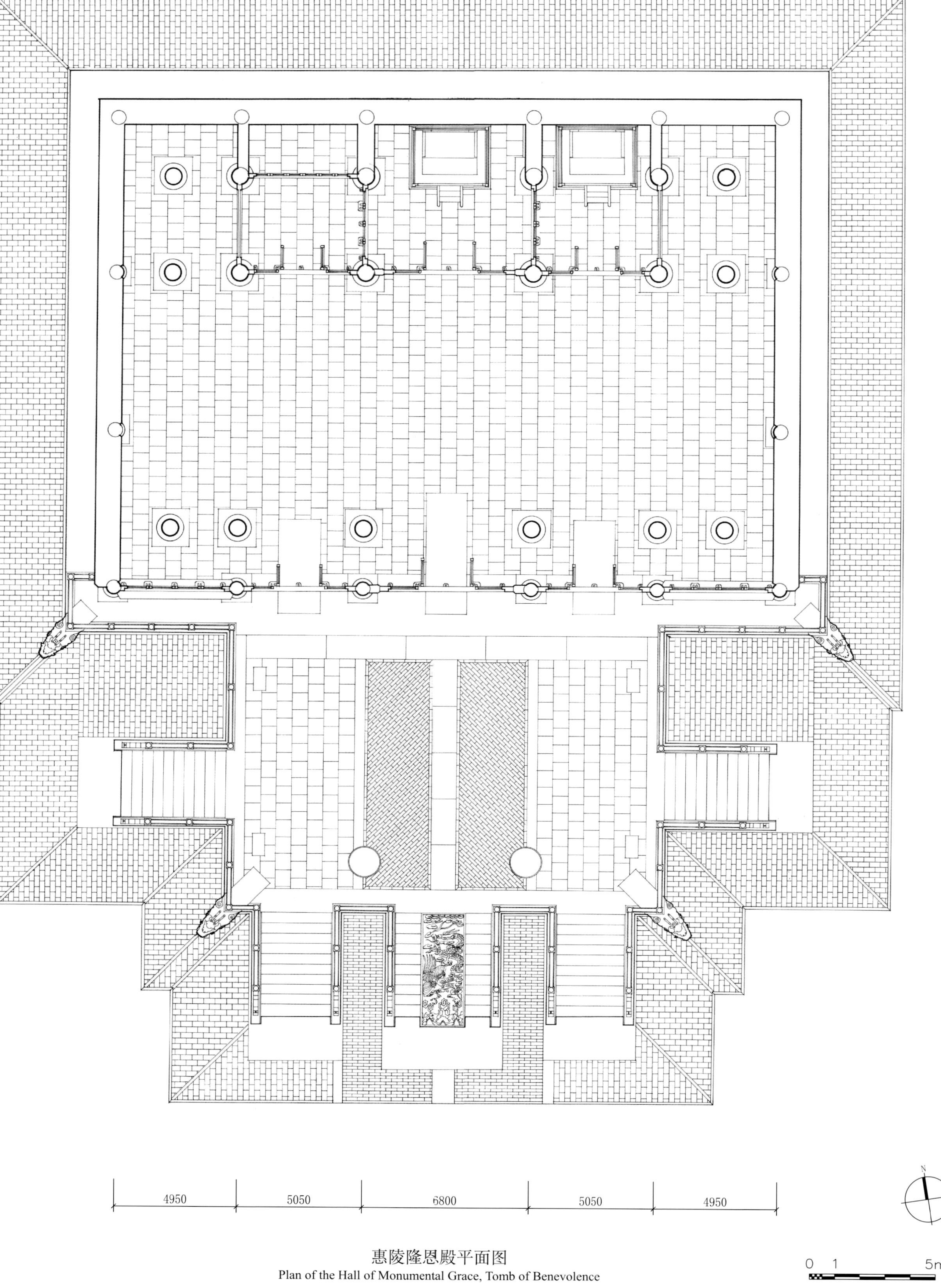

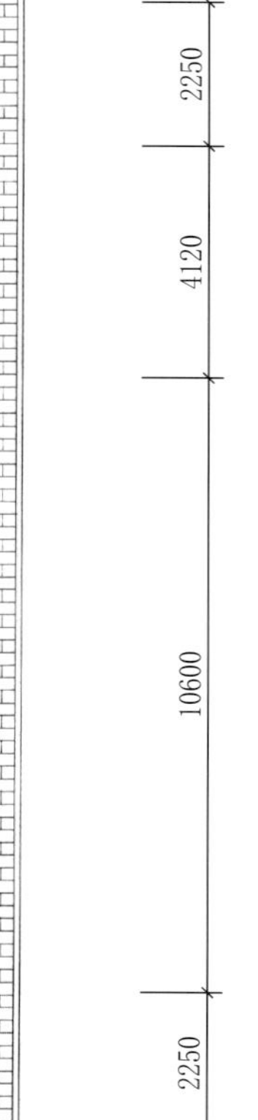

惠陵隆恩殿平面图
Plan of the Hall of Monumental Grace, Tomb of Benevolence

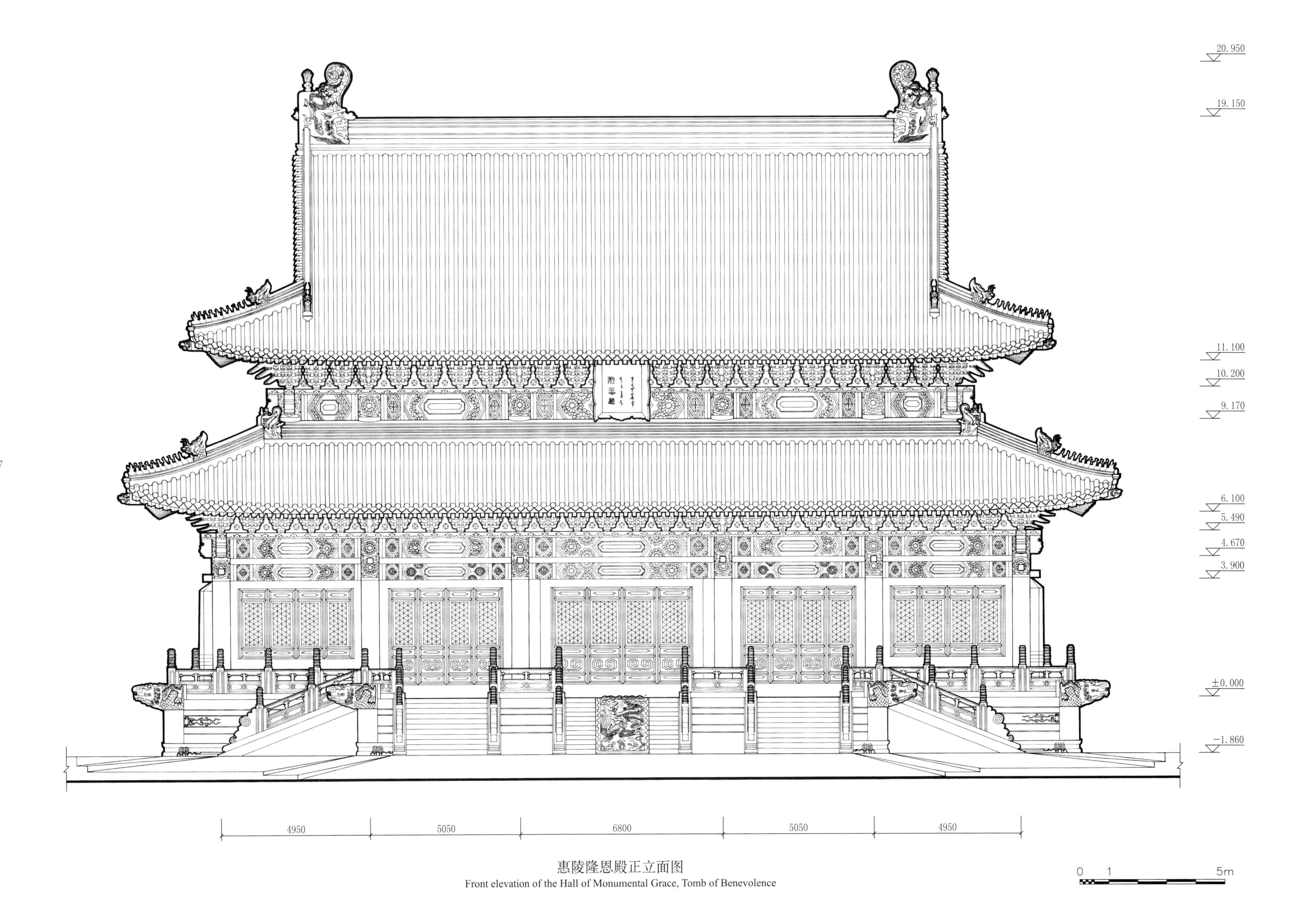

惠陵隆恩殿正立面图

Front elevation of the Hall of Monumental Grace, Tomb of Benevolence

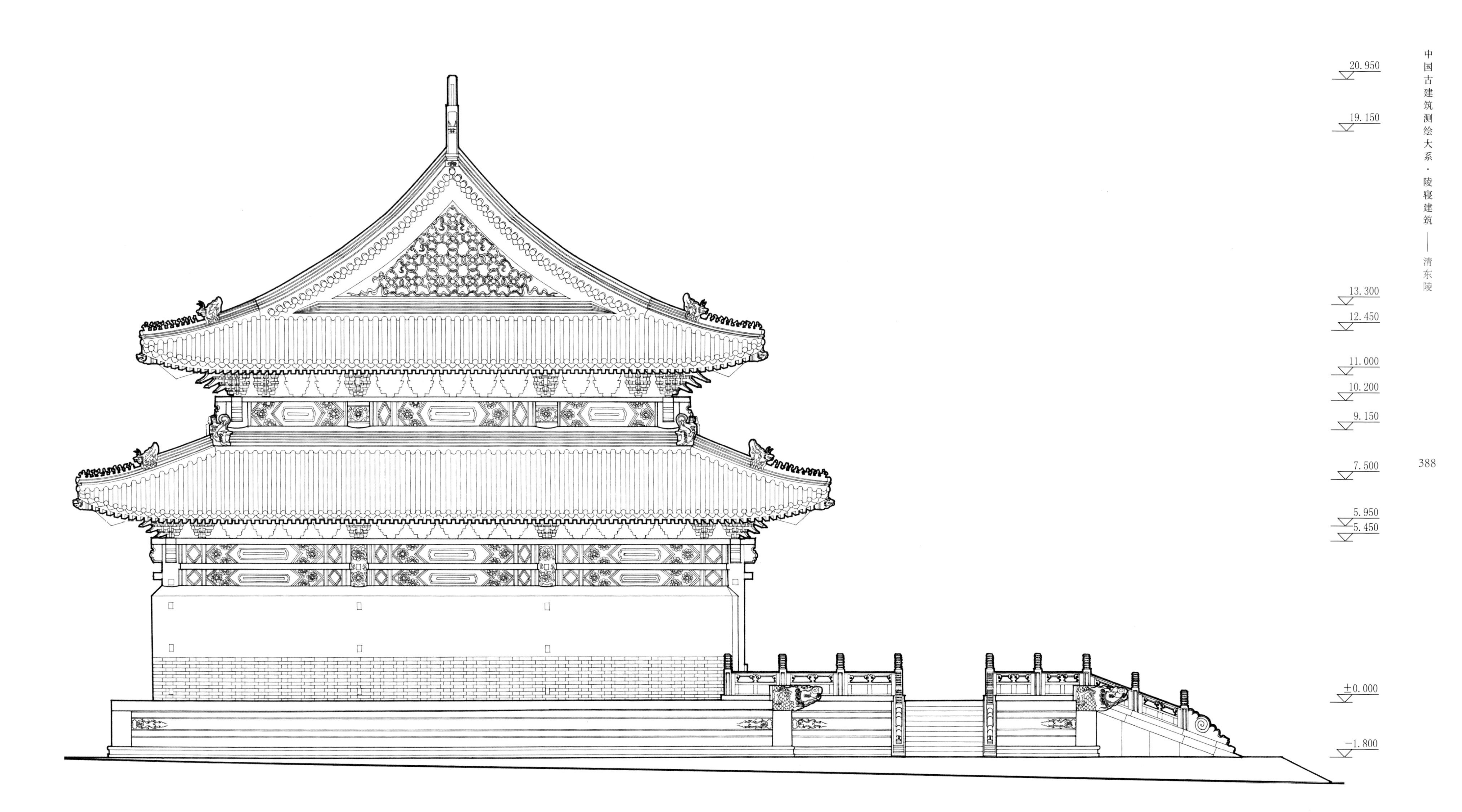

惠陵隆恩殿侧立面图

Side elevation of the Hall of Monumental Grace, Tomb of Benevolence

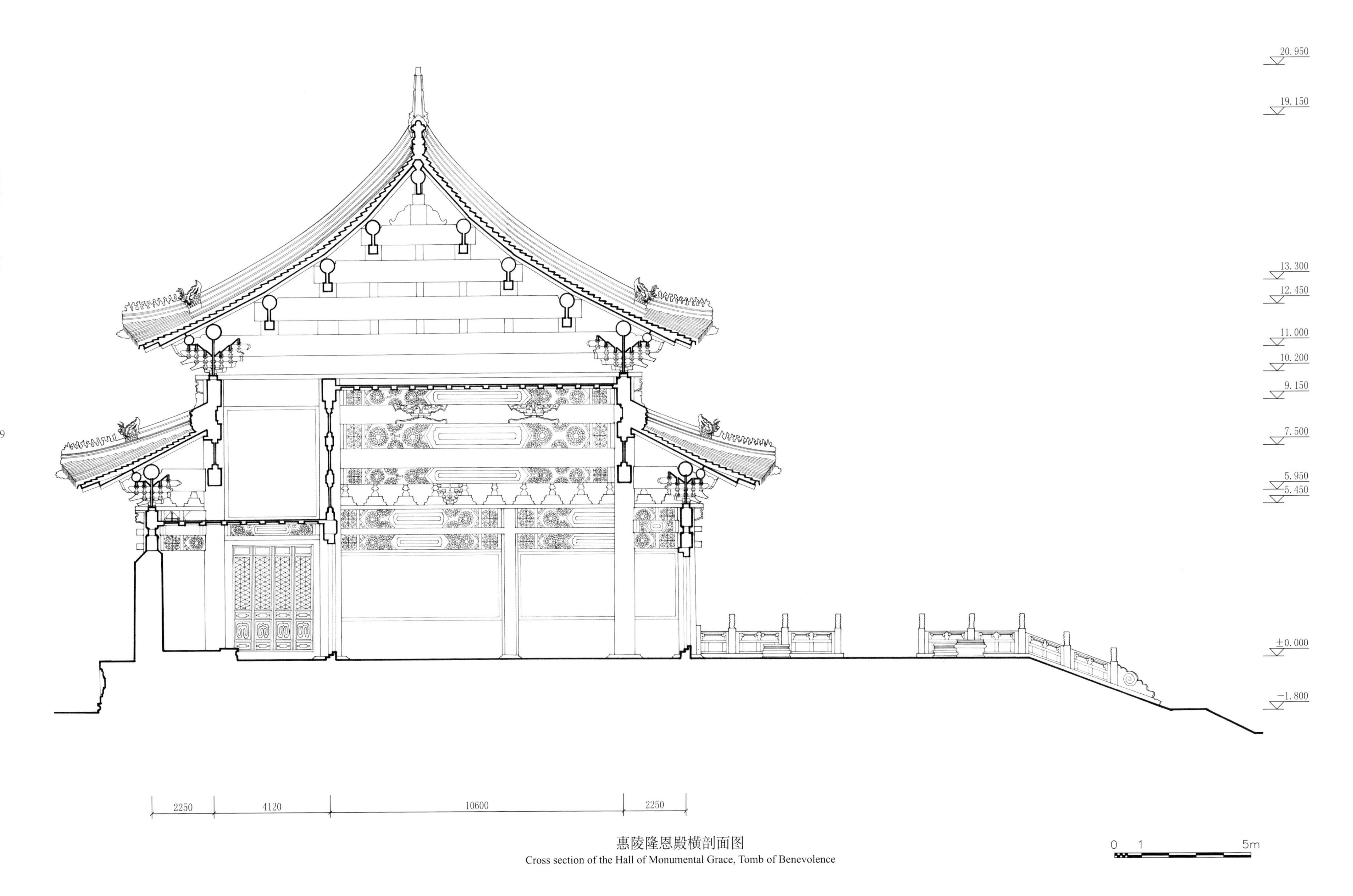

惠陵隆恩殿横剖面图

Cross section of the Hall of Monumental Grace, Tomb of Benevolence

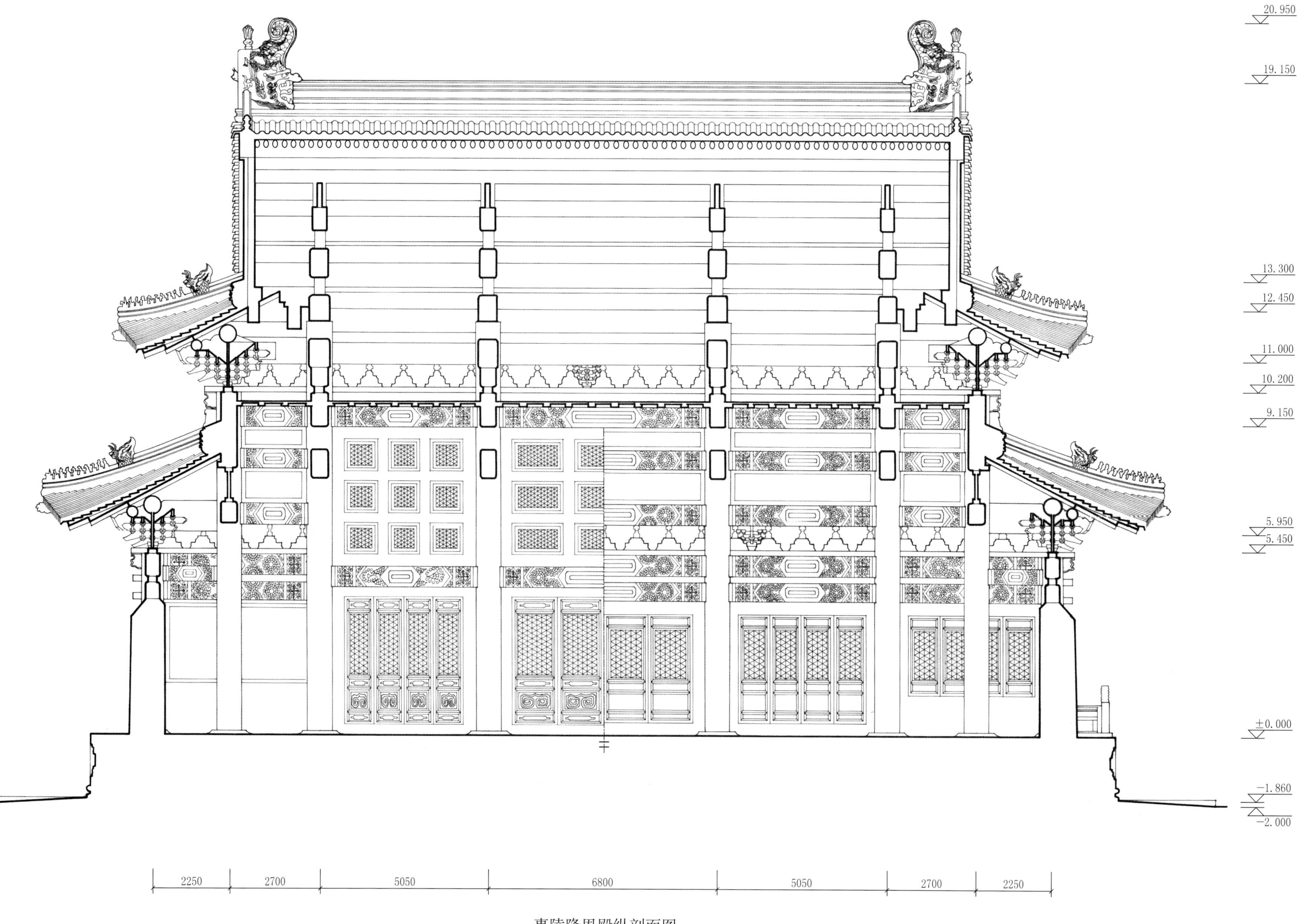

惠陵隆恩殿纵剖面图
Longitudinal section of the Hall of Monumental Grace, Tomb of Benevolence

惠陵隆恩殿螭首大样图
Detailed drawing of the dragon gargoyle (*chishou*) outside the Hall of Monumental Grace, Tomb of Benevolence

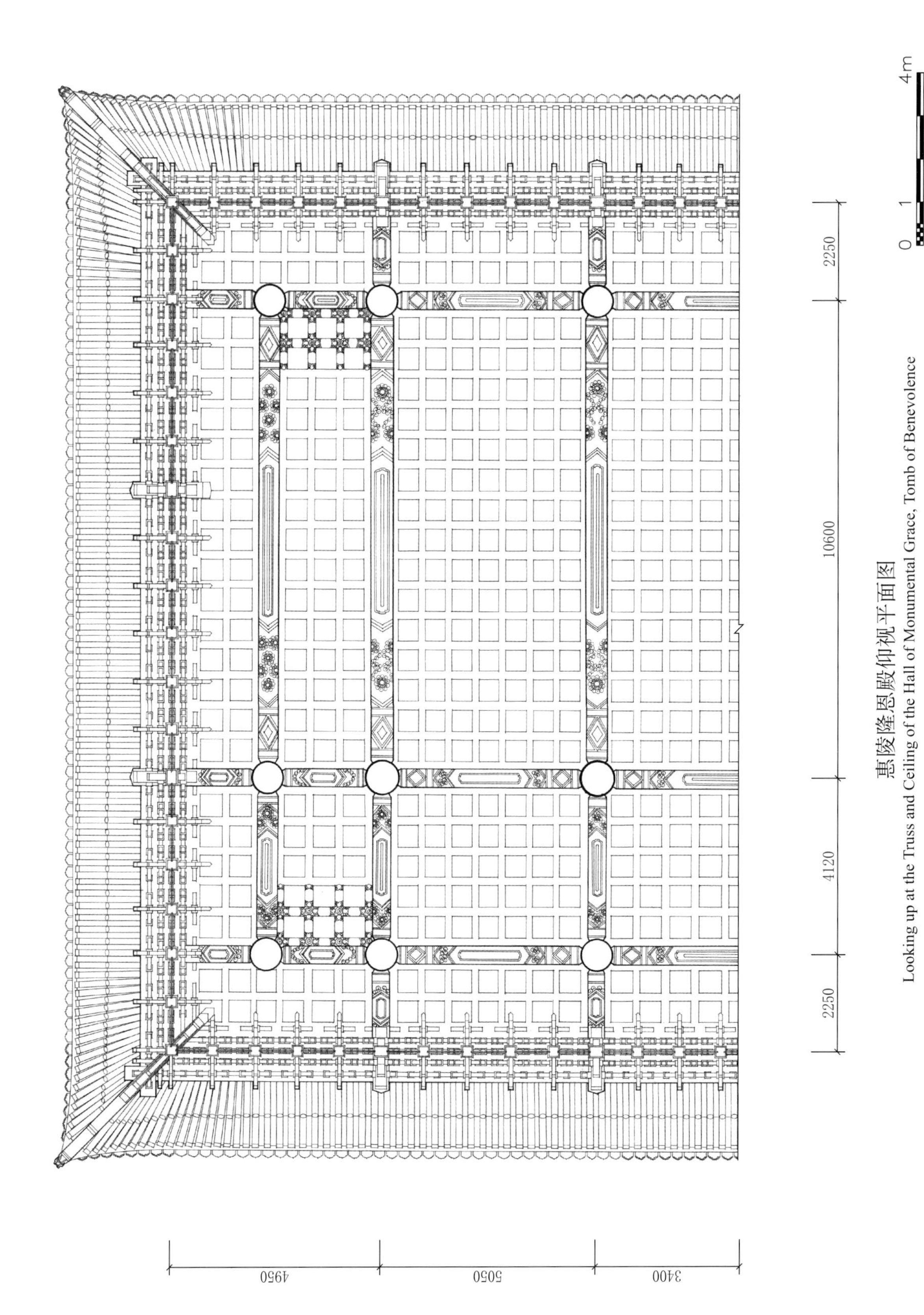

惠陵隆恩殿仰视平面图
Looking up at the Truss and Ceiling of the Hall of Monumental Grace, Tomb of Benevolence

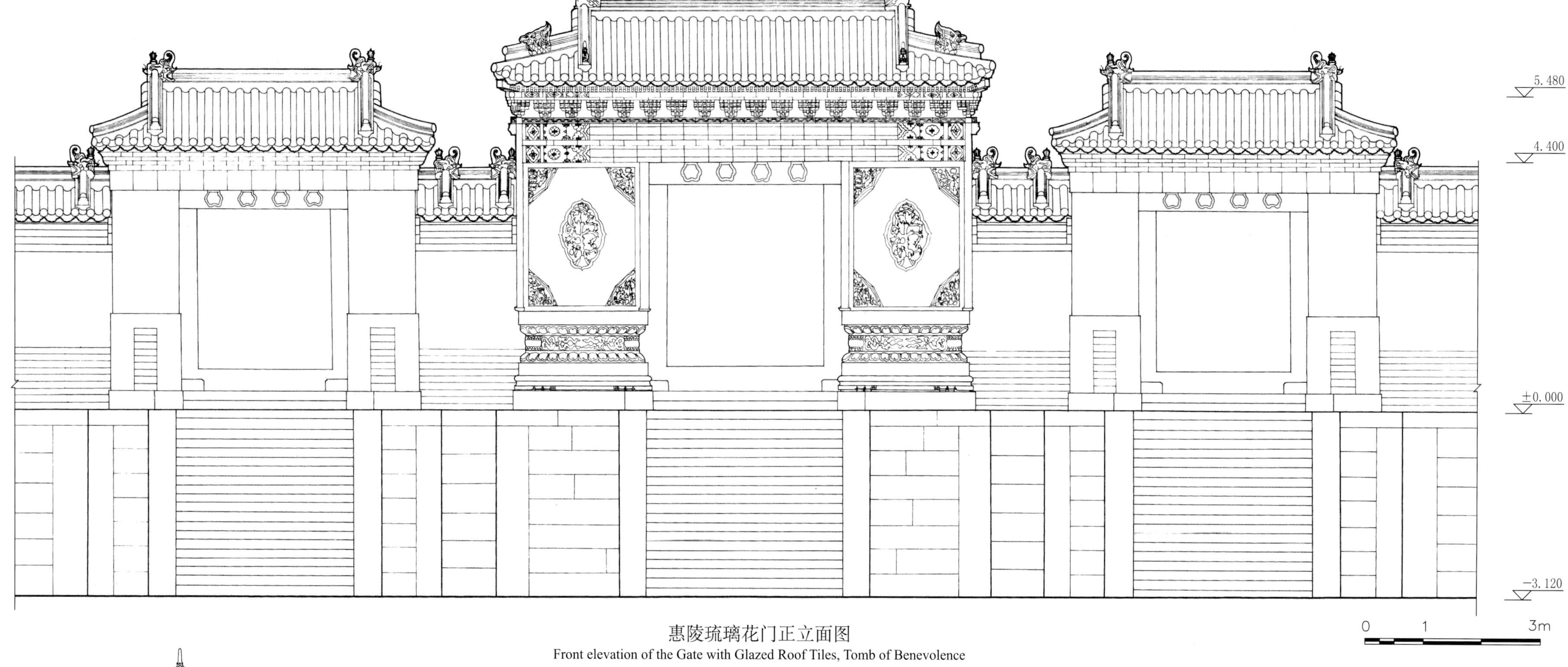

惠陵琉璃花门正立面图

Front elevation of the Gate with Glazed Roof Tiles, Tomb of Benevolence

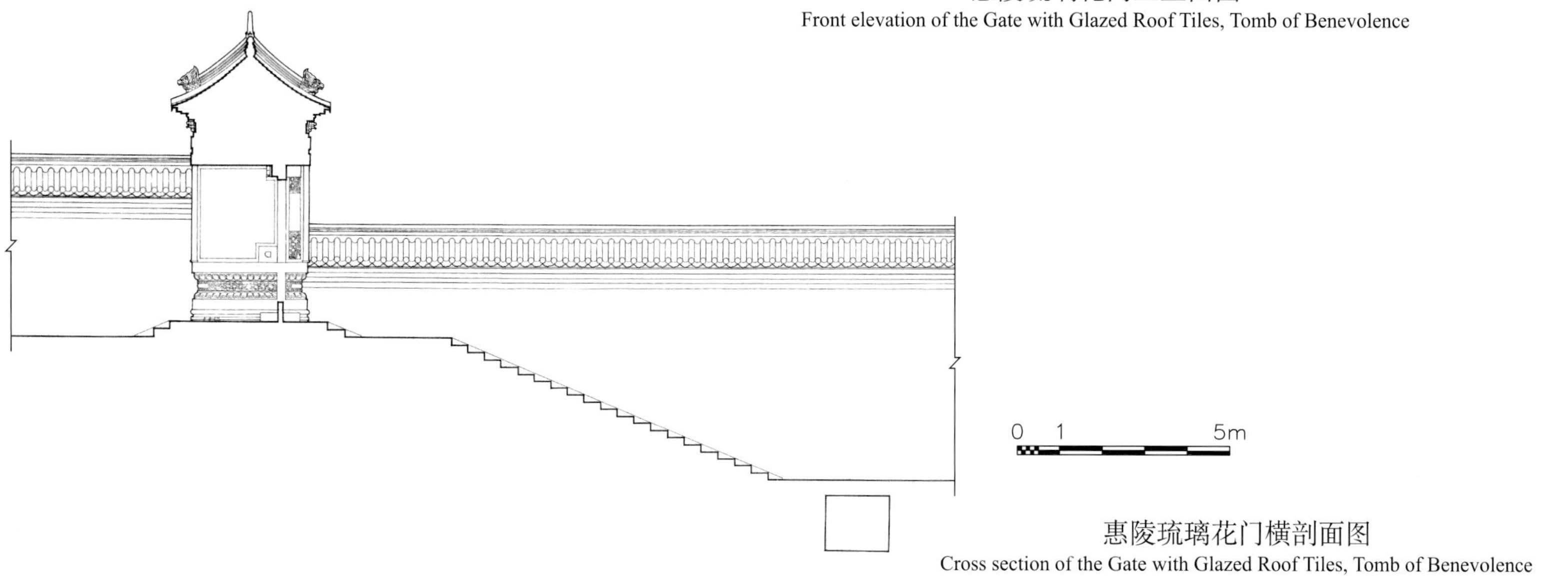

惠陵琉璃花门横剖面图

Cross section of the Gate with Glazed Roof Tiles, Tomb of Benevolence

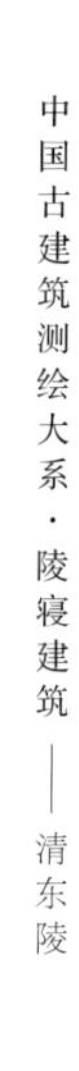

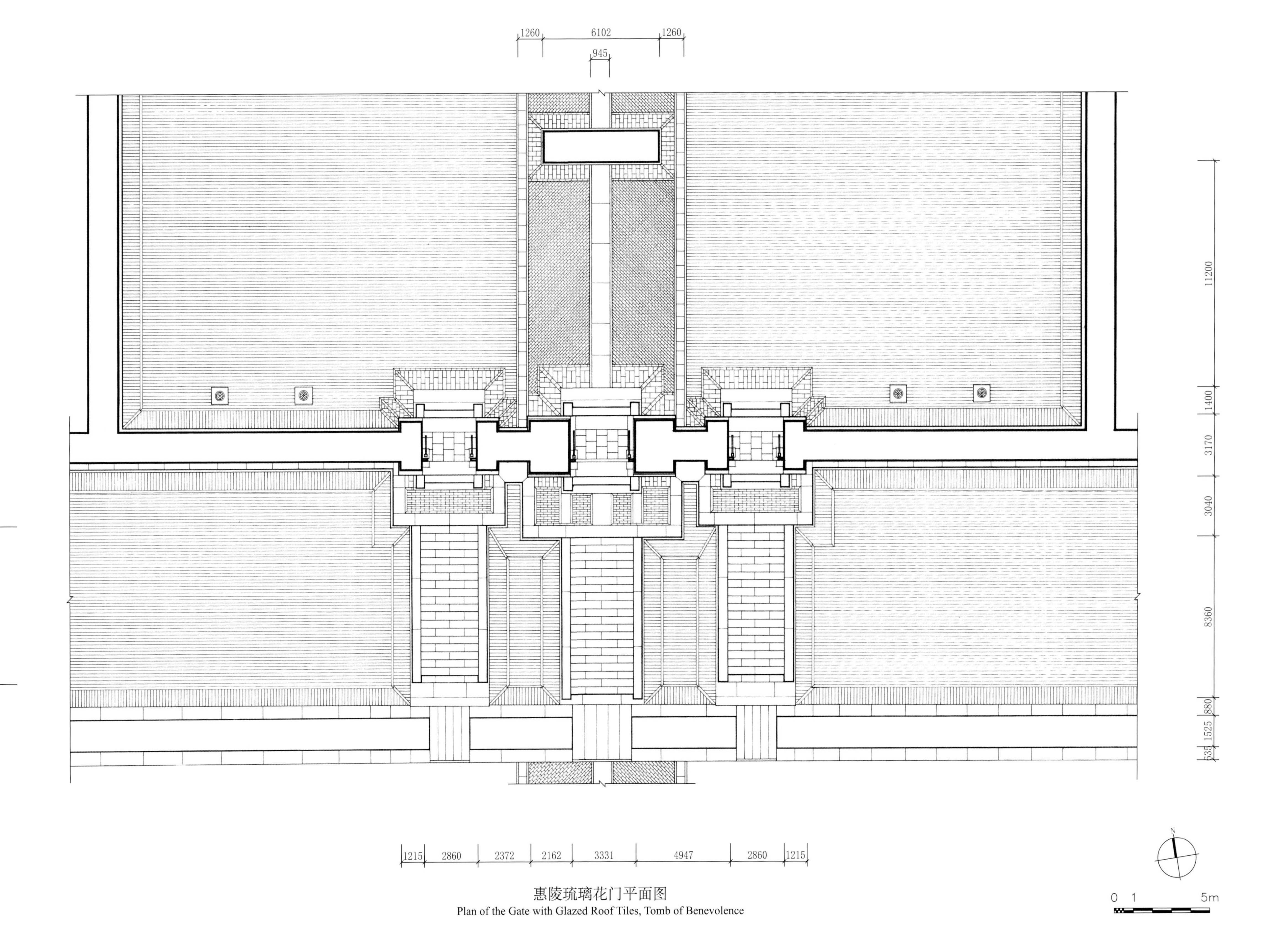

惠陵琉璃花门平面图

Plan of the Gate with Glazed Roof Tiles, Tomb of Benevolence

惠陵石五供正立面图
Front elevation of the Five Stone Ritual Vessels, Tomb of Benevolence

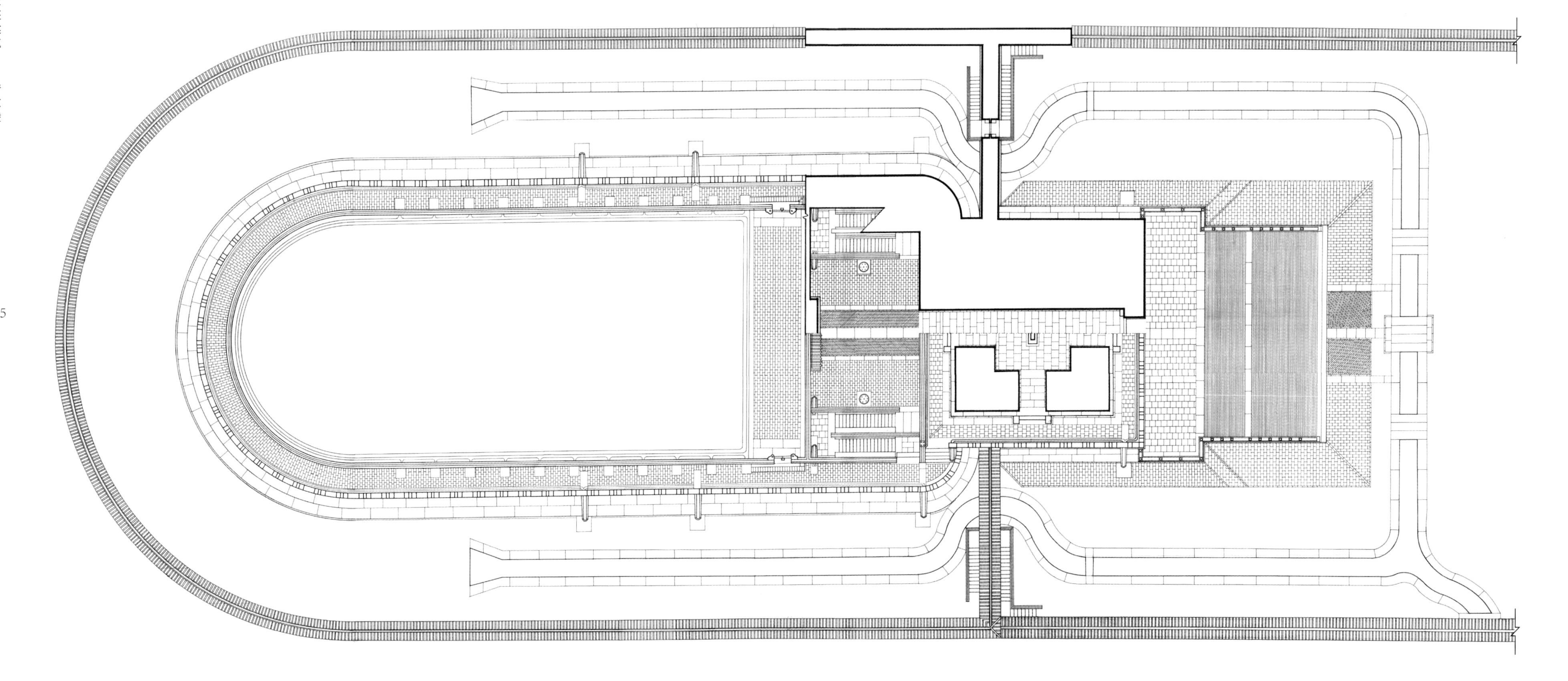

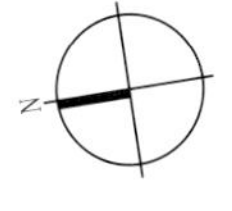

惠陵方城明楼及宝城宝顶平面图

Plan of the Square Walled Terrace and the Memorial Tower as well as the Encircled Realm of Treasure and the Tumulus, Tomb of Benevolence

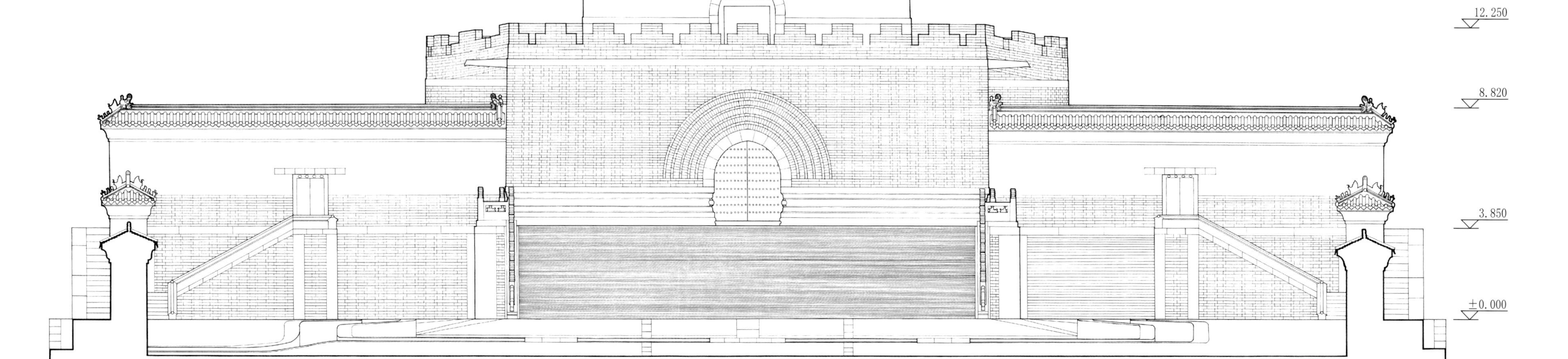

惠陵方城明楼正立面图

Front elevation of the Square Walled Terrace and the Memorial Tower, Tomb of Benevolence

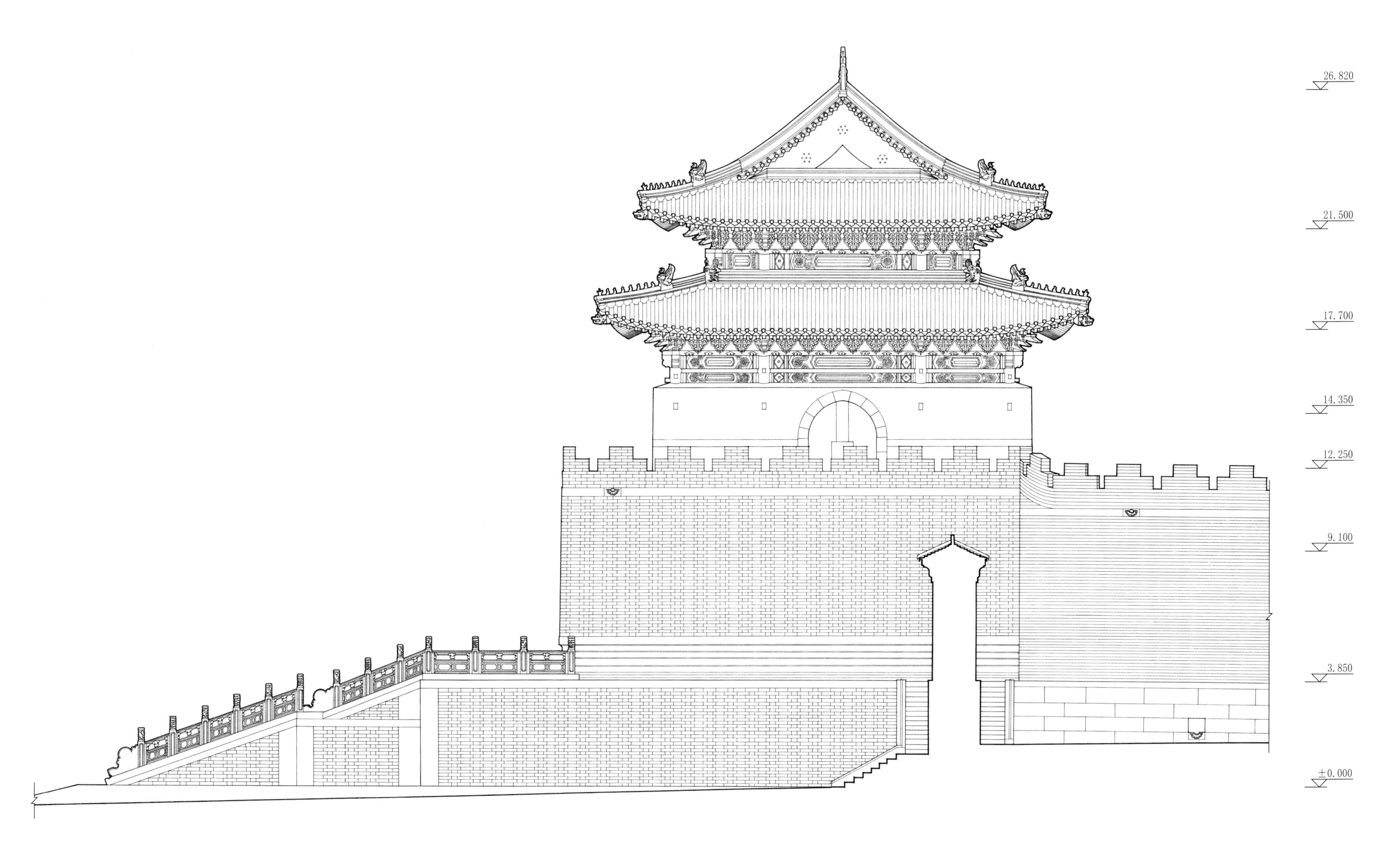

惠陵方城明楼侧立面图

Side elevation of Square Walled Terrace and the Memorial Tower, Tomb of Benevolence

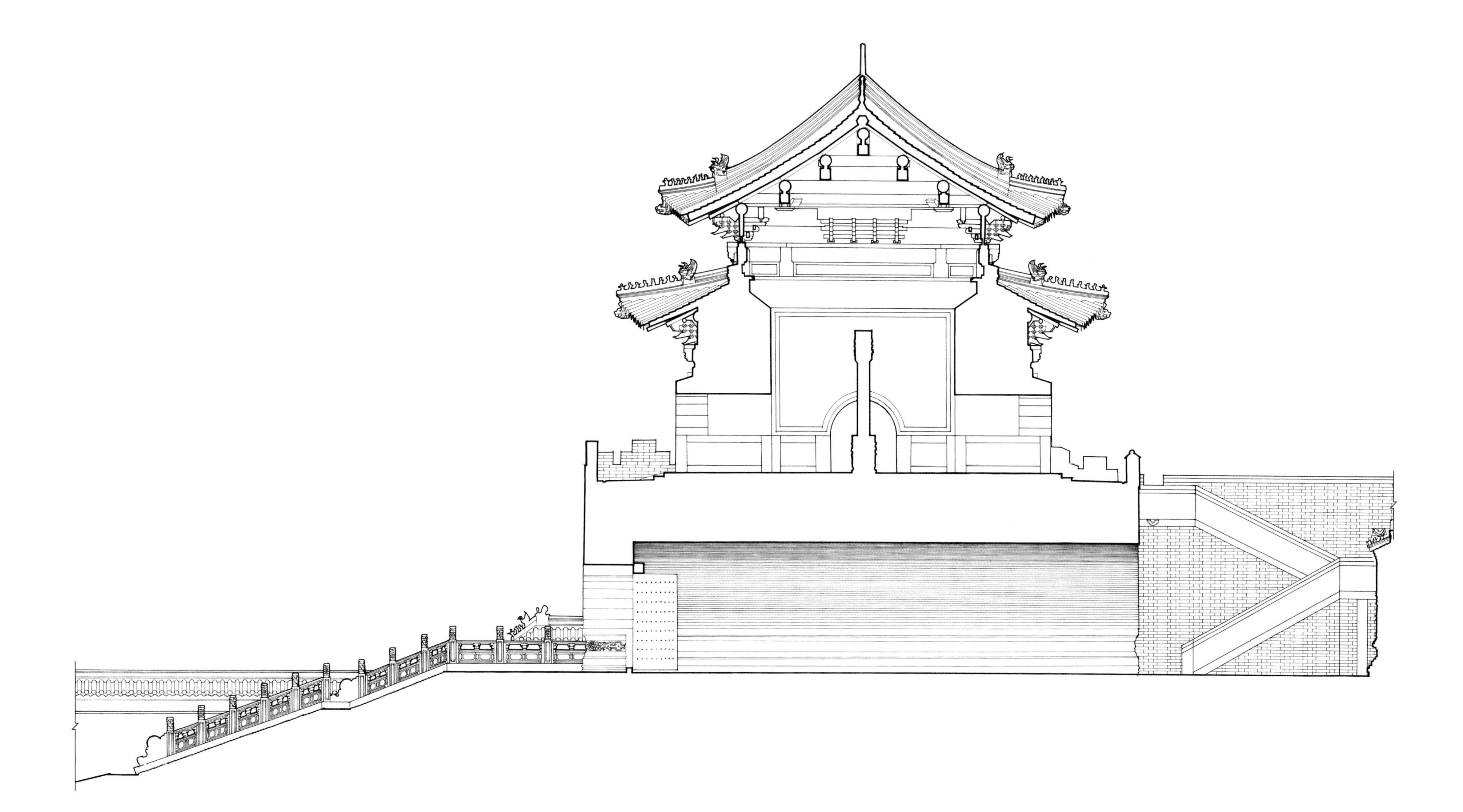

惠陵方城明楼横剖面图

Cross section of the Square Walled Terrace and the Memorial Tower, Tomb of Benevolence

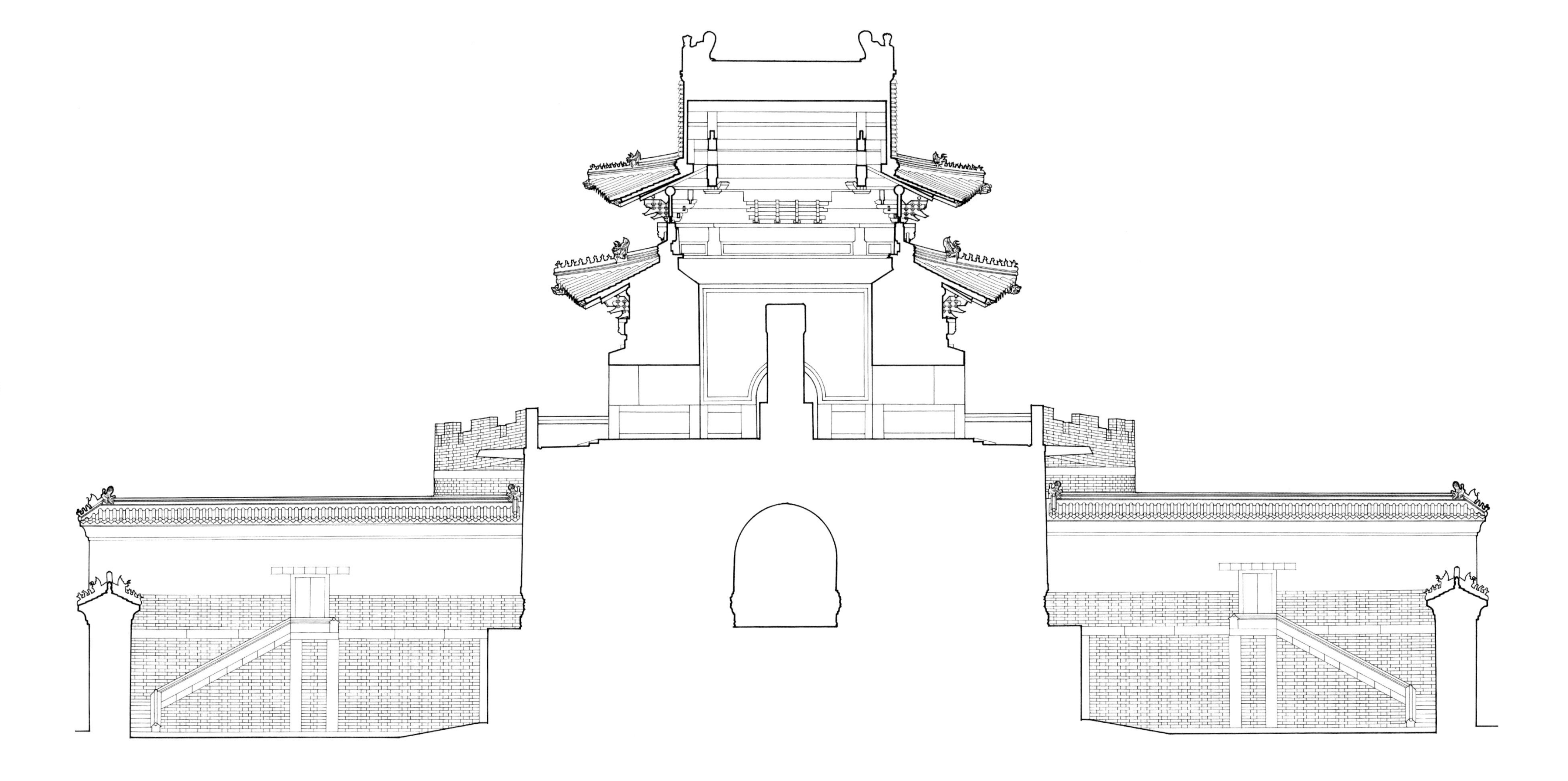

惠陵方城明楼纵剖面图

Longitudinal section of the Square Walled Terrace and the Memorial Tower, Tomb of Benevolence

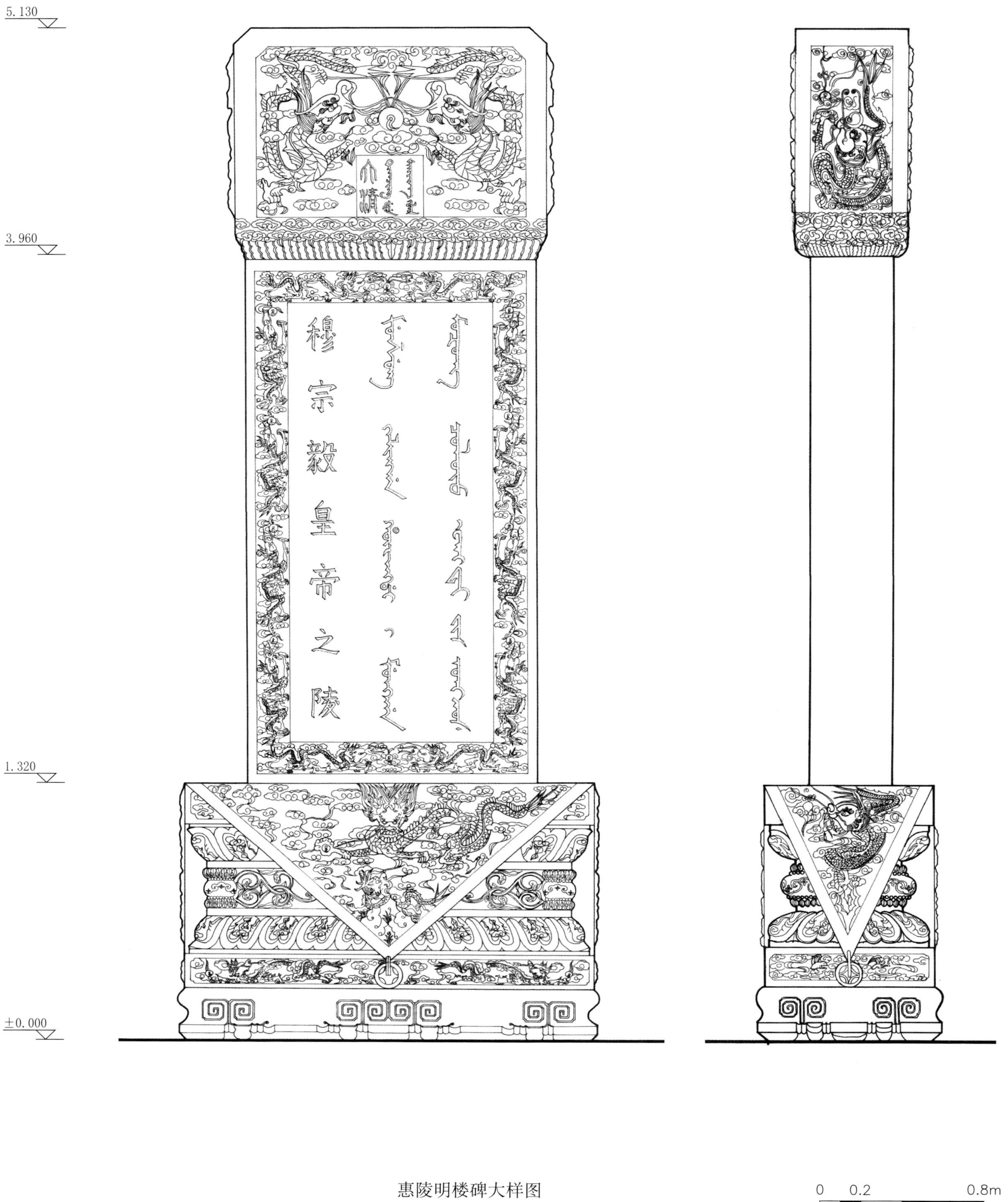

5.130

3.960

大清

穆宗毅皇帝之陵

1.320

±0.000

0 0.2 0.8m

惠陵明楼碑大样图
Detailed drawing of the Marble Stela in the Memorial Tower, Tomb of Benevolence

惠陵妃园寝

Imperial Consorts' Tombs affiliated with the Tomb of Benevolence (Huiling Feiyuanqin)

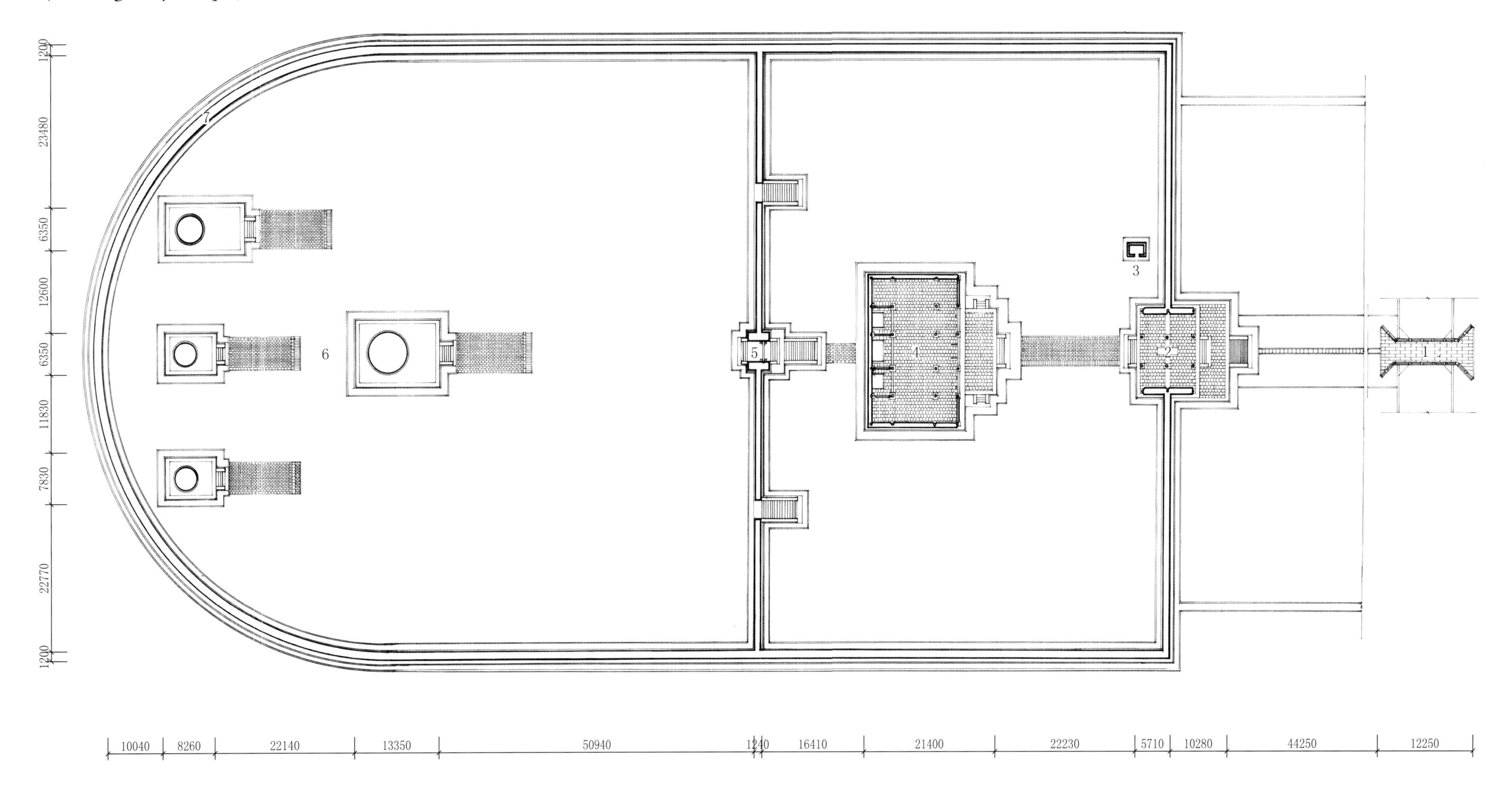

1 单孔桥 Single-Arch Bridge
2 宫门 Main Gate
3 焚帛炉 Sacrificial Burners
4 享殿 Main Sacrificial Hall
5 琉璃花门 Gate with Glazed Roof Tiles
6 妃嫔宝顶 Concubines' Tumulus
7 罗圈墙 Wall Surrounding the Tomb

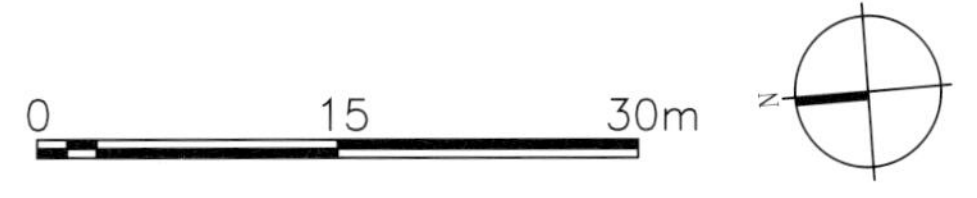

惠陵妃园寝总平面图

Site plan of the Imperial Consorts' Tombs affiliated with the Tomb of Benevolence

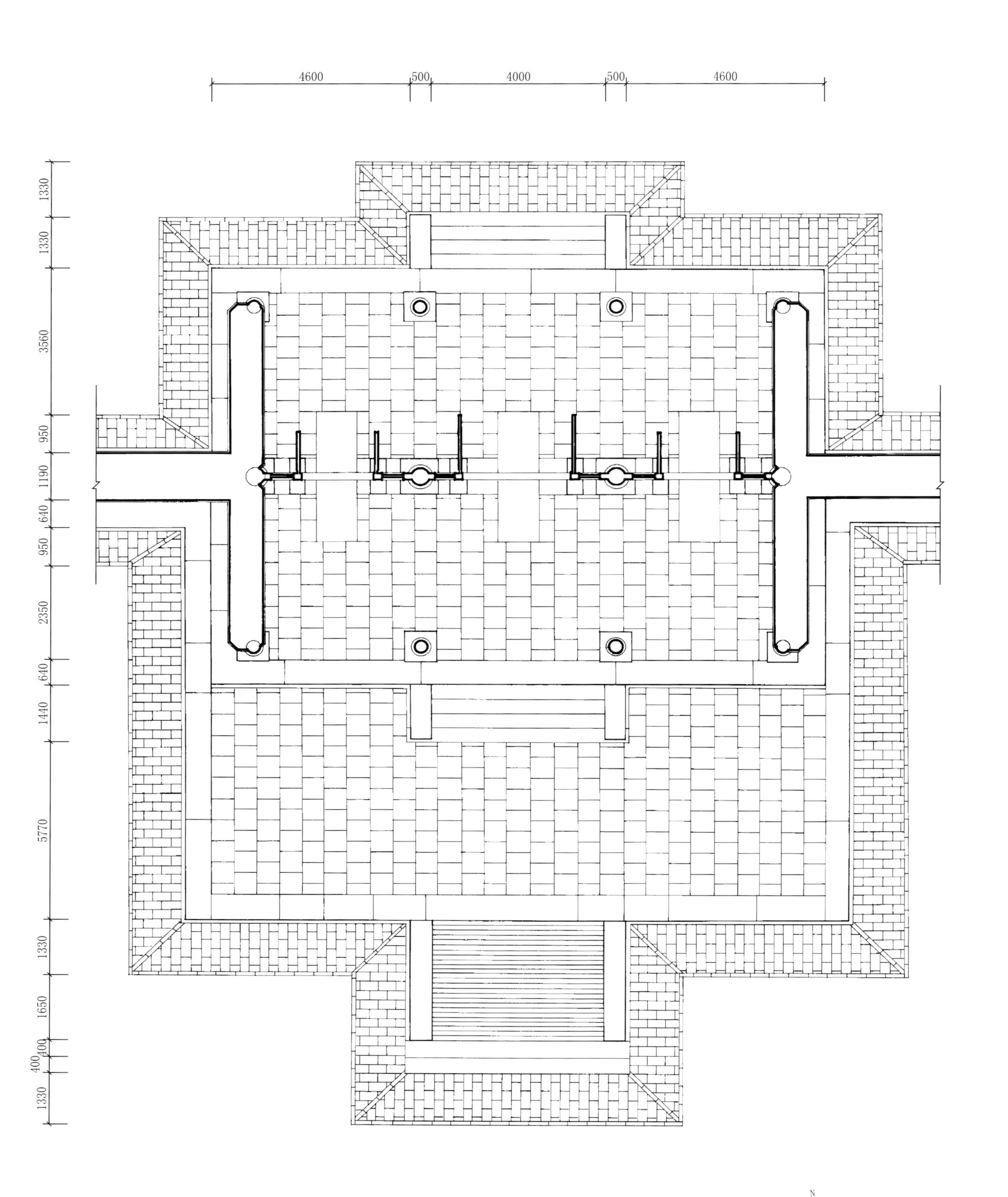

惠陵妃园寝宫门平面图
Plan of the Main Gate, Imperial Consorts' Tombs affiliated with the Tomb of Benevolence

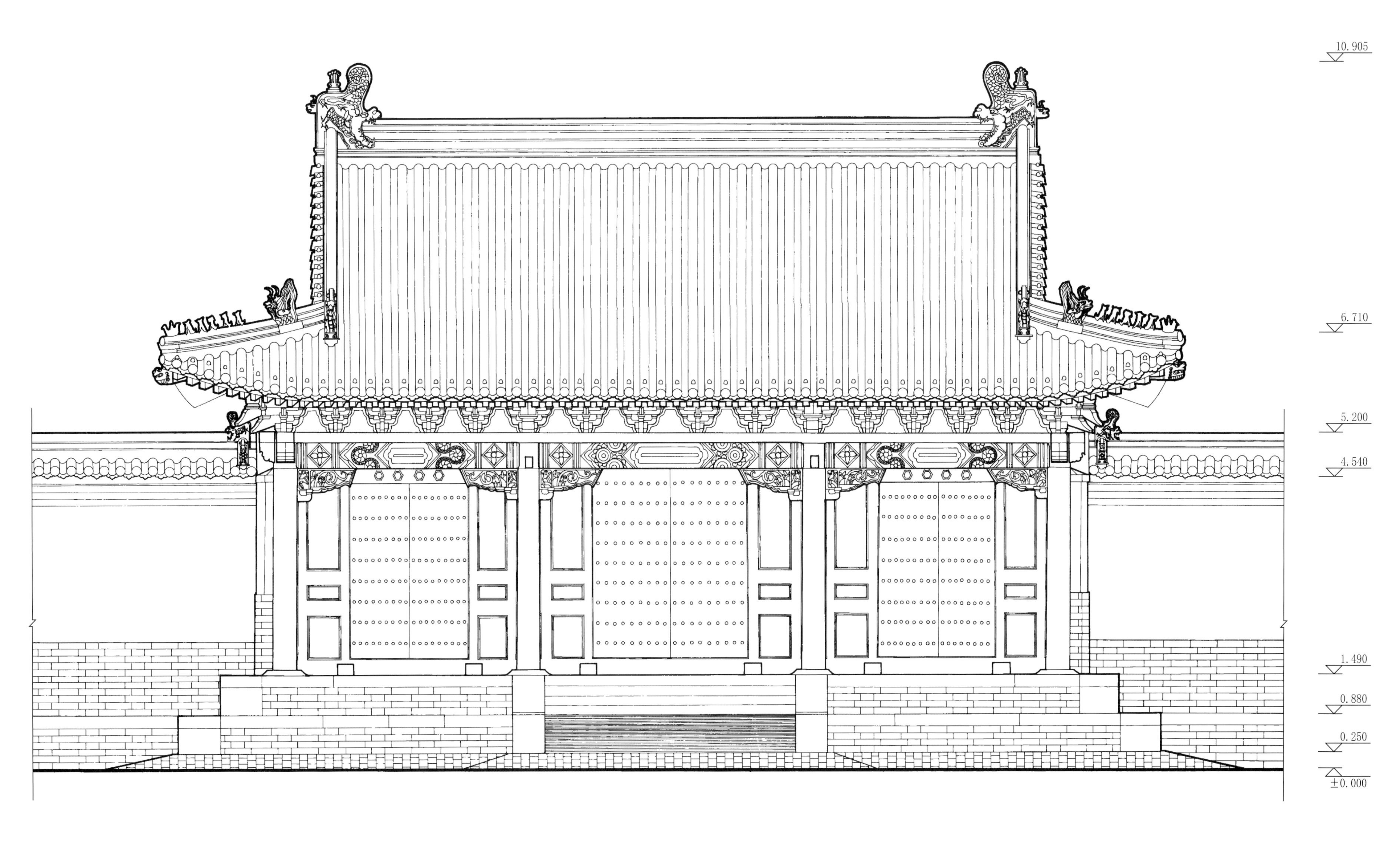

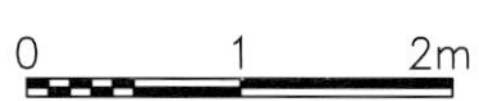

惠陵妃园寝宫门正立面图

Front elevation of the Main Gate, Imperial Consorts' Tombs affiliated with the Tomb of Benevolence

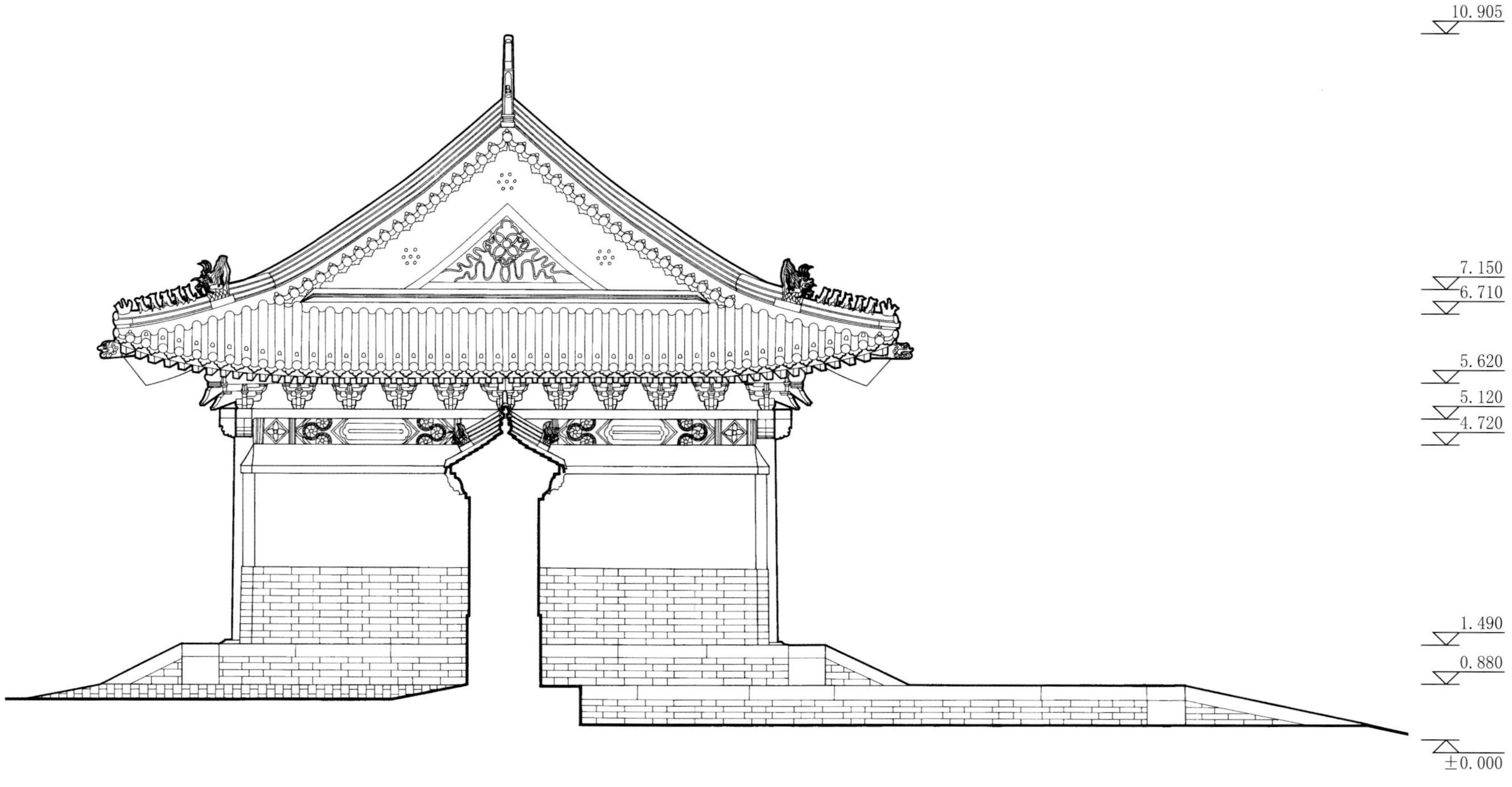

惠陵妃园寝宫门侧立面图

Side elevation of the Main Gate, Imperial Consorts' Tombs affiliated with the Tomb of Benevolence

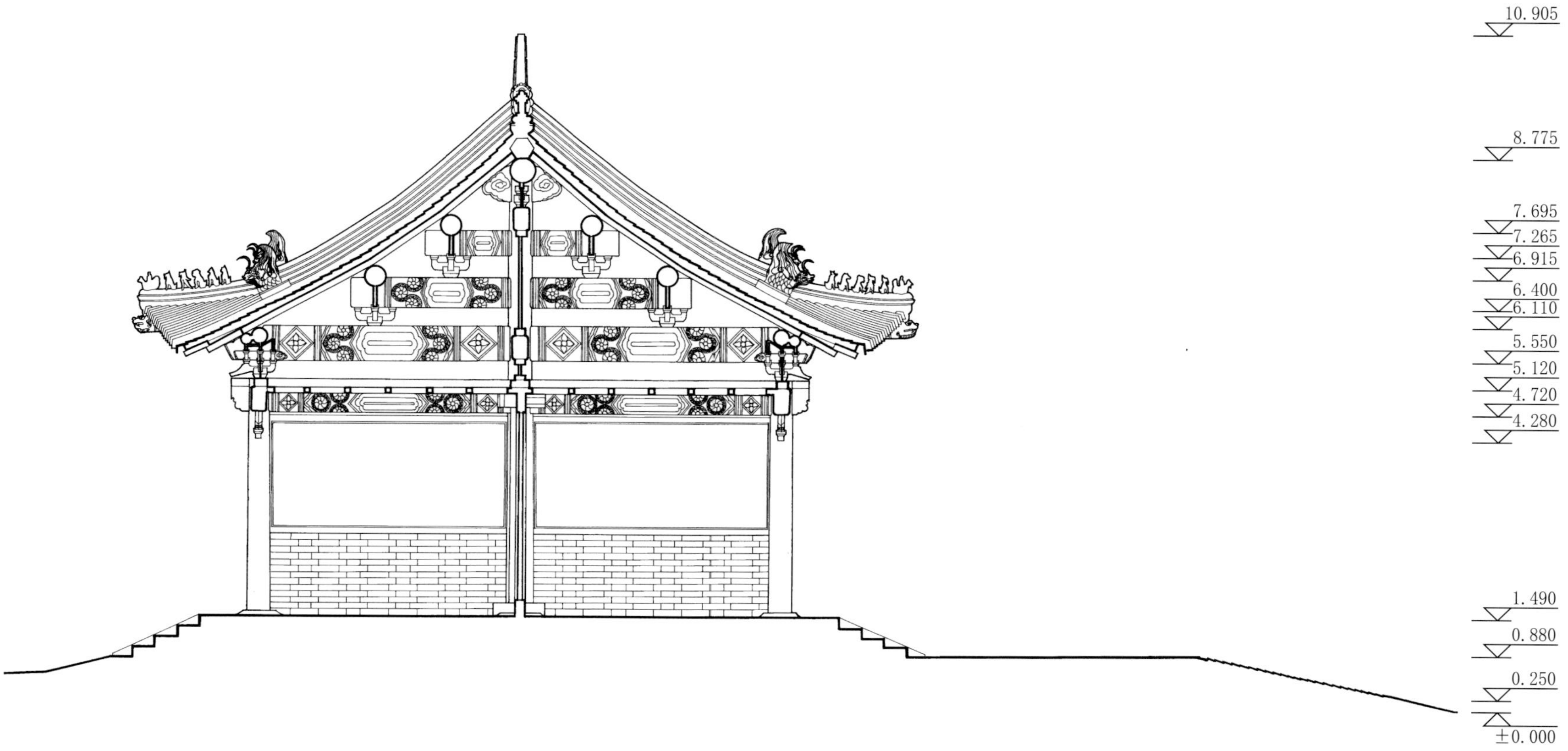

惠陵妃园寝宫门横剖面图

Cross section of the Main Gate, Imperial Consorts' Tombs affiliated with the Tomb of Benevolence

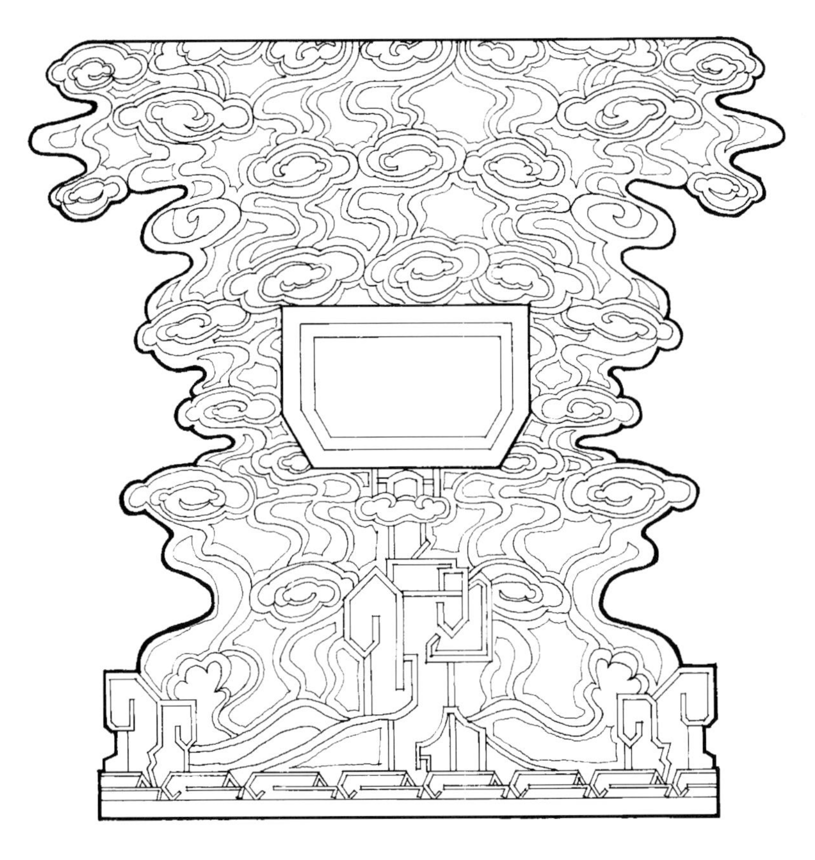

惠陵妃园寝享殿隔架科斗栱大样图

Detailed drawing of the Bracket (*dougong*) between two beams in the Hall of Ritual Sacrifice, Imperial Consorts' Tombs affiliated with the Tomb of Benevolence

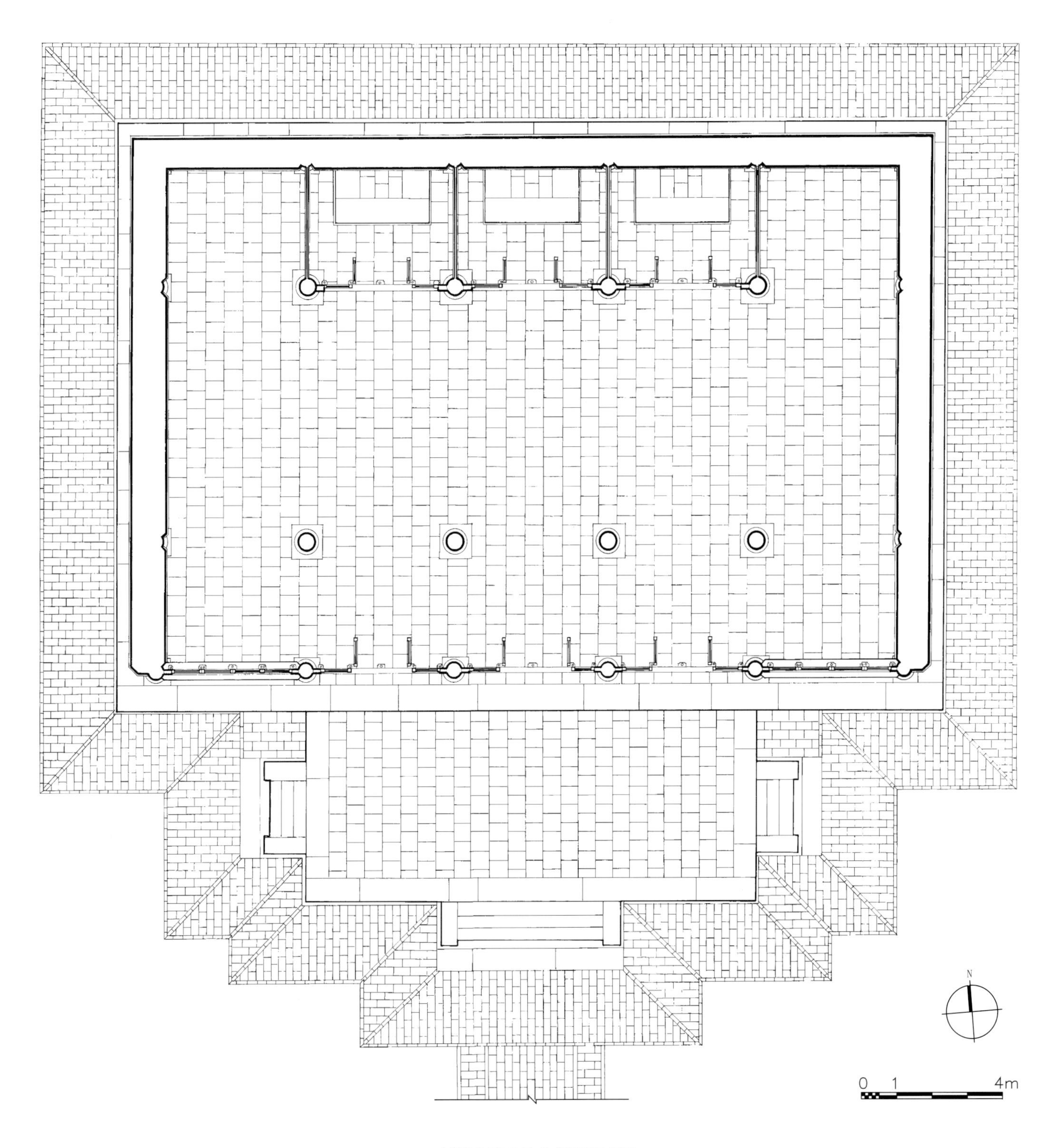

惠陵妃园寝享殿平面图

Plan of the Hall of Ritual Sacrifice, Imperial Consorts' Tombs affiliated with the Tomb of Benevolence

惠陵妃园寝享殿正立面图

Front elevation of the Hall of Ritual Sacrifice, Imperial Consorts' Tombs affiliated with the Tomb of Benevolence

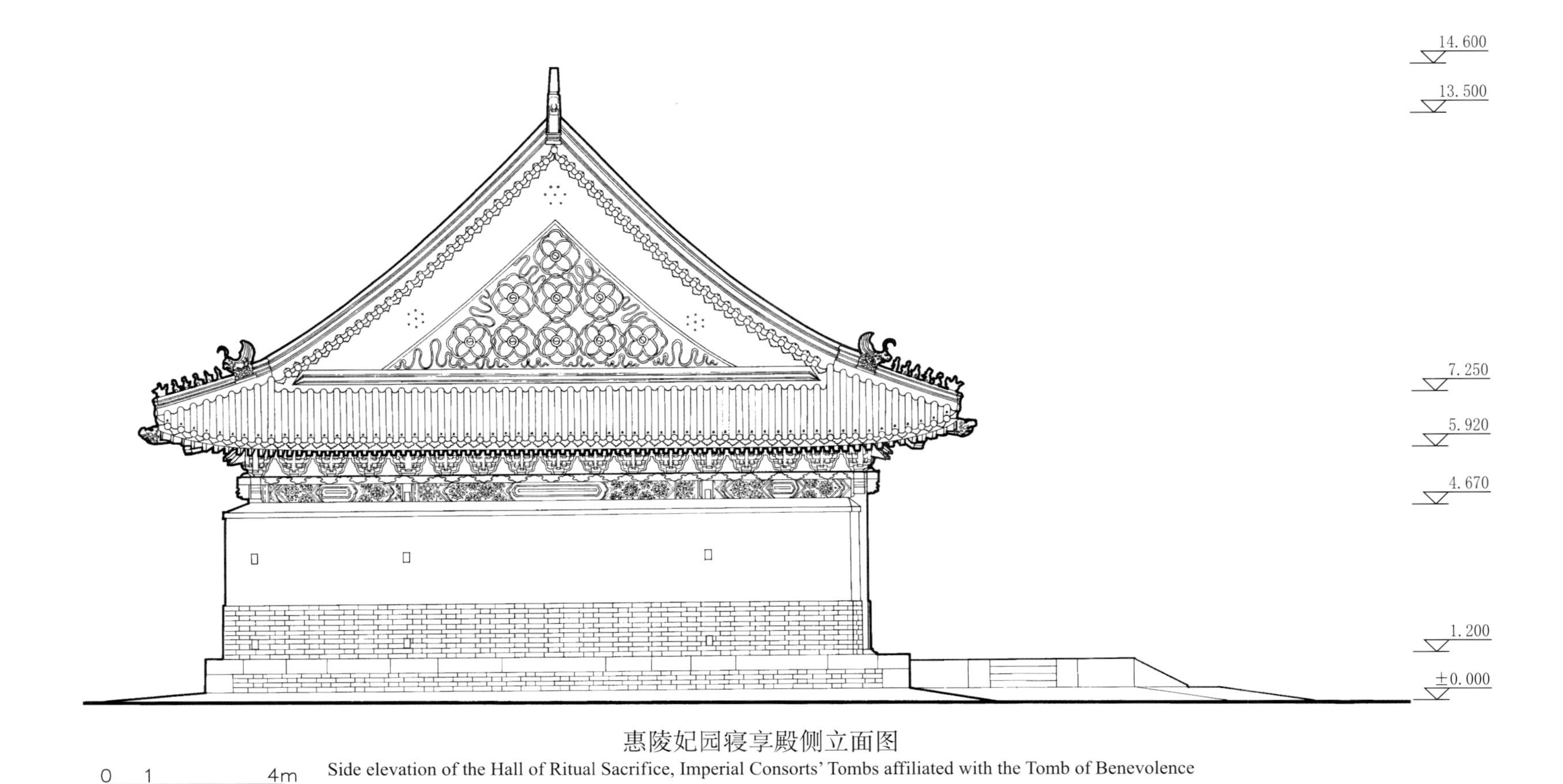

惠陵妃园寝享殿侧立面图

Side elevation of the Hall of Ritual Sacrifice, Imperial Consorts' Tombs affiliated with the Tomb of Benevolence

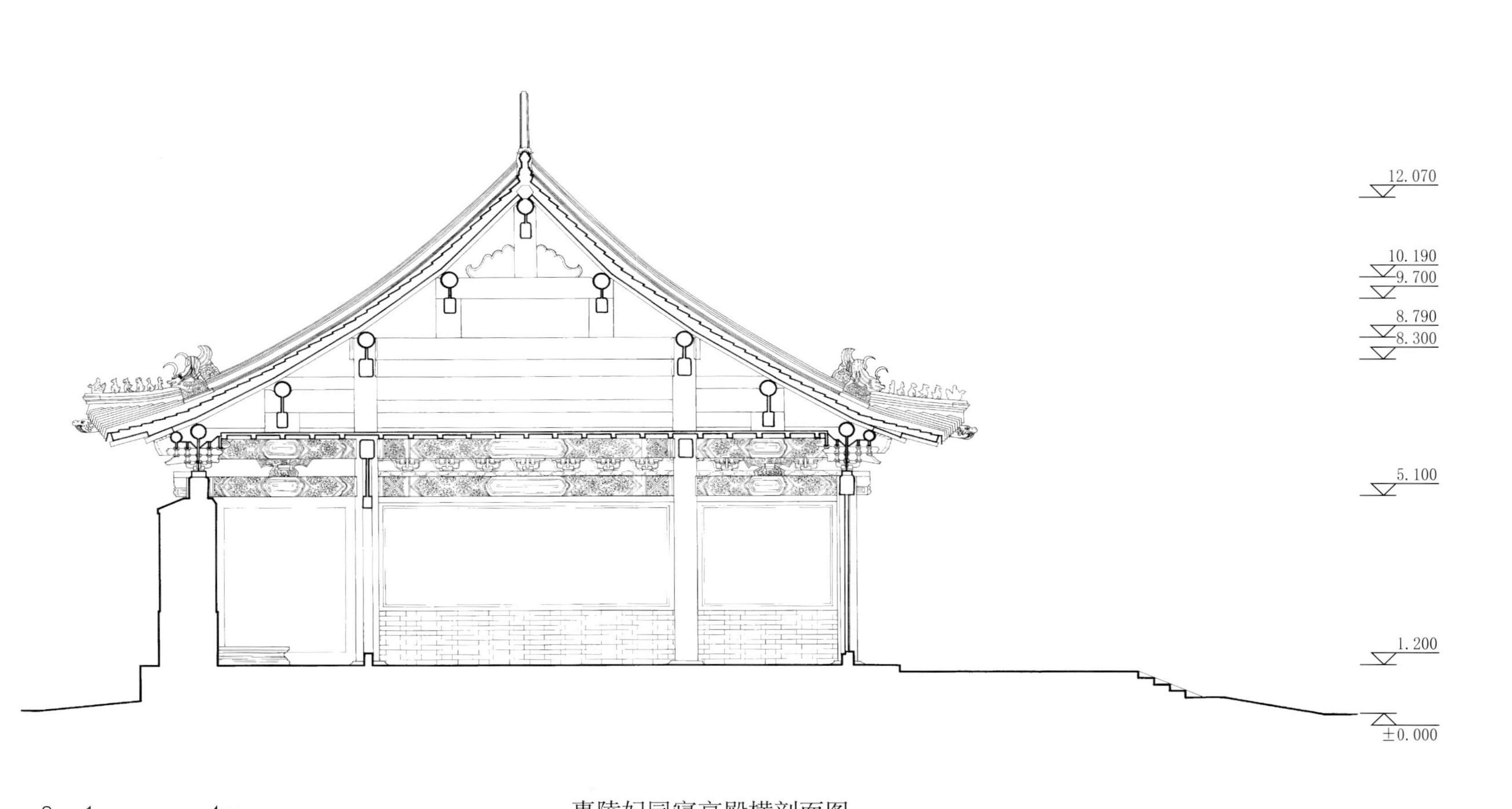

惠陵妃园寝享殿横剖面图

Cross section of the Hall of Ritual Sacrifice, Imperial Consorts' Tombs affiliated with the Tomb of Benevolence

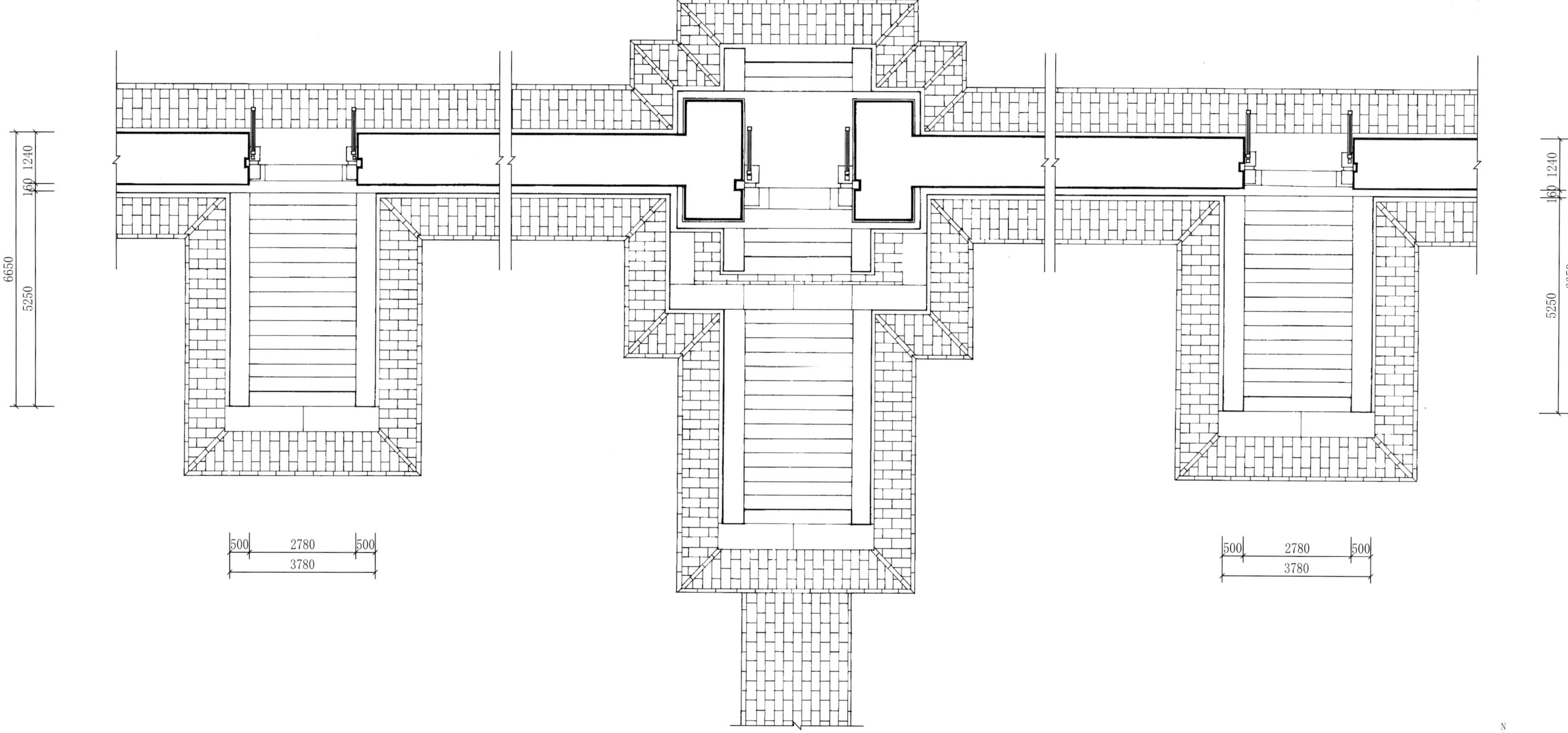

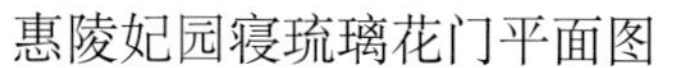

惠陵妃园寝琉璃花门平面图

Plan of the Gate with Glazed Roof Tiles, Imperial Consorts' Tombs affiliated with the Tomb of Benevolence

参与测绘及相关工作的人员名单

一、1980年暑期测绘

孝陵；裕陵；裕陵妃园寝；定陵；定东陵

指导教师：冯建逵　童鹤龄　王玉生　杨道明　慕春暖　曹治政　卢　倓

教师职工：史焕玲

测绘学生

1977级本科：吴宙航　曾大敏　方　元　华　镭　张玉坤　张松山　宫晓玲
徐　燚　张志平　崔　愷　刘均田　杨昌鸣　刘肖林　张繁维
李慧荣　管文海　何洁珩　运迎霞　马万珍　张　华　胡　瑾
刁绍玲　王　岚　戴　月　朱雪琴　张　萍　邓幼莹　蔡　节
李　琳　周湘虎　邱　滨　刘燕辉　吕永泉　覃　力

1978级本科：段　进　张　涛　周贞雄　夏　青　王　蔚　赵国文　赵　冰
聂洪达　吕大力　张　菁　傅之京　余　廲　叶　珉　李　茹
李伟民　张立芳　吴唯佳　唐笑居　王小莉　任俊生　华　超
李金铎　靳瑞珺　王国辉　王延芳　张　伟　路　红

二、1982年暑期测绘

孝陵；孝东陵；裕陵

指导教师：冯建逵　杨道明　张玉坤　刘燕辉　运迎霞

测绘学生

1980级本科：梁　雪　曹建明　常　盾　张　玫　袁逸倩　刘振锋　李雄伟
沈伟洪　洪再生　司小虎　苏惠甫　曾　坚　曹　磊　赵　彤
张　宜　李永玲　付　闽　孙丽萍　朱长青　李瑞林　宫义珍
梁　工　张　军　严　燕　韩玉斌　王　严　陆　平　卫宝岭
徐　凯　徐苏斌　孙　刚　李子萍　赖韶东　许顺法　杜富存
郭建祥　陈志军　刘　平　赵　菁　胡星平　黎建平

三、1982年12月测绘

孝陵石牌坊、大红门、更衣殿；昭西陵；景陵双妃园寝

指导教师：冯建逵　杨道明　曹治政

硕士研究生：王其亨

四、1983年10月测绘

孝陵石牌坊；惠陵

指导教师：冯建逵 杨道明

硕士研究生：王其亨

五、1987年5月测绘

定陵牌楼

指导教师：王其亨

硕士研究生：王西京

六、1990年暑期测绘

景陵；孝陵；定陵；昭西陵；景陵双妃园寝；裕陵妃园寝；定陵妃园寝

指导教师：王其亨 严建伟 曾 坚 刘云月 曾 力 杜海鹰 杨 颖 周 恺

教师职工：苏其盛

硕士研究生：黄 波

测绘学生

1987级本科：周立明

1988级本科：陈晓旭 马江红 张 菁 王海群 张 波 邹 健 林 倩

范 宏 孙惠亭 刘克美 龚清宇 李维纳 陆晓青 白雪茹

黄晓忠 孟 斌 李 浩 夏 磊 冯志强 戴利华 张 擎

李力红 林 淼 徐鹏举 秦 方 雷 建 赵 林 方 明

王 海 杨文岚 陈 蓉 沈勇嵬 潘伟江 张少蓉 张 荔

徐 峻 郭春燕 周海鹰 胡津阳 杨 广 凌 珀 孙忠波

曹金烜 肖 蓝 巢 晖 杜艳哲 刘 征 苏 航 杜 澎

魏青峰 李朝晖 苏小琳 陈 敏 周志宏 田 冰 严 卫

刘 健 盛 健 戴磊涛 张 新 冯 葵 孙其易 袁晨悦

肖宇清

七、1991年12月测绘

孝陵神厨库、宰牲亭

指导教师：王其亨

硕士研究生：黄 波 刘彤彤 王宇慧

八、1992年暑期测绘

景陵妃园寝

指导教师：王其亨

硕士研究生：黄 波

九、2011年暑期测绘

裕陵；景陵双妃园寝；定陵妃园寝

指导教师：曹 鹏 郭华瞻

博士研究生：耿昀

硕士研究生：刘芳 黄兵 王琳 郭满 李天 王方捷 王刚
魏安敏 王原 傅强 刘洪涛

测绘学生

2006级本科：牟文昊

2008级本科：党晟 廖茂羽

2009级本科：

建筑学院

于汉泽 符佳琦 胡伯骥 李默予 朱萌 邱实 郭壮
张志哲 董文乐 侯馨梦 贾彦琪 刘丛 倪旭玮 王韬
赵俊斌 单丹丹 经翔宇 王芸 王姝 于鸿飞 赵鹏
董姝婧 韩志鹏 孙秋莹 王卫童 魏硕璞 郑骥驰 瞿美智
曹睿原 雷玉 胡翔宇 唐婧娴 徐漫辰 陈子钰 黄赫
万福昆 王琛芳 杨馥宁 张薇 张铷航 叶页 王嗣文
胡玥璇 孙钰婷 颜冬 谢奕迪 姜薇 谢怡明 陈君仪
侯宇楠 杨晓昱 韩俊民 胡家源 刘欣然 魏安敏 温亚
郭海阔 白石 孔维媛 刘旸 杨思航 赵丽 王晓颖
程宇 肖宜鹏 苏畅 郝帅 赵熠萌 董烨程 郝璐
郭玉玢 乐桐 胡翔宇 周瀚 张祎 马旭

建筑工程学院

于天昊 赵章泳 王振宇

十、2012年暑期测绘

定东陵

指导教师：王蔚 曹鹏

硕士研究生：王刚 傅强 代朋 刘慧媛 刘洪涛 董瑞曦 苏心
褚安东 石越

2008级本科：党晟

2009级本科：谢怡明 陈君仪

测绘学生

2010级本科：顾嘉禹 石明雨 王庆 雷盼 谢海 朱金运 柴文璞
周鹏威 孙玮 许多锦 王丰亮 曹峻川 毛亚宁 付晓
李舒静 杨丹凝 马思然 金达 刘桐 陈明玉 吴昊
徐海林 李扬淑 游欣 熊毅寒 余啸 张天翔 钱梦菲
刘博 阎晓旭 张文博 赵欣楠 丁宇辰 刘世达 刘芷覃
罗一婷 邓鹤 白文佳 郭婧舒 张文学 赵克仑 袁园
徐玉 张馨文 宋博文 林川人 朱云昕 史颖天 闫振强
郭柳园 汪舒 杨然 张松 邬皓天 刘俐伶 周平
金艺豪 黄浦宽

十一、2013年暑期测绘

孝东陵

指导教师：曹　鹏

教师职工：张建斌

硕士研究生：褚安东　李　超　满兵兵　韩　涛　荣　幸

测绘学生

2011级本科：乔　尚　石晓彤　肖楚琦　孙忠涵　王　禹　黄羽杉　李筱蓬

杨轶乔　刘程明　杨　慧　魏昊楠　杨　莹　耿　佳　李司洋

刘潇雨

十二、2016年暑期测绘

孝陵

指导教师：曹　鹏

硕士研究生：孟晓静　周俊良　田　恬　谢怡明　杨　莹　王婧婷　赵子杰

张文新

2010级本科毕业生：刘未达

2013级本科：张　涛

测绘学生

2014级本科：贝以宁　刘靖旸　王　畅　张晓龙　刘　媛　曹柏青　陈鹏辉

程　婧　刘圆圆　刘麟阅　沈晨思　孙宇桐　王梦薇　王佩璇

王天晓　王旨选　王子良　甄　靓　朱一然　查兴文　苟艇锴

韩工布　胡江晨　黄　豪　黄嘉良　刘纪坤　孟祥瑞　时冬玮

孙亚奇　王铭璐　徐　铭　张　敏　张曦元　梁　毅　邓天怡

李金宗　李雪飞　李　璐　苏　杭　王一帆　许泽坤　张宇程

张奕怡　陈家瑶　陈梦香　周崇曦　李耀达　李　嫣　刘永城

王宇彤　于传孟　张　鹰　周大伟　金炳秀［韩］

十三、2017年暑期测绘

裕陵地宫

指导教师：曹　鹏

博士研究生：曹睿原

硕士毕业生：史　展

测绘学生

2015级本科：蒲洁莹　王舒萱　许宁佳　张梦晓

十四、2018年暑期测绘

定陵；景陵

指导教师：曹　鹏

博士研究生：曹睿原

硕士研究生：李东祖　李文迪　刘　洋　何丽沙　陈　鹏　江林燕　冯亚欣

康博超　王　成

2010级本科毕业生：刘未达

2016级本科：董皓月　金昊争　李美琦　李明远　李阳雨　李寅政　廉佳欣
廖睿妍　林　磊　刘白彬　刘畅翔　刘亨元　刘修岩　刘　岩
刘雨松　刘怡萱　刘　熹　马文超　庞任飞　彭瀚墨　齐　越
任叔龙　邵　仝　沈诗画　孙布尔　郑成禹　费韵霏　黄星宇
雷嵋棵　刘玮琰　卢昱吉　马　聪　倪欣然　聂　月　史德康
史立娟　王苇婧　吴献淋　邢　跃　许瀚文　杨　阳　原　琳
张雪寒　张　真　张　淇　左　桢　闫世航

十五、清东陵测绘图出版整理

指导教师：王其亨　张凤梧

博士研究生：曹睿原　李梦思　周悦煌　何滢洁　吴琳琳　宋　晋　杨元传

硕士毕业生：史　展

硕士研究生：程春艳　孟晓静　李佳璐　王　越　王　晨

2013级本科：朱梦钰　谢　瑾

2015级本科：许宁佳

英文翻译：庄　岳　赵春兰　肖　思

英文校对：[英]布鲁斯·卡瑞

List of Participants Involved in Architectural Survey, Drawing and Work related to the Publication

1. Summer survey and drawing in 1980

Tomb of Filial Piety (Xiaoling), Tomb of Prosperity (Yuling), Imperial Consorts' Tombs affiliated with the Tomb of Prosperity (Yuling Feiyuanqin), Tomb of Stability (Dingling), East Tomb of Stability (Ding Dongling)

Instructors: FENG Jiankui, TONG Heling, WANG Yusheng, YANG Daoming, MU Chunnuan, CAO Zhizheng, LU Yan

Staff: SHI Huanling

Students involved in survey and drawing

1977 undergraduates: WU Zhouhang, ZENG Damin, FANG Yuan, HUA Lei, ZHANG Yukun, ZHANG Songshan, GONG Xiaoling, XU Ying, ZHANG Zhiping, CUI Kai, LIU Juntian, YANG Changming, LIU Xiaolin, ZHANG Fanwei, LI Huirong, GUAN Wenhai, HE Jieheng, YUN Yingxia, MA Wanzhen, ZHANG Hua, HU Jin, DIAO Shaoling, WANG Lan, DAI Yue, ZHU Xueqin, ZHANG Ping, DENG Youying, CAI Jie, LI Lin, ZHOU Xianghu, QIU Bin, LIU Yanhui, LYU Yongquan, QIN Li

1978 undergraduates: DUAN Jing, ZHANG Tao, ZHOU Zhenxiong, XIA Qing, WANG Wei, ZHAO Guoweng, ZHAO Bing, NIE Hongda, LYU Dali, ZHANG Jing, FU Zhijing, YU She, YE Min, LI Ru, LI Weimin, ZHANG Lifang, WU Weijia, TANG Xiaoju, WANG Xiaoli, REN Junsheng, HUA Chao, LI Jinduo, JIN Ruijun, WANG Guohui, WANG Yanfang, ZHANG Wei, LU Hong

2. Summer survey and drawing in 1982

Tomb of Filial Piety (Xiaoling), East Tomb of Filial Piety (Xiao Dongling), Tomb of Prosperity (Yuling)

Instructors: FENG Jiankui, YANG Daoming, ZHANG Yukun, LIU Yanhui, YUN Yingxia

Students involved in survey and drawing

1980 Undergraduates: LIANG Xue, CAO Jianming, CHANG Dun, ZHANG Mei, YUAN Yiqian, LIU Zhenfeng, LI Xiongwei, SHEN Weihong, HONG Zaisheng, SI Xiaohu, SU Huifu, ZENG Jian, CAO Lei, ZHAO Tong, ZHANG Yi, LI Yongling, FU Min, SUN Liping, ZHU Changqing, LI Ruilin, GONG Yizhen, LIANG Gong, ZHANG Jun, YAN Yan, HAN Yubin, WANG Yan, LU Ping, WEI Baoling, XU Kai, XU Subin, SUN Gang, LI Ziping, LAI Shaodong, XU Shunfa, DU Fucun, GUO Jianxiang, CHEN Zhijun, LIU Ping, ZHAO Jing, HU Xingping, LI Jianping

3. Survey and drawing in December 1982

Memorial Stone Archway, Great Red Gate and Dressing Hall affiliated to Tomb of Filial Piety (Xiaoling), West Tomb of Brightness (Zhao Xiling), Two Imperial Consorts' Tombs affiliated with the Tomb of Admiration (Jingling Shuangfeiyuanqin)

Instructors: FENG Jiankui, YANG Daoming, CAO Zhizheng

Graduate student: WANG Qiheng

4. Survey and drawing in October 1983

Memorial Stone Archway affiliated to Tomb of Filial Piety (Xiaoling), Tomb of Benevolence (Huiling)

Instructors: FENG Jiankui, YANG Daoming

Graduate student: WANG Qiheng

5. Survey and drawing in May 1987

Memorial Gateway of Tomb of Stability (Dingling)

Instructor: WANG Qiheng

Graduate student: WANG Xijing

6. Survey and drawing in summer 1990

Tomb of Admiration (Jingling), Tomb of Filial Piety (Xiaoling), Tomb of Stability (Dingling), West Tomb of Brightness (Zhao Xiling), Two Imperial Consorts' Tombs affiliated with the Tomb of Admiration (Jingling Shuangfeiyuanqin), Imperial Consorts' Tombs affiliated with the Tomb of Prosperity (Yuling Feiyuanqin), Imperial Consorts' Tombs affiliated with the Tomb of Stability (Dingling Feiyuanqin)

Instructors: WANG Qiheng, YAN Jianwei, ZENG Jian, LIU Yunyue, ZENG Li, DU Haiying, YANG Yin, ZHOU Kai

Staff: SU Qisheng

Graduate student: HUANG Bo

Students involved in survey and drawing

1987 undergraduate: ZHOU Liming

1988 undergraduates: CHEN Xiaoxu, MA Jianghong, ZHANG Jing, WANG Haiqun, ZHANG Bo, ZOU Jian, LIN Qian, FAN Hong, SUN Huiting, LIU Kemei, GONG Qingyu, LI Weina, LU Xiaoqing, BAI Xueru, HUANG Xiaozhong, MENG Bin, LI Hao, XIA Lei, FENG Zhiqiang, DAI Lihua, ZHANG Qin, LI Lihong, LIN Miao, XU Pengju, QIN Fang, LEI Jian, ZHAO Lin, FANG Ming, WANG Hai, YANG Wenlan, CHEN Rong, SHEN Yongwei, PAN Weijiang, ZHANG Shaorong, ZHANG Li, XU Jun, GUO Chunyan, ZHOU Haiying, HU Jinyang, YANG Guang, LING Po, SUN

Zhongbo, CAO Jinxuan, XIAO Lan, CHAO Hui, DU Yanzhe, LIU Zheng, SU Hang, DU Peng, WEI Qingfeng, LI Zhaohui, SU Xiaolin, CHEN Min, ZHOU Zhihong, TIAN Bing, YAN Wei, LIU Jian, SHENG Jian, DAI Leitao, ZHANG Xin, FENG Kui, SUN Qiyi, YUAN Chenyue, XIAO Yuqing

7. Survey and drawing in December 1991

Imperial Sacrifice Kitchen and Slaughter Pavilion affiliated to Tomb of Filial Piety (Xiaoling)
Instructor: WANG Qiheng
Graduate Students: HUANG Bo, LIU Tongtong, WANG Yuhui

8. Survey and drawing in summer 1992

Imperial Consorts' Tombs affiliated with the Tomb of Admiration (Jingling Feiyuanqin)
Instructor: WANG Qiheng
Graduate student: HUANG Bo

9. Survey and drawing in summer 2011

Tomb of Prosperity (Yuling), Two Imperial Consorts' Tombs affiliated with the Tomb of Admiration (Jingling Shuangfeiyuanqin), Imperial Consorts' Tombs affiliated with the Tomb of Stability (Dingling Feiyuanqin)
Instructors: CAO Peng, GUO Huazhan
Doctoral student: GENG Yun
Graduate students: LIU Fang, HUANG Bing, WANG Lin, GUO Man, LI Tian, WANG Fangjie, WANG Gang, WEI Anmin, WANG Yuan, FU Qiang, LIU Hongtao
2006 Undergraduate students: MOU Wenhao
2008 Undergraduate students: DANG Sheng, LIAO Maoyu
2009 Undergraduate students:
College of Architecture
YU Hanze, FU Jiaqi, HU Boji, LI Moyu, ZHU Meng, QIU Shi, GUO Zhuang, ZHANG Zhizhe, DONG Wenle, HOU Xinmeng, JIA Yanqi, LIU Cong, NI Xuwei, WANG Tao, ZHAO Junbin, SHAN Dandan, JING Xiangyu, WANG Yun, WANG Shu, YU Hongfei, ZHAO Peng, DONG Shujing, HAN Zhipeng, SUN Qiuying, WANG Weitong, WEI Shuopu, ZHENG Jichi, QU Meizhi, CAO Ruiyuan, LEI Yu, HU Xiangyu, TANG Jingxian, XU Manchen, CHEN Ziyu, HUANG He, WAN Fukun, WANG Chenfang, YANG Funing, ZHANG Wei, ZHANG Ruhang, YE Ye, WANG Siwen, HU Yuexuan, SUN Yuting, YAN Dong, XIE Yidi, JIANG Wei, XIE Yiming, CHEN Junyi, HOU Yunan, YANG Xiaoyu, HAN Junmin, HU Jiayuan, LIU Xinran, WEI Anmin, WEN Ya, GUO Haikuo, BAI Shi, KONG Weiyuan, LIU Yang, YANG Sihang, ZHAO Li, WANG Xiaoyin, CHENG Yu, XIAO Yipeng, SU Chang, HAO Shuai, ZHAO Yimeng, DONG Yecheng, HAO Lu, GUO Yubin, LE Tong, HU Xiangyu, ZHOU Han, ZHANG Yi, MA Xu
College of Civil Engineering
YU Tianhao, ZHAO Zhangyong, WANG Zhenyu

10. Survey and drawing in summer 2012

East Tomb of Stability (Ding Dongling)
Instructors: WANG Wei, CAO Peng
Graduate students: WANG Gang, FU Qiang, DAI Peng, LIU Huiyuan, LIU Hongtao, DONG Ruixi, SU Xin, CHU Andong, SHI Yue
2008 Undergraduate student: DANG Sheng
2009 Undergraduate students: XIE Yiming, CHEN Junyi
Students involved in survey and drawing
2010 Undergraduate Students: GU Jiayu, SHI Mingyu, WANG Qing, LEI Pan, XIE Hai, ZHU Jinyun, CHAI Wenpu, ZHOU Pengwei, SUN Wei, XU Duojin, WANG Fengliang, CAO Junchuan, MAO Yaning, FU Xiao, LI Shujing, YANG Danning, MA Siran, JIN Da, LIU Tong, CHEN Mingyu, WU Hao, XU Hailin, LI Yangshu, YOU Xin, XIONG Yihan, YU Xiao, ZHANG Tianxiang, QIAN Mengfei, LIU Bo, YAN Xiaoxu, ZHANG Wenbo, ZHAO Xinnan, DING Yuchen, LIU Shida, LIU Zhiqin, LUO Yiting, DENG He, BAI Wenjia, GUO Jingshu, ZHANG Wenxue, ZHAO Kelun, YUAN Yuan, XU Yu, ZHANG Xinwen, SONG Bowen, LIN Chuanren, ZHU Yunxin, SHI Yintian, YAN Zhenqiang, GUO Liuyuan, WANG Shu, YANG Ran, ZHANG Song, WU Haotian, LIU Liling, ZHOU Ping, JIN Yihao, HUANG Pukuan

11. Summer and drawing Survey in 2013

East Tomb of Filial Piety (Xiao Dongling)
Instructor: CAO Peng
Staff: ZHANG Jianbin

Graduate Students: CHU Andong, LI Chao, MAN Bingbing, HAN Tao, RONG Xing

Students involved in survey and drawing

2011 Undergraduate Students: QIAO Shang, SHI Xiaotong, XIAO Chuqi, SUN Zhonghan, WANG Yu, HUANG Yushan, LI Xiaopeng, YANG Yiqiao, LIU Chengming, YANG Hui, WEI Haonan, YANG Ying, GENG Jia, LI Siyang, LIU Xiaoyu

12. Summer Survey and drawing in 2016

Tomb of Filial Piety (Xiaoling)

Instructor: CAO Peng

Graduate Students: MENG Xiaojing, ZHOU Junliang, TIAN Tian, XIE Yiming, YANG Ying, WANG Jingting, ZHAO Zijie, ZHANG Yixin

2010 undergraduate student (graduated): LIU Weida

2013 undergraduate student: ZHANG Tao

Students involved in survey and drawing

2014 Undergraduate students: BEI Yining, LIU Jingyang, WANG Chang, ZHANG Xiaolong, LIU Yuan, CAO Boqing, CHEN Penghui, CHENG Jing, LIU Yuanyuan, LIU Linyue, SHEN Chensi, SUN Yutong, WANG Mengwei, WANG Peixuan, WANG Tianxiao, WANG Zhixuan, WANG Ziliang, ZHEN Jin, ZHU Yiran, ZHA Xingwen, GOU Tingkai, HAN Gongbu, HUANG Hao, HUANG Jialiang, LIU Jikun, MENG Xiangrui, SHI Dongwei, SUN Yaqi, WANG Minglu, XU Ming, ZHANG Min, ZHANG Xiyuan, LIANG Yi, DENG Tianyi, LI Jinzong, LI Xuefei, LI Lu, SU Hang, WANG Yifan, XU Zekun, ZHANG Yucheng, ZHANG Yiyi, CHEN Jiayao, CHEN Mengxiang, ZHOU Congxi, LI Yaoda, LI Yan, LIU Yongcheng, WANG Yutong, YU Chuanmeng, ZHANG Ying, ZHOU Dawei, JIN Bingxiu (Korea)

13. Summer Survey and drawing in 2017

Underground Palace of Tomb of Prosperity (Yuling)

Instructor: CAO Peng

Doctoral student: CAO Ruiyuan

Graduate student: SHI Zhan

Students involved in survey and drawing

2015 undergraduate students: PU Jieying, WANG Shuxuan, XU Ningjia, ZHANG Mengxiao

14. Summer survey and drawing in 2018

Tomb of Stability (Dingling), Tomb of Admiration (Jingling)

Instructor: CAO Peng

Doctoral student: CAO Ruiyuan

Graduate students: LI Dongzu, LI Wendi, LIU Yang, HE Lisha, CHEN Peng, JIANG Linyan, FENG Yaxin, KANG Bochao, WANG Cheng

2010 Undergraduate student (graduated): LIU Weida

2016 Undergraduate students: DONG Haoyue, JIN Haozheng, LI Meiqi, LI Mingyuan, LI Yangyu, LI Yinzheng, LIAN Jiaxin, LIAO Ruiyan, LIN Lei, LIU Baibin, LIU Changxiang, LIU Hengyuan, LIU Xiuyan, LIU Yan, LIU Yusong, LIU Yixuan, LIU Xi, MA Wenchao, PANG Renfei, PENG Hanmo, QI Yue, REN Shulong, SHAO Tong, SHEN Shihua, SUN Buer, ZHENG Chengyu, FEI Yunfei, HUANG Xingyu, LEI Meike, LIU Weiyan, LU Yuji, MA Cong, NI Xinran, NIE Yue, SHI Dekang, SHI Lijuan, WANG Weijing, WU Xianlin, XING Yue, XU Hanwen, YANG Yang, YUAN Lin, ZHANG Xuehan, ZHANG Zhen, ZHANG Qi, ZUO Zhen, YAN Shihang

15. Participants involved in the compiling of survey drawings of the Eastern Qing Tombs for publication

Instructors: WANG Qiheng, ZHANG Fengwu

Doctoral students: CAO Ruiyuan, LI Mengsi, ZHOU Yuehuang, HE Yingjie, WU Linlin, SONG Jin, YANG Yuanchuan

Graduate master student: SHI Zhan

Graduate students: CHENG Chunyan, MENG Xiaojing, LI Jialu, WANG Yue, WANG Chen

2013 Undergraduate students: ZHU Mengyu, XIE Jin

2015 Undergraduate student: XU Ningjia

English translation: ZHUANG Yue, ZHAO Chunlan, XIAO Si

图书在版编目（CIP）数据

清东陵＝EASTERN QING TOMBS：汉英对照 / 冯建逵，王其亨编著；天津大学建筑学院，清东陵文物管理处编写 . —北京：中国建筑工业出版社，2021.4
（中国古建筑测绘大系 . 陵寝建筑）
ISBN 978-7-112-25745-4

Ⅰ . ①清… Ⅱ . ①冯… ②王… ③天… ④清… Ⅲ . ①陵墓－建筑艺术－遵化－清代－图集 Ⅳ . ① TU251.2-64

中国版本图书馆CIP数据核字（2020）第256169号

丛书策划 / 王莉慧
责任编辑 / 李　鸽　陈海娇
英文审稿 / ［英］布鲁斯・卡瑞（Bruce Currey）庄　岳
书籍设计 / 付金红
责任校对 / 王　烨

中国古建筑测绘大系・陵寝建筑
清东陵
天津大学建筑学院
清东陵文物管理处　编写
冯建逵　王其亨　编著

Traditional Chinese Architecture Surveying and Mapping Series: Tomb Architecture
EASTERN QING TOMBS
Compiled by the School of Architecture, Tianjin University & the Eastern Qing Tombs Cultural Relics Management Office
Edited by FENG Jiankui, WANG Qiheng

*

中国建筑工业出版社出版、发行（北京海淀三里河路 9 号）
各地新华书店、建筑书店经销
北京方舟正佳图文设计有限公司制版
北京雅昌艺术印刷有限公司印刷

*

开本：787 毫米 ×1092 毫米　横 1/8　印张：54½　字数：1444 千字
2022 年 2 月第一版　2022 年 2 月第一次印刷
定价：**428.00** 元
ISBN 978-7-112-25745-4
（35230）